Physics of Active Galactic Nuclei

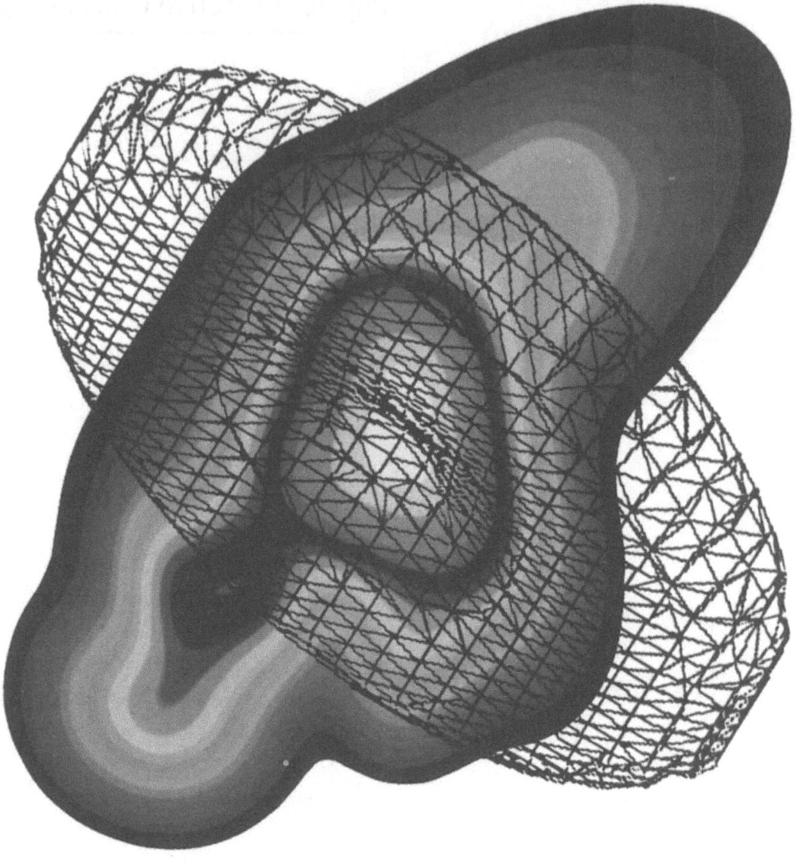

Modern interferometry techniques enable the studies of active galactic nuclei down to scales of a few parsecs. This is comparable to the sizes of the central accretion disks which are supposed to feed all AGN. This now permits detailed comparisons of observations with theory and enables empirical tests of the understanding of the physical mechanisms which drive the nuclear activity. The cover illustrates these achievements, one of the motivations for this conference, by comparing a radio-interferometry (VLBI) map of 3C84 and a computed disk model to scale. The radio map was obtained at 43 GHz at the observing epoch 1988.49 (courtesy T. P. Krichbaum, Max-Planck-Institut für Radioastronomie, Bonn, Germany). The region shown here has a projected diameter of 1.5 miliarcsec. The disk model, shown in the overlay was provided by M. Camenzind (Landessternwarte Heidelberg, Germany).

W. J. Duschl · S. J. Wagner (Eds.)

Physics of Active Galactic Nuclei

Proceedings of the International Conference, Heidelberg, 3–7 June 1991

With 315 Figures

Springer-Verlag
Berlin Heidelberg New York
London Paris Tokyo
Hong Kong Barcelona
Budapest

Dr. Wolfgang J. Duschl
Institut für Theoretische Astrophysik, Universität Heidelberg,
Im Neuenheimer Feld 561, W-6900 Heidelberg, Fed. Rep. of Germany

Dr. Stefan J. Wagner
Landessternwarte Königstuhl, W-6900 Heidelberg, Fed. Rep. of Germany

ISBN-13: 978-3-642-77568-0 e-ISBN-13: 978-3-642-77566-6
DOI: 10.1007/978-3-642-77566-6

Library of Congress Cataloging-in-Publication Data. Physics of active galactic nuclei : proceedings of the international conference. Heidelberg, 3–7 June 1991 / W. J. Duschl, S. J. Wagner (eds.). p. cm. Includes bibliographical references and index. ISBN-13: 978-3-642-77568-0
1. Active galaxies–Congresses. 2. Galactic nuclei–Congresses. I. Duschl. W. J. (Wolfgang J.), 1958–. II. Wagner, S. J. (Stefan J.), 1961–. QB858.3.P49 1992 523.1'12–dc20 92-20770

Softcover reprint of the hardcover 1st edition 1992

Production Editor: A. Kübler
Typesetting: Camera ready by authors/editors

55/3140 - 5 4 3 2 1 0 - Printed on acid-free paper

Preface

The physical processes driving the different manifestations of the phenomenon of active galactic nuclei have been studied extensively during the last decade. A major obstacle in all attempts to understand the relevant processes has always been the wide range of frequencies over which significant fractions of the total power are emitted. During the last decade, orbiting telescopes and instrumental improvements for ground-based instrumentation provided the means for major advancements on the observational side. The organizers felt that it was timely to organize a meeting to discuss the impact of this new situation on the understanding of the relevant physical processes.

More then 400 astrophysicists were interested in participating in the meeting, in spite of the constraints on overseas travel which were imposed in early 1991. Unfortunately only 220 participants could be hosted by the Max-Planck-Haus, the site of the 1991 Heidelberg conference. The meeting was organized by Sonderforschungsbereich 328 "Evolution of Galaxies". During 5 sessions, most of which lasted for one day each, 47 invited and contributed talks and 150 poster papers were given, most, but not all, of which are included in these proceedings. With a few exceptions the order of the written texts follows that of the oral contributions during the meeting. The arrangement of posters into the five sections was not always unambiguous. We hope to have placed them in the most appropriate sections, in which they are listed in alphabetical order.

We started to prepare the conference with Michael Penston as chairman of the Scientific Organizing Committee. Michael has inspired many of the current research programmes on active galaxies and has contributed a lot to our current understanding of the physics of active galactic nuclei. It came as a shock to us and many friends in the AGN community when he passed away in December 1990. The conference and this book are dedicated to his memory.

We are grateful to the speakers and poster authors for sending in the written versions of their contributions. We thank the members of the Scientific Organizing Committee, who set up the scientific program of the meeting. Drs. I. Appenzeller, M. Camenzind, T. Schmutzler, and R. Wehrse helped us to organize this conference. Financial assistance of Sonderforschungsbereich 328 of the Deutsche Forschungsgemeinschaft, the Federal State of Baden-Württemberg, and the European Space Agency (ESA) are gratefully acknowledged. Thanks are due to the many people, who helped with the local organization and technical assistance during the conference and in the course of assembling these proceedings.

Heidelberg, February 1992 W.J. Duschl, S.J. Wagner

Scientific Organizing Committee

M.V. Penston † (United Kingdom)
– Chairman –

R.D. Blandford (USA)
M. Camenzind (Germany)
S. Collin-Souffrin (France)
L. Maraschi (Italy)
M. Malkan (USA)
K. Pounds (United Kingdom)
M. Rees (United Kingdom)
H.-J. Röser (Germany)
S. Wagner (Germany)
A. Wilson (USA)

Local Organizing Committee

S.J. Wagner
– Chairman –

I. Appenzeller
M. Camenzind
W.J. Duschl
T. Schmutzler
R. Wehrse

Contents

POSTER CONTRIBUTIONS

2 – Disks and Broad Line Regions

3 – Extended Narrow Line Regions and Host Galaxies

4 – Magnetic Fields and Jet Formation

5 – Unified Schemes and Evolution

Michael V. Penston

When I was approached to write this foreword to the conference which was dedicated to Michael I accepted gladly. But you should not read this as a normal obituary, for Michael was no normal person. Indeed he is well and alive in the hearts of those of us who were fortunate to know him. In a sense, it was easy to know Michael; sometimes it was just enough to meet him to be pervaded by his lively and kind character. This is significatively reflected by the fact that he was loved by family, friends and collaborators alike; when he was still with us, I witnessed a few occasions when some colleagues were commenting how easy it was to love Michael. Isn't this fact remarkable by itself?

Michael embodied a most extraordinary set of virtues, which have been summarized by Bernard Pagel as "a lively and outgoing personality, a quick mind, a talent for expressing the essentials of a physical problem in simple terms and a strong and generous concern for the good of others." Indeed he was an exceptional person and scientist who "combined to an unusual degree physical insight, technical skills, originality and versatility with a deceptively simple expository style, organizational ability, enthusiasm and leadership."

I first met Michael late in 1982, during the first Anglo-Spanish school on Current Topics in Astrophysics. These were a series of winter schools held in Granada, which he conceived and organized out of his concern for the education of young astronomers; in this case to encourage Spanish students to pursue research in Astronomy, to best obtain the benefits from the new Observatory which was at the time being built in La Palma. He was going to become my thesis supervisor, but this first meeting was simply a reflection of his endeavour to bring Physics and, particularly, Astronomy home to a broader community, paticularly to the young. At the time of his death he was a very active Chairman of the Royal Astronomical Society Education Committee.

Michael had the inquisitive mind which marks a true scientist; his interest running wide not only in the Physical sciences, but in others like Biology and the theory of Evolution. I will not make an appraisal of his scientific career. Others who have known Michael for longer have done this much better than I could (e.g., B. Pagel, 1991: *Quarterly Journal of the R.A.S.*, **32**, 314; J. Wall, 1991: *Gemini, Newsletter of the R.G.O.*, **31**; R. Tayler, 1991: *Gemini, Newsletter of the R.G.O.*, **32**). I will just note that he worked successfully (as a theoretician soon turned into an observer) in several fields of Astronomy, from star formation and early stellar evolution, to the interstellar medium in our and external galaxies, to AGN and cosmology. He was a man of clear thinking, always straight to the heart of the matter. In one of our observing campaings in 1986, while we were watching the signal to noise building up in a spectrum of 3C390.3, he casually commented how similar the profile of the broad Hbeta emission line was to those seen in cataclysmic variable systems. That simple remark led to an article which marked for the astronomical community the begining of the hunt for accretion disk signatures in the profiles of the emission lines in AGN spectra.

Michael had new and creative ideas which he was always happy to share with others, either informally or in collaborations. I remember a comment he made when he published in 1987 almost his last paper as a single author (The FeII/La problem solved by fluorescence?); he was sad because this was the first time in ten years he hadn't found someone to share with as a co-author. That was Michael! He was a master in running international collaborations. It could have only been him who gave to the last big collaboration he established the name of 'Lovers of Active Galaxies', LAG, in a subtle and loving astronomical play on words.

We can't do justice in a few paragraphs to the great personality that Michael was; a full biography would be needed to convey some idea of his ways to those who did not meet him in life. Above all, he was a friend and a master from whom we learned, respected and loved.

Enrique Perez

The Black Hole Model of Active Galactic Nuclei

R.D. Blandford

Theoretical Astrophysics, Caltech, Pasadena, CA 91125, USA.

Abstract. Recent observations have reinforced our belief that AGN are mostly powered by accreting massive black holes, but shaken our confidence in more detailed models of this process. The optical emission seems not to be emitted thermally by a thin accretion disk and reprocessing of the radiation throughout the electromagnetic spectrum occurs from infra-red to X-ray energies. New ideas concerning magnetic transport of angular momentum are briefly discussed and related to a model of the broad emission line region. Prospects for finding more direct observational signatures of black holes are briefly outlined.

1. Introduction

Most astronomers now adopt, at least reluctantly, the hypothesis that the majority of galactic nuclei contain massive black holes in the million to billion $M_{\odot}$ mass range and that these objects are the prime movers for *most* of their energetic activity. It is further believed that these black holes form with their host galaxies, undergo a phase of rapid growth when they are spectacularly luminous, and then continue to exert indirect influence on their surroundings through their gravitational fields. Increasingly attractive though this scheme has become, I should begin with the disclaimer that it is by no means proven that it is correct and that there are alternatives, especially starbursts, which appear to be more important than accretion onto a black hole in many AGN. (Massive back holes and starbursts are not mutually exclusive. Indeed, it is hard to imagine how there can be a rapid rate of gas accretion into the galactic nucleus without there being a considerable amount of star formation and conversely, how a dense star cluster can avoid forming a black hole eventually.) In addition, despite much hard work, there is almost no consensus on the mode of accretion and the dominant means of dissipation and energy release. In this review, I shall try to summarize some current ideas and recent observational developments as they relate to the central black hole.

In an attempt to keep the focus of this conference on the *physics* of Active Galactic Nuclei, (AGN, henceforth), it was decided to try to organise the talks by characteristic lengthscale, starting with the smallest, namely the size of the event horizon. This is like trying to build a house of cards so that each successive story is larger than the one upon which it rests. I hope that I can be forgiven a few anticipatory radial excursions. More detailed discussions and bibliographies for some of these topics can be found in [1] [2] .

2. Arguments for Black Holes

Arguably the strongest reason for AGN being powered by black holes is based upon the efficiency of energy release. In fact, this was one of the first to be given for gravity-, as opposed to nuclear- or atomic-powered quasars, but it has only been in recent years that it has been possible to place it on a firm footing. Its cleanest expression [3] [4] involves estimating the net energy release per unit comoving volume by AGN (dominated by the UV emission of $z \sim 2$ quasars), and then dividing by the density of bright galaxies ($L \gtrsim 0.5L^*$) to obtain the mean energy release per bright galaxy. The answer, good to a factor 3 associated with the uncertain bolometric correction is $\sim 10^6 h^{-3} \mathrm{M}_\odot c^2$, where h is the Hubble constant in units of 100 km s^{-1} Mpc^{-1}. We now compare this mass with the upper limits on central masses associated with local galaxies (3×10^6 $\mathrm{M}_\odot$ in our Galaxy, 8×10^6 $\mathrm{M}_\odot$ in M32, 5×10^7 $\mathrm{M}_\odot$ in M31, 10^9 $\mathrm{M}_\odot$ in M87) [5] . This sample is too small to be quantitative, but it is unlikely to allow an efficiency $\lesssim 10^{-2} h^{-3}$. Such a high efficiency is a strong indication that black holes are involved. A longer list of upper limits would strengthen this deduction. Conversely, a verification that NGC 6240 [6] really does contain a central mass $\gtrsim 10^{11}$ $\mathrm{M}_\odot$, and is not unusual would weaken the argument. This efficiency can be compared with that associated with the faint blue galaxies $2 \times 10^6 h^{-3}$ $\mathrm{M}_\odot c^2$ [7] and the production of metals, $2 \times 10^6 h^{-1}$ $\mathrm{M}_\odot c^2$ [8] It is a curious coincidence that the AGN and galaxy efficiency are similar; the reduced overall efficiency of energy release by stars being compensated by their greater overall mass.

The second strong argument in favour of black holes is the observation of rapid variability, particularly at X-ray wavelengths. The most spectacular example to date is NGC 6814 which varies in a timescale of $\lesssim$ 100s. This is comparable with the light crossing time for the inner disk of a $\sim 10^6$ $\mathrm{M}_\odot$ black hole. Rapid variation appears to be the rule in the majority of a sample of Seyfert galaxies (Pounds, these proceedings) Variation of the Fe line on somewhat longer timescales is reported.

A third argument is furnished by the observation of superluminal expansion in a large fraction of the most luminous compact radio sources. Although the detailed kinematic explanation of this phenomenon remains elusive, it is generally supposed that a relativistic outflow is involved. The most natural way for such high speeds to be produced is from a relativistically deep potential well. (That the jets in extended radio galaxies are able to maintain a fixed direction in space for $\gtrsim 10^7$ yr. also indicates a spinning black hole.)

3. Properties of Black Holes

The properties of black holes are described, with differing emphases on physical rigour and astrophysical relevance, in [2], [9] [10] . Kerr black holes are characterised by a size $m = 1.5 \times 10^{13} M_8$ cm ($= 500 M_8$ s, $= 2 \times 10^{62} M_8$ erg), where M_8 is the mass in units of 10^8 $M_\odot$. The only other variable needed to characterize its gravitational field can be chosen to be the angular velocity Ω. A fraction of the mass of a spinning hole is reducible or extractable. In other words, if we imagine spinning up a non-rotating hole to angular velocity Ω by reversible processes, its total mass will increase to

$$m = \frac{m_0}{(1-\beta^2)^{1/2}}; \qquad \beta = 2\Omega m_0 < 2^{-1/2} \tag{1}$$

where m_0 is the initial irreducible mass. This is reminiscent of the special relativistic formula for the energy of a moving body with the important difference that the "speed" β is limited to $2^{-1/2}$.

The specific angular momentum of the hole can be written as $4m_0^2\Omega$ which is $\sim 5 \times 10^{23} M_8$ cm^2 s^{-1} for a rapidly spinning hole, some six orders of magnitude smaller than the typical specific angular momentum of gas orbiting in the galaxy. For this reason an accretion disk is generally supposed to form. One idiosyncratic property of a black hole is that, when it is sufficiently close to the horizon of a spinning hole, matter must rotate with respect to the distant stars. This is sometimes called the "dragging of inertial frames" by the black hole. This will apply to gas that orbits the black hole in an inclined orbit. Therefore an accretion disk that is inclined will be forced to precess, and in the presence of dissipation, be dragged into the equatorial plane. The manner in which this happens is still not well understood. [11]

4. Black Hole MHD

Magnetic field provides a useful method for extracting the reducible energy of a spinning black hole. The simplest way to see this is to imagine a cloud of plasma threaded by a loop of magnetic field orbiting close to the event horizon of the black hole. Differential frame dragging effects will cause an additional tidal force to act upon the cloud and to do work on the magnetic field that will be frozen into the plasma. The associated increase in energy comes ultimately from the reducible mass of the hole. If it is radiated away before the cloud falls crosses the horizon, then spin energy will have been extracted. [12]

In a more efficient mechanism, a global magnetic field is supposed to thread the hole supported by equatorial currents flowing primarily in an equatorial accretion disk. It seems to be very difficult to avoid producing enough plasma in the magnetosphere to short out any strong electric fields resolved parallel to the magnetic field. In other words, the magnetic surfaces are effective equipotential surfaces. The structure of the magnetosphere may be described by relativistic MHD, or what is to effectively the same for most purposes, force- free electromagnetism. Under these conditions, currents will flow along the magnetic field and can lead to energy extraction. Calculations suggest that roughly half the reducible spin energy will be extracted. It is easy to make an order of magnitude estimate of the strength of magnetic field that is required by treating the black hole as a giant battery. As conditions are essentially electromagnetic close to the hole, we may regard its internal resistance as being roughly the impedance of free space $\sim 100\Omega$. In addition, if the power ultimately escapes the black hole in the form as an electromagnetic wind, then we can also treat the resistive load on the battery as being $\sim 100\Omega$. If the power extracted is $10^{46} L_{46}$ erg s^{-1}, then the current that must flow is $\sim 3 \times 10^{18} L_{46}^{1/2}$ A, and the associated potential difference across the magnetosphere is $\sim 3 \times 10^{20} L_{46}^{1/2}$ V. The magnetospheric field strength is $\sim 10^{4} L_{46}^{1/2} M_{8}^{-1}$ G.

In several recent papers, Punsley and Coroniti [13] have raised some doubts about the viability of this mechanism. In order to appreciate their concern, it is necessary to turn to the relativistic MHD description of the process. This bears some similarity to the more familiar theory of a stellar wind in which gas flows radially passing through a critical point where its speed changes from subsonic to supersonic. In the corresponding hydromagnetic wind, there can be three critical points corresponding to the three possible magnetosonic waves, known as slow, intermediate (or Alfvén) and fast. In a black hole MHD wind the situation is one stage more complicated in that there are both inflowing and outflowing critical points, giving six in general on each field line. Now at a critical point, the behaviour of the characteristics which can carry information about the response of the flow to small changes in the boundary conditions changes. Beyond the outgoing fast mode point, both inwardly and outwardly directed fast mode characteristics travel outward, so the information carried by the fast mode cannot propagate upstream and influence the flow within the critical point. As the fast mode travels the fastest, the flow is unaffected by conditions upstream from the outgoing fast mode point. Conversely, with a stellar source where the flow speed is less than the slow mode speed, all the information can propagate upstream and the wind can respond to small perturbations in the defining conditions at its source. In this sense, the star determines its wind and we can see, at least qualitatively how a particular set of initial velocity and magnetic data can evolve to a stationary flow.

The worry with a black hole MHD wind is that, as the horizon lies within the ingoing fast point, the hole is necessarily causally disconnected from the flow. It has not been demonstrated that a stationary flow is stable; neither that it will be reached from general initial conditions. Although I cannot solve this problem, it is possible to illuminate it a bit by asking what information is being carried by specific characteristics. In the limit when the inertia associated with the plasma is very small, and this is believed to be appropriate to AGN massive black holes, it turns out that the slow waves are essentially responsible for transporting pressure information and are not important for this issue of principle as the pressure is ignorably small. The intermediate waves, however, essentially transport changes in the poloidal field structure. They can propagate both ways through the magnetosphere between the two intermediate wave characteristic surfaces which are essentially just the light surfaces, where a particle corotating with the field would have to travel with the speed of light to remain at constant radius. The location of the outgoing fast point depends upon the boundary conditions at large cylindrical radius, but is typically well removed from the black hole. The ingoing fast mode point is located just outside the black hole event horizon. Fast waves essentially carry information about the toroidal component of magnetic field or, equivalently, about the outflowing poloidal current. This information is not carried by characteristics over a small gap outside the horizon. Therefore, if we were to spin up the hole by a small amount, perhaps by firing a small solid body into it, it might be thought that the magnetosphere would not adjust. However, I would argue that this is not the case because the information about the spin of the hole is actually conveyed well beyond the horizon through the background metric. Any observer on a timelike geodesic at fixed distance from the hole is constrained to rotate with respect to distant stars at an angular velocity very close to that of the hole. Clearly some very difficult calculations will be necessary to settle the matter rigorously.

5. Thin Accretion Disks

It is generally supposed that the optical and UV emission arises fairly close to a massive black hole, perhaps from a thin accretion disk. (Radiation longward of 1μ is now supposed to be re-radiated by hot dust and need not concern us here.) As is well known the formal efficiency of a thin accretion disk increases from 6 to 42 per cent as the angular velocity of the hole increases from 0 to $1/8^{1/2}m_0$. This efficiency represents the binding energy of the smallest stable circular orbit. Unfortunately, this does not tell us how and where the energy is released. The most elementary assumption is that the binding energy is released locally so that the flux F emitted by the disk surface varies as $F \propto r^{-n}$, where $n = 3$ in the Newtonian limit. The simplest assumption to make about the emergent spectrum is that it is emitted locally at the rate prescribed by viscous transport with a black body spectrum so that the characteristic frequency emitted at radius r varies as $\nu \propto F^{1/4} \propto r^{-3/4}$. Integrating over r, gives an observed spectral flux, $S_\nu \propto \nu^{1/3}$. Unfortunately, this model does not describe observed AGN. The spectral flux observed from unpolarized AGN is, typically (but with large variation), $\propto \nu^{-1}$ redward of 1μ, and declining $\propto \nu^{-1/2}$ into the unobservable UV, before falling more rapidly into the X-ray part of spectrum. For a thin accretion disk the peak in spectral energy density, νS_ν, should be found at frequency

$$\nu_{max} \sim 3 \times 10^{15}(h/r)^{1/2} L_{46}^{-1/4} \mathrm{Hz}, \tag{2}$$

where h, by assumption $< r$, is the disk thickness. This is somewhat lower than observed in the most luminous quasars.

Two modifications of these assumptions can be advanced to try to account for the observed spectrum while retaining the hypothesis of a thin disk. Firstly, we can suppose that relatively more power is radiated at larger radius. There are several ways in which this can happen. If angular momentum is transported outward through the disk by internal viscous stress, then energy must also be carried outward. In addition, surface irradiation of the disk can have a similar effect at optical- and UV-emitting radii. Finally, general relativistic effects can suppress the emission from the inner disk. All of these effects can be modelled empirically by adopting an exponent in the range2 $< n < 3$. The second possible modification is to the spectrum which can be quite different from that of a black body. The surface of an optically thick accretion disk should form an atmosphere like that of a low gravity O star. Electron scattering is likely to be very important which will reduce the emissivity relative to a black body ϵ and increase the mean frequency radiated. Suppose that $\epsilon \propto r^\alpha$, the characteristic frequency radiated from radius r will then vary according to $\nu \propto r^{-(n+\alpha)/4}$. The spectral energy distribution will then have the form $\nu L_\nu \propto \nu^{4/(n+\alpha)}$. In order to reproduce the observed exponent ~ 0.5, we require that $\alpha \sim 5-6$, an unreasonably large value. Existing models of thin disks are able to reproduce the observed spectrum over a limited frequency range by invoking steep underlying continuum extending from the far infrared through to X-ray energies. Unfortunately, this continuum is unlikely to exist if the infrared emission is emitted by thermal dust.

There are other difficulties with simple disk models, as discussed here by Collin-Souffrin, Malkan and Peterson. The change in opacity at the Lyman limit almost inevitably creates an edge, though it is not clear whether it is in emission or absorption. Such features are rarely, if ever seen. In addition, the prevalence of electron scattering makes the absence of significant linear polarization in the optical continua of most quasars equally puzzling. In fact the polarization that is observed is oriented parallel to the radio source axis rather than perpendicular as might be expected. A third problem has just been created by the intense monitoring of the Seyfert galaxy NGC 5548, from which it has been concluded that the optical continuum variation follows that of the UV, with no discernible phase lag, not what would be anticipated from a disk with an approximate radius-to-frequency mapping.

Although it is not possible to resolve the innermost parts of an AGN accretion flow directly, gravitational lenses may allow us to do this indirectly. The quasar Q2237+020 has a redshift $z = 1.7$ and is located behind a $z = 0.04$ barred spiral galaxy. Four images are seen of which at least two have been reported to vary independently by up to 0.4^m on a timescale of order a month, long compared with the optical path differences for the different images. This is attributed to microlensing by stars in the nucleus of the galaxy. Unlike most invocations of microlensing, in this case the effect is unavoidable, provided that the source is sufficiently compact [14] . Microlensing arises when there is relative motion of the lens relative to the source-observer line. There are two ways in which this can happen. Firstly, the lens galaxy may move with respect to the microwave background. Secondly, high speed motion of emitting elements within the source may be responsible. (The motion of the earth is essentially along the line of sight, the random motions of the stars responsible for the microlensing are likely to be too slow and bulk motion of the source galaxy is insignificant as it is much more distant than the lens.)

In this case, the upper limit on the compactness of the source actually translates into a requirement that it radiate non-thermally. A microlensing simulation of a variety of thermal accretion disk models shows that they are between four and ten times too large to account for the reported variation [15] . Non-thermal optical emission is indicated.

6. Thick Accretion Disks

There has been a slight renaissance in the study of thick accretion disks. Their initial promise as models of critically accreting gas flows [16] received a severe setback when they were shown to be subject to global dynamical instability [17] . Much work has gone into elucidating the physics of these instabilities both analytically [18] and numerically [19] . In the slender annulus limit, the motions are essentially two dimensional and the modes have negative energy and angular momentum density within the co-rotation radius (where the fluid angular velocity ω equals the quotient of the real part of the wave frequency and the azimuthal quantum number m.) The disks are violently unstable when $-\sqrt{3} > d\ln\omega/d\ln r > -2$ provided that there is efficient reflection at the inner edge of the disk so as to establish a positive feedback cycle. The most rapidly growing mode produces quasi-stationary "planets" in the non-linear regime. Angular velocity variation closer to Keplerian is still dynamically unstable though the growth is less rapid.

The more astrophysically interesting case of a wide torus is more difficult to analyse and three dimensional numerical simulations are necessary. The most recent results [20] indicate that the fastest growing mode has $m = 1$ and that a single spiral arm develops with a saturated amplitude and transports angular momentum globally [21] . If this is the most rapid means of angular momentum transport, then these results indicate that the radial velocity can be as small as $\sim 10^{-3}$ of the tangential velocity which is marginally slow enough to allow the emergent radiation to thermalise.

7. Disk Magnetic Fields

There has also been a resurgence of interest in the influence of magnetic torques upon accretion. The original Shakura-Sunyaev prescription [22] postulated a shear stress either due to fluid turbulence or to magnetic field, proportional to the pressure. In an important recent development, Balbus and Hawley [23] have re-discovered and analysed an MHD instability [24] . In its simplest version, there is a weak vertical field in the accretion disk. A ring of gas is displaced radially outward, convecting the magnetic field with it so that its angular momentum is reduced. Under these conditions, the restoring force of the magnetic tension is overcome by the combined centrifugal and gravitational force provided that the wave vector of the perturbation satisfies $\boldsymbol{k} \cdot \boldsymbol{a} < 3^{1/2}\omega$, where $\boldsymbol{a}$ is the Alfvén speed times a unit vector along the magnetic field. Growth is rapid, in a few radians of rotation and an initially weak field should build up to a pressure comparable with the gas pressure very rapidly, so that the magnitude of the shear stress approaches the pressure, as numerical simulations substantiate. Models of accretion disks relying solely upon fluid dynamics, including the analysis of thick accretion disks outlined in the last section, are probably not internally consistent.

This important discovery, raises several issues, that are under active investigation (*cf.*Heyvaerts, these proceedings). Firstly, it will be important to include the effects of reconnection, which will limit the field growth [25] . Similar remarks apply to buoyancy effects [26] . Secondly, the relationship between this instability and dynamo models of the growth of disk field should be clarified (Camenzind, Lesch, Pudritz, these proceedings). It appears that dynamo processes are always too slow to be of primary importance. Thirdly the growth of non-axisymmetric perturbations must be clarified. The instability is local and so this ought not to be crucial.

8. Hydromagnetic Winds

The prevalence of jets and bipolar outflows associated with disks in AGN (and elsewhere) motivated studies of bipolar outflows. When field lines protrude from a Keplerian accretion disk so that their poloidal components make an angle of less than 60° with the radial direction, conducting plasma, frozen on to these field lines will be flung outward by the action of centrifugal force. When matter is flung out to large enough radius it inertia starts to become important and toroidal field will be created out of the poloidal field. It seems to be generally true [27] [28] that the tension associated with this toroidal field, in combination with the magnetic pressure gradient collimates the outflow towards the axis of symmetry. In this way energy and angular momentum can be extracted from the accretion disk.

Now a key question is does the external torque associated with the wind compete with the internal torque, presumably also magnetic, acting within the disk? To answer this question will undoubtedly require more detailed numerical simulations of the evolution of disk magnetic field.

9. Molecular Disk and Broad Emission Lines Clouds

One of the most significant realizations of recent years has been that molecular gas can be found in abundance close to the innermost regions of an AGN [29] . This gas can be responsible for the equatorial obscuration invoked to account for the distinction between Seyfert 1 and 2 galaxies and some radio galaxies and quasars. It is also probably responsible for the infra red continuum longward of 1μ from Seyfert galaxies and quasars [30] [31] .

Molecular gas can cool in this intense environment because of re-emission by dust. This can only happen while the dust (probably ultimately graphite), becomes so hot that it starts to sublime at a temperature ~ 1800 K and a radius, $r \sim 6 \times 10^{17} L_{46}^{1/2}$ cm, where L_{46} is the isotropic, OUV luminosity in units of 10^{46}erg s^{-1}. An equivalent way of expressing this limit is that the radiation pressure is ~ 0.1 dyne cm^{-2}. Now this is similar to the photo-ionizing radiation pressure inferred in the CIV emitting zone using reverberation mapping in the Seyfert galaxy NGC 5548 (Peterson, these proceedings) and also, less securely, in the more powerful quasars.

These and other considerations have motivated a model of the broad emission line gas in which molecular gas in an accretion disk is flung out along the magnetic field by a combination of radiation pressure and centrifugal force [32] . It is proposed that when the molecular gas is exposed to the full intensity of UV radiation, it will be quickly photoionized. (Shock heating can achieve a similar end.) The gas will then heat to an equilibrium temperature ~ 7000 K and an equilibrium pressure balanced by the surrounding magnetic field. Only one hemisphere will be observed because the disk is opaque and the line velocity profile will reflect a combination of rotational motion and outflow, to give the blue-shifted and blue asymmetric profiles often reported for the highest ionization states. If it is further assumed that the magnetic field extracts a large fraction of the disk angular momentum, then the outflowing ionized gas must have roughly the inferred density and filling factor.

However, the characteristic Keplerian speed associated with the molecular gas, $\sim 1500 M_8^{1/2} L_{46}^{-1/4}$ km s^{-1} is inadequate to account for the broadest line profiles. For this reason, it is also proposed that the lines are broadened by electron scattering in an intercloud medium or by reflection from the disk surface. The requisite temperature of this medium is $\sim 10^6$ K. Under unmagnetized conditions, gas at this temperature is thermally unstable; however magnetic field leads to some stabilization.

This model can be tested by further reverberation and polarization observations. More refined predictions of the general hypothesis that emission line gas is originally molecular are possible using detailed photoionization modelling.

10. Coronal emission

Having argued that in at least one quasar, non-thermal emission is required, we should consider the possible importance of emission from a disk corona. The original notion was that would be a dense, thin accretion disk in Keplerian orbit about the central black hole and that this would be sandwiched between two thicker coronal regions containing much hotter plasma, energised continuously by processes similar to those at work in the solar corona [33] . The coronal gas might well have a relativistic temperature so that electrons and positrons are copiously produced. In this way, a minority of X-ray and γ-ray emission could supplement the dominant thermal UV emission.

However, if magnetic torques are as strong as suggested above, then the inner disk is likely to be optically thin to free-free and bound-free emission and unable to radiate away its binding energy by this process alone. Coulomb heating of the electrons is likely to be ineffectual and the electrons may have to be heated indirectly. In essence the disk itself becomes the corona. Under these conditions, the large bulk kinetic energy associated with the orbiting gas (carried principally by the ions), should be dissipated spasmodically through a combination of shocks and magnetic reconnection events. We know empirically, from observations of young supernova remnants, that such flows are generally efficient non-thermal emitters. It is also quite likely that the sort of layered, two phase medium, inferred independently from X-ray spectroscopy, will be established under these circumstances (Pounds, these proceedings). Indeed, re-processing may be important in the optical and UV as well as in the X-rays and the inner disk may be enshrouded by a quasi-spherical distribution of clouds, with column density large enough to absorb the soft X-rays, but not the rapidly variable hard X-rays.

There is a second source of coronal emission and this is the magnetosphere of a black hole. This is the region where it is easiest to maintain a low particle density, and where, if there is sufficient external magnetic flux (maintained primarily by currents flowing in the accretion disk), intense electrostatic fields will most readily accelerate electrons and positrons.

11. Influence of the Black Hole on the AGN Type

One of the long term goals of research on AGN is to understand how the controlling physical parameters determine the type of source that we observe. At the most basic level, we expect that the mass of the central black hole and the mass accretion rate (normalized by the Eddington rate $4\pi GMm_p/\sigma_T c$)

to be important. If we also believe that spin energy extracted electromagnetically from the black hole can contribute to the total power, then a convenient way to parametrise this is to use the dimensionless quantity $\lambda = 2m\Omega$. These three parameters plus the observer inclination might be enough to describe the gross properties of AGN.

To give one example, suppose that relativistic jets are only formed when there is a rapidly spinning black hole [34] . This may continue to be true even when the mass accretion rate is relatively low. The value of λ may depend in a quite complex manner upon the history of the supply of mass. Radio quiet quasars and their lower luminosity counterparts the Seyfert galaxies would then be associated with objects that are now accreting rapidly but whose central black holes have not been spun up. By contrast, λ would be comparable with its maximal value of unity in radio loud quasars. Broad line FRII radio galaxies are presumably the lower mass and luminosity counterparts of these objects. In this case, AGN would remain radio-quiet or loud for a long time.

Black hole mergers provide a promising way to spin up a black hole. When a large Schwarzschild hole accretes a much smaller companion from a circular orbit that has shrunk as a result of the radiation of gravitational waves, it will be left with a spin angular momentum roughly appropriate to the last stable circular orbit, *i.e.*$\lambda \sim 2.3^{1/2}q$, where $q << 1$ is the mass ratio. This formula indicates that as $q \rightarrow 1$ rapid spinning holes will be produced. It would be of interest to obtain a more accurate estimate for the case $q \sim 1$.

12. Direct Evidence for Black Holes

This subject would rest on a far firmer footing if we had more direct evidence for central black holes. What are the observational prospects for attaining this? As with stellar mass black holes, the most durable results are dynamical. The best hope lies with HST which, if and when repaired, should be able to probe the central regions of galactic nuclei and determine their central velocity dispersions. Determinations of central masses, or least upper bounds upon them would greatly constrain black hole models. It is particularly important to test the idea that Seyfert galaxies harbour low mass black holes as opposed to being defunct quasars with high mass holes. The use of broad emission lines to achieve the same end has always been a bit problematical because of uncertainties in the size of the emission line region and the kinematics of the cloud. If reverberation mapping and photoionization studies of high resolution line profiles rectify these deficiencies, then this promises to be a widely usable technique.

A second possibility [35] is that a star be captured by a black hole and that some periodic emission associated with peribothon or passage through a disk be detectable. It is even possible to imagine measuring orbital precession in this manner. Other more complex scenarios, involving tidal disruption [36] may also be observable. Optical monitoring of nearby galactic nuclei is well motivated by this prospect. The exciting X-ray quasi-periodicity of NGC 6814, discussed here by Abramowicz is especially intriguing in this regard and could even display relativistic precessional effects.

A third way in which black holes might be revealed is as binaries. We have mentioned that galactic mergers are probably quite common especially in elliptical galaxies. If both partners have central holes of mass $\sim 10^6 M_6$ $M_\odot$, these will eventually spiral together under the action of gravitational radiation in a timescale $\sim 40 P_y^{8/3} M_6^{-5/3}$ Myr, where P_y is the period in years. If both holes retain small disks, an eclipsing binary or at least "ellipsoidal" variation might be seen. Now that X-ray lines have been seen from very near to black holes, there is the possibility of finding a spectroscopic binary with a Doppler variation $\sim 3000 M_6^{1/3} P_y^{-1/3}$ km s^{-1}. Observations in low luminosity AGN are most propitious.

Acknowledgements. I am indebted to Ski Antonucci, Steven Balbus, Robert Emmering, John Hawley, Richard Mushotzky, Sterl Phinney, Kevin Rauch, Isaac Shlosman and Martin Rees for collaboration and helpful discussions. I thank the organizers of this workshop for financial assistance. Support under NASA grant NAGW2372 and NSF grant AST89-17765 is gratefully acknowledged.

1. Blandford, R. D., Netzer, H., Woltjer, L.: *Active Galactic Nuclei* (Berlin: Springer-Verlag 1991)
2. Rees, M. J.: Ann. Rev. Astron. Astrophys. **22** 471 (1984)
3. Soltan, A.: Mon. Not. R. astr. Soc. **200** 115 (1982)
4. Padovani, P., Burg, R., Edelson, R. A.: Astrophys. J. **353** 438 (1990)
5. Dressler, A., Richstone, D. O.: Astrophys. J. **324** 701 (1988)
6. Bland-Hawthorn, J., Wilson, A. S., Tully, R. B.: Astrophys. J. Lett. **371** L19 (1991)
7. Tyson. J. A.: Astronom. J. **96** 1 (1988)
8. Cowie, L. L.: In *The Post-recombination Universe* ed. by N. Kaiser, A. Lasenby (NATO Advanced Science Institute Series 1988) p.125
9. Misner, C. W., Thorne, K. S., Wheeler, J. A.: *Gravitation* (San Francisco: Freeman 1973)
10. Blandford, R. D.: In *300 Years of Gravitation* ed. by S. W. Hawking, W. Israel (Cambridge: Cambridge University Press 1987) p.277
11. Kumar, S.: Mon. Not. R. astr. Soc. **233** 33 (1989)
12. Yokosawa, M., Ishizuka, T., Yabuki, Y.: Mon. Not. R. astr. Soc. (1991) (in press)
13. Punsley, B.,, Coroniti, F. V.: Astrophys. J. **350** 518 (1990)
14. Wambsganss, J., Pacszyński, Schneider, P.: Astrophys. J. Lett. **358** L33 (1990)
15. Rauch, K. P., Blandford, R. D.: Astrophys. J. Lett. (1991) (in press)
16. Jaroszyński, M., Abramowicz, M. A., Paczyński, B: Acta Astronomica **30** 1 (1980)
17. Papaloizou, J. C. B., Pringle, J. E.: Mon. Not. R. astr. Soc. **208** 721 (1984)
18. Goldreich, P., Goodman, J., Narayan, R.: Mon. Not. R. astr. Soc. **221** 339 (1986)
19. Hawley, J. F.: Astrophys. J. **356** 580 (1990)

20. Hawley, J. F.: Astrophys. J. (1991) (in press)
21. Spruit, H.: In *Theory of Accretion Disks* ed. by P. Meyer, W. Duschl, J. Frank, E. Meyer-Hofmeister (Dordrecht: Kluwer 1989) p.325
22. Shakura, N. I., Sunyaev, R. A.: Astron. Astrophys. **254** 22 (1973)
23. Balbus, S., Hawley, J. F.: Astrophys. J. (1991) (in press)
24. Chandrasekhar, S.: Proc. Nat. Acad. Sci **46** 253 (1960)
25. Eardley, D. M., Lightman, A. P.: Astrophys. J. **200** 187 (1975)
26. Sakimoto, P., Coroniti, F. V.: Astrophys. J. **247** 19 (1981)
27. Heyvaerts, J., Norman, C. A.: Astrophys. J. **347** 1055 (1989)
28. Norman, M. L., Clarke, D. A., Stone, J. M.: Computers in Physics **Mar/Apr** 138 (1991)
29. Scoville, N. J. Sanders, D. B., Sargent, A. I., Soifer, B. T., Tinney, C. G.: Astrophys. J. Lett. **345** L25 (1989)
30. Rees, M. J., Silk, J. I., Werner, M. W., Wickramasinghe, N. C.: Nature **223** 788 (1969)
31. Sanders, D. B., Phinney, E. S., Neugebauer, G., Soifer, B. T. and Matthews, K.: Astrophys. J. **347** 29 (1989)
32. Emmering, R. T., Blandford, R. D., Shlosman, I.: Astrophys. J. (1991) (in press)
33. Galeev, A., Rosner, R., Vaiana, G. S.: Astrophys. J. **229** 318 (1979)
34. Park, S. J., Vishniac, E. T.: Astrophys. J. **332** 135 (1988)
35. Rees, M. J.: Science **247** 817 (1990)
36. Evans, C. R., Kochanek, C. S.: Astrophys. J. Lett. **346** L13 (1989)

1 – High Energy Emitting Regions

First Results from the ROSAT Survey

Wolfgang Brinkmann

Max-Planck-Institut für Extraterrestrische Physik, D-8046 Garching, Germany

1. Introduction

The German X-ray astronomy satellite ROSAT was launched from Cape Canaveral on June 1 1990, 17:48 local time by a Delta-II rocket into a circular near-Earth orbit of 580 km altitude and 53° inclination. The satellite carries the largest and most precise X-ray telescope built so far with a position sensitive proportional counter (PSPC) and an American High Resolution Imager (HRI) as focal plane instruments. A second telescope, the English Wide Field Camera, extends the energy range of the X-ray telescope from the soft X-ray band into the XUV and EUV. On June 16 the X-ray telescope saw first light during a set of pointings at regions in the Large Magellanic Cloud. After a detailed calibration and verification phase the all-sky survey started with both the X-ray telescope and the Wide Field Camera co-aligned on July 30, 1990.

The survey is performed by scanning the sky along great circles in ecliptic longitude at a rate of about 360° in 96 minutes, the orbital period of the satellite. Thus, the sky is imaged in form of stripes 360° long and 2° wide, corresponding to the field of view of the XRT/PSPC. Through the motion of the Earth around the Sun the great circle is advanced by about 1° per day completing the all-sky coverage in half a year.

Both at the north ecliptic pole and the south ecliptic pole the scan circles from every orbit overlap with each other. Thus, the total exposure of a target ranges between $\sim$ 500 secs for objects near the ecliptical equator up to a few 10^4 secs for circumpolar targets. A source is visible for between typically 3 days up to half a year for objects in the innermost 1° of the polar caps.

After completion of the survey, from February 1991 onwards, the pointing observations of the guest observer programme have been carried out. About 30 % of the observation time will be spend on observations of AGN, indicating the enormous interest of the scientific community and ROSAT's potential for the study of these objects.

In the next chapter I will briefly discuss the properties of the scientific instruments of ROSAT. I will then give a 'preview' of results from the survey, related to AGN, going from a discussion of the large class of previously unknown objects - their number, properties, and contribution to the cosmic X-ray background - to a brief review of detailed studies of well known objects. I will restrict the presentation to a more qualitative form as even now, near the end of the AO-1 phase, many of the obtained results are still preliminary, awaiting the final 'fine tuning' of the data reduction programmes and, in many cases, more detailed follow up observations at other wavelengths.

2. ROSAT: the Scientific Payload and the In-Orbit Performance

With a total mass of the satellite of about 2.4 tons, a maximum length of 4.5m and a diameter of 2.4 m ROSAT is the biggest scientific satellite built so far in Western Europe. The X-ray telescope (XRT) working between 0.1 to 2.4 keV, consists of a 4-fold nested Wolter type I mirror assembly of 2.4 m focal length and a maximum aperture of 835 mm. The angular resolution is estimated to be 3.6 arcsec, and the residual microroughness of the mirror surfaces is 2.8 Å (*Aschenbach* [1]).

The focal plane instrumentation, developed and built at the MPE, consist of two redundant Position Sensitive Proportional Counters (PSPCs) which are operated in gas flow mode (*Pfeffermann et al.* [2]). The PSPC provides spatial and spectral resolution over the full field of view of 2° which vary slightly with photon energy E. The angular resolution is limited by the PSPC to about 25 arcsec on-axis and degrades to a few arcminutes for field angles larger than about 20 arcmin due to the geometric aberrations of the telescope. The energy resolution is $\Delta E/E = 0.41/\sqrt{E_{keV}}$. The effective collecting area including the PSPC efficiency is about 220 cm^2 at 1 keV for on-axis rays and decreases slightly with increasing field angle.

The third detector, the High Resolution Imager (HRI) built at the SAO Cambridge, USA, is a microchannel plate detector with an intrinsic spatial resolution of about 20 μm. With the HRI in the focus of the telescope the angular resolution is dominated by the point spread function of the mirror assembly. A value close to 4 arcsec is expected in orbit.

The X-ray telescope operates in photon counting mode, which means that each recorded count is labeled with detector x-/y-position, pulse height, and arrival time. Using the instantaneous viewing direction of the spacecraft obtained from its Attitude Measurement and Control System the position of each recorded X-ray event is projected back onto a frame of sky coordinates.

Although the intrinsic field of view is only 2° in diameter the size of the effective field is virtually unlimited due to the scan motion of the telescope.

The second telescope is the Wide Field Camera (WFC) furnished by the United Kingdom. The telescope is a 3-fold mirror assembly of Wolter type I arrangement with a focal plane turret from which one of two identical microchannel plate detectors can be selected. A set of 8 filters defines the spectral wavelength passbands which cover the range from about 17 to 210 eV. The telescope has a focal length of 525 mm, a maximum mirror diameter of 576 mm and a field of view of 5° diameter. A more detailed description of the WFC and an account of the first successful observations have been presented by ***Pounds and Wells*** [3].

A comprehensive and more detailed description of ROSAT and the instruments can be found, for instance, in the papers by ***Trümper*** [4].

One major factor, which determines the sensitivity of astronomical instruments is the amount of unwanted background. Fig. 5 of ***Aschenbach*** [5] shows the pulse height spectrum of the charged particle background and the diffuse X-ray background measured with the ROSAT PSPC in orbit during a pointing at β Leonis. The spectrum of the particle background is essentially flat, and the total amount of accepted background events amounts to 3×10^{-5} counts s^{-1} $arcmin^{-2}$. This very low background means that pointed PSPC observations with a typical resolution element of 25 arcsec diameter are not background limited for exposures less than 10^5 sec. For the survey the resolution element is about 2 arcmin in diameter, and PSPC observations become particle background limited only for times exceeding 10^4 sec. Therefore the majority of the ROSAT PSPC observations both in the all-sky survey and the pointing programme are essentially free of particle background contamination, and the majority of the recorded events are true X-ray photons.

Therefore the mapping of the X-ray background by ROSAT is essentially limited by the brightness of the X-ray background emission itself. It is estimated that the morphology of the diffuse emission can be mapped at an angular resolution of about 6 arcmin over the whole sky.

3. The AGN Content of the Survey

Pre-launch studies of number - count relations predicted that about 40 % of all ROSAT sources would be AGN, i.e., about 30 000 to 40 000 objects would be newly detected in X-rays. The identification and subsequent classification of these sources clearly represents a task of really 'astronomical' magnitude.

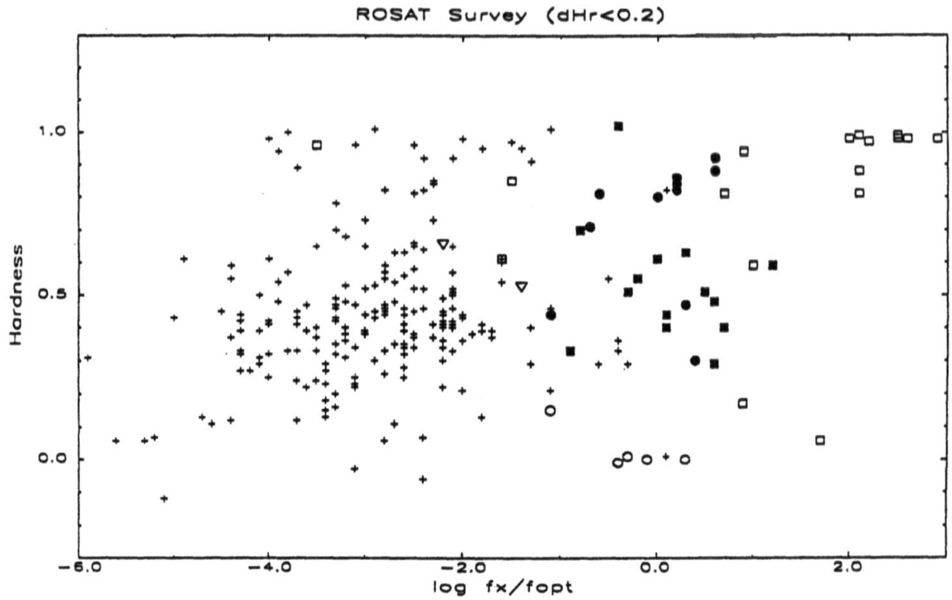

Fig. 1. Hardness ratios as function of X-ray/optical flux: stars (+), white dwarves (open dots), clusters (full dots), LMXB (squares), AGN (blacksquares). (courtesy G. Hasinger)

3.1 The ROSAT properties of AGN

Using the first ROSAT source positions *Hasinger* [6] made a comparative study of these objects with the SIMBAD catalogue. From a total of 1892 ROSAT sources 869 could be identified, the large majority of them (646) with stars, only 32 with AGN.

Plotting the ROSAT count rates of all objects versus their hardness ratios ((400 eV - 2 keV) X-ray flux / total X-ray flux) different classes of objects form clearly distinct branches in the diagram with the accreting neutron stars and clusters of galaxies showing the largest hardness ratio (~ 1) and the white dwarves the smallest (~ 0).

While this purely on X-ray data based classification scheme works in principle it is not a practical way to single out the different object classes: the errors in the hardness ratio determination are small enough only for the small fraction of strong sources, the majority of objects scatter over the plot and cannot be classified unambiguously.

Plotting the hardness ratio against the ratio (X-ray flux / optical flux) results again in a separation of different object classes (see Fig. 1). However, these classes are not clearly distinguished and the available optical information can be insufficient (no optical counterpart or more than one optical objects in the X-ray error box).

A third possibility, plotting the logarithmic ratio of X-ray flux to optical flux against the visual magnitude of the optical counterpart shows a very distinct class separation, very similar to that found from the Einstein EMSS (Stocke et al., 1983 [7]). Clearly, the optically brighter objects are mainly stars whereas, at magnitudes m_v > (17 - 18), the majority of the sources seem to be AGN. This impression is substantiated by the results of the first deep pointed observations.

3.2 First deep surveys

Hasinger et al. [9] analysed several fields around sources observed extensively in the ROSAT calibration and verification phase and at the NEP. All selected fields have limiting sensitivities fainter (up to 3 times) than the Einstein Deep Surveys.

The accumulated flux of the isolated point-like sources makes up a fraction of about 40% of the total flux contained in the deepest exposure image. Further, the cumulative number-count relation shows a significant flattening for X-ray fluxes less than about 3×10^{-14} erg cm^{-2} s^{-1}, which could mean that the number density of sources starts to decrease with increasing distance or that the X-ray luminosity shows a decrease because of evolutionary effects. Cosmological space-time effects have been discussed as well.

The photon spectrum of the resolved sources in the 1 - 2 keV range is parallel to that of the residual background. Both show a power law slope of -2.3, suggesting that the steepening of the X-ray background in this energy range [10] is due to extragalactic sources. The excess diffuse galactic emission starts to dominate the XRB only below 1 keV.

Similar conclusions were reached by Shanks et al. [8]. During the PV phase a 30 ksec exposure of a faint QSO survey field yielded 39 sources in the innermost 40' diameter of the PSPC detector with X-ray fluxes down to $> 2\ 10^{-14} erg cm^{-2} s^{-1}$. This corresponds to a surface density of X-ray sources of $\sim$110 deg^{-2}. From the 30 sources identified in optical follow up observations 24 turned out to be quasars, giving at X-rays a sky density of $\sim$70 QSO / deg^2. This further implies, that at 1 keV 30 - 50 % of the extragalactic X-ray background originates from QSO's.

Finally, the first results from a programme attempting to fully identify all ROSAT sources in certain pre-selected regions of the sky (so called *Study Areas*) confirm the large AGN content of the ROSAT All Sky Survey sources. From the first 126 X-ray sources, observed at La Silla, firm spectroscopic identifications could be made for 30 AGN, 26 stars, 8 clusters of galaxies, 4 normal galaxies, and 6 galactic objects. Most of the 52 as yet unidentified objects are expected to be AGN as well due to the increasing fraction of AGN at fainter optical magnitudes.

These first studies suggest that the AGN content of the ROSAT All Sky Survey is at least as high as the predicted fraction of 40%. However, the enormous amount of data will require quite a time to be analyzed and checked and follow up observation at other wavelengths will very likely take years before a substantial fraction of these sources will be firmly identified.

4. Exploring "the known"

So far, only about 200 of the more than 5000 catalogued AGN are known to emit at X-ray and other wavelengths. A correlation of the ROSAT source list with existing optical and radio catalogues is expected to increase this number considerably.

4.1 The "Master"-catalogue

During the pre-launch phase some efforts were undertaken to collect all available informations on AGN in a "master-catalogue". It is based on the Veron-Cetty - Veron catalogue of quasars [11], supplemented with various other compilations like the Hewitt-Burbidge QSO catalogue, the Kühr catalogue of radio sources, the EMSS, etc., and additional data from current publications.

The source positions from this "master-catalogue" are routinely compared with point sources, found by the Standard Analysis System (SASS, [12]). From the first approximately 22% of the processed ROSAT data one can make the following predictions:

Quasars: We will find $\approx$ 60% of the previously known X-ray QSO. However, $\approx$ 68% of the "ROSAT - QSO" were previously **not** known as X-ray emitters.
We expect a total number of $\approx$ 800 of the catalogued quasars (i.e., about 16% of the sources) to be found in X-rays. This is a factor of $\approx$ 2 improvement in the number of known X-ray quasars.
Unfortunately, about 2/3 of these objects will have total count rates of less than 50 counts, too weak to allow any spectral analysis.

Seyfert 1: The "re-discovery rate" of known X-ray emitting Seyfert 1 galaxies is very high, about 86%. Further, about 58% of the Seyfert 1's in the ROSAT sample were previously not known to emit X-rays. This means, we expect to see finally $\approx$ 92% of all catalogued S 1 galaxies, again a factor of 2 improvement in the number of X-ray emitting objects.
In this sample only 1/3 of the objects have total count rates below 50 counts. For the majority of the Seyfert 1's we will be able to make rough estimates about an underlying power law spectrum or even perform proper spectral fits.

Seyfert 2: Due to their intrinsically low luminosity only very few Seyfert 2 galaxies will be seen in the survey.

BL Lac objects: Less than about 45% of the known BL Lacs will have strong enough emission to be detected.

The spectral properties of many of these sources, their temporal behaviour, and correlations with existing data in other wavelength bands will be subject of various studies in the future (see also chapter 5). However, it is interesting to have a first look on the spectral properties of these objects as the PSPC opened a previously unaccessible energy band.

As mentioned above, the majority of the sources is too weak to allow proper spectral fits, in general only hardness ratios can be given. From these one can, however, determine the slope of an assumed power law and the value of the interstellar absorption. Likewise, one can perform three band fits with assumed galactic absorption.

Figure 2 shows the photon index α for the first ≈ 1000 survey AGN as function of the total count rate. The majority of sources has only few counts in total and the error in the determined power law index is typically of the order of ± 1.0. The 'average' power law index has a value of 2.18 ± 0.95, similar to that found by *Hasinger et al.* [9] in their analysis of the deep field observations.

No obvious differences in the power law index for the different subclasses of AGN could be found yet (Schartl, 1991 [13]). Power laws from spectral fits to the strongest ROSAT sources often show even steeper slopes but it is presently not clear whether selection effects play a role and how the spectral fits are affected by the parameter settings of the SASS.

4.2 Einstein MSS catalogue

An interesting subgroup of the "master catalogue" objects form the 427 AGN from the Einstein Medium Sensitivity Survey ([15], [16]). As Einstein's energy band was rather similar to that of ROSAT, these sources can serve as an important tool for inter-calibrations and comparison of their soft X-ray properties.

In the survey we seem to detect only about 45 % of the EMSS AGN; the reasons for the non-detection of the other objects are not quite clear yet.

To determine the X-ray flux of the EMSS AGN Maccacaro et al. [16] used an energy index of $\alpha = 1.0$. In Fig. 3 we plot the EMSS fluxes versus the ROSAT fluxes using the same index and energy band for the ROSAT detected EMSS sources. It can be seen that there is a systematic offset from the straight line indicating that, in general, the spectrum of the objects in the softer ROSAT band is steeper than previously anticipated.

Fortunately, many of the EMSS objects are strong enough to allow a detailed spectral comparison of the X-ray properties of these sources.

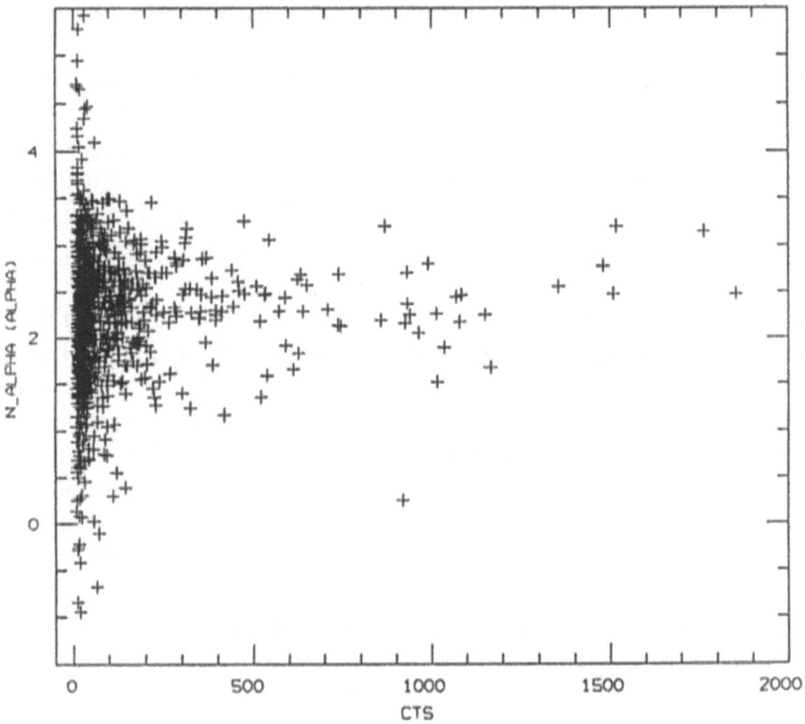

Fig. 2. Photon power law index for the first ≈ 1000 survey AGN, determined from their hardness ratios (Courtesy of N. Schartl)

4.3 The Condon-Broderick-Seielstad 6cm survey

Many AGN show distinctive properties in the radio band, as for example BL Lacs or radio loud quasars. Therefore, a programme was initiated to correlate a point source catalogue generated from the Condon-Broderick-Seielstad 6cm survey of the northern hemisphere [14] with the ROSAT survey. This point source catalogue comprises a total of > 140000 sources with 6cm - fluxes > 20 mJy. For objects found in both energy bands, follow up observations at the 100 m Effelsberg radio telescope were organized, giving radio fluxes at 11 cm, 6 cm, and 2.8 cm. The better radio positions are then used in optical identification programmes to determine the nature of the objects.

More than 1500 sources are expected to be found in the survey data. Figure 4 shows a plot of the 6 cm radio flux versus the ROSAT countrate for the first 501 objects found. 136 of these objects were previously "known", very often only as an entry in another radio survey, i.e., without any optical identification. Those identified are mainly classified as QSO, radio galaxies, BL Lacs, and clusters of galaxies. Only less than a handful of the sources were known previously as X-ray sources.

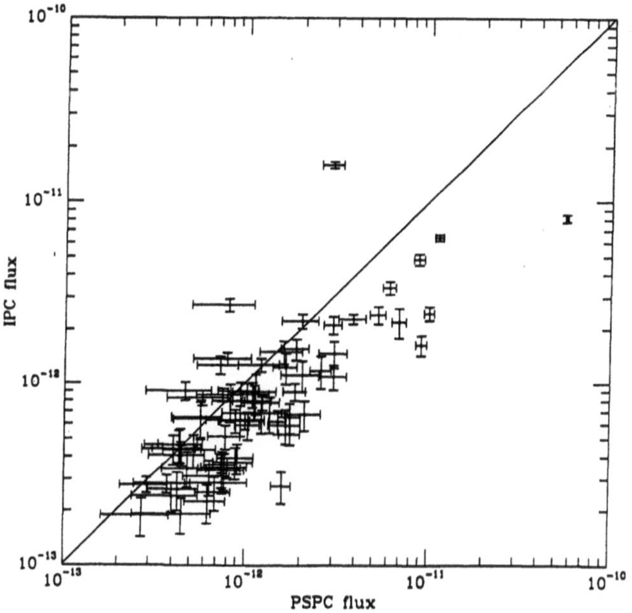

Fig. 3. EMSS flux (from [16]) versus ROSAT flux, calculated for an energy index of $\alpha = 1.0$ for the energy band 0.3 - 3.5 keV

Most of the unknown objects populate the lower, left domain of the diagram, "known" sources are generally either X-ray or radio brighter. The hardness ratios of these two subgroups are slightly different and both are different again from the typical hardness ratios found for the AGN of the "master catalogue". There seems to be a tendency for these radio selected X-ray sources to show harder spectra than the "average" AGN.

However, much more detailed studies invoking the whole data set will be necessary to confirm these first trends and to assess their physical significance.

So far, more about 250 candidates have been re-observed at the Effelsberg telescope and more than 90% of them were positively detected. 44% of them showed a flat or even inverted radio spectrum, 25% had a flat radio spectrum and were polarized, and about 10% had a flat radio spectrum and were extended. So far only about 10 of the candidates could be identified optically.

A detailed discussion of the properties of these sources will be given elsewhere.

4.4 IRAS Point Source Catalogue

Another extended extragalactic database is given by the IRAS Point Source Catalogue. A sample of more than 14500 extragalactic objects was extracted from it employing multivariate statistics [17] and correlated with

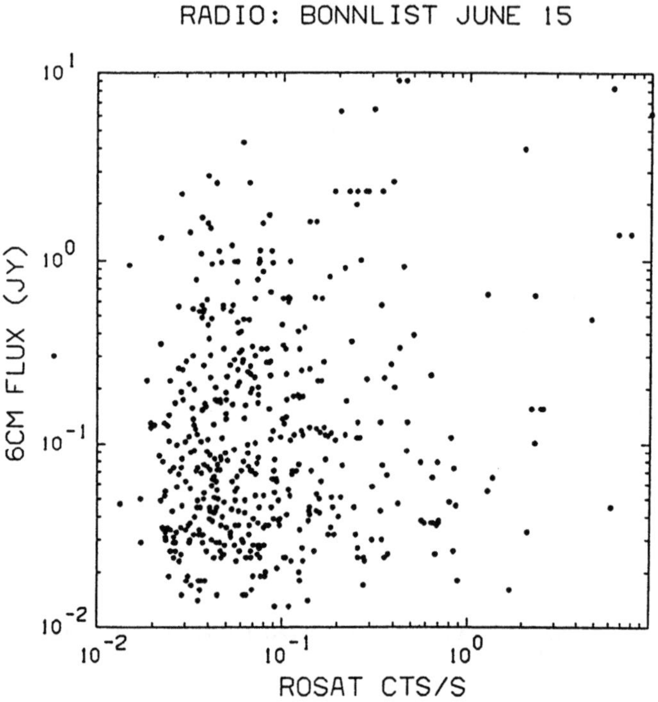

Fig. 4. 6 cm radio flux versus X-ray count rate for the first 501 ROSAT / Condon objects

the ROSAT survey data. This represents the largest homogeneous sample of infrared selected galaxies studied at X-rays so far.

Only a small number of IRAS sources were detected in X-rays (~ 300), $\approx 98\%$ of the objects are emitting below the ROSAT detection limit. There seems to be a grouping of the sources in two populations, spirals and Seyfert 1s and, although the Sy 1 represent only a small fraction of the original sample, about 50% of the X-ray sources are from this class. For more details see the contribution by *Boller et al.* (1991) in these proceedings.

5. Detailed studies of known AGN

Starting already in the pre - launch phase of ROSAT more than 20 international collaborations were set up to study AGN from the survey. Targets were single objects, samples (classes) of objects, certain sky fields (ESO key project, VLA NEP observation, etc.), and simultaneous observations with IUE and Ginga as well as with ground based instruments.

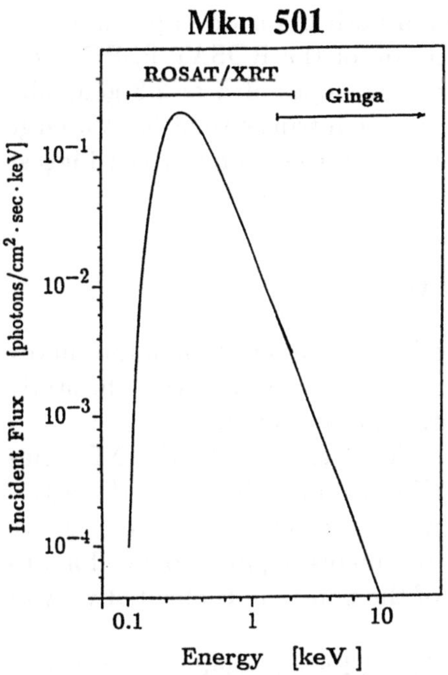

Fig. 5. Combined ROSAT/Ginga power law fits for Mkn 501 (courtesy H.H. Fink, these proceedings)

Furthermore, about 60 MPE internal projects were defined to maximize the scientific output from the survey data.

The primary objectives of these projects are the study of spectral and flux variability for objects near the ecliptical pole (with exposures for up to half a year in the survey), simultaneous multi-wavelength observations from the radio to X-ray range and the investigation of the X-ray properties of well selected samples.

Most of the work in these projects is still in progress; I will only mention two examples presented here in meeting in more detail.

5.1 The ROSAT - IUE/Ginga programmes

In an attempt to study various X-ray sources simultaneously over a wide energy band with ROSAT and Ginga (and, in most of the cases with IUE as well) 24 AGN were selected with long ROSAT exposures and sufficiently high X-ray fluxes.

First results of this programme have been presented elsewhere ([18] and Fink (1991), these proceedings).

Figure 5 shows the combined ROSAT/Ginga spectra for Mkn 501. Both fits are in excellent agreement with each other, the slight spectral break between them at $\sim$ 2 keV is hardly visible.

The close similarity of the parameters in both fits is an important confirmation of the correctness of the calibration of the ROSAT PSPC. This means in particular, that the steep power law slopes of $\alpha \approx 2.3$ generally found for ensembles of extragalactic sources are intrinsic to at least a large class of the objects in the soft X-ray band - perhaps one of the biggest surprises of the ROSAT survey.

5.2 Temporal behaviour of NEP sources

As mentioned above, sources close to the ecliptic poles are seen in the survey up to half a year, 16 times per day for about 30 secs each allowing the study of temporal variations on time scales from $\sim$ hours to months.

A group of 16 objects, with distance closer than 2° to the NEP has been observed simultaneously with ROSAT from the radio to the UV wavelengths. The analysis of spectral and intensity variations is still in progress. Especially the exact calibration of the X-ray fluxes requires some effort, as the sources move through different parts of the non-uniform field of view of the PSPC detector.

First results for Kaz 102 and KUV 18217+64 are presented by *Schaeidt et al.* (1991) in these proceeding.

6. Concluding Remarks

This first review can only give a very preliminary account of the enormous data base available for the study of AGN after the ROSAT All Sky Survey. Most of the data analysis has only just started and will require some time before definite results are available. More observations, in the ROSAT pointed phase, during forthcoming missions, and with ground-based instruments will be necessary to explore the full scientific contents of these data.
Acknowledgement. The ROSAT project is supported by the Bundesministerium für Forschung und Technologie (BMFT). I like to thank my colleagues at the Max-Planck-Institut für Extraterrestrische Physik, B. Aschenbach, T. Boller, H. Fink, G. Hasinger, S. Molendi, N. Schartl, and S. Schaeidt for fruitful discussions and providing parts of the presented material.

References

1. Aschenbach, B., 1988: *Appl. Optics* **27**, 1404.
2. Pfeffermann, E., Briel, U.G., Hippmann, H., Kettenring, G., Metzner, G., Predehl, P., Reger, G., Stephan, K.H., Zombeck, M.V., Chappell, J., Murray, S.S., 1986: *Proc SPIE* **733**, 519.

3. Pounds, K.A., Wells, A.A., 1991: COSPAR Proc. *Adv. Space Research*, The Hague
4. Trümper, J., 1983: *Adv. Space Res.* **2**, No. 4, 241;
Trümper, J., 1984: *Phys. Scripta* **T7**, 209;
Trümper, J., 1990: *Phys. Bl.* **46**, Nr. 5, 137.
5. Aschenbach, B., 1991: *Rev. Mod. Astron.* **4**, 173.
6. Hasinger, G., 1991: private communication.
7. Stocke, J., et al., 1983: *Astrophys. J.*, **273**, 458.
8. Shanks, T., et al., 1991: *Nature*, in press.
9. Hasinger, G., Schmidt, M., and Trümper, J., 1991: *Astron. Astrophys. (Letters)*, **246**, L2.
10. McCammon, D., and Sanders, W.T., 1990: *Ann. Rev. Astron. Astrophys.*, **28**, 657.
11. Veron-Cetty, M.-P., and Veron, P., 1989: A Catalogue of Quasars and Active Nuclei (4th Edition), *ESO Scientific Report* No. 7.
12. Voges, W.: The Standard Analysis Software System for ROSAT, in preparation.
13. Schartl, N., 1991: private communication.
14. Condon, J.J., Broderick, J.J., and Seielstad, G.A., 1989: *Astron. J.*, **97**, 1064.
15. Gioia, I.M., et al., 1990: *Astrophys. J. Suppl.*, **72**, 567.
16. Maccacaro, T., et al., 1991: *Astrophys. J.*, submitted.
17. Boller, T., at al., 1991: *Astron. Astrophys.*, submitted.
18. Makino, F., 1991: paper presented at the *28th Yamada conference "Frontiers of X-ray Astronomy"*, Nagoya.

Discussion

Pustil'nik: In connection with the result which you formulated on bending of quasar X-ray counts at faint fluxes do you see any correlation or may be a trend of X-ray luminosity versus the quasars redshift?

Brinkmann: No trend was seen in the QSO found in the survey. The data of the deep surveys have not yet been analysed to that extend.

Bregman: Could you comment on whether clustering occurs commonly in the ROSAT observations?

Brinkmann: So far it has been seen only in the NEP deep field observations - further deep observations are (or will be) performed in the pointed phase.

Wandel: The ROSAT source-counts suggest that the angular density of the faint sources increases slower than extrapolated from brighter sources (e.g. Maccacaro). Does this present a problem to the hypothesis that the CXRB is contributed by unresolved sources (e.g. AGN) which have the same luminosity function as the presently resolved?

Brinkmann: Theoretical models by Schmidt (1989) predict a flattening of the number-count relation due to cosmological and evolutionary effects in quasars.

Evidence for X-ray Re-processing in Seyfert Galaxies

K.A.Pounds, K.Nandra, and G.C.Stewart

Department of Physics and Astronomy, University of Leicester, UK

Abstract. Our appreciation of the complexity - and the diagnostic potential of the X-ray spectra of active galaxies has been transformed over the past three years. Ginga LAC observations of a large sample of X-ray bright active galaxies, mainly Seyferts, have revealed several new features in the continuum spectrum between ~2 – 30 keV. In particular, an emission line (or line complex) near 6.4 keV and a spectral 'hump' above ~10 keV are common properties of the sample, while a number of sources also show evidence for a 'warm' iron K-absorption edge. The most likely interpretation of the emission line and hump is by re-processing of the primary X-ray power law in near-neutral (or cold) matter within the active nucleus and probably associated with the primary accretion process. The K-absorption edge indicates a substantial column density of highly ionized matter in the line- of-sight to several AGN, including some with otherwise 'bare' Seyfert 1 nuclei. Within the past year further Ginga observations of two 'typical' X-ray-bright Seyferts, MCG–6–30–15 and NGC 5548, have allowed the variability of the above spectral features to be studied, providing the first direct measures of the geometry and scale of the X-ray re-processing matter in these objects.

1. Introduction

Prior to the launch of Ginga, observations from a number of satellites, in particular HEAO-1 and EXOSAT, had shown the hard (2 - 20 keV) X-ray spectra of AGN to be well described by a power-law of photon number index close to 1.7, with negligible low energy absorption, indicating (at least for the largest sub-set, of Seyfert 1 galaxies) a negligible intrinsic column density in line-of-sight to the X-ray nucleus (Turner & Pounds, 1989, and references therein). EXOSAT and Einstein Observatory data had, however, shown a spectral up-turn (or 'soft excess') to be a common feature in many bright AGN, below ~1 keV, widely interpreted as thermal emission from relatively cold ($\sim 10^5$K), dense gas accreting onto the hypothesised central black hole (Arnaud et al., 1985; Pounds et al., 1986; Czerny & Elvis, 1987). Also, the discovery in extended EXOSAT observations of remarkable, large amplitude, rapid and persistent variability in the X-ray emission of several Seyferts lent strong support to the view that both hard and soft X-ray components arise from deep within the active nucleus (McHardy, 1988; Pounds & McHardy, 1988). Nevertheless, prior to Ginga, the ubiquitous (~2 - 20 keV) power law continuum displayed by many AGN offered little prospect for probing this crucial central region by means of X-ray spectroscopy.

2. Early Detections of the Fe K-line in AGN

A strong Fe K-line (E.W. ~200 eV) has long been recognised in the X-ray spectra of NGC 4151 and Cen A, without creating particular interest, since in both cases the large line-of-sight column density ($N_H \sim 10^{23} cm^{-2}$) suggested a straightforward interpretation in terms of fluorescence in a spherical shell of similar density surrounding the X-ray nucleus (Makishima, 1986, and references therein). Fe K-lines at low significance had also been reported at various times, from NGC 5506, MCG–5–23–16 and NGC 2992 (Mushotzky, 1982) and Mkn 509 (Morini et al., 1987).

Heightened interest in searching for Fe K–features arose from theoretical work suggesting matter associated with the primary accretion process may indeed be sufficiently dense to survive in the close proximity of the central hard photon source (Guilbert & Rees, 1988; Lightman & White, 1988). A resultant search of EXOSAT data revealed evidence for Fe K-lines, first from an extended observation of the Seyfert 1 galaxy MCG–6–30–15 (Nandra et al., 1989), and then for several other Seyferts in the EXOSAT survey (Leighly et al., 1989).

About the same time, early Ginga LAC observations were yielding data of sufficient sensitivity to show, quite clearly, both Fe K–emission and K– absorption edge features in three Seyferts, NGC 5548, NGC 3227 and NGC 5506 (Pounds et al., 1989). Furthermore, these new observations provided intriguing detail, in particular a possible blue wing to the Fe line in NGC 3227 and a higher energy K-edge (due to ionized matter?) from NGC 5506. Further Ginga data followed quickly, with reports of similar spectral features for the Seyfert galaxies NGC 7469, IC 4329A (Piro et al., 1990) and MCG–6–30–15 (Nandra et al., 1990, Matsuoka et al., 1990).

3. The Ginga_12 spectrum and its Interpretation

In order to better quantify what appeared to be common features of Seyfert galaxies, a composite spectrum was assembled from 12 Ginga observations of 8 Seyferts (Pounds et al., 1990). The sample was primarily of Seyfert 1 type and was selected from those having small or negligible line-of-sight absorbing column and 'canonical' power law photon number indices, of ~1.7.

Figure 1(a) shows a simple power law fit to the Ginga_12 data, yielding a best fitting photon index of 1.7, close to the weighted mean of the individual spectral slopes. The data-minus-model residuals show very clearly the emission feature at ~6 - 7 keV, (absorption) dip between ~7 - 10 keV and an excess above ~ 12 keV. The addition of an Fe K-line as a free parameter greatly improved the fit, with a mean rest-frame line energy of 6.28 (with 90% confidence limits of 6.07 – 6.49) keV and equivalent width of 110 (75 – 150) eV.

Given that these features (from a sample of Seyferts with very little evidence for cold matter in the line-of-sight) were nevertheless highly suggestive of substantial re-processing of hard X-rays by cold material, the Ginga_12 data were next tested (in Pounds et al., 1990) against a model spectrum, obtained from a Monte Carlo simulation of a power law X-ray continuum irradiating an optically thick, semi-infinite slab of cold, solar abundance matter. A line-of-sight 'warm' absorber was also included in the spectral fit, which was found to be in excellent agreement with the Ginga_12 data (figure 1(b)). The derived parameters of the fit showed the slab to be observed near 'face-on' and to fill 2π steradians (viewed from the X-ray continuum source). The best fit also included a 'warm' Fe K-edge near 8.3 keV (corresponding to Fe XXI), of equivalent (solar abundance) column density $9 \times 10^{22}\,\mathrm{H\,cm^{-2}}$. A further result of this 'reflection model' fit was that the intrinsic power law slope steepened to $\Gamma \sim 1.9$, with the interesting consequence of re-establishing the possible importance of electron-position pair processes in AGN (Zdziarski et al, 1990).

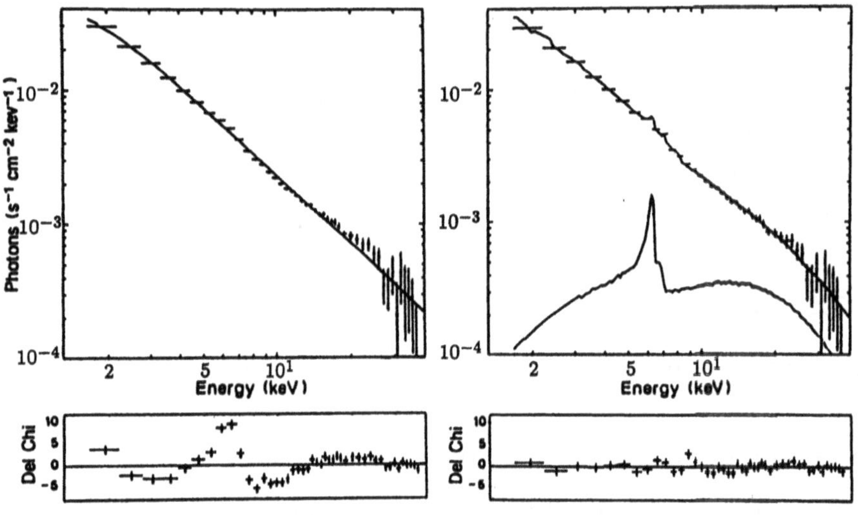

Fig. 1. Model fits to the GINGA_12 spectrum.(a) Simple power law. (b) Power law plus reflection.

In the Ginga_12 analysis, it was also noted that both the reflected spectrum and the warm absorber can give rise to spectral variability. An extensive reflected component is likely to smear out and delay rapid changes in the direct flux, such that the amplitude of variability will be less for the iron line and above ~10 keV, where the continuum reflection becomes significant. Variation in the ionization state of any line-of-sight gas will, of course, affect its X-ray opacity, potentially mimicking a positive Γ v.F_{2-10} correlation.

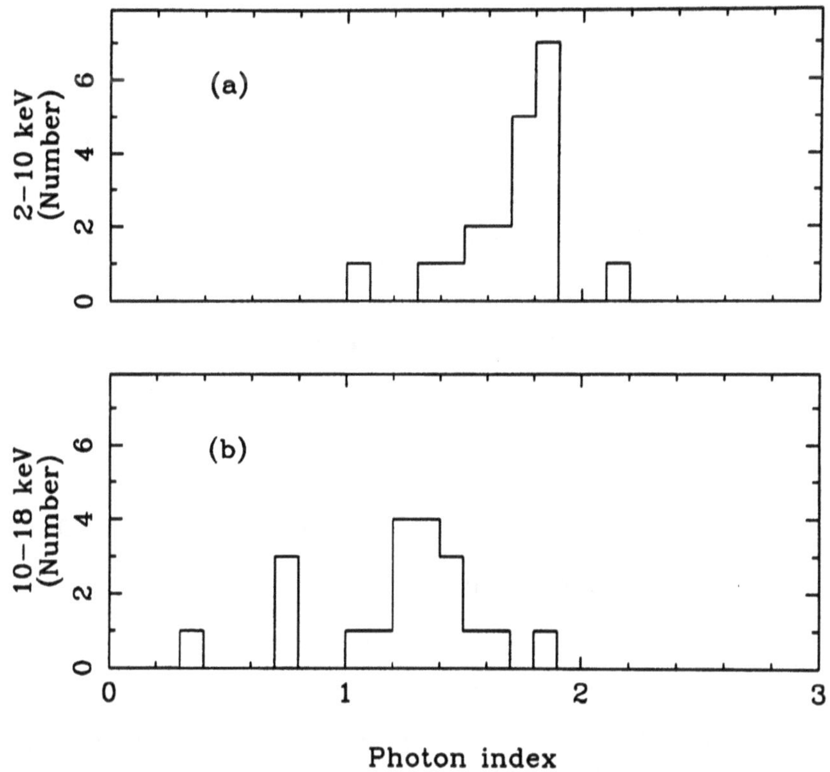

Fig. 2. Histogram of power law indices from the GINGA Seyferts at (a) 2-10keV, (b) 10-18 keV.

4. A Larger Sample of Ginga Spectra of Seyfert Galaxies

In order to assess how commonly and over what range these spectral features occur, a study has now been made using data from the Ginga archive, of 29 observations of 20 individual Seyferts (Nandra, 1991). The sample includes a range of Seyfert classifications; however, the majority of the sources included are again Seyfert 1 galaxies and are generally bright ($>$ 1mCrab), hard-X-ray-selected sources. Most are low luminosity AGN ($L_{2-10} < 10^{45} \mathrm{erg\,s^{-1}}$) and unobscured.

2-10 keV index A histogram of the 2-10 keV spectral indices of the larger Ginga sample is shown in Figure 2(a). The most common value of Γ remains ~1.7, but with individual departures from this. The weighted mean index $\Gamma_{2-10} = 1.722 \pm 0.004$, is consistent with both the EXOSAT (Turner & Pounds, 1989) and HEAO-1 (Mushotzky, 1984) medium energy (2-10 keV) spectral surveys. The standard deviation of this distribution (which measures both the statistical spread and any intrinsic deviation) is 0.25.

10-18 keV: The 'hard tail' Figure 2(b) shows the spectral indices derived from a higher energy band, 10-18 keV. Absorption, negligible in this energy

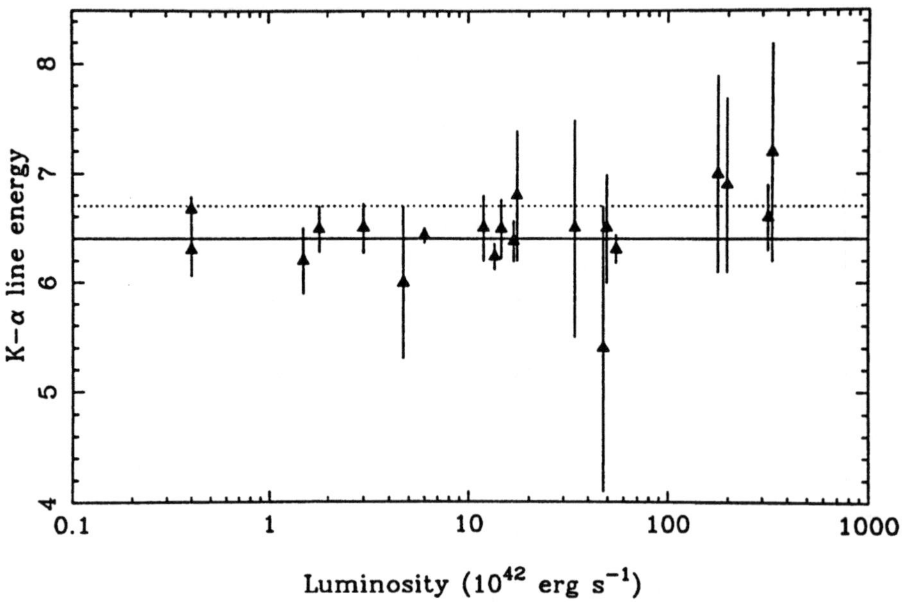

Fig. 3. Fe K-line energies for the larger sample of GINGA Seyferts.

range, was excluded from these fits, which gave a flatter slope, on average $\Gamma_{10-18} = 1.37 \pm 0.03(\sigma = 0.38)$. A KS-test implies that the two distributions are significantly different at > 99 per cent confidence. The flattening at higher energies indicates an additional spectral component, consistent with Compton scattering in cold material.

Although a useful parameterization, most simple power fits gave unacceptable χ^2, with the most prominent deviation from the simple model appearing as an excess close to 6.5 keV. The addition of a narrow gaussian emission line gave substantially improved fits to the data. The F-test supported 26 detections at > 90 percent confidence (most at > 99 percent) in 29 separate spectra and in 18 out of the 20 individual sources, implying that iron K–emission is a very common property of Seyfert galaxies.

Line Energy The average (redshift-corrected) line energy of the sample is 6.43 ± 0.07 keV, with standard deviation $\sigma = 0.38$. Figure 3 shows a plot of the line energy against source luminosity (used, in effect, as a dummy parameter). A test against a constant line energy hypothesis gives $\chi^2_\nu \sim 0.6$, indicating that all of the emission lines are consistent with the mean energy and with fluorescence from cold iron. Thermal line emission from hot plasma, or fluorescence in highly ionized iron, would occur at a higher energy, 6.7–6.9 keV, and is excluded as a unique value by these data ($\chi^2_\nu = 2.7$ for a 6.7 keV line). The apparent trend to a higher line energy suggested by the 4 highest luminosity objects in our sample is intriguing, but not (in these data) statistically significant.

Equivalent width The average narrow-line equivalent width of the sample is 140 ± 10 eV, with a standard deviation, $\sigma = 80$ eV. This substantial mean equivalent width implies columns of $> 10^{23}$ H cm^{-2} for spherical covering, whereas most of the sources in the sample are relatively unobscured (as measured by the small or negligible low energy cut-offs).

Line broadening

Six of the 29 spectral fits showed a significant improvement (at > 95% confidence) when the line width was left as a free parameter, with typical r.m.s. widths of order ~0.7 keV. None of the observations exclude a line width of this order, even in the best determined cases, due to the modest resolution of the LAC detector. Such broadening could arise from Doppler motions, gravity, electron scattering or ionization blends. However, these could also shift the peak energy and such effects must be reconciled with the measured mean energy in any detailed assessment. The equivalent widths implied in these broad-line fits are systematically 50–100% larger than when a narrow line is assumed.

Fe K-absorption Absorption edges at an energy higher than the 7.1 keV for neutral iron have been reported in a few individual spectra (Nandra et al., 1990; Pounds et al., 1990). To test for the presence of such a spectral feature, power law plus absorption edge fits were carried out for the full sample. These gave a significant improvement over the simple power law in 20 out of the 29 spectra, with a mean edge energy ~8 keV. With a narrow emission line (fixed at 6.4 keV) included, 15 of 29 spectra gave a significant improvement over the power law plus-line model when the edge was introduced. Twelve of 20 objects show this evidence for an absorption edge, with typical column densities $N_H > 10^{23}$cm^{-2} (equivalent Hydrogen density). Note that these values are much larger than those obtainable via X-ray reflection (George & Fabian, 1991) and favour a substantial line-of-sight column. The average edge energy is 8.0 ± 0.3 keV implying an ionization state ~ FeXX and temperature ~ 10^7K if collisionally ionized. We conclude that a substantial line-of -sight column of warm material is another common property of Seyfert spectra. (Note, however, the 'cold' Fe K-line energy limits the solid angle subtended by such warm matter at the hard continuum source).

5. Spectral Variability in Seyferts

Rapid variability in both the hard (~2 - 10 keV)and soft (~0.1 - 1 keV) X- ray fluxes of several Seyfert galaxies was established by EXOSAT (Pounds & McHardy, 1988). The variability timescales were found to be generally correlated with luminosity, ranging from minutes (or less) for low luminosity Seyferts such as NGC 4051 and NGC 6814 ($L_{2-10} \sim 10^{42} - 10^{43}$erg s^{-1}) to a few hours for higher luminosity objects, such as NGC 5548 (L_{2-10}

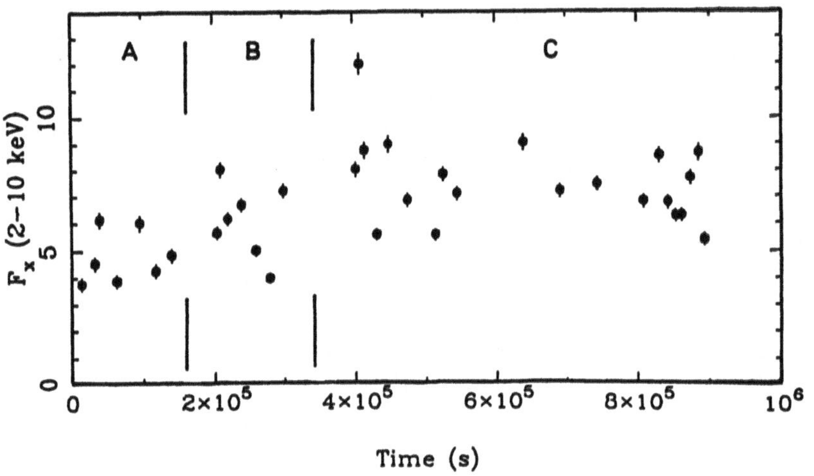

Fig. 4. GINGA X-ray light curve of MCG–6–30–15 from (A) October 1987 (B) June 1989 (C) February 1990.

$\sim 10^{44}$ erg s^{-1}). The Seyfert galaxy MCG–6–30–15 varies on an intermediate timescale, of tens of minutes, and has $L_{2-10} \sim 3 \times 10^{43}$ erg s^{-1}.

Each of the above-named Seyferts has now been observed repeatedly with Ginga, offering the opportunity to study time variability of the complex spectral features described earlier. NGC 6814 has been reported elsewhere (Kunieda et al., 1990)and appears to be atypical of the Seyfert sample as a whole, in showing a flat ($\Gamma \sim 1.3$) and apparently constant power law and an Fe K-line of unusually large equivalent width (~350 eV) which is rapidly variable. MCG–6–30–15 and NGC 5548, which have 'canonical' power law spectra and Fe-line strengths, may be more typical of the sample as a whole. These data are reviewed below.

5.1 Spectral variability of MCG-6-30-15

MCG–6–30–15 was observed by Ginga, in October 1987, June 1989 and February 1990. Figure 4 shows the 2-10 keV light curve in bins of roughly constant flux (typically 10^4 s, sufficient to give adequate statistical quality for a good spectral fit for a source of this strength). Variations over a range of ~3 are seen, with ~50% changes between adjacent bins. Simple power law-plus-absorption fits are poor, with residuals at 6–7 keV and > 10 keV, as found for the main Seyfert galaxy sample. The main interest here, however, is in spectral variability.

Figure 5(a) shows the power law slope against flux, confirming the previously established relationship (Nandra et al., 1990), in which the power law steepens with increasing flux, initially rapidly and then more slowly. As noted earlier, this behaviour is consistent with the reflection model, where the observed slope change is attributed to reduced amplitude changes in the

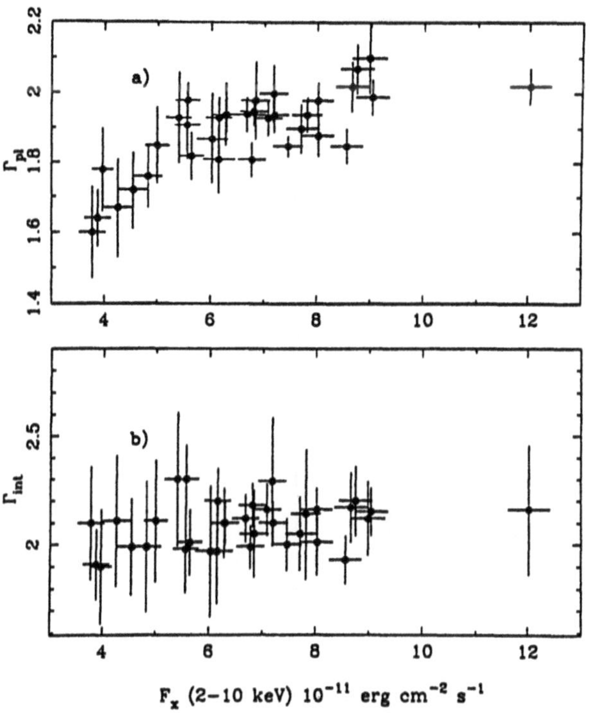

Fig. 5. Continuum slope v. flux of MCG–6–30–15 for (a) single power law (b) reflection model.

re-processed (flatter) continuum component as the intrinsic, or direct, continuum varies. Little or no change is then found to be required in the intrinsic spectral index ($\Gamma_{\rm int} \sim 2.1$) (Figure 5(b). This suggests the re-processing region introduces time lags $> 10^4$ s. The small scatter of individual data points about the mean Γ - flux curve (cf. NGC 5548 in next section) indicates little *absolute* change in the reflected component during each 2-3 day observation and suggests that the re-processing region is $>10^5$ light seconds in extent. Support for this conclusion is provided by the lack of significant variability in of the Fe-line flux.

5.2 Spectral variability of NGC 5548

NGC 5548 is now the most extensively observed Seyfert galaxy in the X-ray band, other than NGC 4151. A series of 12 EXOSAT observations monitored the variability of the soft excess, often dominant below ~1 keV (Branduardi-Raymont, 1988), while a number of Ginga observations have been inspired by the extensive optical - UV monitoring campaigns in 1988/90. A detailed analysis of the EXOSAT data and the 1988/9 Ginga observations has re-

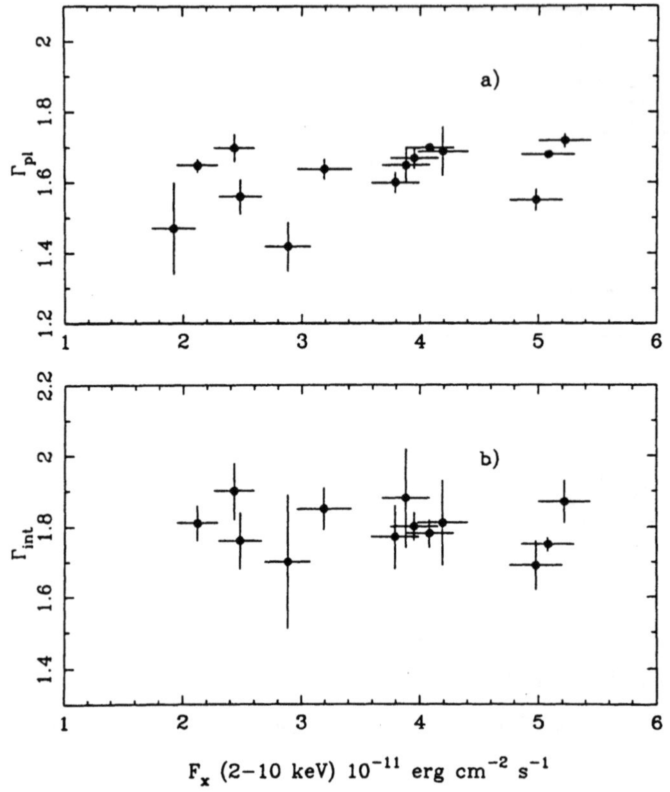

Fig. 6. Continuum slope v. flux of NGC 5548 for (a) simple power law, (b) reflection model.

cently been published (Nandra et al., 1991), revealing a picture consistent with the 'reflection model'. The particular aspect of spectral variability is re-addressed below, adding 9 additional data points from Ginga observations in May – July 1990.

Figure 6(a) shows the distribution of power law index versus flux for all the Ginga data on NGC 5548. Again, a positive trend is apparent, but the individual points are well scattered. Figure 6(b) shows the same data fitted with the reflection model, for which a good fit was obtained to each individual spectrum, with the intrinsic power law index being consistent with a constant value, of $\Gamma_{int} \sim 1.9$. As in the case of MCG–6–30–15, therefore, we conclude that the apparent correlation of power law index with flux is due to predominant changes in the softer, intrinsic spectrum. The much greater scatter of points in the Γ - flux plot for NGC 5548 suggests, in this case, changes also occur in the reflected component between individual spectra, a conclusion drawn in the earlier analysis of Nandra et al. (1991), where an upper limit to the size of the re- processing region of ~135 light days was deduced.

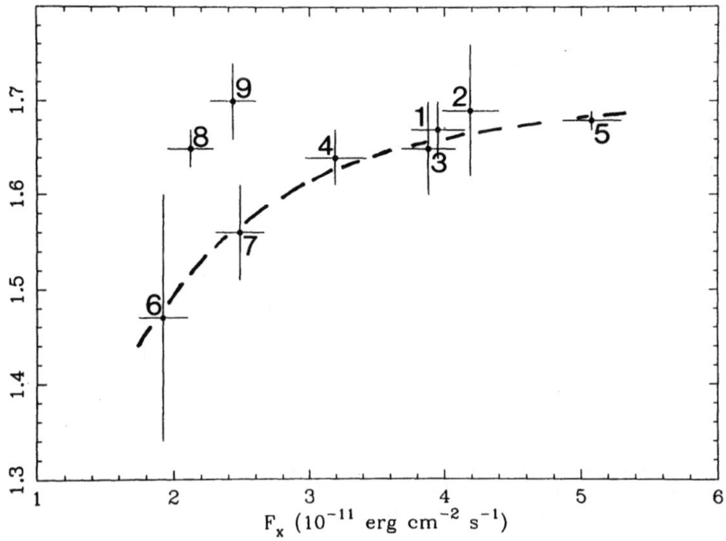

Fig. 7. As 6(a), but only for 1990 observations. Data points refer to time sequence in Figure 8.

To examine this limit in higher resolution we show in Figure 7 the power law v. flux values for the 1990 data alone. Here, the sampling was much more frequent, with 9 spectra covering a 43 day interval. It can be seen that spectra 1 – 7 lie on a single Γ - flux correlation, for $t_{ref} >> t_{int}, \Gamma_{int} = 1.9$ and a face-on disc (see analytical fits in Yaqoob, 1991).

However, from spectral fit 7 to 8, Γ *increases* sharply while F_{2-10} decreases. Between fits 8 and 9, Γ again increases, though this time with increasing continuum flux. Apparently, between spectra 7 and 8/9 the (flatter) reflected component of the continuum must have decreased. Examination of the light curve of the 1990 data (Figure 8) suggests why, since the mean level of F_{2-10} (dominated by the direct component) fell by a factor 2 between data points 1 – 5 and 6 – 9. Smoothing these fluxes, with a time lag of 15 ± 5 days, then leads to the observed spectral behaviour for observations 7, 8, 9.

Our conclusion is that the re-processing region in NGC 5548 is of order 10– 20 light days, not very much larger than the size of the intrinsic power law continuum source (~1-5 light days?), judged on the basis of the typical X- ray flux variability timescales of this Seyfert galaxy.

6. Conclusions

- Iron K-line emission is a common property of Seyfert galaxies. The mean equivalent width is ~140 eV at an energy close to 6.4 keV, with an r.m.s. width up to σ ~0.7 keV. Modelling suggests the bulk of the line emission arises via fluorescence in optically thick material, close

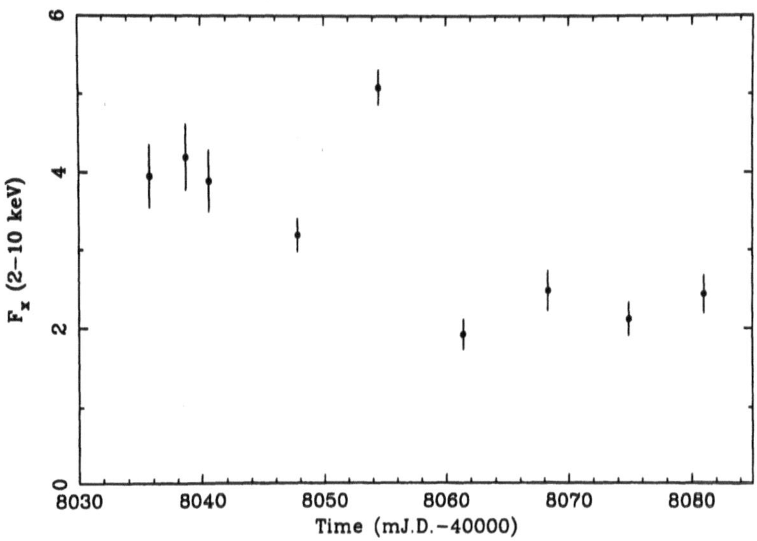

Fig. 8. GINGA X-ray light curve of NGC 5548 from 24 May – July 1990.

to the central black hole. The mean line energy and equivalent width, and the strength of the reflected continuum, suggest low inclination accretion discs are preferentially selected in ~ 2–10 keV X-ray surveys.

- A spectral flattening is observed above ~10 keV. This is consistent with the effects of Compton reflection from such cold material.
- Iron K-shell absorption edges are detected in many sources and may be another common feature. Column densities (equivalent solar abundance of hydrogen) are typically $> 10^{23}\mathrm{cm}^{-2}$ at an energy ~8 keV, implying substantial highly ionized material in the line-of-sight. This may, in some cases, contribute significantly to the line emission, resolving a discrepancy between the observational requirement for the material to lie close to the black hole and the measured line energy (Nandra et al., 1991).
- The intrinsic power law index is steeper than that directly observed, with $\Gamma \sim 2.0. \pm 0.1$. This steeper power law revives some pair models of the emission mechanism (Zdziarski et al., 1990) and is consistent with the Einstein IPC values of Γ ~2.0 in the 0.5–3.5 keV band for a more luminous sample of radio-quiet AGN (Wilkes & Elvis, 1987).
- Soft excesses need to be re-examined, as the deduced luminosity and onset energy will both be reduced as a consequence of the softer underlying power law continuum.
- Spectral variability in NGC 5548 suggests the scale of the bulk of the reflecting material is 15±5 light days.

7. Future Prospects

New Ginga and ROSAT observations, allowing the spectral variability of several bright Seyferts to be regularly monitored, should provide new insights on the origin of the complex X-ray spectrum. In particular, the scale of the re- processing region(s) may be determined in some cases. Further ahead, the availability of X-ray-sensitive CCDs, in the ASTRO-D and Spectrum-X payloads, offers the exciting potential of higher resolution spectroscopy for probing the central power source and its immediate environment in AGN.
Acknowledgements. It is a pleasure to acknowledge the generous support of many colleagues at Leicester and in Japan, especially Martin Turner and Fumiaki Makino. The Ginga project in the UK is funded by the Science and Engineering Research Council, who also provided grants to KN and GCS.

References

Arnaud, K.A., Branduardi-Raymont, G., Culhane, J.L., Fabian, A.C., Hazard, C., McGlynn T.A., Shafer, R.A., Tennant, A.F. & Ward, M.J., 1985, *Monthly Notices Roy. Astron. Soc.*, **217**, 105.
Branduardi-Raymont, G., 1988. 'Active Galactic Nuclei'. Eds. Osterbrock, D.E. and Miller, J.S. Kluwer Acad. Pub., Dordrecht, 177.
Czerny, B. & Elvis, M., 1987, *Astrophys. J.*, **321**, 305.
George, I.M. & Fabian, A.C., 1991, *Monthly Notices Roy. Astron. Soc.*, **249**, 352.
Guilbert, P.W. & Rees, M.J., 1988, *Monthly Notices Roy. Astron. Soc.*, **233**, 475.
Kunieda, H., Turner, T.J., Awaki, H., Mushotzky, R. & Tsusaka, Y., 1990, *Nature*, **345**, 786.
Leighly, K.M., Pounds, K.A. & Turner, T.J., 1989. 'X-ray Astronomy'. Eds. Hunt, J. & Battrick, B. ESA SP–296, 961.
Lightman, A.P. & White, T.R., 1988, *Astrophys. J.*, **335**, 57.
McHardy, I.M., 1988. *Mem. del. Soc. Ast. Ital.*, **59**, 239.
Makishima, K., 1986. 'The physics of accretion onto compact objects.' Eds. Mason, K.O., Watson, M.G. & White, N.E. Springer Verlag, Berlin, 285.
Matsuoka, M., Yamauchi, M., Piro, L. & Murakami, T., 1990, *Astrophys. J.*, **361**, 440.
Mushotzky, R.F., 1982, *Astrophys. J.*, **256**, 92.
Morini M., Lipani, N.A. & Molteni, D., 1987, *Astrophys. J.*, **317**, 145.
Nandra, K., Pounds, K.A., Stewart, G.C., Fabian, A.C. & Rees, M.J., 1989, *Monthly Notices Roy. Astron. Soc.*, **236**, 39P.
Nandra, K., Pounds, K.A. & Stewart, G.C., 1990, *Monthly Notices Roy. Astron. Soc.*, **242**, 660.
Nandra, K., 1991. PhD thesis, University of Leicester.
Nandra, K., Pounds, K.A., Stewart, G.C., George, I.M., Hayashida, K., Makino, F. & Ohashi, T., 1991, *Monthly Notices Roy. Astron. Soc.*, **248**, 760.
Piro, L., Yamauchi, M. & Matsuoka, M., 1990, *Astrophys. J. (Letters)*, **360**, L35.
Pounds, K.A., Stanger, V.J., Turner, T.J., King, A.R. & Czerny, B., 1986, *Monthly Notices Roy. Astron. Soc.*, **224**, 443.
Pounds, K.A. & McHardy, I.M., 1988. 'Physics of Neutron Stars and Black Holes', Ed. Tanaka, Y., Univ. Acad. Press., p. 285.
Pounds, K.A., Nandra, K., Stewart, G.C. & Leighly, K., 1989, *Monthly Notices Roy. Astron. Soc.*, **240**, 769.

Pounds, K.A., Nandra, K., Stewart, G.C., George, I.M. & Fabian, A.C., 1990, *Nature*, **344**, 132.

Turner, T.J. & Pounds, K., 1989, *Monthly Notices Roy. Astron. Soc.*, **240**, 833.

Wilkes, B.J. & Elvis, M., 1987, *Astrophys. J.*, **323**, 243.

Yaqoob, Y., 1991. 'Frontiers of X-ray Astronomy'. Ed. Koyama, K., Univ. Acad. Press, Tokyo.

Zdziarski A., Ghisellini, G., George, I.M., Svensson, R., Fabian, A.C. & Done C., 1990, *Astrophys. J. (Letters)*, **363**, L1.

Iron Lines with Large Equivalent Width and X-Ray Reprocessing from a Thick Medium in Seyfert Galaxies

L. Piro[1], *M. Matsuoka*[2], *and M. Yamauchi*[2]

[1] Istituto di Astrofisica Spaziale, CNR, Frascati, Italy
[2] Institute of Physical and Chemical Research, Wako-shi, Japan

Abstract. Observations by GINGA of the Seyfert galaxies MCG-5-23-16 and NGC 6814 have showed the presence of a strong iron line, with an E.W. of about 400 eV. In analogy with the current model proposed to explain the iron line and the spectral flattening observed in several bright Seyfert galaxies, we have attempted to explain the strong lines observed in NGC 6814 and MCG-5-23-16 in the framework of an optically thick reprocessor. The result is substantially negative, even in the most favourable cases where the line emission is enhanced by increasing the iron abundance or the ionization state of the medium, because the predicted level of the reflected component is not consistent with the data.

1. Introduction

Several bright Seyfert galaxies observed by GINGA have shown the presence of a spectral flattening above 8 keV and an iron line at 6.4 keV with a typical $E.W. \simeq 150$ eV (e.g. Piro et al. 1990, Pounds this conference). These features are currently attributed to the reprocessing of the intrinsic emission by a cold and optically thick medium, producing an iron line by fluorescence and a reflected component by Compton scattering and photoelectric absorption. The latter component is responsible for the observed spectral flattening above 8 keV. Both the equivalent width and the relative intensity of the reflected component to the direct emission are in agreement with those expected by an accretion disk with solar-like abundances seen face on (Matt et al. 1991, Piro 1990)

On the other hand, there are a few objects, like NGC 6814 (Kunieda et al. 1990, Matsuoka et al. 1991), MCG-5-23-16 (Fig.1) and Mkn 841 (Day et al. 1991) where the iron line attains an E.W. of about 400 eV (and the column density derived from the low energy absorption is too small to account for such a line).

The question we have tried to address in this paper is whether the strong lines observed in NGC 6814 and MCG-5-23-16 could be produced by fluorescence by a thick medium ($N_H > 10^{24} cm^{-2}$), in analogy with what proposed for the bright Seyfert galaxies with a 150 eV iron line.

The relevance of this problem comes from the evidence that, at least in NGC 6814, the iron line emitting medium is located very close to the central object (Kunieda et al. 1990), just in the range of distances predicted for an accretion disk (which is, in fact, optically thick). Furthermore both these two objects and Mkn 841 have a spectral photon index ($\alpha \simeq 1.4$) flatter than that observed in bright Seyfert galaxies, that could be explained in terms of a stronger reflected component.

In order to verify the hypothesis that the reprocessing medium is optically thick we go through the following logical steps:

a) compute the spectral shape of the reprocessed component, in particular the ratio of intensities of the iron line to the reflected component. Here is important to point out that this ratio depends on the chemical and physical status of the medium (as discussed in the following) but is only weakly dependent on the geometry (Matt et al. 1991).

b) determine the normalization of the reprocessed component (relative to the direct continuum) in order to have an iron line with E.W.= 400 eV. This relative normalization depends on the geometrical configuration and can be, for example, increased if the central emission is beamed towards the reprocessing medium or if the reprocessor is covering a large solid angle. The point here is that, whatever the geometrical configuration may be, the level of the reflected continuum must be increased by a similar factor as the line. Only in rather peculiar geometries this is not longer true (Matt et al. 1991).

c) Verify if the predicted level of the reflected component is in agreement with the observed spectrum. This is done by fitting the spectrum with the reflection model expressed by the formula:

$$F(E) = [CE^{-\alpha}(1 + rA(E,\alpha)) + I_{Fe}]exp(-\sigma(E)N_H) \qquad (1)$$

where $CE^{-\alpha}$ represents the direct component, $CE^{-\alpha}rA(E,\alpha)$ is the reflected component, $A(E,\alpha)$ is the X-ray albedo of the medium relative to the chemical and physical status under study (computed by using the formulae of Lightman and White 1988 and shown in fig.2) and r is the ratio between the reflected and direct components. The value of r obtained from the fit is then compared with that expected from the E.W. (point b).

2. Thick cold reprocessor with solar-like abundances

Firstly we investigated the case of a thick cold medium with solar-like abundances, which has been successfully applied to bright Seyfert galaxies with an iron line with an E.W. of 150 eV. If an iron line with E.W.=400 eV is to be produced by such a medium, the reflected component should be about

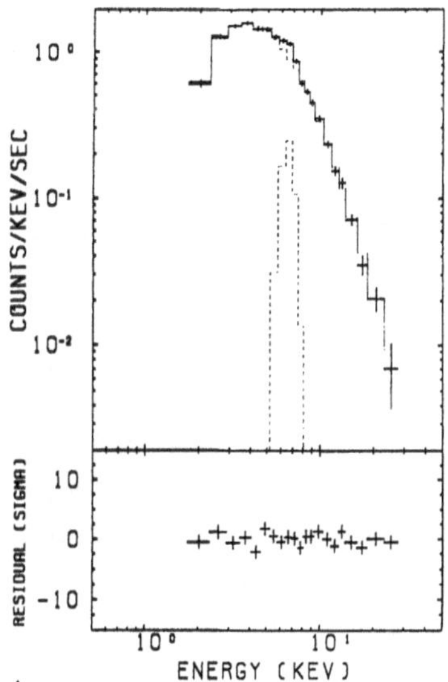

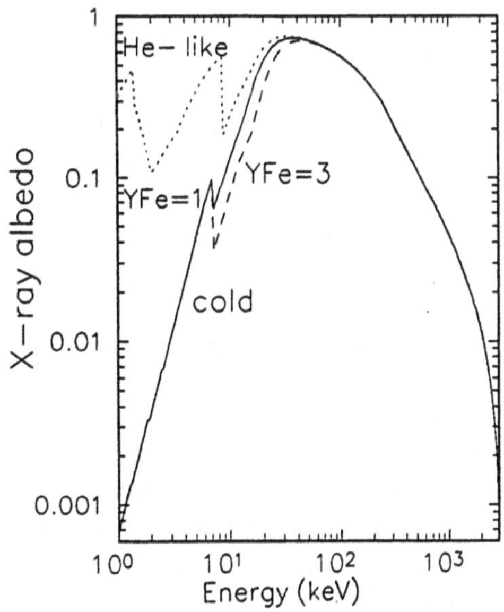

Fig. 1. **The spectrum of MCG-5-23-16 obtained by GINGA is well fitted by a simple power law ($\alpha = 1.42 \pm 0.05$), with an uniform absorber ($N_H = (2.0 \pm 0.3)\ 10^{22}\,cm^{-2}$) and an emission line ($E = (6.4 \pm 0.1)$ keV, $E.W. = (350 \pm 50)$ eV). There is no evidence of an additional iron edge in the range 7-9 keV with an upper limit to the column density $N_{HFe} < 4\ 10^{22}\,cm^{-2}$**

Fig. 2. **The X-ray albedo of a thick medium for different chemical abundances and ionization state**

2.5 times larger than observed in bright Seyfert galaxies. However a fit with the reflection model (1) to the spectra of MCG-5-23-16 and NGC 6814 yields only an upper limit to the level of the reflected component, about 8 times lower than the value predicted from the E.W. of the line.

3. Dependence of the E.W. on physical parameter of reprocessor

The previous result indicates that the problem of a cold thick medium of solar-like abundances is the relatively low value of the ratio of photons reprocessed in the iron line to those in the reflected component, that makes the predicted level of the reflected component much greater than observed.

We have thus explored the possibility of increasing the E.W. by changing the chemical composition or the ionization state of the medium. The dependence of the E.W. on the physical and chemical parameters of the medium can be approximately written in the optically thick case as:

$$E.W. \simeq \omega_K \frac{Y_{Fe}\sigma_{Fe}}{(Y_{Fe}\sigma_{Fe} + Y\sigma_L + 1.2\sigma_T)} exp[\frac{-Y\sigma_{La}}{(Y_{Fe}\sigma_{Fe} + Y\sigma_L + 1.2\sigma_T)}] \quad (2)$$

where ω_K is the fluorescent yield of iron K shell, Y_{Fe} is the iron abundance, σ_{Fe} is the cross section of iron at the K edge, Y , σ_L, σ_{La}, are respectively the abundance, the cross sections at the iron K edge and at the iron line energy of elements lighter than iron, and σ_T is the Thomson cross section. This formula is the product of probabilities of the different processes leading to the emission of a line photon and shows, in fact, that the E.W. is proportional to the fluorescent yield, to the fraction of photons of the incident beam that are ionizing the K shell of iron instead of being absorbed by other atoms or scattered by electrons (second term), and takes into account the absorption of fluorescent photons assuming that they travel a path corresponding to a total optical depth of unity (last term).

Equation (2), although obtained with simple approximations, is indeed useful in showing the trend of the line intensity as a function of the parameters, in particular the non linear dependence on iron and lighter atoms opacities. Note that the dependence on the ionization state is contained in the cross sections. In the following we use the exact solution to the transfer equation of the iron line of Basko (1978).

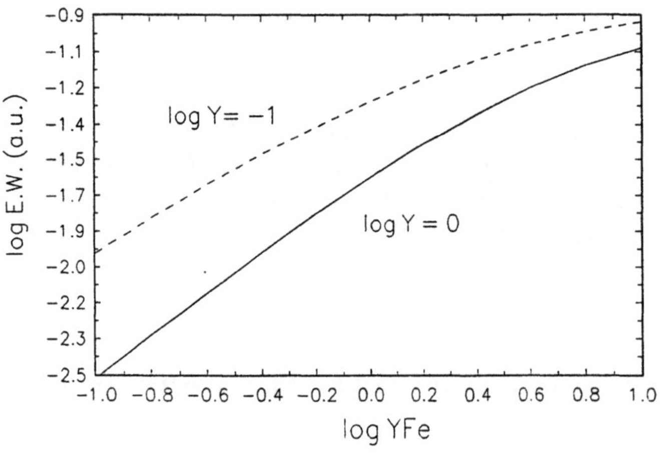

Fig. 3. E.W. vs abundances for a thick cold reprocessor

4. Thick cold medium with non-solar chemical abundances

In the cold case the dependence of the E.W. from Y_{Fe} and Y (fig.3), as also shown by eq. (2), saturates when $Y_{Fe} \gg 1$ and $Y \ll 1$, to a value 4-5 times greater than the one corresponding to solar-like abundances.

We have analyzed in detail the case $Y_{Fe} = 3$, $Y = 1$, that gives an E.W. 2.5 times greater than the case with solar-like abundances.

The corresponding X-ray albedo is shown if fig.2. Note as the increase of the iron abundance produces a deeper iron edge and reduces the albedo in the range 7.1 -20 keV.

A fit with a reflection model gives again an upper limit to the level of the reflected component in both NGC 6814 and MCG-5-23-16 but now, due to the greater E.W. and lower albedo, the upper limit is marginally consistent with the predicted value. A further substantial increase of the E.W. can be obtained only for very large values of Y_{Fe} (fig.3) but this would produce a significant iron edge in the spectra.

5. Highly ionized thick reprocessor

When the medium is photoionized, the E.W. increases, due to the decrease of opacity of elements lighter than iron and the increase of the fluorescent yield of iron K shell (eq.2). In the rather extreme case when the iron atoms are in He-like configuration (corresponding to a photoionization parameter $\Xi = L/(4\pi c r^2 n K T) = 33$ (Krolik and Kallman 1984)) the E.W. is three times larger than the cold case (with solar-like abundances). The line energy is 6.7 keV, significantly greater than observed, but if the medium were located very close to the central black hole (as seems to be in the case of NGC 6814, Kunieda et al. 1990), the line should be gravitationally redshifted.

The X-ray albedo of such a medium (fig.2) is larger than that of a cold reprocessor, particularly below 10 keV, where the photoelectric absorption is reduced. In fig. 4 we show the total (direct+reprocessed) spectrum expected in order to reproduce an iron line with an E.W.=400 eV . Note the presence of a strong reflected component, that gives a deep iron edge (corresponding to $N_{HFe} \simeq 4 \times 10^{23} cm^{-2}$) at 8.5 keV, a feature not observed in the spectra of NGC 6814 and MCG-5-23-16.

A fit with the reflection model of eq.(1) with this albedo gives, in fact, unsatisfactory results. In MCG-5-23-16 the intensity of the reflected component is consistent with zero, with an upper limit 5 times lower than the value expected from the E.W. of the iron line. In NGC 6814 the chi square

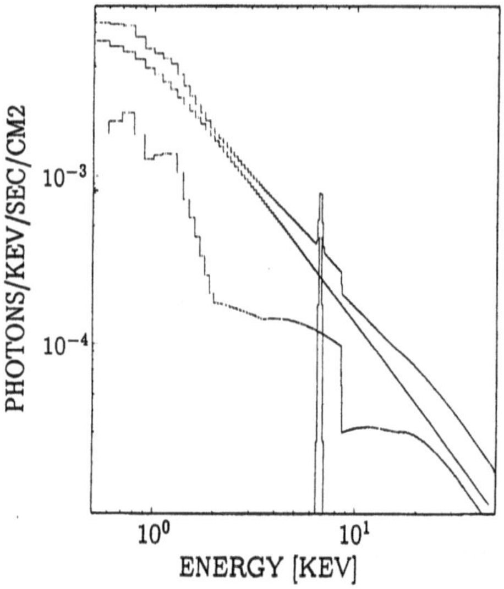

Fig. 4. The X-ray photon spectrum expected in the case of a thick highly ionized reprocessor producing an iron line with an E.W. of 400 eV. The direct and reprocessed components are shown separately

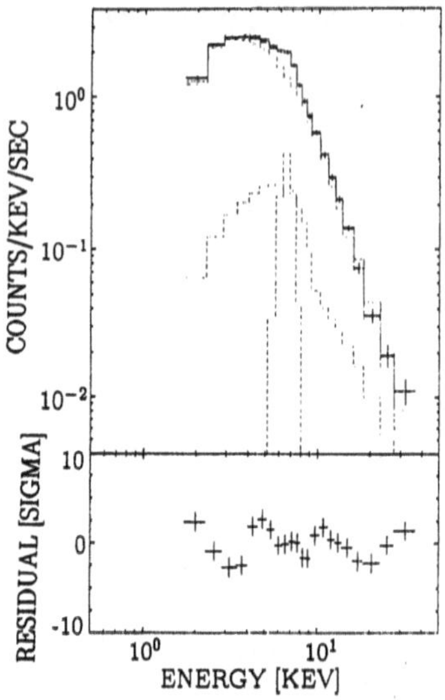

Fig. 5. The spectrum of NGC 6814 fitted with a reflection model from a highly ionized thick reprocessor. The model is clearly unacceptable ($\chi^2_\nu = 3.4$)

value is too large ($\chi^2_\nu = 3.4$) and the level of the reflected component is again too small (fig.5).

6. Conclusions

If the medium producing the strong iron line observed in MCG-6-30-15 and NGC 6814 were optically thick ($N_H > 10^{24} cm^{-2}$) it should unavoidably produce a reflected component, whose main spectral imprints on the total spectrum are a spectral flattening above 8 keV (in the cold case) or a deep iron edge (in the highly photoionized case). Those features are not detected.

Before drawing the conclusion that a substantial fraction of the line should be produced by an optically *thin* medium, we point out that there are some possibilities that leave open an explanation of the line in terms of a thick reprocessor:

a) the *shape* of the intrinsic continuum depends on the direction. For example, if the intrinsic continuum illuminating the reprocessor is steeper than that seen by the observer, the relative weight of the reflected component decreases.

b) in the highly photoionized case (H-He like iron atoms), line emission could be increased by the contribution of recombination (e.g. Ferland and Rees 1988) and of "scattering" of continuum photons by line opacity (Band et al. 1988). Furthermore the detectability of the deep iron edge predicted in this case might be reduced by the spread due to Doppler and gravitational redshift expected in the case of an accretion disk (Matt et al. 1991) and by the filling produced by the Nickel K lines (Mushotzky, private communication).

References

Band D. L. et al.:1989, in *23 ESLAB Symp. on X-Ray Astronomy*, ESA (SP-296), p.885.
Basko, M. M.:1978, *Astrophys. J.*, **223**, 268
Day, C.R.S., Fabian, A.C., George, I.M. and Kunieda, H. 1990, *Monthly Notices Roy. Astron. Soc.*, **247**, 15p
Ferland, G. J. and Rees, M. J. 1988, *Astrophys. J.*, **332**, 141
Krolik, J.H., and Kallman, T.H. 1984, *Astrophys. J.*, **286**, 366
Kunieda, H., et al. 1990, *Nature*, **345**, 786
Lightman, A. P. and White, T. R. 1988, *Astrophys. J.*, **335**, 57
Matsuoka, M. et al., 1991 in *Frontiers of X-Ray Astronomy*, Nagoya, Apr. 8-12 1991.
Matt, G., Perola, G.C. and Piro L. 1991, *Astron. Astrophys.*, , press.
Piro, L., Yamauchi, M., and Matsuoka, M. 1990, *Astron. Astrophys. (Letters)*, **360**, L35
Piro L. 1990, in *Iron Line Diagnostics in X-Ray Sources*, Varenna, Oct. 1990.
Pounds K., *this conference*

Discussion

Krolik: Please comment on a recent preprint by Hayakawa which suggested that the Fe line in NGC 6814 is due to collisional excitation in a hot plasma.

Piro: Hayakawa's work refers to an optically thin medium whereas we explored the possibility of producing the line (by fluorescence) from an optically thick medium. Our negative result (but with the provisions mentioned in the conclusion) would, in fact, suggest the presence of an optically thin medium.

Kazanas: It should be noted that in the case studied by Hayakawa the line appears at 6.4 keV because of gravitational redshift.

EUV Observations of AGN

P.M. Gondhalekar[1], *K.A. Pounds*[2], *S. Sembay*[2], *J. Sokoloski*[3], *C.M. Urry*[3], *L. Matthews*[4], *and J. Quenby*[4]

[1] Rutherford Appleton Laboratory, Astrophysics Group, England.
[2] University of Leicester, X-ray Astronomy Group, England.
[3] Space Telescope Science Institute, Baltimore, USA.
[4] Imperial College of Science, Technology and Medicine, England.

Abstract. The WFC detections of EUV/soft X-ray emission of the BL Lac object PKS 2155-304 and the Seyfert 1 galaxy Mkn 478 have been described. The visible to X-ray continuum energy distribution of PKS 2155-304 is found not to be in agreement with an accretion disk model of this object and alternative models like the synchrotron self-Compton models might have to be considered. It has not been possible to determine the continuum energy distribution of Mkn 478 but the EUV/soft X-ray spectrum of this galaxy determined from the temporal variation of $CIV/Ly\alpha$ ratio is consistent with the WFC observations.

1. Introduction

The radio to X-ray Continuum Energy Distribution (CED) of AGNs has been summarized by Sanders et al. (1989) in their Fig. 1. This figure clearly demonstrates the gap in our knowledge of the CED of AGNs, absolutely nothing is known about the radiation in the crucial EUV/soft X-ray region. A knowledge of radiation in this region is essential for understanding the emission of line radiation from AGNs. After the launch of *IUE*, in 1978, a number of heroic attempts were made to bridge this EUV/soft X-ray gap (Wilson, Carnochan and Gondhalekar 1979, O'Brien, Gondhalekar and Wilson 1989, Reimers et al. 1989) by observations of UV spectra of high redshift quasars. None of these attempts were entirely successful. However, the *IUE* observations established that the continuum flux of low redshift AGNs increases at shorter UV wavelengths. It is not known if this is also true for high redshift AGNs or quasars as the EUV continuum of these objects is absorbed by intervening gas clouds and galactic halos. Similarly, *Einstein* and *EXOSAT* observations demonstrated that there was a "soft excess" in the X-ray spectra of a number of AGNs (Turner and Pounds 1989 and references therein), again indicating an excess flux at EUV/soft X-ray energies.

The British built Wide Field Camera (WFC) on the German satellite *ROSAT* is particularly appropriate for observations of radiation in this

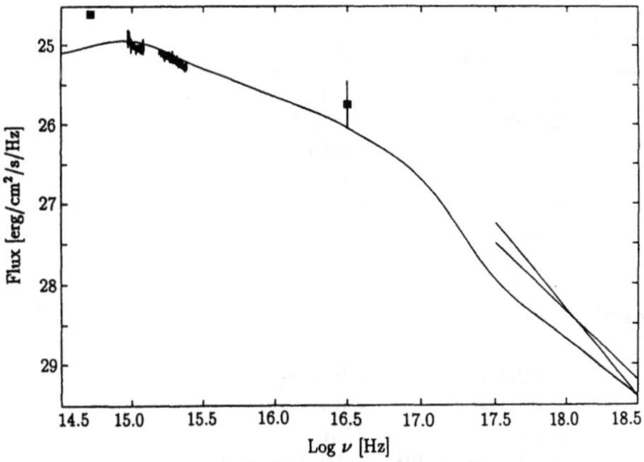

Fig. 1. Continuum energy distribution of PKS 2155-304. The filled squares are IUE/FES and WFC observations respectively. The crossed lines are the *GINGA* data and the continuous curve is the model of Wandel and Urry(1991).

EUV/soft X-ray gap. The prime filters in the WFC, S1a (90-210 eV) and S2a (62-111 eV) are centred close to the expected peak in the EUV/soft X-ray continuum. The *ROSAT* satellite performed an all-sky survey from the end of July 1990 to the end of January 1991. The BL Lac object PKS 2155-304 and the Seyfert 1 galaxy Mkn 478 were detected with the WFC during this survey. These observations and the analysis of these data are described here.

2. PKS 2155-304

PKS 2155-304 belongs to a class of objects known as the BL Lac objects. These objects are characterized by a featureless optical and ultraviolet spectrum and they are variable at all observed energies. PKS 2155-304 (Table 1) is a strong X-ray source and has been observed by both the *Einstein* and *EXOSAT* observatories (Urry, Mushotzky and Holt 1986, Treves et al. 1989, etc.) PKS 2155-304 was in the field of view of WFC from 7 November 1990 to 11 November 1990 and was detected at 30σ level through the S1a filter but was not detectable through the S2a filter. The object was also observed with the *IUE* and the *GINGA* satellites to obtain simultaneous ultraviolet and X-ray spectra respectively. It is not possible to obtain unambiguous flux values in the EUV with just one data point and one upper limit, instead a black-body spectrum was fitted to the count rates. For a *pre-flight calibration* and galactic column of 1.78×10^{20} cm^2 (Stark et al. 1984) a fit within 3σ was obtained for a black-body of 24 ± 4 eV. This is the lowest temperature which gives count rates consistent with observations.

Table 1. Astronomical and Observational Data

	PKS 2155-304	Mkn 478
	BL Lac	Sy 1
α_{2000}	21:58:52	14:42:07
δ_{2000}	-30:13:24	35:25:50
l	72	59
b	-31	65
z	0.117	0.079
m_v	13.09	14.58
M_{abs}	-26.9	-23.8
L_{2kev} $erg\ s^{-1}\ Hz^{-1}$	27.73	
S1a(σ) ct s^{-1}	0.19(30)	0.06(18)
S2a(σ) ct s^{-1}	<0.01(0.1)	<0.01(1.7)
exp(s)	5000	7100
Date	5-11 Nov 1990	3-10 Jan 1991

The visible to X-ray spectrum of PKS 2155-304 is shown in Fig. 1, the data have been corrected for galactic absorption. Also shown in Fig. 1 is the model spectrum of Wandel and Urry (1991). This is an accretion disk model in which a cold optically thin disk is sandwiched between a hot optically thin disk. The X-rays are emitted by the hot disk and the UV-soft X-ray emission is from the cold disk. The cold disk photons radiate through the hot disk and a fraction of these are Compton upscattered to hard X-rays. The model successfully reproduced the *IUE/Einstein* UV to X-ray observations of this AGN (Wandel and Urry 1991). With the present data the model fits the UV region but under estimates the flux in the EUV (*WFC*) and the X-ray (*GINGA*) regions respectively. It is not possible to increase the model flux by a large fraction as the source would then become super Eddington. An alternative to the disk model is the so-called synchrotron self-Compton (SSC) model. This model was proposed for PKS 2155-304 by Urry and Mushotzky (1984), who considered a homogeneous source which requires the the radiation to be relativistically beamed. However, relativistic beaming is not necessary if the source is inhomogeneous (Ghisellini, Maraschi and Treves 1985). The present data seem to suggest that the EUV and X-ray data can be described by a single powerlaw, this would favour a SSC model. But a more detailed comparison is required to see if this model can explain the visible to X-ray CED presented here.

3. Mkn 478

Mkn 478 (Table 1) is a Seyfert 1 galaxy, these galaxies are characterized by a non-thermal continuum, broad permitted and narrow forbidden emission lines. Both the continuum and the permitted emission lines can be highly variable over short time scales. Mkn 478 was in the field-of-view of WFC from 3 January 1991 to 10 January 1991 and was detected at 18σ level through the S1a filter but was not detected through the S2a filter. The EUV flux emitted by Mkn 478 was determined by a procedure similar to that used for PKS 2155-304. For a galactic column of 1.23×10^{20} cm^2 (Stark et al. 1984) and the ***pre-flight calibration*** a 3σ fit is obtained for a black-body of 54 ± 5 eV.

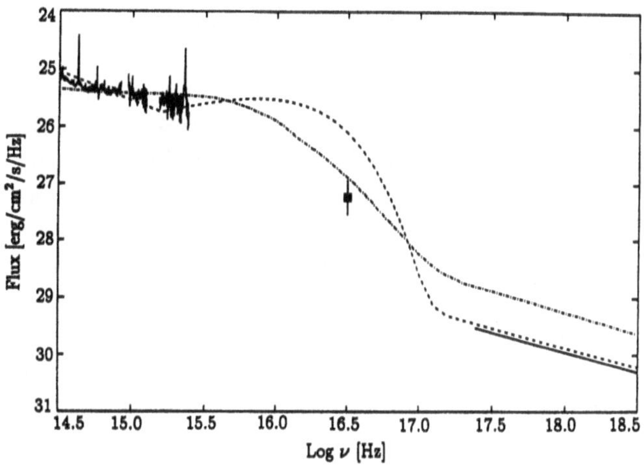

Fig. 2. A representative continuum energy distribution of Mkn 478 (the data in various energy bands have not been obtained simultaneously). The filled square is the WFC observation, the dot-dash curve is an accretion disk model of Laor and Netzer (see text for details) and the dashed curve is a continuum determined from the variability of $CIV/Ly\alpha$ **ratio.**

The visible to X-ray spectrum (corrected for galactic absorption) of this galaxy is shown in Fig. 2 but this is not a true representation of the instantaneous CED of Mkn 478 as the data in different energy bands ware not obtained simultaneously; the optical and ultraviolet data were obtained in May 1991 and the *EXOSAT* data were obtained a few years earlier. The signal to noise ratio of the *EXOSAT* data is rather low and to obtain the X-ray fluxes a power-law with a canonical photon index of 1.7 and a column of 1.0×10^{20} cm^2 were assumed.

In Fig.2 the observed CED of Mkn 478 is compared with the CED of an accretion disk model of Laor and Netzer (1989). It should be noted that Laor and Netzer only model the UV/EUV region of the CED and not the X-ray region. The X-ray CED is added in an *ad hoc* manner to these models. The agreement between observed visible, UV and EUV data is best for a model with a central black-hole of $2.7 \times 10^7\ M_\odot$, accretion rate of $0.3M_\odot$ and a face on disk. No agreement should be expected in the X-ray region. However, it is possible that there are a number of combinations of these parameters which would give an equally good agreement with the data.

It has recently been shown (Binette et al. 1989, Clavel and Santos-Lleo 1990, Gondhalekar et al. 1991) that the temporal change in the ratios of lines (particularly the ratio of CIV and $Ly\alpha$ lines) emitted by AGNs can be modelled to define the EUV/soft X-ray CED seen by the Broad Line Region (BLR) clouds. Observational confirmation of these modelled CEDs has not been possible up to now. It is tempting to perform similar analysis for Mkn 478 and compare the modelled CED with the observed flux. Between 1978 and 1984 Mkn 478 was observed a number of times with *IUE*. The variation of $CIV/Ly\alpha$ was modelled following the procedure described by Gondhalekar et al. 1991. The line ratios could be reproduced for cloud density $n_H \sim 3.0 \times 10^{10}\ cm^{-3}$, column $N_H \sim 10^{23}\ cm^2$ and a BLR radius $r = 20\ lt\ d$. The modelled size of the BLR is consistent with the variability of $Ly\alpha$ and CIV lines in the ultraviolet spectra. An example of modelled CED is shown in Fig. 2. Considering that the observations are not simultaneous, the inflight calibration may differ from the preflight calibration used here and the BLR clouds may not see the CED detected at earth, the agreement between the observed and the modelled CED is acceptable (the agreement at UV and X-ray energies is to be expected as the continuum is forced to agree with observations at these energies). The disagreement between this empirically determined CED and the model of Laor and Netzer (1989) is quite significant, also the CED of Laor and Netzer is unable to reproduce the observed $CIV/Ly\alpha$ ratios of Mkn 478.

4. Summary

The EUV/soft X-ray observations of the BL Lac object PKS 2155-304 and the Seyfert 1 galaxy Mkn 478 made with the WFC on the ROSAT satellite have been described. A visible to X-ray CED of PKS 2155-304 is obtained by simultaneous observations of the ultraviolet, the EUV/soft X-ray and the X-ray emission from this object. The CED is compared with a disk model of this object. The model spectrum is not in agreement with these data and alternative models like the SSC model may have to be considered. It has not been possible to determine the CED of Seyfert galaxy Mkn 478 because the visible, UV, EUV and the X-ray observations were not made simultaneously. An empirically determined EUV/soft X-ray CED of this galaxy is in acceptable agreement with observations.
Acknowledgements. We would like to thank the *ROSAT/WFC* project for the survey data, we would also like to thank the *IUE* and *GINGA* projects

for respectively UV and X-ray observations. PMG would like to thank D Wonnacott and B J Kellett for their assistance with various aspects of data analysis. These data were analysed with the facilities provided by the *STARLINK* project.

References

Binette, L., Prieto, A., Szuszkiewic, E., Zheng, W., 1989: *Astrophys. J.*, **343**, 135.

Clavel, J., Santos-Lleo, M., 1990: *Astron. Astrophys.*

Ghisellini, G., Maraschi, L., Treves, A., 1985: *Astron. Astrophys.*, **146**, 204.

Gondhalekar, P.M., Peterson, B.M., Crenshaw, D.M., Wagner, R.M., 1991: *Astron. Astrophys.*, submitted.

Laor, A., Netzer, H., 1989: *Monthly Notices Roy. Astron. Soc.*, **238**, 897.

O'Brien, P.T., Gondhalekar, P.M., Wilson, R., 1989: *Monthly Notices Roy. Astron. Soc.*, **233**, 801.

Reimers, D., Clavel, J., Groote, D., Engels, D., Hagen, H.J., Naylor, T., Wamsteker, W., Hoop, U., 1989: *Astron. Astrophys.*, **216**, 71.

Sanders, D.B., Phinney, E.S., Neugebauer, G., Soifer, B.T., Mathews, K., 1989: *Astrophys. J.*, **347**, 29.

Stark, A., Heiles, C., Baily, Linke, K., 1984: privately distributed tape.

Treves, A., et al., 1989: *Astrophys. J.*, **341**, 733.

Turner, T.J., Pounds, K.A., 1989: *Monthly Notices Roy. Astron. Soc.*, **240**, 833.

Urry, C.M., Mushotzky, R.F., 1982: *Astrophys. J.*, **253**, 38.

Urry, C.M., Mushotzky, R.F., Holt, S.S., 1986: *Astrophys. J.*, **305**, 369.

Wandel, A., Urry, C.M., 1991: *Astrophys. J.*, **367**, 78.

Wilson, R., Carnochan, D.J., Gondhalekar, P.M., 1979: *Nature* **275**, 457.

Discussion

This paper has benefited from the questions and discussions at this conference. Answers to a number of questions have been included in the text and these questions have not been answered in full here.

Barr: In PKS 2155-304 the high luminosity inferred from the WFC data is due to the assumed spectral index of 1.86. Now, this is the spectral index measured by *GINGA* **above** $2keV$. It is a well established fact - from *EXOSAT, Einstein* and more recently *BBXRT* - that the x-ray spectrum of PKS 2155-304 is NOT a simple power law; it flattens above $2keV$, so the extrapolation of the *GINGA* data to the *WFC* band is invalid, the XUV continuum must be flatter. If you assume a flatter continuum in converting the *WFC* countrate to flux, does that remove the need to assume a XUV excess?

Gondhalekar: The EUV flux from the *WFC* count rate has not been obtained just from extrapolating the *GINGA* power law, the flux has also been obtained by assuming a black-body spectrum. The black-body spectrum provides a better fit and this seems to indicate an EUV excess. *Subsequent analysis has indicated that a flatter power law also does not give a fit as good as the black-body spectrum.*

Mushotzky: The *WFC* flux (comments for PKS 2155-304) is very consistent with the spectral flattening seen in the *BBXRT* spectrum.

Krolik: Do you have any object with upper bounds at an interesting flux level?

Gondhalekar: Analysis of *WFC* data to obtain "interesting" upper bounds has not been completed yet.

Wandel: A standard accretion disk fails to explain hard X-rays. In particular, Laor's model do not claim to explain X-ray, in fact he claims the opposite. You would need an independent model for the X-ray power law. However, an independent model which does fit the UV as well as the soft and hard X-ray spectrum of PKS 2155-304 has been suggested (Wandel and Urry Ap. J. Lett. 367,78). In this model the X-ray power law is produced by an optically thin hot accretion disk which Comptonizes the soft photons from a cool optically thin disk embedded in it. The soft X-ray excess is fitted as well by radiation from inner, electron scattering dominated region from the cool disk.

Gondhalekar: A detailed comparison with this model has been made in this paper.

Zdziarski: It seems to me that the hard X-ray power law you showed in your fits have been added by hand. Thus, they have no value in comparing the data with the disk model.

Gondhalekar: see text for explanation

Variable Soft X-Ray Excesses in AGN from Nonthermal Electron-Positron Pair Cascades

Andrzej A. Zdziarski[1] *and Paolo S. Coppi*[2]

[1] N. Copernicus Astronomical Center Bartycka 18, PL-00-716 Warsaw, Poland
[2] Department of Astronomy and Astrophysics, The University of Chicago 5640 S. Ellis Avenue, Chicago, IL 60637, USA

This is an abstract of a paper published elsewhere (Zdziarski and Coppi, 1991: *Astrophys. J.*, **376**, 480). The paper deals with formation of steep soft X-ray excesses superimposed on flatter hard X-ray power law spectra in nonthermal electron-positron pair cascade sources. Such soft X-ray excesses are similar to those seen below ~ 1 keV in AGN spectra. The soft X-ray excesses in the pair cascade AGN models are due to the Comptonization of UV bump photons by pairs decelerated in the cascade process and thermalized to a nonrelativistic equilibrium temperature. The model parameter space in which an excess is visible is determined. Simple formulae for the spectral form of the excess are found. The soft excess in these models appears as a steep power law superimposed on the tail of the UV bump and the flat nonthermal (hard X-ray) power law.

The time variability of soft excesses in pair cascade models is also examined. It is found that the variability patterns of the soft and hard X-rays may be distinctly different (as often seen in AGNs) despite their common origin in the nonthermal source. The form and the amplitude of the soft excesses depend strongly on the luminosity in soft UV photons. The hard X-ray emission, on the other hand, is roughly independent of the UV luminosity. The soft vs. hard X-ray variability patterns are classified as a function of the relative variability of the UV luminosity and the power supplied to the relativistic electrons.

Finally, it is found that the parameter space in which soft excesses appear encompasses the range of preferred input parameters for a recently developed Compton reflection model of UV and X-ray emission from the central engine of an AGN. In this model, the observed radiation comes from a nonthermal pair cascade source and a cold ($\sim 10^5$ K), optically thick matter distribution which intercepts and reprocesses part of the nonthermal emission. The presence of soft X-ray excesses in AGNs may thus be understood as a consequence of the presence of both cold matter and nonthermal sources in the central engines of AGNs and additional sources of emission at soft X-ray energies may not be required.

Discussion

Lesch: Are your calculations dependent on the isotropy of the energy distribution of the particles?

Zdziarski: Anisotropy would change the results quantitatively to a degree, but it would not affect the conclusions of this work. As we lack the detailed knowledge of the angular distribution of particles, the assumption of isotropy is the simplest one to adopt.

Lesch: What is the confinement mechanism for your pair source?

Zdziarski: The pairs can be confined by tangled magnetic fields, by scattering on plasma waves etc. However, no confinement is actually necessary as there may be a balance between pair production and pair annihilation, advection, and escape, resulting in a steady state pair cloud optically thick to Compton scattering.

Kundt: A comment in support of your model and reply to H. Lesch: The hard component of your pair plasma in the BLR will partially escape through the jets. For our Galactic Center, a typical Lorentz factor of $\gamma \sim 10^4$ (of the pair plasma) has been measured at both Sgr A* and the (much more extended) "chimney" (cf. *Astr. Sp. Sci.*, **172**, 109 [1990]).

Wilkes: Your model predicts that the soft excess should be stronger in low luminosity objects. Contrary to your statement, this has been tested in that *Einstein* observations of quasars show the same incidence of soft excesses ($\sim 50\%$) as *EXOSAT* observations of Seyfert 1s.

Zdziarski: This very interesting result may either be an argument against the present model (and, in fact, against any UV Comptonization model for the origin of the soft X-ray excesses) or against the optically thick thermal origin of the UV bump in AGNs. This origin has been assumed in this work, but it is not an essential feature of the model. If the UV bump is not of the optically thick thermal origin, it could peak at an approximately constant energy regardless of the luminosity, and there would be no correlation of the incidence of soft excesses with the luminosity in our model. In fact, there have been some recent ideas that the bump does *not* originate from optically thick matter (e.g., Ferland *et al.*, *Ap. J. (Letters)*, **363**, L21 [1990]), as well as no trend of the decrease of the energy of the bump with the increasing luminosity (predicted if it came from optically thick thermal matter) has been observed yet.

A Model for the X-ray Variability of the Seyfert Galaxy NGC 6814

M.A. Abramowicz[1,2,3], *G. Bao*[3], *F. Fiore*[2], *A. Lanza*[3,1], *E. Massaro*[2], *G.C. Perola*[2], *E.A. Spiegel*[4,1,3], *and E. Szuszkiewicz*[5,1,3]

[1] NORDITA, Copenhagen,
[2] Universita' di Roma,
[3] SISSA and ICTP, Trieste,
[4] Columbia University, New York,
[5] Queen Mary College, London.

Abstract. We present a model which explains the periodic flares observed in the *EXOSAT* ME 1985/289 X-ray light curve of the Seyfert galaxy NGC 6814 as due to the general relativistic lensing of light emitted by a bright spot closely orbiting a supermassive black hole. For several cases, corresponding to different disks inclinations and spot orbital radii, we calculated the theoretical light curves, including all the relativistic effects. We tested statistically the model by comparing the power spectral densities of the synthetic light curves with that of the observed one and found a good agreement for an inclination angle $i \approx 85°$.

1. Introduction

The X-ray emission of the Seyfert 1 galaxy NGC 6814 is characterized by rapid variations with a large amplitude. Mittaz and Branduardi-Raymont (1989) analyzed the *EXOSAT* X-ray 1985/289 light curve and discovered a series of peaks with a recurrence period P_0 of about 12200 s. The same periodicity is also visible in the *GINGA* data (see *e.g.* Done *et al.* 1990). Figure 1 shows the *EXOSAT* 1985/289 light curve folded on the period P_0. A narrow flare is clearly evident with a duty cycle of about 0.1 and almost symmetric shape, with the signal raising and declining sharply around its maximum. Fiore, Massaro and Barone (1991a, b) calculated the power spectrum density (PSD) of this observation: two components can be distinguished, a series of peaks at frequencies which are harmonic of the fundamental at $1/P_0$, and a low frequency noise (LFN) which can be described by a power law with exponent about -1. The frequencies and the powers of the harmonic peaks $f_0, ..., f_4$ are given in Table 1 (the asymptotic value of the power of the Poisson noise fluctuations is 2 with the adopted normalization).

In this contribution we show that a simple model consisting of a bright spot located at the innermost part of a thin accretion disk which orbits a supermassive black hole is able to represent the observational data when the inclination angle of the system is $\geq 83°$. In the following we explain

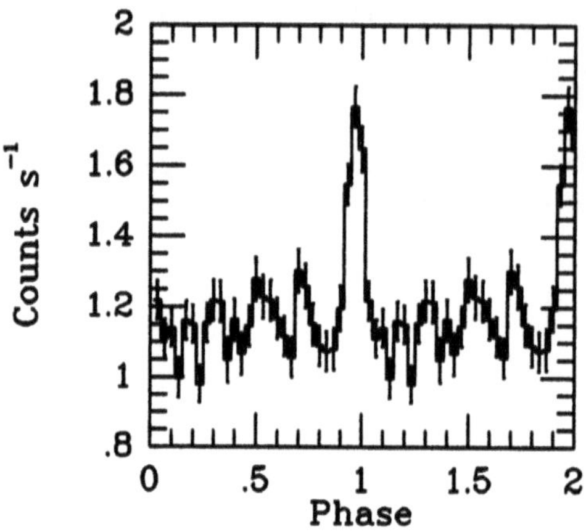

Fig. 1. The *EXOSAT* 1985/289 light curve folded on the period P_0.

the details of the numerical procedures used in the calculations and in the statistical tests supporting this finding. We are aware that the upper limit of 1 keV to the iron line width (Kunieda et al. 1990) could be problematic with our model. We shall discuss briefly this topic in the conclusion.

Table 1. Harmonic peaks in the *EXOSAT* observation 1985/289.

Harmonic N.	f_0	f_1	f_2	f_3	f_4
Freq. (10^{-4}Hz)	0.825	1.625	2.475	3.300	4.125
Norm. Power	27.5	54.3	47.3	39.6	40.7

2. Computation of the light curve

We assume that the disk is very thin. The radius of the orbit of the spot is denoted by r_S and the viewing angle by i. The spot occupies a small area $\Delta r_S \Delta\phi_S$ of the disk surface and in its rest frame isotropically emits radiation with intensity I_S. The orbital angular velocity of the spot is Ω_S and its orbital speed, $v_s = \Omega_S r_S (1 - 2M/r_S)^{-1/2}$. In the special case of Keplerian rotation it is $\Omega_S = (M/r_S)^{3/2}$. Here M is the mass of the central black hole and the units are such that $c = G = 1$. The theoretical light curve, which gives the observed X-ray flux F as a function of the proper time τ of the observer, is fully determined by the model parameters M, r_S, Ω_S, i, $\Delta r_S \Delta\phi_S$, I_S, and the assumed Schwarzschild geometry of the central black hole by the following algorithm (Bao, 1991).

Let ϕ_S denote the orbital phase of the spot which is located at a circular orbit with the dimensionless radius $R_S = r_S/M$. One has

$$\cos\phi_\infty = \sin i \cos\phi_S, \tag{2.1}$$

where ϕ_∞ is given in terms of the Jacobi integrals $J_1 = J(\chi_S/2, k)$, $J_2 = J(\chi_\infty/2, k)$ and the complete elliptic integral $K(k)$ by one of the formulae,

$$\phi_\infty^{(1)} = 2\left(\frac{p}{q}\right)^{1/2}[J_1 - J_2], \quad \phi_\infty^{(2)} = 2\left(\frac{p}{q}\right)^{1/2}[2K(k) - J_1 - J_2], \tag{2.2}$$

with p being the dimensionless perihelion distance of the orbit and $q^2 = (p-2)(p+6)$, $k^2 = (q-p+6)/2q$,

$$\sin^2(\frac{\chi_\infty}{2}) = Q = \frac{qk^2-2}{qk^2}, \quad \sin^2(\frac{\chi_S}{2}) = Q + \frac{2p}{qk^2R_S}. \tag{2.3}$$

Either with $\phi_\infty^{(1)}$ or with $\phi_\infty^{(2)}$ the system of equations (2.1), (2.2), (2.3) has a unique solution $p = p(\phi_S)$. Knowing it one computes the auxiliary quantities

$$B^2 = \frac{p^3}{p-2}, \quad B' = \int_0^{u_S}\sqrt{\frac{2u_S^3 - u_S^2 + 1/B^2}{(2u^3 - u^2 + 1/B^2)^3}}\frac{du}{u_S^2B^3}, \quad u_S = 1/R_S, \tag{2.4}$$

the dimensionless elapsing time $\Delta\tau = \Delta t/M$, and the photon arrival time (measured in terms of the observer proper time) as a function of the orbital phase ϕ_S and the model parameters,

$$\Delta\tau = \int_0^{1/R_S}\frac{du}{u^2(1-2u)\sqrt{1-B^2u^2(1-2u)}}, \quad \tau = \frac{\phi_S}{\Omega_S} - \Delta\tau. \tag{2.5}$$

Next, one computes in function of the orbital phase (and thus in terms of the observer proper time) the photon impact parameter, the redshift factor, and the solid angle $\Delta\Pi$ subtended by the spot at the observer,

$$\lambda = \frac{\sin\phi_S \tan iB}{\sin\phi_\infty}, \quad g = \frac{\sqrt{(1-v_S^2)(1-\frac{2M}{r_S})}}{1-\Omega_S\lambda},$$

$$\Delta\Pi = \frac{\cos iBB'\Delta R_S\Delta\phi_S}{d^2(1-\sin i^2\cos\phi_s^2)} \tag{2.6}$$

where d is the distance from the source to the observer. If the intensity is calculated in arbitrary units, this could be any constant number.

Knowing both the redshift and the solid angle, the observed flux in terms of the proper time of the observer is

$$F(\tau) = I_0\Delta\Pi g^4. \tag{2.7}$$

3. Statistical test of the model

Because of the presence in the source signal of a conspicuous LFN, the possible presence of systematic errors in the background subtraction and the temporal gaps, lasting several hundred seconds, connected with the swaps of the ME array, a direct evaluation of the confidence level for the agreement of the model light curve with the data is difficult and must be discussed with care. We chose therefore to construct, for each orbit in a range of the model parameters, several hundreds simulated time series, taking into account all the effects mentioned above, and to compare their PSDs with that of the observed one. Each synthetic time series was obtained by generating first a noise composed by $N = T/2\Delta T$ (ΔT is the bin size), sinusoidal components $Z_n(t)$:

$$Z_n(t) \propto \sqrt{A} f_n^{-q/2} cos(2\pi f_n t + \phi_n), \tag{3.1}$$

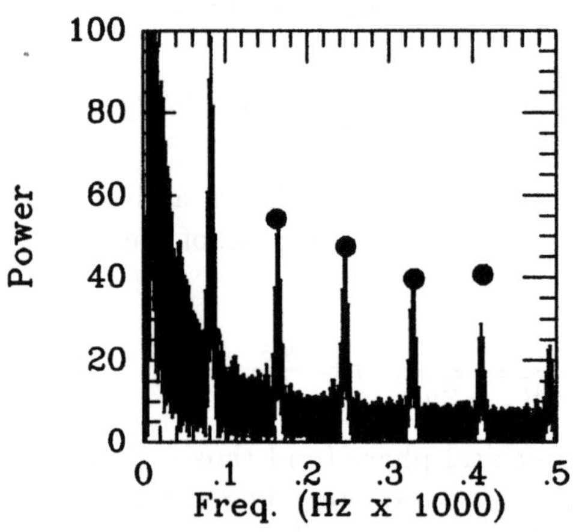

Fig. 2. An example of PSD computed from our model.

with frequencies f_n and phases ϕ_n chosen randomly and where $A = 7.6 \times 10^{-4}$ and $q = -1$ are the best fit values of the observed LFN (Fiore, Massaro and Barone 1991a). We then added to each one of these time series the model light curve and a poisson noise to account for the statistical fluctuations, and sampled them exactly at the same times as the observed light curve. Finally, we considered the effect of systematic errors in the background subtraction, which consists in the possible presence of a step between two consecutive swaps of the ME counters, and added to or subtracted from each segment between two swaps a random level with a mean value equal to 1% of the background count rate. We thus obtained the PSD of each simulated time series and derived the distribution of the power at each frequency.

In Figure 2 we show the PSD obtained for $i = 85°$ and $r_S = 10 r_G$ ($r_G = 2GM/c^2$); the bars repr esent the interval comprising 67 % of the power values in the simulations and the filled circles represent the power actually measured at $f_0, ... f_4$ (and given in Table 1).

We looked first for the existence of a statistical correlation among the powers at the harmonic frequencies and found that it is very small. To evaluate the probability that the observed powers at $f_0, ..., f_4$ are produced by the given model signal, we derived from the simulations performed for each couple of parameter r_S and i the

statistical estimator η^2 defined as:

$$\eta^2 = \sum_{i=0}^{4} \frac{1}{V_i}(P_i - \overline{P}_i)^2, \qquad (3.2)$$

where P_i are the power values at f_i in each synthetic light curve and $\overline{P}_i$ and V_i are the mean values and the variances respectively. We used the distribution of η^2 to estimate the probability of a chance occurrence of $\eta^2 \geq \eta_0^2$, where η_0^2 is the value of the estimator when to the powers P_i in (3.2) are given the observed values.

Table 2. Probability (%) that the observed distribution of power is produced by a model with given i and r_S.

	$r_S/r_G = 8$	$r_S/r_G = 10$	$r_S/r_G = 15$
i = 75°	< 0.25	< 0.25	< 0.25
80°	0.25	< 0.25	0.25
83°	7.75	4.50	6.25
85°	19.75	12.00	11.50

In Table 2 this probability is reported for some representative couples of r_S and i values. We note that it is very small for $i \leq 80°$ and that it rapidly increases for higher inclinations to reach the maximum of about 20% at $i = 85°$ and $r_S = 8r_G$. Because of numerical problems we were not able to investigate inclinations greater than the latter value. We note also that for $i = 85°$ the probability decreases slightly with the radius, but this dependence is so weak that does not allow one to constrain significantly the radius and therefore to reach any significant conclusion about the mass of the black hole. The probability of the model depends only slightly on the radius because at high inclination angles the gravitational lensing is much more effective than the Doppler boosting but it is much less sensitive than the latter to the distance of the spot from the central black hole (Cunningham and Bardeen, 1973).

We would like to note that the above results concern the simple hypothesis of a single spot moving along the orbit. The actual situation could be more complex with several spots with perhaps different intrinsic luminosities moving on the same orbit and the agreement of the orbital model with the data can be improved further by using these two other degrees of freedom.

4. Conclusions

We constructed a detailed quantitative model for the X-ray variability of NGC 6814. The model explains the periodic variability of the source detected in the course of the EXOSAT observation 1985/289 in terms of the orbital motion of a single very bright spot on the accretion disk surface. We found that the data are reproduced quite well for inclinations $i \approx 85°$. A problem with our model seems to be the upper limit of 1 keV to the iron line width (for a gaussian profile) reported by Kunieda et al. (1990). Despite of the fact that the line intensity is enhanced when the spot region is moving almost orthogonally to the line of sight, calculations specific to the "spot model show that the expected width should be considerably larger than 1 keV for distance from the black hole smaller than 15-20 r_G; the expected line profile, however, is complex and very far from gaussian, and a conclusion on what could be the crucial quantity to measure for a direct test of the model, as well as for a precise estimate of the parameter r_S (hence the black hole mass) to which the line width is very sensitive, should await for higher energy resolution experiments. Another interesting possibility is that the spot is occasionally eclipsed for parts of its orbit with the consequent reduction of the line width.

We acknowledge financial support from NORDITA and from the Italian MURST and GIFCO/CNR.

References

Bao, G., 1991: *Astron. Astrophys.*, submitted.

Cunningham, C.T., Bardeen, J.M., 1973: *Astrophys. J.*, **183**, 237.

Done, C., Koyama, K., Kunieda, H., Madejski, G., Mushotzky, R., Turner, T.J., 1990: in: IAU Colloquium No. 129, *Structure and Emission Properties of Accretion Disks*, Paris, France.

Fiore, F., Massaro, E., Barone, P., 1991a: in preparation.

Fiore, F., Massaro, E., Barone, P., 1991b, in: Proceedings of the Heidelberg conference on *Physics of Active Galactic Nuclei*

Kunieda, H., Turner, T.J., Awaki, H., Koyama, K., Mushotzky, R., Tsusaka, Y., 1990: *Nature*, **345**, 786.

Mittaz, J.P.D., Branduardi-Raymont, G., 1989: *Monthly Notices Roy. Astron. Soc.*, **238**, 1029.

Hard X-ray Observations of Active Galactic Nuclei with SIGMA

J. Ballet[1], *F. Lebrun*[1], *J. Paul*[1], *L. Salotti*[1], *E. Jourdain*[2], *L. Bassani*[2], *L. Bouchet*[2], *P. Mandrou*[2], *R. Sunyaev*[3], *E. Churazov*[3], *M. Gilfanov*[3], *A. Finogenov*[3], *S. Grebenev*[3], *M. Pavlinsky*[3], *Yu. Apalkov*[3], *I. Lagunov*[3]

[1] SAp, CEN Saclay, 91191 Gif/Yvette Cedex, France
[2] CESR, 9 av. du C^{el} Roche, BP 4346, 31029 Toulouse Cedex, France
[3] IKI, Profsoyuznaya 84/32, Moscow 117296, Russia

1. Introduction

The main motivation for studying active galactic nuclei (AGN) in hard X-rays is that because of their flat spectra they emit a large fraction of their luminosity in this range ([1] for 3C 273), with a luminosity per decade increasing with energy. Because of the photons' energy and the time scale for variations the hard X-rays are probably emitted close to the central engine, which is another measure of their importance to AGN theory.

The X-ray and hard X-ray radiation is thought to be due to inverse Comptonization of UV seed photons (the "big blue bump) by non thermal electrons ([2] and references therein). It is important to extend X-ray spectra to higher energies, if possible up to the electron rest mass energy and beyond, in order to validate the mechanisms that have been proposed.

Unfortunately, due to the low sensitivity of hard X-ray experiments (compared to the classical X-ray band), only the three brightest AGN (Cen A, NGC 4151, 3C 273) have been substantially studied to date. We present here new results on these three sources.

2. Observations and data analysis

The French SIGMA telescope [3] was launched on December 1, 1989 aboard the Soviet GRANAT space observatory in a highly eccentric orbit (4 days period). It is a coded aperture system designed to provide images of the sky in the 35 keV to 1.3 MeV range with 13 arcmin angular resolution. Its broad band sensitivity is about 20 mCrab at 100 keV, decreasing markedly above 200 keV. Its sensitivity to an annihilation line near 511 keV (FWHM $\leq$ 50 keV) is about $2 \times 10^{-3}\,\mathrm{ph\,cm^{-2}\,s^{-1}}$. This is not enough to constrain severely the electron–positron annihilation line in AGN. These values are obtained for a typical observing time (10^5 s) at the 3σ level. The GRANAT observatory also features the Soviet ART-P telescope [4], sensitive from 3 to 30 keV. ART-P observes the same targets as SIGMA, for the first six hours of a typical one day session. With ART-P and SIGMA, GRANAT spans more than two decades.

The observing logs of Cen A, NGC 4151 and 3C 273 in 1990, as well as images, have been presented elsewhere [5]. We report here on the preliminary analysis of the spectral images. We made two simplifying assumptions leading to systematic errors:

1. The pulse height spectra were transformed into photon (or energy) spectra by dividing the data at each energy by an equivalent effective area, calibrated on the Crab Nebula [6] below 200 keV and on ground calibrations above 200 keV. This procedure doesn't take explicitly into account the Compton contribution inside the detector.
2. The abscissa scale was taken to be the ground channel–energy calibration. The gain has actually changed by about 5% in flight.

Both are smaller than the statistical errors.

The ART-P spectra were taken from *Apalkov et al.* [7]. Because we are interested here in the high energy behaviour, the interstellar absorption was removed before fitting the data (i.e. it was fixed to their best fit value).

Because of the simplifications and uncertainties mentioned above, the quantitative results should be taken with caution, although the main trends are certainly qualitatively correct.

3. Results

3.1 Cen A

SIGMA observed Cen A for only one session (24/07/1990). The ART-P data [7] were summed over two close sessions (19 and 24/07/1990). The spectrum is presented in Fig. 1. It is compatible with an unbroken power law over the whole range. The level of the 100 keV flux is lower than that measured by HEAO 1 in 1978 [8] (also shown in Fig. 1), but comparable to that measured by HEXE in 1987 [9]. What is specific to this observation is the hardness of the spectrum (Cen A usually has an energy index of -0.6 to -0.8), due to the particularly low 10 keV flux. It is also associated with a moderate column density (0.3×10^{23} cm^{-2} whereas 1×10^{23} cm^{-2} is common in this source).

The overall power law index should be taken with some caution because it depends sensitively on the SIGMA/ART-P ratio. They were not obtained exactly at the same time, whereas this source is known to be very rapidly variable [10]. Actually the best fit to the ART-P data alone is at a slope of −0.60 [7]. The ART-P flux was twice lower on the day of the SIGMA observation than on the other session summed in the spectrum. This would hint at an even harder spectrum! However the ART-P session covers only the beginning of the SIGMA session, and the source could have brightened within the observation.

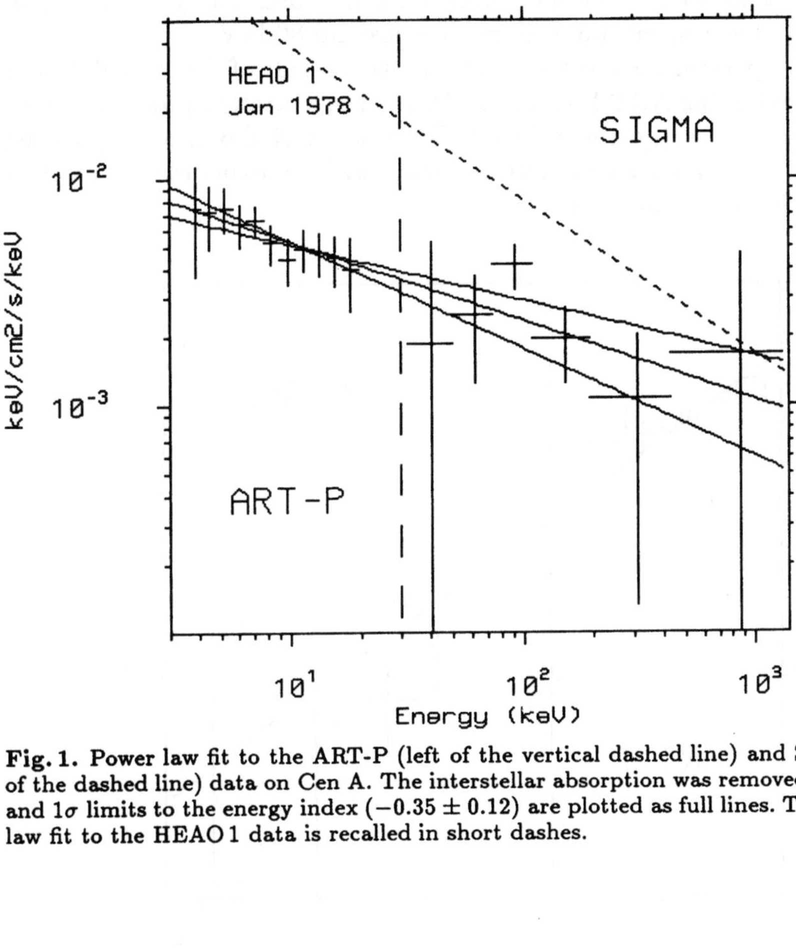

Fig. 1. **Power law fit to the ART-P (left of the vertical dashed line) and SIGMA (right of the dashed line) data on Cen A. The interstellar absorption was removed. The best fit and 1σ limits to the energy index (-0.35 ± 0.12) are plotted as full lines. The best power law fit to the HEAO 1 data is recalled in short dashes.**

3.2 NGC 4151

The SIGMA observations of NGC 4151 in 1990 have been presented in detail elsewhere [11]. The data presented here were restricted to the November 1990 observations, that have ART-P counterparts (actually the ART-P data [7] start four days earlier). There is no significant difference to the average spectrum (the November observations make up more than 75% of the total). The SIGMA spectrum is very steep (the energy index is certainly steeper than −1, and the best fit is closer to −2). This is unusual, since the spectrum of NGC 4151 has mainly been observed as a single power law (with energy index around −0.5) up to the hard X-ray range [12]. One HEAO 1 observation (June 1978) [13] does show a break above 50 keV and could be similar to ours. The ART-P spectrum is flat (−0.44) and at a high level, placing it somewhat outside the flux–index correlation [14]. The absorbing column (removed here) was large ($1.7 \times 10^{23}\,\mathrm{cm}^{-2}$). Note that as the X-ray flux was about twice higher than the usual one, the 100 keV flux was barely below that seen in the flat unbroken state observed by HEAO 1 [13] or HEXE [9]. Their spectra and ours cross around 80 keV.

The joint spectrum is plotted in Fig. 2. As a result of the very different indices derived in the ART-P and SIGMA ranges, a straight power law does not fit the data correctly (the reduced χ^2 is 1.55 for 15 d.o.f., corresponding to 8% confidence in such a model). A model with an exponential cutoff at 80 keV fits the data quite well.

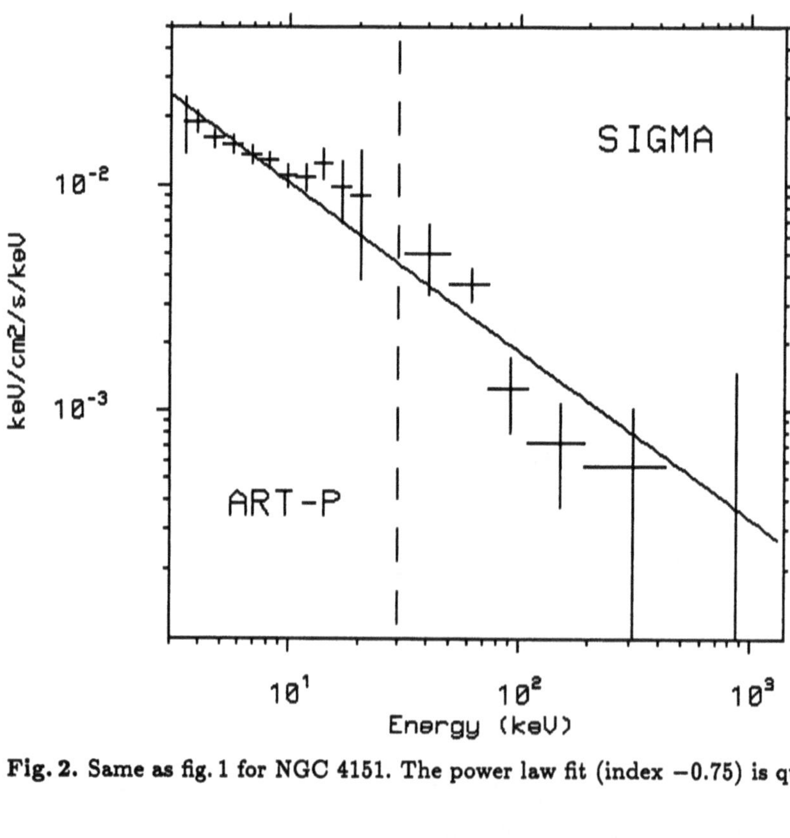

Fig. 2. Same as fig. 1 for NGC 4151. The power law fit (index −0.75) is quite poor.

For this source the break is unlikely to be an artifact of the SIGMA – ART-P comparison:

1. ART-P saw no apparent change in the X-ray flux from one day to the next when SIGMA was on (after November 15) [7].

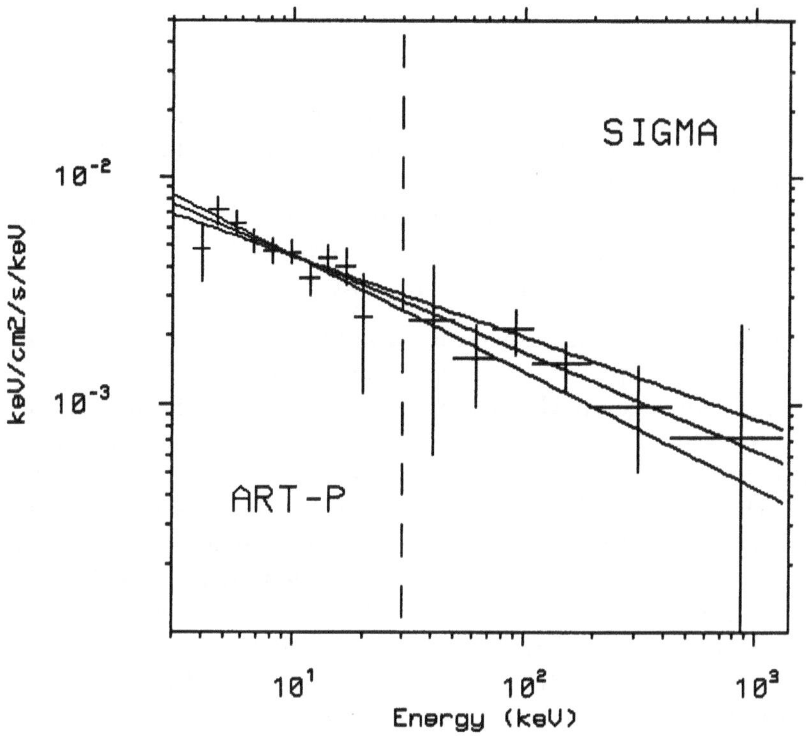

Fig. 3. Same as fig. 1 for 3C 273. The best fit index is -0.43 ± 0.08.

2. The ART-P flux was lower when SIGMA was off, so that the flux averaged only over the sessions with SIGMA on would be even higher.
3. The first two SIGMA points match the last ART-P points well enough.
4. The spectral indices derived in the two ranges are incompatible.

Finally we remark that since this state persisted for one week, and was probably already caught by HEAO 1, it cannot be an extremely rare transient phenomenon, and must be a recurring state, with a smaller probability than the unbroken hard power law.

3.3 3C 273

3C 273 was observed four times by SIGMA in 1990, in June, July and twice in November. The spectrum was summed over all sessions. The ART-P spectrum [7] corresponds only to the June observation. The absorbing column density was negligibly small. The spectrum is plotted in Fig. 3. It is compatible with an unbroken power law over the whole spectrum. The X-ray and hard X-ray fluxes, as well as the spectral slope, agree well with previous measurements [9, 15]. At least one AGN was observed in its usual state !

3.4 Conclusions

We have presented here spectra of the three brightest AGN from 3 keV to 300 keV, obtained with GRANAT during its first year of operation.

1. Cen A was seen in a very low (below 10 keV) and probably very hard (energy index -0.35) state.
2. NGC 4151 had a flat (-0.44) spectrum below 30 keV associated with a high level, broken to a very steep one above 50 keV.
3. 3C 273 was seen in its usual flat (-0.43) state.

It is not clear whether the spectral variability seen in Cen A and NGC 4151 (through comparison with previous observations [8, 9, 12, 13]) is relevant to the "average AGN, since Cen A is very underluminous and NGC 4151 has an X-ray spectrum definitely flatter than average [16]. However in both cases the flux variability seems to be larger in the classical X-ray domain than in hard X-rays. Both 100 keV fluxes are very similar to those observed by HEXE [9], whereas the 10 keV fluxes were stronger (NGC 4151) or weaker (Cen A) than average. **Acknowledgements.** We acknowledge the paramount contribution of the SIGMA project group of the CNES Toulouse space center to the overall success of the SIGMA mission. We thank the staff of the Lavotchin space company, of the Babakin space center, of the Baikonur space center and of the Evpatoria ground station for their unfailing support.

References

1 Courvoisier, T.J.-L., et al., 1987: *Astron. Astrophys.*, **176**, 197.
2 Done, C., Ghisellini, G., Fabian, A.C., 1990: *Monthly Notices Roy. Astron. Soc.*, **245**, 1.
3 Paul, J., et al.. 1991: Proc. 18th COSPAR Symp. (ME.4), *Adv.Sp.Res.*, in press.
4 Sunyaev, R., et al., 1990: *Adv.Sp.Res.* **10**, n° 2, 233.
5 Ballet, J., et al., 1991: Proc. 11th Moriond Astrophysics Meeting on *The early observable universe from diffuse backgrounds*, Editions Frontières, in press.
6 Jung, G.V., 1989: *Astrophys. J.*, **338**, 972.
7 Apalkov, Yu., et al., 1991: Proc. 28th Yamada Conf. *Frontiers of X-ray Astronomy.*
8 Baity, W.A., et al., 1981: *Astrophys. J.*, **244**, 429.
9 Maisack, M., et al., 1989: Proc. 23rd ESLAB Symp., ESA **SP-296**, 975 (1989)
10 Wang, B., et al., 1986: *Publ. Astron. Soc. Japan*, **38**, 685.
11 Jourdain, E., et al., 1991: *Nature* submitted.
12 Perotti, F., et al., 1991: *Astrophys. J.*, **373**, 75.
13 Baity, W.A., et al., 1984: *Astrophys. J.*, **279**, 555.
14 Yaqoob, T., Warwick, R.S., 1991: *Monthly Notices Roy. Astron. Soc.*, **248**, 773.
15 Turner, M.J.L., et al., 1990: *Monthly Notices Roy. Astron. Soc.*, **244**, 310.
16 Turner, T.J., Pounds, K.A., 1989: *Monthly Notices Roy. Astron. Soc.*, **240**, 833.

The X-ray Power Spectrum in the "Hot Spot" Model

Gang Bao

SISSA, Strada Costiera 11, 34014 Trieste, Italy

The non-uniform feature of the innermost part of the accretion disk of AGN, from which most of the X-rays are emitted, can be approximated as a group of independent flares and clumps–hot spots on the surface of the accretion disk. This model is considered to be capable of explaining the short term X-ray AGN variability–timescale independent and multi-peaked features in the variability power spectrum.

Fig. 1 (lower panel) is the light curve with its power spectrum for a spot located at $r = 6r_g$, the viewing angle of the observer being $\theta = 85^o$. The black hole has $10^7 M_\odot$. There is a sharp peak in each period. As we know, the sharp peaks in the light curve result from the gravitational focusing when the spot is exactly behind the black hole. This non-sinusoidal signal makes the high-order harmonics visible in the variability power spectrum (see the upper panel). Two interesting results can be obtained from the power spectrum. First, the high frequency, higher than orbital frequency, contains power comparable to the fundamental, which means that the full relativistic effect, the Doppler and the gravitational focusing effect, will flatten the power spectrum of many spots. Second, since the focusing effect gets stronger when the spots are closer to the black hole, more power goes to the harmonics, which will smooth the predicted "cut off" frequency corresponding to the fastest rotating frequency of a spot.

To produce the power-law-shaped power spectrum, we calculate the light curves and the power spectrum due to many spots. Spots are distributed according to the distribution function $n(R) \sim R$ in the range of $3.5r_g - 25R_g$, and all spots are assumed to be of the same intensity and life time. The black hole has 10^8 solar masses. Fig. 2 shows the power spectrum for $N = 66$ spots, the viewing angle of the observer being $\theta_o = 80^o$. No sharp "cut off" frequency is seen in the power spectrum.

Since AGNs are believed to be a black hole system, our results provide a possible explanation for the difficulty in finding roll-off phenomenon in X-ray variability power spectrum.

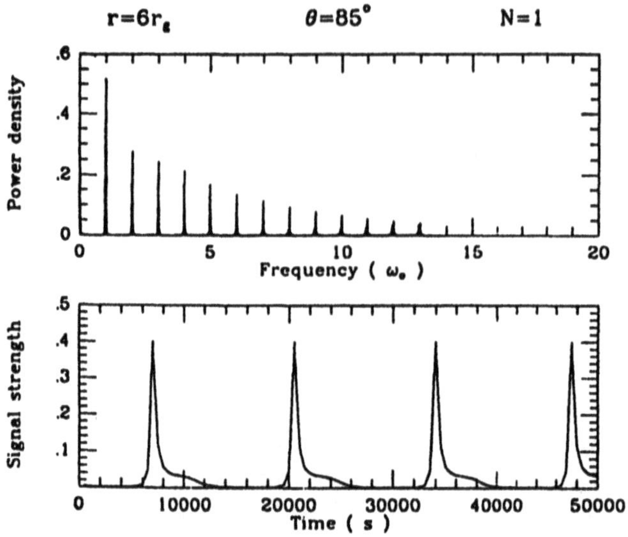

Fig. 1. The light curve (the lower panel) and the power spectrum for one spot. ω_o is the fundamental corresponding to the orbital frequency of the spot.

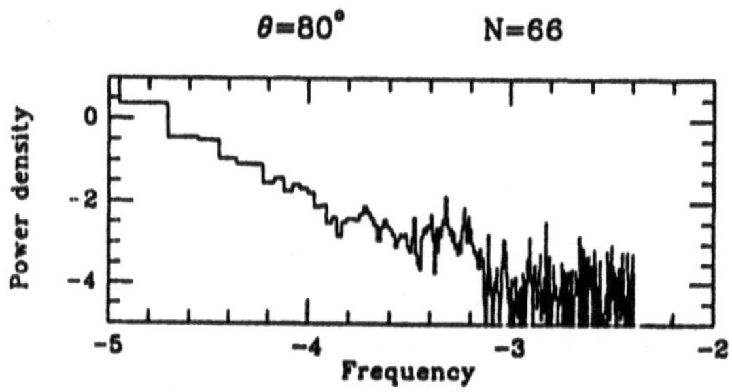

Fig. 2. The power spectrum for $N = 66$ spots.

References

1. Abramowicz, M., Bao, G., Lanza, A., Zhang, X-H., 1991: *Astron. Astrophys.*, 245, 454.
2. Bao, G., 1991: *Astron. Astrophys.*, submitted.

ROSAT All Sky Survey Observations of IRAS Galaxies

Th. Boller[1,2], *E.J.A. Meurs*[2], *W. Brinkmann*[2], *H. Fink*[2], *U. Zimmermann*[2]

[1] Sternwarte Babelsberg, R.-Luxemburg-Str. 17a, D-1591 Potsdam, Germany

[2] Max Planck Institut für extraterrestrische Physik, D-8046 Garching bei München, Germany

We present first results on IRAS galaxies detected in the ROSAT All Sky Survey. The sample of IRAS galaxies, selected from the IRAS PSC employing multivariate statistics [1] consists of 14,315 galaxies. Tests of the quality of the selection [2] on optically confirmed fields (see [3, 4]) and by cross-correlating the sample with the SIMBAD catalogue suggest that the sample is less than 6 % contaminated by galactic sources, allowing to consider statistical correlations with the ROSAT X-ray data.

Some high-flux IRAS galaxies (about 3%) remain unselected using the described procedure, as the classifier rejects objects with infrared fluxes greater than about 10 Jansky at 100 μm. This restriction avoids an even stronger contamination by high-flux galactic objects. We improved the selection by adding a not complete, but likely representative sample of IRAS galaxies, with fluxes greater than 10 Jansky at 100 μm, by cross-correlating the NGC-catalogue with the IRAS PSC. This results in a list of 393 IRAS galaxies with NGC-counterparts. We added these 393 high-flux IRAS galaxies to the previously selected 14,315 IRAS galaxies, resulting in a total number of 14,708 objects to be investigated at ROSAT X-ray wavelengths.

The sample analyzed consists of 75 objects after roughly one quarter of the data have been processed. The work with these data has led to the following conclusions:

1. About 98 % of the galaxies selected from the IRAS PSC exhibit count rates lower than $10^{-2} counts/sec$ in the ROSAT (0.1 - 2.4) keV X-ray band. They remain undetected in the All Sky Survey.
2. The separation of the investigated sample into X-ray detected and X-ray non-detected IRAS galaxies is a consequence of the X-ray activity of the objects and is not a distance effect. The X-ray detected subgroup of IRAS galaxies is dominated by Seyfert galaxies (about 50%). Within the sample of the X-ray non-detected objects their fraction is only 5%.
3. The sample of X-ray detected IRAS galaxies contains nearby galaxies, spiral galaxies, Seyfert galaxies, elliptical galaxies and QSOs. We use scatter diagrams to analyze the luminosity-luminosity correlations for

these subgroups of X-ray detected IRAS galaxies (Fig. 1). We find a common trend that the strongest increase of the X-ray luminosity with respect to the Far-infrared or optical(B)- luminosity is encountered for Seyfert galaxies, followed by spiral galaxies and then nearby galaxies. These different slopes in the scatter diagrams imply constraints for the X-ray production for the subgroups which we investigate in detail.

4. For the mean X-ray luminosities we find the relation $\overline{L_X}(QSO) > \overline{L_X}(Seyfert) > \overline{L_X}(Spirals) > \overline{L_X}(nearby)$.
5. The catalogue of IRAS galaxies detected in the ROSAT All Sky Survey and a detailed interpretation of the results will be published immediately after the survey data have been processed completely.

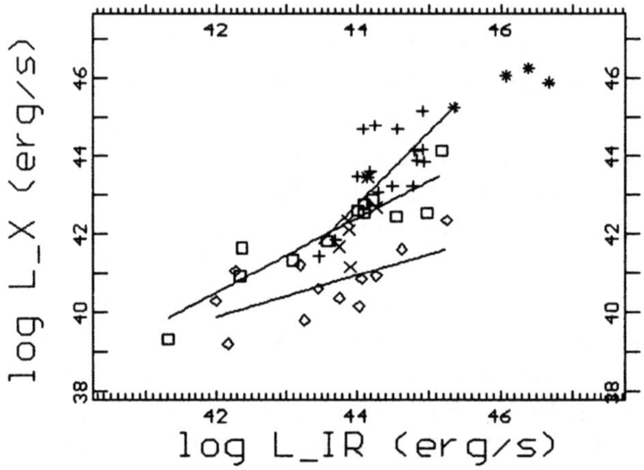

Fig. 1. scatter diagram for Seyfert 1 (+), Seyfert 2 (x), nearby (diamonds), spirals (rectangles) and QSOs (*); best fits from regression analysis are plotted for Seyfert galaxies, spirals and nearby galaxies

References

1 Adorf, H.-M., Meurs, E.J.A., 1988: *Lect. Notes. Phys.* **310**, 315.
2 Boller, Th., Meurs, E.J.A., Adorf, H.-M., 1991: preprint.
3 Leggett, S.K., Clowes, R.G., Kalafi, M., MacGillivray, H.T., Puxley, P.J., Savage, A., Wolstencroft, R.D., 1987: *Monthly Notices Roy. Astron. Soc.*, **227**, 563.
4 Wolstencroft, R.D., Savage, A., Clowes, R.G., MacGillivray, H.T., Leggett, S.K., Kalafi, M., 1986: *Monthly Notices Roy. Astron. Soc.*, **223**, 279.

Soft X-ray Spectra of Radio-loud Quasars: First Results from the ROSAT Survey

H.Brunner[1], *P.Friedrich*[1], *R.Staubert*[1], *H.-U.Zimmermann*[2], *N.Schartel*[2], *and H.Fink*[2]

[1] Astronomisches Institut, D-7400 Tübingen, Germany
[2] MPI für Extraterrestrische Physik, D-8046 Garching, Germany

Results. We have analysed a sample of 98 radio-loud quasars from the Kühr *et al.* [3] catalogue of radio sources, detected in the *ROSAT All Sky X-Ray Survey.* We find a mean energy power law (PL) index between 1.10 to 1.33 (90 % confidence; **Fig. 1**), considerably steeper then the PL spectra measured by the *Einstein IPC* for radio loud quasars (Wilkes and Elvis [6], Canizares and White [2], Brunner *et al.* [1]). We find a similar result for a smaller sample of 33 radio-loud quasars observed both in the *ROSAT Survey* and by the *Einstein IPC*, thus excluding the possibility that the observed difference in X-ray slope might be due to selection effects. We find a mean X-ray spectral index between 1.35 and 1.59 (90 % confidence) in a comparison sample of 30 bright *radio-quiet* quasars (**Fig. 2**). Our preliminary investigation of *ROSAT Survey* data thus at this point does not confirm the large difference in PL index found between radio-loud and radio-quiet quasars observed with the *Einstein IPC*. The result is not changed when we restrict our investigation to quasars with galactic column densities below $2 \times 10^{20} cm^{-2}$, where the effect of low energy absorption on *ROSAT PSPC* spectra is only minor. The difference in X-ray slopes as compared to previous *Einstein IPC* results is also apparent in a small sample of quasars with well defined X-ray spectra (**Fig. 3**). We have performed a KS test on the hardness ratio (hard range; see below) distributions of our radio-loud and radio-quiet quasar samples and find no significant difference between the two distributions. Our result thus should be independent of any remaining uncertainty in the *ROSAT PSPC* energy calibration. As there is no significant difference between the PL index distributions found in the total and hard *ROSAT PSPC* energy ranges (see next section, **Fig. 1** and **Fig. 2**), the difference in X-ray slope as compared to the *Einstein IPC* would have to be due to the *Einstein IPC* sensitivity above 2.4 keV. We are at this point however also investigating to what extent any remaining uncertainties in the *ROSAT* calibration may influence our result.

Method. Our investigation is based on hardness ratios calculated in the 0.1 to 0.4 keV vs. 0.4 to 2.4 keV (total range) and 0.4 to 1.1 keV vs. 1.1 to 2.4

keV (hard range) respectively. When assuming that there is no internal low energy absorption in the quasars investigated, the hardness ratios may be transformed into the corresponding PL indexes using the galactic column density in the line of sight as determined from 21 cm radio measurements (Stark *et al.* [5]), and applying the response of the *ROSAT PSPC* detector. We assess the intrinsic distribution of X-ray PL indexes, which is assumed to be Gaussian, using a maximum likelihood technique described in Maccacaro et al. [4]. Our results are presented in the form of 90 % confidence contours in the ($< \alpha >$,σ) plane, where $< \alpha >$ is the mean energy PL index and σ describes the width of the intrinsic distribution of PL indexes.

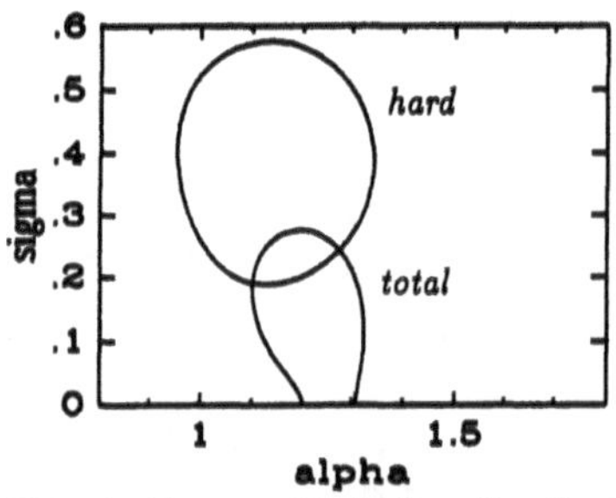

Fig. 1. Energy PL index distribution determined in total and hard energy range for our sample of 98 radio-loud quasars.

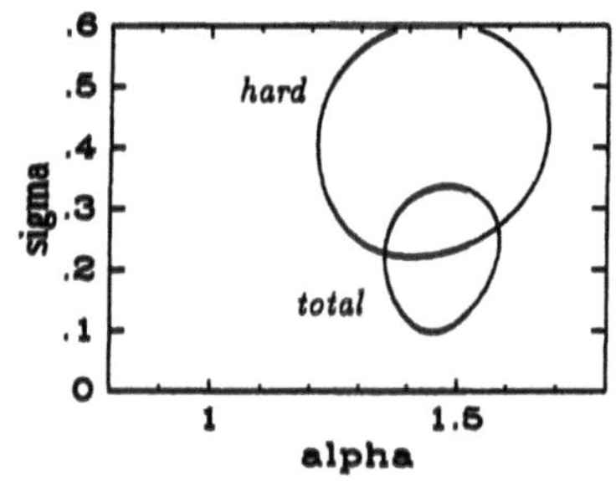

Fig. 2. Same for comparison sample of 30 bright radio-quiet quasars.

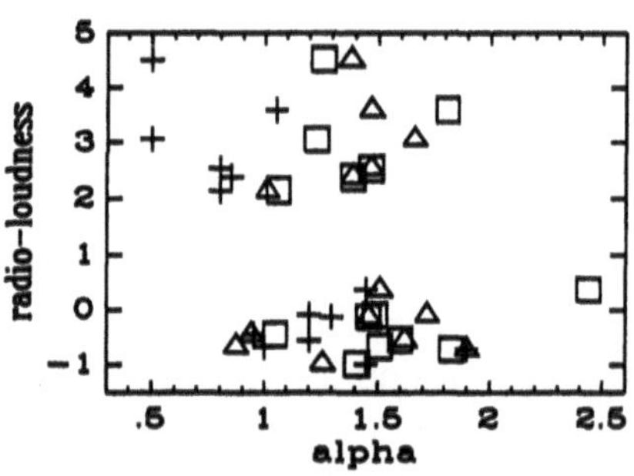

Fig. 3. Radio-loudness (as defined by Wilkes and Elvis [6]) as a function of X-ray PL index for 14 bright quasars. Crosses: *Einstein IPC*; squares: *ROSAT PSPC* total range; triangles: *PSPC* hard range. *IPC* PL indexes were read from the contour plots in Wilkes and Elvis [6] assuming no internal absorption. Note the obvious trend of radio-loudness with X-ray PL index in the *IPC* data. No such trend is apparent in the *ROSAT* data. For the radio-quiet quasars the *IPC* and *ROSAT* slopes agree within statistical errors (*ROSAT* total PL index errors are below 0.2 for all 14 sources).

References

1. Brunner, *et al.*, 1989: Proc.*23rd ESLAB Symp.* **ESA SP-296**, 905.
2. Canizares, White, 1990: *Astrophys. J.*, **326**, 680.
3. Kühr, *et al.*, 1979: *A Catalogue of Radio Sources*, MPIfR preprint
4. Maccacaro, *et al.*, 1988: *Astrophys. J.*, **323**, 243.
5. Stark, *et al.*, 1986: privately distributed computer tape.
6. Wilkes, Elvis, 1987: *Astrophys. J.*, **323**, 243.

EXOSAT X-ray Spectra of Quasars

Andrea Comastri[1,2]

[1] Istituto di Radioastronomia del CNR, Bologna, Italy.
[2] Dipartimento di Astronomia, Universitá di Bologna, Italy.

We have measured the X-ray spectra in the 0.1-10 keV energy range for 17 high luminosity and low redshift AGNs (mostly PG quasars) using the EXOSAT Low and Medium-Energy arrays. The sample has been selected on the basis of soft X-ray flux, and includes **all** the AGNs detected by the *Einstein* IPC with a flux greater than $6 \times 10^{-12} erg\ cm^{-2} s^{-1}$. The most important results of our analysis can be summarized as follow (for a detailed discussion see Comastri *et al.* 1991).

The hard X-ray spectra (above $\sim$ 2 keV) of **all** the AGNs in our sample can be described by a single power law model, with a wide variety of energy indices ($0.4 < \alpha < 1.3$). The average spectral slope is $< \alpha >= 0.89 \pm 0.06$, but the distribution is centered around $\alpha \simeq 1.0$. The mean slope for the objects of our sample is significantly steeper than the one found by Turner and Pounds (1989) for a sample of low luminosity AGNs ($< \alpha >\simeq 0.7$). This difference can be explained by the different selection criteria of the two samples. While our sample, selected in the soft X-ray domain, is likely to be biased against flat spectra, the Turner and Pounds sample, selected in the hard X-ray domain, is biased against steep spectra. Within our limited sample the hard X-ray spectral index does not appear to be correlated with other physical parameters of the sources, such as the intrinsic optical and X-ray luminosities, the redshift, or the radio properties.

The addition of the low energy (LE) data has allowed the analysis of the X-ray spectra over an energy interval spanning about two orders of magnitude (0.1 - 10 keV). Approximately 1/3 (6 out of 17) of the sources in our sample show evidence for soft excess emission, below 1 keV. For these objects a power law plus a soft excess modeled by a steep power law or by the high energy tail of a blackbody is a significantly better description of the data than a single power law over the energy range 0.1-10 keV. In particular, for each of these objects we find that the best fit slopes of the steep power law fits at low energies are in the range $\alpha = 2.7 - 4.2$, while the preferred temperatures of the fitted blackbodies are in the range 40-80 eV.

The "break" energy between the soft and hard component is in the range 0.35-0.75 *keV* at the source.

In order to compare our results with the *Einstein* quasars spectral survey (Wilkes and Elvis 1987), we have performed spectral fits to our data in the 0.1-4.0 *keV* energy range (similar to the IPC range, 0.2-3.5 *keV*). The results show that in this softer energy range there is a clear trend with radio-quiet quasars having steeper slopes than radio-loud quasars, in agreement with the Wilkes and Elvis work. However, while in the Wilkes and Elvis sample the radio-loud objects are grouped around $\alpha \simeq 0.5$ and the radio-quiet ones around $\alpha \simeq 1.0$, our slopes for the two subsample are systematically steeper, with an average $\Delta\alpha$ of the order of 0.2-0.3. It is interesting to note that the mean slope for the radio-quiet objects in the 0.1-4.0 *keV* energy range ($<\alpha>= 1.2 - 1.3$) agrees with the mean slope (1-2 *keV*) of the sources, mostly AGNs, in the ROSAT Medium Sensitivity Survey (Hasinger, Schimdt, and Trümper 1991).

The other apparently significant correlation in our data is between the presence of soft excess and radio loudness. Five out of six objects with detected soft excess emission are radio quiet and four of these are among the last five most radio-quiet objects in the sample.

References

Comastri, A., Setti, G., Zamorani, G., Elvis, M., Giommi, P., Wilkes, B.J., McDowell J. C., 1991: *Astrophys. J.*, in press.
Hasinger, G., Schmidt, M., Trümper, J., 1991: *Astron. Astrophys. (Letters)*, **246**, L2.
Turner, T.J., and Pounds, K.A., 1989: *Monthly Notices Roy. Astron. Soc.*, **240**, 833.
Wilkes, B.J., and Elvis, M., 1987: *Astrophys. J.*, **323**, 243.

X-Ray Observations of Mkn 501 and Mkn 421 with ROSAT/XRT and Ginga

H.H. Fink[1], *F. Makino*[2], *R. Warwick*[3], *N. Kawai*[4], *R. Okayasu*[4], *C. Otani*[2]

[1] Max-Planck-Institut für extraterrestrische Physik, D-8046 Garching, Germany
[2] Institute of Space and Astronautical Science, 3-1-1 Yoshinodai, Sagamihara, Kanagawa, Japan
[3] X-Ray Astronomy Group, University of Leicester, Leicester LE1 7RH, Leicester, Great Britain
[4] The Institute of Physical and Chemical Research, 2-1 Hirosawa, Wako, Saitama 351-01, Japan

The BL Lacertae objects Mkn 501 and Mkn 421 were observed simultaneously with the X-Ray Telescope onboard of ROSAT during the All Sky Survey and the LAC detectors of the *Ginga* satellite in a pointed mode, in order to establish light curves and improved spectra in the (0.1 - 15)keV energy range.

The observation of Mkn 501 took place between Aug 11th and Aug 15th, 1990, with a rather limited common coverage of about 5 hours. Nevertheless, the light curve does show neither in the soft band, (0.2 - 2)keV, nor in the harder bands, (1.2 - 4.6) keV and (4.6 - 10)keV, any significant variation over the entire observational interval. This behaviour of the source agrees well with previous findings that Mkn 501 tends to vary preferentially on longer time scales than days [1]. The spectrum of Mkn 501 was obtained by fitting power law model spectra separately to the ROSAT and Ginga data, respectively. The resulting spectral parameters are given in Tab.1. The spectra of both experiments are consistent in flux in the overlapping region and the combined spectrum can be described by a single power law spectrum with a photon index of -2.63 in the (0.1 - 15)keV range modified by an absorbing column close to the galactic value ($1.73 \cdot 10^{20}$ cm^{-2}, [2]). There is a hint of a depression in the spectrum below 1 keV, the statistical significance of which is, however, too low to establish the existence of an absorption line as recently claimed to be a ubiquitous feature in the X-ray spectra of BL Lac objects [3].

Mkn 421, which was observed in a high state between Nov 13th and Nov 16th, 1990, with a common coverage as long as 1.5 days, showed strong and rapid variability in flux. The soft X-ray flux dropped by more than 20% within about three hours and recovered 30 hours later to its initial level. The Ginga light curve exhibited strong variability, too, the amplitude of

Table 1. Best Fit Spectral Parameters

	MKN501		MKN421	
	ROSAT/XRT	*Ginga*	*ROSAT/XRT*	*Ginga*
Energy Range [*keV*]	0.1 - 2.0	1.5 - 15	0.1 - 2.0	1.5 - 15
Normalization [10^{-2} *phot/cm²sec keV*]	1.76 ± 0.11	1.83 ± 1.00	8.23 ± 0.25	10.7 ± 0.3
Photon Index	2.51 ± 0.15	2.64 ± 0.05	2.41 ± 0.08	2.75 ± 0.02
N_H [10^{20} cm^{-2}]	2.13 ± 0.43	1.44(*fixed*)	1.63 ± 0.17	1.4(*fixed*)
e-fold Energy [*keV*]				25 ± 9
χ^2/dof	37/43	18/26	104/108	22/25

which increased with increasing energy, at least in the covered (1.6 - 20)keV energy range. The time scales of the medium energy variations were again of the order of some hours. Surprisingly, there seemed to be no obvious correlation between the variations in the ROSAT and in the Ginga band, respectively.

The gross spectra taken in the ROSAT band at the higher and lower flux levels do not show significant differences as a function of count rate, although the hardness ratio of the (0.1 - 0.5)keV band to the harder (0.5 - 2.4)keV range decreased slightly as the count rate dropped. As can be seen from Tab.1, the power law indices determined with both experiments are inconsistent, with the Ginga spectrum ,- for which a better fit could be achieved by introducing an high energy cut-off -, being considerably steeper than the ROSAT spectrum. Also the fluxes differ by more than 10% in the overlapping region of the spectra. A possible reason for these discrepancies could be a curvature of the spectrum, - an idea which is supported by a first view onto the overall radio-to-X-ray spectrum, which shows the ROSAT/Ginga spectrum lying directly on the extrapolation of the well established, curved radio-to-UV spectrum, in contrary to previously measured X-ray spectra, [4]. On the other hand, the lack of correlation between the flux variations at soft and harder X-ray wavelengths seems to indicate separate emission mechanisms or emission regions. Further simultaneous observations at soft and harder energies and modelling in the context of inhomogeneous synchrotron and CSS models are necessary to establish the nature of these distinctions.

References

1 Staubert, R., Bazzano, A., Ubertini, P., Brunner, H., Collmar, W., Kendziorra, E., 1986: *Astron. Astrophys.*, **162**, 16.

2 Madejski, G.M., Mushotzky, R.F., Weaver, K.A., Arnaud, K.A., Urry, C.M., 1990: *NASA preprint* No. 90-018.

3 Elvis, M., Lockman, F.J., Wilkes, B., 1989: *Havard-Smithsonian, CfA* No. 2779.

4 Makino, F., et al., 1987: *Astrophys. J.*, **313**, 662.

The X-ray Variability of NGC 6814: A new Analysis of the EXOSAT Observations with Non-Linear Methods

F. Fiore[1], *E. Massaro*[1], *and P. Barone*[2]

[1] Istituto Astronomico, Universita' "La Sapienza", Via G.M. Lancisi 29, I-00161 Roma, Italy
[2] Istituto per le Applicazioni del Calcolo, CNR, Roma, Italy

1. Introduction.

The Seyfert 1 galaxy NGC 6814 was observed by the european X-ray observatory *EXOSAT* five times over a period of about two years (1983/247, 1983/306, 1984/152, 1984/287, and 1985/289). Mittaz and Branduardi-Raymont (1989) discovered in the last observation an intense signal with a period of 12,200 seconds (corresponding to the frequency $f_0 = 0.82 \times 10^{-4}$ Hz). This periodicity is well apparent in the power spectrum computed by means of the usual Fourier Tansform algorithm, where a series of strong peaks at the frequencies of the first four harmonics ($f_1, ..., f_4$) is well evident. Only this observation, however, is long enough ($T \sim 10^5$ s) to obtain by means of standard harmonic tools a good frequency resolution ($= 1/T$); the durations of the other four observations ($\sim$ 25,000 s) do not allow a frequency estimate accurate enough to establish if the above periodic signal were or not present.

To gain more information from these shorter data sets on the stability of this periodic component, we performed a new analysis by means of a non-linear algorithm, the Prony's Method (PM). This method, described in more detail by Barone, Massaro and Polichetti (1989), is particularly suited to analyse series with a small number of points, and, at least in principle, allows a frequency resolution much higher than $1/T$. We apply the PM to time series of 300 s bins, so that having about 72 data points. Because the algorithm can work only with evenly spaced data, it was necessary to interpolate some segments of the original light curves.

2. Results.

The results of the analysis of the four short observations are summarized in Table 1. We reported the frequency values (in 10^{-4} Hz) closest to those of the harmonics detected in the observation 1985/289, the value of $< f_0 >$

estimated by averaging those detected in each individual series and the number of frequencies used in the PM analysis. The largest difference between $< f_0 >$ and the value of the obs. 1989/285 is about 0.5%, much better than the natural resolution ($\sim$ 59%); moreover, the greatest frequency shift found in this analysis occurs at f_4 (1983/306) and corresponds only to 7%, that must be compared with the natural value of 10%. This result is a strong indication that the periodic signal of the obs. 1985/289 does not represent an episodic event but it is a stable characteristic of the X-ray emission of NGC 6814. From the uncertainty on the P_0 estimate (typically $\sim$ 100 s) we can give the upper limit of $\sim 2 \times 10^{-6}$ s/s for the time derivative of the fundamental period. A possible explanation for such a noticeable stability can be given by the orbital motion of some bright source around a massive black hole as discussed by Abramowicz et al. (these proceedings). As a final remark we wish to point out that the fourth harmonic is the only one found in all the observations, while the fundamental frequency was never detected.

Table 1

	1983/247	1983/306	1984/152	1984/287
f_1				1.641
f_2	2.428	2.612		
f_3	3.411		3.282	3.274
f_4	3.967	3.813	4.067	4.043
f_5			4.906	
f_6		5.751		
$< f_0 >$	0.818	0.818	0.818	0.816
Nf	19	18	18	24

References

Barone, P., Massaro, E., Polichetti, A.: 1989, *Astron. Astrophys.*, **209**, 435.
Mittaz, J.P.D., Branduardi-Raymont, G.: 1989 *Monthly Notices Roy. Astron. Soc.*, **238**, 1029.

Are Radio Loud AGN γ-ray Loud?

Demosthenes Kazanas and Peter Giovanoni

NASA/Goddard Space Flight Center, Code 665, Greenbelt, MD 20771

There appears to be a well defined dichotomy of AGN in radio loud and radio quiet. Though the spectral distribution in these two classes appears to be roughly similar between $\sim 10^{12} - 10^{18}$ Hz, they differ drastically in their radio properties: The radio quiet AGN have very little emission below $\simeq 10^{12}$ Hz; they appear to self-absorbed below this frequency. They do exhibit some radio emission at lower frequencies ($\sim 10^9$ Hz) but that comprises only a very small fraction (10^{-6}) of their bolometric luminosity and it is apparently emitted from an extended region. On the other hand, the radio loud AGN, though they can have prominent jets, they have an unresolved compact core which exhibits a continuous smooth spectrum in the range $10^9 - 10^{15}$ Hz; more specifically the intensity of radiation from the unresolved core is "flat" ($F_\nu \propto \nu^\alpha; \alpha \simeq 0$) between $10^9 - 10^{12}$ Hz, while it is roughly $F_\nu \propto \nu^{-1}$ above 10^{12} Hz.

The continuous, smooth spectra of radio loud AGN between $10^9 - 10^{15}$ Hz are suggestive of a single radiation mechanism, thought to be synchrotron. Given that for the typical values of size ($\sim 10^{15}$ cm) and magnetic field (~ 10 G) expected for the "central engine" of AGN their emission should be self-absorbed at $\simeq 10^{12}$ Hz, the continuous "flat" emission between $10^9 - 10^{12}$ Hz indicates that the radiation in this part of the spectrum comes from a source of much larger size. In fact, the form of the spectrum in this range has been modeled as the superposition of a continuum of self-absorbed components extending to distances of several parsecs from the "central engine"; it was also shown (Marscher 1977) that under the assumption of an inhomogeneous distribution of electrons of a power law form in density and energy one could reproduce the observed "flat" spectra by the judicious choice of the appropriate power law indices. However, the mere presence of relativistic electrons at large distances from the compact object poses severe conceptual problems because of their short life times (much shorter than the light crossing time); also, one has to contend with the problem of producing an electron distribution function with a rather specific dependence both in energy and in radius.

These problems can successfully addressed by the presence of relativistic protons in the AGN "central engine". These particles give rise to relativistic neutrons, through pp or $p\gamma$ collisions, which promptly escape their production site to a distance $\simeq 3 \times 10^{13}\gamma_n$ cm, where, upon their decay they can provide locally the requisite relativistic electrons. In addition, because the energy is transported by neutrons, the injection function of the electrons has a low energy cut-off due to the neutron decay which leads to a distribution function $\propto \gamma_e^{-2}$ independent of any details (Kazanas 1984), while the radial dependence of its normalization is $\propto r^{-2}$ (Giovanoni and Kazanas 1990). These functional dependencies (along with a $B \propto r^{-1}$ dependence for the magnetic field) are precisely those needed to give rise to the "flat" spectra of the compact cores of radio loud AGN. This model also suggests that in sources with "flat" radio spectra relativistic electrons and hadrons are present a large distances ($\simeq 1$ pc) from the compact object. These produce γ-rays, either by inverse Compton or π^0 decay, in regions of very small compactness, thereby ensuring their escape without absorption by $\gamma\gamma$ interactions and suggesting that radio loud AGN should also be γ-ray loud. The detection of > 100 MeV γ-rays from 3C 273 and the recent *GRO* detection of 3C 279 in the same energy range (Bertsch et al. 1991) seem to support such a point of view.

The same model also provides a framework for understanding the radio loud - radio quiet AGN dichotomy: For sources of sufficiently high compactness (L/R) the neutrons produced within the central source in the hadronic collisions never get the chance to escape their production region; they are converted back into protons by pn or $p\gamma$ collisions, thus suppressing emission from large radii (and also from frequencies $\nu < 10^{12}$ Hz) thus producing spectra similar to those of radio quiet AGN (Giovanoni and Kazanas 1990) which, by extending the above arguments, should also be γ-ray quiet due to the absorption of high energy γ-rays in the high compactness environment of the compact object.

References

Bertsch, D., et al., 1991: *IAU Circ.* **5311**.
Giovanoni, P.M., Kazanas, D., 1990: *Nature,* **345**, 319.
Kazanas, D., 1984: *Astrophys. J.*, **287**, 112.
Marscher, A.P., 1977: *Astrophys. J.*, **216**, 244.

X-ray Comptonized Spectra of Optically Thick Sources: Softening Versus Hardening as a Sign of Changed Geometry

Iosif I. Lapidus[1,2]

[1] SISSA, International School for Advanced Studies, Strada Costiera 11, 34014 Trieste, Italy
[2] Space Research Institute, Academy of Sciences, Profsoyuznaya 84/32, 117810 Moscow, Russia

The problem of proper treating the radiative transfer in optically thick envelopes at X-ray temperatures appears to be important in various astrophysical circumstances – from compact sources to AGN. X-ray bursters seem to demonstrate most explicitly how the change of geometry of the emitting region results in the drastical change of emergent spectra (Lapidus, 1991, Titarchuk et al., 1991). Similar effects are by any means to be taken into account when constructing detailed AGN models. The spectrum formation is usually dominated by two effects - free-free absorption and emission and Comptonization. The departure from a blackbody spectrum will grow as $l = L/L_{Edd}$ approaches unity. Fitting of emergent spectra by a Planckian-type formula with two parameters - the general multiplicative factor and the fitting temperature - gives T_{fit}/T_0 roughly between 1 and 1.5 at l varying from 0 to 1. Here T_0 is the effective temperature directly related to the luminosity. This is an explanation for the "super"- Eddington luminosities observed for some bursters.

However, in the expansion phase spectra appear to be quite different – flat near to the maximum and with a noticeable soft excess.

The main difference between conditions of spectrum formation in the various phases is one of geometry - the atmosphere becomes essentially spherical in the expansion phase and not plane-parallel as it has been in the hydrodynamically-stable case at $l < 1$. In the latter case the scale-height of the atmosphere (< 1 meter) is much less than the radius of the star and in the former case the relation is entirely the opposite.

Our numerical calculations of the radiative transfer were based on the code enabling one to get both the radiation spectra at all optical depths and the electron temperature in a selfconsistent way. The strategy is that employed in Lapidus et al., 1986 for the determination of influence of Comptonization on the spectra of X-ray bursters at $l < 1$. Two main features were taken into account - spherical geometry instead of plane-parallel and lack of hydrodynamical equilibrium. The first resulted in a different form of the space diffusion operator. The second enforced us to use results of numerical modelling of mass outflows from neutron stars during strong bursts (Ebisuzaki, et al., 1983) as a background for the radiative transfer.

The spectra calculated have a distinct characteristic shape. They may be quite well fitted by a Planckian law in the region near to the maximum and at higher frequencies but at lower frequencies the spectra are very flat showing a noticeable soft excess. Typically T_{fit}/T_0 varies in the region 0.2-0.8 depending on outflow details.

These results make evident the crucial role of geometry on emerging spectra. Namely, Comptonized spectra going out of a plane-parallel semi-infinite medium are always harder than corresponding effective blackbody ones, and those leaving the spherically extended envelopes are always noticeably softer. This main conclusion seems to be independent on details of structure of the atmosphere.

References

Lapidus, I.I. 1991, *Astrophys. J.*, in press..

Titarchuk, L., Lapidus, I., and Haberl, F. 1991, in: *Proceedings of 28th Yamada Conference "Frontiers of X-Ray Astronomy"*, Nagoya, Japan, April 8-12, 1991, in press.

Lapidus, I.I., Sunyaev, R.A., and Titarchuk, L.G. 1986, *Sov.Astron.Lett.*, **12**(6), 383.

Ebisuzaki, T., Hanawa,T., and Sugimoto, D. 1983, *Pub.Astr.Soc.Japan*, **35**, 17

Constraints on Reflection in AGN

M.Maisack[1,2], *T.Yaqoob*[3], *R.Staubert*[1], *and E.Kendziorra*[1]

[1] Astronomisches Institut Tübingen, Germany
[2] George Mason University, Fairfax VA, USA
[3] Institute of Space and Astronautical Science , Kanagawa, Japan

Models including photon-photon pair production (e.g. Lightman and Zdziarski, 1987)and reprocessing by cold material (e.g. Lightman and White, 1988) have recently received wide attention, especially since Ginga spectra of Seyfert galaxies have shown a hard X-ray tail, expected from the latter process.

Zdziarski et al. (1990) have proposed a model for the canonical energy index $\alpha \approx 0.7$ (e.g. Rothschild, R.E. et al., 1983) that includes both effects.

According to this model, the canonical index is the result of the superposition of a steeper ($\alpha \approx 0.9$), pair-dominated spectrum of a source with high compactness and a reprocessed component.

High energy X-ray data have been taken with the HEXE instrument on the Mir space station (Reppin et al., 1983) of three AGN (3C 273, NGC 4151 and Cen A). The original results on spectral fitting and time variability can be found in Maisack et al. 1989 and Maisack, 1991. We have examined these data in the light of the reflection model. The spectra of the three sources were fitted with a power law and a reflection code following White, Lightman and Zdziarski (1988).

One NGC 4151 observation took place on the same days as a Ginga observation of the same source (1987 November 21-24) . The composite spectrum is well described by a power law with energy index 0.37 ± 0.11 and requires no reflected component (Maisack and Yaqoob, 1991). The best-fit value for the reflected component is only $11 \pm 36\%$ of the value expected if the observer sees the full contribution of both the primary and the reflected component as calculated by White et al., 1988. A very small reflected component may be expected if the source is seen edge-on (Maoz, et al. 1991), as has been suggested by some observers (e.g. George and Fabian, 1991).

We find that, while the HEXE data alone cannot constrain the magnitude of a reflected component and are in fact consistent with no reflected component at all, they rule out the above model for 3C 273 (observed 1987 December 3-10) and Cen A (observed 1988 July 23-24) since they require harder primary spectra.

For 3C 273, we find an energy index of α=0.56±0.10 for the power law fit and α=0.54±0.09 for the reflection fit. All errors are given for 90% confidence for joint variation of parameters.

The respective values for Cen A are 0.69±0.07 and 0.68±0.08.

For 3C 273, this means that the spectrum cannot be explained by simple pair cascade models. At an observed compactness of $l \approx 140$, the spectrum should be significantly steeper than the observed 90% confidence limit of α =0.66. Such a hard spectrum at high compactness can be explained by either photon starvation (Zdziarski et al., 1990) or reacceleration (Done, et al., 1989). These effects could occur in a model with spatially separated emission regions, e.g. in the wind and shock model (Camenzind and Courvoisier, 1983 and Courvoisier and Camenzind, 1989).

In Cen A, the situation is less clear since the compactness is poorly determined. As in NGC 4151, a possible accretion disk might be viewed edge-on, consistent with the lack of broad lines (Antonucci and Miller, 1985) and lack of significant UV flux. The HEXE data are not statistically significant enough to constrain the magnitude of a reflected component. However, the slope of the incident power law spectrum cannot be steeper than 0.76 (90% confidence). This implies that either the compactness of this source is low, if simple pair models are a good description, or else photon starvation or reacceleration are suggested.

We note that all three AGN which are bright enough to be studied with present day instruments like HEXE at energies above 30 keV do not show evidence for a reflected component. Although this may be due to a selection effect, it may also be an indication that reflection is not necessarily a universal effect in AGN.

References

Antonucci, R.R.J. & Miller, J.S. 1985. *Astrophys. J.*, **297**, 621 .

Camenzind, M. & Courvoisier, T. J.-L. 1983. *Astrophys. J.*, **266**, L83.

Courvoisier, T. J.-L. & Camenzind, M. 1989. *Astron. Astrophys.*, **224**, 10.

Done, C., Ghisellini, G. & Fabian, A.C. 1989. *Monthly Notices Roy. Astron. Soc.*, **245**, 1.

George, I.M. & Fabian, A.C. 1991. *Monthly Notices Roy. Astron. Soc.*, **249**, 352.

Lightman, A.P. & White, T.R. 1988. *Astrophys. J.*, **335**, 57.

Lightman, A.P. & Zdziarski, A.A. 1987. *Astrophys. J.*, **319**, 643.

Maisack et al. 1989. Proc. 23rd ESLAB Symposium , 975.

Maisack, M. 1991. PhD thesis, University of Tübingen

Maisack, M. & Yaqoob, T. 1991. *Astron. Astrophys.*, in press.

Maoz, D. et al. 1991. *Astrophys. J.*, **367**, 493.

Reppin et al. 1983. in: Non-thermal and Very High Temperature Phenomena in X-Ray Astronomy, 279

Rothschild, R.E. et al. 1983. *Astrophys. J.*, **269**, 423.

White, T.R., Lightman, A.P. & Zdziarski, A.A. 1988. *Astrophys. J.*, **331**, 939.

Zdziarski, A.A. et al. 1990. *Astrophys. J.*, **363**, L1.

Zdziarski, A.A., Coppi, P.S. & Lamb, D.Q. 1990. *Astrophys. J.*, **357**, 149.

X-Ray and FIR Emission from AGN and Normal Galaxies

J. Miguel Mas Hesse[1,2], *I.F. Mirabel*[3], *and H.U. Zimmermann*[1]

[1] Max-Planck-Institut für extraterrestrische Physik
[2] Dpto. de Astrofísica Universidad Complutense de Madrid
[3] Centre d'Etudes Nucleaires de Saclay

We have analysed the soft X-ray and FIR emission in a sample containing 37 "normal" star-forming galaxies (NSFG) (Long & Speybroek 1983, Fabbiano *et al.* 1982) and 64 AGN (Kruper *et al.* 1990, Wilkes & Elvis 1987, Tananbaum *et al.* 1986.) The FIR luminosities have been integrated over the $40 - 120\ \mu m$ IRAS band. The X-ray data are from *Einstein* IPC (range between 0.5 and $3.5 - 4.0$ kev) and have been corrected from interstellar absorption.

We have plotted in Fig. 1 L_x *vs.* L_{FIR}. The solid lines correspond to linear fits performed separately to the Sey 1+QSO and NSFG+Sey 2 subsamples. Both luminosities are correlated for each sample with very similar slopes (around 0.9), but Sey 1+QSO show a significant excess in soft X-ray emission compared to NSFG. The majority of the Sey 2 galaxies appear as the extrapolation to high luminosities of the NSFG. These segregation can be better seen in Fig. 2, where we show the frequency histograms for the ratio XF between X-ray and FIR luminosities. Sey 1 and QSO peak at $log(XF) = -0.25$, around two orders of magnitude higher than NSFG. Sey 2 galaxies clearly overlap with both NSFG and Sey 1+QSO, but peak at significantly lower values than Sey 1 galaxies.

On the other hand, Edelson *et al.* (1987) have shown that there is no segregation between Sey 1 and 2 galaxies when analysing the correlation between FIR and radio luminosities. It seems therefore that they show similar properties at wavelengths weakly affected by extinction, but differ significantly when the extinction is important, as would be expected if Sey 2 galaxies are essentially Sey 1 ones obscured by a molecular torus.

We have used a set of starburst evolutionary synthesis models (Mas-Hesse & Kunth 1991) to derive the values of the XF ratio expected in star-forming galaxies, assuming average luminosities for supernova remnants (SNR), massive stars and compact binaries (Fabbiano *et al.* 1982). The predicted total XF, $log(XF) = -2.4$, is consistent with the values observed in NSFG and Sey 2 galaxies. The soft X-ray radiation would be dominated by compact binaries, being the contribution of SNR only $log(XF) = -4.7$.

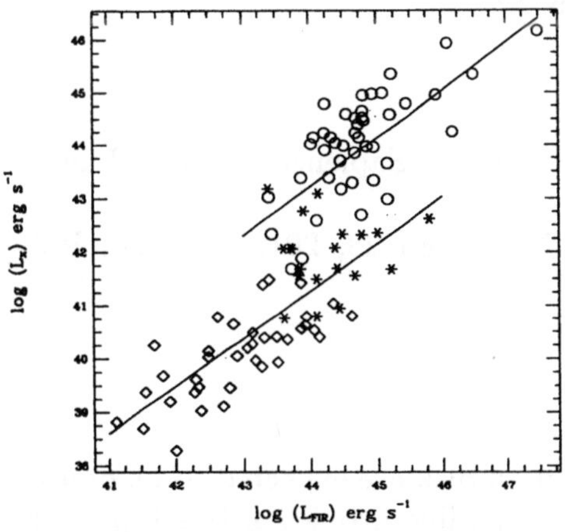

Fig. 1. $log(L_x)$ *vs.* $log(L_{FIR})$. NSFG: ◇; Sey 2: ∗; Sey 1+QSO: ∘. The solid lines are least squares fits $L_x \sim (L_{FIR})^s$, with $s = 0.86$ (Sey 1+QSO) and 0.89 (NSFG+Sey 2).

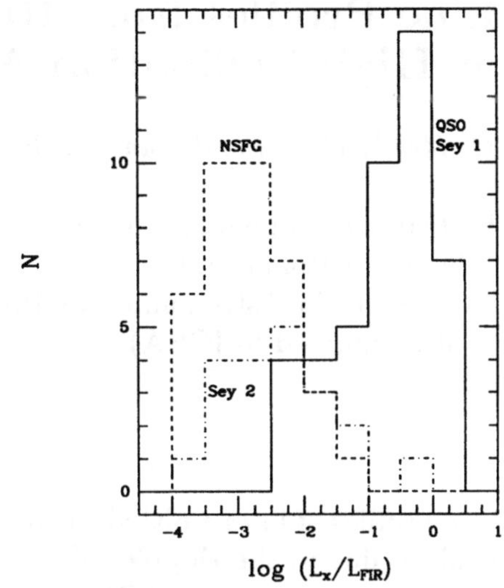

Fig. 2. Frequency histograms for the 3 subsamples.

Our results indicate that the X-ray emission from Sey 1 galaxies can not be accounted for by a star formation episode with average properties, unless the emission from SNR is around 10^4 times higher than observed in Local Group Galaxies, or the abundance of compact binaries relative to OB stars a factor 100 higher than in our Galaxy.

References

Edelson, R.A., Malkan, M.A., Rieke, G.H., 1987, *Astrophys. J.*, **321**, 233.
Fabbiano, G., Feigelson, E., Zamorani, G., 1982, *Astrophys. J.*, **256**, 397.
Kruper, J.S., Urry, C.M., Canizares, C.R., 1990, *Astrophys. J. Suppl.*, **74**, 347.
Long, K.S., Van Speybroek, L.P., 1983, in "Accretion driven stellar x-ray sources", Lewin, W. (ed.), Cambridge Univ. Press, p. 117
Mas-Hesse, J.M., Kunth, D., 1991, *Astron. Astrophys. Suppl.*, **88**, 399.
Tananbaum, H., Avni, Y., Green, R.F, Schmidt, M., Zamorani, G., 1986, *Astrophys. J.*, **305**, 57.
Wilkes, B.J., Elvis, M., 1987, *Astrophys. J.*, **323**, 243.

Line Profiles from Relativistic Disks at High Inclination Angles

G. Matt[1], *G.C. Perola*' *and L. Stella*[2]

[1] Istituto Astronomico, Universitá di Roma "La Sapienza", Via G. Lancisi 29, 00161 Roma, Italy

[2] Osservatorio Astronomico di Brera, Via Brera 28, 20121 Milano, Italy (also affiliated to ICRA)

The profiles of lines emitted by an accretion disk are expected to be doubly peaked, due to the Doppler effect. Hydrogen line profiles of this type have been observed in some AGNs, and in a few of them they are asymmetric, suggesting that relativistic effects are present. We have calculated the profiles from a relativistic disk on the basis of the fully relativistic treatment of the line photon intensity and shifts in the Schwarzschild metric described in Fabian et al. (1989). They apply to the case of a flat, optically thick disk, with inner and outer radii r_i and r_o measured in units of $r_g = GM/c^2$. The main result of our calculations is that at very high inclination angles, due to general relativistic effects, conspicuous structures emerge between the two Doppler horns.

In Figures 1 and 2 profiles are given separately for an isotropic surface emissivity whose radial dependence is parameterized as r^q, and for the iron K_α fluorescence line. In the first instance they were chosen to illustrate the strong dependence on the value of q, and the relevance of the features mentioned above when $i \gtrsim 70°$ and $r_o \lesssim 1000$. The study of a particular line, however, requires a physical model of the angular and radial dependence of its emissivity. We have done so in the second instance, by using the iron K_α emissivity computed in detail by Matt et al. (1991a) for a centrally illuminated, optically thick, "cold" disk. This line is particularly promising for a search of relativistic effects, because the typical values of its integral spectral properties (Nandra et al. 1991) appear to be consistent with those expected from a disk with an inner boundary close to the last stable orbit (Matt et al., 1991b).

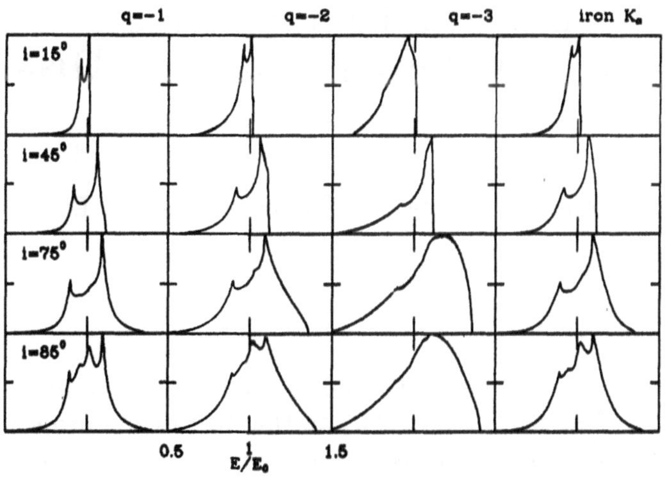

Fig. 1. Line profiles for different i, assuming $r_i=6$ and $r_0=100$.

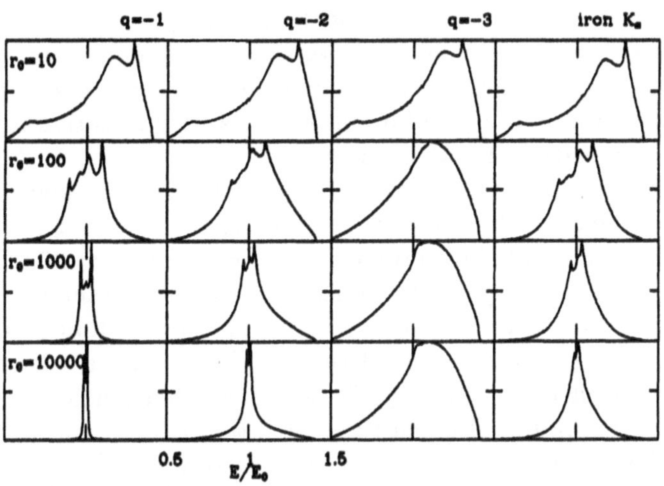

Fig. 2. Line profiles for different r_0, assuming $r_i=6$ and $i=85°$

References

Fabian A.C., Rees M.J., Stella L., White N.E., 1989, *Monthly Notices Roy. Astron. Soc.*, **238**, 739

Matt G., Perola G.C., Piro L., 1991a, *Astron. Astrophys.*, in press.

Matt G., Perola G.C., Piro L., Stella, L., 1991b, *Astron. Astrophys.*, submitted.

Nandra K.,Pounds K.A., Stewart G.C., 1991, to appear in: *Iron Line Diagnostic in X-ray Sources*, eds. A. Treves, G.C. Perola and L. Stella, Springer-Verlag, Berlin

Response Function of the Iron K_α Line from a Centrally Illuminated Accretion Disk

G. Matt and G. C. Perola

Istituto Astronomico, Universitá di Roma "La Sapienza", Via G. Lancisi 29, 00161 Roma, Italy

There is increasing evidence that the iron K_α fluorescence line is a common feature in Seyfert galaxies (Nandra et al., 1991), and that its integral properties are in agreement with the values expected from a centrally illuminated relativistic disk (Matt et al., 1991b). In such a geometrical configuration the temporal evolution of the detailed line profile can yield direct information on the mass of the black hole (Stella, 1990). Here we show that: (a) in the case of a flat disk the temporal behaviour of the intensity I_α, the centroid energy $E_{c\alpha}$ and the line width σ_α of the line, which can be measured also with moderate energy resolution detectors, can be used for the same purpose; (b) the response function of the line intensity is very sensitive to strong deviations from planarity in the disk surface.

We have used the radial and angular dependences of the line emissivity from an optically thick, "cold" disk with cosmic abundances (Matt et al., 1991a, hereafter Paper I), illuminated by a point-like source located at 20 r_g (r_g=GM/c^2) above the disk on its symmetry axis. We define Ψ_α as the line response function, i.e., the light curve of the line in response to a δ-function flash in the ionizing continuum. In the case of the flat disk we have taken into account the relativistic effects, assuming a Schwarzschild black hole (see Paper I for details), and computed also the time evolution of $E_{c\alpha}$ and σ_α for a δ-function flash. Fig. 1 illustrates the three quantities, for a disk with inner radius r_i=10 r_g and outer radius r_o=1000 r_g, for four different values of μ_i, the cosine of the inclination angle. Being the time delay τ expressed in units of r_g/c, comparison between the observed and expected response functions yields an estimate of M. The strong dependence of $E_{c\alpha}$ and σ_α on μ_i can be used to constrain the inclination angle.

The case of a conical or concave configuration of the disk surface has been treated only for a face-on disk (μ_i=1) and without the relativistic corrections. The half-thickness Z is described by the relationship $Z = Ar_o(r/r_o)^\alpha$. When α=1, Ψ_α for different values of of A can mimic a flat disk at different values of μ_i; when $\alpha \gg$1, Ψ_α has a substantially different shape with respect to the flat or conical configurations. As shown in Paper I, to investigate the existence of strong deviations from the planarity, the amplitude of the high energy ($\gtrsim$ 10 keV) bump, produced by the Compton reflection of the primary photons, can be used.

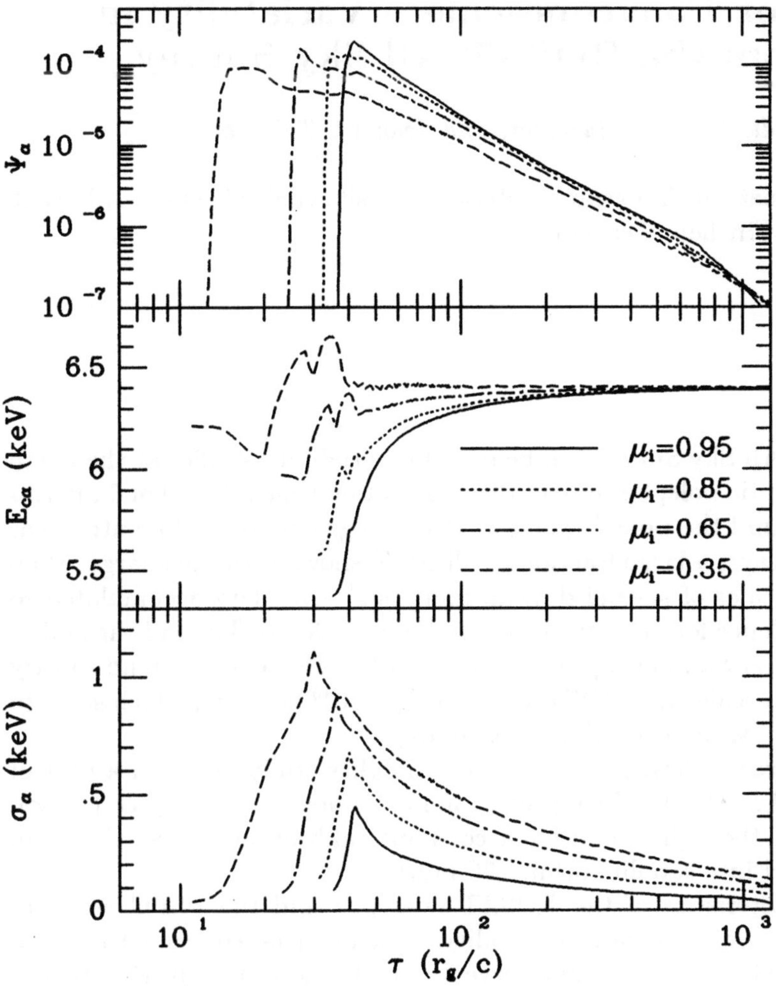

Fig. 1. (see text)

References

Matt G., Perola G.C., Piro L., 1991a, *Astron. Astrophys.*, in press.(Paper I)

Matt G., Perola G.C., Piro L., Stella, L., 1991b, *Astron. Astrophys.*, submitted.

Nandra K., Pounds K.A., Stewart G.C., 1991, to appear in: *Iron Line Diagnostic in X-ray Sources*, eds. A. Treves, G.C. Perola and L. Stella, Springer-Verlag, Berlin

Stella L., 1990, *Nature*, **344**, 747.

Search for Longtimescale – Variability in AGN from the ROSAT All Sky Survey

Stefan Schaeidt, Günther Hasinger, and Joachim Trümper

Max–Plank–Institut für extraterrestrische Physik, Gießenbachstr. 1, D-8046 Garching b. München, Germany

The ROSAT All Sky Survey can be used for variability studies at the north and south ecliptic pole, because of the observation time interval of 6 months within 1° around the pole; beyond 1° it drops gradually to 2 months near 2°. In soft X-rays, a large fraction of all AGN shows variations larger than 10% on timescales of several days to months[1]. The data accumulated so far shows evidence for an anticorrelation between X-ray flux and variability amplitude[2]. Low luminosity AGNs ($Lx < 10^{42} - 10^{43}$ erg s^{-1}) tend to vary on shorter timescales than QSOs ($Lx < 10^{43} - 10^{45}$ erg s^{-1}), which seem to be only weakly variable on timescales of days.

Here we present first results of our variability study, done on a part of the the ROSAT All Sky Survey 6° around the north ecliptic pole (NEP). Fig. 1a shows the light curve for a constant calibration source, the Crab nebula, binned into satellite orbits (96 min).

The characteristics of the ROSAT telescope and the PSPC entrance window introduce systematic variability, which can be connected far to a large degree. The corrected light curve for the Crab is given in Fig. 1b.

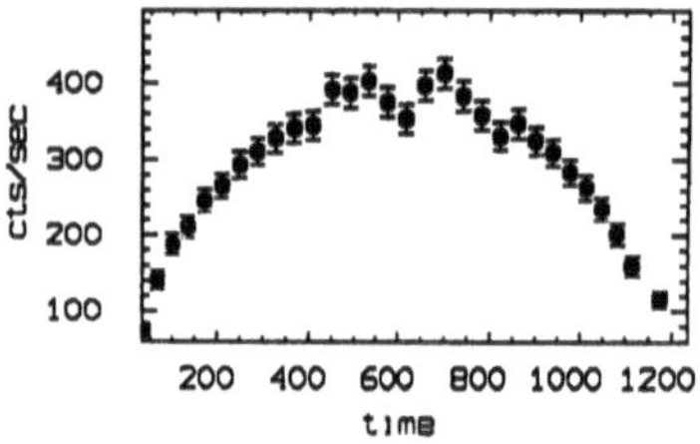

Fig. 1a. Uncorrected light curve for the Crab.

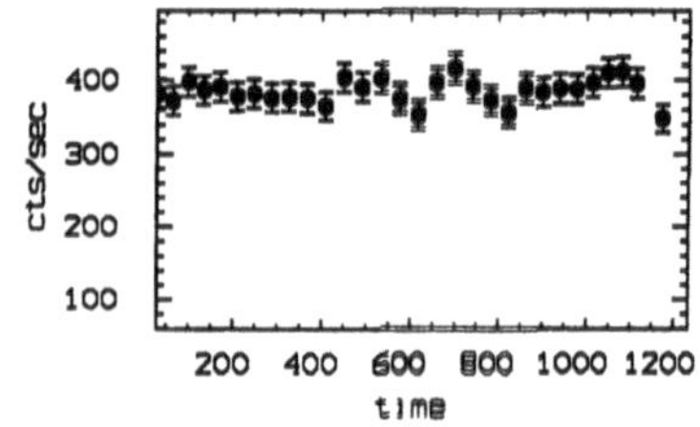

Fig. 1b. Corrected light curve for the Crab.

We binned the lightcurves for all other X-ray sources brighter than 0.2 cts/s on timescale of one satellite Orbit , or one day respectively (see Fig. 2 - 5). Fig. 2 shows the lightcurves of a supposedly constant source (KUV1804+684, a white dwarf). Fig. 3 shows the lightcurve of HR 6927,

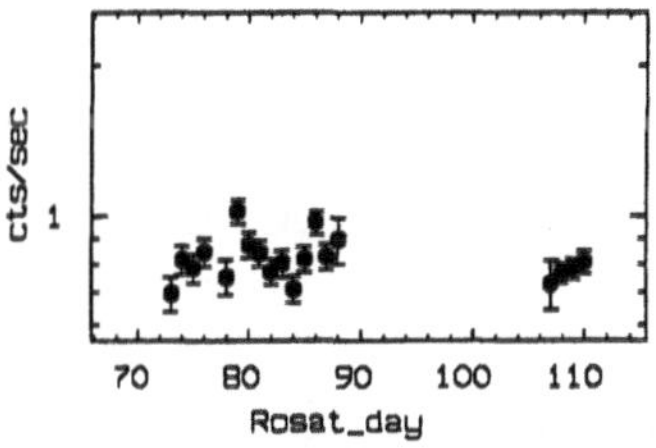

Fig. 2. Light curve of KUV1804+684.

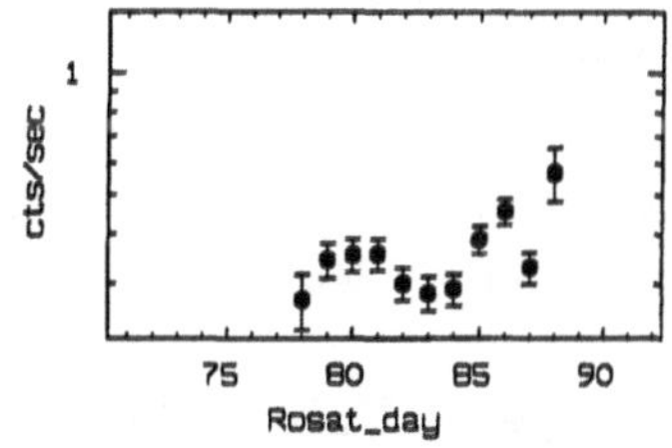

Fig. 3. Light curve of HR 6927, a coronal star.

a coronal star, which displays prominent time variability on a timescale of days.

The lightcurves for the AGNs in the following figures (Fig. 4, 5) are a part of an international multifrequency monitoring campaign in collaboration with H.H. Fink, L. Maraschi, M.-H. Ulrich and M. Malkan. As an example we present the two Seyfert 1 objects KAZ 102 and KUV18217+64. While KAZ 102 shows no significant time variability over the observation, KUV18217+64 dims from a countrate of $\sim$ 1.3 counts/sec to $\sim$ 1.0 counts/sec during the 11 observation days.

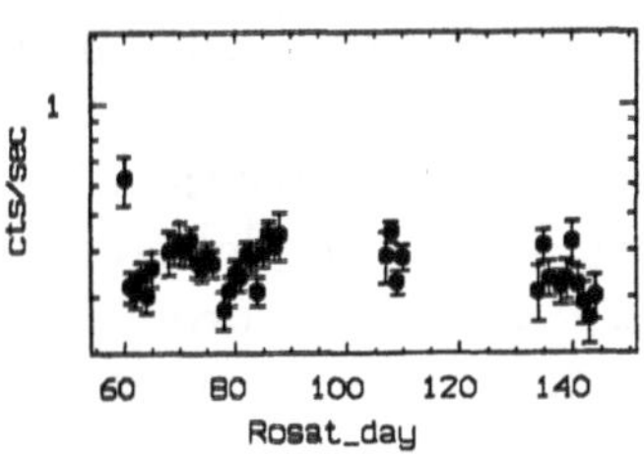

Fig. 4. Light curve of KAZ 102.

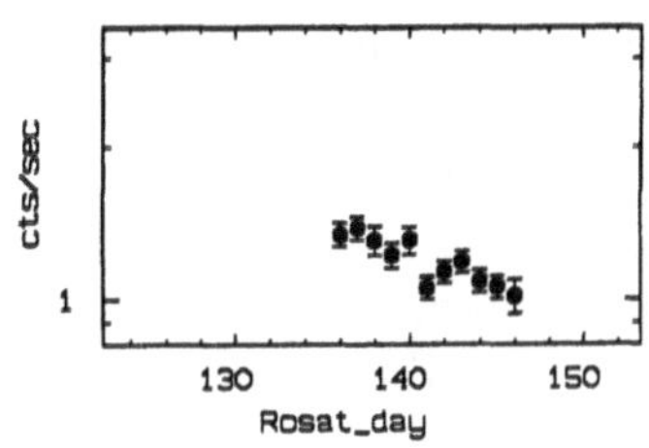

Fig. 5. Light curve of KUV18217+64.

Lightcurves of fainter X-ray sources can be binned in intervals of a few days up to weeks or months, as soon as the full dataset is available. These can also be correlated to the lightcurves obtained in the optical from weekly monitoring observations from La Silla, OHP and Tautenburg using $\sim 5^o$ x 5° Schmidt plates.

References

Stocke *et al.*, 1983, *Astrophys. J.*, **273**, 458.
McHardy, 1989, X- ray Astronomy with EXOSAT, *Mem. S. A. It*, **59**, 239

Recent Observations of PKS 2155–304 by GINGA

S. Sembay[1], *R.S. Warwick*[1], *C. Megan Urry*[2], *J. Sokoloski*[2], *I.M. George*[3], *T. Ohashi*[4], *and F. Makino*[5]

[1] Dept. of Physics, Univ. of Leicester, Leicester LE1 7RH, U.K.
[2] Space Telescope Science Institute, Baltimore, MD 21218, U.S.A.
[3] Inst. of Astronomy, Cambridge CB3 0HA, U.K.
[4] Dept. of Physics, University of Tokyo, Tokyo 113, Japan.
[5] Inst. of Space and Astronautical Science, Kanagawa 229, Japan.

PKS 2155–304 is one of the brightest and most variable BL Lacs known [1], [2]. In *GINGA* observations of this object made during four epochs in 1988 and 1989, the source exhibited extreme variability with a dynamic range in the 2–6 keV band of a factor of 7. The fastest variation was a factor of 2 decline within 4 hours. Spectral fits to the data, integrated in 6400 second time bins, reveal that a single power–law spectral model provides an adequate fit to the data during all intensity states. The spectral index is highly variable and ranges from $\Gamma = 2.6$ to 3.3. The apparent absorption column remained constant at a value of $(3.04 \pm 0.14) \times 10^{21}$ cm^{-2} (compared with the Galactic column of $\sim 1.8 \times 10^{20}$ cm^{-2}), consistent with previous *EXOSAT* observations [2]. This can be explained by the presence of the wide O VIII absorption trough at 0.6–0.7 keV [3]. In Fig. 1 we have plotted the spectral index against the 2–10 keV flux. There is a general anti–correlation of the index with flux: that is the spectrum tends to harden as the intensity increases.

EXOSAT observations of Mkn 501, Mkn 180, PG 1218+304 [4] and Mkn 421 [5] show that the X–ray power law indices follow tracks on the index–flux parameter plane which can be modelled by assuming that the spectrum ***pivots*** about a fixed break point at ~ 0.1 keV. In these sources the UV spectrum is more stable than the rapidly variable X–ray spectrum. In an extension of the standard jet model, this behaviour can be explained if the maximum energy of the synchrotron emitting electrons has a radial dependance [5]. Changes in the efficiency of the re–acceleration process within the X–ray emitting region could provide a plausible physical mechanism for the spectral variability.

Our data are not consistent with a constant, fixed break point. However, the UV continuum of PKS 2155–304 is variable on a time–scale of ≤ 1 day [6]. *IUE* observations [6] simultaneous with our 1989 data have enabled us to estimate the break points for three observations and in Fig. 1 we have also plotted the tracks corresponding to the expected relation between flux and index assuming the spectrum pivots about the derived break points.

The tracks give good fits to the data for the 02–Oct–89 and 10–Oct–89 observations. A closer examination of the data reveals that the index–flux relationship is more complex than this simple model predicts. For example, there are two episodes where the index is *correlated* with intensity in the 02–Oct–89 and 24–Oct–89 data. However, in general the behaviour of the UV to X–ray continuum can be described in terms of a rapidly varying X–ray "tail" and a more slowly varying UV to X–ray continuum.

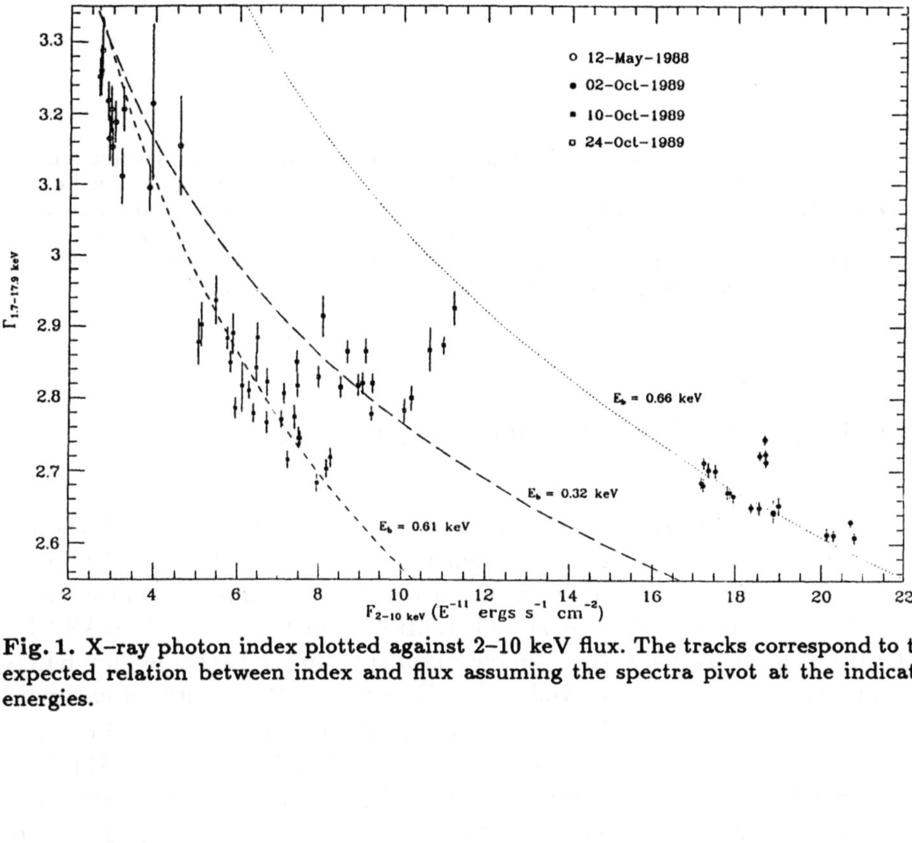

Fig. 1. **X–ray photon index plotted against 2–10 keV flux. The tracks correspond to the expected relation between index and flux assuming the spectra pivot at the indicated energies.**

References

[1] Urry, C.M., *et al.*, 1986: *Astrophys. J.*, **305**, 369.
[2] Treves, A., *et al.*, 1989: *Astrophys. J.*, **341**, 733.
[3] Canizares, C.R., Kruper, J., 1984: *Astrophys. J. (Letters)*, **278**, L99.
[4] George, I.M., *et al.*, 1988: *Monthly Notices Roy. Astron. Soc.*, **232**, 793.
[5] George, I.M., *et al.*, 1988: *Monthly Notices Roy. Astron. Soc.*, **235**, 787.
[6] Edelson, R.A., *et al.*, 1991:*Astrophys. J. (Letters)*, submitted.

Quasar X-Ray Spectra Revisited

P. Shastri[1], *B.J. Wilkes*[2], *M. Elvis*[2], *and J.C. McDowell*[3]

[1] McDonald Observatory, University of Texas, Austin TX, USA
[2] Harvard-Smithsonian Ctr for Astrophysics, Cambridge MA, USA
[3] Marshall Space Flight Center, Huntsville AL, USA

Abstract. We use an expanded sample of 43 *Einstein* IPC quasar spectra (cf. Wilkes and Elvis, 1987, ApJ 323, 243:WE) to re-examine the connection between soft X-ray slope and radio core-dominance, and the much debated relation between X-ray slope and FeII strength. We use published core-dominance values, our new FeII line strengths, and updated X-ray slopes (Masnou et al. 1991, Astr.Ap. in press) that properly account for the ultrasoft excess.

1. X-Rays and Radio Core-Dominance

For the RLQs, X-ray slopes tend to be flatter for larger values of R, the ratio of the nuclear-to-lobe radio flux density, except for 2 discrepant objects (Fig. 1), confirming and extending earlier results (Canizares & White, 1989, ApJ 339, 27; Boroson, 1989, ApJ 343, L9:B89; Shastri, 1991, MNRAS 249, 640:S91). If interpreted in the framework of the "unified scheme" for RLQs (Blandford & Konigl, 1979, ApJ 232, 34), wherein R is a statistical measure of inclination, the "radio-linked" flat-spectrum component of the X-ray emission is anisotropic and is probably relativistically beamed (S91). The anti-correlation in Fig. 1 is significant at >90% level with the discrepant objects, and at >99.5% without them (Spearman rank test). These 2 objects have among the lowest R values; we speculate that at the most edge-on inclinations of the quasar the steep-spectrum X-ray component may be occulted (by an obscuring torus?), leaving the residual flat-spectrum component visible. We also confirm the earlier result of WE, that RLQs have systematically flatter X-ray spectral slopes than RQQs.

2. X-Rays and the FeII Emission

The connection between X-ray slope and optical FeII strength of Wilkes et al. 1987 (ApJ 321, L23) was later disputed. An updated plot (Fig. 2: FWHM<6000km s^{-1} only) using FeII4570 line strengths measured by us for 18 quasars in the sample shows a clear correlation with X-ray slope for RQQs, at the 99% level using the *Ginga* result for 1416-129 ('+' in Fig. 2), and at 97% with the IPC result (that has large uncertainties: '*' in Fig.2). The RLQs show no correlation, and indeed this is consistent with the suggestion that a radio-linked, beamed X-ray component dominates in them. The absence of a correlation in other similar work (B89; Zheng & O'Brien 1990. ApJ 353, 433) is probably due to the inclusion of RLQs. While the correlation for the RQQs implies for the first time a long-sought connection between the ionizing continuum and the broad emission lines, it is in the opposite sense to predictions of standard photoionization models (Krolik 1988, Supermassive Black Holes, ed. Kafatos, M., CUP).

The fact that the X-ray slopes correlate with optical and radio properties of quasars requires that they have a wide intrinsic range, despite the large uncertainties on individual X-ray indices.

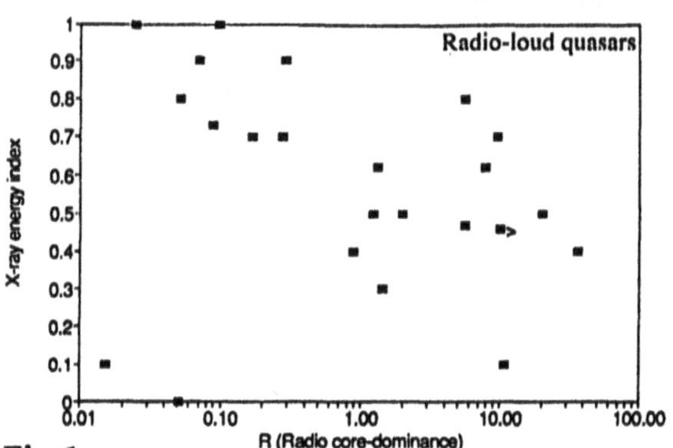

Fig. 1.

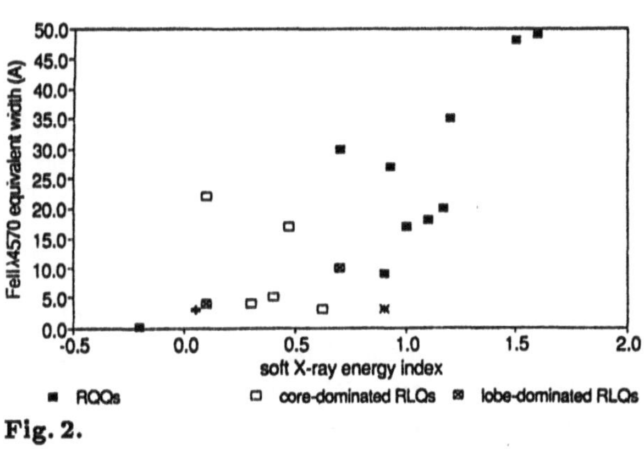

Fig. 2.

The ROSAT X-ray Spectrum of PG 1426+015

Belinda J. Wilkes[1], *Martin Elvis*[1], *Jonathan McDowell*[1], *and Andrew Lawrence*[2]

[1] CfA
[2] QMC

Abstract. We present a preliminary spectral analysis of ROSAT PSPC data for the quasar PG 1426+015. The energy index is 1.37±0.05, steeper than that observed at higher energies.

As part of a ROSAT program to obtain high signal-to-noise (*i.e.* instrument limited) X-ray spectra of quasars, PG 1426+015 (z = 0.086) was observed with the PSPC on 18-19 July 1990. We obtained 10,000 counts in 6.5 ksecs (Open filter) and 3000 counts in 6 ksecs (Boron filter). We report spectral analysis of a preliminary version of the data.

The best fit energy index for a single power law fit to Open and Boron filter data simultaneously, is 1.37±0.05 with $N_H=2.3\pm0.2\times10^{20}$, lower than the Galactic value, 2.64×10^{20} cm^{-2} (Figure 1), which suggests a soft excess. A double power law with the high energy slope fixed at the *Einstein* value of 1.17 yields an equally good fit and a low energy slope of 2.3±0.04. The flux level is uncertain but consistent with *Einstein* and EXOSAT.

Table 1. X-ray slopes for PG 1426+015

Satellite	Date	Energy range keV	Energy Index
EXOSAT (ME)	7/85	2-10	0.38±0.06
Einstein (IPC)	8/80	0.2-3.5	1.19±0.05
Einstein (IPC)	1/81	0.2-3.5	1.16±0.04
ROSAT (PSPC)	7/90	0.1-2.0	1.37±0.05

A comparison with X-ray observations from *Einstein* (Masnou *et al.* 1991) and EXOSAT (Comastri *et al.* 1991) shows that the spectral index increases towards lower energy (Table 1, Figure 2). This may be due to contributions from all three components discussed for quasars, a soft excess,

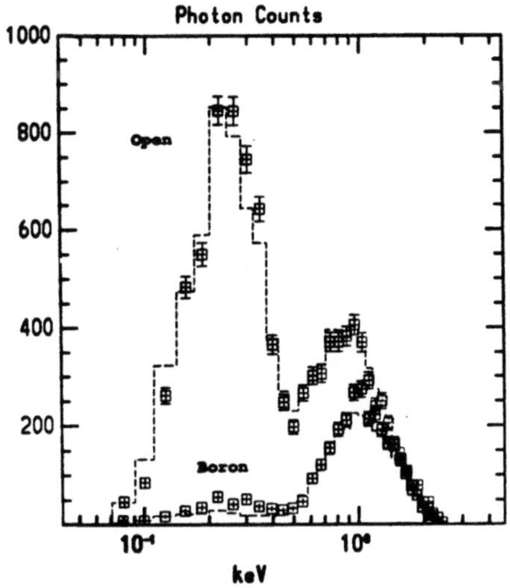

Fig. 1. X-ray spectral fits

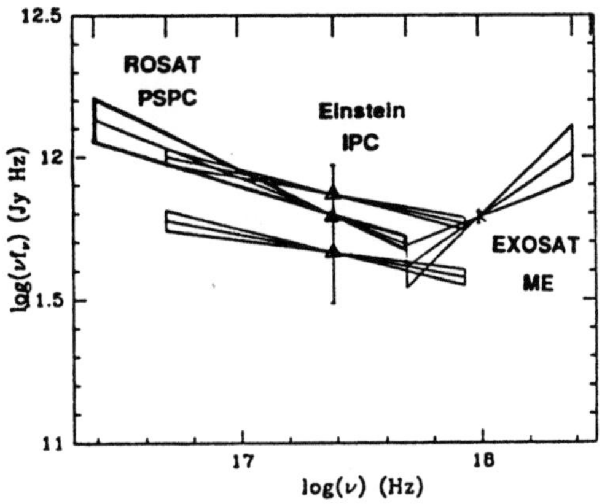

Fig. 2. Comparison to EINSTEIN and EXOSAT

a primary power law and a reflection component (Pounds *et al.* 1990) or to a single, curved component (Schwartz *et al.* 1990).

However, we were unable to obtain a reduced χ^2 value of less than 2. In particular, for both filters the channels around 0.3-0.4 keV are systematically high (Fig.1). Given the PSPC resolution, the feature appears too narrow to be an emission line. The data are currently being investigated via discussion with MPE.

We also obtain optical and infrared data as close as possible to our ROSAT observations to determine the effect of variability on our full energy distributions (Elvis *et al.* 1991). For PG 1426+015, an optical spectrum from the Multiple Mirror Telescope in May 1991 shows no evidence for variation since 1985.

References

Comastri, A., Setti, G., Zamorani, G., Elvis, M., Giommi, P., Wilkes, B.J. and McDowell, J.C. *Astrophys. J.*, submitted.

Elvis, M., *et al.*, 1991: in preparation.

Masnou, J.-L., Wilkes, B.J., Elvis, M., Arnaud, K.A. and McDowell, J.C., 1991: *Astron. Astrophys.*, in press.

Pounds, K.A., Nandra, K., Stewart, G.C., George, I.M., and Fabian, A.C., 1990: *Nature* **344**, 132.

Schwartz, D.A., Qian, Y. and Tucker, W., 1990: *Proceedings of the* 23rd *ESLAB Symposium*, **2**, 1043.

2 – Disks and Broad Line Regions

Examining Alternatives to Accretion Disk Models for the AGN Continuum

M. Malkan

UCLA

The hypothesis that the optical-ultraviolet continuum in AGNs is thermal emission from a black hole accretion disk has received a long string of recent critiques (*e.g.*, Antonucci, Kinney and Ford 1989; Malkan 1990; Courvoisier and Clavel 1991). In fact, there is some danger that the number of articles on the "Big Blue Bump (BBB)" may exceed the number of fresh observational and theoretical insights on it. Since the shortcomings and successes of the disk model have been rather thoroughly summarized, I'll try to minimize the re-hashing in this brief summary.

1. Two Old Problems Which May Not Be Serious

Two objections to thin disk models have been under discussion for at least 8 years (Malkan 1983). Does the observed AGN UV continuum show: a) too little linear polarization, or b) too weak a Lyman limit jump, to be consistent with our expectations for the disk model? The use of the word "or" here implies we have a choice of which problem to worry about, stemming from our fundamental uncertainty about the viscosity in the disk. If the disk has high viscosity (one example is the "alpha" disk, where viscosity is proportional to total pressure), the densities are low, and electron scattering dominates bound-free opacity. In this case a) is likely to be much more of a concern than b) (*e.g.*Sun and Malkan 1987). Conversely, if the disk has low viscosity (as in the "beta" disk, for example), the densities are so much higher that b) is more likely to be a problem than a) (*e.g.*Laor, Netzer and Piran 1990).

The observations appear to indicate that the intrinsic AGN continuum has virtually undetectable polarization and Lyman jump (in emission or absorption). Nonetheless, each of these observational claims could be strengthened by more decisive data. Let's consider them in turn. (a) Some AGNs have observed linear polarizations of a few percent, but in many cases these can be attributed to scattering by an asymmetric distribution of dust grains around the nucleus, which also polarize the broad emission lines (Berriman 1989; Webb et al. 1991). The polarizing mechanism in other quasars, which have unpolarized emission lines, is less clear. The lack of wavelength-dependence of the polarization in these quasars can be attributed to electron scattering, but we do not know whether the electrons are external to the continuum source (but inside the BLR), or mixed in with the continuum-emitting gas. The latter possibility is qualitatively what one might expect from electron scattering in an accretion disk atmosphere. Nonetheless, even if we are observing disk polarization in these quasars, it is still significantly smaller than a simple disk model would seem to predict (Webb and Malkan 1986a and b, but see Laor, Netzer and Piran 1990).

(b) IUE spectra of intermediate-redshift (z=0.4-0.9) quasars show no feature at the Lyman limit, although the typical signal/noise ratio is only 4 or 5 (Koratkar, Kinney, and Bohlin-this conference). Ground-based spectra of high-redshift ($z \sim 3$) quasars have higher S/N, but are complicated by absorptions from intervening intergalactic clouds of neutral hydrogen, which depress the far-UV continuum in a way which is nearly impossible to predict precisely. HST spectroscopy promises to improve the situation soon, and has already settled one related question.

The recently published UV spectra of 3C 273 give the first reliable measure of the optical thickness of the Lyman-alpha forest at low redshift. If one adds up all the intergalactic Lyman-alpha absorption lines detected in the spectrum from a redshift of zero to 0.158, the quasar redshift, the result is that the Ly-alpha "forest" depresses the UV continuum below the Ly-alpha emission line by an average of 0.8%. (Morris *et al.* 1991; Bahcall *et al.* 1991)). (This is not unexpected, from the estimated Ly-alpha discontinuity of 3 to 4% in the continuum of $z_{em} = 1.5$ quasars [Malkan, 1983]). This comforting result means that the UV continuum (at least down to a rest wavelength of 1040Å) in intermediate-redshift quasars is hardly affected by intergalactic absorptions, or more simply, What you see is what you get.

Even if the observational picture can be made decisive, the theoretical expectations remain much vaguer. The chief problem is that predictions of both polarization and spectral features both hinge on accurate knowledge of the vertical structure of the disk. Because the effective gravity increases with z, and energy release is not confined to the midplane, the vertical temperature gradient is likely to be much shallower than in a conventional stellar atmosphere. The most realistic disk/atmosphere models currently available (Wehrse and Störzer-this conference) suggest that the gradients are extremely small. Without a strong temperature gradient (and corresponding vertical gradient in the source function), hardly any polarization *or* Lyman jump is expected.

A related theoretical uncertainty is the probable existence of a hot corona, through which the disk photons must pass before they reach us.

Even a rather modest corona would be expected to have significant electron scattering optical depth (Meier 1982), which will Comptonize the disk photons. Czerny and Zbyszewska (1991) recently calculated how Comptonization would effect the Lyman jump in accretion disk spectra. They showed that for a wide range of coronal optical depths, any flux discontinuity at the Lyman edge would be virtually obliterated (*e.g.*, weaker than 5-10%). And of course this does not even include the Doppler and gravitational wavelength smearing (Sun and Malkan 1987). A more robust signature appears to be a discontinuity in the *slope* on either side of the Lyman limit. That prediction aught to be testable with observations only a little better than are currently available.

2. A New Problem Which May Be Serious: Synchronous Optical-UV Variability

A more serious possible conflict between observations and disk theory is hardly a year old. It was realized as a result of the intensive time-series of quasi-simultaneous optical and UV spectra of NGC 5548 obtained by a combined IUE- and ground-based monitoring campaign (Clavel *et al.* 1991; Peterson *et al.* 1991). The observational result is, again, an upper limit. In NGC 5548, the average lag of the optical continuum behind the UV continuum (as judged from the peak of the Cross-Correlation Function) is no larger than 4 days. A similar limit in the less luminous NGC 4151 is: $Lag(Opt - UV) \leq 3$ days (Clavel *et al.* 1991). There are indications that an optical lag has been *detected* in Fairall 9 (140 days-Clavel *et al.* 1990), and in 3C 273 (40 days- Courvoisier *et al.* 1990). Unfortunately, since the only optical measurements were made with broad bandpasses, which may include some Balmer emission-line contamination, it is possible that these last two values could be considered mere upper limits. Why would the absence of such a lag pose a problem for the thin disk model?

Perhaps the most robust prediction of disk models is that most of the optical thermal continuum is emitted from larger radii than is most of the UV continuum. For example, in the accretion disk model for the average NGC 5548 spectrum, the disk out to $r = 7r_g = 0.15$ light-days produces only $\sim$ 14% of the 5000Å flux (*e.g.*, Malkan 1991a) Since the RMS variability measured by Peterson *et al.* (1991) at 4870Å is 14%, a region at least as large as 0.15 light-days must be changing in some coherent fashion. Suppose the variability is caused by some disk instability (there is no shortage of theoretical instabilities!), which originates at small radii, and moves outward to 7 r_g.[1] As pointed out by Krolik *et al.* (1991), the group velocity of any

[1] A disturbance could equally well propagate from 7 r_g *inwards*, in which case the change in the optical continuum would *lead* the change in the UV. As before, the

disturbance which could cross this region in 4 days would have to be at least 0.04 c. If this is the speed of a sound wave, the disk thickness at $7r_g = 14GM/c^2$ would have to be: $\frac{h}{r} \geq \frac{0.04}{\sqrt{1/14}} = 0.15$. This is far thicker than would be expected from the thin-disk fit which requires an Eddington ratio less than 1%.

The most likely escape for the thin disk model would be to attribute most of the variable optical continuum to reprocessing of UV photons emitted in the inner disk and absorbed (and thermally re-emitted at optical frequencies) in the outer disk (Malkan 1991a, b). That scenario can fit the optical/UV continuum; and since the disturbance is carried by photons, the optical-UV lag would be only several hours. There may also be other motivations for supposing that the cooler outer disk reprocesses some significant fraction ($\geq$ 10%) of the hard radiation from the inner disk, and these are explored in the contribution of Rokaki *et al.* (this conference). Nonetheless, the implication that not all of the optical continuum is described by the standard $T \propto r^{-3/4}$ temperature distribution of the thin disk would be a new complication, which might leave one less satisfied with the entire thin disk hypothesis.

Perhaps the thin accretion disk model will ultimately prove too simple to fit all the observational constraints. In my view, the most strongly motivated alternatives are accretion flows with thicker geometries. Two of the most realistic possibilities are radiation tori (*e.g.*, Madau 1987, Acosta-Pulido *et al.* 1990) or slim accretion disks (Abramowicz, Czerny, Lasota, and Szuszkiewicz 1988; Szuszkiewicz 1990). The calculations are becoming sophisticated enough to allow a detailed confrontation with the observations in the next few years (*e.g.*, Abramowicz, Malkan, and Szuszkiewicz 1991). In these models, the optical-UV continuum is still optically thick thermal emission from a flattened accretion flow. For the sake of objectivity, I now want to consider models which seek to explain the continuum in different ways. We must only insist that these alternate theories be subjected to the same careful scrutiny which has so far been directed toward the thin disk hypothesis.

3. Reconsideration of Alternate Models for the BBB

Several possible physical mechanisms for the BBB were ruled out by comparing their predicted spectral shapes with observed AGN energy distributions (Malkan and Sargent 1982). However, that work assumed that the near-infrared power law ($\alpha \sim -1.2$) should be extrapolated without a cutoff

lag would be the propagation time of the disturbance across the region. The important point is that it would be unlikely for optical and UV variations to be always *synchronized*, given the different radii where they originate.

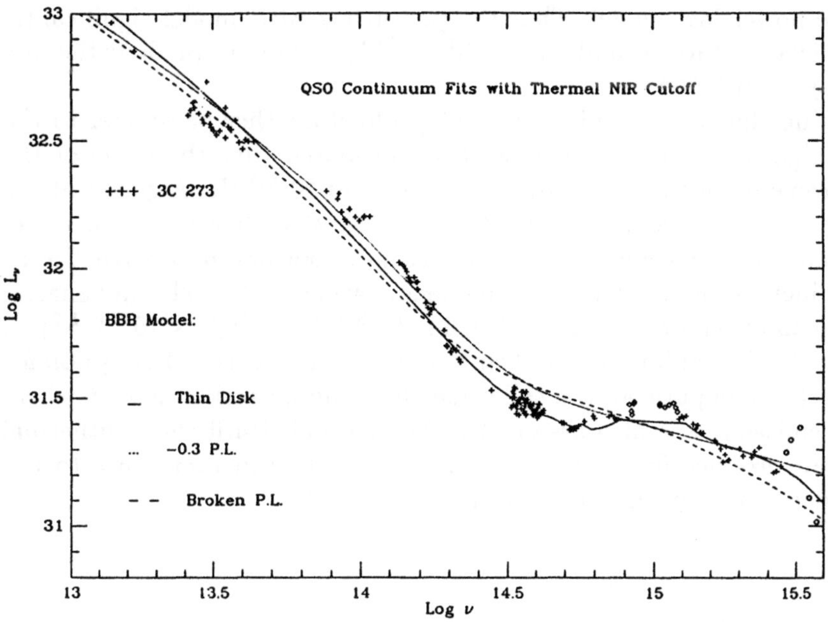

Fig. 1.

to visual wavelengths. If the near-IR emission is actually thermal emission from dust grains, it should cut off exponentially at frequencies above $10^{14} Hz$. This requires that the BBB produce more long-wavelength flux, so that previously rejected models might be revived (Malkan 1989). The two alternate models discussed below do not fit the data if the near-IR is nonthermal, but might have a chance if it is thermal.

4. Nonthermal Synchrotron Emission

The first alternative (which was almost the only one considered before the late '70s and early '80s), is that the optical-UV continuum is nonthermal. Since the *average* AGN continuum slope between $10^{14.85}$ and $10^{15.25}$ Hz is -0.5 (Malkan 1988), it has been interpreted as a nonthermal power law by Puetter *et al.* (1982) and Jones and Stein (1987). In Figure 1, the sum of a thermal near-IR dust component ($L_\nu \propto \nu^{-0.7} \times exp[-\nu/1.1 \times 10^{14}]$) and a $\nu^{-0.3}$ power law (dotted line) is compared with the observed energy distribution of 3C 273. The open diamonds represent data points from 2400 – 4000Å which are contaminated by Balmer continuum and blended Fe II line emission (probably by less than 20%-Malkan 1988). The open circles represent fluxes from Voyager which are particularly uncertain. Even the uncontaminated spectral regions show significantly more downward curvature than the single power law model. Thus the power-law BBB model significantly over-predicts the flux around logν =$10^{14.2-14.8}$, and under-predicts the flux around logν =$10^{14.9-15.2}$.

Realizing this, Jones and Stein (1987) added another parameter to the model-a high-frequency steepening. They proposed that the slope of the synchrotron emission steepens by $\Delta\alpha = 1$ around $10^{15.5}$ Hz. A glance at the failings of the flat power law model demonstrates that they cannot be remedied with a *sudden* steepening (ie with discontinuous first derivative. To try a *smooth* increase in slope to high frequency, we used the following analytical approximation formula: $L_\nu \propto [(\nu/3 \times 10^{15})^{+0.2} + (\nu/3 \times 10^{15})^{+1.2}]^{-1}$. Shown by the dotted line, the broken P. L. is hardly a better description for the BBB. For comparison, the solid line shows an accretion disk fit where the NIR emission was also assumed to be thermal. Until the nonthermal models can produce fits more like that, it would seem premature to use them to derive too much of the physics of the BBB.

5. Free-free Emission

Could the optical-UV continuum be thermal emission from dense, hot optically *thin* gas? The possibility that the continuum is free-free emission was suggested, and quantitatively analyzed 9 years ago (Malkan and Sargent 1982), and more recently by Antonucci and Barvainis (1988), Ferland, Korista and Peterson (1990) and Malkan (1989). Here I'll try to summarize the key quantitative challenges to the thermal bremsstrahlung hypothesis.

It is already known that the emitting gas could not have a density of less than $10^{11} cm^{-3}$, or an electron temperature of less than 50,000 K, to avoid too much bound-bound and bound-free emission, relative to the free-free (*e.g.*, Collin-Souffrin, this conference). *Iff* the near-IR component has a (thermal) high-frequency cut-off above $10^{14} Hz$, the steepening far-UV continuum can be fitted with free-free emission at an electron temperature of 100,000 - 200,000 K (Malkan 1989; although in 3C 273, it would have to be 300,000 K). I'll leave aside the question of whether gas at that temperature can remain thermally stable as its luminosity rises and falls by factors of two or more. Even if it can, it must occupy a significantly larger region than would optically thick gas emitting the same luminosity. This is due to the inherent inefficiency of optically thin bremsstrahlung, compared with blackbody emission.

To take an extreme example, suppose the gas has such a high emission measure ($1.8 \times 10^{37} cm^{-5}$) that it just becoming marginally optically thick at $10^{14} Hz$. [2] In this case-the most favorable possible for the

[2] This upper limit is required because a) the optical continuum is never observed to rise with the steep slope ($L_\nu \propto \nu^{+2}$) expected from the optically thick side of the

free-free hypothesis- the surface intensity for 150,000 K gas is 0.00035 erg/sec/cm^2/ Hz/ sr. At the characteristic frequency where the BBB peaks, $\nu = 2 \times 10^{15} Hz$, the intensity of a 50,000 K blackbody is 0.02 erg/sec/$cm^2/Hz/sr$. (And that estimate is conservative, because most disk models produce substantial UV continuum from even hotter gas). Compared with an accretion disk, the free-free model requires significantly more surface area to produce the same luminosity.

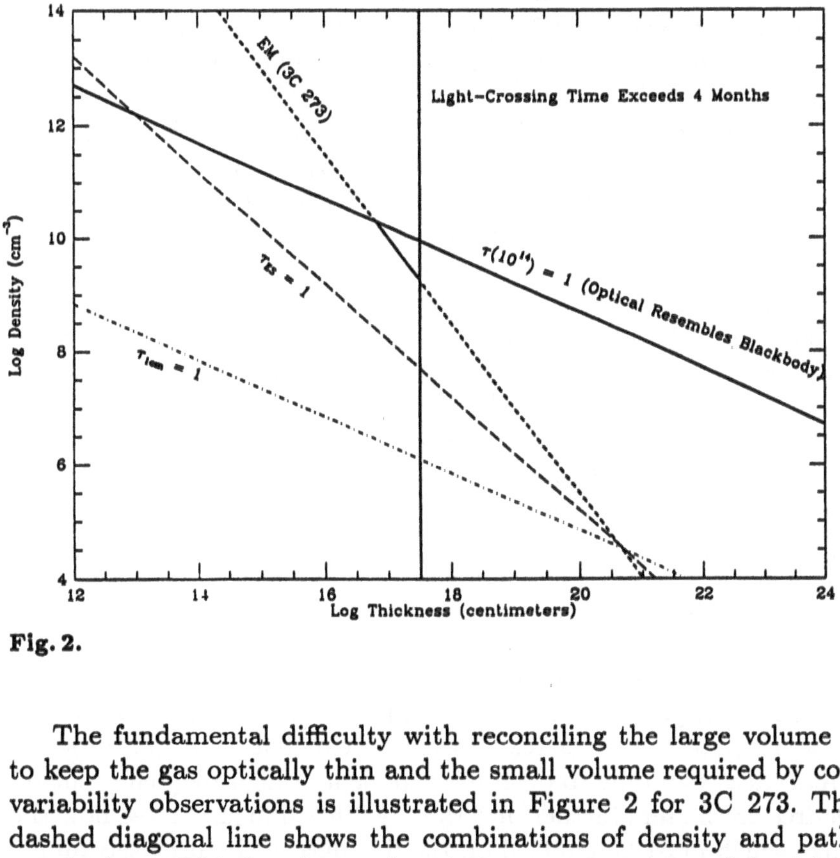

Fig. 2.

The fundamental difficulty with reconciling the large volume required to keep the gas optically thin and the small volume required by continuum variability observations is illustrated in Figure 2 for 3C 273. The heavy dashed diagonal line shows the combinations of density and path length required to yield the observed emission measure to explain this quasar's luminosity, $\sim 10^{71} cm^{-3}$ (Malkan 1989). The region above the sloping solid line is forbidden because it would imply gas which is optically thick at frequencies above $10^{14} Hz$. The most compact optically thin gas which could still satisfy this constraint would have a density of $10^{10} cm^{-3}$ and a size of 0.1 parsec. Unfortunately, this large size is just *marginally* consistent with the observed timescale of strong variations in the BBB flux in 3C 273, while the density may be too low to avoid unreasonably large bound-free emission.

bremsstrahlung spectrum, and b) at higher emission measures, there is no observable difference (or even conceptual distinction?) between the (partially optically thick) bremsstrahlung spectrum and the (completely optically thick) spectrum from an accretion disk.

As illustrated by the vertical solid line, the quasar's UV variability requires the region to be no larger than 4 light months. Thus the free-free model is backed into a tiny triangle of the parameter space bounded by the three solid lines: if more rapid UV variations were detected, or if the frequency where optical depth reaches unity were not fine-tuned to $10^{14} Hz$, it would be untenable. Unfortunately, the gas would have to be extremely optically thick to Thomson scattering, ($\tau_{es} = \sigma n_e dl = 10^3$), raising doubts about how any UV or X-ray variability could be observed.

Antonucci and Barvainis (1988) proposed that the optical-UV *and* the centimeter radio continua of radio-weak quasars are both produced by optically thin free-free emission from the same gas. There are many radio-weak quasars as luminous as 3C 273 (in fact, Antonucci and Barvainis first made the proposal for PG 1634+706, which has M_V = -30.4). To test their hypothesis for luminous, radio-weak quasars in general, I plotted in Figure 2 a dash-dot line showing an optical depth of the free-free emitting gas of 1 at a wavelength of 1 centimeter. Clearly, to allow the optically thin bremsstrahlung to extend to centimeter wavelengths would require an enormous emitting volume. Thus any variations in the BBB flux of any luminous radio-weak quasars (on time-scales shorter than a century) rule out this model decisively. Observations of UV continuum variability in luminous radio-weak quasars seem well established (*e.g.*, Grandi and Tifft 1974)

Even when the free-free hypothesis is applied to low-luminosity AGNs, as in Ferland, Korista and Peterson's (1989) discussion of the Seyfert 1 Mkn 590, variability still raises nagging questions. They proposed that the emission measure of the optically thin gas in Mkn 590 remained constant, while its optical continuum (free-free) flux dropped by a factor of 3. [3] By their assumption, the underlying nonthermal power-law, with $L_\nu \propto \nu^{-0.9}$, makes a minor contribution to the optical continuum, at least when the nucleus is bright. Apparently, the dramatic brightening of the BBB would have to be ascribed to a sharp increase in the electron temperature. However, the principal effect would be to push the exponential high-frequency cut-off of the free-free continuum further into the UV. It is difficult to see how the optical flux (which must be emitted on the flat part of the bremsstrahlung spectrum) could change so much without corresponding changes in the UV emission of several orders of magnitude, which have not been detected by IUE in any normal AGN.

If the free-free emitting volume *could* be made small enough to vary as rapidly as is observed in some AGNs, a more fundamental difficulty remains. All reprocessing schemes have the same trouble explaining the overall energy budget: you can't reprocess more luminosity than was originally there (*i.e.*,

[3] The motivation for this model was their observational claim that the Balmer emission lines in Mkn 590 have broad wings which do not vary. Even if that was the case in the pair of spectra they presented, it is by no means *common* in this, or many other well-observed Seyfert 1 nuclei (*e.g.*, Rosenblatt *et al.* 1991).

produced by the primary power source). The problem is that most AGNs have more of their power in the optical-UV continuum (*i.e.*, in the BBB, which is alleged to come from reprocessing), than at the other wavelengths (*e.g.*X-rays) which are alleged to power the BBB. The discrepancy becomes far worse if the reprocessing gas does not cover 100% of the central engine. (Ferland *et al.* estimated the *volume* filling factor in Mkn 590 would be $10^{-10\ to -2}$). The only way out of this dilemma is to postulate the existence of a primary power source which is hidden from our direct view, and is only seen after it is reprocessed in the BBB. This approach is discussed below.

In any case, the free-free emitting gas would *still* have to be arranged in some special geometry (*e.g.*a flat disk), so that it would allow us to see the rapidly variable soft X-rays emerging from the center.

6. Quasi-spherical Blackbody Emission

If the Big Blue Bump is indeed optically *thick* thermal emission, alternate suggestions about its geometry have been around for over a decade. Why has a flattened disk configuration seemed probable to so many researchers in the field? First, radio (and now optical) observations show that AGNs are consistently bipolar-and it is always tempting to identify this outflow axis with a rotation axis. Furthermore, if much of the AGN energy ultimately derives from accretion, we know that a) even if the accreting gas has only a small net angular momentum, it will still tend to flatten as it nears the center, and b) disk accretion is far more efficient at generating radiant energy than spherical infall. But probably the most airtight argument for a flattened geometry had been the absence of observed absorption in the soft X-ray and UV continua of luminous AGNs. This implied that our line of sight generally reaches the center (where the rapidly varying X-rays must originate) without intercepting any optically thick gas. Note Figure 8 in Ferland and Rees (1988) shows that even small amounts of optically thick gas (with densities up to $10^{17} cm^{-3}$) would produce enormous C and O absorption edges around 0.4 keV (and strong Lyman jumps), which are not observed in quasars.

However, this objection has recently been ingeniously circumvented by invoking the existence of a spherical swarm of very *dense* cloudlets which cloaks the (mostly unseen) central energy source (Rees, this conference). These cloudlets would absorb all of the optical/infrared continuum generated in the center, and thermalize it, so that it re-emerged as blackbody emission in the UV, making the BBB. It is worth considering the ways in which these hypothetical cloudlets would differ from the kind of clouds previously imagined in AGNs.

First, it is essential that the clouds have column densities of *no more* th an $10^{21} cm^{-2}$, so that they are sufficiently transparent to the soft X-rays which must escape through them. To thermalize all of the nuclear optical/infrared radiation, the cloudlets would have densities of at least $10^{18} cm^{-3}$ (Rees, this conference). High pressure is then required to keep them confined. Blandford and Rees (this conference) have proposed that tangled magnetic fields accomplish this. The cloudlets would be extremely small, with a *maximum* size of: d = $\frac{N_H}{n}$ = 10^3 cm. For a density of 10^{18}, their characteristic mass would be a kilogram. Thus I would think of them as hypersonic beachballs, although each one would carry enough kinetic energy to vaporize miles of beachfront.

An aesthetic objection to this model is that it postulates the existence of some unexplained compact, variable source of primary optical/infrared continuum. Although we cannot view it directly, it must usually carry the bulk of the AGN's bolometric luminosity, since the BBB which it powers is energetically dominant in most AGNs. Unfortunately, this "prime mover" has apparently never been observed directly. Could it resemble the violent nonthermal continuum source seen in "blazars"? I think not, because of their different X-ray properties. The problem is that, compared with luminous quasars, blazars are observed to be X-ray "loud", with comparable amounts of energy emitted in the optical-infrared and in the X-rays (*e.g.*, Worrall *et al.* 1987). If a blazar were cloaked by the postulated beachballs, its BBB emission (the reprocessed OIR continuum) would only be energetically comparable to its 0.1-30 keV X-rays (which we would observe directly, since they are unabsorbed). This is quite different from what is actually observed in luminous quasars, in which the total BBB power often exceeds the 0.1-30 keV luminosity by an order of magnitude. Any further differences in the X-ray properties of blazars and normal quasars (*e.g.*the former appear to have steeper X-ray slopes on average) would also be unexplained.

Thus our candidate for the "prime mover" in normal AGNs would be some new sort of nonthermal engine, emitting principally in the optical and infrared, and much less in the X-rays. Why don't we frequently (or infrequently) see these bare nuclei? Evidently they are almost always found with a *complete* cloak of dense cloudlets around them. On the other hand, although the bare nucleus must be well covered, it could not be too enshrouded: if a line-of-sight to its nonthermal engine passes through several beachballs, detectable soft X-ray absorption features would be imprinted, contrary to the observations. Thus we seem pushed into the requirement of a spherical cloak of cloudlets with almost no gaps, but also not too much clumping. How likely is it that the beachballs will usually be arranged in a monolayer, that is, with a characteristic diameter of light days, but a uniform thickness of no more than a few tens of meters?

After invoking circumstances which strike me as rather special to explain the BBB in most AGNs, the model ends up predicting a physical picture which is frustratingly similar to the accretion disk model. This is because the characteristic blackbody temperature is roughly the same function of distance from the central engine for free flying cloudlets as for gas orbiting in a disk.

The conclusion is that no theoretical explanation for the optical-UV continuum in AGNs is free of difficulties. Therefore the relevant judgement for the reader to make is: Which model has the least troubling set of problems?

References

Abramowicz, M.A., Malkan, M. A., and Szuszkiewicz, E., 1991: in preparation.

Abramowicz, M.A., Czerny, B., Lasota, J-P., and Szuszkiewicz, E., 1988: *Astrophys. J.*, **332**, 646.

Acosta-Pulido, J., *et al.*, 1990: *Astrophys. J.*, **365**, 119.

Antonucci, R., Kinney, A., and Ford, H., 1989: *Astrophys. J.*, **342**, 64.

Antonucci, R., and Barvainis, R., 1988: *Astrophys. J. (Letters)*, **332**, L 13.

Bahcall, J., *et al.*, 1991; ApJL 377, 5.

Berriman, G., 1989: *Astrophys. J.*, **345**, 17.

Clavel, J.C., *et al.*, 1991: *Monthly Notices Roy. Astron. Soc.*, in press.

Clavel, J.C., *et al.*, 1991: *Astrophys. J.*, **366**, 64.

Clavel, J. C., Wamsteker, W., and Glass, I., 1989: *Astrophys. J.*, **337**, 236.

Courvoisier, T., *et al.*, 1990: *Astron. Astrophys.*, **234**, 73.

Courvoisier, T., and Clavel, J., 1991: in: *Structure and Emission Properties of Accretion Disks*, eds. C. Bertout, *et al.*, p. 75.

Czerny, B., and Zbyszewska, 1991: *Monthly Notices Roy. Astron. Soc.*, **256**, 352.

Ferland, G., and Rees, M., 1988: *Astrophys. J.*, **332**, 141.

Ferland, G., Korista, K., and Peterson, B., 1990: *Astrophys. J. (Letters)*, **363**, L 21.

Grandi, S., and Tifft, W., 1974: *Publ. Astron. Soc. Pacific*, **86**, 873.

Jones, W., and Stein, W., 1987: *Astrophys. J. (Letters)*, **320**, L 1.

Krolik, J. H., *et al.*, 1991: *Astrophys. J.*, **371**, 541.

Laor, A., *et al.*, 1990: *Monthly Notices Roy. Astron. Soc.*, **238**, 897.

Madau, P., 1988: *Astrophys. J.*, **327**, 116.

Malkan, M.A., and Sargent, W.L.W., 1982: *Astrophys. J.*, **254**, 22.

Malkan, M.A., 1991a: in: *Variability of Active Galactic Nuclei*, eds. H.R. Miller and P.J. Wiita, p. 292.

Malkan, M.A., 1991b: in: *Structure and Emission Properties of Accretion Disks*, eds. C. Bertout, *et al.*, p. 165.

Malkan, M.A., 1990a: in: *Dynamics of Astrophysical Discs*, ed. J. A. Sellwood, p. 89.

Malkan, M.A., 1989: in: *Theory of Accretion Disks*, eds. F. Meyer, *et al.*, p. 19.

Malkan, M.A., 1988: *Adv. Space Res.* **8**, 249.

Malkan, M.A., 1983: *Astrophys. J.*, **268**, 582.

Meier, D., 1982: *Astrophys. J.*, **256**, 706.

Morris, S.L., *et al.*, 1991: *Astrophys. J. (Letters)*, **377**, L 21.

Peterson, B.M., 1991: *Astrophys. J.*, **368**, 119.

Puetter, R., *et al.*, 1982: *Astrophys. J.*, **257**, 487.

Rosenblatt, E., and Malkan, M., 1990: *Astrophys. J.*, **350**, 132.

Sun, W.H., and Malkan, M.A., 1989: *Astrophys. J.*, **346**, 68.

Szuszkiewicz, E., 1990: *Monthly Notices Roy. Astron. Soc.*, **244**, 377.

Webb, W., and Malkan, M.A., 1986a: in: *Continuum Emission in Active Galactic Nuclei*, ed. M. Sitko, p. 71.

Webb, W., and Malkan, M.A., 1986b: in: *The Physics of Accretion onto Compact Objects*, eds. K. Mason, *et al.*, p. 15.

Webb., W., and Malkan, M., 1991: in preparation.

Worrall, D., *et al.*, 1987: *Astrophys. J.*, **313**, 596.

Hybrid Accretion Disk Fits to the UV- to X-ray Spectra of AGN

Amri Wandel

Racah Inst. of Physics, The Hebrew University of Jerusalem, Israel

Abstract. We present new hybrid accretion disk models and techniques of spectral fitting. These models (described in Wandel and Liang 1991) consist of a geometrically thin, optically thick cool disk, and a hot, two-temperature optically thin disk. The dominant radiative process determines the emergent spectrum: inverse scattering of soft photons from the cool disk yields an X-ray power law, while Comptonized bremsstrahlung produces harder X- and gamma rays. These hybrid models are especially suitable for fitting the UV-to-X-ray continuum of AGN, as they can explain two different spectral regimes: the UV bump (thermal modified black body emission from the cool disk) and the hard X-ray power law. This gives a much larger spectral range than standard accretion disk models, which fit only the optical-UV, and can therefore constrain the accretion disk parameters more effectively.

1. Hybrid disk structure

Wandel and Liang (1991, hereafter WL) have demonstrated that the two solution-branches of the accretion disk equations can actually be joined smoothly, and may coexist. The inner, radiation pressure dominated region of the thin alpha-disk is known to be unstable (if the viscous stress is proportional to the total pressure). It is therefore plausible to associate the radius at which the radiation pressure in the cool outer disk becomes comparable to the gas pressure (e.g. Shapiro, Eardley and Lightman 1976, hereafter SLE). At this radius the accretion disk is assumed to switch from the cool, to the hot, 2-temperature solution. This solution is characterized by the ion-electron Coulomb coupling time being longer than the dynamical time, and hence the ions and electrons have different temperatures: while the ions are compression- heated to temperatures that can approach their virial temperature ($kT \sim GMm_p/r$), the electrons remain at a few times 10^9K. The heat reservoir is in the ions, which also provide the pressure to support the vertical structure of the disk, and radiative cooling is done by the electrons, the ion-electron coupling being the energy-transfer bottle neck. At radii close to the transition between the cool and the hot disk, the radiative cooling is dominated by Compton scattering of the soft UV photons from the cool disk off the hot electrons in the hot disk. The solution of the soft-photon Comptonization (SC) 2-T disk is given by SLE and WL. At further in radii the soft photon flux is attenuated by electron scattering outwards of the hot disk, and eventually cooling by self Comptonized bremsstrahlung bremsstrahlung becomes dominant. The solution of the 2-T disk in this case (CB) is given in WL and in White and Lightman (1990). The transition radii depend primarily on the central mass and on the accretion rate. It can be shown (Wandel 1991) that while the 3-region radial structure is quite common in disks around massive black holes, it may be absent or degenerate into a two-region (cool and SC) disk for lower central masses (of solar size), where the transition to the SC hot disk happens at a few schwarzschild radii even for close to Eddington accretion rates, while the CB region may be absent. The geometry of the 2- and 3-region radial disks is schematically shown in fig. 1. An alternative hybrid disk geometry is the stratified disk model (Wandel and Urry 1991; see also Liang and Price 1977), where the hot SC and the cool disk coexist at all radii, but are separated vertically, as the hot disk is geometrically much thicker than the cool disk, and actually engulfs it in a corona-like configuration.

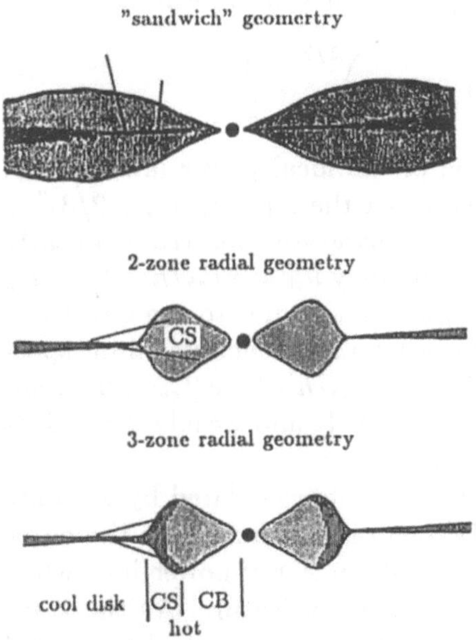

Fig. 1. Hybrid accretion disk configurations: a. Vertically stratified ("sandwich"), b. 2-zone radial, c. 3-zone radial.

2. Spectrum

The spectrum is computed by integrating the local spectra over the radial extent of the whole disk,

$$F_\nu = 2\pi \int I_\nu r^2 dr$$

where I_ν is the local surface emissivity normalized so that

$$\int I_\nu d\nu = \frac{3GM\dot{M}}{8\pi r^3} .$$

The outer cool disk produces a black body spectrum modified by the effect of electron scattering and extended due to the contribution from radii with different temperatures. It has a cutoff at a frequency $\nu_{sc} \sim 2kT_{sc}/h$, where T_{sc} is the highest temperature of the outer disk (at the transition radius to the hot SC disk). Typically $T_{sc} \sim 10^6$K.

The intermediate SC hot disk produces a power law X-ray spectrum $I_\nu \propto \nu^{-\alpha}$ with the spectral index given by the well known formula

$$\alpha = \left(\frac{9}{4} + \frac{\pi^2 m_e c^2}{12kT_e}(\tau_{es} + 2/3)^{-2}\right)^{1/2} - \frac{3}{2} .$$

Actually, the spectrum is a superposition of the local power law spectra, with $T_e \propto r^{1/4}$ and $\tau_{es} \propto r^{-1/4}$, but since $\tau \sim 1$ the term $T_e(\tau_{es} + 2/3)^2$ is weakly dependent on the radius, and also the integrated spectrum is nearly a power law , extending from ν_{sc} to approximately $\nu_{cb} = kT_{cb}/h$, where T_{cb} is the temperature of the SC disk at the transition radius to the CB disk.

The inner region - the CB disk - produces a typical bremsstrahlung spectrum, extending to the frequency $\nu_{cb} \sim (1-3)kT_{cb}/h$ where T_{cb} is the temperature of the hot CB disk (which is of order 10^9K and weakly dependent on radius and on accretion rate ; see WL).

A composite of these three spectral components calculated by a Monte Carlo method is shown in fig. 2. The UV band is dominated by the outer cool disk, the intermediate SC region produces an X-ray power law, while the inner component produces an MeV gamma ray feature. As the transition radii between these three regions depend on the central mass and on the accretion rate , so do the ratios of the three corresponding spectral components.

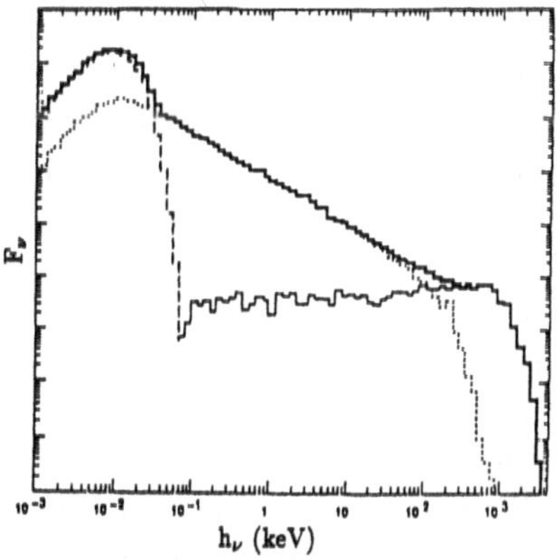

Fig. 2. **Composite spectrum of a 3-zone radial hybrid accretion disk : modified black body component (dashed curve), soft Comptonized photons (dotted) abd Comptonized bremsstrahlung (solid thin). The disk parameters are $\alpha = 1$, $\dot{M} = \dot{M}_{crit}$, $r_{sc} = 100$ and $r_{cb} = 20$ (in units of gravitational radii). The luminosity ratio of the cool:SC:CB components is roughly 1:3:3.**

3. Model fitting of AGN spectra

Much effort has been invested in fitting actual AGN spectra by accretion disk model (Sun and Malkan 1986; Wandel and Petrosian 1989, Laor 1990), motivated by the characteristic blue-UV bump which is nicely reproduced in a modified black body spectrum of an accretion disk around a massive black hole . In principle such a fit may also give an estimate the accretion parameters (accretion rate and central mass). However, the traditional accretion disk models which have been used in previous works have two disadvantages in this respect: first, they fall short of explaining the soft X-ray excesses and hard X-ray power laws commonly observed in AGN spectra, which then require a separate independent explanation with additional parameters. Second, the optical-UV band has a rather small dynamical range, and the absolute luminosity only determines the accretion rate ($\dot{M} \sim L/\epsilon c^2$) not the mass or the Eddington ratio ($= \dot{M}/\dot{M}_{crit} \propto \dot{m} \propto \dot{M}/M$). Without knowledge of the inclination angle, even the best optical-UV data available can determine the accretion parameters (M and $\dot{m}$) only within a factor of $\sim$30, and even if the inclination angle is specified, a range of factor 3 in the accretion parameters is still possible (Laor 1990).

In the framework of the hybrid accretion disk model both difficulties are removed: the model naturally explains the X-ray power law as well as the UV bump with no additional parameters (even soft excesses may be produced in the interface between the cool disk and SC hot region). Because of the much larger spectral dynamical range span between the UV and hard X-ray bands, the accretion parameters are tightly constrained. Roughly there are four observed quantities which characterize the UV-to-X-ray spectrum:

the UV flux and approximate slope, and the hard X-ray flux and spectral index (if the soft X-ray data are included, the number of spectral observables increases to six). These have to be fitted by a hybrid accretion disk model with four free parameters: the central mass, the accretion rate , the viscosity α parameter, and the inclination angle. Eventually, there are two more parameters: the viscosity parameter in the hot disk (which need not be equal to that of the cool disk) and the transition radius from the cool to the hot region. The later may be set to the radius at which the radiation pressure equals some factor times the gas pressure (SLE use $P_{rad} = 3P_{gas}$). This factor can be varied to some extent, giving an additional free parameter. In principle it should always be possible to fit 4-6 observational quantities by a model with the same number of free parameters. In reality however this may be difficult or even impossible, as these parameters are not entirely free and independent. Limiting ourselves to the main spectral properties and model parameters, the number is reduced to 3: the luminosity determines the accretion rate $\dot{M}$, the optical-UV to X-ray luminosity ratio (or the α_{ox} slope), is determined by the transition radius r_{sc}, (which in turn is determined by the Eddington ratio $\dot{m}$), and to some extent by the inclination angle (if the hot disk emits more isotropically than the cool disk, due its larger thickness ($h/r \sim 0.1 - 1$).

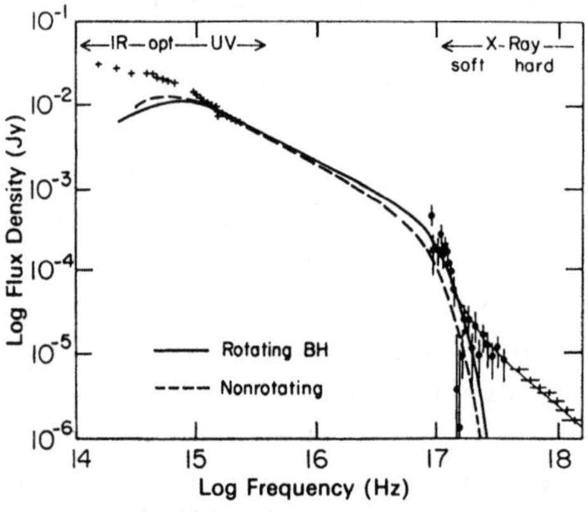

Fig. 3. Stratified hybrid accretion disk model fit to the composite UV-to X-ray spectrum of PKS 2155-304 (from Wandel and Urry 1991). The model parameters for the best fit (a Kerr black hole ; solid line) are $\dot{M}/\dot{M}_{crit} = 0.6$, $M = 710^8 M_\odot$, $\alpha = 0.15$. The dashed line is for a Schwarzschild black hole with somewhat different parameters.

Also the stratified ("sandwich") geometry may be used to fit individual spectra. Since in that configuration the soft photons from the cool disk have to cross the coronal hot disk, a low optical depth of the hot disk is imposed by the rough equality of the UV and X-ray luminosity, hence these model is suitable for fitting AGN with steep X-ray spectra (spectral indices of 1.5-2), such as the BL Lac object PKS 2155-304 whose simultaneous UV, soft X-ray and hard X-ray data were fit by a stratified hybrid accretion disk model

(Wandel and Urry 1991; see fig. 3). In the stratified model the free parameter determining the transition-radius in the radial model is replaced by the ratio of the accretion rates in the cool and hot disks.

References

Laor, A. 1990, *Monthly Notices Roy. Astron. Soc.*, **246**, 369.

Liang, E.P. and Price, R.H. 1977, *Monthly Notices Roy. Astron. Soc.*, **218**, 247.

Sun, W.-H. and Malkan, M.A. 1986, in New Insights in Astrophysics: 8 years of UV Astronomy wit IUE, p. 641.

Shapiro, S.L., Lightman, A.P., and Eardley, D.M. 1976, *Astrophys. J.*, **204**, 187 (SLE).

Wandel, A. 1991 (in preparation).

Wandel, A. and Liang, E.P. 1991, *Astrophys. J.*, **380**, (WL).

Wandel, A. and Petrosian, V. 1988, *Astrophys. J. (Letters)*, **329**, L 11.

Wandel, A. and Urry, C.M. 1991, *Astrophys. J.*, **367**, 78.

White, T.R. and Lightman, A.P. 1990, *Astrophys. J.*, **340**, 1024.

Continuum Variability in 3C 273

Thierry J.-L. Courvoisier

Geneva Observatory, CH-1290 Sauverny, Switzerland

Abstract. Several years of multiwaveband monitoring of the bright quasar 3C 273 reveal a very complex behaviour. The X-ray flux data show no correlations with the other components. The 2-10keV spectral index is, however, related to the blue bump emission. The UV and visible bands show very well correlated emission and the blue bump emission is seemingly related to the radio emission. The near infrared emission is constant for long periods of time, with short and violent outbursts which are also observed in the visible bands, with a smaller amplitude though.

1. Introduction

3C 273 has been the object of an intensive multiwavelength campaign of observations which started with the first EXOSAT observation of this bright quasar. The original aim of the campaign was to test the wind and shock model proposed by Camenzind and Courvoisier (1983). It soon proved that the observed variability was much more complex than the simple predictions of the model. The study of 3C 273 was therefore intensified to probe the the geometrical structure of the nucleus and the relationships between the emission components. Several groups participate in the observations and in their interpretations. The facilities we use include Metsaehovi radio telescope (Finnland), the James Clerk Maxwell Telescope (Hawaii), the Swedish ESO Submillimeter Telescope (ESO, Chile), ESO's infrared and optical telescopes, the 70 cm swiss telescope in Chile, IUE and the X-ray satellites EXOSAT GINGA and ROSAT. The data discussed here has been published in Courvoisier et al. (1987), Courvoisier et al. (1990), Ulrich et al. (1988), Turner et al. (1990) and in the references quoted in these papers.

3C 273 is variable on several timescales. Only a long span of observations covering most of the relevant timescales will allow to properly deconvolve the processes responsible for these variations. One of the problems being that, since the variations linked with different phenomena are deeply meshed together, the sampling must be sufficient to probe the fastest variations. This involves therefore an intensive long duration program.

2. Description of the Variability

Table 1 gives a short summary of the observed variability (amplitude, doubling timescale and characteristics of the duty cycle) in the different wavebands. This table must be read with care, as the results depend on the available sampling, which differs in density in the different wavebands.

Table 1. Amplitudes, timescales and type of variability in the different bands. The different types of duty cycle are explained in the text.

Domain	$\frac{\Delta f_{max}}{t_{average}}$	t_{2x}	Duty cycle
2–10 keV	1.2	30 d	?
1250 Å	1.25	1 a	ℓ
V	0.40		ℓ, s
$J_{outburst}$	1.3	3.4 d	s
$J_{quiescence}$	0.1	—	—
230 GHz	2.0	?	?
90 GHz	1.5		ℓ
37 GHz	0.9	0.5 a	ℓ
22 GHz	0.8		

Several trends can be seen in table 1. The largest amplitude is in the mm domain. It decreases towards longer wavelengths. The amplitude also decreases from the short UV wavelengths to the visible domain. The lack of extreme UV observations prevents to relate the amplitude observed in the X-ray domain to that observed in the UV. The timescale of the variations is the same in the UV and in the visible domains. The amplitude being a decreasing function of the wavelength, the doubling timescale naturally increases from short to long wavelength over the blue bump. The X-ray variability is still so poorly sampled that the light curve cannot be described properly. The fastest X-ray variations we observed had a doubling timescale of approximately 30 days (Courvoisier et al. 1987). Doubling times in table 1 were deduced from significant variations (several 10's of %), they represent thus the low frequency large amplitude variability and not extrapolations from small amplitude fast variations.

The duty cycles observed can be either long (l in table 1) or short (s) or both (as in the visible) or unknown (?). Typical for long duty cycle is the radio variability. In this domain the source is changing most of the time (fig. 1). The same type of variations is observed in the UV and visible domains. In contrast, the infrared light curve (fig. 2) shows long period of constant flux interleaved with short periods of violent activity (Courvoisier

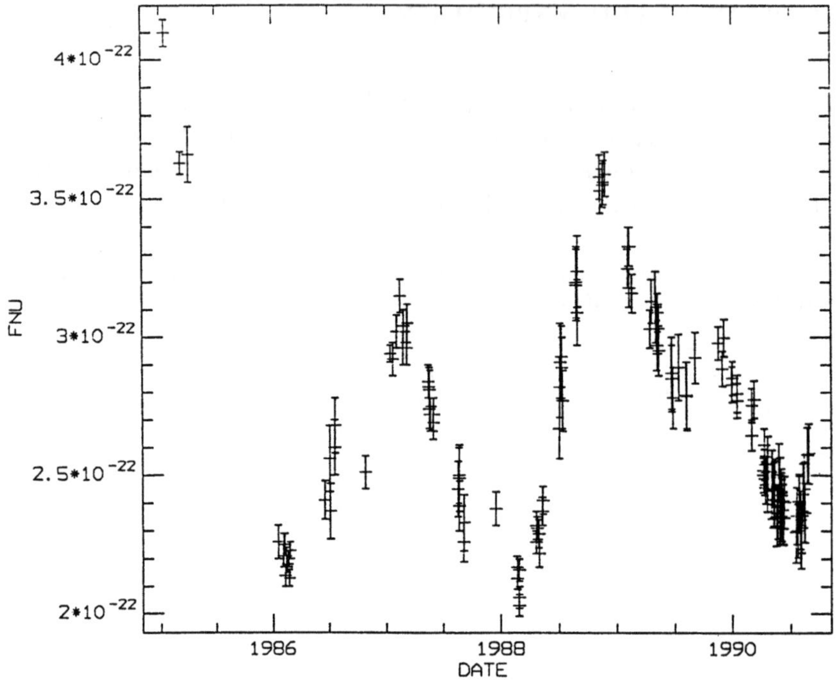

Fig. 1. The light curve (in $ergs s^{-1} cm^{-2} Hz^{-1}$) of 3C 273 at 22GHz from 1985 to 1991

et al. 1988). The amplitude of these short variations decreases towards the short wavelengths, causing in the visible domain a mix between the long duty cycle and the violent variations. The sampling in the mm domain is still insufficient to be able to give a clean description of the very complex variations in this domain (fig. 3).

The variations in the spectral shape of the X-ray and UV domains are described in Turner et al. (1990) and in Ulrich et al. (1988) respectively. Walter and Courvoisier (1991) have analysed together the ultraviolet and X-ray data and in particular the data which was taken quasi simultaneously in both wavebands. They conclude that there is a good correlation between the X-ray spectral index and the logarithm of the ratio of X-ray to UV photon fluxes. This correlation can be explained in terms of thermal inverse Comptonization in a way similar to what had been proposed in the case of NGC 5548 (Walter and Courvoisier 1990, see also the poster by R. Walter in this volume).

In the visible domain, the flux increases associated with the slow variations are of a marked blue colour (i.e. the flux increases more in the U band than in the V band), while the variations associated with the violent activity are very red (the flux increasing more in V than in U). This confirms the associations of these two phenomena with the variations observed in the UV and infrared respectively.

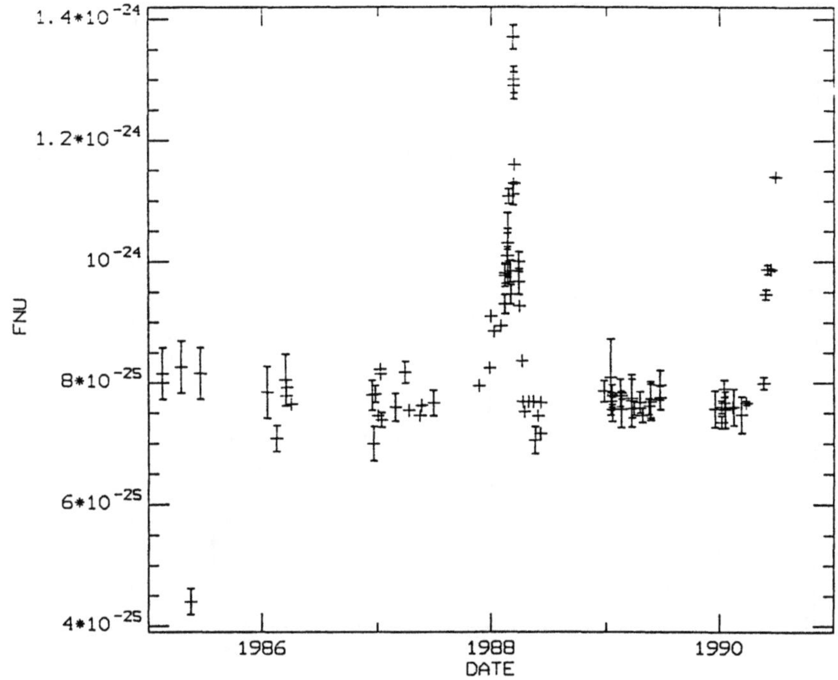

Fig. 2. **The light curve (in** $ergss^{-1}cm^{-2}Hz^{-1}$**) of 3C 273 at 2.2**μ **from 1984 to 1991**

The changes in the spectral shape observed during the period of fast variations in March 1988 (Courvoisier et al. 1988) can be used to study the magnetic field. While in the first analysis we used only the flux information, a deeper study modeling evolution of the flare due to synchrotron cooling indicates that the magnetic field is about $0.3G \pm 0.2G$, slightly less than the value found by looking at the fluxes only ($0.7G$). This analysis will be reported in Robson et al. (in preparation), it assumes that the dominating electron cooling process is synchrotron rather than adiabatic expansion of the source. The timescale of the decrease of the radio flux following the 1988 visible-infrared outburst can be explained by synchrotron cooling with a magnetic field similar to that observed in the near infrared. However, some of the variations seen in the mm domain are much too fast to be understood in this simple way.

3. Correlations

Observational information on the relationships between the emission components is to be found not only in the cross correlations between the different

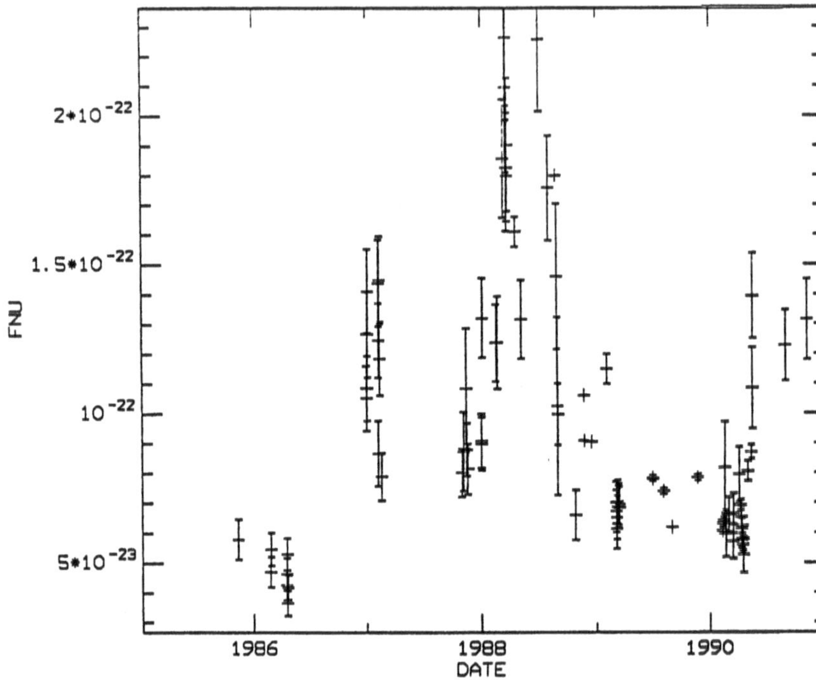

Fig. 3. The light curve (in $ergss^{-1}cm^{-2}Hz^{-1}$) of 3C 273 at 230GHz from 1985 to 1991

fluxes, but also in deeper analysis involving the shape of the continuum energy distribution of each components. For example, the data on the X-ray 2-10 keV flux show no correlations with either the UV flux, or the visible flux or the near infrared flux or the mm and radio fluxes. However, as discussed above, a good correlation can be found between the 2-10 keV spectral index and the log of the ratio of X-ray to UV photon fluxes.

A correlation has been observed at essentially zero lag between the UV and the visible lightcurves (Courvoisier and Clavel 1991). The upper limit on the possible lags is in clear contradiction with the expectations derived from models of optically thick and geometrically thin accretion discs. Since the amplitude of the variations is comparable with the average flux (table 1), we conclude that most of the flux cannot be explained by standard accretion disc models.

A significant correlation is also observed between the UV and the radio light curves. This correlation is also observed when using the visible data (to which the flaring data has been removed). This shows that the flux in the blue bump is correlated with the radio emission. The lag at which this correlation is observed is $0.4 \pm 0.2y$, the UV flux leading the radio flux. Other possible lags are seen at about 2.5y and -1.5y. These maxima in the cross correlation function are related by to a characteristic frequency of approximately $0.5y^{-1}$, which is seen in most of the data. The associated timescale is most probably not related to the observing seasons, as it is seen in data associated with very different celestial constraints. It is, however, too close to a third of the time spanned by our program to be of significance.

The UV-radio correlation is quite unexpected in standard models, it could, however, be explained if magnetic field dominates both the accretion and the the origin of the jets. In this case the lag would give information on the geometry of the synchrotron emission region with respect to the accretion disc. The successive radio flares have different behaviours. A general trend is nonetheless observed when studying the cross correlation between the 37GHz and the 22GHz light curves. Both are very significantly correlated, the lag is of approximately 0.1y, the 37GHZ flux leading, as expected in synchrotron models.

References

Camenzind, M., Courvoisier, T.J.-L., 1983: *Astrophys. J. (Letters)*, **266**, L83.
Courvoisier, T.J.-L., et al., 1987: *Astron. Astrophys.*, **176**, 197.
Courvoisier, T.J.-L., et al., 1988: *Nature* **335**, 330.
Courvoisier, T.J.-L., et al., 1990: *Astron. Astrophys.*, **234**, 73.
Courvoisier, T.J.-L., Clavel J., 1991: *Astron. Astrophys.*, in press.
Turner, M.J.L., et al., 1990: *Monthly Notices Roy. Astron. Soc.*, **244**, 310.
Ulrich, M.-H., Courvoisier, T.J.-L., Wamsteker, W., 1988: *Astron. Astrophys.*, **204**, 21.
Walter, R., Courvoisier, T.J.-L., 1990: *Astron. Astrophys.*, **233**, 40.
Walter, R., Courvoisier, T.J.-L., 1991: *Astron. Astrophys.*, submitted.

Discussion

Blandford: What were the magnitude and direction of the polarisation during the 1988 outburst?

Courvoisier: The polarisation was approximately 10 times the usual value at more than 2%. It was strongly variable both in amplitude and in direction (see Courvoisier et al. 1988). The data come from the last week of February 1988, no data are available for the maximum of the outburst.

Malkan: Were you really so surprised to see the blazar properties in 3C 273. After all, the polarimetry obtained by C. Impey and me (as well as data from some of your collaborators) clearly reveals a highly polarised, violently variable red blazar which rarely makes more than 10 to 20% of the red continuum. Couldn't this component explain most of the blazar properties you identify?

Courvoisier: The blazar properties implied by the data discussed here refer to all or at least most of the flux, rather than to the permanently present low level blazar activity you refer to.

Spherical and/or Disk Accretion in AGN

Suzy Collin-Souffrin[1,2]

[1] DAEC, Observatoire de Paris–Meudon, France
[2] Institut d'Astrophysique, Paris, France

1. Introduction

It is now commonly admitted that AGN are fueled by accretion onto a compact object, most likely a massive black hole. The next question is to whether this accretion is taking place with or without angular momentum (i.e. as spherical accretion or as an accretion disk).

Accretion disks were not always as fashionable as they are presently for explaining AGN fueling. In the seventies spherical accretion was more often envisioned. And it was only after Shields suggested in 1978 that the "big blue bump" observed in the optical continuum of 3C 273 might be the blackbody-like emission of an accretion disk, and after the study of similar features in the spectrum of other quasars and AGN by Malkan and Sargent (1982) and Malkan (1983), that accretion disks became most popular, although they have always been considered by theoreticians as an interesting possibility for AGN.

After a phase of hard work and high production rate (computation of the structure and emission spectrum, hydrodynamical studies...) it became evident that things are not simple. There are actually severe problems with the so-called "standard disks" in AGN. The most conspicuous one is that at the high level of accretion rate required to fuel quasars, these disks should be radiation pressure supported in their central regions and self-gravitating in their external regions, two situations stimulating strong instabilities and leading probably to the destruction of the disk. This, and other reasons, lead Guilbert and Rees (1988) and Lightman and White (1988) to propose another mechanism for explaining the "big blue bump", namely thermal emission of blobs condensed in a hot spherical accretion flow. This suggestion renewed the interest for spherical accretion, in particular because it gives a natural explanation for radiation reprocessing, which is implied by the presence of fluorescence K-shell iron lines in the X-ray spectrum of several AGN.

My approach for trying to choose between these two processes is not a theoretical but a simple phenomenological one. Without prejudice to the solution of the physical problem I will show that spherical accretion is not able to account for the big blue bump. On the contrary the accretion disk is an appealing model, not only because it accounts easily for the optical and UV continuum of AGN, but also as a reprocessor of radiation explaining many observed features, including broad optical lines, X-ray lines, and most results on variability obtained during intensive recent monitoring campaigns. Finally I will suggest an accretion model which agrees with recent results concerning central masses and accretion rates of AGN, and which can circumvent several problems raised with "standard" accretion disks.

If one asks for any "signature" of one or the other type of accretion, obviously things are not comparable. Accretion disks have a privilegious axis, which can lead to several observational tests, while spherical accretion does not. The only signature of pure spherical accretion would be the *absence* of any orientation dependent property, and this is far from being the case. So one is lead immediately to conclude that accretion is not purely spherical, and at least a fraction of the accreted mass should have an angular momentum.

But observational tests for accretion disks based on orientation effects are not clear, and their interpretation is generally model dependent. The most obvious ones, such as the presence of a radio axis or of a "dust molecular torus", concern only a minority of objects and/or regions located far from the central engine. Several orientation effects on line-continuum or continuum-continuum correlations have been predicted, as part of the radiation is expected to be dependent on the direction. However it is not clear which part is orientation dependent. The non-thermal continuum is presumably boosted when the line of sight coincides with the direction of the jet, but this may be a relatively rare event, while the thermal continuum, if mainly emitted by the accretion disk, is also expected to be inclination dependent. So the two effects and their subsequent continuum-continuum correlations cannot be easily disentangled. On the other hand, a fraction of lines may be emitted in a disk structure and be inclination dependent, and another fraction may be emitted more isotropically. So again line-continuum correlations do not prove anything. Finally, one can consider that the only real test is the inverse correlation - actually very loose - found by Wills and Browne (1986) between the FWHM of Hβ and the core-lobe flux density ratio, provided this ratio is taken as an inclination parameter (cf. Browne's talk at this meeting): it may indicate that Hβ kinematics is dominated by rotation.

Predictions concerning the polarization of the continuum are also questionable, since the same dependence on wavelength or on orientation can be obtained with a non-thermal and with a thermal continuum. Perhaps the most promising effect is predicted by Netzer (1991): it is a change in polarization at the Lyman edge, due to a change in the opacity regime for disk emission (atomic versus Thomson).

Emission lines are also a topic of controversy. It is generally believed that double peaked lines are the best signature of a disk emission. This is not clear, since several other effects can lead to this kind of profiles, such as binary black holes, or simply two emitting regions in motion with respect to each other (such as symmetrical jets). Moreover disks do not inevitably lead to double peaked profiles (cf. Boisson's talk at this meeting). So finally, although there are *many suggestive facts in favor of accretion disks, none is a real evidence at the moment.* And we will see that *the most conspicuous evidence for accretion disks is an indirect one, based on the presence of a thick and dense medium giving rise to both the continuum and the lines* (including X-ray lines), which medium is likely to be the accretion disk itself.

2. Radiation regimes

There are two extreme possibilities, according to the value of the ratio t_{cool}/t_{acc}, where t_{cool} is the characteristic time for gas cooling, and t_{acc} the characteristic accretion time.

1) $t_{acc}/t_{cool} \ll 1$: the "Virial regime":

In this case the accretion energy is given directly to radiation, and the temperature is equal to:

$$T_{VIR} = \frac{GMm_p}{kR} \approx 5\ 10^{11}(\frac{R}{10R_G})^{-1}\ K \tag{1}$$

where M is the black hole mass, R is a typical radius of the emitting region, and R_G is the gravitational radius GM/c^2.

This upper limit is generally not reached, but the temperature can stay quite high. For instance, in a strong radial shock, the temperature is equal to 0.2 T_{VIR}. Radiation can be produced through a variety of mechanisms, as non-thermal emission, bremsstrahlung or comptonized free-free emission. Infrared cyclotron radiation can also be emitted if a magnetic field is present, and can be transferred into the X-ray range through Compton interactions. So in any case the bulk of emission will take place in the hard X-ray or gamma-ray range.

2) $t_{acc}/t_{cool} \gg 1$: the "optically thick regime":

In this case the medium cools to a low temperature. The minimum temperature is that of a dense and thick medium radiating as a blackbody:

$$T_{eff} \approx (\frac{GM\dot{M}}{4\pi R^3\sigma})^{1/4} = 10^5(\frac{\eta}{0.1})^{-1/4}\,(\frac{L}{L_{Edd}})^{1/2}\,(\frac{R}{10R_G})^{-3/4}\,L_{47}^{-1/4}K \tag{2}$$

Radiation is then emitted in the UV and soft X-ray range.

Spherical accretion, where the accretion takes place roughly in a free-fall time, leads most naturally to the Virial regime. However a fraction of the hard radiation can be *reprocessed* into a softer range (UV) by dense and thick blobs, if they condensate in the hot flow, as proposed by Guilbert and Rees (1988) and Lightman and White (1988). Accretion with angular momentum leads most naturally to the optically thick regime, since the gas has amply time to cool and a dense and thick disk can be formed. However in this case too, hard X-rays can be emitted, either as Compton *reprocessed* radiation, through a hot corona (cf. Maraschi, this meeting) or through a hot part of the disk (the "two-phase disk", cf. Wandel's talk at this meeting), or they can even be emitted directly through hydromagnetic disk winds for instance (cf. Pudritz talk, this meeting).

A relevant observation would then be to know in which energy band is emitted the bulk of radiation, in the hard X-ray range or in the EUV band. Unfortunately none of these bands is well observed. If the soft X-ray excess extends down to the UV, it can dominate the bolometric luminosity, and if the X-ray continuum extends up to a few hundreds keV with a spectral index 0.7, it will also. So it is not possible to know from direct and simple considerations what is the most likely accretion process. In order to make progress in this problem, we have to discuss in more detail the shape of the UV continuum emitted in both cases.

3. The UV continuum

3.1 Basic concepts

A first immediate result is that the "UV bump" should be emitted by a gas whose temperature is close to 10^5 K. At a higher temperature the bulk of the emission would stay in the X-ray range, and at a smaller temperature, it would stay in the optical or infrared range.

The following basic concept is that a *smooth* UV continuum not displaying strong Balmer or Lyman discontinuities should necessarily be emitted by an *optically thick* medium. In particular the effective optical thickness at *the red side of the Lyman edge* should be much larger than unity. Since the temperature of the gas is of the order of $10^5 K$, the gas should be dense to get a large number of neutral excited hydrogen atoms able to absorb this radiation, and it should have a large column density. Figure 1 displays the continuous spectrum of a shell of gas at T $\sim 10^5 K$ for different values of the column density (Collin-Souffrin, 1991). These computations are made in the same conditions as the Ferland and Rees (1988) model of UV continuum due to reprocessed non-thermal continuum by dense clouds. It shows

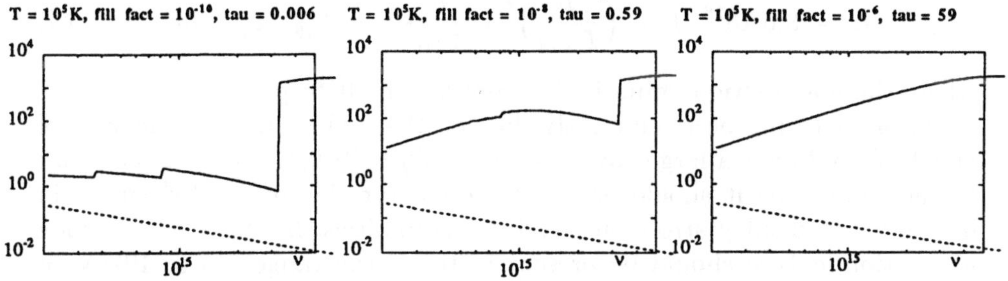

Fig. 1. Non-thermal (dashed line) and thermal reprocessed radiation (full line)

that, even at high densities, a column density of about $10^{25} cm^{-2}$ is required to get a small Lyman discontinuity in agreement with the absence of any observed discontinuity (Antonnucci et al., 1989).

3.2 The spherical model

In this model, dense clouds located in a small volume R $< 100 R_G$, are illuminated by the non-thermal X-ray continuum produced in a more central region. The emitted spectrum is composed of a mixture of directly seen non-thermal radiation and reprocessed thermal radiation (see Figure 8 of Ferland and Rees, 1988).

As shown a decade ago by Krolik, McKee and Tarter (1981) a fundamental parameter governs the reprocessing, namely the ionizing radiation pressure to gas pressure ratio, Ξ, (this parameter is very similar to the "ionization parameter", but physically more meaningful):

$$\Xi = \frac{1}{4\pi R^2} \frac{1}{n_e ckT} \int_{\nu_{Lyman}} L_\nu d\nu \qquad (3)$$

where n_e is the gas density (it is supposed here fully ionized; note that Ξ is in fact about 2.3 the pressure ratio) and L_ν is the differential luminosity.

A gas exposed to a non-thermal spectrum extending in the hard X-ray range displays a bimodal behaviour according to the value of Ξ. For Ξ larger than a few tens, the equilibrium gas temperature is close to the "Compton temperature" (equilibrium between Compton heating and inverse Compton cooling), equal to a few 10^7K for a typical AGN spectral distribution; for Ξ smaller than unity, the equilibrium temperature is governed by photoionization processes and collisional cooling, and is close to 10^4K. A temperature of the order 10^5K is obtained only in a small range of values of Ξ, with $\Xi \sim 10$, and this gives:

$$n \sim 3.6\,10^{16}\ f_{ion}\ \left(\frac{L}{L_{Edd}}\right)^2 \left(\frac{R}{10R_G}\right)^{-2}\ L_{45}^{-1}\ T_5^{-1}\ cm^{-3} \qquad (4)$$

where the bolometric luminosity is expressed in $10^{44} ergs/s$.

In order to reprocess efficiently the non-thermal continuum the shell of clouds should have a large covering factor, $\Omega/4\pi \geq 0.5$. We have seen that at the same time it should have a large column density, $> 10^{24} cm^{-2}$, in order to not display strong spectral discontinuities. In this case a strong absorption feature should be present in the X-ray range, up to 10keV or more, at least in half of the objects. It is in contradiction with observations, which show that the continuum is never heavily absorbed in the hard X-ray range.

A way to circumvent the problem is to assume that the *clouds are embedded in the hot medium* producing the X-ray luminosity. But in this case, not only the density should be very high, of the order $10^{18} cm^{-3}$, but also the UV continuum should display variations in the same time scale as the X-ray continuum, and this is not observed.

Ferland, Korista and Peterson (1990) have recently proposed that the Seyfert galaxy Mkn 590 might be a case where the cloud model applies. In this object the broad wings of Hα and Hβ are constant while the cores of lines vary proportionally to the continuum. The authors suggest that the wings are formed near the central object in a medium covering almost completely the non thermal source (a Very Broad Line Region, or VBLR). This region would be optically thick in the Lyman continuum, but not in the X-ray range. So it would reprocess the Lyman continuum in the optical and near UV, and leave unchanged the harder continuum, which ionizes the classical BLR located further out. This model is quite consistent for the broad line wings, but at the same time Ferland et al. propose that this VBLR is emitting the UV-bump, mainly by free-free process. This last conclusion does not hold because the large value of Ξ necessary to get an equilibrium temperature of 10^5K implies a very small radius of the emitting region, incompatible with the required emission measure (Collin-Souffrin, 1991). *Definitively intense spectral lines and a smooth optical-UV continuum cannot be emitted by the same region.*

3.3 Accretion disk and radiation reprocessing

An accretion disk offers ideal conditions for the emission of a UV-bump, because it is a made of a very dense and thick medium. Its mean density at a radius R is larger than that of a spherical accretion flow by a factor $\sim (Rv_{ff})/(Hv_r) \gg 1$ where H and v_r are the thickness and radial velocity of the disk, and its column density is larger by a factor $v_{ff}/v_r \gg 1$ (assuming that the velocity of the spherical flow is close to the free fall velocity v_{ff}). Several computations of the UV continuum emitted by an accretion disk, either geometrically thin or thick, have shown that the bulk of radiation is emitted as a smooth UV continuum (cf. in particular Madau, 1988, Laor and Netzer, 1990). The UV bump is produced directly by gravitational release in regions of the disk located between 10 and $1000 R_G$, and the hard X-ray continuum comes either from a more central region, or from a region located above the disk, as proposed in several models in this same meeting (Pudritz, Maraschi, Wandel). In both cases the disk will act as a reprocessor of X-ray photons, since it can absorb this radiation and reemit in another wavelength range, and this fact complicates considerably the emission problem.

There are plenty of reasons for the disk being a good radiation reprocessor (cf. Figure 2).

- at small disk radii ($10 R_G \leq R \leq 10^3 R_G$) :

If the X-ray source is located at the very centre and is very small, relativistic bending of light rays allows the disk illumination (Cunningham, 1976). The finite X-ray source size allows also a fraction of the X-ray emission $\sim (R_X/R)^2$ to be absorbed by the disk (R_X is a characteristic radius of the X-ray source).

- at large disk radii ($10^3 R_G \leq R \leq 10^5 R_G$):

"Flaring" of the disk (i.e. increase of the thickness to radius ratio, which occurs in the "standard model" at least) allows a fraction of the X-ray emission $\sim (H/R)^2$ to be absorbed, even in the case of a central X-ray point source. And finally a hot diffuse medium surrounding the disk, whose existence seems to be proved by the observation of polarized broad lines in Seyfert 2 galaxies (Antonnucci and Miller, 1985) Thomson backscatters a fraction of about τ_T towards the disk (τ_T is the Thomson optical thickness of the hot medium).

An important characteristic of disk reprocessing, which is apparently not often realized although it allows to distinguish clearly a spherical medium and a flat structure, is that the illumination flux in this last case does not decrease as R^{-2} but more or less as R^{-3} at large distances. Such a variation gives a quite different emissivity dependence in the region where lines are produced (cf. below).

The problem is then to know to which kind of reprocessed emission this radiation will give rise. As explained above, it depends on the value of the radiation to gas pressure ratio in the region where the bulk of X-rays is absorbed, i.e. in the upper layer of the disk defined by $\tau_T \sim 1$. Detailed computations concerning this problem are not easy to handle, because they depend not only on the radial structure of the disk, which relies strongly on the viscosity mechanism, but also on its vertical structure which is not well known even in a standard model. Qualitative arguments can however help to make some predictions.

The hard X-ray flux will be absorbed under the photosphere if the Thomson opacity is smaller than the atomic opacity in the optical-UV range, i.e. if the temperature of the irradiated layers (and consequently the radiation to gas pressure ratio) is low enough and the gas pressure high enough in the

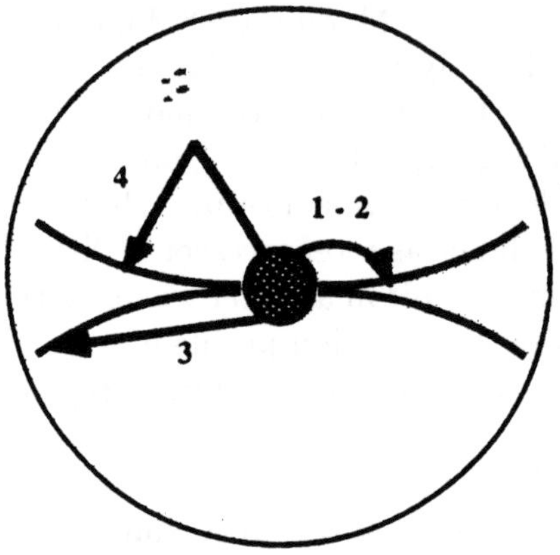

Fig. 2.

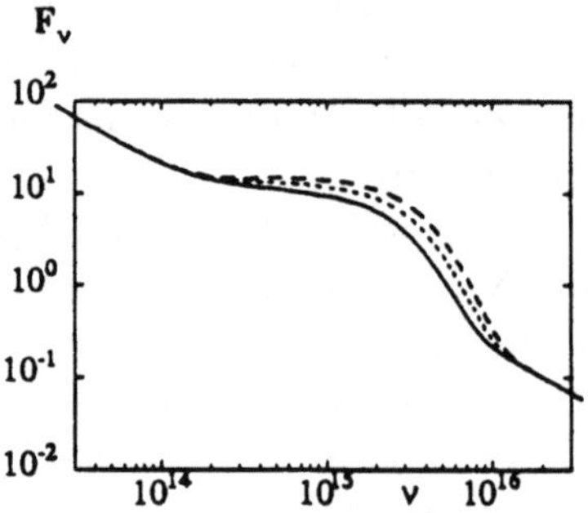

Fig. 3.

upper region of the disk defined by $\tau_T \leq 1$. The X-ray flux leads then simply to an increase of the effective photospheric temperature, and consequently to a change in the optical and/or UV continuum. In the opposite case of a Thomson opacity larger than the atomic opacity, i.e. a large radiation to gas pressure ratio and a small gas pressure, the hard X-ray flux will be absorbed above the photosphere and will create an irradiated chromosphere or corona, which will emit mainly lines in the optical-UV or in the X-ray range. Since the issue depends strongly on the value of the gas pressure, reprocessing in the continuum is more likely to occur in the central regions of the disk where gas pressure is large, while reprocessing in lines occurs in the outer disk regions. Reality is actually more complex, and a fraction of the X-ray flux may also lead to the emission of X-ray lines in the central regions of the disk.

3.4 Continuum and line emission in the accretion disk picture

a. Disk continuum emission Assuming a standard α-disk (Shakura and Sunayev, 1973), one can show that the radiation of a central X-ray source is indeed absorbed under the photosphere in the central regions of the disk $10R_G \leq R \leq 10^3 R_G$ (Collin-Souffrin, 1991), at least in low luminosity objects. It leads to an increase of the optical-UV flux, compared to a pure gravitational release of energy. Note that it may not be the case in high luminosity objects, where this region of the disk is radiation pressure dominated and has a smaller density.

Figure 3 displays the results for an object like NGC 4151. The mass of the black hole is equal to $10^7 M_\odot$ and the bolometric luminosity to 10^{44} $erg\ s^{-1}$

(it corresponds to an Eddington ratio of 0.1, quite common in AGN). A Schwarzschild black hole is assumed. The central X-ray source is spherical and has a radius of $30R_G$. An X-ray luminosity corresponding to 0%, 40% and 80% of the bolometric luminosity is considered (it takes into account the continuum up to the MeV range). A ν^{-1} power law continuum corresponding to a contribution equal to 50% of the total flux at 1μ is added to the disk emission: this does not take into account the starlight which might dominate in the red range.

This figure shows *that the reprocessed radiation can represent a non negligible part of the total flux in the optical-UV range.*

Moreover one can show that the time it takes for the disk to adjust its vertical structure and its effective temperature to a variation of the X-ray flux is very short (20s at $R = 100R_G$ for the case considered above). As a consequence the optical-UV continuum should vary with the X-ray flux. There are two interesting implications:

1. the optical-UV continuum should harden when it brightens, owing to the increase of the surface temperature when the irradiation of the disk is higher. This effect should be amplified in low luminosity AGN, owing to the increasing proportion of starlight in the visible range. The hardening is indeed observed.

2. the time lag between the UV and the optical flux corresponds to light travel time from regions of the disk emitting the UV and the optical continuum (which is about one day in the previous example), and not to the much longer viscous time, as it would be the case if the variations of the optical-UV flux would be due to a dynamical process such as variations of the accretion rate. The small time lag observed between the continuum at different wavelengths in several AGN, in particular in NGC 4151 and in NGC 5848, is therefore easily explained.

b. Disk line emission If the disk extends up to a few $10^3 R_G$ or further out, it will reprocess the X-ray flux into emission lines. This process has been extensively studied by Dumont and Collin-Souffrin (1990) and Rokaki et al. (1991) who have shown that it can account for the bulk of the Low Ionization Lines emission. The amount of High Ionization Lines emitted by the disk depends on the amount of Lyman flux "seen" by the disk, but in any case it is smaller than that of the Low Ionization ones, and another region is certainly required to account for the bulk of this emission.

There are many interesting issues in this process.

First the lines are produced between a few $10^2 R_G$ and a few $10^4 R_G$. For $R <$ a few $10^2 R_G$, the line excitation temperature is smaller than the photospheric temperature, so lines are not seen in

emission, and moreover a major part of the irradiating flux is absorbed under the photosphere as explained above. For $R > 10^4 R_G$ the flux from the central source is too small to heat the disk. This gives *a natural explanation for the line widths*, about equal to the keplerian velocity at R =

$10^3 R_G$, 10000 km s^{-1}. The disk emission matches quite well the observations concerning line intensities and line profiles. In contradistinction with a spherical model a great variety of line profiles is produced (as observed), since they depend strongly on the inclination of the disk on the line of sight.

Second this model solves the long standing problem of the formation and confinement of the emission clouds, at least of those emitting Low Ionization Lines. It solves also a problem similar to that of the continuum emission region, namely the large column density of the emitting medium ($\sim 10^{24} cm^{-2}$), impossible to reconcile in a spherical geometry with the absence of strong absorption in the hard X-ray range (cf. Collin-Souffrin and Lasota, 1988).

4. The accretion disk luminosity

A consistent picture therefore emerges where the optical-UV continuum and a large fraction of the lines, as well as their time variations, can be consistently explained, assuming that accretion is taking place through a disk.

However we know that the standard accretion disk picture suffers from severe problems mentioned at the beginning, in particular if the UV disk luminosity is large and close to the Eddington value. The problem would be reduced if only a small fraction of the gravitational energy is dissipated directly into radiation in the disk. One possibility is a magnetized disk where a large fraction of the gravitational energy is liberated outside the disk through the linkage with a corona or a rotating wind (Heyvaerts, 1991). Another solution is the "two-phase disk" mentioned by Wandel in his talk.

On the other hand we have seen that in order to get the important reprocessed flux required by the observations, the luminosity in the X-ray to gamma-ray range should be comparable to or larger than the disk luminosity, which is mainly emitted in the EUV range. Thus from two different approaches one reaches the same conclusion: the luminosity in the EUV range is likely to be smaller than the non thermal luminosity, in particular in the X-ray to gamma-ray range.

An important question is then to know the respective contributions of the thermal and non thermal luminosity to the bolometric luminosity. Estimations of the bolometric luminosity have been made for a sample of objects by Padovani and Rafanelli (1988) by adding contributions of far infrared, visible, UV and X- ray up to 10 keV. These are underestimations (unless the radiation is strongly collimated), since they do not take into account the EUV bump and the $\geq$ 10keV contribution. Padovani and Rafanelli found an average value of the "observed" bolometric luminosity $L_{bol.obs} \sim 13 L_{opt}$, where L_{opt} is the luminosity in the 3000-6000A range (in the rest frame).

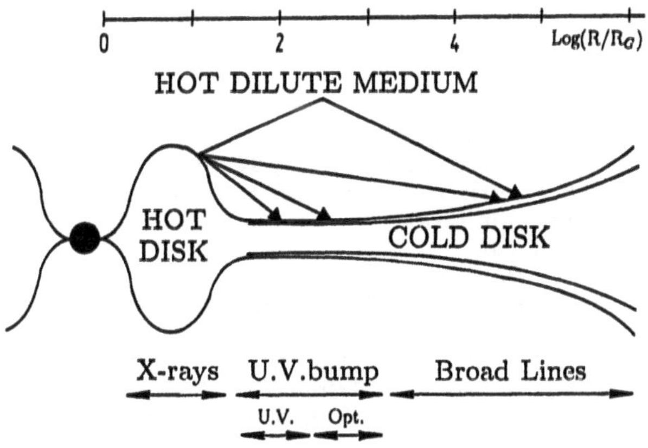

Fig. 4.

Collin-Souffrin and Joly (cf. a poster contribution at this meeting) have estimated the disk luminosity, using a method which links the broad line kinematics to the disk continuum emission. They find that the thermal disk emission is on average not larger than twice L_{opt}, which means that it is smaller than the bolometric luminosity by at least an order of magnitude. However this study is model dependent and it relies on several assumptions which can be questioned: the Hβ emission region is gravitationally bound (it is actually implied by the above model), the disk is radiating like a black body at a given radius, the central black hole is non rotating, and finally an underlying ν^{-1} non thermal component from the infrared to the UV range is assumed. Another determination, where the continuum emission is better fitted but line widths are not taken into account, has been performed by Sun and Malkan (1989) and by Laor (1990). From the results of Sun and Malkan one deduces an average ratio $L_{disk}/L_{opt} \sim$ 4-5, larger than that of Collin- Souffrin and Joly, which nevertheless implies that an important fraction of the bolometric luminosity is non thermal.

One can finally conclude that an accretion disk is certainly present in AGN. It is required to explain the UV bump, and it accounts for many observed other features. However the accretion rate through the disk either is small, and another mechanism energizes the objects (such as spherical accretion), or it is large but an important fraction of the gravitational energy is not evacuated through UV radiation. Figure 4 summarizes what could be the "accretion disk picture" of an AGN according to this model.

References

Antonucci, R.R.J., Miller, J.S., 1985: *Astrophys. J.*, **297**, 621.
Antonnucci, R.R., Kinney, A.L., Ford, H.C., 1989: *Astrophys. J.*, **342**, 64.
Collin-Souffrin, S., Lasota, J.P., 1988: *Publ. Astron. Soc. Pacific*, **100**, 1041.
Collin-Souffrin, S., 1991: *Astron. Astrophys.*, in press.
Cunningham, C., 1976: *Astrophys. J.*, **208**, 534.
Dumont, A.M., Collin-Souffrin, S., 1990: *Astron. Astrophys.*, **229**, 313.
Dumont, A.M., Collin-Souffrin, S., 1990: *Astron. Astrophys. Suppl.*, **83**, 71.
Ferland, G.J., Rees, M.J., 1988: *Astrophys. J.*, **332**, 141.
Ferland, G.J., Korista, K.T., Peterson, B.M., 1990: *Astrophys. J. (Letters)*, **363**, L21.
Guilbert, P.W., Rees, M.J., 1988: *Monthly Notices Roy. Astron. Soc.*, **233**, 475.
Heyvaerts, J., 1991: in Proceedings of IAU Colloquium 129, *Structure and Emission processes in Accretion Disks*, Ed. Frontieres.
Krolik, J.H., McKee, C.F., Tarter, C.B., 1981: *Astron. J.*, **249**, 422.
Laor, A., 1990: *Monthly Notices Roy. Astron. Soc.*, **246**, 369.
Lightman, A.L., White, T., 1988: *Astrophys. J.*, **335**, 57.
Madau, P., 1988: *Astrophys. J.*, **327**, 116.
Malkan, M.A., Sargent, W.C.W., 1982: *Astrophys. J.*, **254**, 122.
Malkan, M.A., 1983: *Astrophys. J.*, **268**, 582.
Netzer, H., 1991, in Proceedings of IAU Colloquium 129, *Structure and Emission processes in Accretion Disks*, Ed. Frontieres.
Padovani, P., Rafanelli, P., 1988: *Astron. Astrophys.*, **205**, 53.
Rokaki, E., Boisson, C., Collin-Souffrin, S., 1991: *Astron. Astrophys.*, in press.
Shakura, N.I., Sunaev, R.A., 1973: *Astron. Astrophys.*, **24**, 337.
Shields, G.A., 1978: *Nature* **272**,423.
Sun, W.H., Malkan, M.A., 1989: *Astrophys. J.*, **346**, 68.
Wills, B., Browne, I.W.A., 1986: *Astrophys. J.*, **302**, 56.

Discussion

Wandel: 1. A comparison of the dynamical (emission line width) and accretion disk modelling has been recently published for a large sample of Seyferts and quasars (Wandel 1991, AA 241,5). It finds a good agreement between the two methods, with $L_{UV}/L_{Edd} \sim 10^{-3}$-$10^{-2}$. Since the efficiency of a thin accretion disk is about 0.1, it indicates that $L_{UV} \ll L_{bol}$, but still an accretion disk could contribute a large part of the bolometric luminosity. 2. You have mentioned the two thermal extremes of accretion, the spherical accretion at the virial temperature and optically thick disk at the BB temperature. I would like to remark that there are also intermediate cases. Spherical flows can be cooler than the virial temperature (e.g. the super adiabatic models discussed by Wandel, Yahil and Milgrom, 1984), while accretion disks can become hotter than $T_{BB} = (F/\sigma)^{1/4}$, e.g. if they are electron scattering dominated, or if they become optically thiner hot 2-phase disks (then the electron temperature is 10^9K).

Collin: 1. I was aware of your work about the comparison of dynamical and accretion disk modelling methods, but I did not know that it was published. I apologize for having not mentioned it. Your results agree with ours. 2. I agree that spherical flows can be cooler than the Virial temperature. Concerning the possibility that the disk itself can become very hot, I mentioned this possibility in my talk (your own 2-phase model). You find that it is the inner part of the disk which becomes hot, so it would give rise to the central X-ray source, which would be dominating the UV disk luminosity. This fits quite well in the present model.

Tagger: I wish to point out that concerning the magnetic stabilization of the disk, we might learn a lot from the center of our own galaxy. The field there is very different from other parts of the galaxy. It is *poloidal and very strong* (1 to a few mGauss). This is precisely the value we computed for the stabilization of spiral waves. Thus I believe that the accreting matter may have carried in its own magnetic flux and compressed it to the point where self- gravity driven wave are nearly stabilized.

Collin: If such a model applies to AGN, it would give a solution to the self-gravity problem, which is very acute in the outer parts of accretion disks.

Lesch: You mentioned a column density of 10^{24}cm^{-2} and a number density of 10^{15}cm^{-3} for the UV-bump emitting plasma. That means the emitting structures have dimensions of 10^9cm. What is the origin of these structures?

Collin: These structures could simply be the external irradiated "skin" a much thicker medium, such as an accretion disk.

Malkan: You argued that in some AGNs, the Big Blue Bump does not produce the majority of the bolometric luminosity. But consider a typical high luminosity QSO (say at z = 2). We know many of them have *typical* X-ray slopes of 1.6 or steeper. Now you can integrate the unobserved extension of hard X-ray however you want. But even if you extrapolate a 0.7 X-ray power law out to a MeV, will you not admit that it still has *far less* luminosity than the UV bump?

Collin: If you assume a spectral index from 2 keV to 1 MeV equal to 0.7, (which is the *canonical* index found near 10keV for AGN, including quasars) you find $L_{10keV-1MeV}/L_{2-10keV} = 8$. Since $L_{2-10keV}$ is in average equal to 10% of the observed bolometric luminosity (limited to 10keV, and not including an EUV bump), it means that the $L_{10keV-1MeV}$ is about equal to the observed bolometric luminosity. Do you think that the EUV bump is larger than the whole observed bolometric luminosity? It is not the result one gets from your own disk accretion fits! Even if the X-ray continuum breaks around 100keV, $L_{10-100keV}$ still represents about 30% of the observed bolometric luminosity, and should be comparable to the luminosity in the UV bump.

Komberg: Why do you not consider a possible influence on the observed picture emission from the relativistic jet?

Collin: I have mentioned the fact that the relativistic jet can be the source of the non thermal radiation (including hard X-rays through inverse Compton process) illuminating the accretion disk.

Pudritz: An excellent method of carrying the gravitational binding energy released during accretion is mechanically through the generation of a hydromagnetic disk wind. Could you compare the mechanical luminosity of jets with that of the UV bump? Note that the mass loss rate in the wind is tiny compared with the accretion rate, $\dot{M}_W \sim 10^{-1} - 10^{-2}\ \dot{M}_a$.

Collin: If the mass loss rate in the wind is indeed small compared to the accretion rate, the mechanical luminosity is clearly smaller than the bolometric luminosity, which is $\sim 0.1\dot{M}_a c^2$, since the velocity in the wind is much smaller than c.

Structure and Spectra of Self-Gravitating Accretion-Disks

Herbert Störzer and Rainer Wehrse

Institut für Theoretische Astrophysik, Universität Heidelberg, Im Neuenheimer Feld 561, 6900 Heidelberg

1. Introduction

Thermal emission of geometrically thin, optically thick accretion disks in the innermost parts of active galaxies is the presently favoured possibility to explain the observed continuous energy distributions in the UV-range (UV-bump) of AGN. The main problem to model the continua is (if it is not assumed that every ring of the disk is locally radiating like a blackbody as it is done for example by Sun *et al.* [1]) that the physical structure of the whole disk has to be calculated. Therefore only very few calculations have been published (c.f. Laor *et al.*, [2]) in which the vertical structure of the disk is explicitly calculated. However, in these calculations the effects of self-gravitation and radiation-pressure are not included or severe far reaching simplifications have been made.

In this contribution we report calculations of models in which the effects of self-gravitation and radiation-pressure are included self-consistently. Self-gravitation can be important because disks in the center of active galaxies have relatively large masses and because they are geometrically very thin. Radiation-pressure can be important because these thin disks have relatively large effective temperatures.

Also, the proper radiative transfer has been solved, which means we take into account the angle dependence and the radial coupling in the radiation field to get the continuous energy distributions. It is shown that, for example, for the modelling of the Balmer-jump these effects have to be included accurately.

In order to limit the required computer time we neglect all general and special relativistic effects. We plan to include them in future investigations.

2. Model Construction

2.1 Basic assumptions

In our calculations we make the following basic assumptions:

- steady state accretion
- axial symmetry
- α-viscosity law of Shakura and Sunyaev [3], $t_{r\varphi} = \alpha P_g$ (α = const.)
- Keplerian rotation law
- the inner edge of the disk is always larger than three Schwarzschild-radii, $r_{\text{in}} \geq 3\,r_s$
- the outer edge of the disk is always smaller than 100 Schwarzschild-radii, $r_{\text{out}} line - emitting\ 100\,r_s$
- we solve the radiative transfer equation in the proper form for accretion disks
- we take into account the effects of self-gravitation and radiation-pressure
- no general relativistic effects
- no magnetic fields

2.2 Basic set of equations

All these assumptions lead to the following set of equations (in a cylindrical system of coordinates r,z,φ), which has to be solved iteratively. The hydrostatic equation defines the vertical gas-pressure stratification P_g

$$\frac{dP_g}{dz} = -\rho\frac{GM_*z}{(r^2+z^2)^{3/2}} + \rho\frac{d\phi_s}{dz} - \frac{dP_{\text{rad}}}{dz}, \tag{1}$$

$$\text{with}\quad \frac{dP_{\text{rad}}}{dz} = \frac{-\chi F_z}{c}.$$

M_* is the mass of the central object, ϕ_s is the gravitational potential of the disk, P_{rad} is the radiation pressure, ρ is the density, χ is the Rosseland average of the total extinction coefficient and F_z is the grey flux in vertical direction. To solve the hydrostatic equation we divide the disk in a certain number of rings and every ring is treated as one dimensional structure. P_g is given by $P_g = (\rho kT)/(\mu m_u)$ (μ is the mean molecular weight, m_u is the atomic mass unit). The Poisson equation yields the self-gravitation potential inside the disk

$$\frac{\partial^2\phi_s}{\partial^2 r} + \frac{1}{r}\frac{\partial\phi_s}{\partial r} + + \frac{\partial^2\phi_s}{\partial^2 z} = 4\pi G\rho. \tag{2}$$

In order to solve the Poisson equation we use a finite element method which allows us to use a two dimensional adaptive grid.

The equation which defines the disk height $z_0(r)$

$$\dot{M}[r^2\Omega - r_{ms}^2\Omega_{ms}] = 4\pi r^2 \int_0^{z_0} \alpha P_g\, dz. \tag{3}$$

$\dot{M}$ is the accretion rate, $\Omega = \sqrt{\frac{GM_*}{r^3}}$, r_{ms} is the radius at the last stable orbit. The energy conservation

$$4\pi\kappa(\frac{\sigma_R}{\pi}T^4 - J) = -\alpha P_g(r\frac{\partial\Omega}{\partial r}). \tag{4}$$

If the mean intensity J is known this equation yields the temperature stratification. κ is the Rosseland mean of the absorption coefficient. To get the mean intensity J in optically thick regions, we solve the 1D diffusion equation

$$F_z = \frac{-4\pi}{3\chi}\frac{\partial J}{\partial z}. \tag{5}$$

For optically thin regions, we solve the radiative-transfer equation to get J (Stenholm *et al.*, [4])

$$\vec{n}\nabla I = \chi(S - I). \tag{6}$$

To calculate the continua we solve the radiative transfer equation for wavelength dependent absorption- (κ_λ) and scattering-coefficients (σ_λ), where we take into account free-free absorption, bound-free absorption and electron scattering.

3. Results

3.1 Effects of radiation-pressure

The accretion disks for which we have found a hydrostatic solution in vertical direction at all radii, are always geometrically very thin and optically thick. In these disks the luminosity is evidently below the Eddington-luminosity. If the luminosity is similar or larger than the Eddington-luminosity, we have problems to find a hydrostatic solution at certain radii. Radiation-pressure blows up the disk, but the disk is still geometrically thin and the height is constantly increasing from the inner edge to the outer one (see Fig. 1). Due to radiation-pressure, gas-pressure (Fig. 2) and temperature (Fig. 3) are decreasing in the symmetry-plane. It can also be realized in figure 2 that when radiation-pressure is taken into account, the gas-pressure is nearly constant for a large fraction of the disk height.

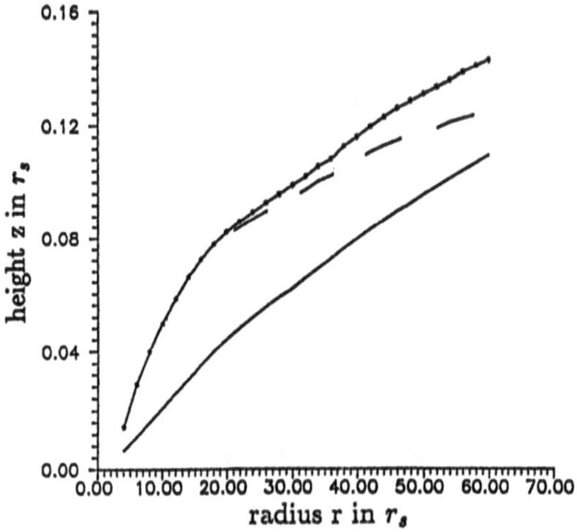

Fig. 1. Height of the disk (we define height as the line where $P_g = 1 \cdot 10^{-3}\mathrm{dyn/cm^2}$) if only gas-pressure is taken into account (full line), if gas-pressure and radiation-pressure are considered (cross marked line) and if gas-pressure, radiation-pressure and self-gravitation are considered (dashed line). The model-parameters are: $M_* = 2.5 \cdot 10^9 M_\odot$, $\dot{M} = 0.25 \cdot M_\odot/yr$, $\alpha = 0.1$, $r_{\rm in} = 3 \cdot r_s$, $r_{\rm out} = 60 \cdot r_s$.

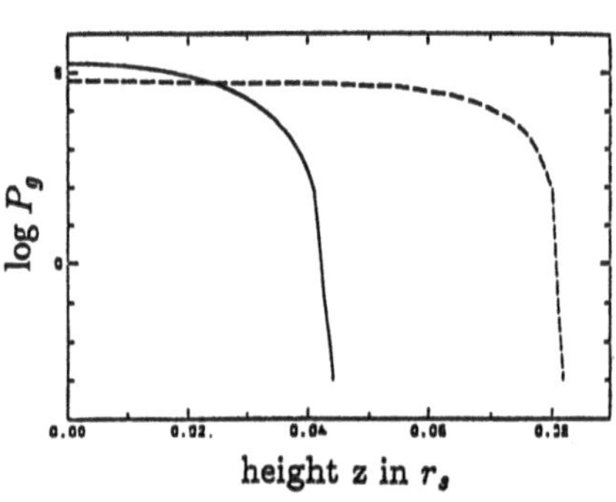

Fig. 2. Vertical gas-pressure P_g stratification at a typical radius of 20 Schwarzschild-radii (r_s). The dashed line represents P_g if radiation-pressure is included, the full line if radiation-pressure is neglected.

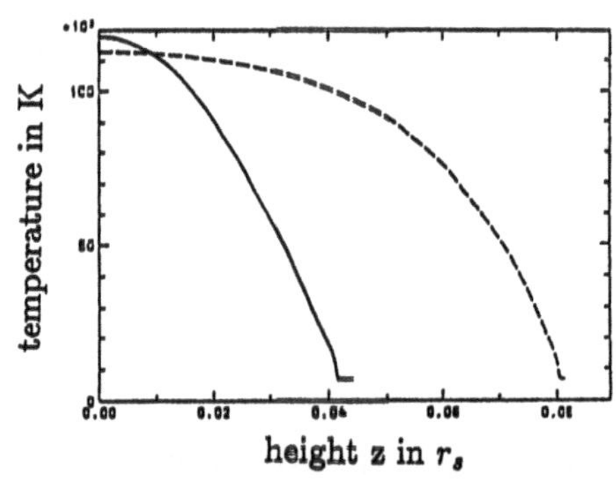

Fig. 3. Vertical temperature stratification at a typical radius of $20 \cdot r_s$. The model-parameters are: $M_* = 2.5 \cdot 10^9 M_\odot$, $\dot{M} = 0.25 \cdot M_\odot/yr$, $\alpha = 0.1$, $r_{\rm in} = 3 \cdot r_s$, $r_{\rm out} = 60 \cdot r_s$.

3.2 Effects of self-gravitation

For the effects of self-gravitation we have to distinguish the effects on the radial disk structure and those on the vertical one. The effects of self-gravitation in radial direction at a given radius are important, if the mass of the disk inside this radius is similar or larger than the mass of the central object. In the disks we have calculated, this is never the case because we have assumed that the outer edge of the disk is always smaller than 100 Schwarzschild-radii. But the effects of self-gravitation on the vertical structure can get important even if the disk mass is much lower than the central mass as the gravity in z-direction is a linear function of the disk height z

$$g_z = \frac{GM_* z}{(r^2 + z^2)^{(3/2)}} \approx \frac{GM_* z}{r^3}. \tag{7}$$

Therefore g_z is getting very low for geometrically very thin disks and the effects of self-gravitation should be taken into account. If self-gravitation is taken into account the geometric height in the outer regions of the disk is decreasing, in the inner parts there are only minor changes (see Fig. 1). For large disk masses ($M_{\rm disk} > 0.005 \cdot M_*$) and large radii ($r > 80 \cdot r_s$) the disk-height is a decreasing function of the radius. This means, that it is very unlikely that the outer regions of the disk can be illuminated by the inner ones.

3.3 Continuous energy distributions

The general run of the continua in our models is similar to a disk which is radiating locally like a blackbody (effective temperature is given by the theory of Shakura and Sunyaev [3]), except for small wavelengths where the flux in our models is higher than the blackbody flux (see Fig. 4). We get only a small Balmer jump and Lyman edge, whereby the Lyman edge can be in emission. That may be due to strong scattering dominance on the long-wavelength side of the edge.

For relatively hot disks, the strongest feature in the spectrum is a Helium I absorption edge at 504 Å(see Fig. 4). If the disk's atmosphere is cooler the Balmer jump is the strongest feature. In Fig. 5 the Balmer edges for different inclination angles i are shown ($i = 0°$ means the disk is being seen face on). If the disk is seen nearly edge on, the Balmer edge vanishes.

4. Discussion

There are several problems to explain the UV-Bump by thermal emission of a stationary, optically thick, geometrically thin disk. First of all it is difficult to explain the raise of the flux for small wavelengths as it is seen, for example, in HS1700+6416 (Reimers *et al.*, 1989 [5]). In all our models where we have found a hydrostatic solution at all radii in the disk, the flux is decreasing for $\lambda < 900$ Å. Furthermore it is difficult to explain luminosities which are similar to or larger than the Eddington-luminosity, because for

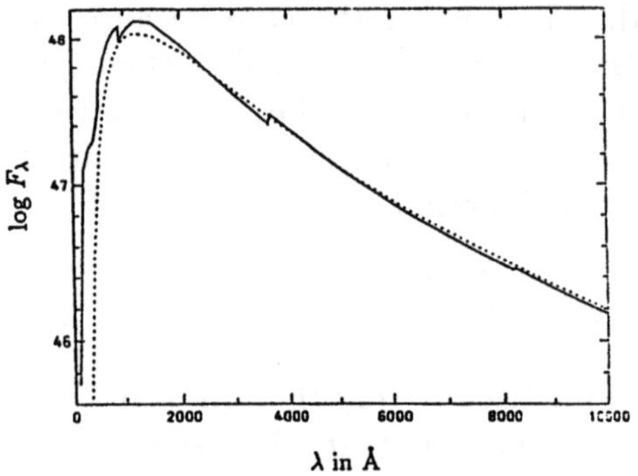

Fig. 4. Continuous energy distribution for a model where the parameters are given by: $M_* = 1.0 \cdot 10^8 M_\odot$, $\dot{M} = 0.1 \cdot M_\odot/yr$, $\alpha = 1.0$, $r_{in} = 18 \cdot r_s$, $r_{out} = 100 \cdot r_s$, $i = 6.25°$. The dotted line represents the flux of a disk which is locally radiating like a blackbody.

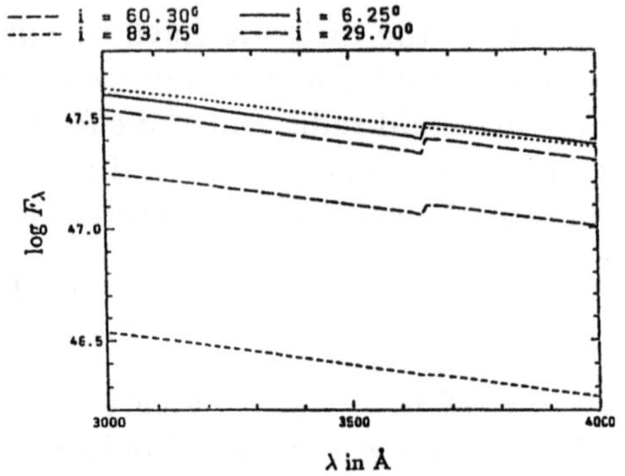

Fig. 5. Balmer edges for different inclination angles. The model-parameters are the same as in figure 4.

such large luminosities we cannot find a hydrostatic vertical solution in a certain radius range.

The most important check of the thin disk approximation is a careful comparison of observed and calculated edges, for which, however, better observations than presently available are needed.

Acknowledgements. This research was partly supported by the German Israeli Foundation (GIF) under number I-94-142.7/88.

References

W.H. Sun, M.A. Malkan: 1989, *Astrophys. J.*, **346**, 68
A. Laor, H. Netzer:1989, *Monthly Notices Roy. Astron. Soc.*, **238**, 897
N.I. Shakura, R.A. Sunyaev: 1973, *Astron. Astrophys.*, **24**, 337
L.G. Stenholm, H. Störzer, R. Wehrse: 1991, JQRST **45**, 47
D. Reimers, J. Clavel, D. Groote, D. Engels, H.J. Hagen, T. Naylor, W. Wamsteker, U. Hopp: 1989, *Astron. Astrophys.*, **218**, 71.

Discussion

Laor: Why does the Balmer edge disappear at large inclination angles ?

Störzer: That is caused by the fact that for large inclination angles the temperature gradients along the lines of sight are smaller than for small inclination angles ?

Siemiginowska: How is the radiation pressure included in your equations ?

Störzer: The radiation pressure itself does not appear in the set of equations. Only the gradient of the radiation pressure is included in the hydrostatic equation.

Abramowicz:: You assume local heat balance and no torque inner boundary but this means that the density is singular at the inner boundary condition. This, in turn, artificially makes self-gravity important at the inner edge of the disk. Is this a problem ?

Störzer: To calculate the potential of self-gravitation inside the disk we neglect the point at the inner boundary of the disk. To assume that the torque is vanishing at the inner boundary is problematic because of the reasons you mentioned. But for larger radii ($r > 30 \cdot r_s$) the vertical disc structure is not very sensitive to the inner boundary condition.

Kundt: You excluded magnetic fields from your considerations. If magnetic fields near the inner edge of the disk exceed 10^6 G, pressure balance with the (diamagnetic ?) disks material exceed your 10^{-3} dyn/cm^2 by more than a factor of 10^{13}.

Störzer:: We have neglected the effects of magnetic fields on the disk structure in our calculations, but I agree that they can be important.

Search for Partial Lyman Edges in Nearby Quasars

A. Koratkar, A. Kinney, and R. Bohlin

Space Telescope Science Institute

Abstract. The Lyman continuum of quasars is expected to be affected by both thermally emitting accretion disks (Sun and Malkan 1989; Laor and Netzer 1989) and broad line region (BLR) clouds (Carswell and Ferland 1988) at the Lyman edge (912 Å). These edge effects range from 10 – 80 % for the disk models and ≤25 % for the BLR models. The IUE archives provide a number of objects to search for the features predicted by these models. A uniform search for the Lyman discontinuity was conducted to search for discontinuities as small as 15% in high signal–to–noise spectra. We found that only 13% of our data showed partial discontinuities which could not be explained by intervening gas extrinsic to the quasar. This fraction is much smaller than predicted by disk models or by BLR models where the clouds are spherically distributed around the central source.

1. Introduction

We have analysed IUE spectra of all Active Galactic Nuclei (AGNs) which contain the waveband around 912 Å ($0.4 \leq z \leq 2.4$) to investigate the nature of the Lyman continuum region. We fit power law continua to the spectral region. The fits are allowed to be discontinuous in a region ±10,000 km/s around the Lyman edge. In this paper we present the preliminary results of this investigation.

A detailed analysis of the continuous spectral region near the Lyman edge (912 Å) is important, because the shape of the continuum in this region is a powerful diagnostic of the emission mechanism of the UV flux in AGNs. Any Lyman continuum absorption at the quasar red-shift is also evidence of absorbing material in the line of sight intrinsic to the quasar and provides clues to the geometry and gas covering factor of the BLR.

The shape at the Lyman edge discriminates between thermal and non-thermal models that are used to explain the UV emission. In thermal models (Sun and Malkan, 1989; Laor and Netzer, 1989) a discontinuity is predicted at 912 Å since there is enormous change in opacity of the gas at this wavelength. In non-thermal models there need not be any discontinuity at all.

Thermal accretion disk models predict 10 – 80% partial discontinuity depending on the inclination angle and thickness of the disk. They also predict that the Lyman edge might be velocity broadened and there might be broad Lyα absorption. If accretion disks are present in all AGNs, then depending on the viewing angle $\approx$ 70 – 80% of the AGNs should show some partial discontinuity. Hence, in thermal models, we expect to see some feature since the gas opacity at 912 Å is changing dramatically.

The absorption signature in the Lyman continuum can be used to get clues about the intervening gas which could be extrinsic to the quasar (for example, host galaxy halo or intergalactic gas) or intrinsic to the quasar. If the absorption is extrinsic to the quasar, i.e., $z_{abs} \leq z_{em}$, the discontinuity is not highly velocity broadened and recovers as ν^{-3}. There are corresponding narrow Lyα and metal (C IV λ1549, Mg II λ2798) absorption lines. Then depending on the red-shift at which the absorption is observed the absorbing material is in the galaxy or intergalactic. If the absorption is intrinsic to the quasars, i.e., by BLR clouds, we have two cases. In the first case, the BLR clouds are larger than the source. In this scenario complete absorption is expected in some lines of sight. Theoretically, the cloud covering factor is predicted to be 0.1, therefore $\approx$10% of the AGNs should show total absorption at $z_{abs} \approx z_{em}$. In the second case, the BLR clouds are smaller than the source. In this scenario, models (Carswell and Ferland, 1988) predict partial Lyman edge depending on the geometry of the BLR. For spherical geometry the discontinuity is $\leq$ 25%, which may be broadened, but will not be larger than the width of the emission lines. Further the continuum does not recover as ν^{-3} and there is no corresponding Lyα absorption. Also most AGNs should show a partial Lyman edge.

2. Analysis

In our analysis we have searched the IUE archival spectra of quasars with red-shifts in the range 0.4 – 2.4. The spectra were optimally extracted and co–added when multiple spectra were available. The improvement in signal-to-noise due to optimal extraction and co-addition made it feasible to investigate partial discontinuities down to the 15% level for the spectra with the best signal-to-noise ratio (5%). 15 of the objects in the list of 76 were deleted because the noise was $\geq$ 60% in the spectrum. For the remaining 61 AGNs, power law continua were fit to regions $\approx$ 250Å on either side of 912 Åby minimizing the χ^2. The discontinuity between the two power law fits was considered to represent the discontinuity at the Lyman edge. See figures 1 – 4 showing examples of typical fits.

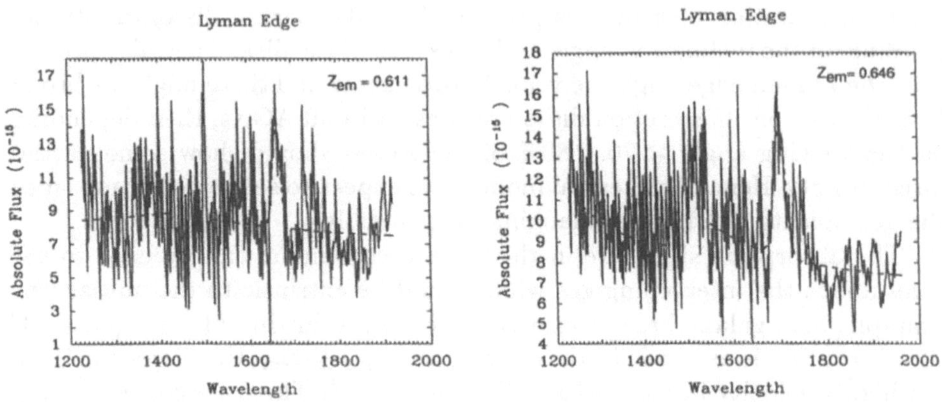

Fig. 1. Two examples of spectra which show no discontinuity. Note that the regions containing the IUE artifacts are excluded from the fit.

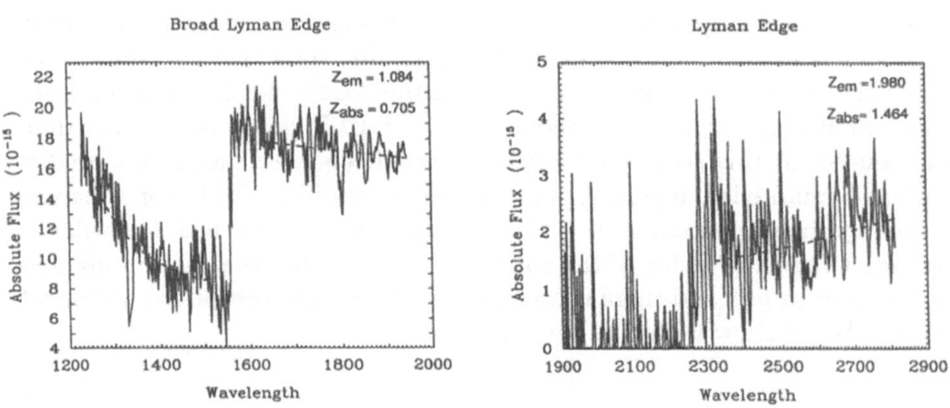

Fig. 2. Two examples of spectra which show discontinuity at $z_{abs} \leq z_{em}$.

3. Results

Our sample can be divided into four groups according to the nature of the discontinuity at the Lyman edge. These groups are as follows:

1. The group of objects which do not show any discontinuity in the IUE spectral range. This group contains 39% of the sample.
2. The group of AGNs which show Lyman continuum absorption at red-shifts smaller than the AGN emission red-shift. This group has 28% of the sample. All the objects in this group, except 1718+481, show complete Lyman absorption edges. These absorption edges are not velocity broadened and have associated Lyα, or metal (C IV λ1549,Mg II λ2798) absorption lines. 1718+481 shows a partial discontinuity which recovers

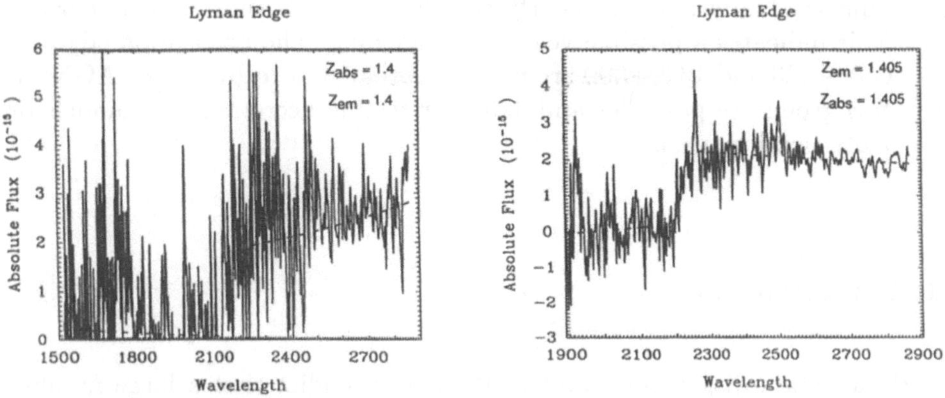

Fig. 3. Two examples of spectra which show discontinuity at $z_{abs} \approx z_{em}$, where the absorption is due to extrinsic intervening gas.

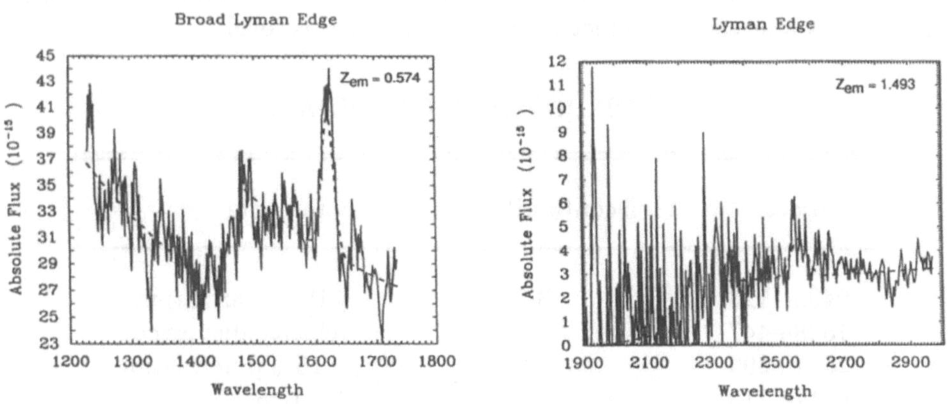

Fig. 4. Two examples of spectra which show discontinuity at $z_{abs} \approx z_{em}$, and belong to our candidate list.

faster than ν^{-3}. The absorption edges seen in this group of objects are due to intervening gas extrinsic to the AGN and its host galaxy.

3. The group of objects which show Lyman continuum absorption at the AGN red-shift with associated Lyα or metal absorption lines. This group has 20% of the sample. Most of the Lyman absorption edges in this group are not velocity broadened and the absorption is down to zero flux for the region shortward of the Lyman edge. These edges are due to gas associated with the AGN or its host galaxy environment.
4. The group of objects which show Lyman continuum absorption at the AGN red-shift with **no observed** associated Lyα or metal absorption lines. This group contains 13% of the sample. The objects which form this group are shown in the table. All objects in this group, except 0405−123 and 1641+399, show discontinuities such that the contin-

uum gradually tends to nearly zero flux towards shorter wavelengths. This indicates a possible velocity broadening. The absorption edges in 0405−123 and 1641+399 are not very broad and recover. The AGNs in this group are possible candidates for testing accretion disk models or BLR cloud models.

4. Conclusions

Both accretion disk models and BLR models predict that a large number of AGNs should show discontinuities at the Lyman edge. We find only a small number (13%) show discontinuities that cannot be accounted for by intervening gas extrinsic to the AGN. If the models are correct then most discontinuities have to be ≤20%. We are limited by the signal-to-noise in the spectra and cannot detect discontinuities at this level.

LIST OF POSSIBLE CANDIDATES

Object	%discontinuity/%noise	comments
0405−123	3.2	≤1% polarization
1538+447	2.6	velocity dispersed
1641+399	2.6	3–15% polarization
0117+213	2.0	flat Lyα
1338+416	1.6	Lyα absorption?
0743−673*	1.1	≤1% polarization
0537−441*	1.0	
2223−052*	0.6	4–17% polarization

References

Carswell, R.F., and Ferland, G.J., 1988: *Monthly Notices Roy. Astron. Soc.*, **235**, 1121.
Koratkar, A., Kinney, A., Bohlin, R., 1991: in preparation.
Laor, A., and Netzer, H., 1989: *Monthly Notices Roy. Astron. Soc.*, **238**, 879.
Sun, W.H., and Malkan, M.A., 1989: in: *Active Galactic Nuclei*, ed. D.E. Osterbrock and J. Miller, p262.

Discussion

Kazanas: Did you find any correlation with the radio properties of the objects?

Koratkar: No I did not. The rate of objects showing discontinuities is very similar to those of the radio-loud and radio-quiet subsets of the sample

Wandel: If a hot disk or corona of low Thomson depth is common in AGN accretion disks, its Compotizating effect could reduce the observed edge, and hence explain the discrepancy between the expected number of objects with an observable discontinuity and the lower number you have actually found.

The Nature of the Broad-Line Clouds

B.M. Peterson[1], *G.J. Ferland*[1], *K. Horne*[2], *W.F. Welsh*[1,2], *and S.N. Nahar*[1]

[1] Department of Astronomy, The Ohio State University, 174 West 18th Avenue, Columbus, Ohio 43210 USA
[2] Space Telescope Science Institute, 3700 San Martin Drive, Baltimore, Maryland 21218 USA

Abstract. Emission-line variability results show that the radiation density in the broad-line region (BLR) is much higher than previously supposed. One consequence of this is that the column density in the ionized zone is very large, so that the BLR clouds are also optically thick in the emission lines. This leads to anisotropic line emission from the clouds, with most of the line emission directed back inwards toward the continuum source. In this case, the observed line transfer functions are consistent with a spherically symmetric distribution of BLR clouds. On this basis, we suggest that the possibility that the BLR clouds originate in stars in the nucleus should be reconsidered.

1. Introduction

The results of the emission-line variability studies of the last few years are fundamentally altering our view about how AGNs work. Massive monitoring efforts such as the International AGN Watch program on NGC 5548 (Clavel *et al.* 1991; Peterson *et al.* 1991) are providing important new constraints on models of both the central engine and the broad-line region (BLR) which are severely straining the current paradigm.

The NGC 5548 monitoring campaign provided reliable continuum – emission-line lags for several lines. The highest ionization lines, such as N V λ1240 and He II λ1640, respond nearly instantaneously to continuum variations, and this has a number of profound implications, as discussed by Ferland. The measured lags for other high-ionization lines, namely Lyα and C IV λ1549, are $\sim$ 8 days, and lower ionization lines such as C III] λ1909 and the hydrogen Balmer lines, respond more slowly. This is taken to be evidence for ionization stratification of the BLR.

More complete information on the BLR geometry can be obtained by solving the emission-line transfer equation (Blandford and McKee 1982). The NGC 5548 monitoring data have been used to obtain the transfer functions for the UV lines by Krolik *et al.* (1991) and for Hβ by Horne, Welsh, and Peterson (1991).

2. Implications for Models

The results for NGC 5548 have at least three important ramifications for the BLR physics:

1. The radiation density in the BLR is more than 100 times higher than previously supposed in "standard" photoionization models. This is a simple consequence of the very small BLR sizes inferred from emission-line variability.
2. The fact that different lines have different lags shows that single-zone photoionization models are inadequate. A direct consequence of this is that diagnostics like the global C III] $\lambda 1909$/C IV $\lambda 1549$ flux ratio are of dubious value, as the respective transfer functions show that these two lines have different radial emissivity distributions.
3. The line transfer functions have maxima displaced from zero lag, so we see little if any emission-line response due to material along our line of sight to the central source. (We note that this statement cannot be made with great certainty for the UV lines with the smallest lags because of the 4-day temporal resolution of the Clavel *et al.* 1991 experiment.)

In this contribution, we show that the first two of these points lead to a natural explanation of the third, which is that the line emission is emitted anisotropically from the ionized clouds. We see no line response along the line of sight to the continuum source because these clouds emit most of their line emission back towards the central source.

3. Anisotropic Line Emission

Photoionization models for this study are based on G.J. Ferland's code "CLOUDY", and our calculations are described in more detail by Ferland *et al.* (1991). Here we outline only the basic results.

The radiation density in the BLR, say at the C IV–Lyα radius of 2×10^{16} cm, is very high. From the mean rest-frame luminosity of NGC 5548 during the 1988–89 campaign, $\nu L_\nu(1315\text{Å}) = 1.8h^{-2} \times 10^{43}$ ergs s^{-1}, we can infer that the ionizing flux in the C IV–Lyα zone is $\phi = 1.9h^{-2} \times 10^{20}$ photons cm^{-2} s^{-1}, where we have adopted the relative spectral continuum energy distribution of Mathews and Ferland (1987), and the quantity h is the Hubble constant in units of 100 km s^{-1} Mpc^{-1}. (We note in passing that the ionizing flux in this continuum does not differ from that used by Krolik *et al.* 1991 by more than $\sim 10\%$.) With such a high ionizing flux, the BLR clouds must have large column densities to be optically thick in the ionizing continuum, which they obviously must be because the recombination lines vary with the continuum. The ionization parameter is

$$U = \frac{\phi}{cN_H} = \frac{0.63h^{-2}}{N_{10}},$$

where N_{10} is the total hydrogen number density N_H in units of 10^{10} cm^{-3}. This corresponds to a Strömgren depth

$$r = \frac{Uc}{N_H \alpha_B} = \frac{7h^{-2} \times 10^{12}}{N_{10}^2} \text{ cm}$$

or a column density through the ionized zone

$$N_H\, r \approx \frac{10^{23}}{N_{10}} \text{ cm}^{-2}.$$

The one poorly constrained parameter is the density; indeed, there is *no reliable* density indicator, as C III] λ1909 and C IV λ1549 have different transfer functions. We can only estimate $N_H \approx 10^{11}$ cm^{-3} from the relative amplitudes of the transfer functions at $\sim$ 8 light days. We note that there is considerable margin for error for several reasons, including the uncertain amplitude of the C III] transfer function at 8 light days; because the C III] transfer function is not well resolved, its true amplitude at 8 days may be overestimated.

The key point is that the column densities in the BLR clouds must be very large, and this means that the clouds also are optically thick in nearly all the important emission lines. A consequence of this is that the emission-line radiation from the clouds is anisotropic, in the sense that most of the line emission escapes from the face of the cloud which is illuminated by the continuum source (i.e., back towards the central source), and relatively little escapes through the outward facing cloud surface (i.e. away from the continuum source). Our representative photoionization models indicate that virtually 100% of the Lyα emission is emitted back towards the central source (although this has been known for over a decade - see Ferland and Netzer 1979), and 60-90% of the Hβ emission as well as 60-70% of the CIVλ1549 emission is directed back towards the central source.

The fact the line emission from the BLR clouds is not emitted isotropically leads to the important conclusion that the transfer functions derived from the monitoring data do not rule out a spherical BLR. Anisotropic line emission can explain the absence of an immediate emission line response due to material along our line of sight to the continuum, where we see only the unilluminated face of the clouds.

We show how the transfer function of a spherical BLR is altered by anisotropic emission in Fig. 1. The illustrative transfer functions shown here were computed by describing the emission from a BLR cloud at a distance r from the central source by

$$F(r,\Theta) = F_\circ \left(\frac{r}{R_{min}}\right)^{-\beta} (1 + A cos\Theta)$$

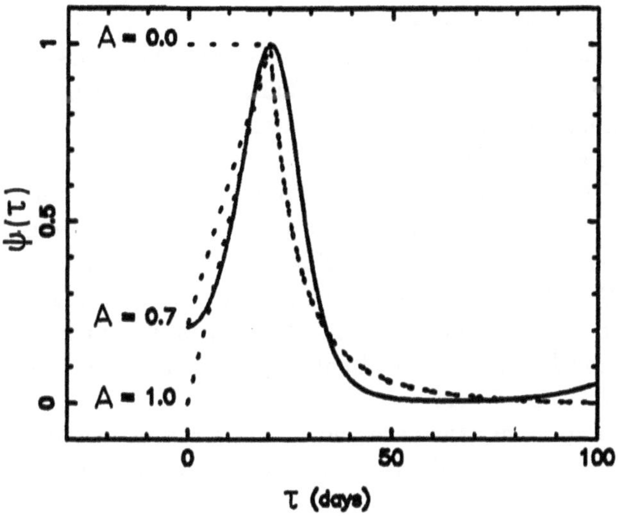

Fig. 1. Sample transfer functions are shown as dotted lines for spherical BLRs with R_{min} = 10 light days, R_{max} = 50 light days, and $\beta = 5$, and three different values for A. For comparison, the transfer function for the Hβ emission line in NGC 5548 (Horne, Welsh, and Peterson 1991) is shown as a solid line.

where Θ is the angular coordinate around the BLR (measured from the observer's line of sight), A is the ratio of anisotropic to total emission (i.e., A = 0 for isotropic emission and A = 1 for complete anisotropy, such as lunar phases), R_{min} is the inner radius of the BLR, and β is a power-law emissivity index. For the examples shown in Fig. 1, we take R_{min} = 10 light days, $\beta = 5$, and use R_{max} = 50 light days for the outer radius of the BLR. We see that for these particular parameters, A = 0.7 provides a plausible description of the Hβ transfer function.

Fig. 1. Sample transfer functions are shown as dotted lines for spherical BLRs with R_{min} = 10 light days, R_{max} = 50 light days, and $\beta = 5$, and three different values of A. For comparison, the transfer function for the Hβ emission line in NGC 5548 (Horne, Welsh, and Peterson 1991) is shown as a solid line.

4. Origin of the BLR Clouds

On the scale of the BLR, there are good physical reasons to believe that any gas should be in a flattened configuration; indeed the transfer functions for NGC 5548 are consistent with a nearly face-on disk-like BLR regardless of whether the line emission is isotropic or anisotropic. The problem with interpretation of the NGC 5548 BLR as face-on disk is that the emission-line widths imply large axial velocities, of order 5000 km s^{-1}. Given this major difficulty, it is reasonable to ask whether or not there exist reasonable scenarios that lead to a spherical BLR.

One possibility is that BLR clouds could be the irradiated atmospheres of red giants, as been suggested in the literature on many occasions (e.g., Edwards 19980; Mathews 1983; Scoville and Norman 1988; Penston 1988; Kazanas 1989; Puetter 1989). The small measured lags imply an even higher radiation field than used in most of the published calculations, which enhances the viability of this mechanism for producing the line emission. This idea has a number of attractive features:

- We know spheroidal component stars are there. They are expected to have a randomized velocity field, which is apparently consistent with the observations.
- Neither cloud confinement nor gas supply is an issue.
- New photoionization models require fewer stars to produce broad profiles, so the nature of emission-line profile structure is less of a problem. The limited profile structure seen in AGN spectra requires something like 10^4 clouds if they are each radiating at their thermal width of $\sim$ 10 km s^{-1} (Capriotti, Foltz, and Byard 1981; Atwood, Baldwin, and Carswell 1982). A simple update of the calculations by Mathews (1983) indicates that the speed at which gas will leave the stars is $\geq$ 100 km s^{-1}. Furthermore, in the high-column density models, the electron scattering optical depth is greater than unity, which will further broaden the lines.
- The emission-line profiles and centroids might show differences as the emissivity distributions and degree to which anisotropic emission is important are likely to be different from line to line.

In view of both the enhanced viability of the mechanism of mass-loss from red giants due to the reevaluation of the radiation density in the BLR, we believe that it is appropriate to reconsider carefully the idea that evolved stars in AGN provide the gas which we see in broad-line emission.

Acknowledgements. This work has been supported by the National Science Foundation under grants AST-8915258 (B.M.P.) and AST-9019692 (G.J.F.) and by NASA under grants NAG5-1366 (B.M.P. and G.J.F.), NAGW-1796 and NAGW-2264 (K.H.). W.F.W. acknowledges support through the STScI Graduate Student Program. We thank the organizers for their kind hospitality at this conference.

References

Atwood, B., Baldwin, J.A., and Carswell, R.F. 1982, *Astrophys. J.*, **257**, 559
Blandford, R.D., and McKee, C.F. 1982, *Astrophys. J.*, **255**, 419
Capriotti, E., Foltz, C., and Byard, P. 1981, *Astrophys. J.*, **245**, 396
Clavel, J., et al. 1991, *Astrophys. J.*, **366**, 64
Edwards, A.C. 1980, *Monthly Notices Roy. Astron. Soc.*, **190**, 757
Ferland, G.J., and Netzer, H. 1979, *Astrophys. J.*, **229**, 274
Ferland, G.J., Peterson, B.M., Horne, K., WElsh, W.F., and Nahar, S., 1991, preprint
Horne, K., Welsh, W.F., and Peterson, B.M. 1991, *Astrophys. J. (Letters)*, **367**, L L5
Kazanas, D. 1989, *Astrophys. J.*, **347**, 74
Krolik, J.H., Horne, K., Kallman, T.R., Malkan, M.A., Edelson, R.A., and Kriss, G.A. 1991, *Astrophys. J.*, **371**, 541
Mathews, W.G., 1983, *Astrophys. J.*, **272**, 390
Mathews, W.G., and Ferland, G.J. 1987, *Astrophys. J.*, **323**, 456.
Penston, M.V. 1988, *Monthly Notices Roy. Astron. Soc.*, **233**, 601
Peterson, B.M., et al. 1991, *Astrophys. J.*, **368**, 119.
Puetter, R.C. 1989, in *Active Galactic Nuclei*, ed. D.E. Osterbrock and J.S. Miller, (Dordrecht: Kluwer), p. 137.
Scoville, N., and Norman, C. 1988, *Astrophys. J.*, **332**, 163.

Discussion

S. Collin-Souffrin: I don't understand, from a simple energetic point of view, since the coverage factor of stars should be of order 10%, how you can account for the BLR emission with a small number of clouds. I would expect that $\sim 10^9$ stars are necessary in luminous AGN.

B.M. Peterson: Yes, for NGC 5548 one requires $\sim 10^8$ stars of radius 10^{12} cm to achieve a covering factor of 10%. The argument for a small number of stars fails spectacularly unless the radii are much larger than for normal giant stars. The whole idea hinges on the atmospheres of the red giant being expanded through interaction with the AGN radiation field.

A.V. Filippenko: Another way to estimate the number of line-emitting clouds is to compare the observed luminosity of broad Hβ with the luminosity emitted per cloud (the latter is derived from the volume and density of a cloud). When this is done for quasars, a very large number of clouds is found. What is the result for NGC 5548, and how does it compare with your independent estimate of the number of line-emitting clouds?

B.M. Peterson: A simpler version of this calculation is to consider Lyα rather than Hβ, since the number of Lyα photons emitted equals the number of ionizing photons ϕ in the absence of Lyα photon destruction (Gary Ferland points out that this is one of the few things that stays simple in the new models). And, of course, we have been considering at this time the Lyα-emitting zone. The number of stars required still works out to be $\sim 10^8$ if they have radii of only 10^{12} cm. The stars must be inflated or develop winds that increase their cross section by a factor of at least 100 to bring the number of clouds required to fewer than 10^4, and 1000 or more to bring the number of stars to under 100.

It should be noted that the radiation density in the BLR is not going to disrupt the stars; as Roger Blandford has pointed out, in NGC 5548 the equilibrium blackbody temperature at 10 light days from the central source is only $\sim$ 2000 K, and a stellar atmosphere can easily deal with this much extra energy. However, this is hard radiation, and there is no doubt that the outer layers will expand and perhaps escape if irradiated by an AGN continuum. In the model of Kazanas (1989), the stars are nearly 100 times larger than normal red giants, even using a radiation density lower than in NGC 5548. This whole issue needs to be reexamined very carefully.

Broad Lines and Continuum Spectrum from Accretion Disc Models

Catherine Boisson

DAEC, Observatoire de Paris–Meudon, France

1. Introduction

There have been several suggestions these last years that the optical–UV bump observed in a number of active galactic nuclei (AGN) can be interpreted as the thermal emission of an accretion disc (Malkan, 1983), but also that this disc can contribute to the broad line emission (Dumont and Collin–Souffrin, 1990).

It is also now well established that the bulk of the high ionization (HIL) and the low ionization (LIL) lines are not emitted by the same physical region. One medium is dilute, optically thin, ionized by the UV continuum and emits mainly the HIL (Lyα,CIV,...). The intensities of the LIL (Balmer lines, MgII, FeII...) are consistent with a denser and thicker medium heated by X–ray radiation (see Collin–Souffrin and Lasota, 1988, for a review).

The structure of these media can be inferred from kinematical arguments. The existence of a weak correlation (Wills and Browne, 1986) between the width of Hβ and the ratio of extended to compact radio emission can naturally be explained if the LIL are confined in a plane perpendicular to the radio axis. The LIL are often asymmetrical with extended red wings, again that is easily reproduced in a disc structure where Doppler transverse effect and gravitational redshifts are taken into account. Also the HIL are blueshifted with respect to the LIL (e.g. Espey, 1989). So a possible model for the broad line region is one where the bulk of the HIL is emitted in an outward flowing medium and the LIL in the irradiated chromosphere of an accretion disc illuminated by an external radiation flux. The part of the HIL emitted by this region strongly depends on the proportion of UV seen by the disc.

In order to assess the legitimity of such an hypothesis, namely the contribution of an accretion disc to the LIL emission , it is necessary to build a consistent model which does not only prescribe line *profiles* but also line *intensities*. The constraints set on the model by fitting *both* line profiles and line intensities, combined with the assumption that a large proportion of the UV continuum is due to disc emission, are indeed so stringent that not only disc illumination can be deduced but also several parameters such as the black–hole mass M_{bh}, the accretion rate $\dot{M}$ and the inclination i of the disc with respect to the line of sight.

Up to now the most used methods to determine black–hole masses are:

1) the "dynamical" method which accounts for the broad line velocities by a gravitationally bound region

2) the "accretion disc" method which consists in fitting the AGN continuum in the near UV with an accretion disc.

Notice that these two methods reach roughly the same conclusions: $R_{Edd} = L_{Bol} / L_{Edd}$ is $\simeq$1 for QSOs and 0.1 for Seyferts. But none of them can precisely determine the mass of an individual object. The first method relies on averaged ionization parameters and densities of the emitting region somehow arbitrarily chosen. The second, once given the working model (thick or thin disc, rotating or non–rotating black–hole) suffers of several uncertainties concerning the disc emission itself such as the inclination, the minimum and maximum radius of emission, etc.

We demonstrate here the power of combining the two methods in determining the physical parameters of individual objects. The lines are assumed to be emitted by the outer region of the accretion disc illuminated by a central X–ray source, and the optical–UV continuum is assumed emitted by the central region of the same accretion disc. The line fluxes are computed combining photoionization models of thick and dense media ($N \simeq 10^{24-25}$ cm^{-2}, $n_e \simeq 10^{11-12}$ cm^{-3}), heated by hard X–rays not the UV bump, with the structure of a stationary α–disc, geometrically thin and axisymmetric (Collin–Souffrin and Dumont, 1989) so that one gets the structure of an irradiated atmosphere (ionization parameter, density...) for a given illumination model.

A full account of this study is given in Rokaki, Boisson and Collin–Souffrin, 1991.

2. The model

Two different illumination models have been distinguished by Collin-Souffrin and Dumont: the Point Source Model (PSM) where the disc is directly illuminated by a source located at a given height above the disc, and the Diffusion Model (DM) in which case the disc is illuminated by radiation backscattered towards the disc by a hot medium. The flux incident on the disc varies in a different way with the distance from the central source (R^{-3} for the PSM, R^{-g} for the DM with $0 \leq g \leq 1$) (Dumont and Collin–Souffrin, 1990).

Line flux computations are performed from the orbit of marginal stability to 10^6 R_G. Lines cannot be emitted at radii smaller than 10^2 R_G as the temperature of the underlying photosphere of the disc is larger than the excitation temperature of the lines, and also because the incident flux becomes larger than the black–body flux in the lines (saturation flux). They are generally not emitted at $R>10^4 R_G$ as the flux from the central source is then too small to heat the disc. The major point of this model is that the region emitting the bulk of the lines is limited to a given range of radii *whatever* the extension of the disc. In a $N \simeq 10^{24-25}$ cm^{-2} and $n_e \simeq 10^{11-12}$ cm^{-3} medium, Balmer lines are nearly in LTE and so do not depend much on the ionization parameter or the spectral distribution of the continuum *but* mainly on the incident flux. The relation between the flux emitted in a line and the incident flux, for thick and dense media at a given radius, has been studied by Collin–Souffrin and Dumont (1989). These relations have been improved for the sake of our study by taking into account the influence of column density (line fluxes at low and high fluxes are modified) and by computing (not assuming) the saturation radius.

Line profiles are computed for each frequency bin by integrating over the allowed radius range for each bin. The emission in each bin is proportional to the line flux weighted by the surface of a disc annulus taking care of relativistic corrections. Line profiles are strongly dependent on the regime where the maximum emission of the disc occurs. This is illustrated in figure 1, for the PSM, keeping same bolometric luminosity, same equivalent width of

$H\alpha$ and same i and varying the height of the source above the disc, R_s. One can see that for large values of R_s the line flux is constant over a large fraction of the disc. Maximum luminosity occurs at a radius of the order of R_s and so the profile is narrow without humps and wings. When R_s is smaller the saturation flux is reached at smaller radius, so is the maximum luminosity in the line and thus the profile is broad with prominent shoulders. The radius of maximum luminosity, R_{max}, thus determines the width of the profile and the size of the line emission region. We have then the constraint that this radius should be smaller than the size given by the characteristic timescale of profile variations.

Although we suppose that a disc is providing the near–UV luminosity, it does not mean necessarily that the whole bolometric luminosity is due to disc accretion. We therefore performed two sets of computation of the near–UV continuum. The first one assuming that the X–ray and non thermal luminosity are not due to the disc so the accretion rate through the disc is related only to the UV luminosity. In the other case, we assume that the whole bolometric luminosity is provided by disc accretion.

The energy distribution from the near–IR to the UV is fitted by adding the contribution of a stellar component, a power–law and the disc contribution (non–rotating black–hole).

3. The fits

Physical parameters of the central source for 6 AGN have been deduced from the Hβ and Hα line fluxes and profiles and their continuum energy distribution.

The adjustment was done assuming first that the whole bolometric luminosity is provided by disc accretion, $\dot{M}(\text{disc}) = L_{Bol} / \eta\ c^2$. Then the inclination is constrained by the humps and asymmetry of the profiles. For a given line flux and inclination, the emission regime and thus the profiles are well determined by R_{Edd}. So, only one value of M_{BH} gives acceptable *intensities* and line *profiles*. Then the estimated mass and accretion rate are used to compute the continuous spectrum. Apart for Arp 102, which in fact is stellar dominated, we cannot fit the continuum with such an accretion rate. $\dot{M}$(disc) has in fact to be decreased. As $\dot{M}$(disc) does not influence much line intensities and profiles, we get the final parameters after a few iterations from line to continuum. This modifies only slightly M_{BH}. Results of such a fit are given for NGC 5548 in figure 2.

The line fits were optimized to match the width and the asymmetries of the profiles. The line fluxes we get are then of the order of 80% of he observed ones, which just means that the bulk of the LIL is emitted by the disk by not the whole intensity.

4. Conclusions

We estimate that the final fits give the mass of the objects within a factor 2 for a given value of the Hubble constant H_0.

Only a small fraction of the bolometric luminosity is reprocessed in the line by the disc, which corresponds to what is expected in the context of AGN. Also R_{Edd} is generally small, apart for 3C 120 where it is one order of magnitude larger suggesting a non thermal radiation both collimated and amplified.

But the new fact from this study is that *the accretion rate through the disc is about one order of magnitude smaller than the total accretion rate.* Such a result could be obtained because we combined both methods (line and continuum) and also because the central mass is strongly constrained by the line intensities and profiles, while the UV continuum depends on both the mass and the disc accretion rate.

Now, some improvement to this study can be envisioned as computing non–axisymmetric disc or taking into account the variability pattern. For a first approach to this last point, see Rokaki's poster.

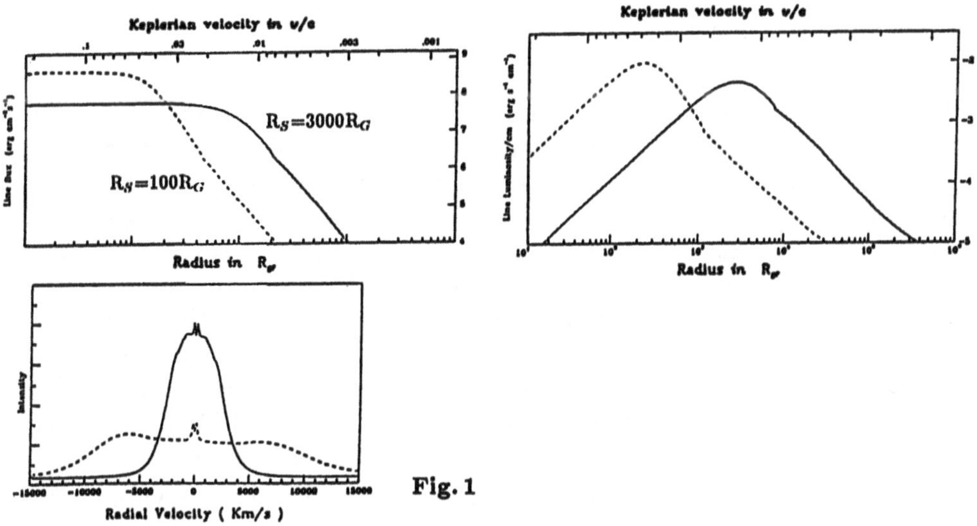

Fig. 1

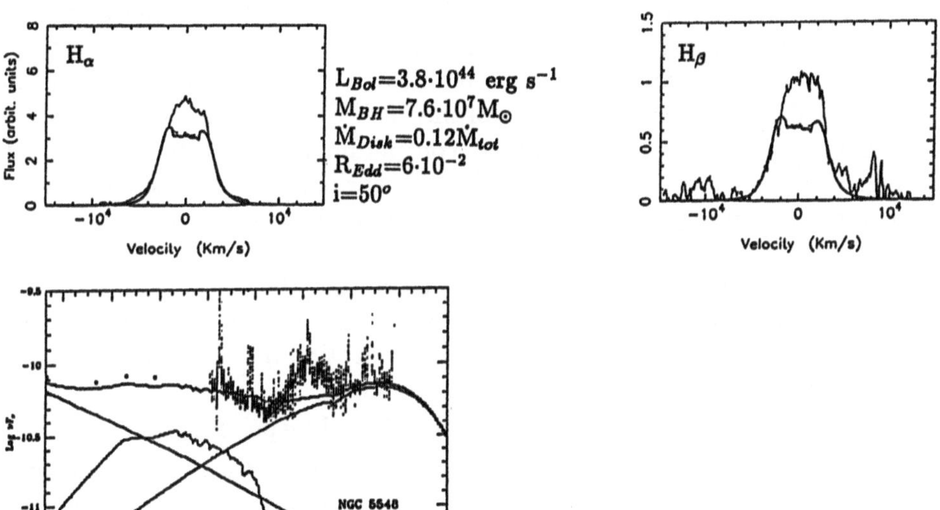

Fig. 2.

References

Collin-Souffrin, S., Dumont, A.M., 1989: *Astron. Astrophys.*, **213**, 29.
Collin-Souffrin, S., Lasota, J.P., 1988: *Publ. Astron. Soc. Pacific*, **100**, 1041.
Dumont, A.M., Collin-Souffrin, S., 1990: *Astron. Astrophys.*, **229**, 313.
Espey, B.R., 1989: *PhD thesis*, Cambridge University.
Malkan, M.A., 1983: *Astrophys. J.*, **268**, 582.
Rokaki, E., Boisson, C., Collin-Souffrin, S., 1991: *Astron. Astrophys.*, in press.
Wills, B., Browne, I.W.A., 1986: *Monthly Notices Roy. Astron. Soc.*, **302**, 56.

Discussion

Wandel: I think this work is very interesting, as it determines the inclination independently, hence allowing a much smaller uncertainty in the acc. disc parameters from the continuum fits than fits which leave the inclination as an open parameter (it should however be cautioned that, although likely, the line emitting gas is not necessarily coaxial with the disc). I have three questions: 1) What is the typical uncertainty in the determination of the inclination angle from the line profiles? 2) Suppose you use the inclination and the optical–UV continuum to fit, what is the uncertainty in the M and $\dot{M}$ found in that way? How do the masses determined by the acc. disc fitting compare with the masses found from the line profiles? 3) To what extent do your results depend on the specific ionization model used?

Boisson: The inclination is fairly well defined when you have asymmetrical profiles. In cases where the profile is symmetric you have a larger range of possible inclinations. But anyway it is known within 15 degrees. For the masses as I said it is then constrained to better than a factor 2. The knowledge of the inclination eliminates already a large cause of uncertainty (a factor 30 for disc around Kerr black–holes). The uncertainty on M and $\dot{M}$ is strongly dependent on the quality of the far UV continuum determination. A change in M shifts the computed continuum from left to right and a change in $\dot{M}$ from top to bottom with some shrinking. So if one has a good observed continuum, once the inclination is determined, there should be almost no uncertainty in the determination of M and $\dot{M}$. To answer your last question, as the Balmer lines are in LTE, they do not depend much on the ionization parameter and the distribution of the continuum but mainly on the incident flux. So in a way it is not "ionization model" dependent but it strongly depends on the illumination model.

Peterson: Can you address the variability of double–peaked profiles?

Boisson: To reproduce variability pattern in the sense that both peaks can alternatively be stronger you need to consider non-axisymmetric disc, unlike the disc model used for our study. Nevertheless for a first element of response, the same model with reprocessing give a good account of the variability line curves (continuum and lines) for NGC 5548, details are given in Rokaki's poster.

Veilleux: Comment to follow up on B. Peterson's question: A recent study by W. Zheng and myself (these proceedings) report the results of 14 years of monitoring of the broad line profile in 3C 390.3 and shows indeed spectacular hump variations. It is shown that a line emitting disc with a co–rotating "hot spot" can explain these variations.

Variability and its Implications for the Structure of the Broad Line Region

Julian H. Krolik

Johns Hopkins University

1. The Goals

Before embarking on a discussion of all that we have learned about the structure of the broad emission line region from variability studies, it is worthwhile spending a little time discussing the long range goals of this enterprise. We will then have a better grasp of the significance of what has been accomplished, and how much more remains to be done.

Painting with a very broad brush, one might say that the real point of the effort is to transform a number of optical and ultraviolet spectra into maps of all the quantities necessary to describe the line-emitting material. If we may make the assumptions that the continuum is radiated isotropically, and that the angle-averaged continuum optical depth of the broad line region is less than unity (two non-trivial assumptions, as I will discuss later), then we immediately obtain a map of one of these quantities, the ionizing intensity. A map of the pressure p would also be desirable, but it cannot be found so easily. Because the emission line material is almost certainly highly clumped, its other physical characteristics must be described by distribution functions which can vary with position. To construct them, we must first take suitably coarsely-grained spatial averages. We can then obtain maps showing the distributions of clump velocity $\boldsymbol{v}$, clump column density N, and clump shape.

This effort is not undertaken for the pure pleasure of cartography. A great deal about the mechanisms of active nuclei could be learned from maps of this sort. For example, if motions inside the line-emitting clumps are subsonic, their internal pressure is a diagnostic of the surrounding pressure, which in turn tells us about one of the forces acting on the volume-filling material, whatever it may be. Similarly, if the velocity distribution function at each location is very narrow, the clump velocity field is probably acting as a tracer of the flow in the inter-clump matter. If, on the other hand, the velocity distribution functions are broad, that is a signal that the dynamics of the line-emitting clumps is essentially "collisionless". The sign of the mean projection of clump velocities on the local radius vector will tell us whether the net flow of line-emitting stuff is inward (in which case it is probably drawn in by the gravity of the central object), or outward (signalling that gravity is overwhelmed by some other force, possibly radiation pressure, or the drag of a fluid wind). Combining the distribution functions of clump thickness with the velocity field will tell us about the mass budget of the region. These examples only scratch the surface of the information available once this program is actually completed.

2. Methodology

Granted such attractive goals, it is natural to ask how we can arrive at them. The crudest vehicle is single-zone photoionization modelling. Using this method, one takes the ratios of observed line fluxes, and uses them to constrain the mean pressure, ionization parameter Ξ (the ratio of radiation to gas pressure), and column density. Applying the two assumptions of continuum isotropy and BLR transparency, we can infer a characteristic radius:

$$\langle r \rangle = \left[\frac{F_{ion}}{c \langle p \rangle \langle \Xi \rangle} \right]^{1/2} D, \tag{2.1}$$

where F_{ion} is the observed ionizing flux. The proportionality between the inferred radius of the region and the distance D from us to the object reveals that this method can never be more accurate than our knowledge of the Hubble constant. In fact, the situation is much worse than that. It is very hard to define a statistic which quantifies the quality of the model fit (see, *e.g.*, Krolik and Kallman 1988), so it is hard to tell the reliability of the mean quantities inferred. The actual values of p and Ξ obtained are also model-dependent in several respects. At a fundamental level, the whole method is based on the assumption that photoionization by the observed ionizing continuum drives the emission lines. Even granted this (by now very well-supported) assumption, the results of photoionization modelling also depend on the shape of the ionizing spectrum (Krolik and Kallman 1988), the specific choices of atomic data made in the photoionization model, and the assumed clump geometry. But our real dissatisfaction with single-zone photoionization models is due to their crudeness: mean quantities are a far cry from maps.

The next simplest method is photoionization modelling of emission line profiles. The profile of a single emission line is a 1-d map of the line emissivity, where the projection is onto a set of isovelocity surfaces. If several emission line profiles are available, photoionization models may be used in a way precisely analogous to the single-zone method, but now applied separately on each isovelocity surface. That would yield a 1-d map of physical conditions averaged over each isovelocity surface. This is now a multi-zone description of the emission line region, and in itself that is an advance. However, this method also suffers from a number of disabilities. First, of course, the quality of the inferences for each surface is subject to the same uncertainties that cloud our confidence in the single-zone models. However, there is an additional problem: we do not know the positions of these surfaces *a priori*. Moreover, we do not even know if these surfaces are well-defined. If the local velocity distribution functions are broad, these surfaces acquire finite thickness. It is quite possible that they badly overlap. For these reasons, this method has seen comparatively little use, though some recent results show that it can yield interesting new information (Kallman *et al.* 1991).

Much more work is required to make use of the next method, though it also offers proportionately greater rewards. This is the method of total flux reverberation mapping. Though first qualitatively proposed many years ago (Bahcall, Kozlovsky, and Salpeter 1972), it has only come into its own in the last few years. Just as for photoionization modelling, this method also requires the assumption that the ionizing continuum powers the emission lines, but it relies on a different consequence of this assumption, namely that the line luminosity from a clump generally rises when the ionizing flux striking it does. Given the further assumptions that no other quantities (*e.g.,* the pressure) change, and that changes in the continuum intensity occur slowly with respect to the thermal and ionization equilibration timescales of the clumps, the response of an individual line-emitting clump to a small change in the continuum luminosity can be written as $\delta j_l \simeq (\partial j_l/\partial J_{ion})\delta J_{ion}$.

This last relation is purely *local.* To find the change in the total line flux when the continuum changes, one must employ values of δJ_{ion} at *retarded* times, *i.e.* one must allow for the fact that the BLR can be large enough that the time for light to cross it is long compared to the time for the continuum to change significantly. Making use of the assumption that the continuum is radiated isotropically, we may write the time-variation of the line flux as

$$F_l(t) = \langle F_l \rangle + \frac{1}{4\pi D^2} \int d^3r \, \frac{\delta L_{ion}\left[t - \tau(\boldsymbol{r})\right]}{4\pi c r^2} \frac{\partial j_l}{\partial J_{ion}}, \qquad (2.2)$$

where $\tau(\boldsymbol{r})$ is the extra time required for light to travel to position $\boldsymbol{r}$ and then to us, compared to the direct path. Fortunately, we know $\tau(\boldsymbol{r})$ exactly: it is $\tau = (r/c)(1 - \cos\theta)$, where θ is the polar angle defined with respect to our line of sight. Equation 2.2 can then be thought of as a convolution:

$$F_l(t) = \langle F_l \rangle + \int d\tau \, \delta F_c(t - \tau)\Psi(\tau), \qquad (2.3)$$

where all the details of the emissivity distribution are bundled into $\Psi(\tau)$, the marginal emissivity integrated over the paraboloid which responds to the continuum with lag τ. The method of total flux reverberation mapping then boils down to inverting the convolution of equation 2.3. This provides a 1-d map projected onto the isodelay surfaces.

The principal advantage of this method is that, unlike the case of profile mapping, the form of the projection surfaces is known *a priori*. Its greatest disadvantage is that inverting the convolution requires large amounts of data. It is this last fact which accounts for the long delay between suggestion and implementation for this method.

The most complex method, though also the one offering the greatest potential, is that of profile reverberation mapping. First suggested by Blandford and McKee (1982), it potentially gives a 2-d map: the marginal emissivity is projected onto the curves which are the intersections of the isodelay paraboloids and the isovelocity surfaces. As one might imagine, the problems of data quantity and quality are even more severe than for the total flux reverberation method. Nonetheless, a first attempt has been made at using this method, and I will report on the results at the end of this talk.

3. Current Status

Not surprisingly, the simplest methods have been developed the most. A great many single-zone photoionization models have been constructed (Bahcall and Kozlovsky 1969; Davidson 1972; MacAlpine 1972; Weisheit, Shields, and Tarter 1981; Kwan and Krolik 1981; Mushotzky and Ferland 1984; Krolik and Kallman 1988; Rees, Ferland, and Netzer 1990), and they have achieved significant success. The relative strengths of the stronger lines (Lα, CIV 1549, CIII] 1909, Hα, Hβ, NV 1240, MgII 2800) can usually be reproduced to within a factor of two, and different modellers generally find similar parameters (to within factors of three). However, there are some examples in which no single-zone model works (for example, NGC 5548: Krolik, *et al.* 1991), there is considerable difficulty in explaining (not to mention measuring) the strength of the FeII lines (Wills, Netzer, and Wills 1985), and it seems difficult to do better than factor of two agreement, particularly with the Balmer lines.

Surprisingly little effort has been given to profile analysis. The undetermined character of the isovelocity surfaces has deterred many people from attempting this problem. Perhaps the most thorough attempt to date is the work by Kallman, *et al.* (1991), who found that the simplest imaginable dynamical picture (radial free-fall) is consistent with the profiles of the stronger UV lines in a sample of a dozen high-redshift quasars.

A great deal of effort has been given to total flux reverberation in the last year or so. Two major efforts have been mounted, one tracking the Balmer lines and optical continuum in NGC 4151 (Maoz *et al.* 1991), the other monitoring the UV continuum, UV lines, and optical lines in NGC 5548 (Clavel *et al.* 1991; Peterson, *et al.* 1991; Krolik, *et al.* 1991). In both cases, ~ 60 observations were obtained with a mean interval of a few days, enough to permit inversion of the convolution relation.

However, it is worth noting that in both instances it was found that direct inversion (*e.g.,* by Fourier transforms) was impractical. The reason is that the power spectrum of continuum fluctuations in AGN is so "red" that signal dominates noise only at the lowest frequencies. As a result, direct inversion gives a response function which is severely corrupted by noise on short timescales.

Both efforts hit upon the same solution to this problem: replacing direct inversion with maximum entropy model-fitting. Using this method, one searches for that $\Psi(\tau)$ and those values of $F_c(t)$ at unobserved times which result in a line light curve consistent with the one measured, and which are optimal with respect to a criterion chosen in advance (for example, that they should be maximally smooth). I will discuss the results obtained for NGC 5548 in some detail.

4. The BLR in NGC 5548

This galaxy was selected at the Segovia meeting in October, 1987 as the best candidate for the first attempt at reverberation mapping. It is bright ($\bar{F}_c \simeq 5 \times 10^{-11}$ erg cm^{-2} s^{-1} in the UV), nearby ($z = 0.0173$), and has strong emission lines (their summed flux is roughly a quarter of the UV continuum flux). It was also recommended by its history of substantial variation. The only missing element, as always, is information about the shape of its EUV spectrum; but it has been well-studied by both *IUE* and a number of X-ray satellites.

4.1 Total flux reverberation mapping

One of the major goals of the reverberation mapping effort was to test the characteristic size estimated on the basis of single-zone photoionization models (see equation 2.1). Interestingly, Krolik *et al.* (1991) found that *no* single-zone model was adequate. Instead, even before attempting the reverberation solution they were driven to a two-zone picture. In this model the high ionization lines (which also carry most of the flux) come from material $\simeq 13h^{-1}$ lt-d from the central continuum source, while the lower ionization lines are generated at a radius roughly twenty times greater.

The results of the total flux reverberation mapping are shown in Figure 1, taken from Krolik, *et al.* (1991). The first thing to observe is that the reconstructed light curves are, indeed, good fits to the observed light curves, *i.e.,* the reduced χ^2 really is unity. However, the most important part of the figure to examine is the eight panels on the left showing the response functions for the eight measured lines. They are ordered roughly in decreasing ionization sequence; this almost perfectly coincides with the sequence of characteristic delay. That is, there is an excellent correlation between degree of ionization and quick response.

The next point to notice is that the first five lines, He IIλ1640, NVλ1240, Lα, CIVλ1549, and SiIVλ1400 (actually a blend with OIV]λ1400) all peak either at or very near zero lag. This is in contrast with CIII]λ1909, which has a definite minimum at zero lag and peaks between 20 and 30d. MgIIλ2800 and OIλ1304 also peak at lags well away from zero, but the relatively small effective signal/noise for these lines (*i.e.,* the ratio of *rms* fluctuation to measurement error) is so small that their response functions are too sensitive to model assumptions to be reliable. HeII, CIV, and CIII] all also seem to have a secondary peak around 175d, but, as I will discuss later, these peaks may not be trustworthy.

Because these results are the product of model-fitting, and not true direct solutions of equation 2.3, it is wise to examine how model-dependent they are. To this end, we found the maximum entropy solution under a number of different assumptions. We shortened the range of lags for which we would permit Ψ_l to be non-zero; we changed defaults from maximal smoothness to maximal uniformity; we let the reconstructed continuum be different for different lines; and we arbitrarily set the reference levels $\bar{F}_l$ and $\bar{F}_c$ to zero. Any feature which persisted throughout these changes we deemed robust enough to believe. Our primary result, that the five lines HeII, NV, CIV, Lα, and SiIV all peak near zero and fall to low levels by $\simeq$ 20d, while CIII] is low near zero and peaks between 20 and 30d, clearly passes this test.

These solutions for the line response functions amount to a 1-d mapping of the marginal line emissivity response. To proceed farther, to obtain 3-d information and a mapping of physical conditions, requires inserting some physical assumptions. The most important of these is a geometrical symmetry assumption. Fortunately, the shapes of the response functions guide these assumptions.

To illustrate how they do, consider some idealized examples. Suppose that the line-emitting gas were arranged on a series of nested spherical shells, each of them radiating isotropically. Then the shell at radius r would contribute a square-wave shaped piece to the response function. The amplitude of this contribution is proportional to the marginal line emissivity response and the amount of material in that shell, and the square wave is non-zero from zero lag to $2r/c$. Summing the contributions from all shells, the response function would always have a maximum at zero lag, and drop smoothly to zero amplitude at a lag corresponding to the diameter-crossing time of the largest shell. On the other hand, individual annuli produce

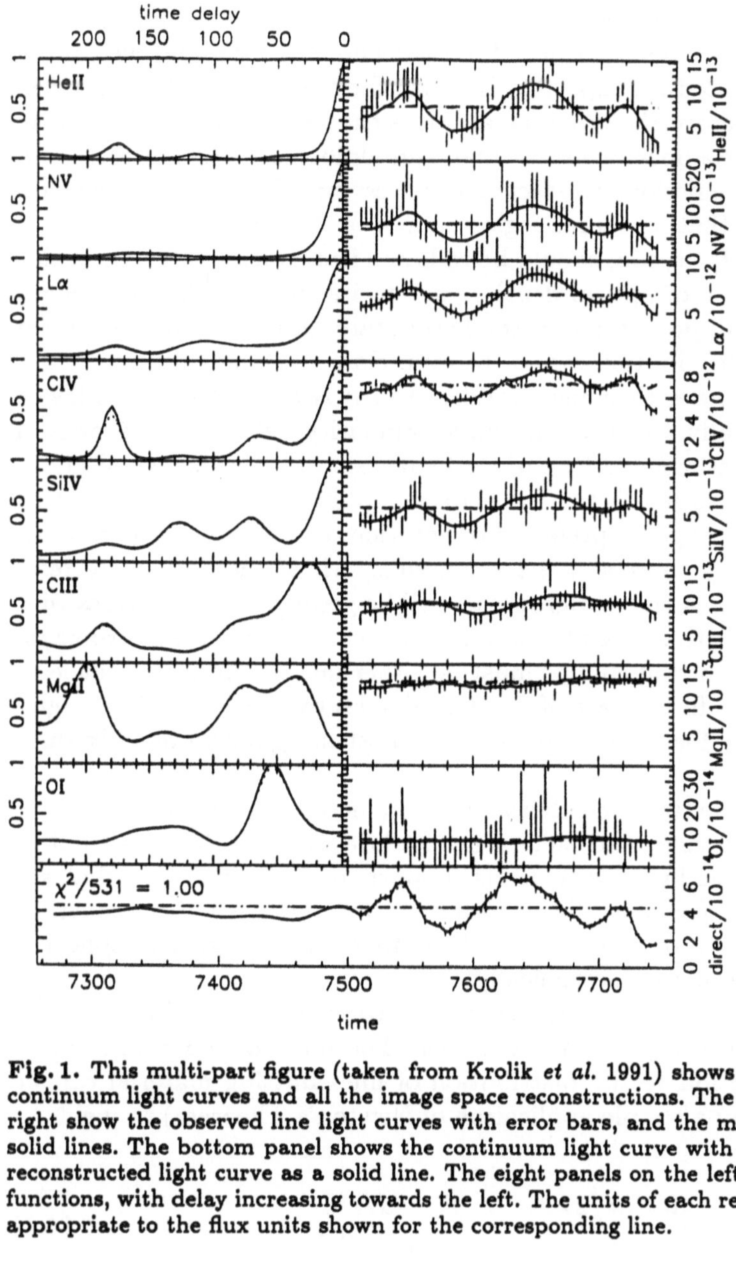

Fig. 1. This multi-part figure (taken from Krolik *et al.* 1991) shows the actual line and continuum light curves and all the image space reconstructions. The eight panels on the right show the observed line light curves with error bars, and the model light curves as solid lines. The bottom panel shows the continuum light curve with error bars, and the reconstructed light curve as a solid line. The eight panels on the left show the response functions, with delay increasing towards the left. The units of each response function are appropriate to the flux units shown for the corresponding line.

double-peaked response functions divergent at $\tau = (r/c)(1 \pm \sin i)$. Sums of these shapes always peak away from zero (unless $\sin i = 1$). Anisotropic radiation, which generally tends to favor the illuminated side of line-emitting regions, would emphasize the large τ end of these curves; distributed obscuration emphasizes the small τ side.

With these idealized pictures in mind, it is fairly easy to choose which symmetry best describes each line. First consider the two lines we can be confident are radiated isotropically. Because the lower state of HeII $\lambda1640$ is high in energy above the ground state (it is the HeII analog of Hα), there should be very little opacity in this line, and almost any configuration will produce isotropic emission. The fact that its response function clearly peaks at zero lag means that its source region cannot be flattened unless $\sin i \simeq 1$. Next consider CIII]. It, too, should be radiated isotropically, though in this case the suppression of optical depth is due to its small oscillator strength rather than a small population in the lower state. Since it clearly peaks at non-zero lag, its distribution cannot be even close to spherical.

We do not know with equal confidence that NV, Lα, CIV, or SiIV is radiated isotropically, but the qualitative similarity between the shapes of their response functions and that of HeII suggests that this is a good guess. Physically, this guess entails two further assumptions: that the line-emitting regions have on average no preferred orientation; and that they be at most marginally optically thick at the Lyman edge. All of these lines are resonance lines, so their surface brightnesses will be greatest on those sides with the least optical depth to the photon creation zones; to preserve isotropic radiation, the line-emitting clumps must be either quasi-spherical, or at least oriented randomly. In the case of Lα, for all sides to be comparably bright, it is also necessary for the clumps to be at most marginally optically thick at 1 Rydberg.

The fact that the two-zone photoionization model produces most of these lines in the inner zone, and that zone needs a covering factor of 0.3 to produce enough line luminosity, lends support to the assumption that the region responsible for these lines cannot be too flattened. Consequently, we model this group of five lines as all coming from a region of spherical symmetry which radiates isotropically.

Once we make that assumption, it is easy to transform the line response functions into maps of physical conditions. The integral equation relating the response function to the distribution of line-emitting material can be trivially solved (by a simple application of Newton's Theorem) to yield

$$\frac{dC\,(c\tau/2)}{dr} = -\frac{\tau}{c}\frac{F_c}{F_{ion}}\frac{d\Psi_l/d\tau}{\partial S_l/\partial F_{ion}}. \tag{4.1}$$

The LHS of equation 4.1 is a unique function of r for any given object, while the RHS depends on the particular line l for which it is evaluated. If they are to be truly equal, the ratio of line-dependent quantities on the RHS must be the same for all lines at any particular value of τ. Thus, we can obtain the physical conditions as a function of r by searching (*via* photoionization models which compute $\partial S_l/\partial F_{ion}$) for that set which best achieves this goal.

Table 1 (from Krolik, *et al.* 1991) shows the results of this exercise. The first column is the lag at which equation 4.1 is evaluated; the next two show the best-fit values of Ξ and p_H for that lag; Φ_2 is a measure of quality of fit for which values well less than unity are acceptable; r_{pred} is the distance from the continuum source inferred using that lag's values of Ξ and p_H (h is the Hubble constant in units of 100km s^{-1} Mpc^{-1}); and the final column is the integrated covering factor interior to that radius.

Table 1 The Radial Run of Physical Conditions in the Broad Emission Line Region

Lag (d)	Ξ	p_H (dyne cm^{-2})	Φ_2	r_{pred} (lt-d)	$C(\leq r)$
4	0.56	0.027	1.1	$32h^{-1}$	0.00080
8	0.56	0.027	0.73	$32h^{-1}$	0.0026
12	0.42	0.27	0.59	$12h^{-1}$	0.12
16	0.24	0.27	0.48	$15h^{-1}$	0.26
20	0.24	0.27	0.37	$15h^{-1}$	0.41
24	0.10	0.27	0.42	$24h^{-1}$	0.50
28	0.10	0.27	0.85	$24h^{-1}$	0.62
32	0.10	0.27	1.09	$24h^{-1}$	0.67

The fits at small lags are not well-determined because the slope of Ψ_l is not well-determined for several of the lines, but from lags of 8d to 28d, reasonable fits can be obtained. Surprisingly, they suggest nearly constant (and fairly high) pressure—the implied density is $n_H \sim 10^{11}$cm^{-3}. The predicted radius should always be half the lag (*cf.* equation 6), but the inferred distances instead tend to be equal to the lag times h^{-1}. Assuming that $1/2 \leq h \leq 1$, that means the photoionization fits to the response functions predict radii a factor of 2 – 4 too large. Given the uncertainties in the multiple stages of analysis, we regard this degree of match as excellent. Finally, we see that there is significant high ionization material at all radii from 4 to 14 lt-d, though the peak in dC/dr occurs near $r =$ 8lt-d.

Finally, it is time to turn to the low ionization lines. Unfortunately, the response functions for Mg II and OI were too ill-determined to be useful, so the only low ionization lines available for analysis are CIII] and Hβ. These two lines have very similarly shaped response functions: both are small at zero lag, rise to peaks between 20 and 30d, and fall at larger lags (CIII] from Krolik *et al.* 1991; Hβ from Horne, Welsh, and Peterson 1991). Both are therefore better described by flattened geometries than by spherical. However, at least two different flattened geometries are consistent with these response patterns. In the first, the flattened region is nearly face-on, and our maximal smoothness assumption has wiped out the sharp features that might otherwise occur; in this case, the radius of the annular region would be $\simeq$ 25d. The second possibility is that the flattened region is nearly edge-on, and the second peak is at such large lags that we have missed it; in this case, to determine the radius of the ring, we must locate the large lag peak. In fact, as mentioned before, there are weak, ill-defined peaks in the response functions of CIII], CIV, and HeII near 175d (though none is visible in the response function derived from the longer train of data on Hβ). If these are the missing peaks, the ring's radius would be $\simeq$ 100lt-d.

In terms of photoionization physics, the second alternative would be highly desirable. If the CIII] region is really as close as 25 lt-d from the source of the ionizing continuum, production of λ1909 would be very hard to understand. Because we know the ionizing flux there to within a factor of a few, the only free parameter is the pressure, but a pressure high enough to keep CIII in the ionization balance would lead to strong deexcitation of the λ1909 line, while pressure low enough to permit efficient radiation of λ1909 would lead to an ionization balance in which CIII had very low abundance. At $\simeq$ 100 lt-d, however, as suggested by the two-zone line ratio model, these problems would disappear.

The problem with the second alternative is the strong possibility that the peaks near 175d are artificial. Because the lightcurves for both the continuum and the lines were quasi-periodic, there is an intrinsic ambiguity in the determination of the response function which can artificially produce repeated peaks in Ψ_l at intervals of the quasi-period. Because the difference in time between the positions of these questionable peaks and the positions of the definite peaks is not far from two quasi-periods, it may be that they are artifacts due to this effect. On the other hand, because they are two quasi-periods away, and no features appear at one quasi-period, one might argue that they are real. For the time being they must be regarded with skepticism.

4.2 Velocity-resolved reverberation mapping

The same *IUE* data set from which the preceding results were obtained also contained well-resolved emission line profiles. Although the signal/noise does not always permit using such fine resolution, in the case of the strongest line, CIV 1549, the wavelength scale of the instrument allows up to 56 velocity bins across the profile. This data therefore permits, in principle, analysis of the most detailed kind possible.

Obtaining the velocity-dependent response functions from this data is more complicated than obtaining the total flux response functions, but the method is not fundamentally different. Just as in the simpler case, for each velocity bin one searches for the response function which is both consistent with the data and maximizes the entropy function. In practise, all bins

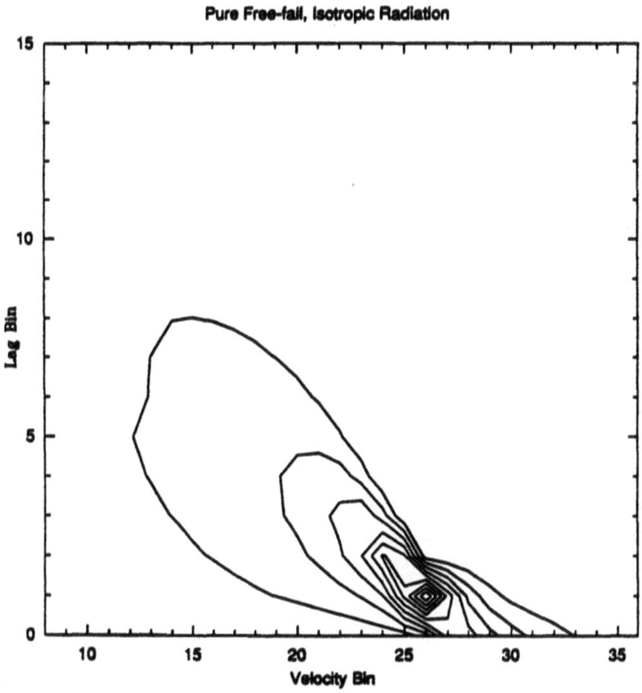

Fig. 2. Contours of $\Psi(v, \tau)$ for a model in which the lines are emitted from spherically distributed clouds falling freely towards a point mass and radiating isotropically. The zero of velocity is in bin 22.

within a given line are solved for at once, so that consistency with the data is defined by requiring χ^2 for the entire data set to be unity.

Unfortunately, the results to date are so much more complicated than those found from the total flux analysis that no satisfactory model has yet been found. To illustrate just how complex this problem is, I will first show a sample diagram displaying what would be expected from a very simple model (see Welsh and Horne 1991 for more detail on how to construct models like this one, and a fuller set of examples). Figure 2 shows the response function from what is arguably the simplest possible model: spherically symmetric free-fall with isotropic radiation. The peak at small lag and positive velocity is a cusp characteristic of any radial flow model in which $v_r(dv_r/dr) < 0$. There is also a trend for negative velocities to have larger average lags. This is, of course, the immediate consequence of motion which is away from us (and towards the nucleus) on the near side, and motion towards us on the far side. Any organized radial flow model, even much more complicated ones, can be expected to retain this qualitative characteristic, though it can be considerably weakened by, for example, anisotropic radiation.

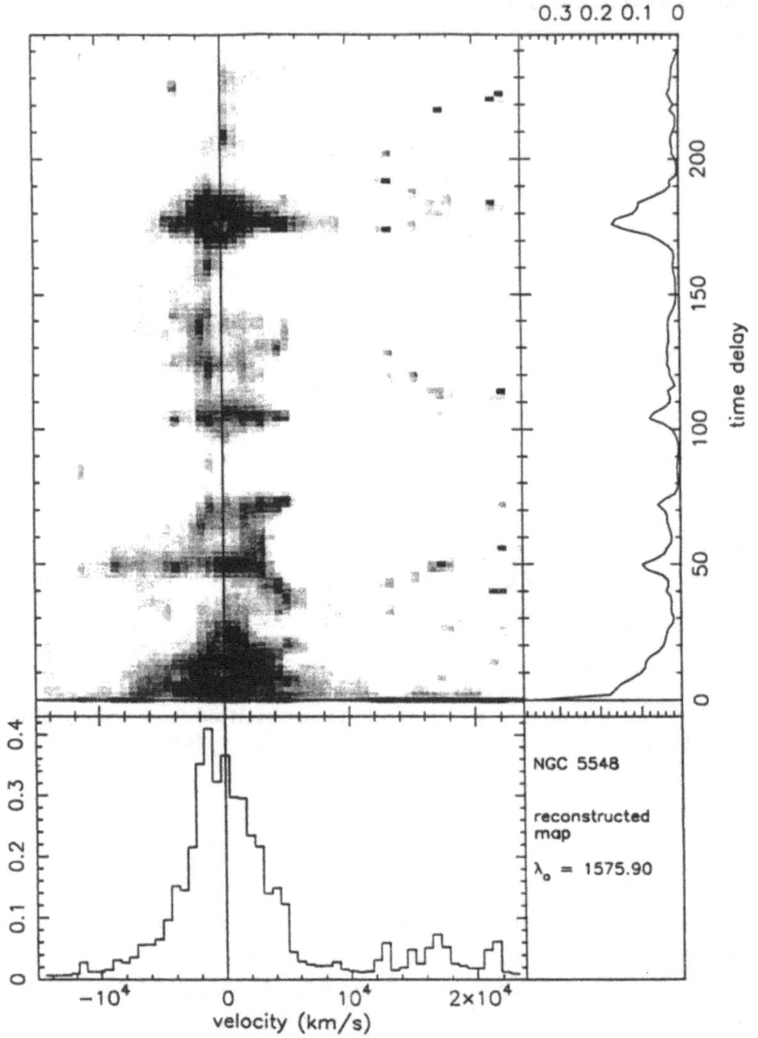

Fig. 3. A grayscale representation of $\Psi(v,\tau)$ for the CIV 1549 line in the NGC 5548 data taken with *IUE*.

Figure 3 (taken from Horne and Krolik 1992) shows what the real response function looks like for CIV 1549. It is clearly *not* a result of any sort of simple spherical, radial flow model. In fact, its near symmetry between positive and negative velocities suggests that we consider "collisionless" models, *i.e.*, models in which at each location in the broad line region the emitting material has a substantial spread in velocities. However, I stress that these results are still in a very early stage, and I expect to have a much fuller story to tell when these results are finally published.

5. Conclusions

The monitoring programs of the past couple of years have generated significant progress in our effort to infer the structure of the broad line region in active galactic nuclei. Total flux reverberation mapping has shown that (at least in the case of NGC 5548), the region is well described by a two component model: an inner spherical shell in which the material exists in a state of fairly high ionization, and an outer flattened region of lower ionization whose structure is not so clearly determined.

Results from profile reverberation mapping are near at hand. Preliminary work on the NGC 5548 data has produced a very complicated response function whose interpretation is just beginning. Because there may be signal/noise problems even in the cleanest line, CIV 1549, it may be very difficult to derive the sort of ionization state diagnostics that were possible with the total flux method. At this point, about the most that can be said is that the results are most naturally consistent with models containing substantial random velocities, but there is almost no reasonable class of model they clearly rule out.

Beyond these efforts, more work certainly awaits us. In particular, there are a number of what I call "skeletons in the closet" whose implications for this program need to be better understood before our confidence level will rise much. These include:

• Time-variability in the geometry on scales shorter than or comparable to the duration of the monitoring. Although this suggestion has already appeared in the literature (Netzer and Maoz 1990; Sparke 1992), its statistical base is weak and the measured time for emission line material to cross the broad line region is quite long compared to the relevant timescales.

• Anisotropic continuum radiation, so that the fluctuations in the continuum we see are not the same as those exciting emission regions on other lines of sight. The fact that the cross-correlation functions between observed line and continuum fluctuations peak at such large amplitude (Clavel, *et al.* 1991) always this fear to some degree.

• It is possible that fluctuations in the ionizing continuum are not simply proportional to the fluctuations in the sub-Rydberg ultraviolet which we observe.

• A geometrically dispersed source of ionizing radiation would lead to a thickening of the isodelay surfaces in addition to exacerbating worries about whether the continuum we observe is a good indicator of what the continuum radiated in other directions is doing (Wanders and van Groningen 1992).

• If the continuum fluctuations are large enough, or the emitting material is sensitive enough, the linear approximation may be inadequate (Maoz 1991; Sparke 1992). In the case of NGC 5548, the ratio of *rms* continuum fluctuation to mean continuum flux was $\simeq 0.3$, so this is probably not a major concern here, but it could certainly arise in other cases.

• Finally, and most important, the effort involved in mounting monitoring campaigns and analyzing their results is so large that it is unlikely more than a few will ever be done. How confident can we be that the results from this very small sample can be extrapolated to other active galaxies?

References

Bahcall, J.N., and Kozlovsky, B.-Z., 1969: *Astrophys. J.*, **155**, 1077.
Bahcall, J.N., Kozlovsky, B.-Z., and Salpeter, E.E., 1972: *Astrophys. J.*, **171**, 467.
Blandford, R.D., and McKee, C.F., 1982: *Astrophys. J.*, **255**, 419.
Clavel, J., *et al.* , 1991: *Astrophys. J.*, **366**, 64.
Davidson, K., 1972: *Astrophys. J.*, **171**, 213.
Edelson, R.A., and Krolik, J.H., 1988: *Astrophys. J.*, **333**, 646.
Horne, K., and Krolik, J.H., 1992: in preparation.
Horne, K., Welsh, W., and Peterson, B.M., 1991: *Astrophys. J. (Letters)*, **367**, L 5.
Kallman, T.R., Wilkes, B., Krolik, J.H., and Green, R.F., 1991: in preparation.
Krolik, J.H., and Kallman, T.R., 1988: *Astrophys. J.*, **324**, 714.
Krolik, J.H., Horne, K., Kallman, T.R., Malkan, M.A., Edelson, R.A., and Kriss, G.A., 1991: *Astrophys. J.*, **371**, 541.
Kwan, J.Y., and Krolik, J.H., 1981: *Astrophys. J.*, **250**, 478.
MacAlpine, G.M., 1972: *Astrophys. J.*, **175**, 11.
Maoz, D., 1991: in preparation.
Maoz, D., *et al.* , 1991: *Astrophys. J.*, **367**, 493.
Mushotzky, R.F., and Ferland, G.A., 1984: *Astrophys. J.*, **278**, 558.
Netzer, H., and Maoz, D., 1990: *Astrophys. J. (Letters)*, **365**, L 5.
Peterson, B.M., *et al.* , 1991: *Astrophys. J.*, **368**, 119.
Rees, M.J., Netzer, H., and Ferland, G.A., 1989: *Astrophys. J.*, **347**, 640.
Sparke, L., 1992: these proceedings.
Wanders, O., and van Groningen, E., 1992: these proceedings.
Weisheit, J.C., Shields, G.A., and Tarter, C.B., 1981: *Astrophys. J.*, **245**, 406.
Welsh, W., and Horne, K., 1991: *Astrophys. J.*, in press.
Wills, B., Netzer, H., and Wills, D., 1985: *Astrophys. J.*, **288**, 94.

NGC 4151: A Test for the Stellar Origin of Variability in AGN

Itziar Aretxaga[1], *Roberto Terlevich*[2], *Angeles I. Díaz*[1], *and Elena Terlevich*[2]

[1] Dpto. de Física Teórica, C-XI. Universidad Autónoma de Madrid. Cantoblanco. 28049 Madrid. Spain
[2] Royal Greenwich Observatory. Madingley Rd. Cambridge. CB3 0EZ. U.K.

In the Starburst Model, the variability observed in Seyfert galaxies and optically selected quasars is thought to be produced by the supernova explosions (SN) resulting from the evolution of a massive stellar cluster in the characteristic high density and high metallicity medium of the nuclei of early galaxies (Terlevich *et al.*, 1987).

The observed nuclear optical light curve of NGC 4151 is represented in Figure 1 ($H_o = 50$ Km s^{-1} / Mpc). The arrow in the graph indicates the date when Stratoscope II observed the nucleus of the galaxy, March 26–27, 1970, setting $0''.08$ as an upper limit for the size of the galaxy nucleus (Schwarzschild, 1973).

Assuming that the observed luminosity at minimum light is dominated by the young stellar cluster, the SN rate resulting from the evolution of this cluster has been calculated. For a $V = 12.7$ mag cluster, we find values between 0.1 and 0.8 SN/yr, depending on the assumed Initial Mass Function and age. The time evolution of a supernova remnant in a high density medium is adopted from that of Wheeler and collaborators (1980) for a total energy $\epsilon = 3 \times 10^{51}$ erg and circumstellar density $n = 10^7$ cm^{-3} . Figure 2 shows the computed light curve of a young cluster with a SN rate of 0.3 SN/yr, where gaussian errors of 0.1 mag have been introduced to simulate the observational errors. Due to the stochastic nature of the problem, it is very difficult to distinguish solutions between the predicted rates, 0.1 to 0.8 SN/yr.

From the comparison of Figures 1 and 2, it can be seen that the simulations reproduce both the observed number of peaks of the light curve (SN events) and the mean luminosity.

The observation of NGC 4151 reported by Stratoscope II on March 26–27, 1970, could correspond to a SN explosion. At the time of the observation, the luminosity of the nucleus of the galaxy could have been dominated by the SN light and therefore would be unresolved.

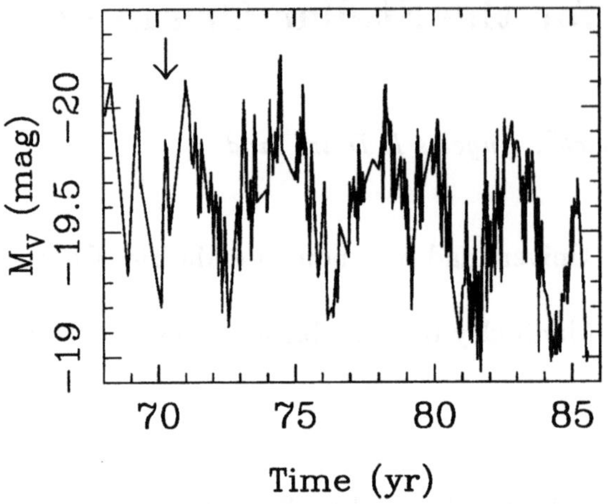

Fig. 1. Light curve of NGC 4151 in absolute V magnitudes, derived from those collected by M.V. Penston (private communication).

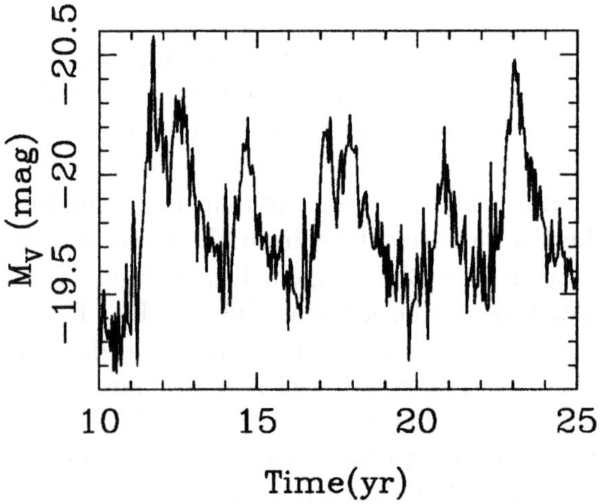

Fig. 2. Simulated light curve with a supernova rate of 0.3 SN/yr

References

Schwarzschild, M., 1973: *Astrophys. J.*182,357.

Terlevich, R., Melnick, J., Moles, M., 1987: IAU Symp # 121: *Observational Evidence for Activity in Galaxies*, eds. E.Ye. Khachikyan, K.J. Fricke, J. Melnick. Reidel, Dordrecht.

Wheeler, J.C., Mazurek, T.J., Sivaramakrishnan, A., 1980: *Astrophys. J.*237,781.

A Study of Emission-line Variability of the Nucleus in the Seyfert Galaxy NGC 4151 in 1986-1989

N.G. Bochkarev[1], *A.I. Shapovalova*[2], *and S.A. Zhekov*[3]

[1] Sternberg State Astronomical Institute, Moscow, Russia
[2] Special Astrophysical Observatory, Nizhnij Arkhyz
[3] Institute Space Research, Bulgarian Academy of Sciences

1. Observation Data

The NGC 4151 nucleus was observed in dense sets (each night, when possible) during individual periods in March - April, 1986 (17 nights), in March and May, 1987 (18 nights), and in February - March, 1988 (8nights), using the 1000-channel TV spectral scanner of the 6-m telescope (Drabek et al., 1987). The spectral region from $< 3900Å$ to $4960Å$ were recorded. Most of the spectra were obtained with a $2 - 3Å$ spectral resolution and a signal-to-noise ratio $S/N = 10 - 25$ for individual spectra. The contribution of the stellar component of nucleus was subtracted from the average spectra. The standard galaxy was selected to be NGC 4339 for which Filippenko and Sargent (1988) obtained a spectrum at a high signal-to-noise ratio ($S/N = 100$).

2. Results

1) he stellar component contribution in the NGC 4151 nucleus was found to be $\sim 40\%$ of continuum near the H_β line.
2) The narrow emission lines in the NGC 4151 spectrum show a blue shift (the radial velocity of the line centers is $V_r = +970 \pm 16km/s$ relative to the systemic velocity of the host galaxy ($V_r = 1000 \pm 3km/s$; see Table 2 in Schulz, 1987)).
3) In 1986-1988 all broad lines were asymmetric and shifted bluewards with respect to the narrow-line centers by $1000km/s$. Within 1986-1988 the $HeII$ 4686 line asymmetry inverted.
4) In 1987 the broad component of the H_β and H_γ lines decreased strongly as compared with 1986. In 1988 it increased up to approximately the level of 1986.
5) Satellites of the strong H_β, H_β and H_δ lines and of the $HeII$ 4686 line were suspected in the NGC 4151 nucleus spectrum at radial velocities of $+9500km/s$ and $-7500km/s$ (Fig. 1, 2). In February-March, 1988 the

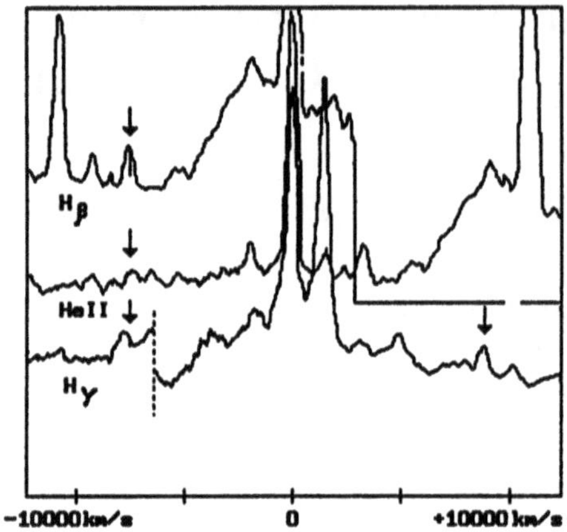

Fig. 1. The fractions of the average NGC 4151 nucleus spectrum of February and March, 1988 in the scale of radial velocities. The arrows indicate the positions of satellites.

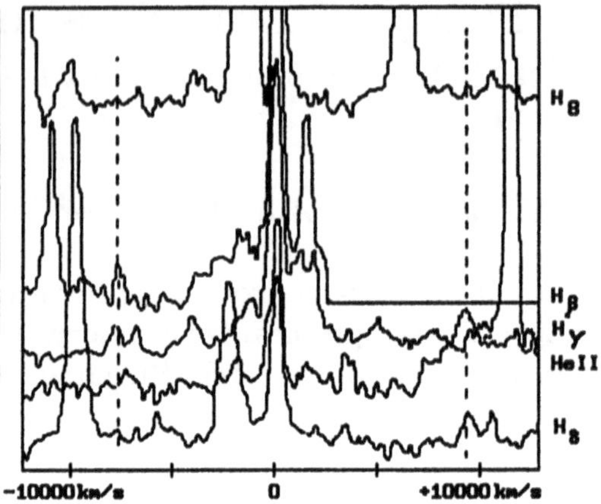

Fig. 2. The fractions of the NGC 4151 nucleus spectrum of March 6, 1987 in the scale of radial velocities. The dashed lines indicate the positions of the satellites.

satellites seemed to exhibit variability. An attractive – but speculative – possibility is that the satellites are produced in jets.

6) In 1988 the spectrum of the nucleus exhibited variability on time scales smaller than 30 days. The $HeII$ 4686 line broad component and the H_β line blue wing enhanced early in March. A burst in the NGC 4151 nucleus occurred in February, 1988 and the continuum in the H_γ - H_δ region increased by 40 - 50 % within not more than 20 days.

References

Drabek, S.V., Kopylov, I.M., Somov, N.N., Somova, T.A., 1987: *Astrofiz. Issled (Izv.SAO)* **22**, 54 (in russian).
Filippenko, A.V., Sargent, W.L.W., 1988: *Astrophys. J.*, **291**, 677.
Schulz, H., 1988: *Astron. Astrophys.*, **203**, 333.

Self Gravitating Thick Accretion Disks

Gianluigi Bodo and Anna Curir

Osservatorio Astronomico di Torino, Pino Torinese, Italy

A problem which has not yet received much attention in the modelling of thick accretion disks is the effect of their self-gravity (see however Hachisu, 1986; Hachisu, Eriguchi and Nomoto, 1986). Here we study the equilibrium structure of self–gravitating, geometrically thick accretion disks by solving autoconsistently the Poisson equation for determining the potential of self–gravity and the hydrostatic equation (Bodo and Curir, 1991). The forces acting in the hydrostatic equation are the gravitational force of the central black hole, the centrifugal force and the gravitational force of the disk itself. By assuming a politropic equation of state, the centrifugal force can be represented as the gradient of a potential and the hydrostatic equation can be easily integrated:

$$\rho = \left[(B - B_s)\frac{\rho_c^{1+1/n}}{P_c(n+1)}\right]^n \quad , \tag{1}$$

where the subscript s refers to the surface of the torus and the subscript c to a circle in the equatorial plane where density and pressure attain their maximum value. Eq. (1) and the Poisson equation are solved alternatively in an iterative way until convergence is reached (self consistent field method). For the potential of the central object we assume a pseudonewtonian form.

Here we want to discuss the effects of self gravity on the size of the disk, on its temperature distribution and on its luminosity. In Fig. 1 we show the outer radius of the disk as a function of the mass ratio between the disk and the central object. We can see that at intermediate values of the mass ratio the outer radius reaches a minimum and then increases again. Fig. 2 and Fig.3 represent the temperature distribution obtained assuming that each element of the disk radiates like a blackbody at the critical limit. We see an increase of the surface temperature due to the effect of self-gravity. Finally from Fig. 4 we can see that the total luminosity of the disk increases with the mass ratio.

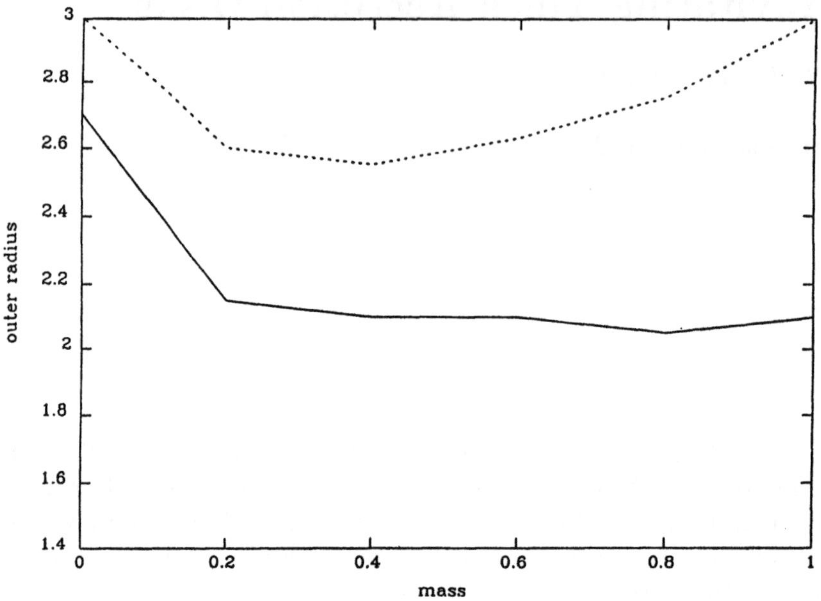

Fig. 1. Outer radius of pseudonewtonian accreting disks as a function of the mass ratio. The solid line is for $\beta = 0.01$, the dashed line is for $\beta = 0.4$.

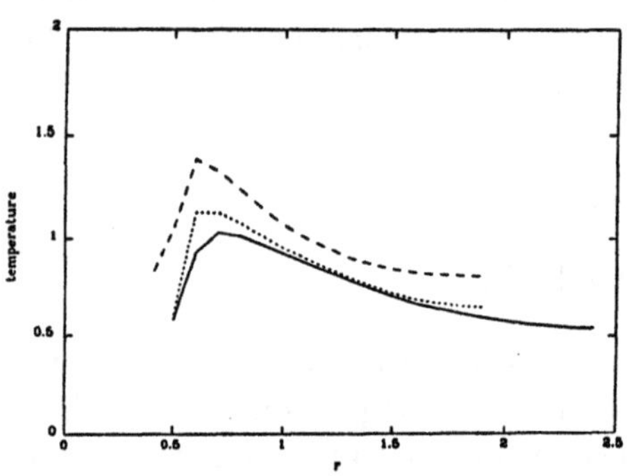

Fig. 2. $T_* = T(a^2\sigma_B/L_e)^{1/4}$ as a function of radius for a sequence of pseudonewtonian accreting disks. The solid line refers to a mass ratio 10^{-3}, the short dashed line to a mass ratio 0.2 and the long dashed line to a mass ratio 1, for $\beta = 0.4$.

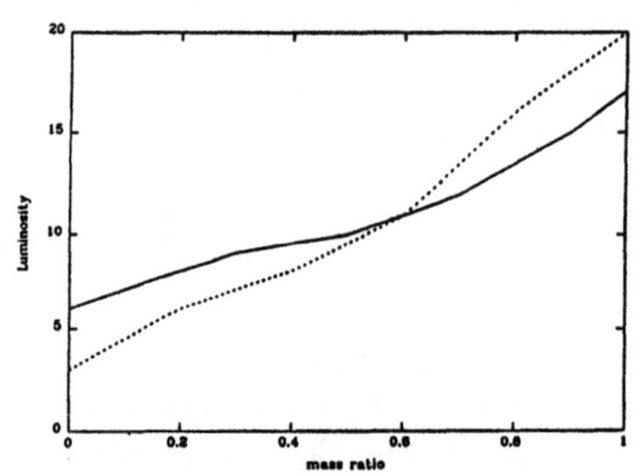

Fig. 3. The total luminosity of accreting pseudonewtonian disks as a function of mass ratio. The solid line refers to $\beta = 0.01$, the dashed line refers to $\beta = 0.4$.

References

Bodo, G., Curir, A., 1991: *Astron. Astrophys.*, in press.
Hachisu, I., Eriguchi, Y., Nomoto, K., 1986: *Astrophys. J.*, **308**, 161.
Hachisu, I., Eriguchi, Y., Nomoto, K., 1986: *Astrophys. J.*, **311**, 214.

First Results from the 500–7000 Å Monitoring Campaign of Markarian 509

Timothy E. Carone

Experimental Astrophysics Group, Space Sciences Laboratory, University of California, Berkeley, California 94720

Abstract. The Seyfert galaxy Mkn 509 was monitored from September through mid–November, 1990. We were able to monitor this active galaxy from 500 Å to Hα for most of this period. The emission and continuum light curves of Mkn 509 decreased and then began to brighten at the end of this campaign. Mkn 509 was detected shortward of 1200 Å but no emission shortward of 912 Å was seen. The Ly α λ1216 and C IV λ1549 emission lines showed a time lag of between 13–19 days relative to the continuum at 1350 Å. No substantial variations were seen in Mg II λ2798. The amplitude of the continuum variations increased with decreasing wavelength. The continuum became more red as it decreased in brightness and then became bluer as the brightness increased. However, it is not clear that the continuum flux distribution shortward of 1200 Å changed in shape.

1. Introduction

Variability in the continua and broad lines of Seyfert galaxies has been known for over 20 years (Peterson 1988). Recently, multispectral monitoring campaigns have been undertaken for several active galaxies, including an intense study of NGC 5548 (Clavel *et al.* 1991; Peterson *et al.* 1991). This and other studies have documented the temporal spectral behavior of the nuclear regions of active galaxies and have clearly shown the need for further similar efforts.

As part of the international effort to monitor the temporal behavior of active galaxies, the Seyfert galaxy Mkn 509 was monitored for over two months starting in September, 1990. This monitoring campaign used the *Voyager* Ultraviolet Spectrometers (**VUVS**) to cover the 500–1700 Å spectral region (Broadfoot *et al.* 1977), **IUE** to cover from 1200–3300 Å and six ground–based observatories that observed longward of 4200 Å. The **VUVS** monitored Mkn 509 continuously during the month of September. 18 **IUE** SWP images were taken and 16 LWP images were taken during one–half of a US1 shift every week for nine weeks. The ground–based observatories contributed 86 optical spectra.

The **VUVS** reduction was done under IRAF (Carone *et al.* 1992) and involves determining an aspect solution on a large number of high–rate spectra in which the back–scattered solar emission and the RTG particle backgrounds have been removed. The **IUE** data was reduced using the GEX software (Clavel *et al.* 1991) available at the Goddard Space Flight Center RDAF.

The optical data processing was done in an identical manner as that done for the NGC 5548 campaign (Peterson *et al.* 1991).

2. Results and Discussion

The continuum emission and line emission light curves clearly show variability over the two month period. The results are qualitatively similar to the light curves for NGC 5548. However, the Mkn 509 light curves show that this active galaxy was sampled as it became dimmer and then began to brighten. Therefore, we did not sample an entire period.

Preliminary analysis of the **VUVS** observations show that Mkn 509 was detected shortward of 1200 Å during the first half of the monitoring campaign. However, after Mkn 509 began to decrease in brightness, no emission longward of 1200 Å was found. In addition, no emission was detected shortward of 912 Å in any of the **VUVS** spectra. There is a correlation between the variations of the continuum at 1350 Å (in the rest frame of Mkn 509) and the Ly α λ1216, CIV λ1549 emission lines. The time lag between the variation at 1350 Å and the variations of Ly α λ1216, CIV λ1549 emission lines is between 13 and 19 days. No significant variability was seen in Mg II λ2798. The amplitude of the variations became stronger with decreasing wavelength. At the end of the monitoring campaign, Mkn 509 was definitely starting to brighten in the continuum. The continuum became redder with decreasing brightness and more blue with increasing brightness longward of 1200 Å. The **VUVS** spectra do not clearly show this same behavior shortward of 1200 Å.

Acknowledgements. I would like to thank Baber Ali, R. Bertram, Karsten Bischoff, Michael Crenshaw, Matthias Dietrich, Peter Garnavich, John Huchra, Wolfram Kollatschny, Kirk Korista, Matthew Malkan, Nancy Lame, Bruce McCollum, Brad Peterson, Richard Pogge, Ron Polidan, Gail Reichert, Alberto Sadun, Jack Sulentic, R. Wagner, Belinda Wilkes, and Wei Zheng for their help in this project. I would also like to thank O.H.W. Siegmund for his support. This work was funded under NASA Grant NAG 5–1142.

References

Broadfoot, A.L., *et al.* , 1977: *Space Sci. Rev.* **21**, 183.
Carone, T.E., *et al.* , 1992: in preparation.
Clavel, J., *et al.* , 1991: *Astrophys. J.*, **366**, 64.
Peterson, B.M., 1988: *Publ. Astron. Soc. Pacific*, **100**, 18.
Peterson, B.M., *et al.* , 1991: *Astrophys. J.*, **368**, 119.

Preliminary Results on the Baldwin Effect and Correlation Between Disc Luminosity and Mass

Suzy Collin-Souffrin and Monique Joly

DAEC, Observatoire de Meudon

Abstract. Since the discovery of the Baldwin effect, the luminosity dependence of line intensities and equivalent widths in quasars has been intensively studied and several interpretations have been proposed. The origin of the Baldwin effect in AGN is however still controversed. Our purpose here is to discuss the origin of line luminosities on overall luminosity dependence, and to determine the relation between disc luminosity and overall luminosity.

1. Hypothesis

We try to give an account of the correlations between line luminosities and continuum luminosities in the framework of classical photoionization models without invoking any luminosity dependent parameter.

We assume that the continuum of AGN is made of the superposition of 2 components:

-a non thermal power law component with a spectral index -1, which dominates the near IR and the far UV;

-a thermal component from a thin accretion disc which dominates in the UV range, it is described by a thin relativistic Schwarzschild disc with an inclination of 40°.

We consider only objects in which the stellar component is negligible ($L_{1\mu} > 10^{29.5}$).

The Hα and Hβ lines are emitted in the outer part of this disc, i.e. their emission region is gravitationally bound. An estimation of the dispersion velocity is obtained from the FWHM of Hβ. This assumption allows to determine the "dynamical mass" of the central object:

$$\mathrm{M} = 3/\mathrm{G} \times (\mathrm{FWHM(H}\beta)/2)^2 \times (\mathrm{Q}/4\pi \ \mathrm{c\ Un})^{0.5}$$

where Q is the number of ionizing photons per sec., U the ionization parameter ($U=Q/4\pi c r^2 n$) and n the density. We assume **Un** $= 10^9$, and a **constant covering factor**. The computed spectral distribution is normalized to the observed luminosity at 1450Å. It provides Q and the accretion rate.

2. Observational data

We use a data base on QSO and Seyfert 1 nuclei, first described in Joly et al. (1985). It provides a sample of about 150 objects for which line fluxes, equivalent widths and FWHM of Lα , CIV, Hβ or Hα are available, together with the monochromatic continuum flux at 1450Å or 1μ. For a subsample of 20 objects all these data are available.

Fig. 1 shows the observed correlation between L(CIV) and the continuum luminosity at 1450Å for the two samples. The slope of the regression line is 0.77±0.05 with a correlation coefficient of 0.94.

3. Results

The mass of the central body, the luminosity of the disc and the number of ionizing photons have been computed for each object of the subsample, using an iterative process.

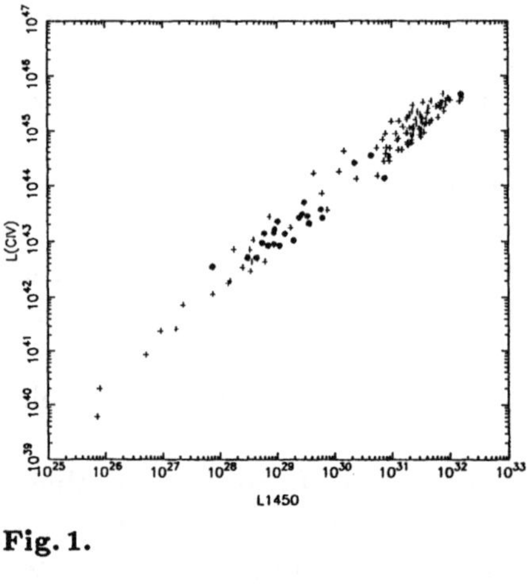

Fig. 1.

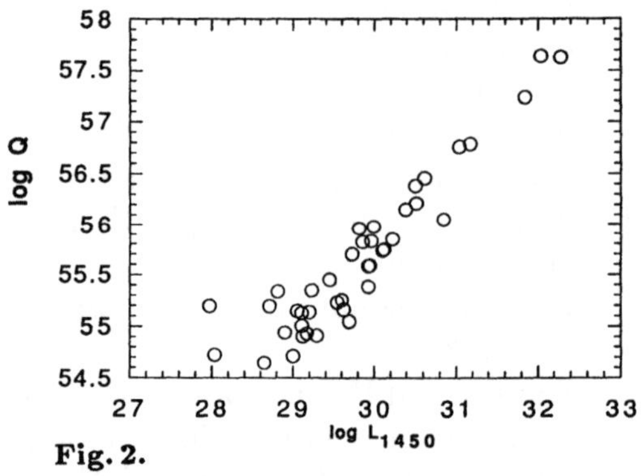

Fig. 2.

Fig. 2 shows the number of ionizing photons, Q, as a function of $L_{1450\AA}$. The slope of the relation is 0.79±0.06 implying a straightforward relation between L(CIV) and Q: **the luminosity relation L(CIV) versus $L_{1450\AA}$ is explained without variation of U or $\Omega/4\pi$.**

Fig. 3 shows the relation between the mass M of the central body and $L_{1450\AA}$ and Fig. 4, the relation between the luminosity of the disc and M:

$$\log M \propto 0.6 \pm 0.2 \log L_{1450\AA}; \qquad \log L_{disc} \propto 1.8 \pm 0.6 \log M$$

Note that **the disc luminosity is clearly below the Eddington luminosity.**

Fig. 5 gives the ratio of the disc luminosity to the optical luminosity as a function of $L_{1450\AA}$. It shows that, the more luminous the object, the more luminous the disc and the more it dominates the overall luminosity.

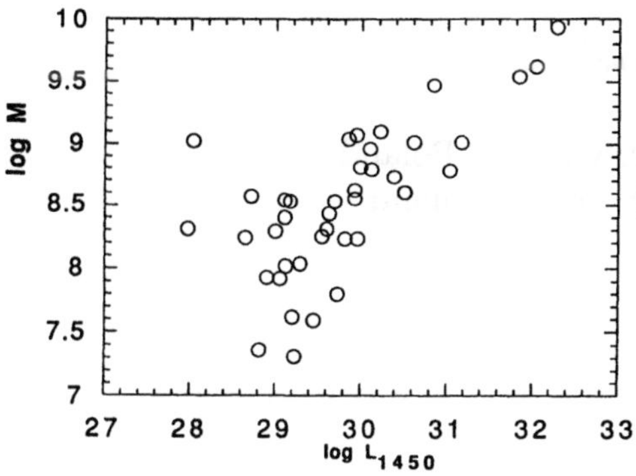

Fig. 3.

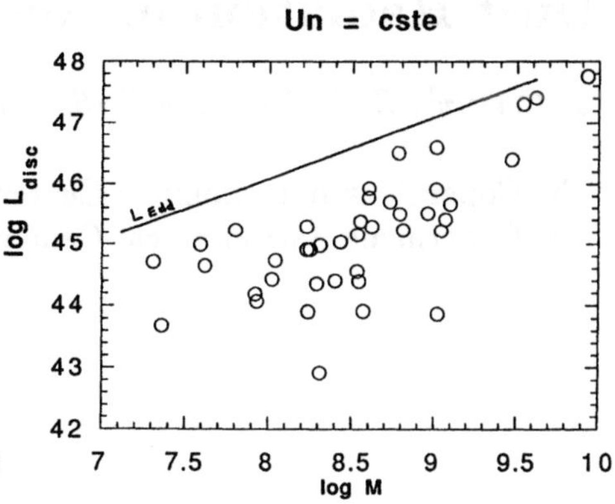

Fig. 4.

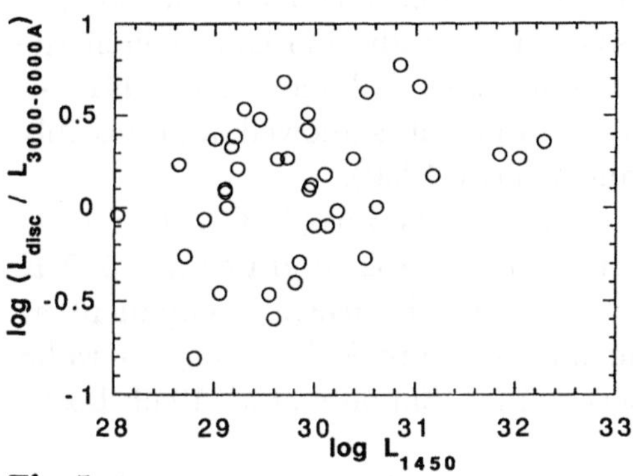

Fig. 5.

References

Joly, M., Collin, S., Masnou, J.L., Nottale, L., 1985: *Astron. Astrophys.*, **152**, 282.

Dust Radiation in Active Galactic Nuclei

B. Czerny[1], *Z. Loska*[2], *and R. Szczerba*[2]

[1] N. Copernicus Astronomical Center, Warsaw, Poland
[2] N. Copernicus Astronomical Center, Toruń, Poland

1. Introduction

Radiation from the dust has been recently recognized as one of the components of the continuum emission in AGN. Dust radiation can explain the infrared (IR) AGN spectrum (e.g. Barvainis 1990; Sitko and Zhu 1991) although a contribution of a non–thermal component is allowed, and possibly required (e.g. Berriman 1990; Loska and Czerny 1990).

Simple model of dust radiation was used by Barvainis (1987). His results show that the evaporation temperature of the graphite grains at 1500 K provides a high frequency cut–off for the emitted spectral component at about $\log(\nu)=14.0$. This explains the universal break frequency in radio quiet quasar spectra seen at about $\log(\nu)=14.5$ (Sanders *et al.* 1989; Loska and Czerny 1990).

Here we present very preliminary results of radiative transfer calculations for case of spherical dust distribution and anisotropic uv source.

An exact numerical code for axisymmetric dust distributions was recently developed by Efstathiou and Rowan–Robinson (1990). However, because of the complexity, this code is very time consuming. Therefore, we rather follow simpler approach of Barvainis (1987) and Sanders *et al.* (1989), with some modifications.

2. Method of solution

The numerical method is adopted from the computer program developed by one of us (see Siemiginowska, Szczerba & Sitko 1991) for the case of GQ Comae.

In these computations we neglect multiple scatterings (Barvainis 1987; Sanders *et al.* 1989) as the spherical dust cloud cannot be optically very thick at uv frequencies.

We assume power law distribution for the dust density profile (index $-\beta$). Both silicate and graphite grains are assumed to be present, with the relative ratio treated as a free parameter. The distribution of grain sizes was usually adopted after Mathis, Rumpl & Nordsieck (1977).

The temperature of a silicate or graphite grain of given size a is determined from the energy balance between absorption of the local uv flux and the reemission. Adopted evaporation temperature: 1000 K for silicate grains and 1500 K for graphite grains.

The dust contribution to the spectrum is calculated through simple integration, including contribution of scattered photons.

The intrinsic spectrum of an AGN is modelled as a sum of two components: spherically symmetric power law representing the nonthermal radiation of relativistic electrons, and thermal emission from optically thick Keplerian accretion disc, with cosine dependence on the inclination angle (Loska and Czerny 1990).

3. Results

Two examples of spectra models are shown in Fig. 1, and exemplary fit to the data is shown in Fig. 2. Reemision of uv by graphite dust well represents the observation whilst the data does not indicate any features characteristic for silicate emission.

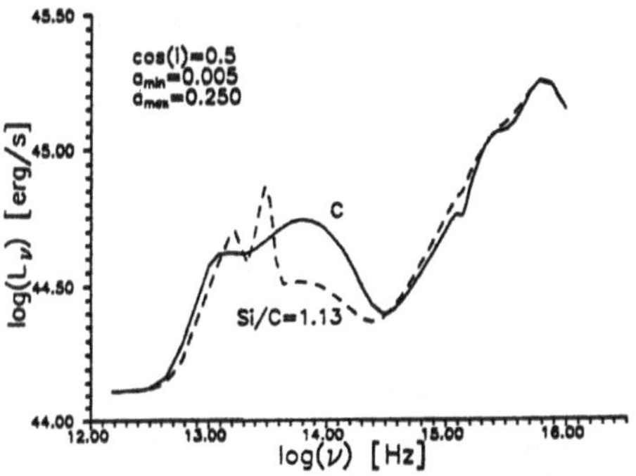

Fig. 1. Two examples of chemical composition

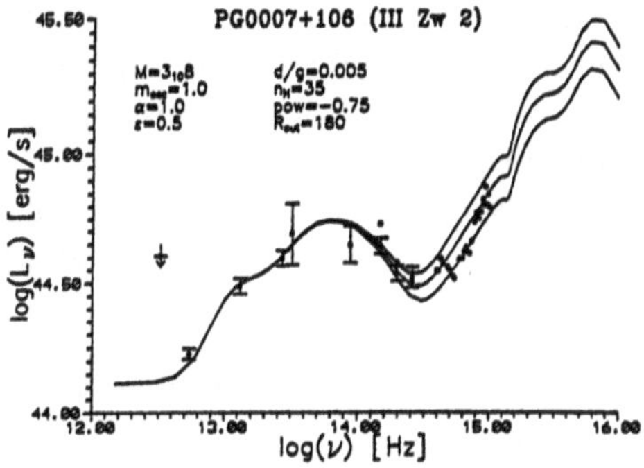

Fig. 2. Three values of disk inclination angle

References

Barvainis, R., 1987: *Astrophys. J.*, **320**, 537.
Barvainis, R., 1990; *Astrophys. J.*, **353**, 419.
Berriman, G., 1990: *Astrophys. J.*, **354**, 148.
Efstathiou, A., Rowan-Robinson, R., 1990: *Monthly Notices Roy. Astron. Soc.*, **245**, 275.
Loska, Z., Czerny, B., 1990: *Monthly Notices Roy. Astron. Soc.*, **244**, 43.
Mathis, J.S., Rumpl, W., Nordsieck, K.H., 1977, *Astrophys. J.*, **217**, 425.
Siemiginowska, A., Szczerba, R., Sitko, M.L., 1991: preprint.
Sitko, M.L., Zhu, Y., 1991: *Astrophys. J.*, **369**, 106.

Variability of the Broad Emission Line Profiles in NGC 5548

Matthias Dietrich[1] *and Wolfram Kollatschny*[1,2]

[1] Universitätssternwarte, Geismarlandstr. 11, D-3400 Göttingen, Germany
[2] Dr.Remeis-Sternwarte Bamberg, D-8600 Bamberg, Germany

Abstract. Light curves of Hα, HeI 5876, and Hβ are presented. The difference spectra of these three optical emission lines reveal the existence of at least two individual components showing different response to the continuum variations of NGC 5548.

The variations of the Hα, HeI 5876, and Hβ emission line fluxes scaled to the narrow forbidden lines are shown in Figure 1a-c. The light curves of these three emission lines show the same variability pattern as the UV continuum (Clavel et al., 1991, Ap.J **366**, 64) but shifted with respect to the UV continuum. For estimating the extent of the BLR one usually assumes a linear response of the emission lines with respect to the variation of the central continuum source. The CCF of the Hα, HeI 5876, and Hβ lines were calculated using methods similar to those of Gaskell and Peterson, 1987, Ap.J. Supp **65**, 1. An ionization stratification of the BLR is clearly present.

line	delay [d]	line	delay [d]	line	delay [d]
F_{1350} F_{1840}	0.0	CIV 1548	9.0	Hβ 4861	17.0
HeII 1640+OIII]	2.0	HeI 5876	11.0	Hα 6563	19.0
Lyα + NV	9.0	Hγ 4340	12.0	CIII] 1909	26.0

The variations of the broad emission line profiles give us informations about the velocity field in the BLR. Furthermore, we detected individual subcomponents in the line profiles varying in a different way. These components are seen more clearly in the difference spectra. In Figure 2a-c the difference spectra of Hα, HeI5876, and Hβ, are shown for several epochs with respect to a reference spectrum. The difference spectra of the three emission lines show the same pattern; the FWZI amounts to 8000 km s^{-1} in these lines. Furthermore, the existence of a blueshifted component, $v_{bl} = -1800$ km s^{-1}, is clearly visible. A detailed investigation of the temporal evolution of the difference spectra shows that the blue component varies in a different

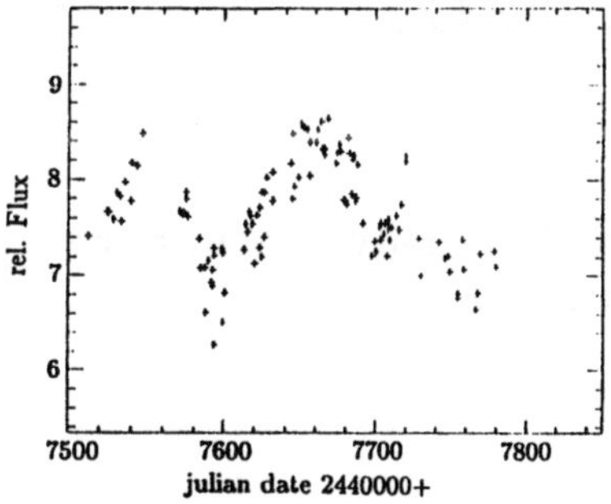

Fig. 1a. Hα light curve scaled to the flux of [NII]6548,6584 and $H\alpha_{small}$

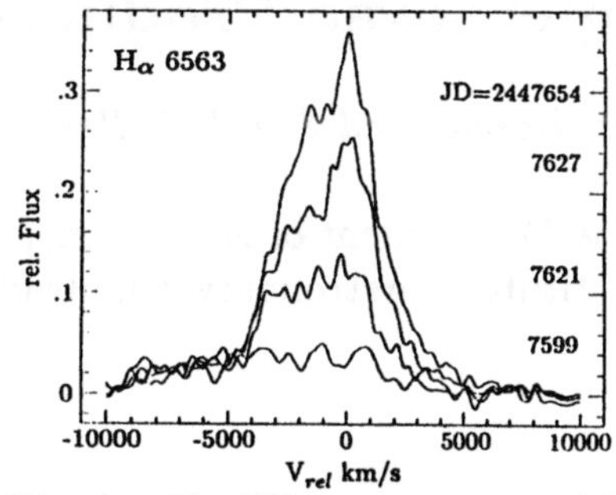

Fig. 2a. Hα difference spectra with respect to the epoch JD=2447749

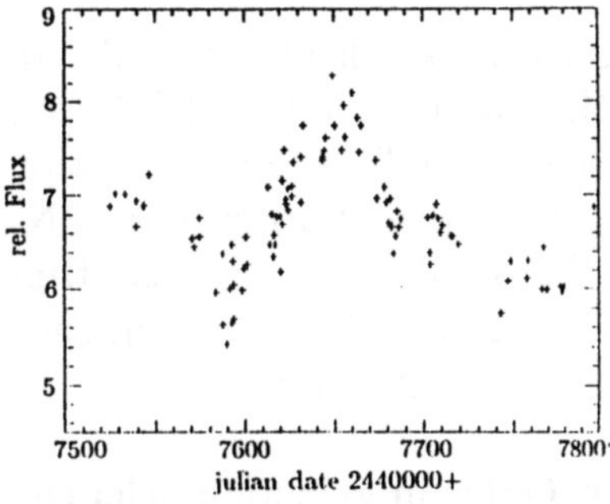

Fig. 1b. HeI 5876 light curve scaled to the flux of [FeVII]5721,6087

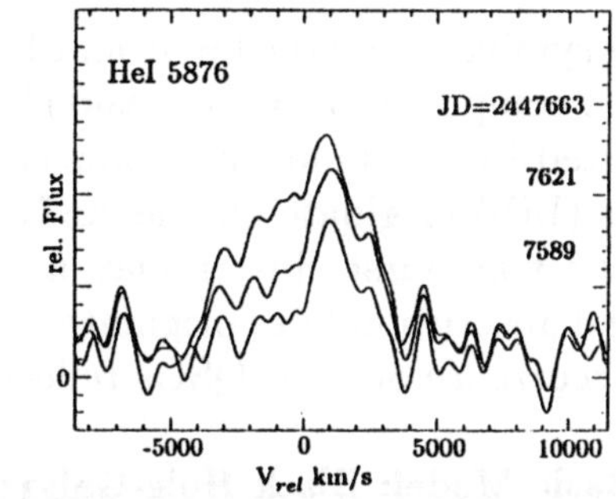

Fig. 2b. HeI 5876 difference spectra with respect to the epoch JD=2447746

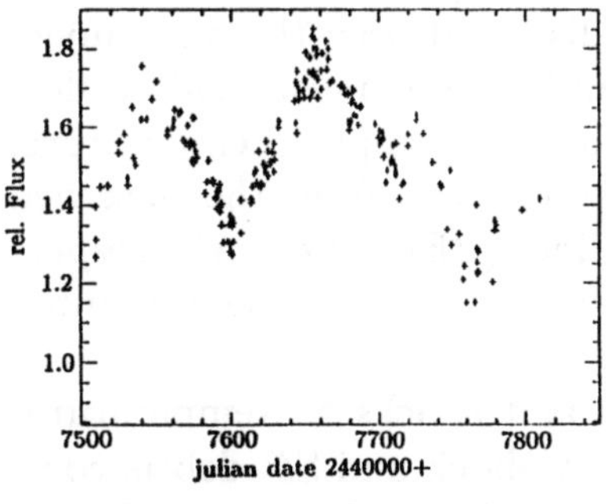

Fig. 1c. Hβ light curve scaled to the flux of [OIII]4959,5007

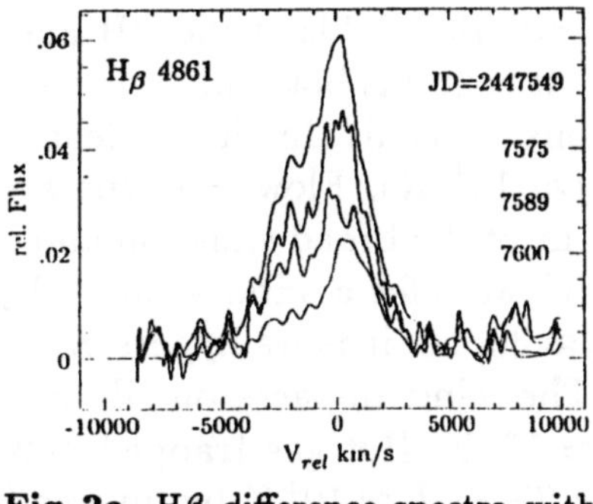

Fig. 2c. Hβ difference spectra with respect to the epoch JD=2447767

way than the dominant central component (Kollatschny an Dietrich, 1991, in: *Variability of Active Galaxies*, eds. W.J. Duschl et al., Springer Verlag, p. 57); therefore they must originate in different spatial regions.

This work has partly been supported by BMFT grant Verbundforschung Astronomie FKZ 05 5G042A 1

Supernovae Produced BAL QSOs

John Dyson[1] *and Judith J. Perry*[2]

[1] The Department of Astronomy, The Victoria University of Manchester
[2] Institute of Astronomy, Cambridge, CB3 0HA, UK.

The symbiosis of star-burst and black-hole AGN models leads naturally to a two component broad emission line region. High ionization lines (HIL) are emitted by continuously produced but evanescent cloudlets; low ionization lines (LIL) originate on the surface of an accretion disc. Supernovae (SN) in the star burst cluster play a key role in this model. We propose that broad absorption lines originate in supernova debris and that the frequency of occurrence of BAL QSOs reflects the debris covering factor.

1. Basic Model: Black Hole-Galaxy Symbiosis. QSOs in galactic nuclei can be modelled using a massive black hole (BH), a nuclear stellar cluster, an accretion disc and the interstellar medium (ISM). The ISM is created by stellar mass loss from the cluster and receives energy and momentum from the radiation fields of the BH and accretion disc, and from the mass injection. It contains both inflow (fueling the BH and accretion disc) and outflow regions (a wind). In the nuclear region it is at the Compton temperature, T_C, ($\approx 10^7$ K). Flow patterns are quite complex, but from the observed widths of the broad emission and absorption lines (BELs and BALs respectively) we infer a wind velocity $V_w \approx 10^4\,\mathrm{km\,s^{-1}}$. The wind Mach number $M_w \approx 30$, i.e. it flows hypersonically.

The wind impacts on SN in the cluster and shocks to temperatures $T_S \approx 10^9$ K. Hot gas trapped between the bow shock and SN debris cools from T_S to $T \approx 10^4$ K by inverse Compton scattering and bremsstrahlung, followed by heavy element cooling. In the latter phase fragmentation into fast moving, optically thin, photoionized BEL clouds occurs. These clouds produce the HIL. They are evanescent and continuously replenished. The BEL line widths reflect V_w. The process works at $r_{HIL} \lesssim L_{47}^{1/2}$ pc. (Perry & Dyson, 1985, *MNRAS* **213**, 665.)

The post-shocked gas backscatters the central continuum and emits radiation while cooling; both illuminate the accretion disc. Its illuminated surface radiates the LIL. Their line widths reflect the disc rotational velocity. (Collin-Souffrin, Dyson, McDowell & Perry, 1988, *MNRAS* **232**, 539.)

2. A Simple Model for Supernovae produced BALs. A natural connection between the BEL model and the BAL exists through the cluster SN. Their ejecta velocities are similar to the observed absorption line widths; the frequency of occurrence of BAL QSOs agrees with calculated SN covering factors for BEL production, $\approx 10\%$.

When a SN explodes in the ISM a two shock flow is set up: a fast forward shock initially moves out at about the ejecta velocity, V_e; a reverse shock in the ejecta accelerates back towards the explosion centre. The shocked ISM is always very hot since it shocks at $\sim V_e$ or V_w; the shocked ejecta is initially cool ($\sim 10^4$K) but heats up as the relative velocity between the ejecta and shock increases. The ejecta is stopped in a time, $t_s \approx 1.4\,(M_e/n_o)^{1/3}\,V_e^{-1}$ yr, at a radius, $R_s \approx V_e t_s \approx 0.04\,(M_e/n_o)^{1/3}$ pc. [M_e is in $M_\odot$, V_e in 0.1c and n_o in 10^5 cm^{-3}.] As expected, ejecta moving in toward the source are slowed down far more quickly, and reach a much smaller radius, than ejecta moving out. Consider a momentum conserving solution for a SN at ≈ 1 pc from a source of bolometric luminosity 10^{47} erg s^{-1}. For $M_e = 5$, $V_e = 0.5$, and $n_o = 0.1$, $t_s \approx 10$ yr and $R_s \approx 0.15$ pc. Red shifted ejecta effectively have stopped at $t \approx 1.7 t_s$ when blue shifted ejecta still are travelling at $V_b \approx 0.75 V_e$. The ratio of blue:red covering factors is then $\approx$ 4:1. Thus only blue shifted absorption is observed. $N_H \approx 8\times10^{19}$cm^{-2} and $n_e \approx 3\times10^9$cm^{-3}. All BAL fragments are decelerating, as inferred from observations.

The only cool fast moving material is the $\approx 10\%$ of the ejecta which enters the reverse shock at low relative velocity. Since the forward shock velocity $V_F \approx V_e \approx V_W$, the cooled ejecta has a pressure and ionization parameter comparable to that of the HIL. BALs exist for $r_{BAL} \gtrsim r_{HIL}$ and so overlap, but lie outside, the HIL BELR. Their covering factor $\propto$ (Size of the Absorber $\times$ Number of SN) $\propto P_{stag}(r)\rho_*(r) \propto r^{-\beta}$, $\beta \approx 5 - 7$. It falls off rapidly with r, defining an outer "edge of the region. The nuclear environment influences the range of BAL properties, so second-order correlations, such as that with NV line strength, are likely.

Absorption occurs in fragments produced by Rayleigh-Taylor instabilities and in gas ablated from their edges as they move relative to the hot shocked ambient gas. Thus complex structure is inevitable. Attachment or detachment from the emission lines depends on the relative velocities of the dominant emitting and absorbing material. Smoothness and velocity extent depend on the number of fragments and SN on the line of sight.

Velocity variations should be observed on the slowing down timescale t_s^o in the observers frame, $t_s^o = (1+z)t_s$. For high redshift QSOs with $z \approx 2$ say, $t_s^o \approx 30$ yr. Therefore velocity variations should not be common. However since Rayleigh-Taylor instability growth times can be $\ll t_s$ column density and thus line strength variations can be observed over times $\ll t_s^o$.

The existence of BALs may discriminate starburst models with and without black holes, since the BAL clouds must cover the continuum source.

Magnetic Accretion Disks in AGN

G.B. Field and R.D. Rogers

Harvard-Smithsonian Center for Astrophysics

Abstract. We propose that both the X-ray and gamma-ray components of the 3 keV to 10 MeV high-energy background (HEB) originate in magnetic coronae of AGN accretion disks (Rogers and Field, Ap.J. Lett., 378, L17). This model gives an excellent fit to the spectrum of the HEB (Figure 2) and is consistent with observations of individual AGNs.

1. Introduction

The numbers and luminosities of AGNs can account for the intensity of the XRB, but their spectra in the few keV range until recently appeared to be too steep. Now 1-30 keV *Ginga* spectra of AGNs indicate that this "spectral paradox" can be resolved by the presence of a Compton reflected component in AGN spectra (Figure 1). As shown by Fabian *et al.* (1990 MNRAS, **242**,14P), if a Compton reflected component of ~ 90 % is superimposed on a power law, then AGNs at various redshifts may be superimposed to fit the XRB. We have shown that if the X-ray spectrum is inverse-Compton scattering by relativistic electrons, then scattering of the X rays produces gamma rays, which we identify with the gamma-ray background (Figure 1).

2. Magnetic Accretion Disks

The observations require that relativistic electrons are confined close to the disk and that they be reaccelerated to compensate for inverse-Compton losses. A strong ($\sim 10^3$ G) magnetic field embedded in an accretion disk serves both purposes. Magnetic buoyancy forms coronal loops about 10^{12} cm across with B$\sim 10^3$ G, and Keplerian rotation continually brings fields of opposite polarity into radial contact. Magnetic reconnection between loops of opposite polarity at adjacent radii dissipates energy, allowing accretion and causing particle acceleration. The electrons, accelerated with an energy index ~ 3, inverse-Compton scatter thermal photons into X rays, and X rays into gamma rays. The heat absorbed by the disk is reradiated thermally at $T \sim 30,000$ K, accounting for the blue bump in AGN spectra. The electrons

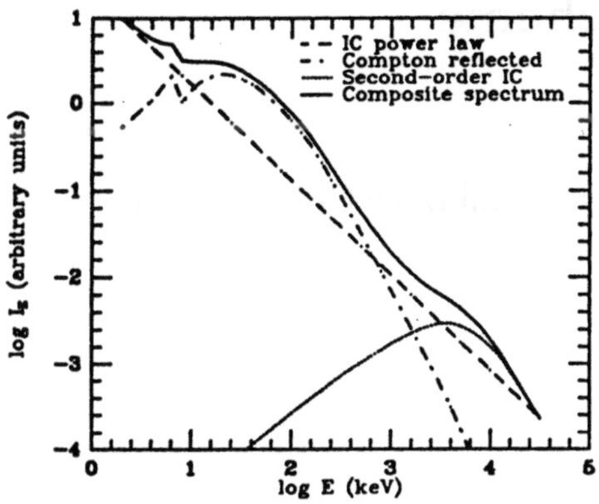

Fig. 1. Three components originate in the magnetic corona and in the cold gas of an accretion disk around a $10^8 M_\odot$ black hole: (i) Dashed line: Inverse-Compton scattering of thermal radiation from the disk by magnetically trapped relativistic electrons with $n \sim \gamma^{-3.2}$ and low-energy cutoff at $\gamma_1 = 28$, gives an energy index of 1.1, (ii) Dot-dash line: 89 % of component (i) interacts with the disk giving a Compton-reflected spectrum, (iii) Dotted line: Inverse-Compton X rays scatter from the relativistic electrons, giving γ rays whose spectrum is shown for electron optical depth $\tau_T = 3 \times 10^{-4}$. (iv) Solid line: The sum of components (i) - (iii) above.

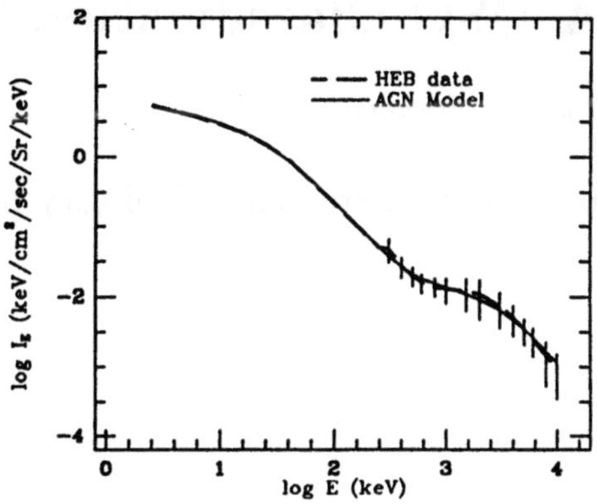

Fig. 2. The data are compared with the emission predicted by superposing spectra of Figure 1 with various redshifts. This best fit was obtained with comoving luminosity varying like $(1+z)^{2.8}$, out to $z_m = 4.6$.

also emit synchrotron radiation with spectral index ~ 1; self-absorption cuts off the spectrum in the far infrared, consistent with observation. With $L \leq 10^{44}$ erg sec^{-1}, collisions between X and gamma rays produce pairs mostly outside of magnetic loops, where they are free to escape.

3. Conclusions

The high-energy background from 3 keV to 10 MeV can be explained by AGNs in which particles are accelerated by reconnection of magnetic loops above an accretion disk. Our model is consistent with *ROSAT* observations of the background below 3 keV, but it requires a larger Compton reflected component than is observed by *Ginga* for individual AGN.

A QSO Classification Scheme

Paul J. Francis

Institute of Astronomy, Madingley Road, Cambridge CB3 0HA, UK.

The analysis of QSO spectra has long been dominated by morphological techniques – features are identified by shape ("if it sticks up, it must be a line"), and are deblended by model fitting. I present a quite different approach, which identifies features not by shape, but by correlations. Using Principal Components Analysis (PCA, see Mittaz *et al.* 1990), applied directly to the spectra, rest-frame wavelength regions that correlate together are identified as components.

I have applied PCA to a large and carefully selected QSO survey (the Large Bright QSO Survey, Hewett *et al.* 1991 and references therein). The results are startling – with only three linear components, over 80% of the diversity of QSO spectra can be reproduced. These components are shown in figure 1.

The first principal component is by definition the largest source of diversity in the survey spectra. It shows correlating bumps at the wavelengths of the emission lines, telling us unsurprisingly that the lines vary in equivalent width, and that they correlate with each other. Note however that the bumps are narrow, and that the Lyman-α bump is larger than the CIV bump, when compared to the average fluxes. Thus in objects where this component is strong, we will see strong narrow "Eiffel Tower" shaped lines with a high ration of Lyman-α to CIV, but when it is weak, the lines will have low equivalent widths and show broad flat topped profiles.

The second principal component is the largest way in which QSO spectra differ after the first component has been removed. This shows a slope from red to blue, telling us that the continuum slope is variable. Several correlations with line strengths are identified.

The third principal component, showing correlated troughs blueward of high ionization emission lines, is only strong in a minority of QSOs. These are Broad Absorption Line QSOs, and this work identifies a continuum of properties between them and normal QSOs, and also demonstrates that they show excess NV, and to a lesser extent CIV, emission.

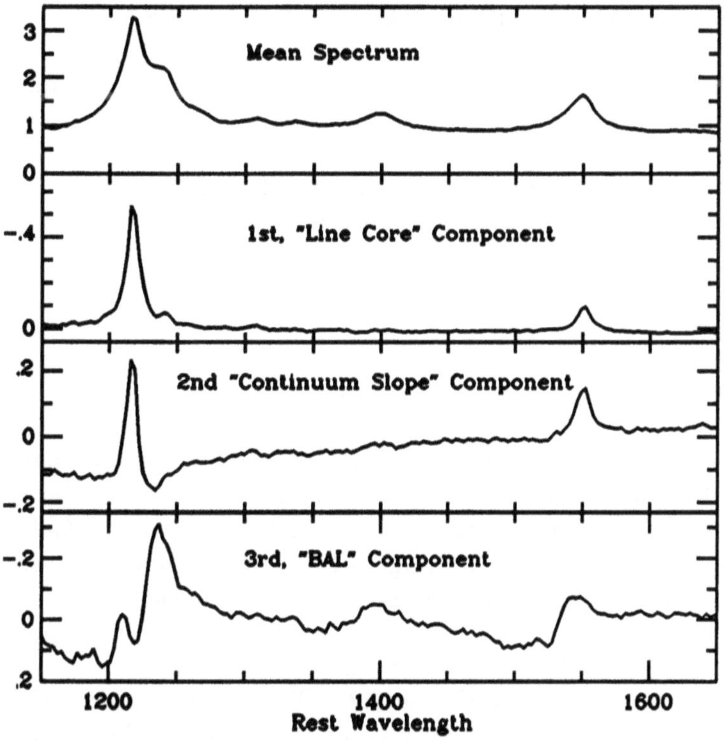

Fig. 1. Mean spectrum, and the first three principal components.

PCA thus produces a remarkably simple description of the diversity of QSO spectra, identifies correlations in a model-independent way, and quantifies the range of QSO spectra.

This work has been carried out in collaboration with Paul Hewett (IoA Cambridge), Craig Foltz and Fred Chaffee (MMT Observatory), and Ray Weymann (Carnegie Observatories).

References

Hewett, P.C., Foltz, C.B., Chaffee, F.H., Francis, P.J., Weymann, R.J., Morris, S.L., Anderson, S.F., MacAlpine, G.M., 1991: *Astron. J.*, **101**, 1991.

Mittaz, J.P.D., Penston, M.V., Snijders, M.A.J., 1990: *Monthly Notices Roy. Astron. Soc.*, **242**, 370.

UCM2257+2438: A new Seyfert Galaxy

J. Gallego[1], *J. Zamorano*[1], *A. Vitores*[1,3], *M. Rego*[1], *and R. González-Riestra*[2]

[1] Dpto. de Astrofísica. Universidad Complutense de Madrid
[2] Tracking Satellite Station, European Space Agency
[3] EUIT Industrial. Universidad Politécnica de Madrid

We are currently carrying out a survey in selected fields of the Northern Hemisphere with the aim of finding low-metallicity galaxies (Zamorano et al., 1990). The survey is performed with the Schmidt Telescope at the Calar Alto German–Spanish Observatory. Although a large proportion of the extragalactic objects found in the survey are star-forming galaxies, there are some objects that show some evidences of activity. In this paper we report the identification of UCM2257+2438 as a new Seyfert 1.5 galaxy.

UCM2257+2438 was observed with the 2.2m Telescope at Calar Alto German-Spanish Observatory through an r Gunn-Thuan filter. The image of the galaxy fills only a small portion of the frames and the sky in the neighborhood is flat within a 1-2 %. Sky level is 21.0 magnitudes per square arcsecond. The spectroscopic observations were carried out with the 3.5m Telescope at Calar Alto Observatory, using the Cassegrain Twin Spectrograph and an exposure time was 2700 s.

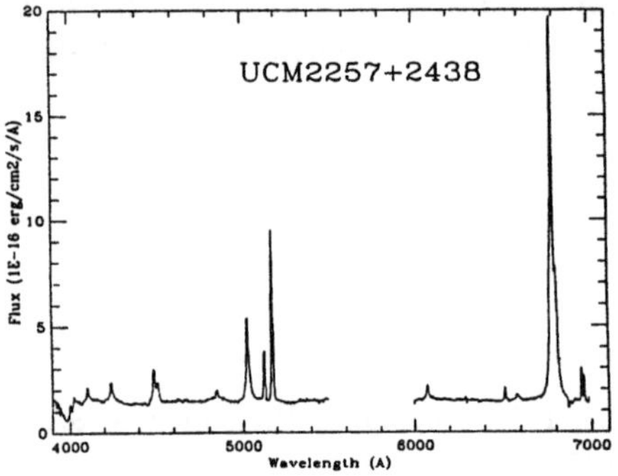

Fig. 1.

The brightness profile of UCM2257+2438 is plotted in figure 1 against the mean radius, i.e. $r = \sqrt{ab}$. Also showed is the best fit obtained of the summation of a $r^{1/4}$ law and a exponential disk law. We have followed the procedure outlined in Schombert & Bothun, 1987, using the Downhill

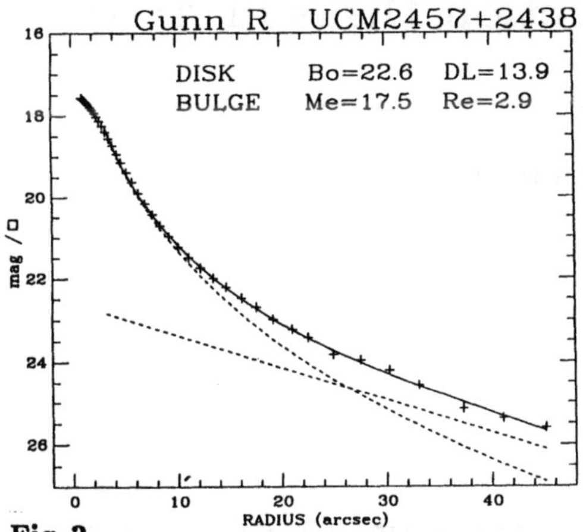

Fig. 2.

Simplex Method, to perform the minimization. No attempt has been made to fit the central part of the galaxy (radius $\leq$ 3 arcsec) were seeing smears the intrinsic light distribution. The Bulge-to-Disk ratio is B/D = 0.12 and corresponds to morphological type 4, i.e. Sb-Sc (Simien & de Vaucouleurs, 1986).

The spectrum (figure 2) is typical of Seyfert 1.5 galaxies with both broad and narrow Balmer HI emission lines and narrow forbidden lines. Balmer lines of Hydrogen show two components: The broad one with FWHM = $2300 km/s$, FWZI = $6800 km/s$, appears clearly blueward. [OIII] lines are asymmetric.

UCM2257+2438 is a spiral galaxy seen almost face on, with a diameter of 14 kpc which lies 200 Mpc from us ($H_o = 50 kms^{-1} Mpc^{-1}$). Inside a radius of 10 kpc from the nucleus the hard radiation field ionizes the interstellar medium producing an extended narrow line region (ENLR). The presence of lines of high ionized species (FeX, NeIII) and the strong asymmetries in the [OIII] lines suggest the existence of a high radial flow of turbulent material and a broad line region (BLR). Assuming a density Ne=10000 cm^{-3} in the region where the [OIII] lines are emitted in preference, the [OIII]4959+5007/[OIII] 4363 ratio is consistent with a temperature T=45000K, typical of objects with high ionization.

References

Kent, S.M., 1985: *Publ. Astron. Soc. Pacific*, **97**, 165.
Kunth, D., Sargent, W.L.W., 1983: *Astrophys. J.*, **273**, 81.
Simien, F., de Vaucouleurs, G., 1986: *Astrophys. J.*, **302**, 564.
Schombert, J.M., Bothun, G.D., 1987: *Astron. J.*, **93**, 60.
Zamorano, J., Rego, M., González, R., 1989: *Astron. Astrophys. Suppl.*, **79**, 443.

Magnetized Thin Accretion Disks in AGN

Ramon Khanna and Max Camenzind

Landessternwarte Königstuhl, W-6900 Heidelberg, Germany

Abstract. We study the magnetic field topology and the magnetic field strength in a thin accretion disk surrounding a supermassive rotating black hole. In a Newtonian treatment, the existence of a boundary layer between the black hole and the disk is taken into account in the particular form of the rotation law $\Omega(R)$ and of the accretion velocity $u(R)$ within the marginal stable orbit. Turbulence in the disk plasma and boundary conditions for the magnetic field at the horizon of the hole require a radially dependent magnetic diffusivity $\eta(R)$. Neglecting the dynamo effect as a source for the poloidal magnetic flux $\Psi(R, z)$, we present steady state solutions for $\Psi(R, z)$ and the toroidal magnetic field $B_\phi(R, z)$ for even and odd symmetry of $\Psi(z)$.

The magnetic fields

In the stationary case the thin disk induction equations are

$$\frac{\partial \Psi}{\partial R} + D(R)\frac{\partial^2 \Psi}{\partial \zeta^2} = 0 \qquad (1.a)$$

for the poloidal flux, and

$$\frac{\partial}{\partial R}\left[(u + \frac{\eta}{R})B_\phi\right] + \frac{\partial \eta}{\partial R}\frac{\partial B_\phi}{\partial R} + u(R)D(R)\frac{\partial^2 B_\phi}{\partial \zeta^2} = + \frac{d\Omega}{dR}\frac{1}{H(R)}\frac{\partial \Psi}{\partial \zeta} \qquad (1.b)$$

for B_ϕ, where $u(R)D(R) = \eta(R)/H^2(R) = t^{-1}_{\mathrm{diff},z}(R)$ is the inverse characteristic timescale for vertical diffusion at the radius R. $\zeta(R) = z/H(R)$ is a dimensionless vertical coordinate.

For a jet to be produced by centrifugally accelerating particles along magnetic field lines, the disk solution has to fit to a wind solution. Figure 1 shows the calculated dipole solution for $\mathbf{B_p}$ and $\mathbf{j_p}$ in combination with a qualitatively possible jet solution. In the disk solution itself we find $\mathbf{j_p} || \mathbf{B_p}$ at the disk surface, so that a force-free jet solution fits intriguingly well. The current field lines turning in the disk make it possible to have currents flowing out in the jet envelope and coming back along the jet axis. The

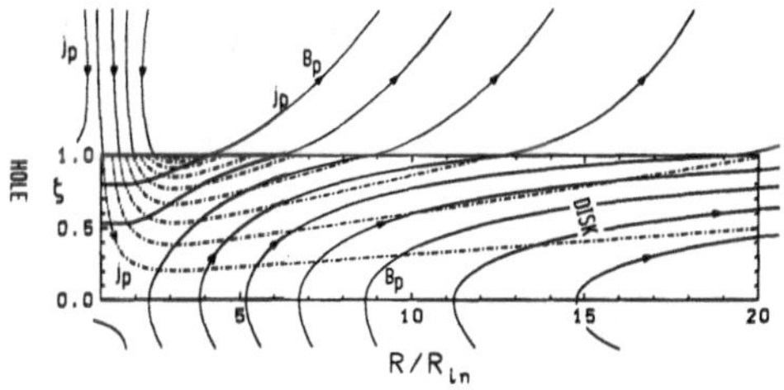

Fig. 1. $\mathbf{B_p}$ **and** $\mathbf{j_p}$ **- topology in disk (calculated) and jet (qualitatively) for a dipole.**

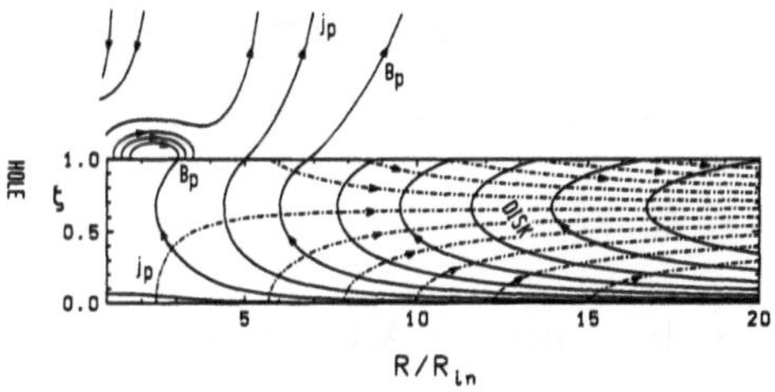

Fig. 2. $\mathbf{B_p}$ **and** $\mathbf{j_p}$ **- topology in disk (calculated) and jet (qualitatively) for a quadrupole.**

current can be closed in the disk at one end and somewhere in the jet at the other end (in the hot spot e.g.). The turning of the currents in the disk is an effect of the boundary layer, where $\frac{d\Omega}{dR}$ changes its sign, so that the source term for B_ϕ has an extremum. The fact that the turning point lies very close to the horizon allows to produce fast jets, since plasma is ejected from the disk at high rotation velocity. Furthermore, the Newtonian solution suggests that the current can be closed in a region where relativistic effects are important. This result gives support to the idea of Blandford and Znajek (1977) to extract rotational energy from the black hole by some kind of magnetic braking.

The quadrupole topology is not suitable for the production of a force-free jet (Fig. 2). Moreover, since the currents turn quite far away from the hole, where the rotation velocity is low, a jet produced in this topology could not be fast. However, quadrupolar field topologies can also produce jets (Haehnelt and Camenzind, 1991). They have a maximum of Ψ and $|RB_\phi|$ near the hole. In Khanna and Camenzind (1991) we suggest a possible disk solution.

References

Blandford, R.D., Znajek, R.L., 1977: *Monthly Notices Roy. Astron. Soc.*, **179**, 433.
Haehnelt, M., Camenzind, M., 1991: in preparation.
Khanna, R., Camenzind, M., 1991: *Astron. Astrophys.*, submitted.

On the Isotropy of the IR/Optical Emission in AGN

Ari Laor

The Institute for Advanced Study, Princeton NJ 08540, USA

Thin accretion disks have been proposed as the source for the optical-UV emission in AGN. The optical flux of thin accretion disks is highly non-isotropic, a face-on disk appears more than a factor of 10 brighter than a nearly edge on one. The observed near IR continuum cannot be emitted by a simple thin accretion disk. A possible alternative source is a very hot dust, in which case the IR emission is likely to be isotropic. The IR/optical flux correlation of a sample of objects viewed at random inclinations should therefore display a significant scatter even if the luminosities of the intrinsic IR and the Disk components are perfectly correlated.

This hypothesis is tested by comparing the scatter in the IR ($1.69\mu m$) *vs.* optical (4200Å) correlation of the PG sample (Neugebauer *et al.* 1987) to the one deduced from a sample of disk models.

The theoretical sample consists of thin accretion disk models (see Laor and Netzer 1989) + IR power law component with a flux proportional to the Disk bolometric luminosity. The IR component is likely to be dust emission, however a power law form was used since accretion disk fittings of the optical/UV continuum of AGN generally require an additional component around $1.0 - 0.5\mu m$ which cannot be due to dust and is usually described as an IR power law (Sun and Malkan 1990, Laor 1990).

The disk model parameters were drawn at random from a uniform distribution of inclinations, log of black hole mass and log of the accretion rate in Eddington units. The models were further subject to other selection criteria which were designed to mimic selection effects present in the PG sample.

The scatter in the IR/optical correlation was measured as the deviation in the Y direction of each data point from a best fit power law to the IR *vs.* optical fluxes. An F-test was used to test the significance of the difference between the widths (variances) of the observed and predicted distribution of deviations (see Fig.1), and a K-S test was used to test the difference in the shapes of the cumulative probabilty distributions. Both tests indicate that **the observed distribution is significantly narrower than the predicted**

one. Note that the model distributions do not take into account other effects such as optical variability or a dispersion in the intrinsic IR/Disk luminosity ratio, which would further increase the width of the predicted distributions.

This result implies that the angular distributions of the near IR and optical emissions in AGN do not differ as much as assumed above, either the near IR emission is non-isotropic and has an angular distribution similar to that of the optical emission, or possibly the optical flux is more isotropic than expected from a thin accretion disk. The last possibility might occur in case of a thick accretion disk, or in the case of a thin disk which is obscured over a certain inclination range.

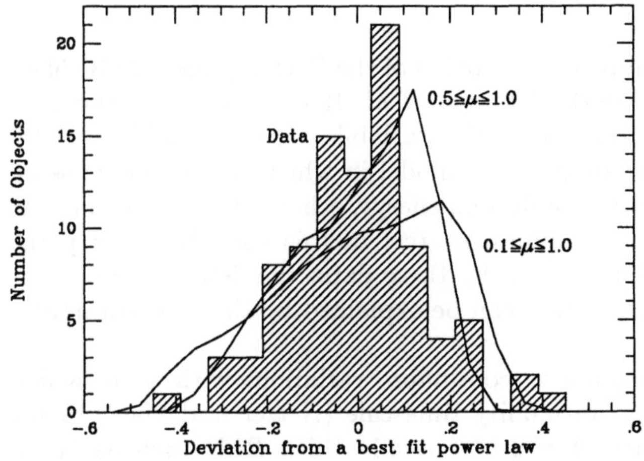

Fig. 1. The distributions of deviations from a power law fitted to the IR *vs.* Optical fluxes. The predicted distribution is significantly broader than the observed one. It can be made consistent with the data if the disk inclination range is reduced, as shown above.

References

Laor, A. 1990, *Monthly Notices Roy. Astron. Soc.*, **246**, 369
Laor, A. and Netzer, H. 1989, *Monthly Notices Roy. Astron. Soc.*, **238**, 897
Malkan, M. A. 1983, *Astrophys. J.*, **268**, 582
Neugebauer,G. *et. al.* 1987. *Astron. Astrophys. Suppl.*, **63**, 615
Sun, W. H. and Malkan, M. A. 1989, *Astrophys. J.*, **346**, 68.

Non-Linear Emission-Line Response and the Structure of the Broad-Line Region in NGC 5548

Dan Maoz

Institute for Advanced Study, Princeton, NJ 08540, USA

The highly successful monitoring campaign of the Seyfert galaxy NGC 5548 in 1988-89 (Clavel *et al.* 1991, Peterson *et al.* 1991) resulted in the best sampled and largest spectral range AGN variability data available to date. Overall, the data strongly support the model in which the broad emission lines are reprocessed ionizing continuum energy, echoing from a region that is large compared to the continuum source. In this contribution I point out, however, some problems raised by these data: the details of the variability are more complicated than can be explained by the simplest analyses/models.

The emission lines respond to continuum variations with a lag which depends on the continuum variability timescale (Netzer and Maoz 1990). This requires the presence of a geometrically thick BLR such as is, in fact, indicated by the deconvolutions which assume linear response (Krolik *et al.* 1991, Krolik, these proceedings). However, these deconvolutions do not reproduce well the relative amplitudes of the outbursts in the high-ionization line light curves; the similar amplitudes of response to continuum bursts with very different total energy is strongly suggestive of a non-linear emission-line response (Fig. 1.).

Instead of the usually assumed linear response of a BLR cloud to the ionizing continuum ($F_{line} \propto F_{ion}$), I have developed and applied to the data a maximum entropy deconvolution algorithm which allows and solves for a non-linear response ($F_{line} = f(F_{ion})$), where f is a non-linear function approximated by a polynomial in F_{ion}. Fig. 2 shows that the relative amplitudes of the variations in the *CIV* λ1549 lightcurve are much better reproduced when non-linearity is allowed, without requiring a complicated transfer function. A fairly large degree of non-linearity is indicated ($a_2/a_1 = -0.3$ in the polynomial), suggesting the *CIV* λ1549 emission reaches saturation. Such behavior can arise either because the ionization front of the species reaches the rear side of the line-emitting cloud, or due to a softening of the ionizing continuum when it is in a high state.

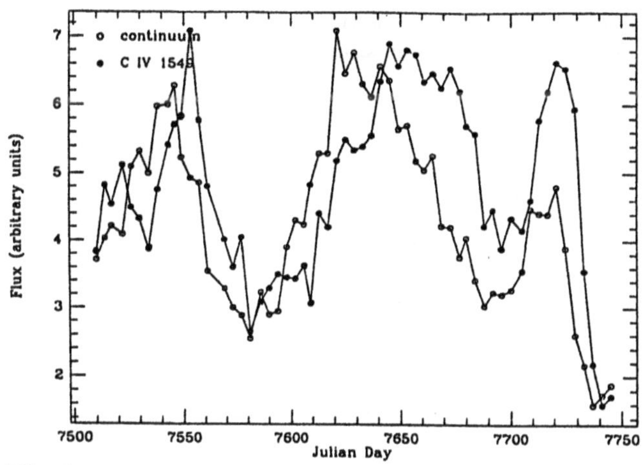

Fig. 1.

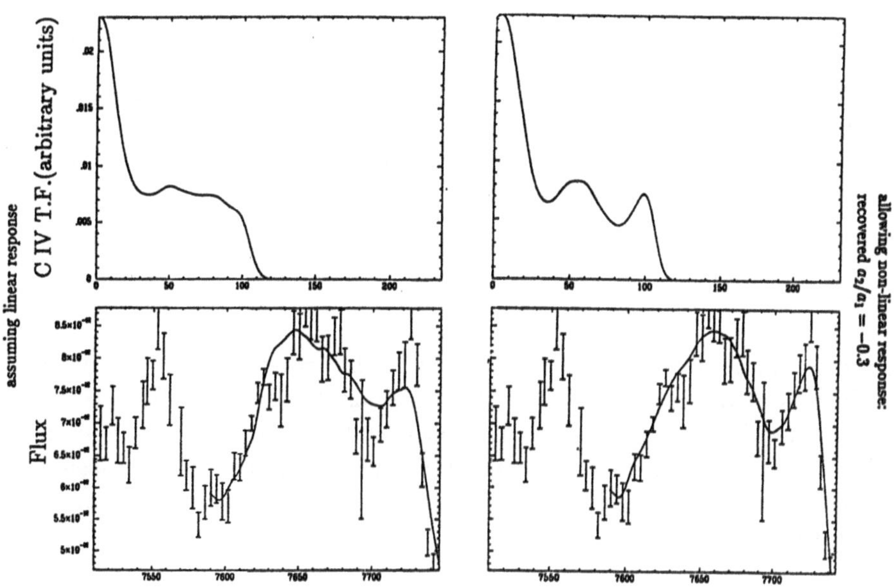

Fig. 2.

While the high-ionization line light curves are not well reproduced by the simplest models, the light curves of $H\beta$ and the low-ionization UV lines are too noisy to constrain well the geometry of their emission region. Horne, Welsh, and Peterson (1991, HWP) deconvolved the $H\beta$ light curve from the optical continuum light curve, and obtained a transfer function peaked at 20-days lag. Fig. 3, top-left panel, shows the transfer function resulting from a similar maximum-entropy calculation. The top right panel shows the $H\beta$ transfer function which results when the UV continuum light curve is used in the calculation. The transfer function is now peaked at 0 lag, and de-

Hβ transfer function obtained from maximum entropy deconvolution using:

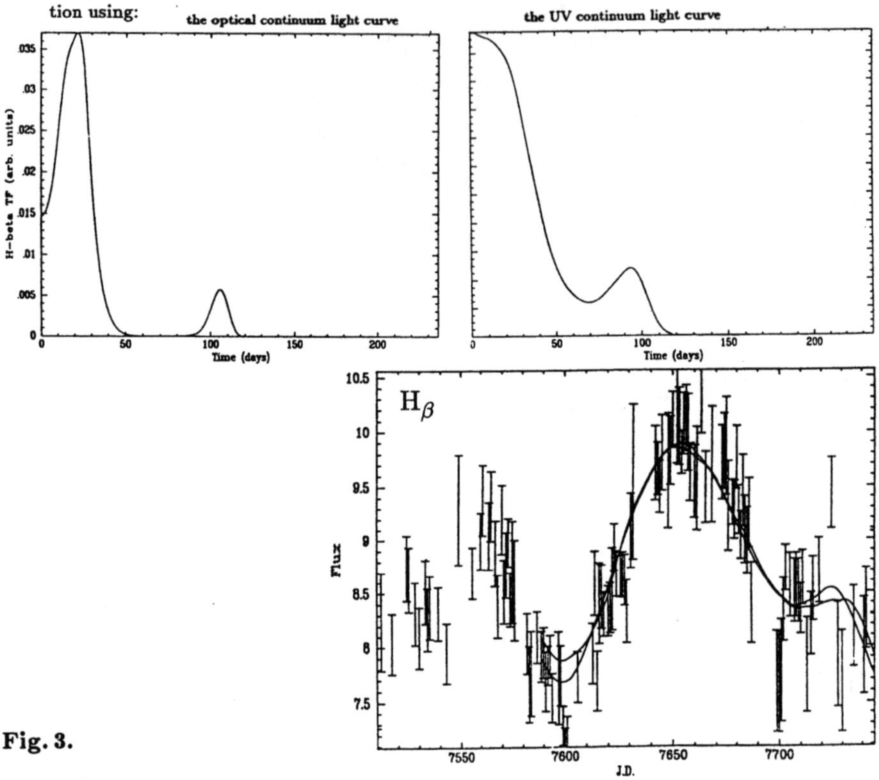

Fig. 3.

creases monotonically, and is quite similar to the transfer functions derived for the high-ionization UV lines. The bottom panel shows that both transfer functions can faithfully reproduce the observed light curve. The suggestions that $H\beta$ comes from a face-on disk (HWP) or that the $H\beta$ emission is anisotropic (Peterson, these proceedings) may therefore constitute an over-interpretation of the data.

References

Clavel, J., et al. 1991, *Astrophys. J.*, **366**, 64.

Horne, K., Welsh, W.F., and Peterson, B.M. 1991, *Astrophys. J. (Letters)*, **367**, L5.

Krolik, J.H., Horne, K., Kallman, T.R., Malkan, M.A., Edelson, R.A., and Kriss, G.A., 1991, *Astrophys. J.*, **371**, 541.

Netzer, H., and Maoz, D. 1990, *Astrophys. J. (Letters)*, **365**, L5.

Peterson, B.M., et al. 1991, *Astrophys. J.*, **368**, 119.

Observations of AGN with Peculiar or Very Broad Emission Line Profiles: OQ 208

P. Marziani[1], *M. Calvani*[2], *and J. W. Sulentic*[3]

[1] International School for Advanced Studies, Trieste, Italy
[2] Osservatorio Astronomico di Padova, Italy
[3] Department of Astronomy, University of Alabama, Tuscaloosa, USA

Abstract. Good prospect for finding evidence of line emission from an accretion disk around a supermassive black hole in AGN are Sey–1 galaxies and BLRG with the broadest and most irregular Balmer line profiles (*e.g.* Arp 102B, 3C 332). We are carrying out a survey aimed to obtain high resolution ($\approx$ 2 Å), high S/N spectra of these objects. Here we present some result on the BLRG OQ 208 ($\equiv$ 1404+28) observed on Febr. 19, 1991 with the Kitt Peak 2.1m telescope (Hβ, Fig. 1a), and on Apr. 4, 1990 with the ESO 1.5m telescope (Hα, Fig.1b).

1. Results

A strong variation (a factor $\approx$ 3) has occurred in the continuum level in the time elapsed between our observations of Hα in Apr. 1990 and of Hβ in Feb. 1991. A comparison with the spectra published by Osterbrock and Cohen (1979) and with those of Perez (1987, P87) confirms the variability and suggests that this object goes through a *high* and a *low* luminosity phase. The broad Hα profile appears double–peaked: two broad *shelfs* are present on the blue and the red side of the narrow component of Hα. The shelf on the red side is by far the most prominent. The radial velocity of A with respect to the Narrow Line Regions ($v_{r,A} \approx 3000$ km s^{-1}) is in good agreement with that measured by Cohen and Osterbrock (1979). In the Hβ profile, the *blue* shelf appears to be depressed with respect to the red one. Emission line profiles are strongly variable in shape. The red shelf is much stronger in our spectrum than in those shown by P87, when lines and continuum were fainter. Furthermore, one gets the impression that it has been displaced toward the red (the position of the of the centroid of Hβ measured on the spectra of P87 is marked in Fig.1b).

2. Discussion

The observation of a *red* shelf much stronger than the *blue* one disagrees with the predictions of simple relativistic disk models (*e.g.* Chen and Halpern, 1989). If *only* the *blue* shelf of Hα is interpreted as *pure* emission by an accretion disk, a relativistic disk model profile ($q = 3.2, R_{in} = 200R_g, R_{out} =$

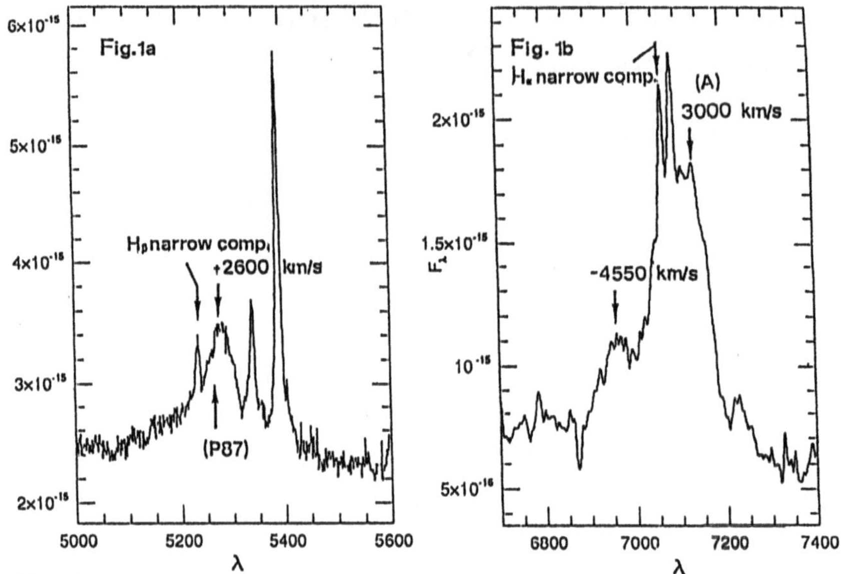

Fig. 1.

$1250R_g, i = 37°$) plus an additional redshifted component (needed to explain part of the *red* shelf) provides a satisfactory reproduction of the broad component of Hα. The enhancement of the Hβ red shelf and the increase of the radial velocity of its centroid with time and with continuum luminosity can be explained in several conflicting ways. First, the increase of v_r may be used as a supporting evidence for *infall* motion of the emitting gas. Second, it may be due to the orbital motion of a binary black hole; this hypothesis however is unlikely, since v_r increases with continuum luminosity. Third, the increase of v_r is suggestive of bipolar outflow driven by radiation pressure, but strongly against ballistic outflow. Concluding, we suggest that the most likely explanation for the complex profiles of OQ 208 involves either a rotating disk (heavily obscured) plus an infalling component or bipolar outflow of the gas (probably radiatively accelerated).

References

Chen K., Halpern J.P.: 1989, *Astrophys. J.*, **344**, 115
Perez E.: 1987, *Ph.D. Thesis* (P87)
Osterbrock D.E., Cohen R.: 1979, *Monthly Notices Roy. Astron. Soc.*, **187**, 61p.

LAG Optical Monitoring of QSO 1217+023

Paul T. O'Brien and Mike Goad

Department of Physics and Astronomy, University College London

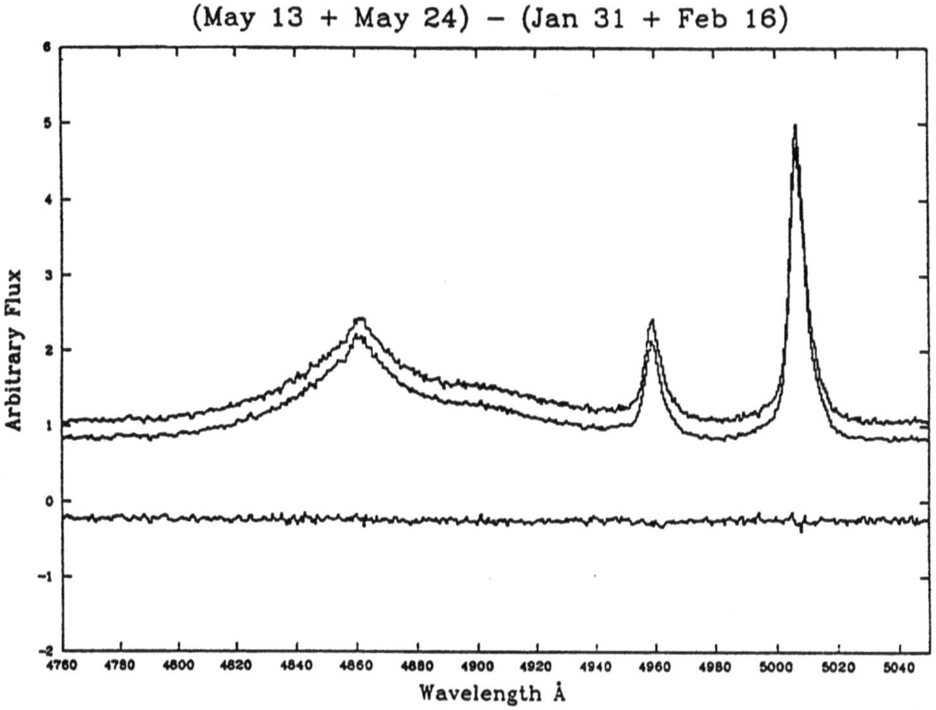

Fig. 1. Top: The merged Jan 31 and Feb 16 WHT rest-frame spectrum. Middle: The merged May 13 and May 24 WHT rest-frame spectrum. Bottom: The difference spectrum ((May 13 + May 24) - (Jan 31 + Feb 16)). The continuum level has decreased by ≈ 20 percent. There are no discernable changes in the emission lines.

As part of the LAG (Lovers of Active Galaxies) international time monitoring programme the QSO 1217+023 was observed at 16 epochs between January and June 1990 using the INT (12 epochs) and WHT (4 epochs) telescopes on La Palma. The spectra were extracted using a simple optimal extraction routine, put onto a relative flux scale using the narrow [OIII] $\lambda\lambda$4959, 5007 lines, and corrected to the rest-frame. We find no evidence for a significant change (> 4 - 8 percent) in the flux of Hβ during the monitoring period. The internal flux calibration is accurate to ≈ 4 percent. There is an ≈ 20 percent drop in the continuum level during late March/early April 1990, as illustrated in Figure 1. There is a broad red component in the Hβ line profile. The LAG spectra and previous published spectra show no evidence for significant amounts of FeII emission from this quasar, and the red feature is displaced too far from the narrow [OIII] lines to be associated with broad [OIII] emission. We note that the narrow Hβ line is at the same redshift and has the same profile as the broad Hβ line. The nature of this feature is under investigation.

Synthetic Photometry of AGN in the Range 50-200 Å

Astrid Orr and Thierry J.-L. Courvoisier

Observatoire de Genève, CH-1290 Sauverny, Switzerland

Abstract. Different models of AGN soft excess emission are used to compute flux ratios between 8 wavebands corresponding to the future detectors of the Extreme UV Imaging Telescope Array (EUVITA). Detector characteristics and interstellar photoelectric absorption are taken into account. The simulations show that ratios of fluxes in the different spectral bands allow the measurement of physical parameters such as temperatures and absorbtion column densities. Effects due to line emission are considered in the case of optically thin plasma models.

1. EUVITA

Up to now few observations have been made in the extreme UV and soft X-ray bands due to high interstellar absorption and lack of instrumentation. The EUVITA experiment will be flown on the Soviet Spectrum-X-G satellite. The current design consists of 8 normal-incidence-reflection multilayer coated mirrors of 20 cm ⊘. Each mirror is sensitive in a different waveband between 50 and 175 Å, with $\Delta\lambda/\lambda$ equal to a few %. The detectors are microchannel plates with position sensitive readout. Each telescope has an effective area of a few square centimeters. The spatial resolution will be ~ 10 arcsec in a field of view of $\sim 1°$ (for a description of the instruments see Zehnder et al., 1990). Typical exposures will be of $\sim 10^5$ seconds. This experiment is planned to cover a wide range of scientific objectives, including study of the "soft excess" component observed in several AGN (Turner and Pounds, 1989). We have calculated the expected count rates in the different EUVITA wavebands for a minimal soft X-ray emission model of 3C 273 based on Courvoisier et al. (1990), ($z = 0.158$, $H_0 = 50\,\mathrm{km\,s^{-1}}$). They are between $2\,10^{-5}$ (at 130 Å) and $1.2\,10^{-3}$ counts/sec (at 70 Å) for $N_H = 1.8\,10^{20}\ \mathrm{cm}^{-2}$. We therefore expect a clear detection in all bands for observing times larger than 10^5 seconds.

2. AGN spectrum models

We present simulations of the EUVITA instruments, for different soft excess emission models possibly applicable to AGN (black body, power law and optically thin plasma models). We include instrumental parameters and interstellar absorption according to photoelectric cross sections given by Morrison & McCammon (1983). For each model we obtain a contour plot giving flux ratios between EUVITA wavebands for different source parameters (see Fig. 1 for an example). These ratios prove to be good indicators of **both** the source parameters and the amount of absorbing galactic hydrogen.

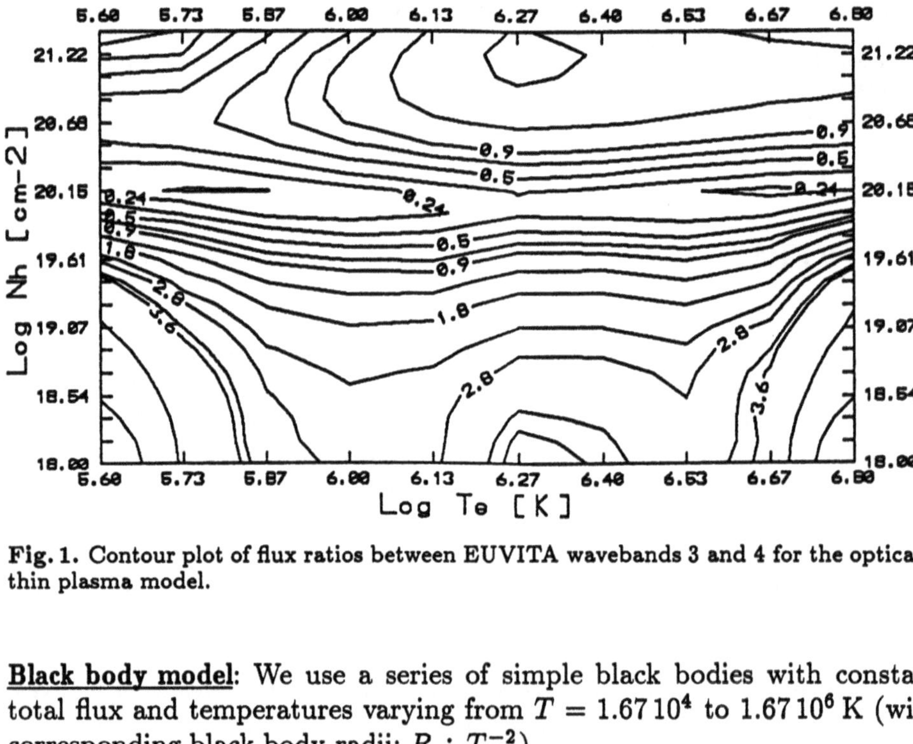

Fig. 1. Contour plot of flux ratios between EUVITA wavebands 3 and 4 for the optically thin plasma model.

Black body model: We use a series of simple black bodies with constant total flux and temperatures varying from $T = 1.67\,10^4$ to $1.67\,10^6$ K (with corresponding black body radii: $R \div T^{-2}$).

Power law model: We consider power laws described by: $f(\nu) \sim \nu^{-\alpha}$ with α indices between 0.1 and 1.5 . Flux ratios of EUVITA detectors give a poor measurement of the index, but are sensitive to interstellar absorption.

Optically thin plasma model: R. Mewe et al. (1985) have computed continuum emission and line power output for an optically thin plasma in ionization equilibrium for several electron temperatures T_e. This type of plasma is found in stellar coronae. In the case of AGN the soft excess emission process is unknown, an important contribution might however be expected from a hot corona surrounding the accretion disc (L. Maraschi, 1990). Line profiles were not taken into account since natural broadening and thermal Doppler line broadening are small compared to the bandwidths ($\Delta\lambda \sim 10^{-4}$ [Å] and $\Delta\lambda/\lambda \sim 10^{-4}$ respectively). However, a macroscopic velocity field of 3000 [km/s] would produce a Doppler broadening of $\Delta\lambda/\lambda \sim 10^{-2}$. This is no longer negligable, especially for the narrowest waveband at 50 Å. Flux ratios between pairs of EUVITA detectors were calculated for 13 electron temperatures between $\log(T) = 5.6$ and 6.8.

References

Courvoisier, T.J.-L., et al., 1990: *Astron. Astrophys.*, **234**, 73.
Maraschi, L., 1990: *IAU Coll.*, **129**,283.
Mewe, R., et al., 1985: *Astron. Astrophys. Suppl.*, **62**, 197.
Morrison, R., McCammon, D., 1983: *Astrophys. J.*, **270**, 119.
Turner, T.J., Pounds, K.A., 1989: *Monthly Notices Roy. Astron. Soc.*, **240**, 833.
Zehnder, A., et al., 1990: *Proc. SPIE*, 132-143.

Composite Models of the Broad Line Region

P. Rafanelli and M. Radovich

Department of Astronomy, University of Padova, Italy

The *Broad Line Region* of AGN is often described as an ensemble of gaseous clouds photoionized by a strong central source. Sophisticated codes have been developed in order to discuss the physical processes involved and to compute the emission line intensities using the so–called 'photoionization models' (Kwan & Krolik 1981). A single–valued hydrogen density $n_H \sim 10^{9.5}$ cm^{-3} is generally adopted, because of the absence of any 'broad' forbidden line and the presence of the 'broad' semi–forbidden line CIII] λ1909 ('standard' *single–cloud* models).

We present here some preliminary results we obtained considering a *multi–density* BLR, characterized by the presence of a high density component in addition to the 'standard' one. The density of such component was chosen in the range $n_H \sim 10^{11} - 10^{12}$ cm^{-3}, since for higher values the correct handling of many physical processes is not assured by the available photoionization codes. We assumed that the two cloud populations are located at the same distance from the central source, and that they are illuminated by the same continuum: this is acceptable, if the covering factor of the clouds is small. With such assumptions, if $U = Q_H/4\pi R^2 n_H c$ is the ionization parameter of a cloud (where Q_H is the number of hydrogen ionizing photons emitted per unit time, R the distance between the cloud and the central source), we have that $U_2 = U_1 n_{H_1}/n_{H_2}$. Other multi–density photoionization models of the BLR, with different assumptions about the cloud configuration, have been computed by Netzer (1987), Rees, Netzer & Ferland (1989), and Collin–Souffrin & Dumont (1990).

In fig. 1 we compare the observed emission line ratios given by Wu, Boggess & Gull (1983) with those produced by both single–cloud and composite models; in our computations we adopted the continuum shape discussed by Mathews & Ferland (1987) and Ferland & Persson (1989). We see that the **single–cloud** models do not give a good fit with the observed Lyα/Hβ and MgIIλ2798/CIVλ1549 ratios; however, a better fit is obtained when we add some reddening by dust, with $E_{B-V} = 0.2$ mag (DeZotti & Gaskell 1985). In the **composite** models, the presence of the very high density component leads to an enhancement of the Balmer lines and of the low ionization lines as MgIIλ2798. We have now a significative improvement in the fitting of the Lyα/Hβ and MgII/CIV observed ratios, while the agreement is not so good for the MgII/Hβ ratio: this could mean that our models

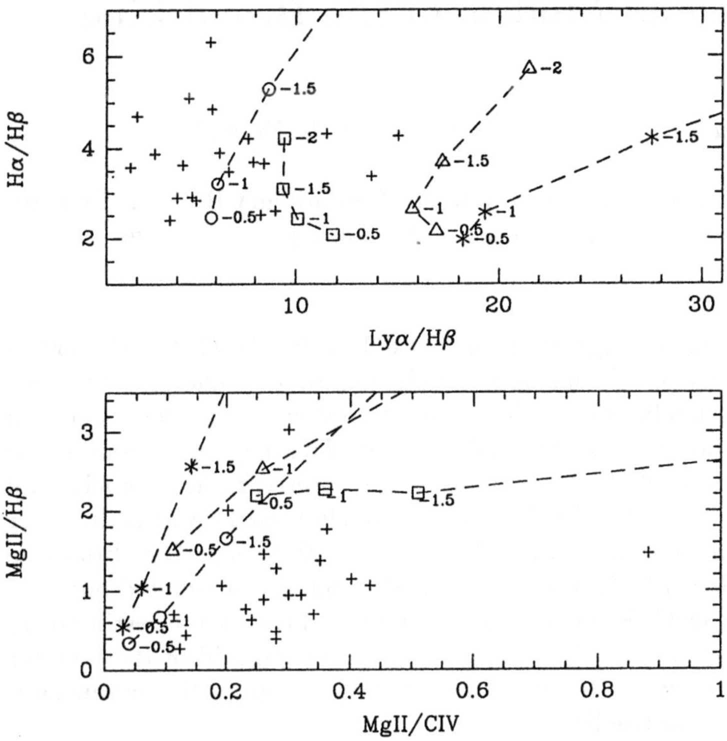

Fig. 1. Emission line ratios. The logs of the ionization parameters are indicated for each model; for the multi-density models, they are those of the lower density component. +: Emission line ratios given by Wu, Boggess & Gull (1983); $*$: Not reddened single-cloud models ($n_H = 10^{9.5}$ cm^{-3}); o: Reddened ($E_{B-V} = 0.2$ mag) single-cloud models; Δ: Multi-density models ($n_{H_1} = 10^{9.5}$, $n_{H_2} = 10^{11}$ cm^{-3}); ⊔: Multi-density models ($n_{H_1} = 10^{9.5}$, $n_{H_2} = 10^{12}$ cm^{-3})

are still too simple, but better UV data are required. We wish to thank Dr. Gary Ferland, who kindly sent us his photoionization code **Cloudy 76**.

References

Collin-Souffrin S., Dumont S., 1990, *Astron. Astrophys.*, **229**, 313.
De Zotti G., Gaskell C.M., 1985, *Astron. Astrophys.*, **147**, 1.
Ferland G.J., Persson S.E., 1989, *Astrophys. J.*, **347**, 656.
Kwan J., Krolik J., 1981, *Astrophys. J.*, **250**, 478.
Mathews W.G., Ferland G.J., 1987, *Astrophys. J.*, **323**, 456.
Netzer H., 1987, *Monthly Notices Roy. Astron. Soc.*, **225**, 55.
Rees M.J., Netzer H., Ferland G.J., 1989, *Astrophys. J.*, **347**, 640.
Wu C.C., Boggess A., Gull T.R., 1983, *Astrophys. J.*, **266**, 28.

Discrete Cloud Models of the Broad Line Region

Andrew Robinson[1], *Enrique Pérez*[2], *and Luis de la Fuente*[2]

[1] Institute of Astronomy, Madingley Road, Cambridge CB3OHA, England
[2] Instituto de Astrofisica de Canarias, 38200 La Laguna, Tenerife

Following the original suggestion of Gaskell & Sparke (1986) the cross-correlation function (CCF) has been widely employed to analyse variations of the broad emission lines in AGN, usually to determine the delay between the continuum and emission line light curves. This lag, corresponding to the shift in the peak of the CCF, is often interpreted rather vaguely as a measure of the "size" of the BLR but it is strongly biased towards the inner regions if the BLR is radially extended (Edelson & Krolik 1988; Robinson & Perez 1990). Recently, Koratkar & Gaskell (1991) have pointed out that the centroid of the CCF should yield a more representative estimate of the BLR size since, in principle, it is equal to the centroid of the transfer function which, in turn (Robinson & Perez 1990), gives the luminosity-weighted radius, R, of the BLR.

In this note, we summarize an investigation (Perez, Robinson & de la Fuente 1991) of the reliability of the CCF as a means of measuring the BLR size. This is based on numerical simulations of the response of an ensemble of photoionized clouds to a 'realistic' continuum light curve. The clouds are linear reprocessors, radiate isotropically and are distributed within a spherical shell centred on the continuum source. The ionizing continuum light curve is based on B-band measurements of NGC 4151 compiled by Gill et al. (1984).

Response light curves were computed for a series of BLR models characterized by different values of R (see Robinson, Perez & de la Fuente 1991). A 'windowing analysis' was then applied in which the light curves (each of total length 4000d) are scanned with a sampling window of a given width. At each timestep, the CCF is computed for the sections of the line and continuum light curves which fall within the window and the displacement of the peak (lag, Δt) and centroid (τ) of the function are determined. Frequency distributions for these quantities are thus obtained for several different window widths.

The frequency distributions are presented and discussed in detail by (Perez, Robinson & de la Fuente 1991). Here we highlight the most important points which emerged from the analysis.

* For narrow sampling windows (~100d) broad distributions of Δt and t are obtained whose median values are considerably smaller than the expected values (i.e., the light crossing time, t_0, at the inner radius of the

cloud shell and R/c, respectively). The probability of either Δt or t coinciding with the expected values in any given segment of the light curve is very small.

* There is substantial overlap, for narrow windows, between the distributions produced for different values of R and hence a significant probability that the value of Δt or τ obtained for a larger BLR (e.g., R=200ld) will coincide with the value obtained for the R=25 model.

* As the window width is increased, the distributions gradually converge to the expected values. The convergence is more rapid for small R; in the R=25ld case, a sampling window of ~1000d is required for a 50% chance of obtaining $\Delta t = t_0$ or τ=R/c, respectively.

Note that the scatter in either the lag or centroid is not caused by CCF's of low significance. Only CCF's whose peak exceeds a significance threshold of 75% were included in the frequency distributions shown in Fig.1 but in fact, the choice of threshold does not significantly alter the results of the analysis.

We conclude that whereas, in principle, the displacement of the CCF peak corresponds to the inner radius of the cloud system while its centroid corresponds to the luminosity-weighted radius, in practice, the CCF is sensitive to the size of the sampling window and its location within the light curve. In particular, for small sampling windows (in this context, ~100d), there is a significant probability that the CCF lag and centroid will deviate substantially from the *expected* values. Thus, these results imply that BLR size scales determined from the cross-correlation function may be subject to a substantial uncertainty which would be difficult to quantify from most current data-sets.

References

Edelson, R. & Krolik, J., 1988: *Astrophys. J.*, **333**, 646.
Gaskell, C.M. & Sparke, L.S., 1986: *Astrophys. J.*, **305**, 175.
Gaskell, C.M. & Peterson, B.M., 1987: *Astrophys. J. Suppl.*, **65**, 1.
Gill, T.R., Lloyd, C., Penston, M.V. & Snijders, M.A.J., 1984: *Monthly Notices Roy. Astron. Soc.*, **211**, 31.
Koratkar, A.P. & Gaskell, C.M., 1991: *Astrophys. J. Suppl.*, **75**, 719.
Perez, E., Robinson, A. & de la Fuente, L., 1991: *Monthly Notices Roy. Astron. Soc.*, submitted.
Robinson, A. & Perez, E., 1990: *Monthly Notices Roy. Astron. Soc.*, **244**, 138.
Robinson, A., Perez, E. & de la Fuente, L., 1991: in: *Variability of Active Galaxies*, Duschl, W.J., Wagner, S.J. & Camenzind, M. (Eds.), Springer-Verlag.

A Model for the Correlated Line and Continuum Variations in NGC 5548

Evlabia Rokaki[1,2] *and Christian Magnan*[1,3]

[1] Institut d'Astrophysique de Paris, France
[2] DAEC, Observatoire de Paris-Meudon, Meudon, France
[3] Collège de France, Paris

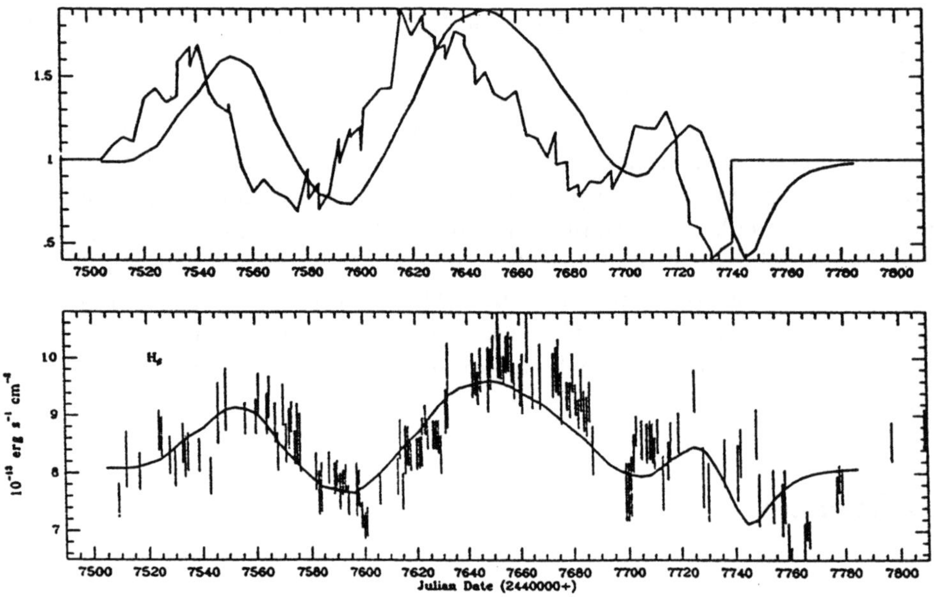

Fig. 1. 1a: Normalized X-ray light curve taken as input and the computed time delayed Hβ line fluxes; 1b: Theoretical Hβ light curve compared to the observations [8]. A constant value of 4.3 10^{-13} erg s^{-1} cm^{-2} has been added to the computed fluxes.

We consider the accretion disc model developed by Collin-Souffrin and Dumont [1-4] based on the theory of an α-disc [5].This model assumes that the disc fueling the central massive black hole is illuminated by a X-ray source located at a given height (A) above the disc center. The reprocessing of the X-ray in the inner part of the disc ($\sim 100R_g$, where R_g=2GM_{BH}/c^2, M_{BH} the mass of the black hole) gives rise to a portion of the UV and visible continuum radiation [6], while the outer region emits the characteristic lines of a photo-ionized medium. The line emission responds to changes in the ionizing radiation with a certain time-lag due to the light-travel delay.

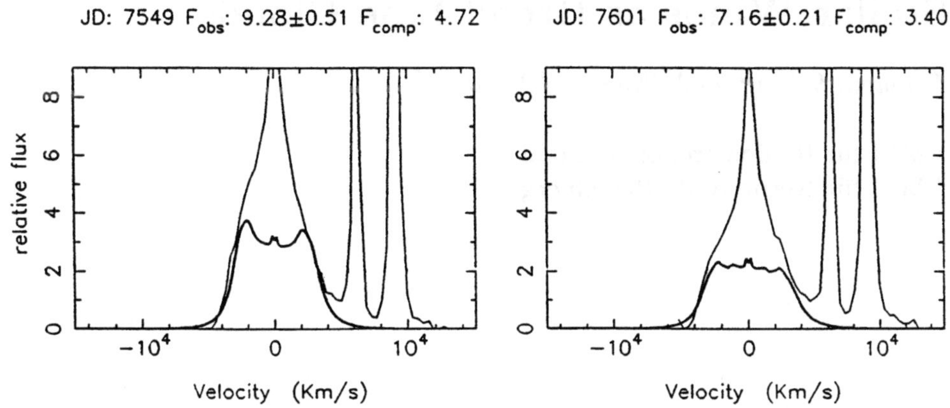

Fig. 2. Hβ calculated (thick line) and observed (thin line) profiles from [9] at two Julian Dates.The integrated fluxes are expressed in units of 10^{-13} erg s^{-1} cm^{-2}.

This time of response is a direct measurement of the spatial scale of the line emitting region. We infer the X-ray variations from the observed continuum variations at 1350 Å [7] by supposing that the X-ray and the continuum follow the same variation pattern, i.e. are linearly dependent, but are shifted in time with a 5.5 day lag (Fig. 1a). The X-ray luminosity seen by the disc varies between 0.4 and 1.9 10^{44} erg s^{-1} corresponding respectively to the low and high state. By taking into account the geometrical factors governing the paths of the continuum and the line radiation reaching the observer, we compute the Hβ line intensities as a function of time (fig.1b). One sees that the fit between the observational data [8] and the theoretical modelling is quite satisfactory. In fig. 2 we show Hβ computed line profiles for A = 500 R_g, $M_{BH} = 5.5\ 10^7\ M_\odot$ and an inclination of the disc, i=35°. The line intensities account for about 50 % of the observed values, so that a part of the line emission should come from another region.

References

[1] Collin-Souffrin, S., Dumont, A.M., 1990: *Astron. Astrophys.*, **229**, 292.
[2] Dumont, A.M., Collin-Souffrin, S., 1990: *Astron. Astrophys.*, **229**, 302.
[3] Dumont, A.M., Collin-Souffrin, S., 1990: *Astron. Astrophys.*, **229**, 313.
[4] Dumont, A.M., Collin-Souffrin, S., 1990: *Astron. Astrophys. Suppl.*, **83**, 71.
[5] Shakura, R.I. and Sunyaev, R.A., 1973: *Astron. Astrophys.*, **24**, 337.
[6] Collin-Souffrin, S., 1991: *Astron. Astrophys.*, in press.
[7] Clavel, J. et al., 1991: *Astrophys. J.*, **366**, 64.
[8] Peterson et al., 1991: *Astrophys. J.*, **368**, 119
[9] Kollatschny, W., Dietrich, M., 1991: in: *Variability of Active Galactic Nuclei*, ed. H.R. Miller & P.J. Wiita, Cambridge University Press, p. 325

Stellar Winds as Broad Line Clouds

Reinhold Schaaf[1] *and Thomas Schmutzler*[2]

[1] MPI für Radioastronomie, Bonn
[2] MPI für Kernphysik, Heidelberg

Gas in stellar winds near the central engines of AGN was repeatedly proposed to be the origin of the broad emission lines (e.g. Edwards 1980, Kazanas 1989). This concept avoids several major difficulties of other models for the broad line clouds, e.g. the problems of formation and confinement of the emitting material. We want to address the questions, whether the intense continuum radiation of an AGN will drive winds from stars inside the broad line region, and whether the observed emission lines are formed in these winds.

For this purpose we solved the hydrodynamical equations for a spherical symmetric, stationary stellar wind that is heated by an AGN continuum. We neglected effects of optical depth and of radiation forces, but treated heating and cooling of the gas by the illuminating radiation in great detail, using a time-independent version of the thermodynamic program described by Schmutzler and Tscharnuter (this volume). The mass-loss rate of the star was treated as a free parameter at the present stage of the calculations.

Figure 1 shows the temperature and the Mach number for winds with a range of mass-loss rates from a $1\,M_\odot$ star, which is illuminated by a 3C 273-like continuum (adopted from Bezler *et al.* 1984) of energy flux $8\times10^8\,\mathrm{erg/cm^2\,s}$. Close to the star the gas is dense enough to cool efficiently, hence its temperature is relatively low ($T < 10^5\,\mathrm{K}$). In this region, the flow is mainly subsonic ($v \approx 1\,\mathrm{km/s}$). When the gas becomes sufficiently tenuous, cooling becomes inefficient and the material is heated up to temperatures $T = 10^6 \ldots 10^7\,\mathrm{K}$ on a timescale which is very short compared with the dynamical timescale of the flow. At the same instance the flow is accelerated to supersonic velocities ($v \approx 100 \ldots 1000\,\mathrm{km/s}$).

The physical conditions in the dense part of the wind are favourable for the formation of AGN emission lines: The temperature is of the order $10^4\,\mathrm{K}$ and the density is higher than $10^8\,\mathrm{cm^{-3}}$, increasing towards the star. The column density of the material with $T < 3\times10^4\,\mathrm{K}$ and $n_H < 10^{10}\,\mathrm{cm^{-3}}$ is $10^{22} \ldots 10^{23}\,\mathrm{cm^{-2}}$ for all mass-loss rates considered here, consistent with photoionization models of the BLR.

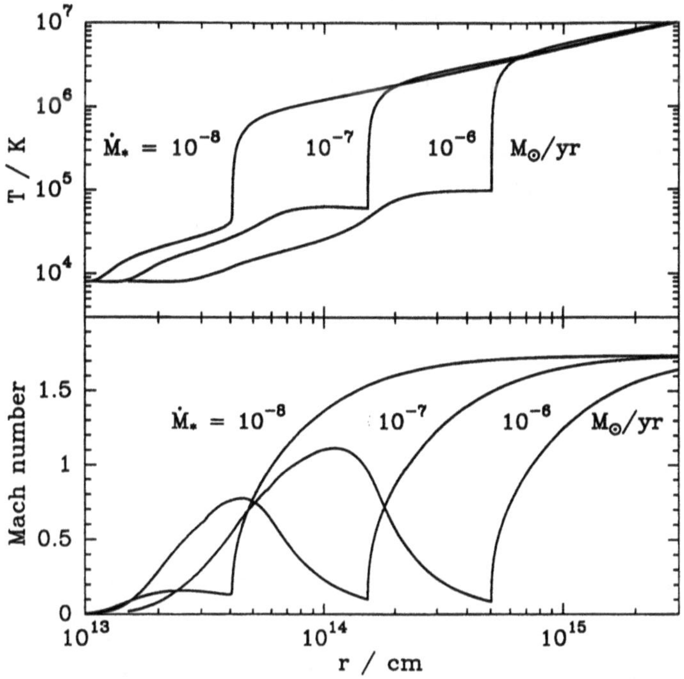

Fig. 1. Temperature and Mach number in the stellar wind from a $1\,M_\odot$ star which is illuminated by a 3C 273-like continuum of energy flux $8 \times 10^8\,erg/cm^2\,s$. Solutions are labelled with the mass-loss rate, which is a free parameter of the model.

Because both the density ($n_H = 10^{10} \ldots 10^{12}\,cm^{-3}$) and the low velocity ($v < 1\,km/s$) of the wind at $r \approx 10^{13}$ cm allow to match the wind solutions to stationary atmospheres of red giants, a consistent picture of broad line clouds emerges: The continuum radiation of the AGN heats the atmospheres of red giants, *driving thereby the winds, in which the emission lines are formed.* The number of stars necessary to account for the observed line fluxes turn out to be moderate: For an AGN with luminosity 10^{44} erg/s, covering factor of the broad line clouds 1%, and radius of the broad line region of 10^{17} cm 1600 stars, each with mass-loss rate $10^{-6}\,M_\odot$/yr, or 2.5×10^5 stars with $10^{-8}\,M_\odot$/yr are needed.

References

Bezler, M., *et al.* 1984, *Astron. Astrophys.*, **136**, 351.
Edwards, A.C., 1980: *Monthly Notices Roy. Astron. Soc.*, **190**, 757.
Kazanas, D., 1989: *Astrophys. J.*, **347**, 74.

Timescales of Phase Transitions in BLRs

Thomas Schmutzler[1,2] *and Werner M. Tscharnuter*[1,2,3]

[1] Max-Planck-Institut für Kernphysik, Heidelberg
[2] SFB 328 "Entwicklung von Galaxien", Heidelberg
[3] Institut für Theoretische Astrophysik der Universität Heidelberg

1. Scenario

We discuss the relevant timescales of phase transitions within the frame work of the two-phase model for the BLR/NLR of AGNs and study the required model parameters for BLR cloud formation in hot AGN winds due to bow shocks induced by obstacles like SNRs or stars with strong winds. Effective cooling in the downstream region of the shock with a high stagnation pressure causes phase transitions from hot ($T \simeq 10^{6-9}$ K), thin ($N \leq 10^6\,\mathrm{cm}^{-3}$) gas to cool ($T \simeq 10^4$ K), dense ($N \geq 10^8\,\mathrm{cm}^{-3}$) filaments (cf. Perry and Dyson 1985).

Physics. Our calculations include all relevant atomic processes in a gas composed of the 10 most abundant chemical elements. These processes are collisional ionization incl. autoionization, charge exchange, radiative and dielectronic recombination; photoionization incl. Compton ionization, Auger effect and ionization by suprathermal electrons. Heating and cooling are due to ionization, Compton effect, recombination, thermal bremsstrahlung, line excitation and two-photon emission (Schmutzler 1987).

2. Method

The set of equations to be solved comprises energy balance, time dependent state of ionization, charge conservation, and the equation of state. Moreover, we have to prescribe the time variations of the pressure (or density or temperature) according to the timescales of the scenario briefly indicated above. This system of 103 ordinary differential equations (together with 14 algebraic relations) is well known to be a stiff one and requires an implicit method for an efficient solution.

3. Results

We use different shapes for the illuminating continuum, but assume a total luminosity of 10^{47} erg s^{-1} in all models and a distance of the obstacle from the centre of the AGN of 1 pc. The wind at this distance is characterized by its hydrogen density $N_H/[cm^{-3}]$, equilibrium temperature $T_{eq}/[K]$ and wind velocity $V/[cm\,s^{-1}]$. These parameters and the size ($d_0/[pc]$) of the obstacle determine the time interval for the shocked gas staying at high pressure and cooling radiatively (Perry and Dyson 1985). Afterwards we assume an exponential pressure decrease until the thermal pressure of the wind has been reached. We present the results of four models:

1) Continuum: "standard broken power law" (Krolik *et al.* 1981).

a) $N_H = 3.9\,10^3$, $T_{eq} = 1.2\,10^8$, $V = 5.0\,10^8$, this corresponds to a Mach number $M_a \simeq 3.0$ and a mass loss rate $\dot{M} \simeq 8.4\,M_\odot\,yr^{-1}$ for a spherically outflowing wind. With $d_0 = 0.1$ we find that the flow stays $2.5\,10^9$ s at high pressure. Within this time the shocked gas can cool only to $T = 2.0\,10^8$; thus a phase transition is impossible.

b) $N_H = 3.0\,10^4$, $T_{eq} = 1.2\,10^8$, $V = 1.5\,10^9$, ($M_a \simeq 9.2$ and $\dot{M} \simeq 194\,M_\odot\,yr^{-1}$(!)). With $d_0 = 1.0$(!) the flow stays $8.3\,10^9$ s at high pressure. The shocked gas cools to $T = 1.0\,10^4$ within $5.0\,10^9$ s; thus a phase transition occurs.

2) Continuum: broken power law fitted to the observed spectrum of 3C 273 (Bezler *et al.* 1984). $N_H = 1.0\,10^4$, $T_{eq} = 7.5\,10^8$, $V = 1.5\,10^9$, ($M_a \simeq 3.6$ and $\dot{M} \simeq 65\,M_\odot\,yr^{-1}$). With $d_0 = 0.1$ the flow stays $8.3\,10^8$ s at high pressure. Within this time the shocked gas can only cool to $T = 2.0\,10^9$; thus no phase transition occurs. Here we also have calculated the timescale of evaporation starting at $T = 1.0\,10^4$ and $N_H = 6.5\,10^7$ assuming isobaric heating after a small rarefaction. $T = 1.0\,10^5$ has been achieved within $1.4\,10^5$ s, $T = 1.0\,10^6$ within $2.0\,10^6$ s, and $T = 1.0\,10^7$ within $2.2\,10^7$ s.

3) Continuum: broken power law fitted to the average continuum for radio-loud quasars (Sanders *et al.* 1989). $N_H = 1.0\,10^4$, $T_{eq} = 1.3\,10^6$, $V = 8.0\,10^8$ ($M_a \simeq 47$ and $\dot{M} \simeq 35\,M_\odot\,yr^{-1}$). With $d_0 = 0.1$ the flow stays $1.6\,10^9$ s at high pressure. Although the shocked gas can cool only to $T = 2.6\,10^8$ within this time, a phase transition occurs because the gas can cool furthermore to $T = 2.0\,10^4$ within the next $3.4\,10^9$ s whilst pressure decreases.

Conclusion. It is easy to evaporate filaments, however, it is hard to find convincing model parameters to produce line emitting filaments by phase transitions: Compton heating reduces the effective cooling rate at temperatures near the hot phase. Therefore the cooling time is longer than Perry and Dyson (1985) estimated.

References

Bezler *et al.*: 1984, *Astron. Astrophys.*, **136**, 351
Krolik *et al.*: 1981, *Astron. J.*, **249**, 422
Perry, J.J., Dyson, J.E.: 1985, *Monthly Notices Roy. Astron. Soc.*, **213**, 665
Sanders *et al.*: 1989, *Astrophys. J.*, **347**, 29.
Schmutzler, T.: 1987, Ph.D. thesis, Bonn University.

The Baldwin Effect and the Shape of the Ionizing Continuum

Hartmut Schulz

Astronomical Institute, Ruhr University Bochum, Postfach 102148, D-4630 Bochum, Germany

1. Introduction

Albeit with large scatter, the equivalent width W(CIV) of the emission line CIV1550 is known to decrease with increasing continuum luminosity at the line center (Baldwin 1977, Baldwin et al. 1978, Kinney et al. 1990). There is no generally accepted explanation for this so-called "Baldwin-effect".

Another crude luminosity dependency is given by the apparent gradient between the optical and x-ray region: high-luminosity objects (QSOs) are x-ray fainter than low-luminosity objects (Seyferts) (Zamorani et al. 1981).

This raises the question: are both luminosity indicators related, or more precisely: Is W(CIV) weakened in high-luminosity objects because the intensity of CIV1550 is diminished due to a relative lack of C^{3+} generating photons?

Lacking suitable data, rather than directly addressing the Baldwin effect, we investigate the presumably equivalent luminosity dependence of the emission-line intensity ratio CIV/Lyα.

2. Method

UV - x ray gradients α(UX) (interpolated power-law slope between 145nm and 0.5 kev), CIV/Lyα and CIV/CIII are determined for a sample of Seyfert galaxies and a QSO sample of intermediate redshifts (Kinney et al. 1985, 1987; Kriss et al. 1980, Padovani and Rafanelli 1988, Tananbaum et al. 1986, Wu et al. 1983). Mean values are -1.39±0.08 and -1.78±0.04 for α(UX), 0.75±0.12 and 0.47±0.04 for CIV/Lyα, and 5.7±0.8 and 2.7±0.5 for CIV/CIII, respectively.

Single-component broad-line region photoionization models employing the observed UV-x ray slopes are computed to predict CIV/Lyα and

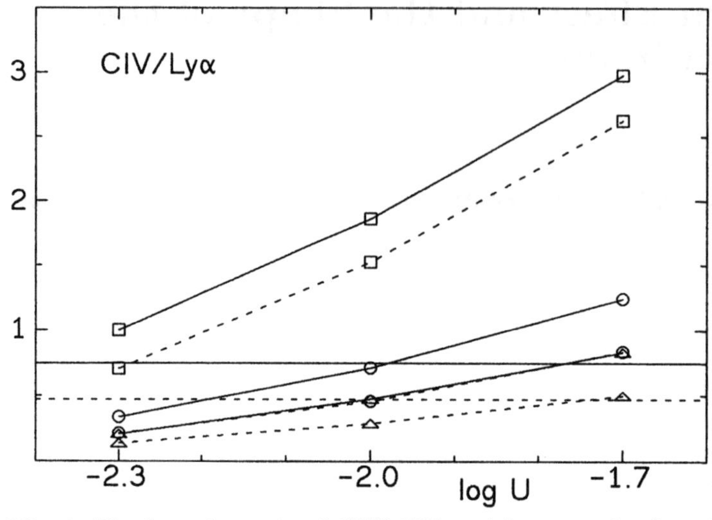

Fig. 1. The intensity ratio of CIVλ1550 and Lyman α is plotted versus log(ionization parameter)(see text). Continuous lines: Seyferts; dashed lines QSOs. Horizontal lines represent empirical line ratios.

CIV/CIII. We assume a column density (in CGS units) of log N = 23, hydrogen number densities log n = 9, 10 and 11 (triangles, circles and squares in Fig.1) and ionization parameters (Ferland 1990) log U = -2.3, -2.0 and -1.7.

Computed and observed line ratios are compared.

3. Results

Plain power-law interpolations between the UV and x-ray give consistent fits in CIV/Lyα (shown in fig.1) and CIV/CIII (not shown) with log U = -2 and log n = 10 for both Seyferts and QSOs.

A small UV-bump (peak at 3.5 ryd) does fit as well, but an extended UV bump (peak at 20 ryd) would require different ionization parameters for Seyferts and QSOs.

Changes of the continuum in the x-ray between 0.5 and 4.5 kev do not significantly affect the line ratios considered here (harder photons are even less important because the cloud is transparent beyond 4.5 kev).

4. Conclusion

Our results support the view that the Baldwin effect may be simply caused by the steepening of the ionizing continuum with luminosity.

References

Baldwin, J.A., 1977, *Astrophys. J.*, **214**, 679

Baldwin, J.A., Burke, W.L., Gaskell, C.M., Wampler, E.J., 1978, *Nature*, **273**, 431

Ferland, G.J., 1990, CLOUDY 76.03, OSU Internal Report 90-02

Kinney, A.L., Huggins, P.J., Bregman, J.N., Glassgold, A.E., 1985, *Astrophys. J.*, **291**, 128

Kinney, A.L., Huggins, P.J., Glassgold, A.E., Bregman, J.N., 1987, *Astrophys. J.*, **314**, 145

Kinney, A.L., Rivolo, A.R., Koratkar, A.P. 1990, *Astrophys. J.*, **357**, 338

Kriss, G.A. et al., 1980, *Astrophys. J.*, **305**, 57

Padovani, P., Rafanelli, P., 1988, *Astron. Astrophys.*, **205**, 53

Wu, C.-C., Boggess, A., Gull, T.R., 1983, *Astrophys. J.*, **266**, 28

Zamorani, G. et al., 1981, *Astrophys. J.*, **245**, 357.

A Thermal Model of the Continuum Variability of GQ Comae (PG 1202+281)

Aneta Siemiginowska[1] *and Michael L. Sitko*[2]

[1] Harvard-Smithonian Center for Astrophysics, Cambridge
[2] Univ. of Cincinnati

GQ Comae (PG 1202+281) is a radio-quiet quasar with observed variability in the optical/UV/IR part of the spectrum in the 1984–1987 (Sitko 1986, Proc of a workshop in Tuscon; Sitko et al. in preparation). We present (Fig. 1) the sequence of the 7 different observational nights on the color–color diagram (1 year of observations). During this time an increase and decrease of the UV luminosity was observed. The color–color variations show an almost straight line on the $log(IR/V)$ vs. $log(V/UV)$ plot. The differences from that line are visible when the time step between the observational nights is longer then two months.

The maximum luminosity of GQ Comae in the UV was observed in May 19, 1985. We assumed that the continuum emission is due to thermal emission from a thin accretion disk around a supermassive black hole with $M = 1.5 \times 10^8 M_\odot$ and accretion rate equal to $\dot{M} = 1.0\, M_\odot/yr$. We also add a power law component with a spectral index equal to 0.97.

We assumed a simple variability model. In this model the variability is originated in the inner, radiation pressure dominated regions of the accretion disk. These regions are thermally unstable. Thermal instabilities can develop on the characteristic thermal time-scale throughout the disk causing luminosity variations. We assumed that the disk is divided into two parts: a hot, optically thin, inner part which emits the non-thermal X-rays and does not contribute to the optical/UV part of the spectrum; and a cold, optically thick outer part which is thermally stable and radiates locally as a blackbody contributing to the optical/UV band of the spectrum. Therefore, the opt/UV variability depends on the dimensions of the outer region.

The maximum luminosity in the opt/UV range is observed when the whole accretion disk is optically thick and the instability is not present in the inner part of the disk.

We draw two theoretical variability lines on the color – color diagram (Fig. 1). One represents a pure accretion disk with an underlying power law component, and the second includes accretion disk, underlying power law and absorption due to cold dust (E(B–V)= 0.04) in a host galaxy.

We present results of our model calculations on the Fig. 2. The change in the UV part of the spectrum is well reproduced. However, the optical variations are not. The optical part of the spectrum originates in the region which is situated outside the unstable, radiation pressure dominated parts

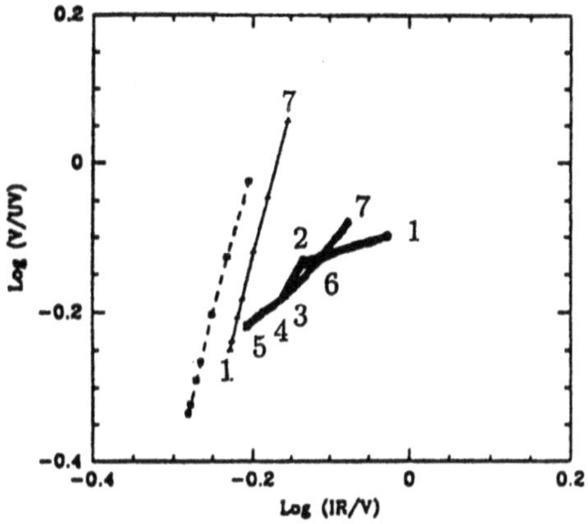

Fig. 1. The color-color diagram for the observational data of GQ Comae. The frequency bands: (log ν) (14.45 – 14.75); (14.75 – 15.05); (15.15 – 15.45). The solid line connects the observational points. Observational dates (open triangles); 1) 11/30/84; 2) 02/15/85; 3) 03/25/85; 4) 04/28/85; 5) 05/19/85; 6) 11/15/85; 7) 02/14/85. The dashed lines represent the theoretical variability models. The dotted line represents power law variations.

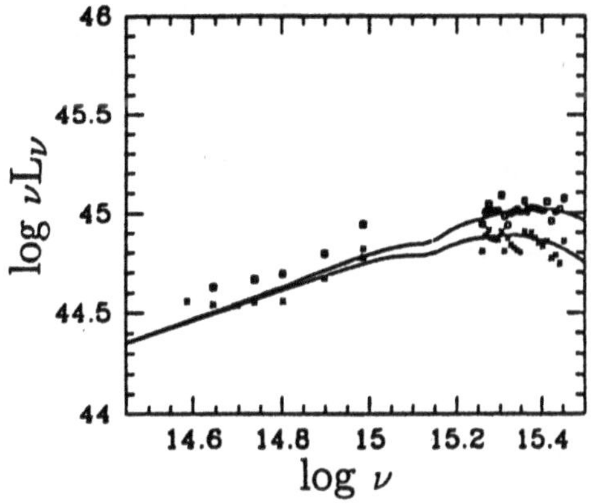

Fig. 2. The observational data data from May 19, 1985 are indicated with open squares, and the data from March 25, 1985 are indicated with crossed points. The solid lines shows the theoretical spectrum in two different epochs. We assumed that the cold dust in the host galaxy ($E(B-V) = 0.04$) caused the absorption.

(Siemiginowska & Czerny 1989, MNRAS 239, 289). A different, more detailed model has to be used to describe the optical variability. Irradiation and inclination effects could be more important here.

Conclusions

1/ The color – color diagram for the observational data of GQ Comae shows an almost the straight line. It means that the observed variability does not have a chaotic character, but may be caused by one physical process.
2/ We assumed that a thin accretion disk is a source of the thermal continuum emission. Then the variability may originate in the inner, unstable, radiation pressure dominated regions. A simple model of variability explains the observed UV variations.
3/ The theoretical model of variability is represent by different lines than the observed variations on the color – color diagram. The optical variations could be originated outside the unstable regions. So, a more detailed model is needed to explain the optical variation. Then, also the theoretical lines on the color – color diagram have to be different.

Optical Variability from a Thermally Unstable Region in AGN Accretion Disks

Aneta Siemiginowska, Martin Elvis, Olga Kuhn, and Belinda Wilkes

Harvard-Smithsonian Center for Astrophysics, Cambridge, USA

Hydrogen ionization causes viscous and thermal instabilities in thin accretion disks. This leads to vertical expansion or contraction of the disk. If the accretion disk surrounds a supermassive black hole ($> 10^6 M_\odot$) the vertical expansion will be significant, (Fig. 1), leading to a 'wall' with a contrast of ~ 5 in the disk structure at a characteristic radius. The 'wall' will exist during the hot phase (~ 0.3 of whole cycle time). We discuss some theoretical and observational consequences:

1/ The 'wall' will be exposed to direct irradiation by the central source.

2/ Part of the central radiation will be absorbed by the optically thick matter of the unstable regions. Self-blocking by the 'wall' may explain the existence of a group of 'Weak Bump Quasars' (McDowell et al. 1990, Ap.J.Lett. 345, 19).

3/ The UV/optical radiation is emitted from the inner regions situated close to the wall (Fig. 1). This will lead to the blocking of the optical light for all quasars seen more edge-on so there should be a population of quasars that are UV bright yet optically less dramatic.

4/ Color–color diagrams for quasars (Kuhn et al. 1991 in preparation) show an complete absence of objects in the edge–on region. Disk self–blocking due to the 'wall' provides a natural explanation for this unpopulated region of the diagram.

5/Where the X-ray optical depth through the 'wall' is close to unity variability in the soft–X-ray spectrum due to photoelectric absorption might be observed (Done et al. 1991).

Instabilities which develop in the partially ionized region can cause spectral variations up to a factor 2 in the visible band. The temperature of the unstable region is fixed by physics (at $\sim 10^4$K) and so it is always seen in the optical band. The whole optical continuum (from B–band to R-band) varies together in a quasi–gray shift. The timescales depend on the dimension of the unstable region and the velocity of fronts (similar to the cooling and heating fronts in cataclysmic variables; Meyer 1984 A&A, 131, 303; Clarke

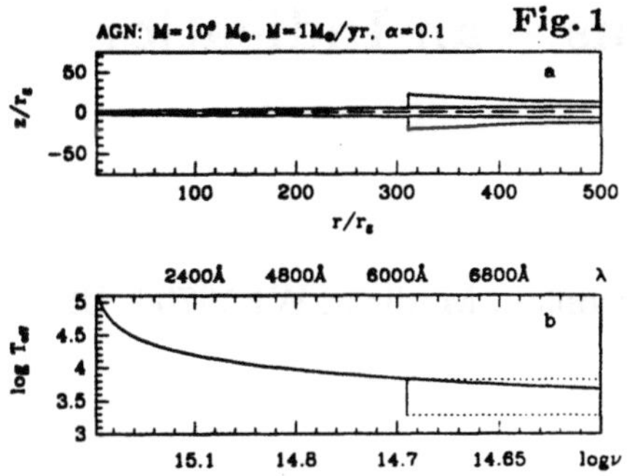

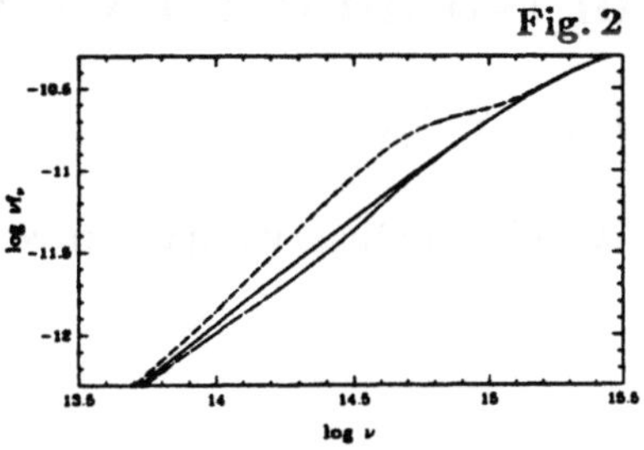

Fig. 1. (a) Height vs. radius for the accretion disk around $10^8 M_\odot$, $\alpha = 0.1$, and $\dot{M} = 1 M_\odot/yr$. Profiles for the extreme high and low cases in the unstable region are shown. (b) The peak emission frequency (wavelength) vs. effective temperature relation. The marked log (frequency) labels the spectral region in (νf_ν) where the largest contribution comes from the disk with a given effective temperature ($h\nu = 3.9\, kT_{eff}$).

Fig. 2. Theoretical spectrum emitted from the disk ($10^8 M_\odot$, $1 M_\odot/yr$, $\alpha = 0.1$). The solid line represents the spectrum emitted by the stable accretion disk. The lower curve is the spectrum emitted by the cold disk and the upper, dashed curve by the hot disk.

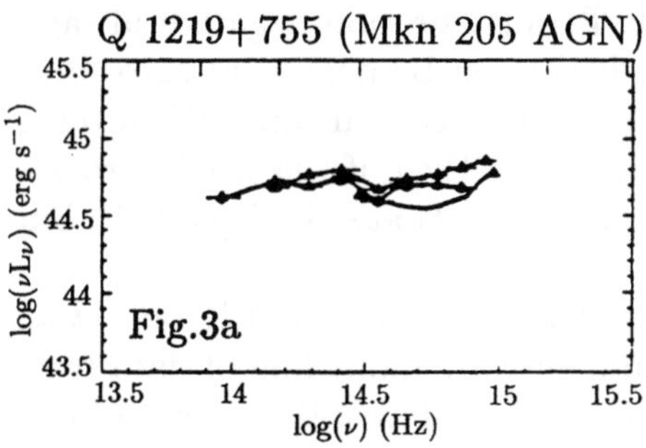

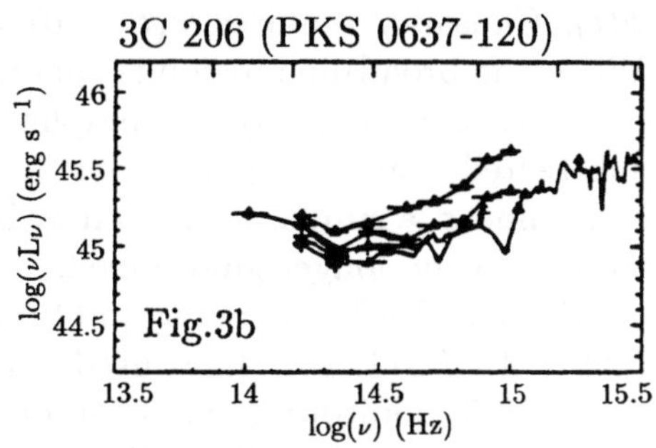

& Shields 1989, Ap.J. 338, 32) $t_F = r/v_F \sim (r/H)\, t_{thermal}$ (H – height of the disk).

Quasars in the PG quasar sample show ~10% variability in the optical range of the spectrum (Allen & Keel, private communication) with the largest variations being factor ~2. Two quasars out of our sample of 38 (Elvis et al. 1991 in preparation) show variations: Mkn 205 gives a good fit to this model if hot dust emission is included near $1\mu m$ (Fig.3a); 3C 206 shows a well–sampled rise and fall over $\sim 12yrs$.

However, there is a problem with the timescale of observed variability. The theoretical time scale expected is too long by 10^3. Several effects may decrease this timescale: irradiation, different viscosity prescriptions, reflection of transitions fronts; all of which need to be investigated to see if the observed variability can be explained by the 'wall'.

Veilleux & Zheng (these proceedings) show the H_α profile variability of 3C 390.3 over 10.4yrs. They argue that a hot region on an accretion disk can reproduce the detailed profile variability. We note that the distance at which they require a hot region corresponds to the partial ionization zone (the 'wall').

Line Correlation Functions – an Indication of Nonlinear Response

Linda S. Sparke

Washburn Observatory, University of Wisconsin, Madison WI 53706, USA.

Standard analysis of broad-line variability assumes that at any time, the line radiation from each gas cloud is linearly proportional to the continuum flux striking the cloud at that instant. The line flux $L(t)$ is given by the continuum flux $C(t)$ at earlier times, convolved with a transfer function $\Psi(t)$. Causality requires $\Psi(t < 0) = 0$; Ψ must be non-negative always. When the broad-line region is spherical, $\Psi(t)$ must be peaked at zero, then decrease monotonically. A complication: in practice a mean 'DC' level is subtracted from the light curves before analysis. Even if the whole process is not linear, a similar relation may hold for small *changes* in $C(t), L(t)$; but $\Psi(t) > 0$ is no longer guaranteed.

Horne, Welsh and Peterson (1991; *Ap J Lett* 367, L5) and Krolik *et al.* (1991; *Ap J* 371, 541) have used maximum entropy modelling of data from the 1988-9 monitoring campaign on NGC 5548 to derive $\Psi(t)$: for the Hβ, SiIV, CIII] and Mg II lines, the transfer function does not peak at zero lag, but between 5 and 40 days of positive lag. So the BLR is not spherical, or does not respond in a uniform linear way to the continuum flux (or both).

The **cross-correlation** $CCF(t)$ between the continuum flux at any time and the line flux at a time t later is related to the continuum autocorrelation function C_{acf} by

$$CCF(t) \equiv \int_{-\infty}^{\infty} C(\tau)L(t+\tau)d\tau = \int_{-\infty}^{\infty} C_{acf}(t-\tau)\Psi(\tau)d\tau.$$

Whatever the geometry, if the transfer function Ψ is always positive, the cross-correlation $CCF(t)$ is a 'fuzzier' version of the continuum autocorrelation, and should have a gentler slope. But the cross-correlations of the Lyα, NV and CIV fluxes with the continuum of NGC 5548 are *steeper* on the negative-lag side than the continuum autocorrelation.

Similarly, if Ψ is always positive, the **autocorrelation** L_{acf} of the line flux $L(t)$ should be broader than the continuum autocorrelation, by an amount comparable to the displacement of the cross-correlation peak away from zero

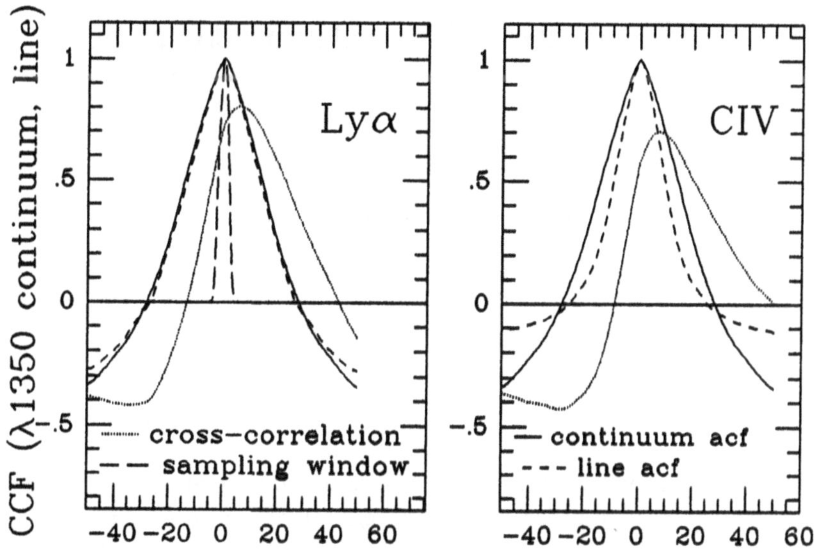

Fig. 1.

time. But the autocorrelation of the flux in these lines is actually narrower than that of the continuum.

Large line flux errors might make the line autocorrelation too peaked. Could errors be causing the narrowness? No: the 'sampling window ACF' is much sharper than the L_{acf}. The sharply peaked line correlation functions are not consistent with *any* model in which the line flux is strictly linear with the continuum brightness.

Physical conditions vary with position in the BLR. If the flux from the innermost part of the region which is responsible for a given line begins to drop at high levels of the continuum flux, this could explain both the lag in $\Psi(t)$ and the steep correlation functions. Photoionization calculations (Netzer, in *'Variability of Active Galaxies'*, Heidelberg, 1991) show that deep within the BLR where the ionization parameter is high, the flux in the Mg II and CIV lines decreases as the continuum source brightens: the inner BLR has effectively a *negative* transfer function. Since this matter is closest to the central source, its negative response would decrease $\Psi(t)$ most strongly at early times. If enough of the material responsible for producing a given line is in this regime, then the total transfer function Ψ may be negative near zero lag, and the line correlation functions could be more peaked than the continuum autocorrelation. Current maximum-entropy methods assume that Ψ is everywhere positive, and would miss negative values.

Gas Release Through Tidal Disruption of Stars

Hanno Sponholz

Inst. f. Theor. Astrophysik d. Univ. Heidelberg, Im Neuenheimer Feld 561, D-6900 Heidelberg

Capture and tidal break-up of stars by a massive black hole and subsequent accretion of the dispersed matter onto the central object may provide a crucial mechanism for the fueling of AGNs. To reinvestigate the tidal action of a black hole on a compressible polytrope, especially to include additional effects by a Kerr black hole, we apply a Smoothed Particle Hydrodynamics code for a fixed background metric. (Figs. 1–5) show a purely Newtonian treatment. We model a $1M_\odot$ star with 200 particles as an $n = 3/2$ polytrope. It approaches a $10^4 M_\odot$ black hole on a parabolic orbit. The relevant quantities are: The black hole mass M_{BH}, the radius r_* and the mass M_* of the secondary star, the characteristic 'Roche'-radius $r_r = r_* (M_{BH}/M_*)^{\frac{1}{3}}$, the distance of closest approach of the star to the hole r_p, and the penetration factor $\beta = r_r/r_p$. The Roche limit lies between $\beta = 0.55$ ($r_p = 1.2r_r$), when the star starts to lose some outer layers (Fig.3), and $\beta = 0.8$ ($r_p = 1.8r_r$), when it is completely disrupted. For very close encounters (Figs. 4, 5), the polytrope is stretched along the orbit and severely compressed in the z-direction (Fig.1) before being dissolved. This is accompanied by an increase in the central density just after the closest approach (Fig.2). The star does not keep an ellipsoidal form as the tidal forces overwhelm the internal stress. (This 'affine star' assumption has been discussed in detail in the frame of the 'nuclear runaway' scenario e.g. Luminet and Carter, 1986.

Relativistic context:

For a small main-sequence star orbiting in the equatorial plane around a Kerr Black Hole, the trajectory of the centre-of-mass of the star will be given by Carter's integrals of motion. The relative (tidal) acceleration of a nearby test-particle may be obtained by the equation of geodesic deviation. An orthonormal tetrad, locally defined and parallel-propagated along the centre-of-mass trajectory allows to derive a tidal tensor in terms of the Riemann tensor from the equation of geodesic deviation. The separability properties of a Kerr Black Hole make it possible to construct such a tetrad analytically and to write down the components of the tidal tensor (Marck, 1983). Fig.6 illustrates the tidal disruption (Figs. 6a, b) for a 'parabolic' (E=1) corotating orbit around an extreme (a=0.999) $10^9 M_\odot$ Kerr-BH. The figs. are scaled in units of the half Schwarzschildradius. Fig.7 shows a 'parabolic' trajectory for a very close encounter (in terms of the Roche radius) around a $10^6 M_\odot$ Schwarzschild-BH; Fig.7a confirms the appearance of the multiple double-points as described in Luminet and Marck, 1985 – the 'vertical' axis (solid line) suffers at least three strong flattenings in the vicinity of the periastron. Due to disruption, a multiple density peak cannot be confirmed (Fig.7b).

References

J.-P. Luminet and B. Carter, 1986, *Astron. Astrophys. Suppl.*, **61**, 219
J.-A. Marck, 1983, Proc R. Soc. London **A 385**, 431
J.-P. Luminet and J.-A. Marck, 1985, *Monthly Notices Roy. Astron. Soc.*, **212**, 57.

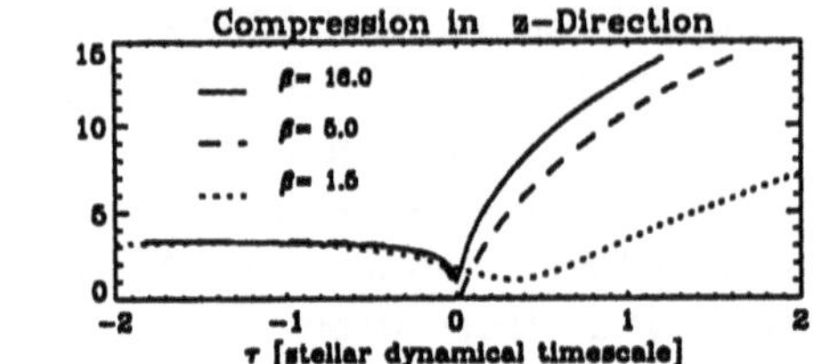

Fig.1 Time from Closest Approach

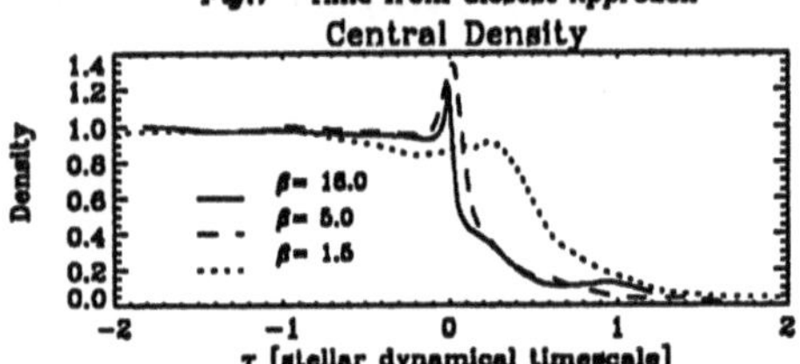

Fig.2 Central density vs. τ

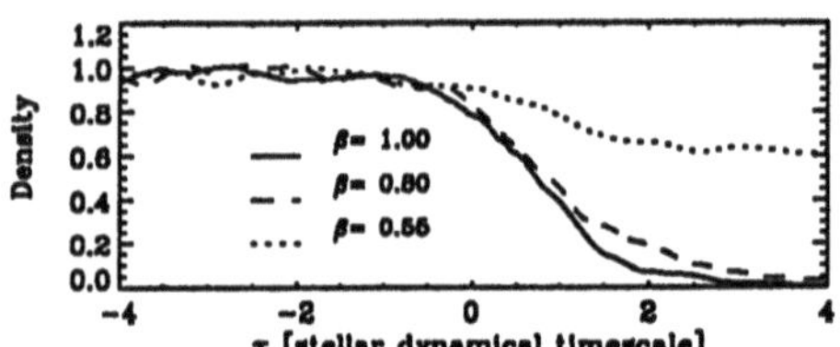

Fig.3 Effective disruption limit

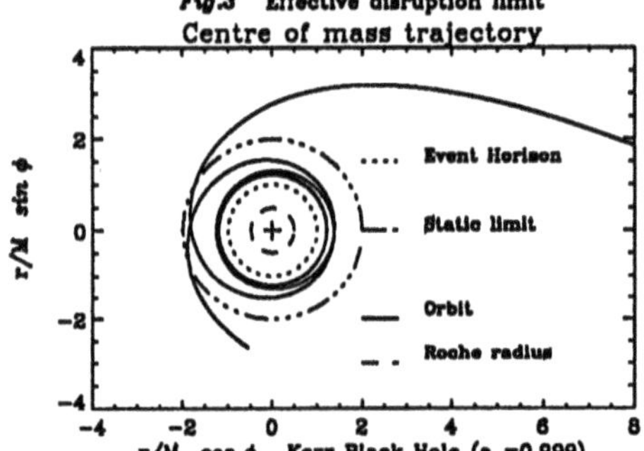

Fig.6 $\beta = 0.40$, $r_p = 1.2$, $r_{ROCHE} = 0.5$, $M_{BH} = 1000.M_\odot$

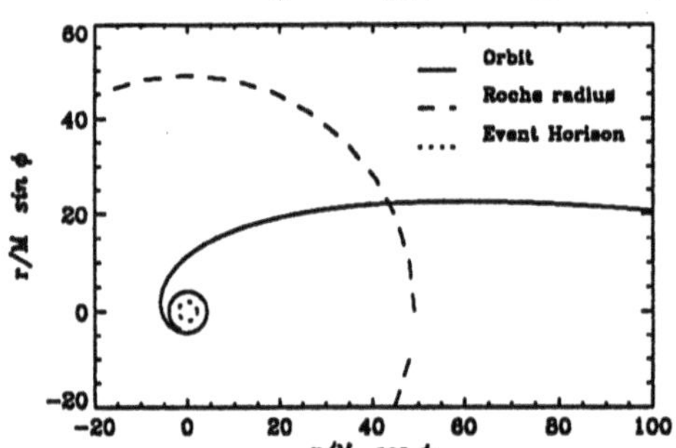

Fig.7 $\beta = 12.0$, $r_p = 4.1$, $r_{ROCHE} = 49.0$, $M_{BH} = 1.M_\odot$

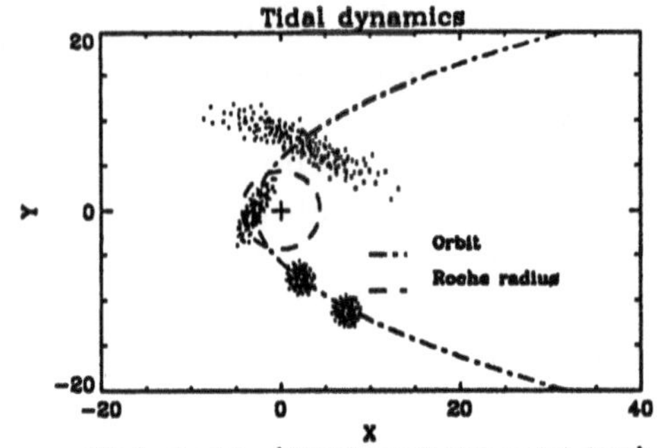

Fig.4 $\beta = 1.5$ (Orbital coordinates scaled down)

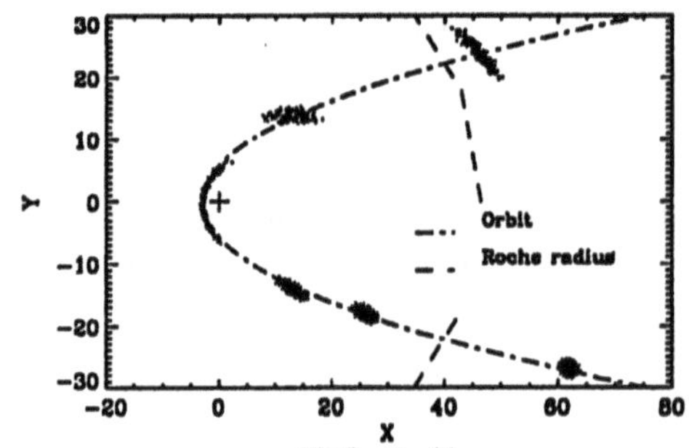

Fig.5 $\beta = 16$

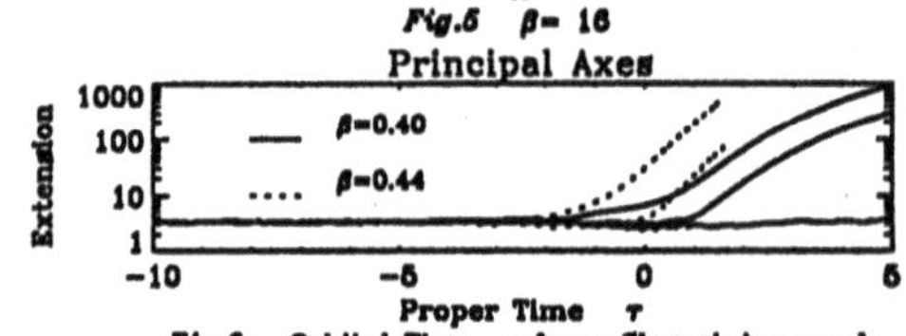

Fig.6a Orbital Time τ from Closest Approach

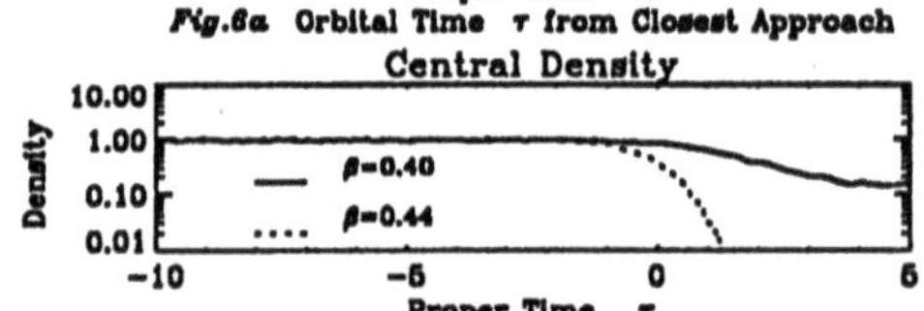

Fig.6b

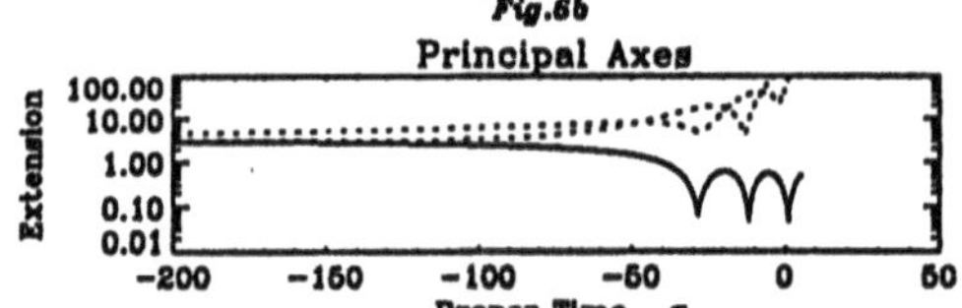

Fig.7a Orbital Time τ from Closest Approach

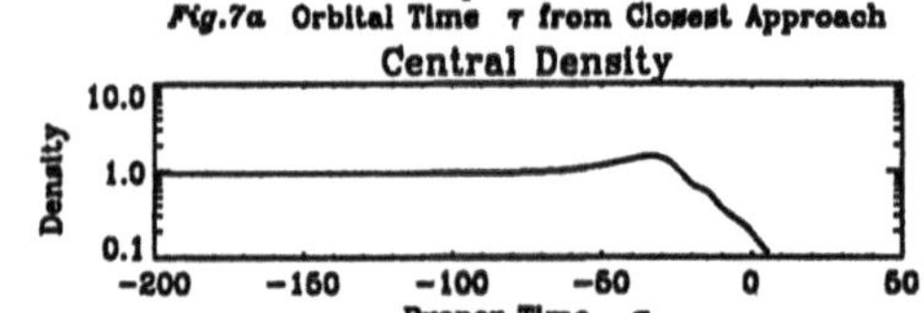

Fig.7b

Dust Emission from Extended AGN Disks

Lars G. Stenholm[1,2]

L. Stenholm

[1] Astronomical Observatory, Uppsala University, Box 515, S-75120 Uppsala, Sweden
[2] Swedish Defence Research Establishment, S-17290 Sundbyberg, Sweden

1. Motivation

Quasars emits substantial energies over wide ranges of the wavelength spectrum (from far-IR to X-ray). The optical and X-ray emission is produced within less than 1 pc from the central black hole (see Sanders at al., 1989). The IR radiation can be either non-thermal radiation from close to the central engine or thermal reradiation from dust outside the central parsec. This work intends to prove or disprove the suggestion of thermal dust emission from an extended possibly warped disk.

2. Radiative transfer

The radiative transfer method used has been described in detail by Stenholm et al. (1991) and I will only summarize its main characteristics here. Our method solves the transfer equation iteratively for the case of unpolarized radiation in a 3-dimensional media.

We consider the transfer equation for unpolarized radiation

$$\mathbf{N} \cdot \nabla I = \kappa (S - I) \tag{1}$$

$\mathbf{n}$ is a unit vector in the ray direction, I is the specific intensity, κ the extinction coefficient and S the source function. A cartesian coordinate system is employed. Eq. (1) can thus be written

$$\sin\theta\cos\phi\frac{\partial I}{\partial x} + \sin\theta\sin\phi\frac{\partial I}{\partial y} + \cos\theta\frac{\partial I}{\partial z} = \kappa (S - I) \tag{2}$$

where θ being the angle between ray direction and positive z axis and ϕ being the angle between the positive x axis and projection of the ray on the $x - y$ plane.

Since we assume that the infalling intensities are given as boundary values for all surfaces, we now discretize eq. (2) in the ray directions, i.e. for positive n_x, n_y, n_z and solve for

$$I_{i+1,j+1,k+1} = \frac{\frac{\sin\theta\cos\phi}{\delta x} I_{i,j+1,k+1} + \frac{\sin\theta\sin\phi}{\delta y} I_{i+1,j+1,k} + \frac{\cos\theta}{\delta z} I_{i+1,j+1,k} + \kappa_{i+1,j+1,k+1} S_{i+1,j+1,k+1}}{\frac{\sin\theta\cos\phi}{\delta x} + \frac{\sin\theta\sin\phi}{\delta y} + \frac{\cos\theta}{\delta z} + \kappa_{i+1,j+1,k+1}} \tag{3}$$

We use the method if Ng (1974) to accelerate the convergence of the basic method.

The radiation spectrum and the radius of the central source is assumed to be known.

The energy equation is written in the gray form

$$\epsilon\kappa\,(B - I) = 0 \tag{4}$$

where κ is the frequency integrated extinction coefficient, B the frequency integrated Planck function, J the frequency integrated mean intensity. The temperature distribution is obtained from eq. (4) solved for B.

The spectrum finally is calculated fully frequency dependent.

3. The Model

The spectral distribution of the radiation from the central source is based on the average energy distribution derived from Sanders et al. (1989) but the IR component is not included in the source since we assume that component is emitted from the extended disk.

The standard radius and luminosity of the central source is resp. 0.1 pc & 10^{12} solar luminosities.

The inner radius of the dust disk is 1 pc and the outer radius is varied between 100 and 1000 pc.

Various radial density distributions are tested.

The shape of the disk is the most important parameter in this study and disks with a constant opening angle as well as disks with convex and concave shapes are tested.

The opacity characteristics of the silicate dust used is taken from Draine (1985) and extended to shorter wavelengths according to the suggestions of Mathis (1990).

4. Results

The project is still in an early stage but figure 1 is an example of the results. The figure shows the radial temperature structure in the central plane for three models. The concave disk has $z_{max} = 0.5 r_{max}$. The optical depth given in the figure is based on the Planck mean opacity for the source

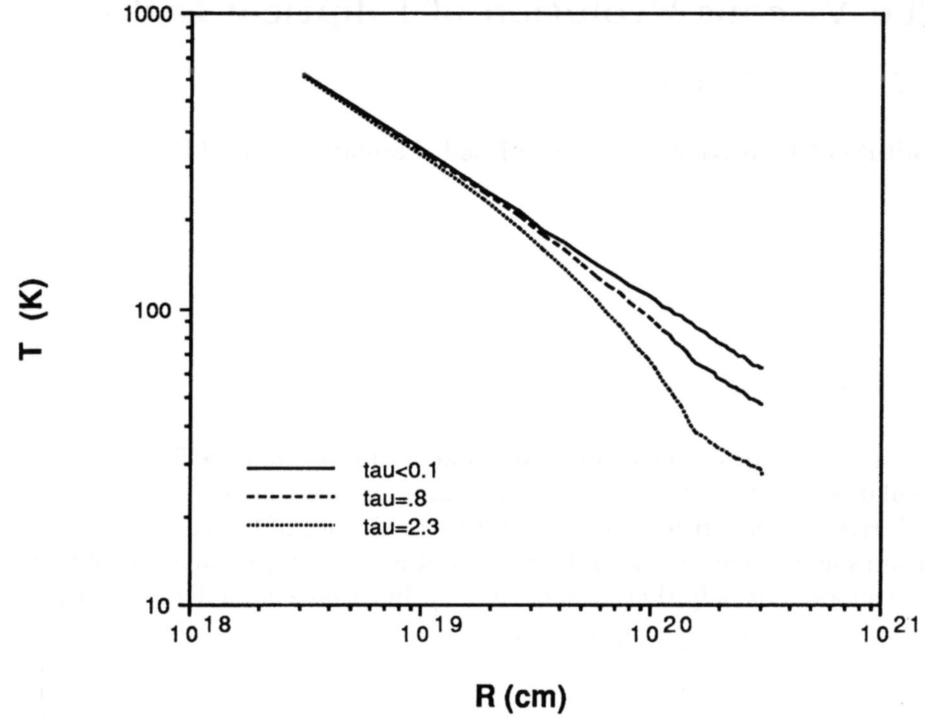

Fig. 1.

emission (i.e. UV optical depth), all three models are optically thin in the IR.

References

Draine, B.T., 1985: *Astrophys. J. Suppl.*, **57**, 587.

Mathis, J.S., 1990: *Ann. Rev. Astron. Astrophys.*, **28**, 37.

Ng, K.C., 1974: *J.Chem.Phys.* **61**, 2680.

Sanders, D.B., Phinney, E.S., Neugebauer, G., Soifer, B.T., Matthews, K., 1989: *Astrophys. J.*, **347**, 29.

Stenholm, L.G., Störzer, H., Wehrse, R., 1991: *J. Quant. Spectrosc.Radiat.Transfer* **45**, 47.

The Viscous Evolution of Elliptical Disks

D. Syer and C.J. Clarke

Institute of Astronomy, Madingley Road, Cambridge, CB3 0HA

We have examined the assumption that accretion flows will, in general, circularize in a time t_e much shorter than a viscous time t_ν.

Firstly, we constructed the constant e analogues of the circular disks of Shakura and Sunyaev (1977). Carrying out a perturbation analysis of these disks shows, not only that $t_e \gtrsim t_\nu$ (Fig. 1), but that e actually *increases*.

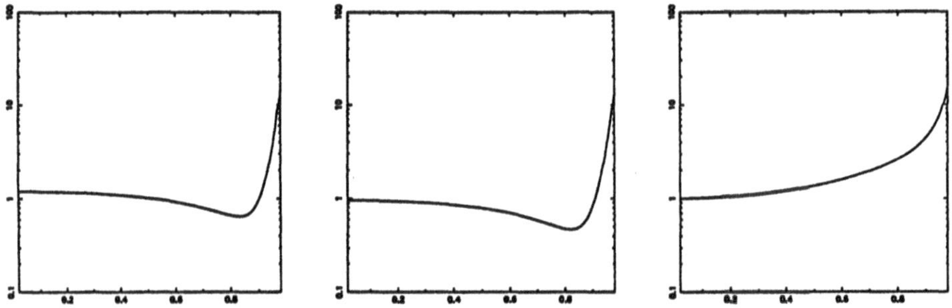

Fig. 1. **The ratio of circularization to viscous timescales t_c/t_ν for constant eccentricity disks in local equilibrium (left) 'outer disk' regime of Shakura and Sunyaev (1977), 'middle disk', and (right) 'inner disk' with viscous stress $\propto$ gaspressure.**

We have also carried out 'sticky particle' (Lin and Pringle, 1976) simulations of eccentric disks with three different initial conditions:

Model A: was a constant e disk in a $1/r$ potential and broadly confirmed the perturbation analysis.

Model B: was the same but with general relativistic effects included. This lead to differential precession of elliptical streamlines and consequent circularization out to a radius determined by the ratio of precession timescale and viscous timescale.

Model C: had a $1/r$ potential but assumed an initial condition where the particles are bunched together at pericentre – reminiscent of the debris from a tidally disrupted star. The final state after about a viscous time is shown in figure 2, which confirms that eccentricity is a long lived property of the resultant disk.

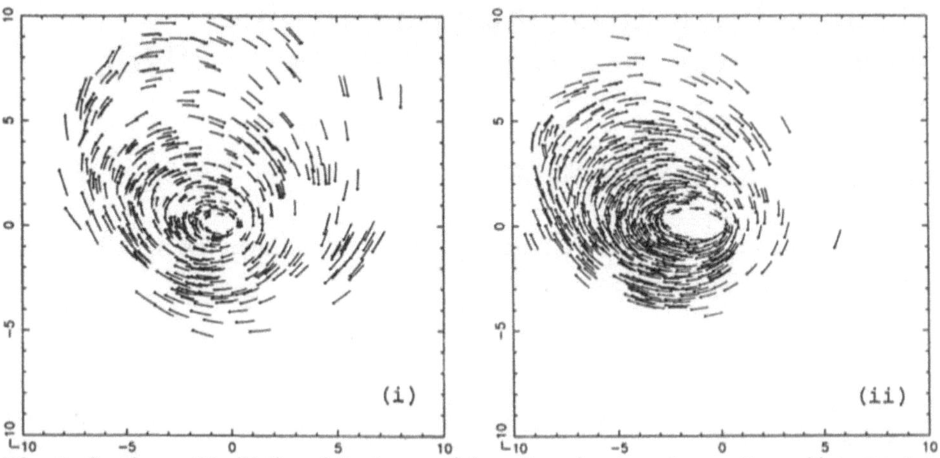

Fig. 2. Sections of ballistic orbits for particles after about a viscous time: (i) in Model A, (ii) in Model C.

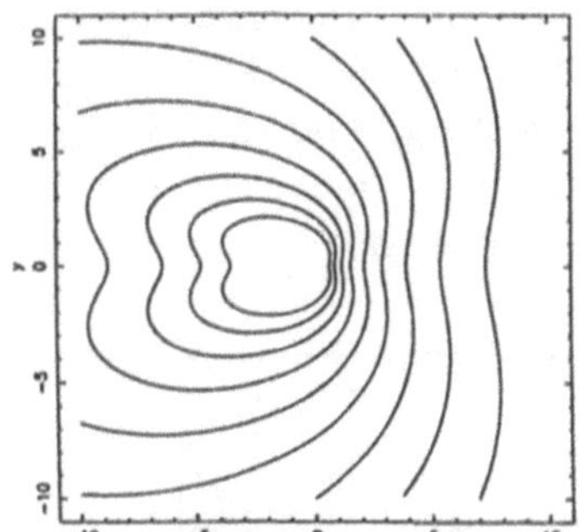

Fig. 3. Logarithmic isophotal contours of an $e = 0.8$ disk in the 'inner disk' regime with viscous stress $\propto$ gas pressure.

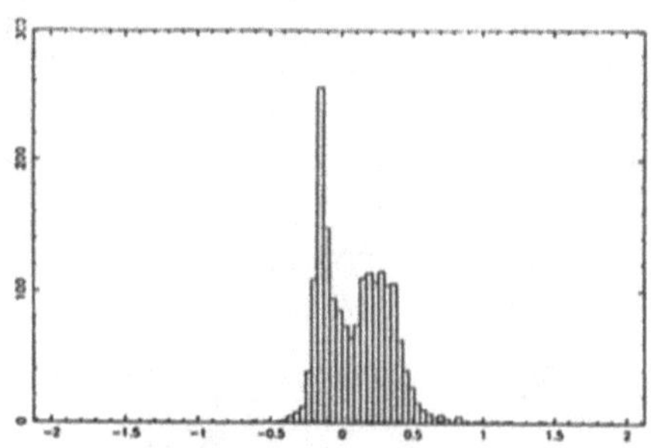

Fig. 4. Line profile of integrated light from an $e = 0.8$ disk with constant surface brightness.

Observational signatures of eccentric disks are isophotal contours (Fig. 3) and line profiles (Fig. 4) which are asymmetric, even for low e.

Thus we conclude that initially elliptical streams of gas could remain so, either in a Kepler potential or where the accretion rate is high compared to the rate of differential precession of streamlines. In fact, a high accretion rate actually *implies* radiatively inefficient accretion, because the flow remains elliptical. This has important implications for the observability of tidal flares from galactic nuclei (Rees, 1988).

References

Shakura, N.I., Sunyaev, R.A., 1977: *Astron. Astrophys.*, **24**, 337.
Lin, D.N.C., Pringle, J.E., 1976: in: *IAU Symposium #73*, ed. Eggleton *et al.*.
Rees, M.J., 1988: *Nature* **333**,523.

Double-Peaked Broad-Line Profiles: Evidence for Accretion Disks with "Hot Spots"?

Sylvain Veilleux[1] *and Wei Zheng*[2]

[1] U. Hawaii, Institute for Astronomy, 2680 Woodlawn Drive, Honolulu, Hawaii 96822
[2] U. Alabama

1. Observations

We have collected high signal-to-noise ratio spectra of 3C 390.3 obtained at 39 epochs during the years 1974-1988 with the Lick Observatory 3-m telescope. IUE spectra of 3C 390.3 covering the years from 1978 to 1989 supplemented the optical spectra. The analysis of the Lyman and Balmer line profile history of 3C 390.3 has shown the following facts that clearly support a complex model for its broad line-emitting region:

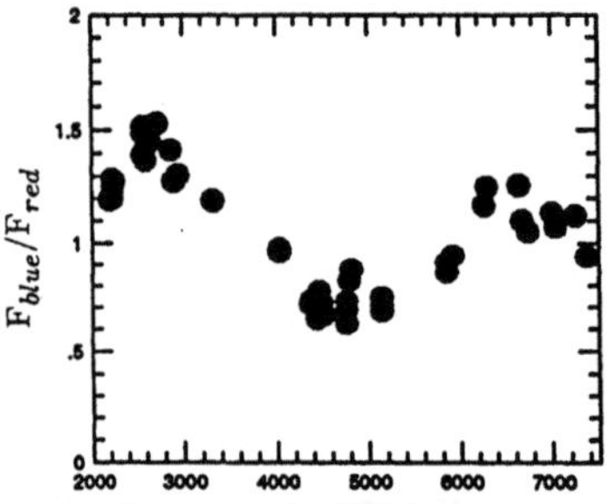

Fig. 1.

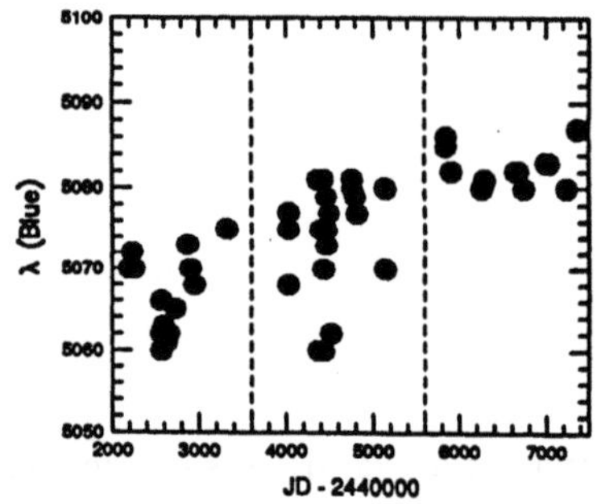

Fig. 2.

(1) A blue "bump", extending over $v \sim (-7000, 0)$ km s^{-1} relative to systemic velocity, is visible during the years 1974 – 1980 and 1985 – 1988. A red "bump" is also visible most of the time at $v \sim (0, +3800)$ km s^{-1}. Figure 1 shows that the ratio of the blue-to-red bump fluxes follows a fairly smooth, nearly sinusoidal pattern with a time interval of 10.4 years between the two maxima.

(2) Figure 2 shows that the peak of the blue bump has shifted by about +900 km s^{-1} over the 14-year interval covered by the data. In this figure the values of the peak positions within the vertical dashed lines are very uncertain because the blue bump was essentially absent during that period.

(3) A third, weaker broad-line feature, partially blended with the blue wing of [O III] λ4959, is visible at $v \sim +4600$ km s^{-1}.

This feature appears to be present only in the 1974 – 1975 scans.

(4) The values of the Lyα/Hα and Hα/Hβ ratios are lower in the humps than in the rest of the profiles, perhaps indicating a higher density in the gas producing these humps.

2. Models

A: Accretion Disk + Hot Patch According to this hypothesis, the emission from an accretion disk makes the typical double-hump profile. In addition, there is a hot spot on the disk which co-rotates with the disk and produces a moving component in the emission line profile. This component swings between the two humps and changes their relative intensity ratio. To fit the data of 3C 390.3, we used the disk prescription of Chen & Halpern (1989, ApJ, 344, 115) and required that the hot spot extends along both the radial and angular directions. The moving component in this case is a hump with a considerable line width which remains fixed for quite a long period of time before moving quickly to the other profile end. During the transition period, the hump appears very broad and is not easily distinguishable from the underlying disk emission.

B: Double-Stream BLR Displaced double humps could also be the result of a bipolar BLR (Zheng, Binette, & Sulentic 1990, ApJ, 365, 115). In this case, the opening angle of the cones greatly affects the profile shape: too small an angle produces a pair of detached peaks, which are not present in the data. Unfortunately, a wide bipolar BLR can produce only a flat profile without prominent hump, and an additional term is therefore needed to produce the humps. Such a region could be a lump of gas, or a region of limited size so that the velocity dispersion is smaller than the wide bipolar BLR. The simple model adopted in our calculations is a narrow stream along the cone's axis which contributes 20% of the total Hα emission.

Models involving an accretion disk with a co-rotating enhanced region double-stream BLR appear equally consistent with our data. Long-term monitoring of 3C 390.3 will be needed to differentiate between these two models. These results are discussed in more detail in Veilleux & Zheng (1991, ApJ, 377) and Zheng, Veilleux, & Grandi (1991, ApJ, 381).

Propagation of Light from Accretion Disks Around Kerr Black Holes

Sven Viergutz and Rainer Wehrse

Institut für Theoretische Astrophysik, Im Neuenheimer Feld 561, 6900 Heidelberg

1. Introduction

Since supermassive rotating black holes are supposed to be located in the centers of active galactic nuclei, general relativity plays an essential role for a correct transformation of spectra emitted from accretion disk surfaces to those received by observers.A transformation scheme developed by Cunningham [1] was restricted to plane parallel atmospheres of disks located exactly in the equatorial plane and to observers at infinity. Since it is our aim to synthesize spectra from radiative transfer calculations in Kerr metric without these restrictions it is necessary to devise a more flexible and general scheme.To demonstrate the importance of general relativistic effects we show the optical appearance of a simple disk to a distant observer as well as visualizations of corresponding lightlike geodesics in Kerr geometry.

2. Method of calculation

The geodesic equations in the Kerr metric are integrable once to yield four constants of motion [2],where in case of light only two independent impact parameters remain. After separation of variables and some substitutions, it is possible to express all relevant coordinates by means of elliptic integrals and Jacobian elliptic functions. In order to connect emitter and observer location correctly, two relationships have to be fulfilled for the impact parameters which—when found—can be transformed to local observer angles.

Results

Some results are given in figures 1 to 3. In order to simplify the presentation, we restrict ourselves here to flat disks, but we want to stress that our method is capable of handling any arbitary disk shape. The hole is assumed to be maximally rotating counterclockwise with mass M and specific

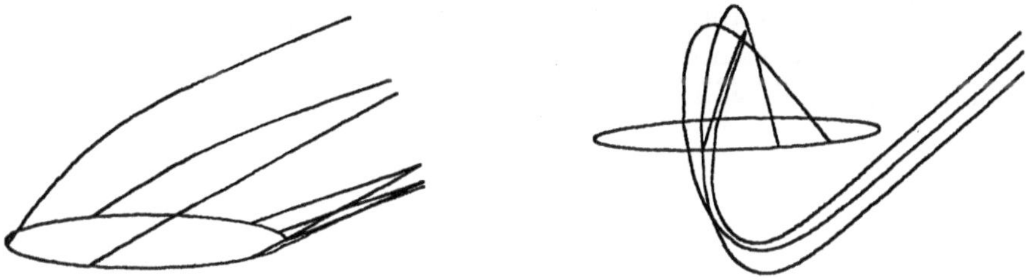

Fig. 1,2. Lightlike geodesics emitted at 10 M (see text).

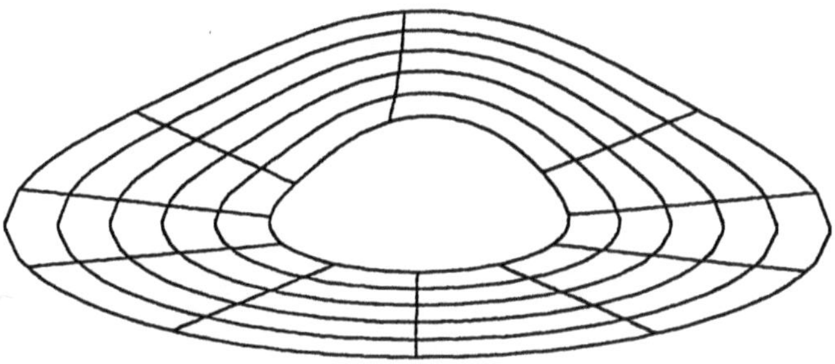

Fig. 3. Distant image of a simple accretion disk (see text).

angular momentum $a = 0.998\,\mathrm{M}$. As a simple model for the disk, we take six concentric rings equally spaced and located at radii of 10 M to 30 M. All rings are situated in the equatorial plane and connected by ten lines of constant azimuth. The observer is located at coordinate radius $r = 1000\,\mathrm{M}$ and at a polar angle of 18° above the equatorial plane.

Fig. 1 shows some light rays emitted from the ring $r = 10\,\mathrm{M}$ which all meet at the chosen observer. These direct rays do not traverse the equatorial plane and in their totality produce the innermost ring in Fig. 3. Only the region up to $r = 20\,\mathrm{M}$ is plotted.

Fig. 2 is a similar plot to Fig. 1, but this time the light traverses the equatorial plane twice and is drastically dragged by the hole Additional images covering parts of the ring are seen by the observer resulting from such indirect light.

Fig. 3 shows the direct distant image of the disk. Light coming from the front part is not significantly deflected, the mapping is as one would classically expect. On the contrary, the view of the back part is drastically distinct from usual expectation because the light must pass the vincinity of the hole and is deflected towards the hole. A typical Kerr effect can be seen at those two azimuth lines at the central backside which deviate counterclockwise when going inwards: the dragging produced by the rotating hole increases strongly with decreasing distance to the hole.

Mathematical details will be published elsewhere.

References

C.T.Cunningham 1975, *Astrophys. J.*, **202** , 788

S.Chandrasekhar 1983, The Mathematical Theory of Black Holes, Clarendon Press, Oxford

Thermal Comptonization in AGN

R. Walter

Observatoire de Genève, CH-1290 Sauverny, Switzerland

Abstract. X-ray and ultraviolet observations of quasars and Seyfert galaxies for which a sufficient number of observations exist in both wavebands are used to study the relation between the optical-ultraviolet big blue bump and the X-ray powerlaw component. Two groups of objects with different modes of spectral variations and with different ultraviolet to X-ray flux ratios are found. In the first group ($L_{UV} \gtrsim L_X$), the spectral variations are characteristic of thermal Comptonization where the blue bump photons are Comptonized into the X-ray band by hot electrons. In the second group ($L_{UV} \lesssim L_X$), the powerlaw index increases with the X-ray flux and is not coupled with the ultraviolet component.

1. Thermal Comptonization

We have plotted the positions of several emission line type AGN in a diagram showing the 2-10 keV powerlaw index α as a function of the logarithm of the ratio of the count rates measured between 2 and 10 keV and in the ultraviolet (*IUE*) or in the soft X-rays (*EXOSAT 3Lx*). When the ultraviolet to X-ray spectral shape varies for an object, its position changes in this diagram. If the spectra is well represented with a single powerlaw component in the considered spectral range, we expect to see an obvious anti-correlation in this diagram when the spectral slope varies. Such an anti-correlation may be seen between the X-ray powerlaw index and the ratio of X-ray and soft X-ray fluxes in 3C 120 and MR 2251-178 indicating that there is no soft excess or variable absorber in these objects (Walter and Courvoisier, 1991).

However, in 3C 273 (figure 1) and 3C 120, the X-ray powerlaw index was anti-correlated to the ratio of X-ray to ultraviolet count rates. In these objects, the spectra are not well fitted with a single powerlaw, the X-ray and ultraviolet fluxes are not correlated and the powerlaw index is not correlated to either of them. In NGC 5548 (Walter and Courvoisier, 1990) and PG 1211+143 this anti-correlation has been observed between the X-ray powerlaw index and the ratio of X-ray to soft X-ray count rates. In these objects the contribution of the X-ray powerlaw component is negligible in the ultraviolet and less than 20% in the soft X-rays. The soft X-ray photons are emitted by an independent emission component and their variations cannot be interpreted with a variable cold or hot absorber.

We notice that such spectral variations are not compatible with standard pairs models. We have interpreted these significant correlations as the signature of thermal inverse Comptonization of small and variable optical depth and approximately constant covering factor and temperature. The slopes of the observed correlations give a temperature of a few $m_e c^2$ for the Comptonizing electrons.

2. Two populations ?

Using a sample of objects observed several times with *EXOSAT* and showing spectral variations, we noticed that all objects showing spectral variations characteristic of thermal Comptonization have a strong blue bump in comparison with the X-ray luminosity. On the other hand, objects with a weaker ultraviolet component show a correlation between the X-tray powerlaw slope and flux. We conclude that there may be two different emission processes for the X-ray component in these two groups of objects (Walter and Courvoisier, 1991).

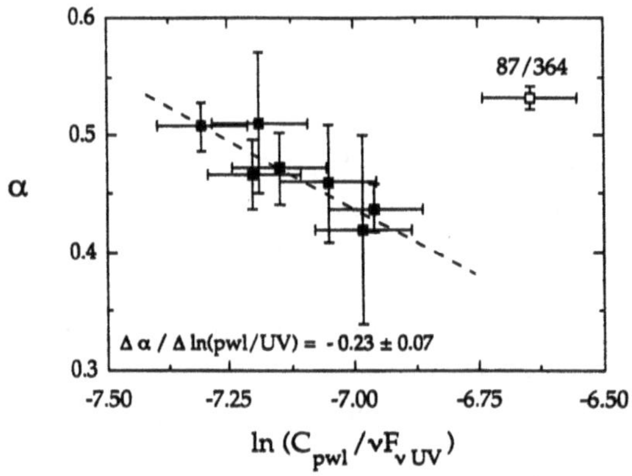

Fig. 1. 2-10 keV spectral index as a function of the ratio of the 2-10 keV to ultraviolet photon fluxes in 3C 273. Data are from *IUE*, *EXOSAT* and *Ginga* observations. The white point was observed during an X-ray flare corresponding to the synchrotron flares occuring in March 1988 (Courvoisier et al., 1990).

References

Courvoisier, T.J.-L. *et al.* : 1990, *Astron. Astrophys.*, **234**, 73
Walter, R. and Courvoisier, T.J.-L. : 1990, *Astron. Astrophys.*, **233**, 40
Walter, R. and Courvoisier, T.J.-L. : 1991, *submitted to A.& A.*

Rapid Variations in the BLR of NGC 3516

Ignaz Wanders and Ernst van Groningen

Astronomiska Observatoriet, Box 515, S-751 20 Uppsala, Sweden

1. Introduction

During the first five months of 1990 the Seyfert 1 galaxy NGC 3516 was observed spectrometrically as part of the LAG monitoring subproject (LAG stands for Lovers of Active Galaxies). The observations were carried out on the 2.5m INT and the 4.2m WHT on La Palma. The time coverage was quite good at the beginning of the campaign but due to bad weather serious gaps in the time series exist in the latter half of the observing period. NGC 3516 is a SB0 type galaxy at a redshift of 0.009. The narrow lines were used as internal calibrators to scale the spectra to each other. We checked that this procedure is not affected by the extended narrow emission line region since the latter is weak compared to the central unresolved NLR and since we can clearly separate it from the unresolved NLR. The stellar contribution to the nuclear spectra is as high as 50 - 70 % but can be corrected for in the Hβ spectra by subtracting a template spectrum (of the bulge of NGC 3516 itself) from the observed spectra such that, amongst others, the Mg b absorption lines completely disappear. In this way we can find both the continuum and the line (profile) light curves. We developed a new method for scaling spectra to each other when total emission or absorption line fluxes in the spectra can be assumed constant in time. This method (van Groningen and Wanders, in preparation) obtains an accuracy of better than 5 %, typically 2 - 3 %.

2. Results

The lightcurves and difference spectra all show that significant variations in both the continuum and the broad emission line fluxes are present at time scales of less than 2 - 3 days. The Hα and Hβ lines lag the continuum variations by zero days which simply indicates that we are not only still undersampling but also that the BLR of NGC 3516 is extremely small. Profile variations do also occur but at a longer time scale than total flux variations.

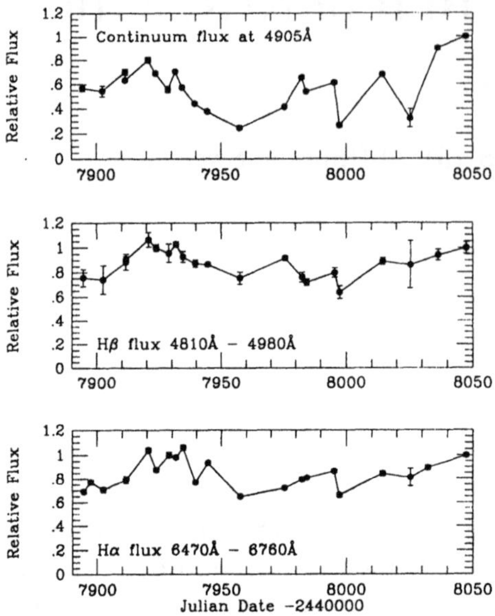

Fig. 1. Continuum, Hβ & Hα lightcurves

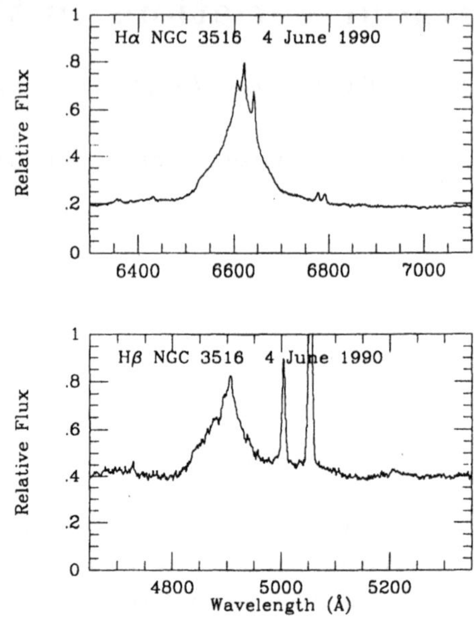

Fig. 2. Example spectra of 4 June 1990

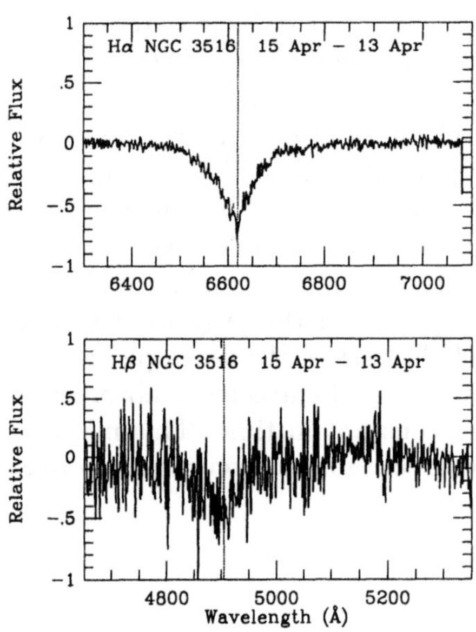

Fig. 3. Difference spectra of 15 April - 13 April 1990

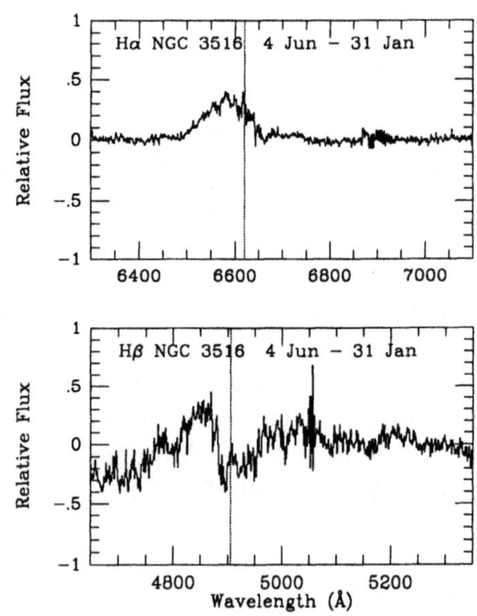

Fig. 4. Difference spectra of 4 June - 31 January

AGN with Strong SiIV/OIV] 1400Å Emission

Wang Tinggui, Jean Clavel and Willem Wamsteker

ESA IUE Observatory, VILSPA, Madrid, Spain

1. Characterizations of Strong 1400 Å Feature Objects

The emission at 1400Å is a common feature in high redshift quasars. But it has seldom been discussed in low redshift AGNs. This is partly due to the relative weakness and the associated difficulty to measure and study it in low S/N spectra taken with IUE. However in the IUE Archive we found that a number of AGNs have very strong 1400Å features: the intensity ratio the (1400ÅCIV 1549Å) ranges from 0.4 to 1.0 (fig 1). While it is 0.16 as predicted by the standard model (Kwan and Krolik, 1981), 0.25±0.16 found for a radio-loud dominated high redshift quasars sample (Wilkes 1986).

Table 2 gives the far UV continuum flux, integrated line fluxes and, UV spectral slope, as well as the ratio Fe II (multiple 37,38) and Hβ, and the optical to X-ray spectral index whenever available. The error in the line fluxes is less than 20%. The strong-1400Å objects show the following characteristics: (a) Strong narrow (FWHM less than 2300 km.sec^{-1}) Lyα emission, frequently containing half flux the total flux in Lyα +NV. (b) CIV 1549Å is weak relative to Lyα: CIV/Lyα = 0.29 ± 0.09 for the strong-1400Å objects, while CIV/Lyα= 0.75±0.12 for average Seyfert galaxies (Wu *et al.* 1983). (c) For strong-1400Å objects, the peak wavelength of the 1400Å feature is at 1399.1 ± 0.66Å , while for the weak-1400Å objects the peak of the feature is found at 1401.5 ±0.76Å . This wavelength difference between strong and weak 1400Å is statistically significant at the 98% confidence level. (d) These objects also show strikingly strong optical FeII emission and possible also strong CaII triplet emission. The six (out of ten) objects for which optical Fe II measurements are available showed FeII/Hβ>0.5 ($\langle$ FeII/Hβ $\rangle$ = 0.85 ± 0.27) and they belong thus to the class of strong FeII emitters as defined by Joly (1991). The galaxies I ZW1, II ZW 136 and Mkn 478 also have strong CaII triplet emission (Persson, 1985). I ZW1, Mkn 478 and PG 0844+34, show at the same time the strongest 1400Å /CIV as well as the strongest FeII/ Hβ ratio of the objectsin our sample.

Table 1: UV Line and continuum fluxes, (FeII4570)/Hβ and αox

source	Z	Lya	CIV	1400	F_{1350}	σ	αuv	FeII/Hβ	αox	ref.
I ZW 1	0.061	299	67	60	1.02	0.29		1.34	1.36	a,b
Mkn 975	0.050	149	39	31	0.35	0.12				
II ZW 1	0.054	104	49	39	0.64	0.18	1.25	0.53		a
PG 0804+76	0.100	510	213	105	2.73	0.27	1.10		1.41	
PG 0844+34	0.064	474	101	85	2.95	0.67	0.83	0.92		c
Mkn 734	0.059	428	91	53	1.74	0.33	1.02		1.41	
Mkn 734	0.059	622	112	92	3.45	0.76	1.35		1.41	
Mkn 771	0.064	566	184	89	3.64	0.34	0.40	0.66	1.37	c
PG 1415+45	0.114	121	37	24	0.49	0.16	1.19			
Mkn 478	0.064	429	74	73	2.59	0.37	1.23	1.00	1.31	a,b
II ZW 136	0.065	623	190	120	3.65	0.54	0.99	0.59	1.32	a
II ZW 136	0.065	746	221	108	4.85	0.69	0.37	0.59	1.32	a
II ZW 136	0.065	523	182	66	2.63	0.42	0.95	0.59	1.32	a
F 9 high	0.054	3055	860	435	17.86	0.96	0.64	–	–	
F 9 low	0.054	926	454	117	4.03	0.36	0.64	–	–	

(1) αox from Mushotzky and Wandel(1989) (2) FeII/Hβ ratio a. Joly 1991, b. Philips 1977, c.Wilkes *et al.* (1987) (3) Line and continuum flux in 10^{-14}ergs/s/cm^{-2}. UV slope of F9 from Clavel et al. (1988).

(e) For the 6 objects, for which α_{ox} is available in the literature, the mean α_{ox}=1.36 $\pm$0.04, indistiguishable from the canonical value of 1.36 $\pm$ 0.23 for AGN (Mushotzky and Wandel, 1989).

2. Discussion

Wills and Netzer(1979) argued that 1400Å /Lyα > 0.2 is difficult for the photoionization model. Recent high density model (Rees,*et al.* 1989, Ferland *et al.* 1991) can predicted 1400Å /Lα as high as 0.30. To our knowledge, yet no AGNs violate this limit, but when we compared the SiIV/CIV, the problem turns out. These lines came from the same region, and less affected by reddening, or by continuum shape (similar ionization potential). The standard model predicted the SiIV+OIV]/CIV = 0.16, while the high density model of Rees *et al.* (1989) predicted SiIV+OIV]/CIV = 0.35. This cannot covered most o bjects in our strong SiIV sample. Detail photoionization calculation, covered a wide range of U and ne, failed to produce the 1400Å /CIV larger than 0.6. This suggested that chemical composition different from solar is needed.

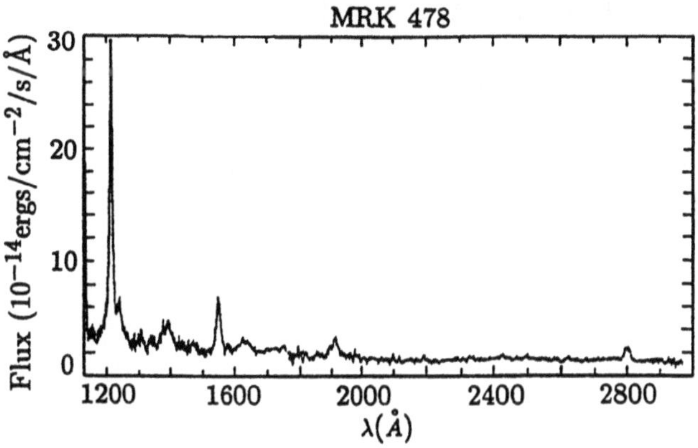

Fig. 1. IUE spectra of Mkn 478

Acknowledgement. TGW would like to thank F.H.Cheng for discussions, M.Santos for help in photoionization calculations and M. Malkan and G.Ferland for comments and suggestions. WTG wishes to thank ESA for suporting his stay in VILSPA.

References

Clavel, J., Wamsteker, W. and Glass I.:1988, *Astrophys. J.*, **337**, 236.
Kwan, J. and Krolik, J.H.: 1981, *Astrophys. J.*, **250**, 478.
Joly, M.: 1991, *Astron. Astrophys.*, **242**, 49.
Mushotzky, R.F. and Wandel, A.: 1989, *Astrophys. J.*, **339**, 674.
Netzer, H. and Wills, B.J.: 1983, *Astrophys. J.*, **275**, 445.
Persson, S.E.: 1988, *Astrophys. J.*, **330**, 751.
Phillips, M.M.: 1978, *Astrophys. J.*, **215**, 746
Rees, M.J., Netzer, H. and Ferland, G.: 1989, *Astrophys. J.*, **347**, 640.
Wills, B.J. and Netzer H.: 1979, *Astrophys. J.*, **233**, 1.
Wilkes, B.J.: 1986,*Monthly Notices Roy. Astron. Soc.*, **218**, 331.
Wilkes, B.J., Elvis, M. and McHardy I.: 1987, *Astrophys. J. (Letters)*, **321**, L23.
Wu, C.C. *et al.* : 1983, *Astrophys. J.*, **266**, 28.

Kinematic Axisymmetry in QSOs

Beverley J. Wills, Dongping Fang, and Michael S. Brotherton

McDonald Observatory and University of Texas at Austin, USA

We present new correlations among the widths and strengths of broad emission lines in radio-loud and radio-quiet quasars, and use these to test a simple axisymmetric model for the continuum emission and BLR kinematics.

The similarity in relative broad line strengths from one QSO to another suggests similar physical conditions, so the wide range in observed line width may be a result of geometrical projection (Osterbrock 1977). This interpretation is supported by the observed inverse correlation between Hβ broad-line widths and R, the ratio of radio core-to-lobe flux density; the lines are narrower when the system is viewed closer to the jet direction representing the kinematic axis of the quasars' central engine. This suggests that broad emission line gas is moving predominantly in a plane perpendicular to the radio jets (Wills and Browne 1986). The polarized optical synchrotron continuum in radio-loud quasars is also axisymmetric – beamed in the jet direction (Impey *et al.* 1991, Wills 1989, and references therein), giving rise to the anti-correlation of [OIII] equivalent width, EW, with R (Jackson *et al.* 1989). In radio-quiet quasars and lobe-dominated (low R) quasars the synchrotron continuum is unimportant and the Big Blue Bump dominates in the UV and optical. This is true even in some core-dominant quasars. If the Big Blue Bump is 'thermal' emission from an accretion disk it too should appear brighter when viewed at smaller angles to the disk axis (for $\lambda > 912$ Å).

In investigating this picture further we have found a strong *inverse* correlation between CIV λ1549 EW and FWHM (full width at half of the peak line intensity), for both radio-loud and radio-quiet quasars – contrary to that predicted. Perhaps the motion of *high ionization* gas is predominantly parallel to the kinematic axis. Here we test the predictions of this modified picture.

Table 1 shows results of rank correlations among EW, FWHM, and R for various samples of radio loud and radio-quiet quasars, and high-ionization (CIV) and lower-ionization lines (Hβ and Mg II). The selection of some samples (*e.g.*, for flat or steep radio spectrum) and quality of some data may bias

Table 1. Rank Correlation Results

	High Ionization				Low Ionization			
	r	n	σ	ref.	r	n	σ	ref.
EW vs. FWHM	−0.58	61	5.5	1	+0.29	54	2.2	2
(radio quiet)					+0.34	13	1.2	3
					+0.31	31	1.8	4
EW vs. FWHM	−0.49	26	2.8	5	+0.41	112	4.7	7
(radio loud)	−0.22	36	1.3	6	+0.25	45	1.7	8
					+0.92	10	6.7	9
					+0.28	33	1.6	10
					+0.87	12	5.6	11
FWHM vs. R	−0.37	33	2.2	12	−0.41	84	4.1	13
(radio loud)								
EW vs. R	−0.32	36	2.0	14	−0.53	46	4.2	15
(radio loud)					−0.40	39	2.6	16

References

1. CIV, Sargent, W.L.W., Boksenberg, A. and Steidel, C.C. 1988 Ap.J.Suppl., 68, 539; Sargent, W.L.W., Steidel, C.C. and Boksenberg, A. 1989 Ap.J.Suppl., 69, 703.
2. Hβ, Stephens, S.A. 1989 A.J. 97, 10; QSOs and Seyfert 1 nuclei.
3. Hβ, Stephens (1989), QSOs alone.
4. MgII 2798, Baldwin, J.A. *et al.* 1989 Ap.J. 338, 630, the BQS.
5. CIV, our measurements from data of Sargent *et al.* (1988, 1989).
6. CIV, our measurements from data of Barthel, P.D., Tytler, D.R., and Thomson, B. 1990 A.A.Suppl. 82, 339.
7. Hβ, some MgII, Wills, B. and Browne, I.W. 1986 Ap.J., 302, 56.
8. Netzer, H. 1987 MNRAS, 225, 55, using data from Wilkes (1986), predominantly a radio-loud sample.
9. Mg II 2798, Espey B.R., Carswell, R.F., Bailey, J.A., Smith, M.A., and Ward, M.J. 1989 Ap.J. 342, 666.
10. Mg II 2798, Smith, H.E. *et al.* 1977 Ap.J., 215, 427; Baldwin, J.A. *et al.* 1973 Ap.J., 185, 739.
11. Hβ, Molonglo sample, as in 10.
12. CIV, high redshift data (Barthel 1990), low redshift IUE data.
13. Hβ mostly, Wills and Browne (1986), and other data.
14. CIV, Barthel *et al.* (1990), our measurements.
15. [OIII], Wills and Browne (1986).
16. [OIII], Jackson, N., Browne, I.W.A., Murphy, D.W., and Saikia, D.J. 1989 Nature, 343, 43.

the results. Luminosity dependences may affect results for low-redshift samples, and the results of comparison of high-ionization lines in high redshift samples with low ionization lines easily measurable in low redshift samples. However, the signs of the correlations follow expected trends, except for the lack off a positive correlation of FWHM with R for the radio-loud CIV sample.

We conclude that, if the Big Blue Bump arises from a standard accretion disk, and the strength of CIV emission is isotropic, velocities of high-ionization gas are greater parallel to the disk axis than in the disk plane. The large scatter in most relations suggests that geometry is not the only or even the major factor determining the observed line widths but better samples need to be investigated. Perhaps the high-ionization lines arise in a wind blowing parallel to the axis of the flattened lower-ionization region. Interestingly, this kinematic interpretation is along the lines suggested by angle-dependent photoionization models (Netzer 1987); if radiation pressure is important a high-ionization wind may be identified with an accretion disk corona or ablated stellar atmospheres (Ferland and Persson 1989).

We thank Drs W.L.W. Sargent, C.C. Steidel and A. Boksenberg for sending us their data in digital form, and acknowledge the support of NSF grant AST-8714937.

References

Ferland, G.J., and Persson, S.E.: 1989, *Astrophys. J.*, **347**, 656.
Impey, C.D., Lawrence, C.R., and Tapia, S.: 1991, *Astrophys. J.*, in press.
Netzer, H.: 1987, *Monthly Notices Roy. Astron. Soc.*, **225**, 55.
Osterbrock, D.E.: 1977, *Astrophys. J.*, **215**, 733.
Wills, B.J.: 1989, in *'BL Lac Objects'* ed., L. Maraschi, T. Maccacaro, M.H.-Ulrich (Berlin: Springer-Verlag), p.108.

Determining the Extinction, Luminosity and Flux Distribution of Seyfert Galaxies using Broad-band Photometry

Hartmut Winkler and Ian Glass

South African Astronomical Observatory, P.O. Box 9, 7935 Observatory, Cape Town, South Africa

We investigate the data obtained during a multi-aperture $UBV(RI)_c$ variability study of a sample of 35 Seyfert galaxies during the period 1986-1990 (Winkler et al., 1991). After converting the magnitudes to fluxes we made F_{ν_1} vs F_{ν_2} plots for each object. The example in Fig. 1 demonstrates a linear relationship between the U and B fluxes in NGC 3783. The same effect is observed in most other objects when comparing the fluxes of these or other filters. We call the constants of proportionality Γ. The simplest way to interpret this result is to postulate that the nuclear colours are invariate, as proposed by Choloniewski (1981). These colours can then be derived directly from the Γ-values.

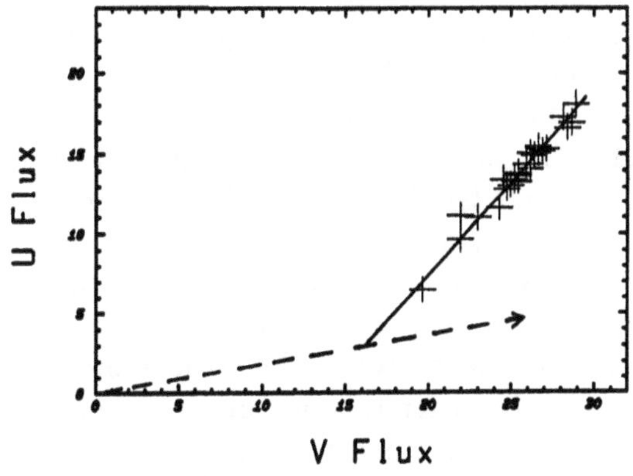

Fig. 1. The F_U vs F_V diagram of NGC 3783. The broken line corresponds to the extra-nuclear component and the solid line describes the nuclear (variable) component.

The bluest nuclei in our sample have $U - B = -1.1$ and $B - V = 0.0$. We propose that these are the intrinsic colours of Seyfert nuclei. Reddenings and extinctions were then estimated using Galactic reddening relationships. It was found that for most cases $E(U - B)/E(B - V)$ is larger than in our galaxy, suggesting that the reddening law in Seyfert galaxies is anomalous,

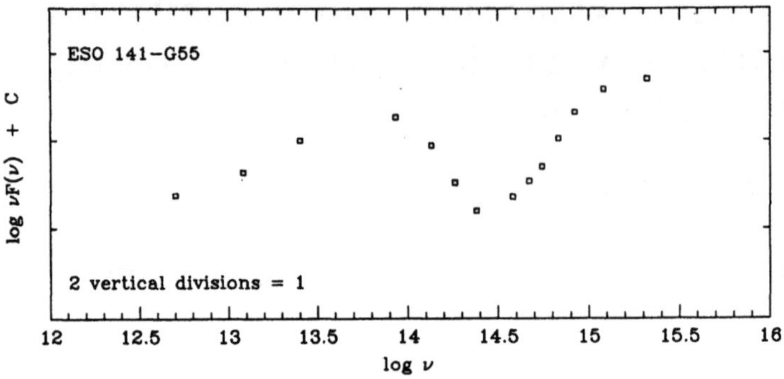

Fig. 2. The flux distribution of ESO 141-G55

as suspected by Martin et al. (1982). Despite this, we believe that the extinction values calculated with our method are at least as good as those found using currently employed methods, such as measuring the Balmer decrement and fitting X-ray spectra.

It is further possible to calculate the strengths of the nuclear and galactic components of the total flux, as shown graphically in Fig. 1, if we know the colours of the host galaxy. These were estimated by taking the difference between large and small aperture measurements of each Seyfert. Note that the procedure assumes that the shape of the flux distribution remains constant. Studies of the flux distribution of Seyfert galaxies in the ultraviolet have instead found that the spectrum is harder at maximum brightness (Barr et al., 1983, Clavel et al., 1991). A relationship proposed by Krolik et al. (1991) to descibe this behaviour in the UV is $\Delta F_\nu / F_{\nu,average} \propto \nu^{0.8}$. This would also lead to a linear relationship between F_{ν_1} and F_{ν_2}. It is, however, inreconcilable with the fact that $\Delta F_\nu / F_{\nu,average}$ can be quite large in the optical. For example, $\Delta F_U / F_{U,average}$ for NGC 3783 is ~ 0.8 (see Fig. 1.). In order to keep ΔF_U smaller than $F_{U,average}$ at higher frequencies we need $\Delta F_\nu / F_{\nu,average}$ to have a much smaller frequency dependence in the optical than suggested by Krolik et al. Hence we propose that the shape of the flux distribution in the optical is constant and that therefore the optical and UV fluxes do not vary in the same manner. In order to find the origin of the optical flux we constructed $0.1 - 100\mu$m nuclear flux distribution, combining our optical results with IUE and IRAS satellite data as well as $JHKL$ fluxes corrected for starlight contamination using the method of Glass (1981). The derived flux distribution of ESO 141-G55 is illustrated in Fig. 2. The other objects' flux distributions, while sometimes not as well defined, appear to have essentially the same shape. Fig. 2 shows that the optical continuum is dominated by the 'blue bump', which is presumed to arise from thermal emission from an accretion disk.

Acknowledgements: We thank Gerhard Bauer, John Innis, Giovanna Stirpe and Andrew Stoddart for their help in the preparation of this paper.

References

Barr, P., Willis, A.J., Wilson, R., 1983: *Monthly Notices Roy. Astron. Soc.*, **202**, 453.
Choloniewski, J., 1981: Acta Astr., 31, 293.
Clavel, J., et al., 1991:*Astrophys. J.*, **366**, 64.
Glass, I.S., 1981: *Monthly Notices Roy. Astron. Soc.*, **197**, 1067.
Krolik, J.H., Horne, K., Kallmann, T.R., Malkan, M.A., Edelson, R.A., Kriss, G.A., 1991: *Astrophys. J.*, in press.
Malkan, M.A., 1983: *Astrophys. J.*, **268**, 582.
Martin, P.G., Stockman, H.S., Angel, J.P.R., Maza, J., Beaver, E.A., 1982: *Astrophys. J.*, **255**, 65.
Winkler, H., Glass, I.S., Van Wyk, F., Marang, F., Spencer Jones, J.H., Buckley, D.A.H., Sekiguchi, K., 1991: *Monthly Notices Roy. Astron. Soc.*, submitted.

3 – Extended Narrow Line Regions and Host Galaxies

Stellar Dynamics of Galactic Nuclei

Ortwin E. Gerhard

Landessternwarte, Königstuhl, 6900 Heidelberg, Germany

1. Introduction

The energy density in the integrated background light from quasars amounts to $\sim 3000\,\mathrm{M}_\odot c^2\,\mathrm{Mpc}^{-3}$, and with the known number density of galaxies this corresponds to about $10^6 h^{-3}\,\mathrm{M}_\odot c^2$ per bright ($L > 0.5L_*$) galaxy[1] (Soltan 1982, Rees 1990). Thus a number of now quiescent galaxy nuclei may have gone through an early active phase. If indeed active galactic nuclei (AGNs) occur in the cores of ordinary galaxies, three equally fascinating questions arise: (i) Can one relate the dynamical structure of the inner galaxy to some of the properties of the embedded AGN? (ii) How does the stellar system react to the gravitational field of the supermassive black hole (SBH) that is believed to power the AGN, and must continue to influence the dynamics of gas and stars in its vicinity after the nuclear activity has faded away? Finally, (iii) how, when, and in what order do the inner galaxy and the SBH form? — This review will concentrate on some aspects of (i) and (ii); the third subject is discussed by Rees in this volume.

The observational evidence relating to the connection between AGNs and galaxy properties, as well as galaxy interactions and environment, is summarized in, e.g., Fricke & Kollatschny (1989) and Woltjer (1990). Conversely, several studies of the stellar kinematics in the nuclei of nearby *normal* galaxies have provided evidence for an increase of the mass-to-light ratio at the smallest observable radii, and have interpreted this as the signature of a dormant SBH (Tonry 1987; Kormendy 1988a,b; Dressler & Richstone 1988). — In the following sections I discuss the dynamical conditions in the cores of normal galaxies, a number of characteristic processes and scales for stellar systems with embedded SBH's, the present dynamical evidence for SBH's in nearby galaxies, and finally some recent work on gas motions and gas inflow in the centers of our own and other galaxies.

---

1 h is Hubble's constant in units of $100\,\mathrm{km\,s^{-1}\,Mpc^{-1}}$. L_* is the scale luminosity in Schechter's luminosity function of galaxies.

2. The Central Parts of Normal Galaxies

Two key issues addressed here are (i) the central stellar density in the nuclei of galaxies, and (ii) the question as to whether cold gas can find stable orbit configurations in the central gravitational potential.

2.1 Core Luminosity Profiles

In most bright galaxies the central light distribution is dominated by a spheroidal component. For a typical bulge or elliptical galaxy the steep outer surface brightness (SB) profile usually flattens inside the inner few 100 pc (hereafter ≡ 'core region'); the question is whether it flattens to constant SB, or keeps rising inwards to pc scales, reaching very high central densities. Seeing always ensures that the innermost part of an observed SB profile is flat; observers then use the concept of an 'apparent core radius' $r_{c,app}$, defined as that radius at which the SB has dropped to half of its central value. $r_{c,app}$ measures a structural property of the intrinsic light profile only when the core is 'well-resolved'.[2] Otherwise the intrinsic inner profile can be close to a pure power law despite the apparent near-constant-SB core.

High-resolution observations show that only some very bright elliptical galaxies have intrinsic King profile cores with constant SB (Kormendy 1985; Lauer 1985). Fainter ellipticals have profiles with less curvature at $r_{c,app}$ which do not flatten completely to constant SB. Nieto *et al.* (1991b) have recently reported that the group of 'disky ellipticals' (probably related to S0 galaxies) remain unresolved. Bulges of spiral galaxies have still steeper inner profiles. Besides M31 (Kent 1983) there are now high internal resolution data on NGC 3115 (Kormendy & Richstone 1991) and NGC 7457 (Lauer *et al.* 1991). In both cases the inner luminosity density profiles are consistent with power laws, with upper limits to the intrinsic core radii r_c of 30 pc and 5.4 pc, respectively (at $h = 0.5$). The density profile of the bulge of our Galaxy is inferred to be $\propto r^{-1.8}$ for $1\,\mathrm{pc} < r < 1200\,\mathrm{pc}$ on dynamical grounds (Genzel & Townes 1987, Binney *et al.* 1991). Finally, the nearby small elliptical M32 also has a power law profile in to $r < 1.1\,\mathrm{pc}$ (Michard & Nieto 1991). The central stellar densities in the spheroids of our Galaxy, NGC 7457, NGC 3115, and M32 are all at least of order 10^5–$10^6\,\mathrm{M_\odot\,pc^{-3}}$, very much more than in the core of a resolved giant elliptical ($\sim 100\,\mathrm{M_\odot\,pc^{-3}}$).

2.2 Nuclei and Nuclear Disks

A nucleus is a dense, dynamically distinct star cluster superposed on a much larger core region. The best-known case is that of M31 (Light *et al.*

[2] Requires $r_{c,app} \gtrsim 5\sigma$ of seeing PSF; e.g., Kormendy 1985.

1974); this nucleus has a mass of several $10^7\,M_\odot$ within $\sim 2\,$pc, luminosity $L_r \sim 5\times10^6\,L_\odot$, and is (2-3) mag brighter than the underlying nuclear bulge. Nuclei are also indicated by inflection points in the inner bulge profiles of a few other nearby spirals (Kormendy 1985), but none are known in ellipticals. (The spike in the center of M87 is now believed to be non-thermal; Dressler & Richstone 1990). Many dwarf spheroidal galaxies contain separate nuclei (Caldwell 1983; van den Bergh 1986), as does M33, a late-type spiral without bulge (Gallagher *et al.* 1982).

The large ratio of rotation velocity to dispersion (v/σ) suggests that the nucleus of M31 is disk-like (Kormendy 1988a), and the misalignment of the nuclear isophotes with those of the bulge that it is barred (Light *et al.* 1974; Gerhard 1988). A nuclear disk with scale $\sim$ 300 pc is found in NGC 4594 (Kormendy 1988b), and probably NGC 3115 contains one (Kormendy & Richstone 1991). Recently, it has emerged that some elliptical galaxies contain disks on scales somewhat larger than the 'core region' (e.g. NGC 3610, Scorza & Bender 1990). These as well as the nuclear disk in NGC 4594 are dominated by the spheroid potential, so do not strictly qualify for a nucleus as defined above. The colours of the disk in NGC 3610 indicate an old stellar population, while the M33 nucleus appears to contain stars of a broad age range. The nuclear densities are $\sim 10^6\,M_\odot/\,pc^3$ in M31, and $\sim 3\times10^4\,M_\odot/\,pc^3$ in M33.

2.3 Core Kinematics and Dynamics

High-resolution kinematic data of galaxy cores are still scarce. Some large resolved cores appear to be anisotropic. This is clearest in NGC 1600, for which the core radius after deconvolution (Lauer 1985) is $r_{c,dec} = 3''.7$ and $(v/\sigma)_c \simeq 0$ (Jedrzejewski & Schechter 1988a). Probably similar are the cores of NGC 720, NGC 4472, NGC 4636 ($r_{c,dec} = 5''.0, 4.''2, 4.''1$, $v/\sigma \lesssim 0.1$), although the kinematic data (Davies & Birkinshaw 1988, Binney *et al.* 1990) only just resolve the cores. An isotropic and rotationally flattened oblate stellar system has $v/\sigma \simeq \sqrt{\epsilon/(1-\epsilon)}$, where ϵ is the apparent ellipticity (Binney 1978; approximate expression from Kormendy 1982). Certainly the main bodies of these ellipticals are strongly anisotropic. Axisymmetry of an anisotropic non-rotating stellar system requires that the mean kinetic energies in two directions are exactly equal; this is generally considered unlikely and a triaxial shape is assumed.

A number of elliptical galaxies have been found to contain kinematically distinct subcomponents in their outer 'core regions': NGC 3608, NGC 4365, NGC 4406, NGC 4494, NGC 5322, NGC 7626, IC 1459 (Bender 1988, Franx & Illingworth 1988, Jedrzejewski & Schechter 1988b). In three of these the 'core' counterrotates relative to the outer galaxy, in two cases 'core' and mainbody rotate around perpendicular axes. Both photometry and absorption line profiles indicate that these are disk-like structures with masses up to 10^{10} $M_\odot$. These observations are generally interpreted as evidence for a past merger, during which the remaining gas settled in a disk according to its own net angular momentum. Because of lack of resolution, little is known about whether the inferred subcomponents actually extend inside the core radius.

Contrary to many bright ellipticals, bulges of disk galaxies rotate as rapidly as is required for an oblate isotropic configuration (Kormendy & Illingworth 1982) and some are fitted well in their detailed kinematics by distribution functions of the form $f(E, L_z)$ (Jarvis & Freeman 1985). This is also true for at least the inner 200 pc of the bulge of M31 (Kent 1989), although this may be triaxial (Stark 1977). Similarly, the ionized gas kinematics in the bulge region of the Sa spiral NGC 4845 is well fitted by the streamlines expected in a triaxial potential (Bertola *et al.* 1989; Gerhard *et al.* 1989), and a study of bulge-disk misalignments (Bertola *et al.* 1991) has recently been used to argue that *the majority* of bulges may not be axisymmetric. In addition, barred structures in the inner disks of spiral galaxies appear to be more common than assumed sofar (Section 5); it will be difficult to disentangle these different components in a dynamical analysis.

2.4 Stellar Orbits in the Core Region

Consider the possible orbits for a stellar or gaseous disk in the core region of a triaxial galaxy. Gas would quickly settle to some family of planar, non-intersecting closed orbits (Section 5), and also a stellar disk with large v/σ must have a distribution function strongly peaked around these closed orbits. In the central parts of a slowly rotating triaxial potential two families of planar loop orbits exist, one in the plane with the two longest, the other in that with the two shortest axes of the potential (e.g., Merritt & de Zeeuw 1983). Both are near-elliptical and elongated along the shorter axis in their plane. Fig. 1 schematicaly shows the equatorial family for various triaxial density profiles. The family extends inwards as long as the density keeps rising; the orbits are the more elongated the shallower the density power-law. In a King-type core, however, all motions are harmonic, and loop orbits are not possible. These appear only by bifurcation of an axial orbit as the latter becomes unstable, which happens at some critical energy E_c corresponding to apocenter near r_c. Thus a rotating disk cannot exist inside a flat triaxial core, and hence the disk-like distortion of the isophotes Nieto *et al.* (1991a) find well inside the anisotropic core of NGC 1600 probably does not signify the presence of a rotating disk.

If gas clouds in a disk outside a triaxial core dissipate enough energy to fall below E_c, they will then fall quickly to the center, where they will loose their orbital energy by collisions with other clouds. A new stable orbit configuration can then be reached only when the gas can cool and find an axisymmetric or sufficiently anharmonic potential at very small radii. On

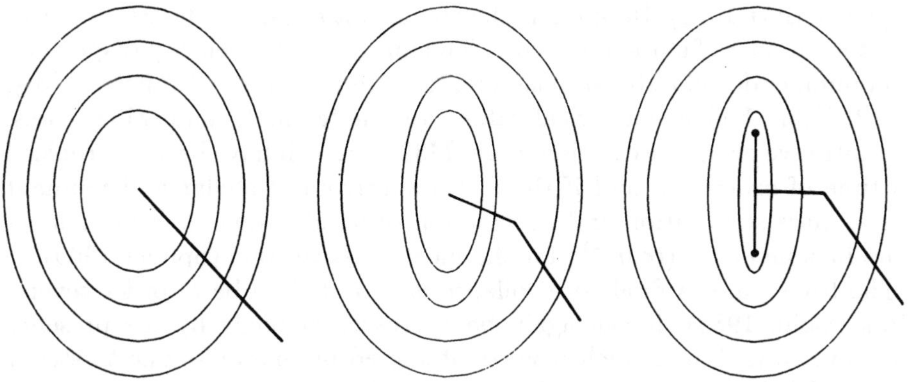

Fig. 1. Closed loop orbits in the equatorial plane of a non-rotating triaxial potential (schematic). From left to right for: pure power-law density profile, double power-law, profile with flat core.

the other hand, Fig. 1 shows that in those spiral bulges with an unresolved power-law density profile, axisymmetric or not, gas can track down to very small radii along a sequence of closed orbits more slowly, giving it more time to form stars on its way. In Section 5 we discuss how this picture is modified if the potential contains a rapidly rotating component.

3. Some Characteristic Scales and Processes

3.1 Core Parameters and Collisions between Stars

For an isothermal core of radius r_c and central one-dimensional velocity dispersion σ_0, the central density is $\rho_0 = 9\sigma_0^2/(4\pi G r_c^2)$. The circular velocity at r_c is $v_c(r_c) = (5/4)\sigma_0$, and hence $M(r_c) \equiv r_c v_c^2(r_c)/G = 1.56 r_c \sigma_0^2/G$ (e.g., Binney & Tremaine 1987). Table 1 gives numerical values for these quantities and the central dynamical time $t_d \equiv 2r_c/\sqrt{3}\sigma_0 \simeq (G\rho_0)^{-1/2}$ in the core of a giant elliptical and a core with the parameters of M32. The numbers for the central parsec of the Galaxy or the nucleus of M31 are not too different from those of M32.

On account of their high densities, some galactic nuclei evolve dynamically by two-body scattering between individual stars. This occurs on the central relaxation time scale

$$t_{r,0} = 0.34 \frac{\sigma_0^3}{G^2 m_* \rho_0 \ln \Lambda} = 1.5 \times 10^9 \; \sigma_{0,100}^3 n_6^{-1} m_{*,\odot}^{-2} (\ln \Lambda)_{12}^{-1} \; \text{yr} \qquad (1)$$

(Spitzer & Hart 1971). Here m_* is the stellar mass and $\Lambda = 0.4 M_c/m_*$ is of order the number of stars in the core. The units used for velocity dispersion, stellar number density, stellar mass, and $\ln \Lambda$ are $100\,\mathrm{km\,s^{-1}}$, $10^6\,\mathrm{pc^{-3}}$, $1\,\mathrm{M_\odot}$, and 12. Table 1 gives numerical values for the 'giant E' and 'M32' cores. The central velocity distribution in an 'M32' core will have been isotropized in a time of order $t_{r,0}$, and if there is a spectrum of stellar and remnant masses, mass segregation and contraction of a subsystem of heavier stars occurs in time $\sim t_{r,0}$ ($\propto m_*^{-1}$) for the most massive stars (Spitzer 1969).

The time-scale to final core collapse is expected to be of order several $100 t_{r,0}$ (Cohn 1980). It can again be decreased strongly by the presence of massive stars, but this effect is counteracted by mass loss due to stellar evolution (Weinberg & Chernoff 1989), the relaxation time being long compared to the main sequence time of the more massive stars. So core collapse should not have occurred in M32 unless the observational limits underestimate its central density by an order of magnitude or more. Finally, the time-scale for direct collisions between stars is given by

$$t_{\mathrm{coll}} \equiv \left[n < \sigma_{\mathrm{coll}} v_{rel} >\right]^{-1} = \frac{\sigma_0}{4\sqrt{\pi} R_*^2 n (v_*^2 + 4\sigma_0^2)} = \frac{0.1 \ln \Lambda \;\, v_*^4}{\sigma_0^2 (v_*^2 + 4\sigma_0^2)}\, t_{r,0}. \quad (2)$$

Here we have assumed a Gaussian distribution of relative velocities with dispersion $\sqrt{2}\sigma_0$ to average over the cross section for stars to come within $2R_*$ of each other (e.g., Spitzer 1987). For solar-type stars in M32 the escape velocity from the stellar surface $v_* \sim 10\sigma_0$, and the gravitational focussing term in (2) dominates. Then the time-scale for direct stellar collisions and tidal formation of binaries (for which the cross section is also of order σ_{coll}; Lee & Ostriker 1986; Benz & Hills 1987) is $\sim 100 t_{r,0}$. Again this time-scale may become interesting for M32 if the current observational limits significantly underestimate the central density.

Table 1. Characteristic numbers for the core of a giant elliptical and the center of M32.

	r_c/pc	$\sigma_0/(\mathrm{km\,s^{-1}})$	$\rho_0/(\mathrm{M_\odot/pc^3})$	$M(r_c)/\mathrm{M_\odot}$	t_d/yr	$t_{r,0}$/yr
giant E	240	240	200	5×10^9	10^6	6×10^{13}
M32	$\lesssim 1$	60	6×10^5	1×10^6	2×10^4	5×10^8

3.2 Cusps around Central Black Holes

How does the stellar system respond to the presence of a massive central BH? Some insight can be obtained by first considering several characteristic radii of the problem. Assume a stellar density profile of the form

$$\rho(r) = \rho_0 \left(\frac{r}{r_0}\right)^{-\alpha}, \quad M(r) = \frac{4\pi}{3-\alpha}\, \rho_0 r_0^3 \left(\frac{r}{r_0}\right)^{3-\alpha} \equiv M_0 \left(\frac{r}{r_0}\right)^{3-\alpha}. \quad (3)$$

If the hole has mass M_h, it will dominate the gravitational force and deflect stars from their orbits within the gravitational radius

$$r_g = (M_h/M_0)^{1/(3-\alpha)} r_0. \tag{4}$$

For the 'giant E' core in Table 1, $r_g \simeq 65 M_{h,8}^{1/3}$ pc ($M_{h,8} \equiv M_h/10^8\,\mathrm{M}_\odot$; α=0). For M32 the result depends on whether the density increases inwards of 1 pc: 0.1 pc $< r_g <$ 0.7 pc for $0 < \alpha < 2.2$ and $M_h = 10^6\,\mathrm{M}_\odot$.

Closer to the hole, stars become bound at radius

$$r_h \simeq GM_h/\sigma_0^2 = 43 M_{h,8}\sigma_{0,100}^{-2}\,\mathrm{pc} \tag{5}$$

and move on essentially Keplerian orbits. In this region a cusp of stars forms in which the density increases inwards and $\sigma \propto r^{-1/2}$. At some point the velocity dispersion in the cusp will be comparable to the escape velocity from the surface of a star. When this happens, stars can no longer deflect each other without physically colliding. This is therefore called the collision radius (Frank & Rees 1976); for solar-type stars

$$r_{\mathrm{coll}} = (M_h/m_*)R_* \simeq 7 \times 10^{18} M_{h,8}\,\mathrm{cm}. \tag{6}$$

Finally, stars entering a radius

$$r_t = 2\,(M_h/m_*)^{1/3} R_* \simeq 6.5 \times 10^{13} M_{h,8}^{1/3}\,\mathrm{cm} \tag{7}$$

will be disrupted by the tidal field of the BH, and some of the debris accreted by the hole. The details of tidal capture and diffusion of stars into the loss-cone will not be discussed further here; the reader is referred to a recent review by Rees (1988).

Inside r_h and outside r_t or r_{coll} a cusp will form around the BH. If the hole grows adiabatically by accretion of loose gas, the structure of the cusp is determined by the adiabatic invariance of the action integrals $\boldsymbol{J}$ of the stars drawn into the cusp. The density and mean-square velocity profiles are given by integrals over velocities of the phase-space distribution function (DF) at small $\boldsymbol{J}$. This results in (Binney & Petit 1989).

$$\rho \propto r^{-3/2} \qquad \bar{v^2} \propto r^{-1} \qquad t_r \propto r^0. \tag{8}$$

The velocity distribution in such a cusp is isotropic near its inner boundary and increasingly tangentially anisotropic further out (Goodman & Binney 1984). From equation (1) one also sees that t_r is proportional to the inverse of the mean phase-space density, so that it is approximately adiabatically invariant for stars drawn into the cusp. Such cusps may therefore form in cores with long relaxation times such as of the 'giant E' in Table 1.

On the other hand, when the relaxation time is short compared to the age of the cusp (as in the fiducial 'M32' core) the structure of the cusp is determined by relaxation effects. In this case the inner boundary is r_{coll}.

For stars with $r_{coll} \ll r \ll r_h$ one might expect near-isotropic velocities and look for a scale-free steady-state DF $\propto E^p$. In a Keplerian potential $E \propto r^{-1}$ and $\bar{v^2} \propto r^{-1}$, so integration over velocities gives $\rho(r) \propto r^{-p-3/2}$. The exponent is $p = 1/4$, as first derived by Bahcall & Wolf (1976) from one-dimensional Fokker-Planck calculations for $f(E)$. This remains true for more general solutions $f(E, L)$ (Murphy *et al.* 1991). Thus in these cusps

$$\rho(r) \propto r^{-7/4} \qquad \bar{v^2} \propto r^{-1} \qquad t_r \propto r^{1/4}. \tag{9}$$

One can derive $p = 1/4$ by a simple scaling argument due to Shapiro & Lightman (1976), which balances inward mass flow and outward energy flow at the tidal radius. For further discussion of dense star clusters containing massive BH's and their evolution see the review by Shapiro (1985).

3.3 Box Orbit Scattering in Triaxial Potentials

In a triaxial potential angular momentum is not conserved, and the influence of a central compact object on the macroscopic dynamics is more profound than in cases with higher symmetry. A triaxial galaxy model depends critically on the existence and selective population of the *box orbits*, in particular the so-called 'thin' box orbits (Schwarzschild 1979; de Zeeuw 1985; Statler 1987). These orbits go arbitrarily close to the centre after sufficient time. Different box orbits of the same energy differ by the sizes of their 'waists', and at $\boldsymbol{r}$=0 by the direction of their velocity vectors. A large enough central mass can scatter stars on box orbits, and by redistributing them in terms of the angular actions, dissolve the heart of the system.

Numerical orbit integrations (Gerhard & Binney 1985) show that most box orbits are destroyed and become stochastic when a BH with mass $M_h = 0.02 M(r_c)$ is introduced at the center of a triaxial galaxy. Similarly, Hasan & Norman (1990) have shown that a BH at the center of a rotating bar destroys a significant fraction of the bar-supporting orbits when its mass becomes greater than several percent of the mass of the bar. In the case of an elliptical galaxy, stochastic orbits at intermediate energies can be described as segments of regular box orbits, separated by discrete scattering events. Denote $\mu \equiv M_h/M(r_c)$; then in a flat core the gravitational radius (4) becomes $r_g \simeq \mu^{1/3} r_c \simeq 0.3 r_c$ for $\mu \simeq 0.02$. Assume that stars from the galaxy enter r_g with a velocity v_g, and that inside r_g they move on hyperbolic Keplerian orbits with asymptotic velocity v_g. Then large-angle scattering occurs for impact parameters $b \lesssim b_l \equiv GM_h/v_g^2$, and empirically one finds that scattering by a point mass becomes effective in destroying the regularity of a box orbit when

$$b \lesssim 3b_l = \frac{3GM_h}{v_g^2} = 3\mu r_c \left(\frac{v_c}{v_g}\right)^2. \tag{10}$$

Here v_c is the circular velocity at r_c. A sufficiently compact nuclear star cluster has a similar effect; in terms of half-mass radius r_n or characteristic velocity v_n an approximate criterion for the necessary compactness is

$$r_n \lesssim 2b_l, \quad \text{or} \quad v_n^2/v_g^2 \lesssim 5. \tag{11}$$

According to equation (10) even a small central point mass will ultimately destroy the box orbits of the *entire* system, so we must next estimate the time-scale. Assuming that a box orbit has uniform probability of going through any point in its waist (see Gerhard & Binney 1985) we have $N(< b) \propto b^2$. Then a star will be scattered after about (area of the waist)/$9\pi b_l^2$ passages through the centre, or after a time

$$\tau_l \simeq \frac{(2r_c)^2}{\pi 9 b_l^2} t_d \propto \mu^{-2} \left(\frac{v_g}{v_c}\right)^4 t_d, \tag{12}$$

which is proportional to M_h^{-2}. In equation (12) both the dimensionless velocity factor and the dynamical time increase with the energy of the orbit; thus for any value of the mass ratio μ, there is a maximum radius out to which box orbit scattering is effective, and, conversely, for any radius there is a minimum BH mass required for scattering in a Hubble time. A central BH of $\sim 10^8 M_\odot$ in a giant elliptical core (Table 1) will scatter box orbits out to ~ 1 kpc. For the nucleus of M31 the maximum radius for box orbit scattering is not determined by the time-scale τ_l, but through the compactness condition (11); it is numerically found to be $\sim 500pc$ (Gerhard 1986).

The scattering by a central BH in a triaxial galaxy will tend to establish a uniform distribution of box orbits in angular actions. Thus one expects that the inner parts of the system must ultimately become axisymmetric. Preliminary calculations indicate that an isotropic distribution of box orbits in the core favours evolution towards a prolate shape. However, Pfenniger & de Zeeuw (1989) have raised the possibility that the evolution might cease when the core encounters a resonant frequency ratio. Outside the core, triaxiality may in principle be maintained by the two loop orbit families, but some roundening is expected (Norman *et al.* 1985). To follow the self-consistent evolution of a triaxial system due to box orbit scattering by a central BH is a difficult problem; even qualitative aspects remain unclear.

4. Kinematic Evidence for Massive Black Holes in Nearby Normal Galaxies

Detecting dormant black holes in normal galaxy nuclei and determining their masses and frequencies can provide important clues for quasar models and lifetimes. So what would be the dynamical signatures of a remnant SBH? The clearest evidence for a very deep potential well would be the observation of stellar velocities far above those normally seen in galaxies. A $10^8\,\mathrm{M}_\odot$ SBH at the center of M31 would result in $\sim 500\,\mathrm{km\,s^{-1}}$ at 0.″5 resolution and this has not been seen. In more distant galaxies the footprints of the hole are more subtle for any reasonable mass, and one must rely on comparing high-resolution measurements of surface brightness, stellar kinematics, and their gradients with stellar-dynamical models.

4.1 Dynamics

(i) Cold Nuclear Disks Observing the rotation velocities of gas or stars on near-circular orbits sufficiently close to the hole would give fairly unambiguous results:

$$M(r) = v_c^2 r/G \tag{13}$$

and one would hope for $M(r) \to \text{constant} = M_h$ as $r \to 0$.

(ii) Spherical Models The next simplest case is that in which the surrounding star cluster is spherical. Then we can use the Jeans equation (e.g., Binney & Tremaine 1987)

$$M(r) = -\frac{r\sigma_r^2}{G}\left(\frac{\mathrm{d}\ln\rho}{\mathrm{d}\ln r} + \frac{\mathrm{d}\ln\sigma_r^2}{\mathrm{d}\ln r} + 2\beta\right), \tag{14}$$

where ρ is the density of a tracer population that need not be self-consistent with the potential, $\sigma_r(r)$ its radial velocity dispersion, and $\beta(r) \equiv 1 - \sigma_\phi^2/\sigma_r^2 < 1$ the anisotropy parameter. A SBH would be indicated by negative logarithmic gradients of ρ or σ_r such that $M(r) \to M_h$ for $r \to 0$. However, radial variations of β introduce an intrinsic complication that often renders the results ambiguous, and a confident detection is possible only if reasonable mass-to-light ratios require $\beta > 1$ somewhere. Equation (14) is a moment of the Vlasov equation and as such represents, for fixed potential $\Phi(r)$, a constraint that every DF $f(E,L)$ must satisfy. However, there is no guarantee that a solution of (14) with constant M/L, say, actually corresponds to a smooth and everywhere non-negative DF. Thus it is important to construct model DF's for a set of observations.

(iii) Axisymmetric Models These models are the simplest that allow one to incorporate the observed flattening and rapid rotation in most of the

BH candidate nuclei discussed below. Unfortunately, the extra freedom introduces further complications. The Jeans equation in the equatorial plane $z = 0$ of the system in cylindrical coordinates now reads

$$\begin{aligned}\left(\frac{\partial \Phi}{\partial R}\right)_{z=0} = \frac{\bar{v}_\phi^2}{R} - \frac{\sigma_R^2}{R}\left[\frac{\partial \ln \rho}{\partial \ln R} + \frac{\partial \ln \sigma_R^2}{\partial \ln R}\right. \\ \left. +1 - \frac{\sigma_\phi^2}{\sigma_R^2} + \frac{R}{\sigma_R^2}\frac{\partial \overline{v_R v_z}}{\partial z}\right]_{z=0}.\end{aligned} \tag{15}$$

Here $\bar{v}_\phi$ is the intrinsic rotational streaming velocity. The application of this equation requires assumptions about the kinematic structure of the model, as the last derivative term depends on the orientation of the stellar velocity ellipsoid. This term vanishes for a flattened system with velocities isotropic in the meridional plane, and also when the velocity ellipsoid separates in cylindrical coordinates. Equation (15) is also valid for stars in the equatorial plane of a spherical potential, in which case $\partial \overline{v_R v_z}/\partial z = (\sigma_R^2 - \sigma_z^2)/R$. The equatorial potential gradient (15) may be associated with a 'mass' $M(R) = R^2(\partial \Phi/\partial R)_{z=0}/G$, which overestimates the real mass within R by a factor of order unity except in this spherical case.

Assuming $\sigma_R^2 = \sigma_z^2$ restricts the distribution function to depend only on the two classical integrals E, L_z but not on the third, non-classical I_3. Complete isotropy (assuming also $\sigma_\phi^2 = \sigma_R^2$) further restricts the DF. Therefore in the axisymmetric case it is even more necessary to explore the full range of *distribution functions* that are consistent with the observed kinematics. Techniques for doing this are now available (Dejonghe & de Zeeuw 1988 using Stäckel potentials when M/L = constant, McGill & Binney 1990 using numerical torus construction, and Gerhard & Saha 1991 using perturbation theory to approximate the third integral), but no detailed such models have yet been constructed for any of the BH candidate nuclei.

(iv) Triaxial and Rotating Bar Models Even less is known about triaxial models that could answer whether a given velocity dispersion profile is consistent with constant M/L, and yet these may well be the relevant models since Merritt (1987) has shown that Newton & Binney's (1984) anisotropic spherical model for M87 is unstable to bar formation. Binney & Petit (1989) have constructed approximate DF's by using a spherical model as starting point and shifting stars on energy surfaces until the density distributions along the principal axes were approximately reproduced. From the kinematics of these models they conclude that it will be more difficult to produce centrally peaked velocity dispersion profiles with a triaxial model than with an anisotropic spherical model of the same density structure. Statler (1987) has constructed self-consistent triaxial Stäckel models for which three integrals of motions are known. However, these have not yet been applied to any galaxy with a suspected central SBH.

Some of the nuclei discussed below rotate very rapidly and are probably disk-like. Especially for the nucleus of M31 both the existing photometry and the fact that it is self-gravitating suggests that it may be a rotating, possibly fat bar. Self-consistent models of such systems to date are mostly N-body. Some orbit distribution functions have been constructed by Pfenniger (1984). These models rely to large degree on the rapidly corotating long-axial family of orbits. The observed streaming motions depend strongly on aspect angle. Only a heuristic application to M31's nuclear kinematics has been attempted, (Gerhard 1988), but no dynamically self-consistent such model exists yet.

4.2 Candidate Black Holes

Table 2 below lists present candidate BH galaxies and inferred BH masses. The notes give details about the type of model used for the dynamical analysis and the respective assumptions and uncertainties. In all cases the primary indication for a SBH stems from the measured stellar kinematics (major and often minor axis; approximate velocity field from several slits in M32). No cases are known with a distinct central light cusp or large nuclear gas rotation. In the IR galaxy NGC 6240 (Bland-Hawthorn *et al.* 1991) one of two interacting disks shows very large streaming velocities. The ongoing interaction makes the interpretation difficult; so this galaxy is not included in Table 2. Only recent data and modelling based on it are included. For earlier work see references in the papers quoted.

In M87, a strong central dip in line strength indicates that the central light cusp is non-thermal. The measured dispersion σ_p levels off at $360\,\mathrm{km\,s^{-1}}$ inside 1". The new data are consistent with M_h=0.

In M32, rotation is unresolved at $v_m = 60\,\mathrm{km\,s^{-1}}$, while the intrinsic $\sigma = 56\,\mathrm{km\,s^{-1}}$ is constant. Assuming that the galaxy is less inclined will increase velocities and masses. The central rise in M/L is modest. The assumption of isotropy is reasonable as the central relaxation time is short, but this may also have caused some mass segregation as discussed in Section 3.

In M31, the photometric center of the nucleus in both the optical and near-infrared is offset from its kinematic center, and the nuclear isophotes are not aligned with the bulge (Light *et al.* 1974; Mould *et al.* 1989). The cause of this is unclear at present, as is its effect on the existing dynamical models.

NGC 4594 contains a nuclear disk. The same is probably true for NGC 3115; isophote analysis (Nieto *et al.* 1991a) indicates that the disk may be a non-negligible part of the luminous mass at 1". If so, it may increase the predicted rotation velocity in the radial range within which the zero BH mass maximum entropy model deviates most from the observations; but whether this is a significant effect is unknown.

Table 2. Galaxies with black hole candidates. Columns give galaxy, resolution of kinematic observations, typical model values for the cumulative mass-to-light ratios ζ^V inside 0."5 and 1."0, bulge value, range of derived black hole masses, code for type of dynamical model employed, references.

Galaxy	Resol^n	$\zeta^V_{<0.''5}$	$\zeta^V_{<1.''0}$	ζ^V_b	$M_h/M_\odot$	Model	Ref.
M87	1."0	10	10	10	0-2 $\times 10^9$	1	1
M32	1."0	7	4	3	3-10$\times 10^6$	2	2
	1."0				0-3 $\times 10^6$	3	3,4
M31	1."1	40	20	5	0.4-5 $\times 10^7$	4	5
	1."0				4-5 $\times 10^7$	3	3,4
	—	8	8		0	5	6
N 4594	0."9	25	10	4	0.5-2 $\times 10^9$	4	7
	2."0				1 $\times 10^9$	6	8
N 3115	1."0	35	20	4	0.3-3 $\times 10^9$	4	9
	1."0				0.1-1.4 $\times 10^9$	7	9

Notes: The number given under 'resolution' is the seeing FWHM (in one case the larger slit width) during the best spectroscopic observation. Models used to fit the data and to determine masses and mass-to-light ratios are denoted as follows:

[1] Spherical DF $f(E, L)$, by maximum entropy method from orbit library. Assumptions: Isotropic or anisotropic. Uncertainties: ragged $\sigma_{r,\theta}(r)$.

[2] Jeans equation in spherical potential (obs. $\epsilon \simeq 0.2 - 0.3$). Assumptions: Isotropic, edge-on for kinematic model.

[3] Spherical DF $f(E, L)$ as for M87. Maximum streaming, tangentially anisotropic.

[4] Assumes an intrinsic bulge velocity and dispersion field, adding Keplerian and possibly disk components. Masses from spherical Jeans equation. Isotropic.

[5] Heuristic model of flat nuclear bar, assumed in galaxy plane; orientation from isophote deprojection. Uncertainties: streaming motions in self-consistent bars.

[6] v^2r/G at 3."5.

[7] As [3] but includes flattening correction not present in other models.

References are: [1]Dressler & Richstone (1990); [2]Tonry (1987); [3]Dressler & Richstone (1988); [4]Richstone *et al.* (1990); [5]Kormendy (1988a); [6]Gerhard (1988); [7]Kormendy (1988b); [8]Jarvis & Dubath (1988); [9]Kormendy & Richstone (1991). Quoted mass-to-light ratios for M87 stem from Richstone & Tremaine (1985), and value for bulge of M31 from Kent (1989).

Compact nuclear star clusters cannot yet be ruled out as an explanation for the observed mass concentrations (Goodman & Lee 1989). A small admixture of young stars in a nucleus, or an unusually large fraction of dark remnants with star-burst origin might both modify the expected M/L of the central stellar population (in opposite ways).

In conclusion: There are considerable uncertainties in individual cases, as shown by the range of derived BH masses. In most galaxies, these are only a fraction of the total mass inside 1", so careful work is required. However, not all possible anisotropies and other ways out need to be realized in nature; the weight of the evidence seems to point to the existence of compact central mass concentrations in nearby normal galaxies. In order to constrain the implied masses more tightly, future modelling should involve constructing self-consistent DF's in axisymmetric and triaxial potentials and make use of information in the absorption line profiles besides rotation velocity and dispersion.

5. Gas Dynamics at the Galactic Center

In the past year a group of us (Stark *et al.* 1991; Binney *et al.* 1991) have been working on the dynamics of cold gas at the Galactic center (GC). I include the results of this work here, not only because of my recent interest in this topic, but also because these results may have some relevance for angular momentum loss and gas motions in AGNs on scales of order 100 pc.

Interferometric observations of molecular emission at a resolution of several arcsec have shown that nuclear disks, rings, and bars of molecular gas are common in spiral galaxies (Sofue 1991). Non-circular motions appear to play an important role in driving the present evolution in the starburst galaxy IC 342 (Ishzuki *et al.* 1990). The classical Seyfert galaxy NGC 4151 shows signs of a non-axisymmetric velocity field – albeit on kpc scales (Bosma 1977); it is tempting to speculate that there also barred structures are driving gas inwards.

5.1 Kinematical Model of Galactic Center Gas

In summary, the kinematics of the GC gas indicate a rapidly rotating barred potential. Our model is based mostly on the mm-line survey of Bally *et al.* (1987, 1988), but also on HI data by Burton & Liszt (1978) and Sinha (1979). The mm-survey provides high-resolution data cubes of intensity as a function of longitude, latitude, and radial velocity (l, b, v) in ^{12}CO, ^{13}CO, and CS, tracing molecular gas of densities from $\sim 200\,\mathrm{cm}^{-3}$ to $\sim 3\times10^4\,\mathrm{cm}^{-3}$. The data are fully sampled and typical resolution is $1' = 2.5\,\mathrm{pc}$. They show a number of different components (see Binney *et al.* 1991), of which only the 'Galactic Center Disk Material' is modelled. This comprises about 70% of the molecular emission, including a few very massive clouds, and is highly flattened to the plane. It is lumpy, strongly asymmetric both in l and v, and shows large non-circular motions.

The key to the model is a distinct feature in the (l, v)-diagram, which we have called the molecular parallelogram (also known as 'expanding molecular ring'). This feature is produced in the model by closed orbits in a barred potential in the vicinity of the Inner Lindblad Resonance (ILR). Closed orbits provide the basis to understanding the orbital structure of galaxy potentials: most other orbits can be described as epicyclic oscillations about an underlying stable closed orbit. A large cloud of gas released into a potential will shear and then settle onto closed orbits, because collisions between subclouds dissipate the kinetic energy stored in epicyclic librations. Thereafter, dissipation will cause the gas to track down slowly along a sequence of closed orbits of decreasing energy.

In a barred potential, this picture is complicated by the fact that, at most energies, more than one closed orbit exists. Furthermore, the sequence of closed prograde long-axis orbits (also called 'x_1') becomes self-intersecting near the ILR and is thus not suitable for a smooth gas flow at lower energies. Near such intersections pressure forces become important and the description in terms of simple ballistic orbits must be modified. Hydrodynamic simulations (Mulder & Liem 1986; Athanassoula 1988) indicate that when gas reaches the highest-energy self-intersecting x_1-orbit, a shock forms which causes it to transfer to another family of closed orbits, the prograde short axis or 'x_2' family. These orbits, deeper in the potential well, are roughly elliptical and less elongated than the x_1-orbits. Fig. 2 shows the sequence of closed orbits and their projection into the (l, v)-diagram for an observer near the long axis of the bar, as in the model of Binney *et al.*

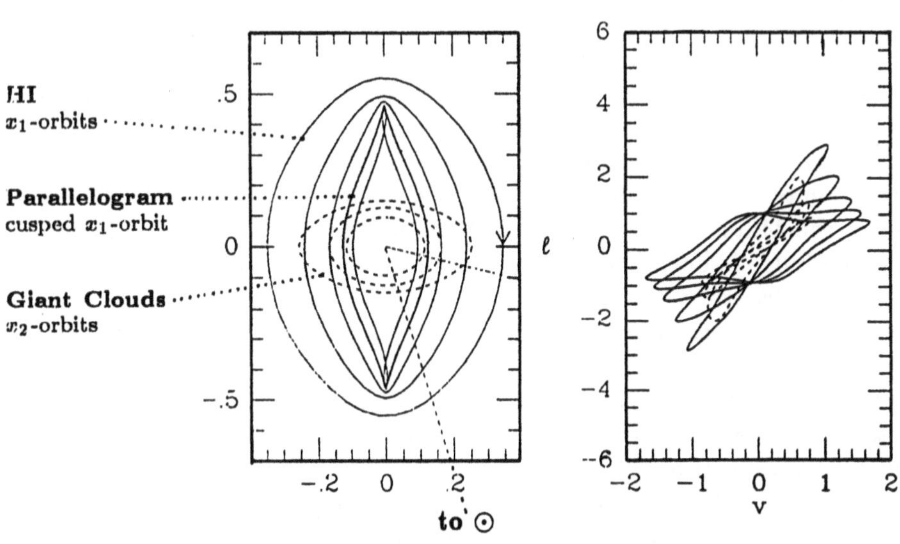

Fig. 2. Closed orbits in a rotating barred potential and their projections into the (l, v)-diagram for an observer near the long axis of the bar

The model parameters are as follows. The GC bar has density profile

$$\rho(\boldsymbol{r}) = a^{-\alpha}, \qquad a = \left(x^2 + \frac{y^2 + z^2}{q^2}\right)^{1/2} \qquad \alpha = \begin{cases} 1.7, & \text{if } a < 1.2\,\text{kpc}; \\ 3.5, & \text{if } a > 1.2\,\text{kpc}. \end{cases} \tag{16}$$

The axial ratio of the bar is constrained only weakly, but the aspect angle of the bar for an observer near the Sun, measured from the bar major axis, is determined well:

$$q \gtrsim 0.75, \qquad \theta_{maj} = 16 \pm 2\,\text{deg}. \tag{17}$$

The pattern speed of the bar must be such that the ILR is near the parallelogram, and its M/L is fixed by the maximum velocity in the HI (l, v)-diagram occurring just outside the ILR. Corotation radius and M/L in the model are

$$R_{corot} \simeq 2.4\,\text{kpc}, \qquad M/L_B \simeq 4\,\mathrm{M}_\odot/L_{\odot,B}. \tag{18}$$

The clearest indications that a model of this form approximates the dynamical conditions near the GC are:

- The molecular parallelogram, which is interpreted as the (l, v)-trace of almost a single orbit. This is the 'cusped' orbit in Fig. 2, modified by hydrodynamic effects. The absence of CO on larger x_1-orbits suggests that molecular material is formed in the strong shocks near this orbit.
- The rapid fall-off in the HI terminal velocity observed between $l = 2$ deg and $l = 12$ deg. That this decline occurs over a considerable range in l indicates that corotation and ILR are well-separated, and this in turn implies that the circular velocity increases *outwards* through this range.
- The observed left-right asymmetry of the parallelogram (~ 0.3 deg to $l > 0$). This is at least partly explained as a perspective effect, the nearer side of the bar lying at positive l. The sense of the predicted effect is consistent with one's impression from NASA's COBE/DIRBE near-infrared fotograph of the Galactic bulge, and with Blitz & Spergel's (1991) analysis of the Matsumoto *et al.* (1982) balloon data, where they concluded that the bulge is both brighter and more extended in the first than in the fourth quadrant.

Several other features seem to fit well into this general scheme. A few of these are:

- The absence of cold gas between $R \sim 1.5$ kpc and $R \sim 3.5$ kpc, since in the neighbourhood of corotation there are no stable closed orbits on which it could settle.
- The molecular ring observed at $R \sim 3.5$ kpc may be associated with gas accumulating near the outer Lindblad resonance of the bar.
- The giant molecular clouds at the GC like Sag B and Sag C are predicted to move presently on x_2-orbits inside the ILR .

For details of the comparison to the observations see Binney *et al.* (1991).

5.2 Dynamical Friction on Galactic Center Clouds

In this rapidly rotating barred potential molecular gas is moving inwards through the cusped orbit shock and then presumably settles on the stable near-elliptical x_2-orbits. In the Galaxy this is a region of $\sim 200\,\mathrm{pc}$ radius. Gas distributed on x_2-orbits could again set up a smooth flow without much dissipation and angular momentum loss. However, a large fraction of the Galactic gas in this region is in the form of massive, dense molecular clouds, which in turn loose angular momentum to the bulge stars by *dynamical friction.*

When a massive cloud moves through a background of light particles (here the bulge stars) it deflects these particles preferentially behind itself. The resulting overdensity is proportional to the mass of the cloud m and causes a decelerating ('friction') force $\propto m^2$. If the could moves on a circular orbit in an isothermal sphere at initial radius R, this 'friction' causes it to spiral to the center in time

$$\begin{aligned} t_{\mathrm{fric}} &= \frac{1.17}{2\pi \ln \Lambda}\left(\frac{M(R)}{m}\right) t_{\mathrm{orb}} \\ &= 0.4\left(\frac{m}{5\times 10^6\,\mathrm{M}_\odot}\right)^{-1}\left(\frac{R}{500\,\mathrm{pc}}\right)^2\left(\frac{\ln \Lambda}{6}\right)^{-1}\,\mathrm{Gyr}, \end{aligned} \tag{19}$$

where $\ln \Lambda$ is the Coulomb logarithm, $t_{\mathrm{orb}} \equiv 2\pi R/v_c$ is the initial orbital period, and a circular velocity of $v_c = 180\,\mathrm{km\,s^{-1}}$ is assumed.

The masses of the GC cloud complexes were estimated by Stark *et al.* (1991) in several independent ways; for Sgr B a mass of $m \gtrsim 5 \times 10^6\,\mathrm{M}_\odot$ resulted. This correspond to a spiraling-in time $t_{\mathrm{fric}} \lesssim 0.4\,\mathrm{Gyr}$. The masses of the other clouds are smaller, but they still spiral to the center within a Gyr. The conclusion is that the present population of Galactic centre clouds must be transient, and – if the present rate of angular momentum loss is typical – that material is accreting onto the GC at an average rate of $\sim 0.01 - 0.1\,\mathrm{M}_\odot/\,\mathrm{yr}$. At a galactocentric radius of $\gtrsim 30\,\mathrm{pc}$ (the present projected radius of Sgr A) the tidal field of the nuclear bulge becomes strong enough to overwhelm the internal gravity of a typical cloud; the fate of the gas after the ensuing disruption is uncertain.

Star formation might disrupt a cloud complex. But because the friction wakes from individual sublumps add extensively, it is not sufficient to expand a complex by merely a factor of a few; the minimum requirement for halting the infall of a cloud is to disperse its material into an annulus. Then it could only reform in several dynamical times. The observed configuration of clouds suggests that the material remains lumpy on average, and continues its irreversible angular momentum loss, presumably making stars on its way. But it does not seem likely that a cloud's star formation rate would make it disappear just before it reached the center. As the cloud sinks in the potential well, its rate of descent increases, until finally the remaining mass is dumped rapidly onto the GC, leading presumably to a period of enhanced star formation. Integrated over the age of the Galaxy, the mass added to the inner bulge in this way must be significant.

I am grateful to my collaborators, J. Bally, J. Binney, A. Stark, K. Uchida; to J. Kormendy for sending material on NGC 3115 in advance of publication; to R. Bender and J. Binney for comments on the manuscript; and to the Heisenberg foundation for support.

References

Athanassoula, E., 1988. In: Proceedings of the joint Varenna-Abastumani International School & Workshop on Plasma Physics, ESA SP-285, vol. I, eds. Guyenne, T.D., Hunt, J.J. (ESA, Paris) p.341
Bahcall, J.N., Wolf, R.A., 1976. Astrophys. J. **209**, 214.
Bally, J., Stark, A.A., Wilson, R.W., Henkel, C., 1987. Astrophys. J. Supp. Ser. **65**, 13.
Bally, J., Stark, A.A., Wilson, R.W., Henkel, C., 1988. Astrophys. J. **324**, 223.
Bender, R., 1988. Astron. Astrophys. **202**, L5.
Benz, W., Hills, J.G., 1987. Astrophys. J. **323**, 614.
Bertola, F., Rubin, V.C. Zeilinger, W., 1989. Astrophys. J. (Letters) **345**, L29.
Bertola, F., Vietri, M., Zeilinger, W., 1991. Astrophys. J. (Letters) **374**, L13.
Binney, J.J., 1978. Mon. Not. R. astr. Soc. **183**, 501.
Binney, J.J., Davies, R.L., Illingworth, G.D., 1990. Astrophys. J. **361**, 78.
Binney, J.J., Gerhard, O.E., Stark, A.A., Bally, J., Uchida, K.I., 1991. Mon. Not. R. astr. Soc. , in press.
Binney, J.J. & Petit, J.-M., 1989. In: Dynamics of Dense Stellar Systems, ed. D. Merritt (Cambridge Univ. Press, Cambridge) p.43
Binney, J.J. & Tremaine, S.D., 1987. Galactic Dynamics, Princeton University Press, Princeton.
Bland-Hawthorn, J., Wilson, A.S., Tully, R.B., 1991. Astrophys. J. (Letters) **371**, L19.
Blitz, L., Spergel, D.N., 1991. Preprint.
Bosma, A., 1977. Astron. Astrophys. **57**, 97.
Burton, W.B., Liszt, H.S., 1978. Astrophys. J. **225**, 815.
Caldwell, N., 1983. Astron. J. **88**, 804.
Cohn, H., 1980. Astrophys. J. **242**, 765.
Davies, R.L., Birkinshaw, M., 1988. Astrophys. J. Supp. Ser. **68**, 409.
Dejonghe, H., de Zeeuw, P.T., 1988. Astrophys. J. **333**, 90.
de Zeeuw, P.T., 1985. Mon. Not. R. astr. Soc. **216**, 273.
Dressler, A., Richstone, D.O., 1988. Astrophys. J. **324**, 701.
Dressler, A., Richstone, D.O., 1990. Astrophys. J. **348**, 120.
Frank, J., Rees, M.J., 1976. Mon. Not. R. astr. Soc. **176**, 633.
Franx, M., Illingworth, G.D., 1988. Astrophys. J. (Letters) **327**, L55.
Fricke, K.J., Kollatschny, W., 1989. In: Active Galactic Nuclei, IAU Symp. No. 134, eds. D.E. Osterbrock & J.S. Miller (Kluwer, Dordrecht) p.425.
Gallagher, J.S., Goad, J.W., Mould, J., 1982. Astrophys. J. **263**, 401.
Genzel, R., Townes, C.H., 1987. Ann. Rev. Astr. Astrophys. **25**, 377.
Gerhard, O.E., 1986. Mon. Not. R. astr. Soc. **219**, 373.
Gerhard, O.E., 1988. Mon. Not. R. astr. Soc. **232**, 13P.
Gerhard, O.E., Binney, J.J., 1985. Mon. Not. R. astr. Soc. **216**, 467.
Gerhard, O.E., Saha, P., 1991. Mon. Not. R. astr. Soc. **251**, 449.
Gerhard, O.E., Vietri, M., Kent, S.M., 1989. Astrophys. J. (Letters) **345**, L33.
Goodman, J., Binney, J.J., 1984. Mon. Not. R. astr. Soc. **207**, 511.
Goodman, J., Lee, H.M., 1989. Astrophys. J. **337**, 84.
Hasan, H., Norman, C., 1990. Astrophys. J. **361**, 69.
Ishizuki, S., Kawabe, R., Ishiguro, M., Okumura, S. K., Morita, K-I., Chikada, Y., Kasuga, T., 1990. Nature **344**, 224.
Jarvis, B.J., Dubath, P., 1988. Astron. Astrophys. **201**, L33.

Jarvis, B.J., Freeman, K.C., 1985. Astrophys. J. **295**, 324.
Jedrzejewski, R., Schechter, P.L., 1988a. Astron. J. **98**, 147.
Jedrzejewski, R., Schechter, P.L., 1988b. Astrophys. J. (Letters) **330**, L87.
Kent, S.M., 1983. Astrophys. J. **266**, 562.
Kent, S.M., 1989. Astron. J. **97**, 1614.
Kormendy, J., 1982. In: Morphology and Dynamics of Galaxies, 12th Adv. Course Swiss Soc. Astr. Astroph., eds. L. Martinet, M. Mayor (Geneva Obs.) p.115.
Kormendy, J., 1985. Astrophys. J. (Letters) **292**, L9.
Kormendy, J., 1988a. Astrophys. J. **325**, 128.
Kormendy, J., 1988b. Astrophys. J. **335**, 40.
Kormendy, J., Illingworth, G.D., 1982. Astrophys. J. **256**, 460.
Kormendy, J., Richstone, D.O., 1991. Preprint.
Lauer, T.R., 1985. Astrophys. J. **292**, 104.
Lauer, T.R. etal., 1991. Astrophys. J. (Letters) **369**, L41.
Lee, H.M., Ostriker, J.P., 1986. Astrophys. J. **310**, 176.
Light, E.S., Danielson, R.E., Schwarzschild, M., 1974. Astrophys. J. **194**, 257.
Matsumoto, T.ëtal., 1982. In: The Galactic Center, AIP Conf. No. 83, eds. G.R. Riegler & R.D. Blandford (American Institute of Physics, New York) p. 48.
McGill, C., Binney, J.J., 1990. Mon. Not. R. astr. Soc. **244**, 634.
Merritt, D. 1987. Astrophys. J. **319**, 55.
Merritt, D., de Zeeuw, P.T., 1983. Astrophys. J. (Letters) **267**, L19.
Michard, R., Nieto, J.-L., 1991. Astron. Astrophys. **243**, L17.
Mould, J., Graham, J., Matthews, K., Soifer, B.T., Phinney, E.S., 1989. Astrophys. J. (Letters) **339**, L21.
Mulder, W.A., Liem, B.T., 1986. Astron. Astrophys. **157**, 148.
Murphy, B.W., Cohn, H.N., Durisen, R.H., 1991. Astrophys. J. **370**, 60.
Newton, A.J., Binney, J.J., 1984. Mon. Not. R. astr. Soc. **210**, 711.
Nieto, J.-L., Bender, R., Arnaud, J., Surma, P., 1991a. Astron. Astrophys. **244**, L25.
Nieto, J.-L., Bender, R., Surma, P., 1991b. Astron. Astrophys. **244**, L37.
Norman, C.A., May, A., van Albada, T.S., 1985. Astrophys. J. **296**, 20.
Pfenniger, D., 1984. Astron. Astrophys. **134**, 373.
Pfenniger, D., de Zeeuw, P.T., 1989. In: Dynamics of Dense Stellar Systems, ed. D. Merritt (Cambridge Univ. Press, Cambridge) p.81
Rees, M.J., 1988. Nature **333**, 523.
Rees, M.J., 1990. Science **247**, 817.
Richstone, D.O., Bower, G., Dressler, A., 1990. Astrophys. J. **353**, 118.
Richstone, D.O., Tremaine, S., 1985. Astrophys. J. **296**, 370.
Schwarzschild, M., 1979. Astrophys. J. **232**, 236.
Scorza, C., Bender, R., 1990. Astron. Astrophys. **235**, 49.
Shapiro, S.L., 1985. In: Dynamics of Star Clusters, IAU Symp. No. 113, eds. J. Goodman & P. Hut (Kluwer, Dordrecht) p.373.
Shapiro, S.L., Lightman, A.P., 1976. Nature **262**, 743.
Sinha, R.P., 1979. Astron. Astrophys. Supp. **37**, 403.
Sofue, Y., 1991. In: Dynamics of Galaxies and their Molecular Gas Distributions, IAU Symp. No. 146, eds. F. Combes & F. Casoli (Kluwer, Dordrecht) p.287.
Soltan, A., 1982. Mon. Not. R. astr. Soc. **200**, 115.
Spitzer, L., 1969. Astrophys. J. (Letters) **158**, L139.
Spitzer, L., 1987. Dynamical Evolution of Globular Clusters, Princeton University Press, Princeton.
Spitzer, L., Hart, M.H., 1971. Astrophys. J. **164**, 399.
Stark, A.A., 1977. Astrophys. J. **213**, 368.
Stark, A.A., Gerhard, O.E., Binney, J.J., Bally, J., 1991. Mon. Not. R. astr. Soc. **248**, 14P.
Statler, T.S., 1987. Astrophys. J. **321**, 113.
Tonry, J.L., 1987. Astrophys. J. **322**, 632.
van den Bergh, S., 1986. Astron. J. **91**, 271.
Weinberg, M.D., Chernov, D.F., 1989. In: Dynamics of Dense Stellar Systems, ed. D. Merritt (Cambridge Univ. Press, Cambridge) p.221

Woltjer, L., 1990. In: Active Galactic Nuclei, R.D. Blandford, H. Netzer, L. Woltjer (Springer, Berlin) p.1.

Discussion

Bland-Hawthorn: It would seem that proper motions from VLBI observations over the next few years could help to check your model. How confident are you that the Galactic Center potential is smooth?

Gerhard: Fluctuations will be present at least on the level corresponding to the masses and sizes of the observed giant molecular clouds.

Kundt: An estimate of gas spiral-in rates via magnetic torques for our and similar galaxies yields slightly higher rates than your $0.1\,M_\odot/\,yr$ (for the inner part; *Astrophys. Sp. Sci.* **172**, 285, 1990).

Lesch: Why are the observed molecular gas ring like structures not destroyed by dynamical friction?

Gerhard: Stationary, smooth rings in a symmetry plane gain no net momentum or energy from scattered stars. The more a ring breaks up in sublumps and the less stationary it is, the more energy exchange (friction) with the bulge stars occurs.

Pudritz: Your interesting time-scale for dynamical friction $t_{df} = 0.4\,\text{Gyr}$ is still much larger than the lifetime of a molecular cloud. Massive star formation in such a cloud would likely disperse it long before it reached the center. Could you comment on this.

Gerhard: Yes, star formation might disperse a cloud. But, as I have argued, to stop angular momentum loss requires to disperse it on *large* scales, into an annulus. This requires a starburst of energy $\sim {}^3/_2 m\sigma^2 \simeq 10^{53}$erg; $\gtrsim 100$ supernovae would in principle suffice (using $m = 5 \times 10^6\,M_\odot$ for Sgr B). However, unless we live at a special time, the fact that we see large lumps now suggests that the annulus would lump up again in a few dynamical times, and angular momentum loss would continue. — Also notice that Sgr A is now about as close to the center as it can come without being disrupted by the tidal field of the nuclear bulge.

Internal Dynamics of NGC 6240: Evidence for a Massive, Dark Compact Core?

Jonathan Bland-Hawthorn[1] *and Andrew S. Wilson*[2]

[1] Dept. of Space Physics and Astronomy, Rice University, Houston, TX 77251-1892, USA
[2] Astronomy Program, University of Maryland, College Park, MD 20742, USA

In a recent paper, we presented an imaging Fabry-Perot Hα study of the luminous infrared galaxy NGC 6240. The velocity field from these spectrophotometric observations may be interpreted in terms of two overlapping "spider diagrams" whose centers are closely spaced in velocity and in projection. One of these systems (disk 1) exhibits flat rotation centered on the double nucleus and coincides with the main body of near-infrared continuum. The other system (disk 2) has an unresolved, or marginally resolved, central velocity gradient with a projected peak-to-peak amplitude of at least 800 km s^{-1} within a radius of $\approx$1.2 kpc. The peak "rotation" implies a compact mass greater than $4.5\times10^{10}M_\odot/\sin^2 i_2$ although, currently, there are no obvious detections of a luminous counterpart at any wavelength. Here, we report on new observations and their implications for the internal dynamics of NGC 6240.

1. Context

The Infrared Astronomical Satellite (IRAS) discovered a class of galaxies that emit the bulk of their radiation at far-infrared wavelengths. The importance of understanding the origin of these extremely luminous ($\geq 10^{11}L_\odot$) far-infrared galaxies (ELFs) is hard to overstate. Not only are they amongst the most luminous objects in the universe, with bolometric luminosities higher than most quasars, but they dominate the galaxy luminosity function beyond $10^{12}L_\odot$ (Scoville & Soifer 1990). In these galaxies, the molecular gas is centrally concentrated, constitutes a large fraction of the dynamical mass, and is thought to fuel the source of activity giving rise to the far-infrared radiation. The ELF prototypes, Arp 220 and NGC 6240, are two of the most luminous galaxies within 100 Mpc, with bolometric luminosities that are almost two orders of magnitude higher than for ordinary spiral galaxies (de Jong *et al.* 1984). However, the source of the far-infrared radiation in both objects remains highly controversial. Opinions differ on the relative importance of large-scale shocks (Harwit *et al.* 1987), massive star formation or a "buried" active nucleus in heating the dust (Soifer *et al.* 1984). It is now known that almost all ELF systems have highly disturbed appearances suggesting that they result from galactic collisions. Because axes of symmetry are rare within these objects, the imaging Fabry-Perot interferometer has a major advantage over conventional long-slit techniques in studying ELF galaxies, which we now demonstrate.

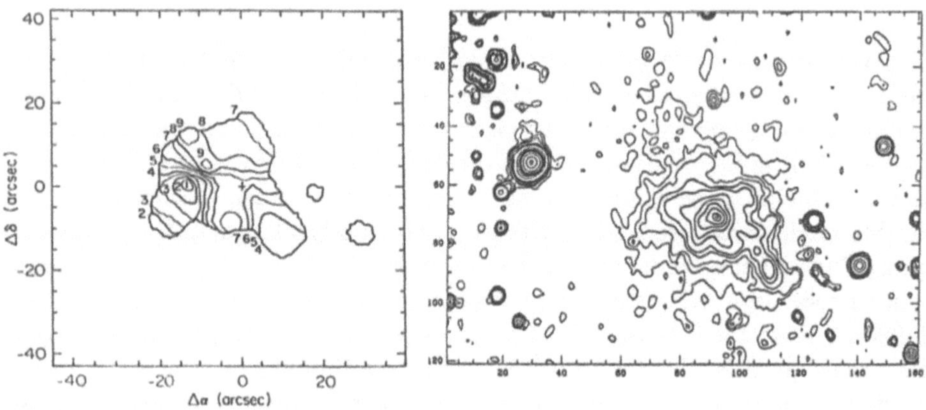

Fig. 1. *a.*– **The Hα velocity field for an 85″×85″ field of view (north at top). The cross marks the position of the double radio source. The velocity contours in km s^{-1} are: (1) 7120, (2) 7170, (3) 7220, (4) 7270, (5) 7320, (6) 7370 (systemic for disk 1), (7) 7420, (8) 7470, (9) 7520.** *b.*– **The Hα line flux distribution within a 90″×70″ field of view. The outer boundary is where the Hα surface brightness falls below ~10^{-17} erg cm^{-2} s^{-1} arcsec^{-2} Å^{-1}. The innermost contour marks the position of the double nucleus; the two bright compact objects to the west are HII regions.**

2. Earlier Observations

Three years ago, we observed NGC 6240 with the Hawaii Imaging Fabry-Perot Interferometer (HIFI) at the UH 2.2m telescope and obtained the detailed Hα velocity field shown in Fig. 1*a* (Bland-Hawthorn, Wilson & Tully 1991, hereafter BHWT). We showed that the line flux distribution is characterized by two axes that run SW-NE and SE-NW; these axes cross north of the nucleus where the Hα lines are commonly found to have two components. We argued that, in the simplest interpretation, the two axes correspond to two separate dynamical systems (disks 1 and 2). These have radically different rotation and are closely spaced in velocity ($\approx$70 km s^{-1}) and in position ($\approx$12″, $\approx$5.7 kpc). Disk 1 (P.A. $\approx$ 45°) is roughly aligned with the major axis of the near-infrared continuum. In Fig. 2*a*, the disk rotation is flat out to at least 20″ in radius, centered on the double nucleus seen at many wavelengths (e.g., Condon *et al.* 1982). The rotation turns over at $\approx$7.2″ with a peak-to-peak velocity amplitude $\approx 280/\sin i_1$ km s^{-1}, where i_1 is the disk inclination.

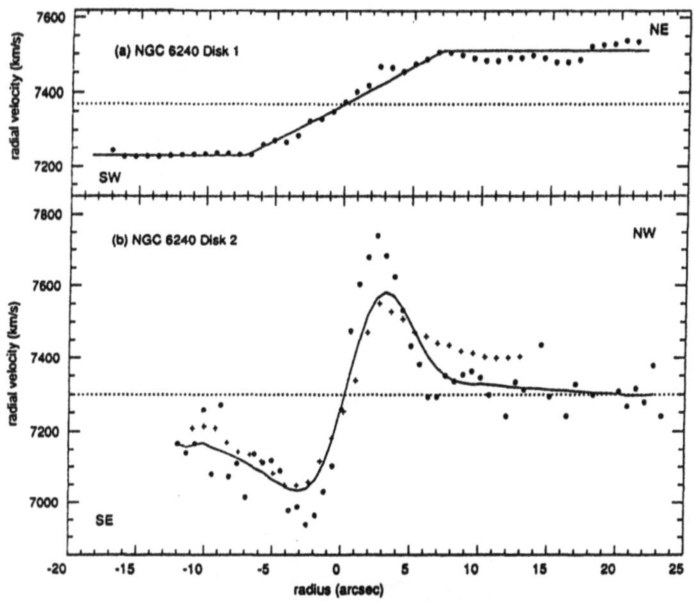

Fig. 2. .– **The steepest gradients through the HIFI Hα velocity field are found, *a*.– along P.A. = 45° through the double nucleus (disk 1) and, *b*.– along P.A. = 155° offset by (11.5″E, 2.5″N) from the origin in Fig. 1*a*. Beyond the turnover radius, the curves exhibit very different behavior. The total, projected velocity amplitude in *a* is $\approx$280 km s^{-1}. In *b*, the combined Hα+[NII]λ6583 measurements from the long-slit data are shown as filled circles: these have a total amplitude of $\approx$800 km s^{-1} within a turnover radius < 2.5″.**

The properties of the second dominant axis (disk 2) are more remarkable. In Fig. 2*b*, the observed radial velocities along P.A. = 155° have been plotted, centered on a point $\approx$12″ almost due east of the double nucleus. The associated spider diagram in Fig. 1*a* is most easily understood in terms of differential rotation. However, the "rotation curve" comprises an unresolved, or marginally resolved, central velocity gradient with a peak-to-peak velocity amplitude of $\approx 800/\sin i_2$ km s^{-1} within a turnover radius < 2.5″ (1.2 kpc). If the kinematics reflect a system close to dynamical equilibrium, the peak velocity implies a compact mass greater than $4.5\times10^{10}\mathrm{M}_\odot/\sin^2 i_2$. Presently, there are no observations of a luminous, compact counterpart to the core of disk 2 at radio, millimeter, far-infrared or X-ray wavelengths.

3. New Observations

In March 1991, we used HIFI at the CFHT 3.6m telescope to obtain another Hα data set of NGC 6240 with finer spatial sampling ($\approx$0.57″), higher spectral resolution ($\approx$40 km s^{-1} FWHM), and wider velocity coverage ($\sim$1100 km s^{-1}) than presented in BHWT. A preliminary total Hα flux map is presented in Fig. 1*b*. Four months later, we used the Rutgers Fabry-Perot (RFP) at the CTIO 4.0m telescope to obtain a [NII]λ6583 data set at lower spectral resolution ($\approx$110 km s^{-1} FWHM) over a similar field of view ($\sim$4′). A cursory inspection of the RFP data shows that the strength of [NII]λ6583 with respect to Hα throughout disk 2 is everywhere higher than within disk 1. The new HIFI data broadly confirm the earlier HIFI velocity field shown in Fig. 1*a* and even indicate that the projected velocity amplitude in disk 2 may be greater than 800 km s^{-1}. In Fig. 1*b*, the elongated axes NW-SE ($\sim$30″) and NE-SW ($\sim$45″) are well pronounced: these are in the same sense as the steepest gradients in Fig. 1*a*.

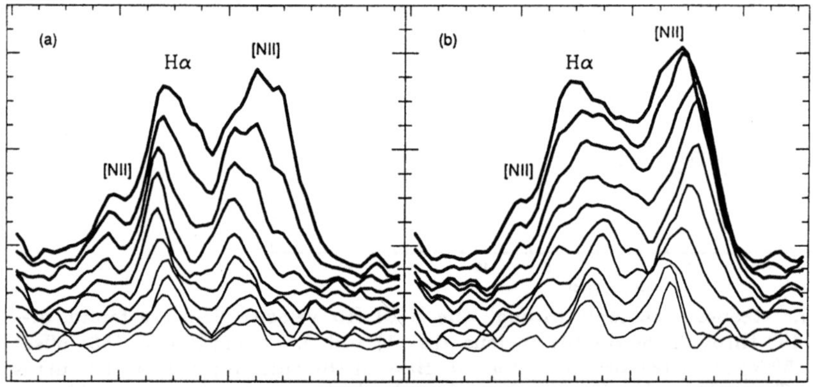

Fig. 3. .– The relative spectral intensity of the [NII] and Hα lines approaching the centre of disk 2 along P.A. = 155°, *a*.– from the SE and, *b*.– from the NW. The data are taken from the spectrogram presented in BHWT (Fig. 3) where successive spectra are separated by 1.6″ along the slit and are offset by 2 units along the y axis.

It is hard to look at Fig. 3 (BHWT) and deny the impression of a real center in disk 2, where the well defined, radial Hα distribution is confirmed by Fig. 1*b*. This impression is further compounded by Fig. 3 above which plots the relative intensity of the Hα and [NII]$\lambda\lambda$6548,6583 lines towards the center of disk 2. On approaching the center from both sides, the lines increase systematically in total line flux and dispersion reaching a peak line width of $\sim$700 km s^{-1} FWHM. A similar behavior is observed in disk 1 along P.A. = 45° although the line widths are even broader ($\sim$1000 km s^{-1} FWHM) towards the double nucleus. In disk 2, the [NII]λ6583/Hα line ratio increases systematically in the range 0.6 to 1.6 towards the center. Notice that the Hα and [NII] lines are essentially symmetric everywhere except that the Hα line develops a red wing NW of the nucleus of disk 2. This component arises from the overlap with disk 1 where the [NII]λ6583/Hα ratios are considerably smaller ($\lesssim$0.5) than in disk 2 such that only the redshifted Hα component from disk 1 is clearly seen.

Recent multiband (BVRIzK) high resolution imaging supplied by W.C. Keel and R.B. Tully shows evidence of continuum emission that peaks near the R band at the core of disk 2. After subtracting the emission line contribution, a provisional upper limit on the total optical and infra-red luminosity (L_{tot}) in disk 2 is $\sim 1.5 \times 10^9 L_\odot$. The implied lower limit of $M/L_{tot} > 30$ makes the core about as dark as the nucleus of M87 (Sargent *et al.* 1978) although much more massive. This does not include a possible contribution at far-infrared wavelengths. Conceivably, the large IRAS beam ($0.76' \times 4.6'$) could hide a bright, compact source at the center of disk 2, although the listed IRAS position is very close to the center of disk 1 and the core of disk 2 lies close to the 95% confidence position ellipse (Fullmer and Lonsdale 1989). However, beyond the turnover radius, the rotation has a faster than Keplerian drop-off which calls into question the specific nature of the kinematics in Fig. 2*b*.

4. Concluding Remarks

With particular reference to our work, Binney (1991) notes from N-body simulations that strongly interacting galaxies tend to develop stellar bars. This induces elliptic streaming into the gas kinematics which could explain the fast drop-off in Fig. 2*b*, thereby reducing the required mass to roughly $1.5 \times 10^{10} M_\odot / \sin^2 i_2$. Blandford (1991) suggests that we may be seeing complex hydrodynamics in a strong galactic collision. However, Barnes & Hernquist's (1991) simulations demonstrate that the gas is susceptible to strong radiative cooling in shocks and, consequently, it may be difficult to generate such large velocity amplitudes in the gaseous component. The observed [NII]λ6583/Hα ratios are high enough to argue against any explanation involving non-radiative shocks. If higher resolution observations confirm that the peak-to-peak velocity amplitude in disk 2 is even larger than $\sim$800 km s^{-1}, an interpretation explicitly in terms of the gas motions within two colliding systems becomes even less attractive.

Rees (1991) notes that one or more high-velocity outflows, e.g., the double radio jets in 3C 75 (Owen *et al.* 1985), could possibly explain the observed kinematics. The post-shock regions of rapid outflows – radio jets from Seyfert nuclei (Cecil 1988), Herbig-Haro systems, etc. – are characterized by line profiles that display complex structure and very strong wings irrespective of viewing angle. Presently, the large-scale distribution of complex line profiles within the HIFI data does not suggest an obvious morphological signature. While more contrived explanations – deflected high-velocity outflows (Rudnick & Anderson 1990), multiple tidal tails (Borne & Richstone 1990), etc. – are even less appealing, they can only be addressed by better data. If the observed gas motions do reflect dynamical equilibrium, they provide substantial evidence for the existence of a massive, compact dark object at the center of disk 2. Indeed, the systematic behavior of Hα and [NII] in Fig. 3 is commonly observed in Seyfert and LINER galaxies. Finally, we anticipate that newly proposed N-body simulations of galactic interactions (in collaboration with R.D. Blandford and L. Hernquist), when combined with our Fabry-Perot data, will help to constrain the large-scale mass distribution within NGC 6240.

Acknowledgments. J.B.H. is supported in part by NSF grant AST 88-18900. We wish to thank Roger Blandford and Martin Rees for interesting exchanges, and W.C. Keel, R.B. Tully and S. Veilleux for contributed observations. As a personal valediction, it is fitting that this conference has been dedicated to Michael V. Penston. He was a constant source of encouragement and inspiration to the graduate students and postdocs who passed through the Royal Greenwich Observatory during the 1970's and 1980's. Michael's youthful effervescence will long be remembered by those who knew him.

References

Barnes, J.E., and Hernquist, L., 1991: *Ap. J. (Letters)*, **370**,,L65.
Binney, J., 1991: *Nature*, **351**,,102.
Blandford, R.D., 1991: this conference.
Bland-Hawthorn, J., Wilson, A.S., and Tully, R.B., 1991: *Ap. J. (Letters)*, **371**,,L19.
Borne, K., and Richstone, D., 1991: *Ap. J.*, **369**,,111.
Condon, J.J., *et al.*, 1982: *Ap. J.*, **252**,,102.
Cecil, G., 1988: *Ap. J.*, **329**,,38.
de Jong, *et al.*, 1984: *Ap. J. (Letters)*, **278**,,L67.
Fullmer, L., and Lonsdale, C.J., 1989: *Joint IRAS Sci. Work. Group*, JPL D-1932.
Owen, F.N., *et al.*, 1985: *Ap. J. (Letters)*, **294**,,85.
Rees, M.J., 1991: this conference.
Rudnick, L., and Anderson, M., 1990: *Ap. J.*, **355**,,427.
Sanders, D.B., *et al.*, 1988: *Ap. J.*, **325**,,74.
Sargent, W.L.W., *et al.*, 1978: *Ap. J.*, **221**,,731.
Scoville, N.Z., and Soifer, B.T., 1990: in: *Massive Stars in Starbursts*, eds. C. Leitherer & N. Walborn, Baltimore.
Soifer, B.T., *et al.*, 1984: *Ap. J. (Letters)*, **278**,,71.

Discussion

Burbidge: I would like to think that NGC 6240 is a galaxy that is young in evolutionary terms and not a merger. Can you interpret the velocity field in terms of non-circular motions, rather than effects due to the rotations of two inclined disks?

Bland-Hawthorn: Yes, I am familiar with your 1986 Astronomical Journal paper on "young galaxies". Presently, it is very difficult to see how bipolar winds or jet termini can explain the line flux distribution when considered in conjunction with the kinematics and (mostly) symmetric line profiles. Martin Rees (private communication) suggests something like the double radio jets in 3C 75. One might imagine something rather exotic like a deflected outflow (e.g. Rudnick & Anderson) or shearing/shocking in a galactic collision. But these are rather contrived unless related directly to the system as a whole. Taniguchi and Noguchi's recent characterization of "wing galaxies" from strong tidal interactions offers some hope in understanding NGC 6240. In their simulations, the outer envelope, the inner disks and many of the tidal features bear close resemblance with NGC 6240 although this is a highly transient phase in their simulations. In any event, if HST demonstrates that the central velocity amplitude in "disk 2" increases much beyond 800 km $s^{-1}/\sin i_2$ at higher resolution, this would be very difficult to reconcile with the hydrodynamics of a disk-disk collision.

High Spatial Resolution NIR Imaging Spectroscopy

A. Krabbe[1], *V. Rotaciuc*[1], *S. Drapatz*[1], *R. Genzel*[1] *and J.W.V. Storey*[2]

[1] Max-Planck-Institut für extraterrestrische Physik, Garching, Germany
[2] University of New South Wales, Kensington, Australia

Abstract. We report subarcsec/300 km s^{-1} resolution imaging spectroscopy of HI Brγ 2.17μm and H_2 1-0 S(1) 2.12μm line emission towards the nuclei of NGC 1068, NGC 1808, and in the 30 Dor Complex. For 30 Dor in addition we present far infrared imaging spectroscopy of C II $^2P_{3/2}-^2P_{1/2}$ 157.7μm line emission. In the Seyfert 2 galaxy NGC 1068 both lines originate within 2" (170pc) of the peak of radio, near-infrared, and the visible continuum radiation. The H_2 emission has a double-lobed shape with a minimum towards the nucleus in contrast to the Bγ emission, which is peaked on the nucleus similar to the narrow line region (NLR) in the visible. The H_2 emission probably comes from a system of molecular clouds in the vicinity of the NLR. Possible excitations mechanisms for both lines are discussed. In the vicinity of the nucleus of the hot spot galaxy NGC 1808 more than 10 condensations of Brγ emission were detected. They are mostly super HII regions and supernova remnants (SNR) but differ in flux and diameter. H_2 was discovered to form an elongated extended structure close to the nucleus. The Brγ emission in the 30 Dor complex forms a cavity with a diameter of about 2' around the R 136 cluster. Outside the cavity diffuse H_2 emission was discovered north of R136 in a region of enhanced far infrared activity and CII emission. The abundance of CII compared to CO is explained in terms of a destruction mechanism for CO due to UV and X-rays emitted by the R136 cluster.

1. Introduction and Observations

Many galaxies with nuclear activity or starbursts are heavily obscured in their central parts. Near infrared (NIR) observations are therefore vital in any attempt to unravel the morphology and excitation conditions close to the nuclei. The MPE imaging NIR Fabry-Perot spectrometer FAST was used to investigate a sample of bright southern active and starburst galaxies, among them NGC 1068 and NGC 1808. The 30Dor complex in the Large Magellanic Cloud (LMC) was chosen as well because it represents a region of enhanced activity in our neighbouring galaxy, which can therefore be studied in much greater detail than distant galaxies.

The measurements with FAST (Krabbe et al. 1992) were obtained in January on the 2.2m MPI/ESO telescope in Chile, and in September 1990 on the 3.9m Anglo-Australian telescope. FAST uses a 62x58 pixel InSb array detector manufactured by Santa Barbara Research Center behind a Queensgate scanning Fabry-Perot spectrometer, resulting in a resolving power of 10^3 in the K/band. The filterwheel in the beam was equipped with broad band J, H, K filters and an order sorting circular variable filter (CVF). For the ESO/Chile observations an image scale of 0.8"/pixel was chosen while the scale at the AAT/Australia was 0.95"/pixel. The diameter of the field of view was about 40".

2. Results and Discussion

2.1 NGC 1068

NGC 1068 is one of the most luminous galaxies at infrared wavelengths and is the nearest galaxy with a Seyfert 2 nucleus. Its relative proximity enables spatially resolved study of the nucleus and its surroundings, which should provide clues to the origin of the nuclear activity.

Figure 1 presents images of the H_2 and the Brγ lines, along with a continuum image taken through the CVF alone. The continuum shows a bright nuclear source (FWHM 2.7") embedded in a stellar cluster with a prominent NE-SW bar component in good agreement with Scoville et al. (1988) and Thronson et al. (1989). The compact and slightly elongated Brγ emission source is centered on the 2μm continuum source. The H_2 emission is offset from the continuum source by $\approx$ 0.9" east and is more extended than the Brγ emission. The H_2 source has a FWHM diameter of about 2.3" and the lower contours extend out to about 8" (along p.a. 70°). By selecting data with the best seeing and optimum focussing, it is possible to obtain even better spatial resolution. Then the H_2 emission possesses an intensity minimum at the location of the compact continuum source with two emission peaks of different size $\approx$1.3" on either side of it. Our data show that the molecular and ionized emission components are spatially distinct.

The H_2 emission probably arises in a system of massive molecular clouds in the vicinity of the NLR. The total H_2 mass of $3 \cdot 10^7 M_\odot$ leads to a mean H_2 density of $<n>=50\text{cm}^{-3}$ in the emitting region. Since the density required for thermal population of the H_2 vibrational levels is a $\geq 5 \cdot 10^4 \text{cm}^{-3}$, so the molecular emission region is probably sheetlike or very clumpy with a cloud filling factor less than 10^{-3}. Observations of Oliva & Moorwood (1989) with 6" spatial resolution confirm that the intensity of the H_2 2–1 S(1) transition is more than a factor of 7 weaker than the 1-0 S(1) line. This implies that the molecular emission is probably produced in hot dense gas which has been shocked or heated by external UV and X-ray radiation

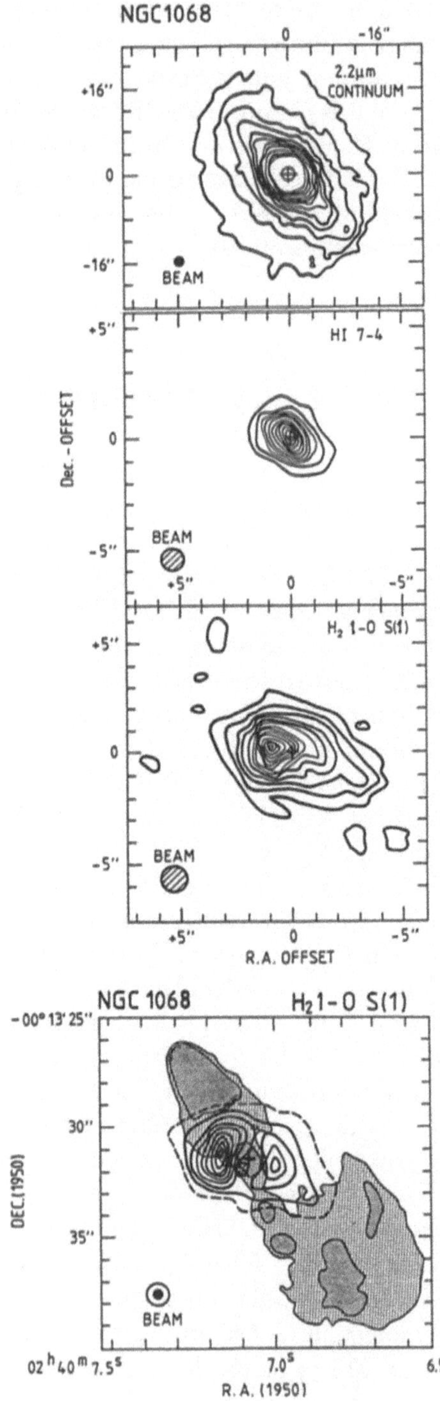

Fig. 1. *Top*: 2.2μm continuum map of NGC 1068, smoothed to a resolution of 1.6". Integrated fluxes in the central 6" and 16" diameter regions are 9.3 and $16 \cdot 10^{-16} W/m^2$. *Upper middle*: Brγ map of the line core with a velocity width of 316 km/s. The contour unit is $5.9 \cdot 10^{-19} W/m^2$. The cross denotes the position of the 2μm continuum peak. *Lower Middle*: H_2 1-0 S(1) map of the line core with a velocity width of 300 Km/s centered at v_{LSR}=1200 km/s. The contour unit is $3.9 \cdot 10^{-19} W/m^2$. *Bottom*: Map of the highest quality H_2 data (best seeing and focussing), superposed on a schematic of the 5 GHz radio continuum map of Wilson & Ulvestad (1983). The contour unit is 4.9 $\cdot 10^{-19} W/m^2$. The cross denotes the position of the 2μm continuum peak.

Fig. 2. Hot spot galaxy NGC 1808. *Right:* Continuum image at 2.2μm taken with the CVF. *Bottom left:* Brγ line emission at 2.17μm. About 10 condensations can be identified. *Bottom right:* H_2 line emission at 2.12μm.

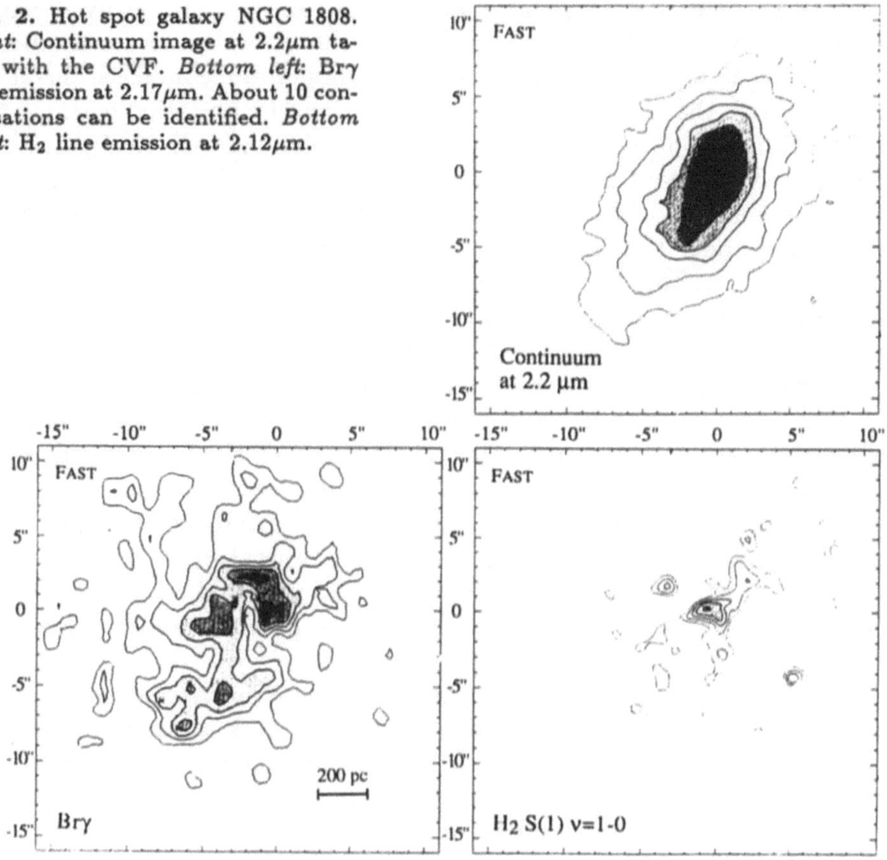

fields which are present in the same nuclear region that produces the Brγ emission. The conditions at the nucleus of NGC 1068 are discussed in more detail in Rotaciuc et al. (1991).

2.2 NGC 1808

NGC 1808, a southern starburst galaxy, contains a bright Seyfert nucleus and several nuclear 'hot spots' (Sersic & Pastoriza 1965, Veron-Cetty & Veron 1983). Parts of it also suffer from heavy extinction of the order of $A_v = 16 \pm 9$ (Philips et al. 1984).

The 2.2μm CVF continuum image in figure 2 looks remarkably different from the optical image. At an extinction of K$\leq$1 mag, it displays a prominent center and a more normal shaped galaxy instead of hot spots. This may be interpreted as very patchy extinction towards the nucleus which in turn means that instead of 'hot spots' the nucleus of NGC 1808 is obscured in the optical by 'anti hot spots'.

The Brγ emission line shows more than ten spatially resolved condensations towards the SW of the nucleus linked by bridges of lower intensity which tend to form spiral like structures. Though the condensations do not coincide with the hot spots there is a good overall correlation with the 6 cm map of Saikia et al. 1990. Integrating the Brγ flux yields $2 \cdot 10^6 L_\odot$ which can be emitted by about 10^5 O/B stars. Such a population would imply a SN rate of about 0.1 SN/year, in good agreement with the estimates from the 6 cm map under the assumption that this radiation is mostly of nonthermal origin. The Brγ line emission then indicates that NGC 1808 possesses a moderate extranuclear starburst region. It consists of several super HII condensations (containing the SNR's) differing in size and activity which are driven by intense, but not dramatic, recent star formation.

The 10μm dust continuum flux predicted from our Brγ measurements using the method of Ho et al. 1990 leads to 4 to 6 times lower values than those observed by Telesco et al. 1987. This 10μm excess, which has also been observed in other galaxies (e.g. NGC 2903) could indicate non-equilibrium heating of very small dust grains to high temperatures by visible or UV photons. In this case the 10μm excess radiation would be an indicator for either the local radiation field or the dust abundance around these active regions.

The shape of the H_2 emission is very clumpy and is extended over 9" in SE/NW direction with the emission peak being located ≈1" east of the continuum and Brγ emission peak. The clumpy emission tends to fill the regions between the Brγ emission. Again, as in NGC 1068, the atomic and molecular emission seems to be excited in spatially distinct regions.

2.3 LMC/30 Doradus

The 30 Doradus complex is the most luminous spot in the Large Magellanic Cloud at almost all wavelengths. Due to its relative proximity (55kpc) it is very well suited as a case study for an example of an extragalactic activity centre.

Our Brγ emission line mosaic image in Figure 3 is made of 27 single frames and covers an area of 3.5'x3'. The circular structure with extended filaments suggests that we are looking at the inside of a shell, which surrounds the R136 star cluster and which was probably created by the intense UV radiation and stellar winds of the cluster members. The bright knots would then indicate regions of stronger interaction of the expanding shell with the adjacent clouds of dust and molecular gas. This view is supported by the discovery of weak diffuse emission in the H_2 line beyond the northern Brγ peak as indicated in figure 3.

Contrary to what one might expect, no or only very little CO emission has been seen so far behind the rim, delineated by the Brγ emission. Instead C II line emission at 157μm, measured with MPE imaging far infrared

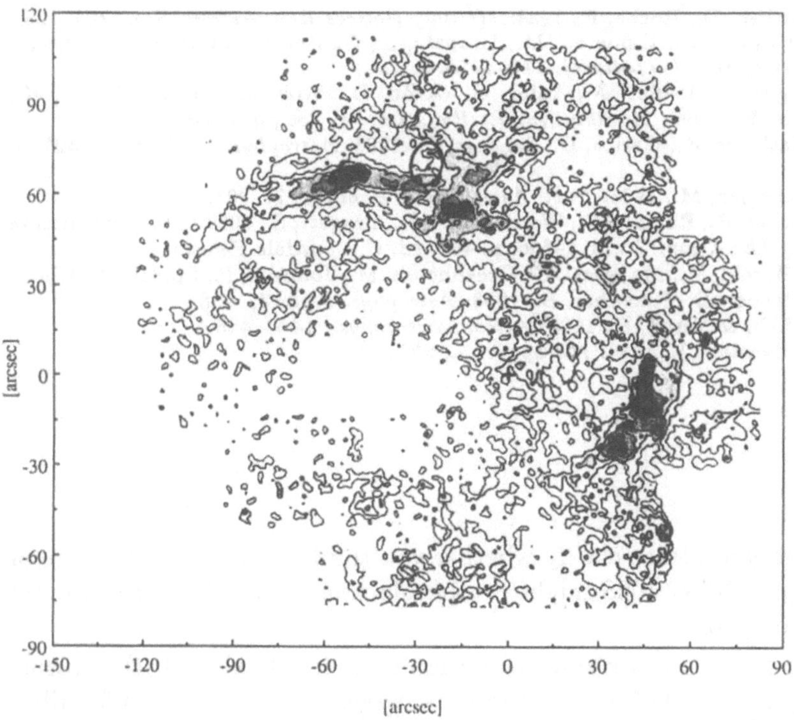

Fig. 3. Brγ line emission in the 30 Dor complex. The lowest of the linearly spaced contours is at $4 \cdot 10^{-19} W/m^2/arcsec^2$ and the highest at $1 \cdot 10^{-17} W/m^2$ /arcsec^2. The position of the R136 cluster is indicated by a cross (right of the center). Diffuse weak H_2 emission has been discovered within the ellipse(north of the center).

Fabry-Perot spectrometer (Krabbe et al. 1991) was detected in this complex. Peaks in the C II emission lay adjacent to those observed in Brγ. As the R136 cluster is emitting strong UV radiation, these photons tend to dissociate the CO molecules and thus produce the C II. The H_2 molecules can survive much longer in this radiation field because of "self shielding", whereby a substantial fraction of them reemit the absorbed photons instead of being dissociated. Our measurements can best be explained as the result of the interaction of the intense UV radiation field and wind with the cloud system that surrounds R136.

References

Krabbe, A., et al., 1991: in preparation.
Krabbe, A., et al., 1992: in preparation.
Oliva, E., Moorwood, A., 1990: *Astrophys. J. (Letters)*, **348**, L 5.

Phillips, M., Aitken, D., Roche, P., 1984: *Monthly Notices Roy. Astron. Soc.*, **207**, 25.
Rotaciuc, V., Krabbe, A., Cameron, M., Drapatz, S., Genzel, R., Sternberg, A., Story, J., 1991: *Astrophys. J. (Letters)*, **370**, L 23.
Saikia, D., Unger, S., Pedlar, A., Yates, G., Axon, D., Wolstencroft, R., Taylor, K., Gyldenkerne, K., 1990: *Monthly Notices Roy. Astron. Soc.*, in press.
Scoville, N., Mathews, K., Carico, D., Sanders, D., 1988: *Astrophys. J. (Letters)*, **327**, L 61.
Sérsic, J. & Pastoriza, M., 1965: *Publ. Astron. Soc. Pacific*, **77**, 287.
Telesco, C., Decher, R., Ramsey, B., Wolstencroft, R., Leggett, S., 1987: in: *Starformation in Galaxies* NASA conf. Publ. 24666, p. 497, ed. C. Lonsdale Persson
Thronson, H., Hereld, M., Majewski, S., Greenhouse, M., Johnson, P., Spillar, E., Woodward, C., Harper, D., Rauscher, B., 1988: *Astrophys. J.*, **343**, 158.
Veron-Cetty, M.-P. & Veron, P., 1985: *Astron. Astrophys.*, **145**, 425.
Wilson, A. & Ulvestad, J., 1983: *Astrophys. J.*, **275**, 8.

Discussion

Wilson: Could you please clarify whether you think the double structure of your H_2 map of NGC 1068 might be the blocking molecular torus viewed more or less edge on.

Krabbe: No, I don't think our H_2 structure is identical with this torus, because the postulated obscuring torus, as suggested by Antonucci & Miller (1985), should be only a few parsecs in radius, whereas our H_2 source is a few hundred parsecs in extent.

Extended Emission-Line Regions as Probes of Nuclear Activity in Galaxies

Andrew S. Wilson

Astronomy Program, University of Maryland, College Park, MD 20742, USA

Abstract. This paper reviews recent observations of extended emission-line regions and their rôle in probing the nature of active nuclei. There is currently strong support for a "unified scheme" in which Seyfert 2's, Seyfert 1's and radio-quiet quasars represent one class of object (radio-quiet AGNs) and narrow-line radio galaxies, broad-line radio galaxies and radio-loud quasars represent another class (radio-loud AGNs). Three main lines of evidence - geometric, energetic and spectropolarimetric - indicate that optical and ionizing radiations escape anisotropically from the nuclei of both classes, being "beamed" into oppositely directed, wide cones, with axes close to the radio axis. The geometric and energetic arguments are discussed in detail. The evidence for this "beaming" is stronger for the radio-quiet than for the radio-loud objects. There is a strong correlation in both classes between the power of ionizing radiation which escapes to large distances from the nucleus and ionizes low density gas, and the kinetic power of the outflow (e.g. as "jets") which fuels the steep-spectrum radio lobes. There are no powerful radio sources which radiate much less luminosity in ionizing radiation than in radio emission. If the uv radiation in radio-loud objects comes directly from an accretion disk, this disk is probably not an ion torus for such objects may radiate too little thermal emission. The extended emission-line nebulosities in both classes of object are photoionized by the nucleus. However, the electron temperatures inferred from observations are higher than predicted by photoionization models. Either the metal abundances are sub-solar or additional heating sources are present. The presence of radiative shock waves in a photoionized medium may account for the elevated temperatures and a possible correlation between excitation and surface brightness. Mechanisms for collimating the ionizing photons include relativistic beaming, shadowing by a dense molecular torus, shadowing by broad-line region clouds and the intrinsic anisotropy of a thick accretion disk. Observational tests for distinguishing these four processes are discussed. While the currently available evidence is inconclusive, shadowing by a torus or the intrinsic anisotropy of a radiation torus are tentatively favored. Radio-loud objects are never found in spiral host galaxies, have much lower space densities than radio-quiet objects, and seem to be associated with galaxies which have undergone mergers. On the basis of these well-known results, I speculate that radio-loud objects represent the small fraction of merger products for which *both* merging galaxies contained a supermassive black hole, thus resulting in a supermassive binary.

1. Why Study Extended Emission-Line Regions?

The existence of an Extended Emission Line Region (EELR) requires the presence of both gas and an energy source for ionization and excitation. For the types of active galaxies with which I am concerned in this review (Seyfert galaxies, radio galaxies and quasars), the energy source is usually the active nucleus itself, so the study of EELR's can provide valuable information on *both* the active galactic nucleus (AGN) and the interstellar gas in the host galaxy or the intergalactic gas around it. In this paper, I shall emphasize the value of EELR's for learning about the AGN.

EELR's are, by definition, remote from the "prime mover" of the AGN. Their properties may be strongly influenced by entities other than the nucleus, such as the Hubble type and absolute magnitude of the galaxy, dust and photoelectrically absorbing gas between the EELR and the nucleus, nearby (interacting) galaxies which gravitationally perturb the gas in the AGN-host galaxy, and so forth. The possible presence of these effects can lead to ambiguities in deductions about the AGN. Components of the AGN which are closer to the central object (broad line region [BLR], non-thermal continuum sources etc.) may be expected to be much less influenced by environmental effects and their properties dominated by the central object. Nevertheless, the study of EELR's has recently played an increasingly important rôle in AGN research for three reasons: (i) the widespread current availability of high quantum efficiency, low read noise CCD's permits detection and spectroscopy of the often faint nebulosities; (ii) the physics of EELR's is much simpler than that of BLR's or accretion disks, primarily because of their lower gas densities ($n_e \leq 10^3$ cm^{-3}), radiation densities and magnetic field strengths; and (iii) EELR's are spatially resolved, so important parameters (electron density, ionization parameter, kinematics etc.) can be directly mapped out, rather than having to be inferred by indirect means (e.g. reverberation mapping of the BLR). The distance of the gas from the nuclear ionizing source is also directly determined (apart from the projection factor), and emission at individual off-nuclear locations is usually dominated by gas with a relatively narrow range of density. This situation may be contrasted with that for single aperture *nuclear* spectra, which include contributions to the forbidden line emission from gas with wide ranges of density ($10^2 \leq n_e \leq 10^7$ cm^{-3}), distance from the ionizing source (few pc to 100 pc for a nearby Seyfert galaxy), and velocity (100 - 1000 km s^{-1}). An example of a unique benefit of off-nuclear spectroscopy is the direct determination of the number of ionizing photons emitted by the nucleus from the measured density, ionization parameter and distance of the gas from the nucleus. A catalogue of extended ionized nebulosities around AGN is given in ref. 16.

Excluded from this review are objects with low level nuclear activity (e.g. LINER's), high-redshift radio galaxies (which seem to be fundamen-

Radio Quiet	Radio Intermediate	Radio Very Loud	Direction of View
Seyfert type 2's	Radio galaxies with weak emission lines, FR I	Narrow Line Radio Galaxies (NLRG), FR II	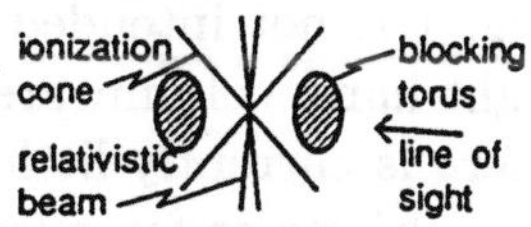
Seyfert 1.5's, Seyfert 1's, Radio-quiet quasars	Radio galaxies with weak emission lines, FR I	Broad Line Radio Galaxies (BLRG), FR II, Radio-loud steep-spectrum quasars	
Seyferts with flat spectrum core radio source?	BL Lac objects	"Blazars"/core dominated quasars	

Fig. 1. The "Unified Scheme" for AGN. The three columns represent three intrinsically different classes of object. The different rows refer to the angle of view with respect to a "blocking torus", "ionization cone", or radio ejection axis, as illustrated in the right-hand column

tally different from their low-redshift countersparts, e.g. [33]), emission-line gas powered by hot stars (HII regions, "starburst" galaxies, and Far-Infrared Galaxies e.g. [25]), and any explicit discussion of the relation between cooling flows and nuclear activity. Also, because of the large quantity of recent literature in this area (cf. ref. 36), the discussion is necessarily selective.

2. Relationship to "Unified Schemes"

A summary of the currently popular "Unified Scheme" for AGN is given in Fig. 1. The left three columns divide the population by radio power. Objects with very high radio powers (Fanaroff-Riley [FR] class II radio galaxies, radio-loud steep spectrum quasars and "blazars" or "core-dominated" quasars) are in the first column, objects with intermediate radio powers (FR class I radio galaxies and BL Lac objects) are shown in the second column, while radio-quiet AGNs (Seyfert galaxies and radio-quiet quasars) are listed in the third. These three columns segregate *intrinsically different* classes of objects. The different rows refer to the supposed *angle of view* with respect to a "blocking torus", "ionization cone", or radio ejection axis, as illustrated in the right-hand column. There is considerable evidence for "blocking tori", and that their axes align well with that of both the ionization cone and radio axis. For example, it is likely that narrow-line radio galaxies are intrinsically similar to broad-line radio galaxies and radio-loud quasars, but are viewed from such a direction that the broad line region is blocked by an optically thick (molecular?) torus. I should emphasise that Fig. 1 is not intended to imply that all the components illustrated in the right-hand column are present in all the classes of object. In particular, there is currently little evidence for blocking tori in FR I galaxies and BL Lac objects or for *relativistic* jets in Seyfert galaxies.

In this review, the two intrinsically different "parallel tracks" listed in columns 1 and 3 of Fig. 1 are discussed separately (Sections 3 and 4). Despite their greatly different radio powers, a strong symbiosis exists in both classes between their radio and emission-line properties, including evidence for anisotropic escape of ionizing photons. The nature of this anisotropy is discussed in Section 5, while Section 6 contains speculations on the difference between radio-quiet and radio-loud objects. The objects in column 2 of Fig. 1 are excluded from this paper because they have only very weak nuclear and extended line emission.

3. Radio-Quiet AGNs

I shall concentrate here on Seyfert galaxies, since more detailed information is available on their EELR's than is the case for radio-quiet quasars. A schematic diagram of a Seyfert nucleus, emphasising our current understanding of the large-scale features, is shown in Fig. 2.

3.1 Radio Sources and Narrow Line Regions in Seyfert Galaxies

The relationship between radio sources and the NLR's of Seyfert galaxies has been reviewed in detail recently [68,73], so I shall confine myself to a brief summary. Strong correlations between the integrated radio power, the (narrow) emission line width and the (narrow) emission line luminosity of Seyfert galaxies have been known for many years. Further, there is very good alignment and coincidence of spatial scales between the "linear" (i.e. double, triple or jet-like) radio sources and the NLR's in those Seyferts known to have such radio sources [13,17,22,40]. At least in Seyferts with high power ($P_{1415 MHz} \geq 10^{22.5}$ W Hz^{-1}) "linear" radio sources, most of the line width is a consequence of "stirring" by the radio ejecta [69,70,72]. There seems to be a general consensus [42,62,74] that these relationships arise through the ejected radio plasma driving radiative shocks into the ambient interstellar medium. The resulting cool condensations are then maintained in an ionized state by the luminous ultraviolet emission from the nucleus (Fig. 2). For Seyferts with lower radio luminosity, the width of [OIII] λ5007 appears to be dominated by virial (gravitational) motions [69]. The correlation between radio power, [OIII] line width, and [OIII] luminosity may then arise

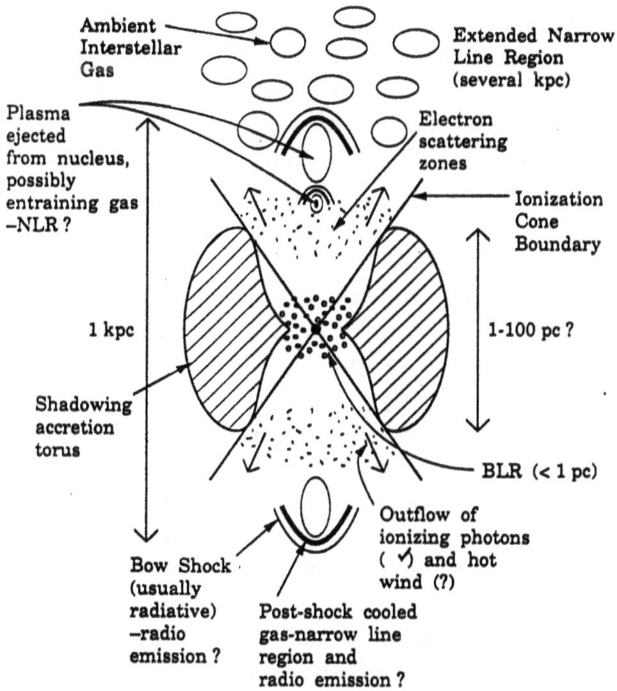

Fig. 2. A schematic diagram of a Seyfert nucleus, showing currently recognised or probable components, and their typical size scales

indirectly, through mutual correlations with another parameter, most likely the mass of the galaxy bulge [35,67,69].

3.2 Anisotropic Ionizing Radiation in Seyfert Galaxies

There are three main kinds of argument indicating that optical and ultraviolet continuum radiations escape anisotropically from the nuclei of at least some Seyfert galaxies - geometric, energetic and spectropolarimetric. The last of these is outside the scope of this article and will not be discussed; two recent papers in this area are those by Miller and Goodrich [38] and Miller, Goodrich and Mathews [39]. I now summarize the evidence from the first two types of argument.

3.2.1 Geometric Evidence. The alignment described above between the radio sources and the NLR's does not, in itself, show that the ionizing radiation is partially beamed along the radio axis because the enhanced line emission near the radio components could result from density enhancements in an isotropic radiation field. It has, however, been known for many years that some Seyfert galaxies exhibit highly excited gas which lies *farther* from the nucleus than either the radio source or the broadened emission lines (NLR). Using long slit spectroscopy of a sample of seven Seyferts, Unger et al.

[64] showed that this gas extends farther along the extension of the radio axis than perpendicular to it. The gas radiates narrow (≤ 45 km s^{-1}) lines and appears to follow normal rotational motion. Unger et al. called this component the "Extended Narrow Line Region (ENLR)" and interpreted it in terms of ambient interstellar gas ionized by uv radiation from the nucleus which is beamed along the radio axis. Schulz [53] found similar evidence for a "beamed" ionizing source from studies of the string of high excitation, extranuclear clouds *outside* the NLR of NGC 4151.

In some galaxies, the ENLR/NLR has a clear conical or bi-conical form, with the cone apices at the nucleus and opening angles several tens of degrees. These cones may show up directly in images taken in high excitation lines, but often stand out more clearly in flux ratio maps, such as F([OIII]λ5007)/F(Hβ) or F([OIII]λ5007)/F(Hα+[NII]$\lambda\lambda$6548,6583), which effectively separate the Seyfert-excited gas from HII regions. Only about eight galaxies with clearly defined or probable cones have been discovered [9,15,23,45,46,47,54,58,59,60], so inferences of a statistical nature are limited. Nevertheless, the axis of the cone always seems to align with the radio source (to within $\sim 5°$), and is often misaligned w.r.t. the axis of the galaxy disk. There may, however, be some trend for cone axes to align perpendicular to the galaxy disk in *late* type spirals; alternatively this could be a selection effect, since the highly excited gas is most easily distinguished from HII regions when it lies above or below the plane of an edge-on system. Wagner and Anton [66] find that radio *jets*, when seen, are often one-sided, but the high excitation gas is usually two-sided.

The "best" examples of extended, high excitation gas associated with Seyfert nuclei tend to be found when the host galaxy is of early type – S0 or S0/a. This trend may result from a number of properties of early type systems [23,34]: a lower mean interstellar gas density compared with late type spirals (which allows the ionizing photons to penetrate further from the nucleus); an absence of confusing line emission from HII regions; and the possibility that the interstellar gas has been accreted and is spread over multiple planes, thereby increasing the chances of intersection with a radiation cone emitted from the nucleus. A similar preference for early morphological types is exhibited by those Seyfert galaxies known to contain "linear" radio sources.

In the type 1 Seyfert galaxy NGC 3516, there is an elongated gas cloud which "points" at the nucleus and lies directly along the present nuclear radio axis [40]. This gas may be ionized by a narrow "blazar beam", as has also proposed for an elongated system of filaments in Cen A (see paper by R. Morganti).

3.2.2 Evidence from Energetics. A simple test for anisotropy of ionizing photons is illustrated in Fig. 3. A cloud of gas is illuminated by a nuclear point source of ionizing radiation. For case B conditions, the number of Balmer recombination photons at Earth, $N_{rec} = \alpha_B F(H\beta)/\alpha^{eff}_{H\beta} h\nu(H\beta) =$

$8.6N(H\beta)$ photons cm^{-2} s^{-1}, where the α's are, as usual, the recombination coefficients and $\nu(H\beta)$ is the frequency of Hβ. Hβ is preferred over Hα or Lyα because the last two lines may have collisional contributions to their excitation [20,29]. The number of ionizing photons at Earth is $N_{ion} = (F_0 - F_{912})/\alpha h$ photons cm^{-2} s^{-1}, where F_0 and F_{912} are the flux per unit frequency interval at zero wavelength and 912 Å, respectively, α is the continuum slope assuming a power law spectrum ($F \propto \nu^{\alpha}$), and h is Planck's constant. In a model in which the nuclear continuum radiates isotropically, $N_{rec}/N_{ion} = C \leq \Omega/4\pi$ (C is the covering factor of the ionized gas). In the occultation plus reflection picture [3], $N_{rec}/N_{ion} = C/f \leq \Omega/4\pi f$, where f is the fraction of ionizing radiation scattered into the line of sight. In general, values of $N_{rec}/N_{ion} > 1$ indicate anisotropy (i.e. the gas is illuminated by more intense ionizing radiation than we infer by direct extrapolation of the observed far uv or X-ray continua).

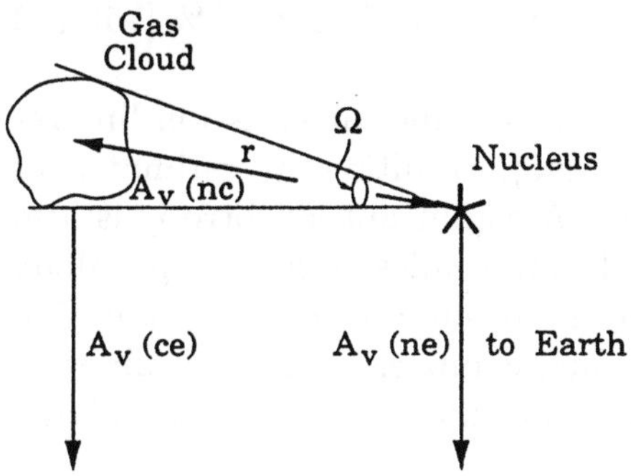

Fig. 3. A gas cloud ionized by a point nuclear ultraviolet source. The three values of A_v discussed in the text are indicated

A second, related approach is to estimate the ionization parameter, $U(= Q/4\pi nr^2 c$, where Q is the number of ionizing photons emitted by the source per second), and density, n, of the gas by comparing the emission line spectrum with photoionization models and by measuring the [SII]$\lambda\lambda$6717, 6731 doublet ratio, respectively. The number of ionizing photons at Earth derived from the emission line spectrum is then $N_{ion,lines} = Unr^2c/R^2$ photons cm^{-2} s^{-1}, where R is the distance to the galaxy. This value may then be compared with N_{ion}, derived from the observed continuum spectrum (see above). In an isotropic model, $N_{ion,lines}/N_{ion} = 1$, while in the occultation plus reflection model $N_{ion,lines}/N_{ion} = 1/f$.

It is important to allow for the effects of obscuration in making these tests; three values of A_v are needed (Fig. 3). One assumes that the (unknown) obscuration between the nucleus and the gas cloud $A_v(nc) = 0$; this is a conservative assumption in the sense that it *maximises* the number of nuclear continuum photons incident on the cloud and thereby *minimises* any indication of beaming. The obscuration between the cloud and the Earth,

$A_v(ce)$, is estimated from the observed Balmer decrement. Lastly, the obscuration between the nucleus and the Earth, $A_v(ne)$, is obtained from the form of the uv continuum (see below) and the strength of the 2200 Å absorption feature.

The first approach described above ("photon counting") has been applied recently to Seyfert 2's [29,75]. In particular, Kinney et al. [29] have measured the uv continua of 15 Seyfert 2's down to 1200 Å using IUE, providing the first study of the uv spectral properties of a significant sample of this class of object. Extrapolation of these spectra to ionizing frequencies introduces little uncertainty. In general, the slopes of these continua are similar to those of Seyfert 1's (favoring electron over dust scattering in the reflection zone) and no 2200 Å feature is seen. Both of these results indicate that $A_v(ne)$ is very small. Both refs 29 and 75 find $N_{rec}/N_{ion} > 1$ for a significant fraction of the galaxies studied, indicating anisotropy even if $C = 1$ is assumed. In some cases, the emission- line images may be used to set direct upper limits on C. For example, in Mkn 78, $C \leq 17\%$ [75] and the ratio $N_{rec}/N_{ion}C \geq 100$.

The second approach has been applied to off-nuclear emission-line regions in NGC 1068 [5] and NGC 4151 [43], resulting in estimates of $N_{ion,lines}/N_{ion} = 200$ and 13, respectively. A significant anisotropy is also found in Mkn 573 [63]. Again, it appears that the emission-line clouds along or near the radio axis are illuminated by more intense ionizing radiation than is emitted towards Earth. In principle, comparison of the values of N_{rec}/N_{ion} and $N_{ion,lines}/N_{ion}$ determined from the two approaches allows an estimate of C to be made.

There are three possible alternative explanations for the high values found for N_{rec}/N_{ion} and $N_{ion,lines}/N_{ion}$, but all face difficulties, as I now discuss. 1) The first alternative is that N_{ion} is higher than estimated by power-law extrapolations, perhaps because of the presence of a thermal component which peaks at ionizing energies. This option is unattractive now that spectra are available for Seyfert 2's down to 1200 Å [29]. 2) There are sources of ionization local to the gas. However, continuum images generally reveal little or no evidence for any unusually blue continuum light near the high excitation gas. Also, collisional ionization in shock waves cannot enhance $H\beta$ enough and the predicted line ratios disagree with observations. 3) The nucleus was much brighter in the past when the photons which are ionizing the gas "now" were emitted. Because of the trend for $N_{rec}/N_{ion} > 1$, Seyfert 2's would have to be galaxies in which the ionizing continuum has decreased by a large factor in the last 10^{2-4} yrs. This idea can account for the energetic, but not the geometrical (Section 3.2.1), evidence for anisotropy. On balance, I conclude that the evidence for anisotropy in the ionizing continuum is very strong, at least for Seyfert 2's.

4. Radio-Loud AGNs

This section reviews recent work on the EELR's of radio galaxies with only passing reference to radio-loud quasars (see refs 26 and 57 for discussions of extended line emission in quasars). I shall draw heavily on recent work by the Baltimore/Maryland and Munich/Cambridge groups.

4.1 Statistical Results

Baum and Heckman [6,7] have recently reported optical broad-band (R or V) and narrow-band ([OIII]λ5007 or Hα+[NII]$\lambda\lambda$6548,6583) imaging of a radio flux-density limited sample of 38 radio galaxies plus another 5 radio galaxies known from previous work to be optically interesting. 31 of the 38 and 4 of the 5 have $L_{radio} \geq 10^{42}$ erg s^{-1} and some 70% of the galaxies are in the FRII class. Some of the chief results of this survey are as follows.

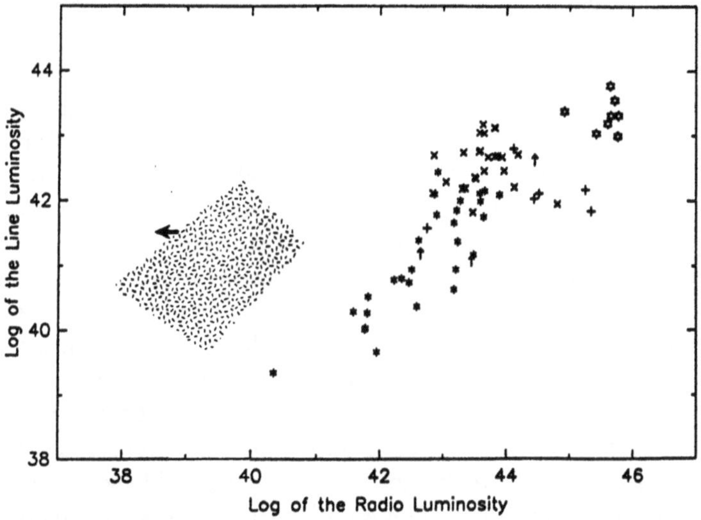

Fig. 4. (from [7]). Plot of the log of the Hα+[NII]$\lambda\lambda$6548,6583 luminosity against the log of the total radio luminosity (both in erg s^{-1}). Asterisks and crosses represent radio galaxies, $\times$'s represent radio-loud, steep spectrum quasars [57], and the stars in the upper right represent high-redshift radio galaxies. The region of the diagram occupied by Seyfert galaxies is shown hatched

1. Spatially extended emission-line gas is detected in 85% of the sample.
2. The median nebulosity extent is $\simeq$ 10 kpc (which is $\simeq$ 10 times larger than for normal early-type galaxies), with a range of 1 - 96 kpc. The median line luminosity is $\simeq 3\times10^{41}$ erg s^{-1} (which is $\simeq$ 10 times greater than is characteristic of normal early-type galaxies of similar absolute magnitude), with a range of $2.2\times10^{39} - 1.3\times10^{43}$ erg s^{-1}.

3. The narrow emission line luminosity is strongly correlated, over a range of ≥ 4 orders of magnitude, with both the total radio luminosity (see Fig. 4 and also [48]; cf. the similar correlation for Seyferts) and the radio core power, the latter relation showing more scatter (cf. earlier discussions in refs 27 and 71). There is a better correlation with the *narrow* line luminosity than with the sum of narrow and broad (as also found for Seyferts). This last result may be influenced by possible aspect angle or variability effects in the broad line fluxes. Lastly, a related result is found for *quasars*: quasars with steep-spectrum, extended radio lobes tend to have much more luminous nuclear and extended *narrow* line emission than do quasars which are radio-quiet or have radio emission dominated by a flat-spectrum compact radio core [11,12,57].
4. The luminosity ratio $L_{total\ narrow\ line}/L_{radio}$ ranges between $\simeq 0.1$ and $\simeq 3$ which may be compared with the values of 10^{2-4} found for Seyferts. Presuming the gas is photoionized by the nucleus (see below), this result implies that the power available in ionizing photons is $\approx$ or $\geq$ the power output in radio emission for powerful radio galaxies. *There are NO powerful radio galaxies which radiate much less luminosity in ionizing radiation than in radio emission.*
5. A distinction may be drawn between small (<150 kpc) and large (>150 kpc) radio sources with extended gas. Members of the former group show very distorted radio morphologies and low fractional radio polarization, are often associated with the dominant galaxy at the center of a rich cluster of galaxies and commonly have emission-line gas co-spatial with the radio emission. These small radio sources seem similar to (but much more luminous than) Seyfert galaxy radio sources, in that the radio morphology, physical extent and polarization properties are strongly affected by a gas-rich environment. The large radio sources, on the other hand, show undistorted FRII morphologies, have "normal" fractional radio polarization, tend to be associated with galaxies which are either isolated or in a small group, and only occasionally have emission-line gas adjacent to bright radio features. These differences may be interpreted as follows. The small radio sources are likely confined by a dense interstellar medium and the line emission is associated with radiative shocks driven into this gas by the radio ejecta (as appears to be the case for Seyferts). The large radio sources are propagating through an intergalactic medium of much lower density, so the radiative cooling time associated with ejecta-driven shocks may be longer than the dynamical time. The distribution of line emission is then determined primarily by the pre-existing density distribution of cool, ambient or accreted gas and any anisotropy (see below) in the escape of ionizing photons from the nucleus.

4.2 Ionization and Excitation of the Emission-Line Gas in Radio Galaxies

4.2.1 Evidence for Photoionization by the Nucleus. The lines of evidence indicating the emission-line gas is photoionized by the nucleus are as follows. 1) A strong correlation is found between N_{ion} and N_{rec} [6], similar to that for Seyferts and quasars. 2) EELR's in radio galaxies have similar spectra to nuclear NLR's, ranging from low excitation LINER-type ([OIII]λ5007/H$\beta \leq 1$) to high excitation Seyfert-type ([OIII]λ5007/H$\beta \geq 3$) [50]. 3) Photoionization models with power-law and hot ($[1-2] \times 10^5$ K) black-body continua provide equally good fits to the line ratios [50]. It is then plausible to interpret variations in excitation as a result of variations in the ionization parameter, U.

At this point, it is worth recalling that for *quasars*, there is a strong correlation between the narrow emission-line luminosity in the *nuclear* region and the presence of strong, extended line emission [57], which again argues that the extended gas is photoionized by the nucleus.

4.2.2 Evidence for Anisotropic Ionizing Radiation. The evidence for anisotropic ionizing radiation in radio galaxies is similar to, but less strong than, the evidence for this behavior in Seyfert galaxies (Section 3.2). Consideration of the ratio N_{rec}/N_{ion} implies that $C \simeq 1$ for reasonable extrapolations of an assumed isotropic, featureless, nuclear optical continuum [6,50] (note that very few uv continuum spectra are available for radio galaxies). However, images show that the nebulosities are very patchy, which suggests $C << 1$. Evidence that the gas is highly clumped is also found from the much higher densities (10^{2-3} cm^{-3}) inferred from the measured [SII] ratio than the lower limits ($\geq$0.01 - 0.1 cm^{-3}) obtained from the Hβ surface brightness [50]. The extended gas in both radio galaxies and radio-loud steep-spectrum quasars is found preferentially, although not exclusively, in the two radio quadrants [7,21]. Other possible sources of ionization include cosmic rays, shock waves, and "in situ" photoionization. Detailed coincidences between EELR's and radio hot spots are the exception and some emission-line features are further from the nucleus than the radio lobes; both of these observations argue against cosmic rays as the dominant source of ionization [7,50]. The electron temperatures derived from the [OIII] ratio are much lower than expected from collisional ionization in shocks [50]. Lastly, extranuclear sources of non-stellar blue continuum light are only occasionally seen, and in some cases result from scattering of nuclear light [55]. The evidence thus tends to support the notion that the EELR's in radio-loud AGN are photoionized by a nuclear uv source which is partially "beamed" along the radio axis, but further observational data, such as uv continuum spectra of NLRG's from HST (to provide more accurate estimates of N_{ion}), are required for a definitive conclusion.

4.2.3 Possible Problems with "Pure" Photoionization Models. The most significant "problem" is the discovery [61] that the electron temperature derived from the [OIII]λ4363/$\lambda\lambda$4959,5007 ratio is typically 12,800 < T_e < 22,000 K, whereas the otherwise successful photoionization models predict T_e < 11,000 K. Possible explanations include [61]: 1) The clouds are optically thin, so we only sample the hottest, fully ionized zone. The problem with this idea is that lines which originate in the partially neutral zone (e.g. [OII], [SII]) in optically thick models are then much weaker than observed. 2) Additional heating sources, such as shock waves [10,65] or relativistic particles [19], elevate the electron temperature. 3) The metal abundances are reduced w.r.t. solar, thereby making cooling by forbidden lines less effective, and increasing the electron temperature. Of these possibilities, heating by shocks and sub-solar abundances appear the most promising [61].

A curious result found for the nebulosity associated with 3C 227 is a possible correlation between the $H\beta$ surface brightness and the excitation level as indicated by the [OIII]λ5007/Hβ ratio [50]. Further data are needed to establish whether this sort of relation is ubiquitous or rare. By way of explanation, one possible scenario would involve driving radiative shocks into a photoionized nebula [37]. The bright knots would be associated with the dense, cooling gas behind the shock, and thus would have a higher [OIII] temperature than unshocked gas, which would be at the "ambient" photoionization temperature. Because of their higher temperature, the [OIII]/Hβ ratio would be higher in the knots than in the purely photoionized gas. In this way, a correlation between Hβ surface brightness and [OIII]/Hβ ratio might result. This notion predicts that the brighter knots should have higher temperatures than the more diffusely spread nebulosity.

5. Origin of the Collimation of the Ionizing Photons

We have seen that ionizing photons are collimated into cones of wide opening angle (several tens of degrees) in at least some AGNs. Possible processes responsible for this collimation include [e.g. 43,44]:

1. Beaming by a mildly relativistic jet. In favor of this possibility is the good alignment between the cone and radio axes, but there is no evidence yet for bulk relativistic motion in radio-quiet AGNs. The wide opening angles may present a problem, for the bulk motions would have to be consistently *mildly* relativistic, a "fine tuning" that may be difficult to achieve in jet models.
2. Shadowing by a dense molecular torus. There is considerable evidence for "blocking" material in Seyfert 2's, perhaps in the form of a thick torus [3] or warped disk [52]. This evidence comes from their optical polarization [3,38], strong, fluorescent Fe K X-ray features [30,31], high gas columns as indicated by X-ray spectra [4], nuclear molecular hydrogen [28,51], and strong nuclear reddening found in some objects [24,59].

In NGC 4151, however, the ENLR subtends only 1-2% of the sky at the nucleus. Thus, if the central optical-uv source radiates isotropically, almost all of its radiation must be absorbed by the putative torus and reradiated in the infrared. This expected large infrared signature is not observed - the infrared luminosity is similar to the optical-uv-soft X-ray luminosity (L(2-100 μm)$\simeq$1.2L(4000 Å-2 kev). This result presents a problem for any model of NGC 4151 in which an unseen, initially isotropic continuum of the strength deduced from the ionization parameter is reprocessed into the infrared [43]. This argument is a powerful one and should be applied to more galaxies (cf. also ref 39).

3. Shadowing by the BLR clouds. Ferland and Mushotzky [18] interpreted the strong soft X-ray variability of NGC 4151 in terms of obscuration of the nuclear continuum by broad line clouds. To account for the collimation, the BLR would have to be disk shaped or possibly more spherically symmetric with holes "punched" by the radio ejecta.
4. Intrinsic anisotropy of a thick accretion disk. Radiation pressure supported accretion tori collimate soft X-rays into wide cones by means of their hot inner funnels [1,32]. Unlike the dust shadowing case, no strong reprocessed far infrared emission is expected. Photoionization models using the expected ionizing continua of radiation tori can account for the emission-line intensities and ratios of the ENLR [2]. These objects radiate near or somewhat above the Eddington limit; relatively low mass black holes would be implied, in possible conflict with the higher masses needed if the BLR motions are gravitational in origin [14,43]. Another worry concerns the stability of these objects. Ionized cones are seen to extend up to tens of kpc from the nuclei, so if a radiation torus is responsible, it must be stable over 10^5 yrs, which is many torus dynamical timescales.

6. Speculations on the Difference between Radio-Quiet and Radio-Loud Objects

I have noted the remarkable correlation between the luminosity of the narrow emission lines (a measure of the power of ionizing photons reaching the NLR/ENLR) and the luminosity of the *steep spectrum* radio emission (a measure of the kinetic power of collimated plasma ejecta which fuel radio lobes). Such a correlation is found (Fig. 4) for *both* radio-loud objects (including radio galaxies and quasars) and radio-quiet objects (Seyfert galaxies; I expect radio-quiet quasars will be found to lie to the upper right of the hatched area, which represents Seyferts, in Fig. 4). However, the *ratio* of ionizing luminosity to radio luminosity is typically $\approx 0.1 - 1$ for radio-loud objects and $\approx 10^{2-4}$ for radio-quiet ones. This difference appears to be dominated by differences in the power of collimated plasma ejecta in the two classes of object. If the ionizing photons are radiated directly from an accretion disk, the fact that radio-loud objects generate similar ionizing photon luminosities to radio-quiet ones (Fig. 4) is consistent with there being similar disks in the two classes. In particular, this result argues against the idea [49] that the central disk is an ion torus in the radio-loud objects; such ion tori are expected to be very inefficient radiators of thermal emission. More generally, this line of argument suggests that the process of *thermal uv* emission is similar in radio-loud and radio-quiet objects.

One key to the difference between radio-loud and radio-quiet objects has been known for 30 years: *radio-loud objects are never associated with spiral galaxies.* A second is that *radio-loud objects have much lower (typically by a factor of 100) space densities than radio-quiet ones.* Approximate space densities at the present epoch are [41]: 10^{-4} Mpc^{-3} for Seyfert galaxies; 10^{-6} Mpc^{-3} for radio galaxies; 10^{-7} Mpc^{-3} for radio-quiet quasars; and 10^{-9} Mpc^{-3} for radio-loud quasars. Third, there is increasing evidence that elliptical galaxies, in general, and radio galaxies, in particular [56], may be *the products of galaxy mergers.*

Suppose that some fraction, f, of spiral galaxies are born with, or acquire in infancy, the "central engine" (say a supermassive black hole) required for a radio-quiet AGN. Suppose further that a substantial fraction of these galaxies subsequently merge, the result of which is an elliptical or elliptical-like galaxy. Then only a fraction $\approx f^2$ of galaxies will contain two "central engines" (i.e. a binary black hole). I speculate that the radio-loud objects are just this small fraction of merger products which contain two "central engines". The longest evolutionary phase of a black hole binary is the period after the binary has become "hard" but before it becomes so tightly bound that gravitational radiation shrinks the orbit [8]. There is a wide uncertainty in the length of this phase, but both radio galaxy timescales ($\approx 10^8$ yrs) and the Hubble time can be accomodated [8]. Precession effects associated with such a binary can account for inversion symmetric large scale jets and bending of VLBI jets [8]. The process of generating a large-scale luminous jet (radio-loud object) would be connected with the existence of the binary; I don't know how this might work. However, the thermal emission would be associated with the individual holes (e.g. one or both accretion disks) and scale with the mass in the traditional manner. In this scheme, single holes are supposed to be responsible for radio-quiet objects and generate their thermal emission in the same manner; again, as usually supposed, the thermal luminosity would depend primarily on the hole mass, accounting for the wide range of optical-uv luminosities observed. At the present epoch, $\approx 1\%$ of galaxies are Seyferts, so if $f \approx 0.01$, as suggested by the relative number of radio-quiet and radio-loud objects, the active phase would be comparable to the Hubble time. Relatively low mass holes would be more common and account for low luminosity activity (e.g. LINER's). Cosmological evolution would be associated with higher fueling rates at earlier epochs.

I thank Dr. J.P. Harrington for drawing my attention to ref 37 and for a valuable discussion.

References

1. M.A. Abramowicz, M. Calvani and L. Nobili: ApJ **242** 772 (1980)
2. J.A. Acosta-Pulido, I. Pérez-Fournon, M. Calvani and A.S. Wilson: ApJ **365** 119 (1990)
3. R.R.J. Antonucci and J.S. Miller: ApJ **297** 621 (1985)
4. H. Awaki, K. Koyama, H. Kunieda and Y. Tawara: Nature **346** 544 (1990)
5. J.A. Baldwin, A.S. Wilson and M. Whittle: ApJ **319** 84 (1987)
6. S.A. Baum and T.M. Heckman: ApJ **336** 681 (1989)
7. S.A. Baum and T.M. Heckman: ApJ **336** 702 (1989)
8. M.C. Begelman, R.D. Blandford and M.J. Rees: Nature **287** 307 (1980)
9. J. Bergeron, P. Petitjean and F. Durret: A&A **213** 61 (1989)
10. L. Binette, M.A. Dopita and I.R. Tuohy: ApJ **297** 476 (1985)
11. T.A. Boroson and J.B. Oke: ApJ **281** 535 (1984)
12. T.A. Boroson, S.E. Persson and J.B. Oke: ApJ **293** 120 (1985)
13. G. Cecil, J. Bland, and R.B. Tully: ApJ **355** 70 (1990)
14. J. Clavel et al.: ApJ **321** 251 (1987)
15. M.A. Corbin, J.A. Baldwin and A.S. Wilson: ApJ **334** 584 (1988)
16. F. Durret: A&A Suppl **81** 253 (1989)
17. I.N. Evans, H.C. Ford, A.L. Kinney, R.R.J. Antonucci, L. Armus and S. Caganoff: ApJ **369** L27 (1991)
18. G.J. Ferland and R.F. Mushotzky: ApJ **262** 564 (1982)
19. G.J. Ferland and R.F. Mushotzky: ApJ **286** 42 (1984)
20. G.J. Ferland and D.E. Osterbrock: ApJ **289** 105 (1985)
21. R.A.E. Fosbury: In [36], p169 (1989)
22. C.A. Haniff, A.S. Wilson, and M.J. Ward: ApJ **334** 104 (1988)
23. C.A. Haniff, M.J. Ward and A.S. Wilson: ApJ **368** 167 (1991)
24. C.A. Haniff, M.J. Ward and A.S. Wilson: In preparation.
25. T.M. Heckman, L. Armus, and G.K. Miley: ApJS **74** 833 (1990)
26. T.M. Heckman, M.D. Lehnert, W. van Breugel and G.K. Miley: ApJ **370** 78 (1991)
27. R.G. Hine and M.S. Longair: MNRAS **188** 111 (1979)
28. K. Kawara, M. Nishida and B. Gregory: ApJ **342** L55 (1990)
29. A.L. Kinney, R.R.J. Antonucci, M.J. Ward, A.S. Wilson and M. Whittle: ApJ **377** 100 (1991)
30. K. Koyama et al.: PASJ **41** 731 (1989)
31. J.H. Krolik and T.R. Kallman: ApJ **320** L5 (1987)
32. P. Madau: ApJ **327** 116 (1988)
33. P.J. McCarthy, W. van Breugel, H.Spinrad, S. Djorgovski: ApJ **321** L29 (1987)
34. E.J.A. Meurs: In [36], p 405 (1989)
35. E.J.A. Meurs and A.S. Wilson: A&A **136** 206 (1984)
36. E.J.A. Meurs and R.A.E. Fosbury (Editors): "Extranuclear Activity in Galaxies", ESO Conference and Workshop Proceedings No. 32 (1989)
37. D. Middlemass, R.E.S. Clegg, J.R. Walsh and J.P. Harrington: MNRAS **251** 284 (1991)
38. J.S. Miller and R.W. Goodrich: ApJ **355** 456 (1990)
39. J.S. Miller, R.W. Goodrich and W.G. Mathews: ApJ **378** 47 (1991)
40. T. Miyaji, A.S. Wilson, and I. Pérez-Fournon: ApJ (in press) (1991)
41. D.E. Osterbrock: Rep. Prog. Phys. **54** 579 (1991)

42. A. Pedlar, J.E. Dyson and Unger, S.W. MNRAS **214** 463 (1985)
43. M.V. Penston et al.: A&A **236** 53 (1990)
44. I. Pérez-Fournon and A.S. Wilson: ApJ **356** 456 (1990)
45. R.W. Pogge: ApJ **328** 519 (1988)
46. R.W. Pogge: ApJ **332** 702 (1988)
47. R.W. Pogge: ApJ **345** 730 (1989)
48. S. Rawlings and R. Saunders: Nature **349** 138 (1991)
49. M.J. Rees, M.C. Begelman, R.D. Blandford and E.S. Phinney: Nature **295** 17 (1981)
50. A. Robinson: In [36], p259 (1989)
51. V. Rotaciuc et al.: ApJ **370** L23 (1991)
52. D.B. Sanders, E.S. Phinney, G. Neugebauer, B.T. Soifer and K. Matthews: ApJ **347** 29 (1989)
53. H. Schulz: A&A **203** 233 (1988)
54. H. Schulz: AJ **99** 1442 (1990)
55. S. di Serego Alighieri et al.: Nature **334** 591 (1988)
56. E.P. Smith and T.M. Heckman: ApJ **341** 658 (1989)
57. A. Stockton and J.W. MacKenty: ApJ **316** 584 (1987)
58. T. Storchi-Bergmann and C.J. Bonatto: MNRAS **250** 138 (1991)
59. T. Storchi-Bergmann, A.S. Wilson and J.A. Baldwin: Proceedings of this conference.
60. C.N. Tadhunter and Z. Tsvetanov: Nature **341** 422 (1989)
61. C.N. Tadhunter, A. Robinson and R. Morganti: In [36], p 293 (1989)
62. D. Taylor, J.E. Dyson, D.J. Axon, and A. Pedlar: MNRAS **240** 487 (1989)
63. Z.I. Tsvetanov and J.R. Walsh: ApJ in press (1991)
64. S.W. Unger, A. Pedlar, D.J. Axon, M. Whittle, E.J.A. Meurs and M.J. Ward: MNRAS **228** 671 (1987)
65. S.M. Viegas-Aldrovandi and M. Contini: ApJ **339** 689 (1989)
66. S.J. Wagner and K. Anton: In [36], p 195 (1989)
67. M. Whittle: MNRAS **213** 33 (1985)
68. M. Whittle: In [36], p 199 (1989)
69. M. Whittle: ApJ (in press) (1991)
70. M. Whittle, A. Pedlar, E.J.A. Meurs, S.W. Unger, D.J. Axon, and M.J. Ward: ApJ **326** 125 (1988)
71. A. Wilkinson, R.G. Hine and W.L.W. Sargent: MNRAS **196** 669 (1981)
72. A.S. Wilson: In "Extragalactic Radio Sources", IAU Symposium No. 97, Eds D.S. Heeschen and C.M. Wade, p 179 (1982)
73. A.S. Wilson: In "The Interpretation of Modern Synthesis Observations of Spiral Galaxies", Astronomical Society of the Pacific Conference Series Vol 18, Eds. N. Duric and P.C. Crane, p 227 (1991)
74. A.S. Wilson and J.S. Ulvestad: ApJ **319** 105 (1987)
75. A.S. Wilson, M.J. Ward and C.A. Haniff: ApJ **334** 121 (1988)

Discussion

Bregman: 1) How does the opening angle deduced from the observations of narrow lines compare to the opening angle required for the unified models? 2) For the high power radio sources, where jets are seen only on one side, do you see emission lines associated at the presumed location of the counter jet?

Wilson: 1) The typical observed full cone opening angles are $\sim 60° - 80°$, which is consistent with the commonly quoted space density ratio of N(Seyfert 2)/N(Seyfert 1) $\simeq$ 4 to 6. 2) Yes.

Filippenko: Could you please clarify the [OIII] temperature problem for me? It seems to me that low abundances in the extended narrow-line region could reasonably account for the somewhat high electron temperature. Has a similar problem been found in HII regions or in the nuclei of those Seyferts having relatively low electron densities?

Wilson: I agree that low abundances are probably a viable explanation. In the *nuclei* of Seyferts, it is likely that high densities render temperatures from [OIII] unreliable.

Robinson: 1) In reply to Filippenko, the situation in the NLR differs from that in the ENLR. In the NLR, densities may get as high as 10^6 cm^{-3} and we can explain the high observed values of [OIII]4363/4959+5007 in terms of collisional suppression of [OIII] 4959+5007. In the ENLR, on the other hand, we know the density is $\leq 10^3$ cm^{-3}, so presumably the [OIII] ratio gives a reliable temperature determination. 2) Comment to speaker. I would like to add another object to your list of cases where a "blazar beam" may be illuminating a cloud. This may even be the best example. The object is PKS 2152-69, which has a very highly ionized emission line knot situated about 8 kpc from the nucleus. There are two main pieces of evidence which suggest the knot is ionized by a radiation beam: the knot has a transverse ionization gradient which peaks near the centre of the leading edge, and it has polarized optical continuum rising to the blue, which we believe is light scattered out of the beam by dust.

Wilson: Thank you for reminding me of this object.

Abramowicz: I would like to point out another possibility for ionizing cones. It could be that there are many (hundreds) of small-scale vorticies on the surface on a standard *thin* (or slim) accretion disk. They are scaled-down versions of the gigantic double vortex postulated for thick radiation pressure supported disks. Many vortices do collimate radiation equally well (This possibility was suggested in a paper by myself, A. Lanza, E. Spiegel and E. Szuszkiewicz, Nature 1991, submitted).

Bland-Hawthorn: It is difficult to believe that dust scattering can dominate the polarization in AGNs. We calculate the scattering phase functions of Mathis-Rumpl-Nordsieck grains (Sokolowski, Bland-Hawthorn and Cecil,

ApJ July 10 1991) and show that dust is almost entirely forward scattering beyond the Lyman limit (see also Martin and Rouleau 1990).

Wilson: I think you are referring to the question of whether *scattered nuclear* radiation is a significant source of ionization of disk gas away from the radio axis of NGC 1068. I have some doubts about this model on energetic/covering factor grounds, but do agree that the scattering particles in the nucleus are electrons. Nevertheless, there is certainly *some* contribution to the optical polarization of Seyfert 2 nuclei from dust scattering; this process is seen directly for an off-nuclear cloud in NGC 1068 [39] and is strongly suspected in more distant galaxies [38].

Blandford: Do you see a good way to discriminate ionization cones defined by emission from those delineated by absorption, and, in the latter case, can you distinguish an obscuring torus from a highly warped disk?

Wilson: 1) For cones delineated by absorption (e.g. dusty molecular tori), the absorbed energy is radiated in the infrared, and allows an energetic test (see text). 2) The inner edge of a molecular torus cannot be too close to the nucleus, since X-rays will heat this edge and evaporate the gas (Krolik and Begelman ApJ 308, L55 1986). Therefore if collimation were observed on still smaller scales, an absorbing torus could not be responsible. Unfortunately, such a test may have to await optical interferometers, since HST's resolution is 7 pc at NGC 1068, too large to make this check. 3) One can try to map out the "polar diagram" of a cone using emission line diagnostics (see e.g. paper by Tsvetanov). Unfortunately, I think radiation tori models are too poorly developed to give useable comparisons with the results. 4) Perhaps there are continuum. polarization signatures to be expected when viewing a radiation torus directly. 5) As for the question of thick absorbing tori versus warped thin disks, I hope that direct maps of the molecular gas (e.g. in the near infrared H_2 lines - see paper by Krabbe - or in CO with millimeter interferometers, which will achieve 0.1 arc sec resolution in the medium term future) will delineate the geometry directly.

HST Observations of Jets and AGN

F. Macchetto[1]

[1] On assignment from the Space Science Department of ESA

1. Introduction

The Hubble Space Telescope (HST) was launched on the 24th of April 1990. The main mission of the HST is to make available to the international scientific community a long-lived space observatory. It carries a 2.4 m diameter Ritchey-Chretien telescope, with a set of five scientific instruments, the Wide Field Planetary Camera (WFPC), the Faint Object Camera (FOC), the Faint Object Spectrograph (FOS), the Goddard High Resolution Spectrograph (GHRS) and the High Speed Photometer (HSP).

The long lifetime will be assured by regular visits by the astronauts, who will replace defective or worn-out equipment (solar arrays, gyroscopes, computers, etc.) and will upgrade the scientific capabilities by deploying new up-to-date instruments.

The HST was designed to provide three main advantages over ground-based telescopes. 1) An extended wavelength range covering from 1200Å in the UV to a few microns in the IR; 2) a decreased background, because of its location above the atmospheric scattering and 3) much improved spatial resolution capability, the result of excellent pointing accuracy and superb optical quality.

The latter has not been achieved! As is well known, as a result of the defective construction of the primary mirror, the telescope suffers from spherical aberration.

This has been well documented (Burrows, et al. 1991) and results in both a reduction in contrast and ultimate resolution.

Plans to correct the problem are well underway and it is expected that in early 1994 a WFPC II and a Corrective Optics Structural Axial Replacement (COSTAR) instrument will be deployed to finally achieve the HST performance goals.

In the intervening period, HST will remain a unique facility, unmatched for observations requiring high spatial resolution or ultraviolet wavelength coverage.

In the last few months, the HST observatory has begun to carry out its scientific mission. At the time of writing about 60% of the observing time is used on scientific observations, while the remainder is dedicated to ongoing calibrations and engineering tests of the telescope and scientific instruments.

The observing program of HST includes a considerable number of studies dedicated to the investigation of the active galaxies. So far, only a few have been started, but very interesting results have already been obtained.

Imaging observations of NGC 1068, with the WFPC, have been reported by Evans, et al. (1991) and by Lynds, et al. (1991). These were followed by spectroscopic observations with the FOS, (Caganoff, et al. 1991).

The FOS Team also studied the quasar UM675, (Beaver, et al. 1991); the low-luminosity Seyfert 1 galaxy NGC 1566, (Kriss, et al. 1991). The quasar 3C 273 was observed with both the FOS, (Bahcall, et al. 1991) and the GHRS, (Morris et al. 1991).

Finally, the Faint Object Camera has carried out observations at high-spatial resolution of the optical counterparts to three radio jets.

The study of the optical counterparts to radio jets will be the subject of an intensive investigation program with the European Space Agency's Faint Object Camera first proposed 10 years ago (Macchetto 1981, Miley 1981).

Since the optical emission appears to be the continuation of the radio synchrotron spectrum, optical observations are essential to define the physical parameters and constrain the jet emission models.

High spatial resolution observations in the visible and ultraviolet, offer the possibility to determine the precise location where particle acceleration occurs, since the electron lifetime for optical emitting electrons is extremely short. Comparison of radio and optical morphologies obtained with similar resolution will allow the investigation of confinement mechanisms and diffusion processes within the relativistic plasma.

The following sections will describe the results obtained for PKS 0521-36, (Macchetto et al., 1991a) for 3C 66B (Macchetto, et al., 1991b) and for M87.

2. PKS 0521-36

One of the most prominent radio and optical jets is that found in the elliptical galaxy PKS 0521-36, a relatively isolated radio galaxy at a redshift $z = 0.055$ which also harbors a bright $V = 16$ BL Lac nucleus and extended optical line emission. Detailed spectroscopic and morphological studies have been carried out by Danziger et al. (1979, 1983); Cayatte & Sol (1987); and Boisson, Cayatte, & Sol (1989). Recently Sparks, Miley, & Macchetto (1990) reported optical polarization measurements of the jet and nucleus,

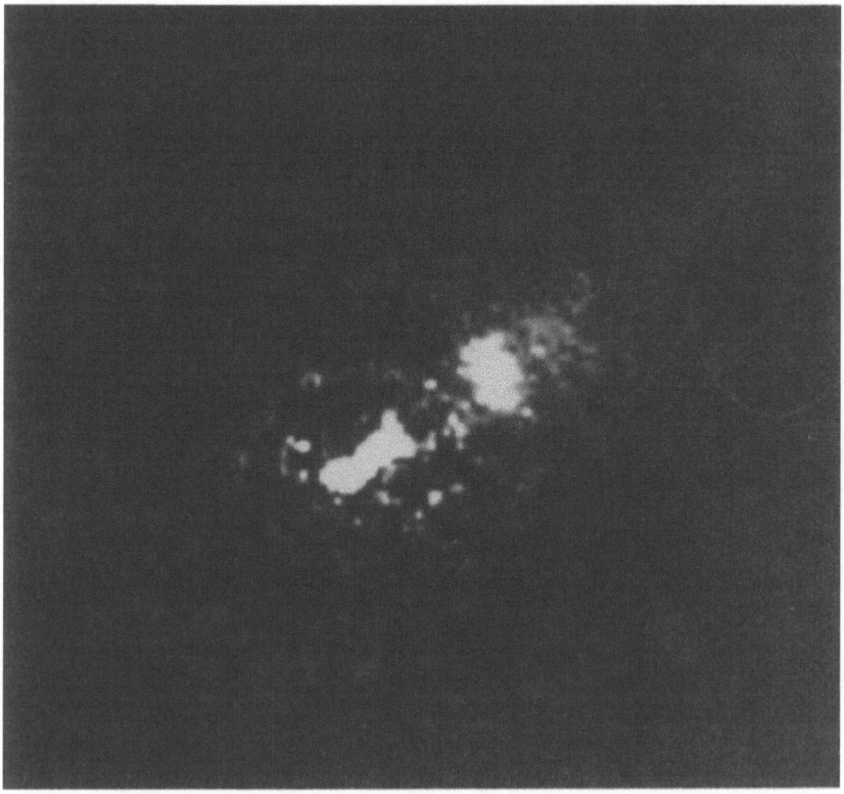

Fig. 1. The jet in PKS 0521-36 as observed by the FOC. An elliptical model of the galaxy has been subtracted. The jet is approximately $6''.5$ long.

which confirmed the expected high polarization if the emission is due to synchrotron radiation.

Images of PKS 0521-36 were obtained on 1990 August 25 with the FOC using the F430W and F320W filters and f/96, 512 × 512 mode (Paresce 1990) with a corresponding pixel size approximately 0".022. A galaxy subtracted deconvolved image is shown in Figure 1.

The VLA contour data as published by Keel (1986) is shown superposed on the FOC data in Figure 2. It is immediately obvious that the FOC data has a resolution very similar to that of the VLA data, but shows considerably more morphological information that the ground-based optical data.

The FOC image shows a bright knot located $\approx 1''.8$ to the NE and clearly resolved as in the VLA data. The width of the knot is $\approx 0''.8$. Beyond this bright knot, the jet has approximately constant surface brightness and a morphology similar to the VLA image with a total length of $6''.5$. The jet is also resolved in width, $0''.6$ wide in the fainter regions of the jet, with little or no evidence of structure on a scale of $\leq 0''.1$. The FOC data appears to

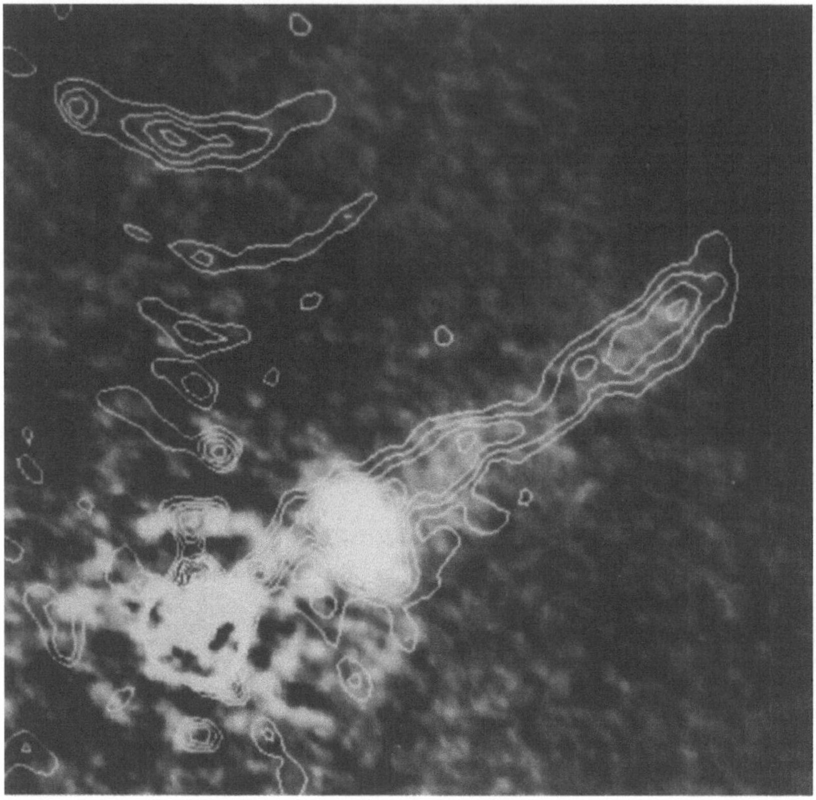

Fig. 2. The VLA contour data of Keel (1986) is shown superposed on the FOC data. Note the excellent agreement of the two data sets.

show more flux than the VLA data in the region at slightly larger radius from the nucleus but close to the southern tip of the knot.

There is a second component seen in the deconvolution at $0''.36$ from the nucleus, Figure 2. It is visible in the raw data, although by no means as clearly as in the deconvolved image. Such a source would be completely within the core of the highest resolution VLA contour map published by Keel. We believe we have detected a previously unseen inner jet structure at a distance corresponding to 300 pc, $H_o = 75$ km s^{-1} Mpc^{-1}. Presumably, the optical polarization measured by Sparks, Miley & Macchetto (1990) comes from both these components. The fact that the nuclear polarization is large and perpendicular to the jet direction (Sparks et al. 1990; Angel & Stockman 1980; Bailey, Hough & Axon 1983; Bridle et al. 1986) is consistent with what has been observed in other quiescent blazars and quasars. The second component close to the nucleus may be an inner extension of the jet, since it lies on the geometrical projection of the jet toward the nucleus. However, since this extension falls entirely within the VLA radio core, there is no independent way of confirming its existence. If real, it could be a site of electron acceleration along the jet due to transverse shocks (Drury 1983; Blandford & Eichler 1987).

The large bright knot further along the jet is a clear counterpart to the radio knot. The radio and optical polarization position angles (Keel 1986; Sparks et al. 1990) suggest a magnetic field aligned along the jet direction. This knot is unresolved in the optical polarimetric measurements, but the general sense of the magnetic field is still along the jet direction at that position. The bright knot is clearly an important site where particle acceleration is occurring. The general optical morphology of the jet does not exhibit any significant degree of clumpiness even at FOC resolution.

Using standard formulae and values for the relevant parameters, we derive a mean lifetime for the electrons of $t_{1/2} \approx 600$ yrs.

This implies that there must be continuous acceleration along the jet of the electrons responsible for the optical emission, since electron diffusion from the bright knot could not account for the observed optical extent.

3. 3C 66B

3C 66B is a relatively nearby bright radio source associated with a 13th magnitude galaxy at a redshift of 0.0215 (Matthews, Morgan, & Schmidt 1964). The galaxy lies in a small group close to the cluster Abell 347, at a distance of 86 Mpc. At this distance an angular scale of $0''.1$ corresponds to a projected linear size of 41 pc.

The radio source has been the subject of several comprehensive studies (e.g., Northover 1973; Miley & van der Laan 1973; van Breugel & Jägers 1982; Leahy, Jägers, & Pooley 1986). Its structure is intermediate between that of an edge-darkened double and a head-tail morphology, indicating that the morphology may be affected by motion of the parent galaxy through an ambient medium (Miley et al. 1972). When mapped at sufficient resolution, the radio emission from the nuclear "head" apears jetlike and one-sided. An optical counterpart of the jet in 3C 66B was detected by Butcher et al. (1980) and further studied by Fraix-Burnet et al. (1989b).

Images of 3C 66B were obtained with the Faint Object Camera (FOC) on 1990 September 23 using the F320W filter in the f/96 512 × 512 mode (Paresce 1990) and reduced as in Macchetto et al. (1991a). The filter bandpass is about 900 Å FWHM centered at 3360 Å. The pixel size after correction for geometric distortion is approximately $0''.022$, $\approx$ 10 pc, and the field size approximately 11" $\approx$ 4 kpc. Two exposures of 1500 s duration were made with this filter, both in fine lock, resulting in a tracking accuracy of $0''.007$ rms.

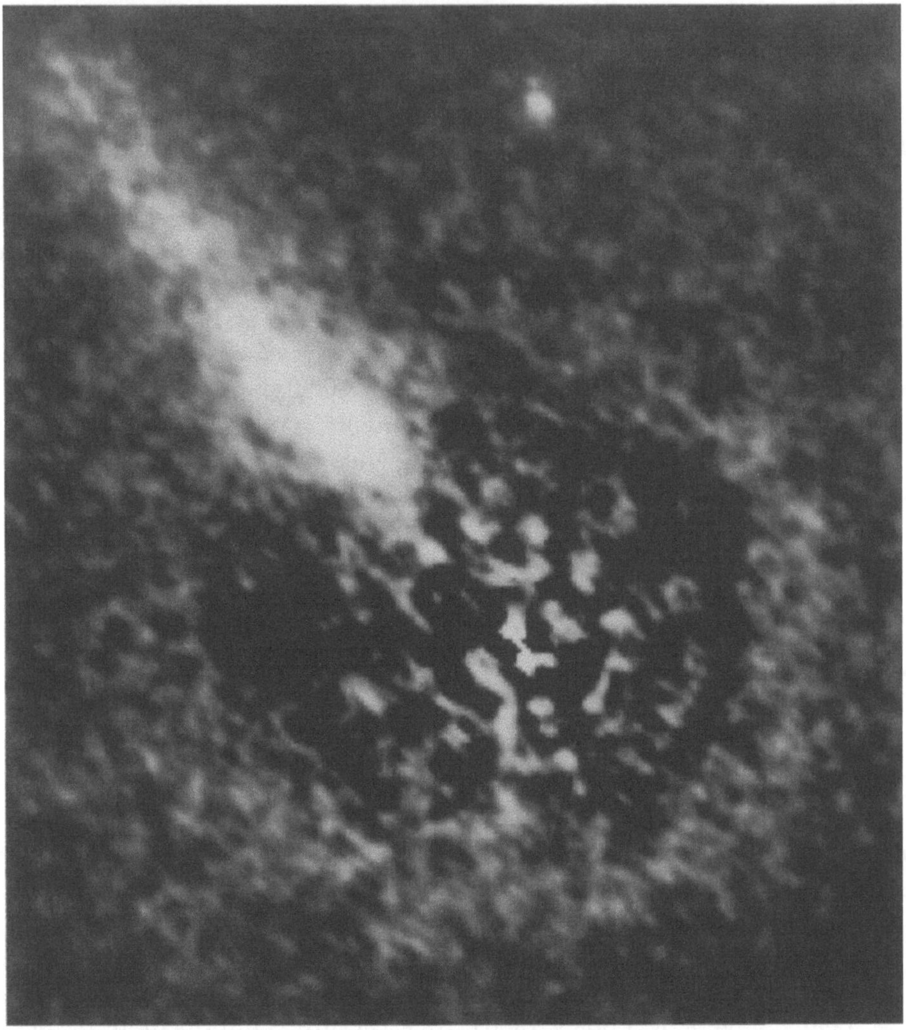

Fig. 3. Deconvolved and galaxy-subtracted image of the jet of 3C 66B. The jet is 7″.6 (3 kpc) long. The FOC resolution of 0″.1 corresponds to 40 pc at the distance of the galaxy.

The galaxy was clearly visible in the raw data, but in order to investigate the jet alone, a model of the underlying galaxy light was subtracted from the deconvolved image (Fig. 3).

In order to compare our HST image with the best VLA map of 3C 66B, we smoothed the deconvolved and galaxy-subtracted optical image with a circular Gaussian function having $\sigma = 3.1$ binned pixels (FWHM $\approx 0''.35$; Fig. 4a). Figure 4b shows a contour map of the "A-configuration" VLA map of 3C 66B obtained at 6 cm wavelength by Leahy et al. (1986).

Several conclusions can be drawn from these figures.

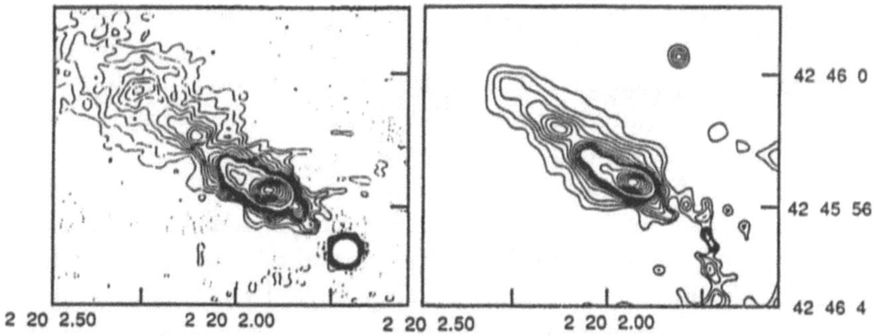

Fig. 4. Contour map of 3C 66B jet smoothed with a Gaussian "beam" with $\sigma = 3.1$ pixels ($0''.44$ each). The contours are at intervals of 1.12 counts from 1.12 to 11.2 counts and intervals of 3.73 counts from 14.92 to 37.3 counts, corresponding to the contours of Leahy et al. 1986 for a nominal flux conversion and spectral index $\alpha = 1.1$. Note that the spatial registration and scale are only approximate for the HST data. The coordinates' R.A. and Decl. axes are from Leahy et al. (left). The 5 GHz VLA map at comparable resolution to the optical image of (right) (from Leahy et al. 1986).

1. The similarities between the smoothed optical image and the radio map imply that on a scale of $0''.3$ the radio and optical emission have very much the same radio-to-optical spectral index all along the jet.

2. On the scale of the HST resolution, the jet of 3C 66B is filamentary. Filamentary structure has not previously been seen for optical jets, although it has been observed on a scale of $0''.1$ for the M87 jet in the radio (Owen et al. 1989), (and now with the FOC, see Section 4)

3. Two distinct "strands" appear in the HST image. These can be traced from $\geq 3''.7$ (1.5 kpc) from the nucleus out to a distance of $7''.6$ (3 kpc), where they disappear into the noise. The separation between the strands varies between about $0''.3$ and $0''.4$, that is, about 150 pc and they appear to undergo sharp "kinks" at distances of $2''.5$ (1.0 kpc) and $6''.2$ (2.5 kpc) from the nucleus.

It is interesting to compare our observations of 3C 66B with the high-resolution radio data of Owen, Hardee & Cornwell (1989) and our FOC high resolution observations of M87 (next section).

Although, unlike M87, 3C 66B is not a cD galaxy at the center of a rich cluster, there are several similarities between M87 and 3C 66B. Their total radio luminosities are comparable, but the jet in 3C 66B is slightly longer and can be traced out to about 3 kpc from the nucleus, a factor of 2 farther than in M87. Both jets appear filamentary and edge brightened. However, in 3C 66B, the double-stranded filaments appear to be embedded in a broader structure, which is not accurately colinear with the filaments. Also, although the 3C 66B jet is filamentary, the pitch angle of the filaments is quite different from that in the filaments in M87 - we do not see the tightly wound helix that the M87 shows.

A very basic question posed by these observations concerns constraints on the regions where the synchrotron-radiating electrons are accelerated. Does the presence of optical radiation along the jet imply that localized particle acceleration is required, or could the electrons be accelerated in the nucleus and transported to the sites of the radiation?

The average age of the radiating synchrotron electrons, with typical values for the magnetic field strength and cut-off frequency, is less than 1000 yr for the optical electrons, while the light travel time from the nucleus to the end of the observed optical jet is about 10^4 yr. Hence, at first sight, localized particle acceleration would be required, however, as discussed later for M87, in situ acceleration is not needed in a two fluid model, (Pelletier and Roland 1989).

The other new features present in our observations are the sharp bends or "kinks" in the double-stranded filaments. The origin of these kinks is unclear. The fact that they are mimicked in more than one filament suggests that they are not due to an instability mode in an individual filament. They may well trace out irregularities in the ISM of the galaxy and/or be due to time-dependent variations in the power of the nuclear machine responsible for producing the jet.

Another possibility is that these bends and kinks are the result of observing a jet which contains magnetic field modes which are at an angle to the jet. Models for force-free magnetized jets have been calculated by Königl & Choudhuri (1985) containing two magnetic field orientations, one parallel to the jet axis and the other a double helix twisted field. Both our observed filamentary edge-brightened configuration and the presence of bends are at least consistent with their model.

4. M87

The EO galaxy M87 harbours the prototypical and most studied example of an optical jet. First observed by Curtis in 1918, it remained little more than a curiosity, until Baade and Minkowski studied it in 1954 and first used the term "jet" to describe the sequence of optical knots extending to about 20" from the nucleus. Since then, the jet has been observed at radio (Owen, Hardee & Cornwell, 1989, Biretta, Stern & Harris 1991), optical (de Vaucouleurs & Nieto 1979, Keel 1988, Fraix-Burnet, Le Borgne & Nieto 1989) and X-rays wavelengths (Schreier, Gorenstein & Feigelson, 1980).

The radio and optical morphologies and polarization structure of the jet are similar (Schlötelburg et al., 1988) to within the resolution limits of the ground based observations. These results are best explained by emission from synchrotron radiation. In addition, the emission detected at x-ray wavelengths in the jet region also suggests that the synchrotron spectrum extends to high frequencies.

Images of M87 centred on the nucleus and positions along the jet were obtained with the FOC, utilizing the following filters in the f/96 512 × 512 modes: F120M, F140W, F220W (direct and with polarizers POL0,POL60,POL120), F430W (with polarizers only) and F501N. Additionally, a zoomed acquisition image was made with the F372M filter. The exposures were made in fine lock, with an expected tracking accuracy of 0″.007.

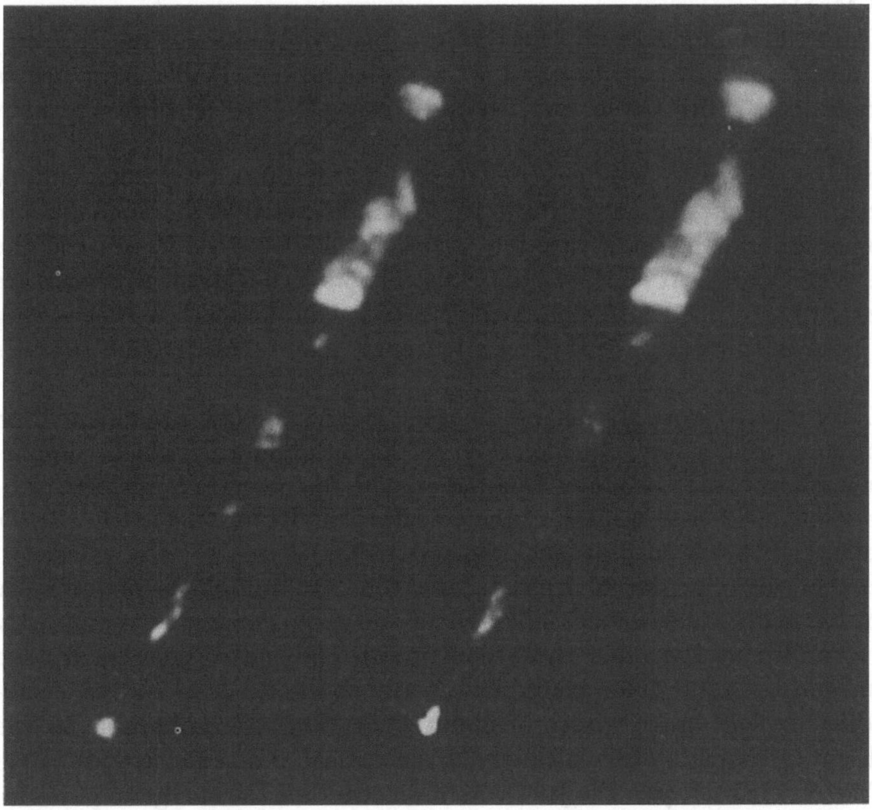

Fig. 5. Left. Deconvolved FOC image of the M87 jet. Two FOC images are needed to cover the length of the jet (≃ 20"), hence the small gap in the spatial coverage. Right. Deconvolved VLA image kindly supplied by F. Owen and J. Biretta. The images show the remarkable agreement of the ultraviolet and radio observations.

The F220W exposures were corrected using the Lucy (1974) deconvolution technique to produce the final deconvolved images with minimum beam size 0″.2 FWHM. The results of this restoration are shown in Figure 5. This figure shows side-by-side the deconvolved FOC and VLA data. As is easily observed, two FOC frames (each of 11" × 11") are needed to cover the jet which extends for over 20" in length.

The complex structure and the wide range of intensities make it very difficult to display all the features in a single picture. Furthermore, the value of the relative intensity of the ultraviolet and radio data is arbitrary and was chosen only to show the most prominent features. Detailed intensity comparisons and determination of spectral indices as a function of position along the jet will be carried out in the near future.

The FOC observations demonstrate for the first time that the radio and ultraviolet brightness distribution is generally the same over a scale of about $0''.1$ or about 10 pc (M87 presents 78 pc arcsec^{-1} at an assumed distance of 16 Mpc).

The FOC data shows that all the prominent optical knots (A,B,C, etc) have now been fully resolved and show the same remarkable structure as the radio data. (See Owen, Hardee and Cornwell, 1989, also for the knot notation). The jet is limb brightened, shows very well-defined edges along its conical structure with an opening angle of $\sim 6.5^o$ and has a tight filamentary structure. The appearance of this structure indicates that the filaments are wrapped around the jet with pitch angles of about 30° - 40° between the nucleus and knot I. At knot A, and between knots A and B, the filaments are more tightly wrapped with a pitch angle between 80° and 90° and decreasing from knot A towards knot B. The pitch angle may increase again towards knot C.

Slices taken across the jet at different locations show prominent limb brightening similar, but not identical, to the radio data. Detailed comparisons are beyond the scope of this paper, but this general agreement provides strong evidence in favor of the two fluid model for this jet. (Pelletier & Roland, 1988, Owen, Hardee & Cornwell 1988).

In this scenario, the jet consists of a cone around which one or more bright filaments are wrapped. The optical and radio emission comes mostly from a surface layer in which these filaments are embedded. The synchrotron lifetimes in the ultraviolet are typically only of the order of 100 yr, corresponding to light travel times of about 30pc. This is comparable to the width and presumably the thickness of the optical and radio strands.

The low emissivity in the jet's core indicates that the energetic particles are not suffering significant synchrotron losses. This is also compatible with the model in which the jet's core has a relatively low magnetic field. In this case, the high-energy particles can be produced in the central black-hole and propagate along the jet's interior in a low magnetic field region, thereby suffering only modest synchrotron losses. As they diffuse across the jet and into the high magnetic field boundary layer, they can produce the optical emission without the need for in situ acceleration.

Several mechanisms can be invoked to explain how the emission from this boundary layer is produced. Non-linear evolution of synchrotron instabilities has been proposed by Bodo et al., 1991, to explain the formation of filaments in jets out of equipartition, such as that in M87, where the energy of the relativistic electrons exceeds that of the magnetic field. They find that in a plasma subject to constant heating, after an initial phase in which the instability growth rate follows the linear model, the instability reaches a quasi equilibrium state on timescales of the order of several synchrotron timescales. This mechanism can explain the formation of filaments of enhanced emission observed in the lobes and jet of M87.

Optical synchrotron emission can also be produced through a diffusive shock acceleration mechanism (Fraix-Burnet, 1991). This process is so efficient that requires the magnetic field turbulence to be quite low. The source of energy of this turbulence could be the kinetic energy of the jet which can be transferred to the magnetic field (or to the plasma) through the interaction of the jet with the interstellar medium.

The answer to which of the competing mechanisms and scenarios are at work and the determination of the relevant physical parameters will have to wait further detailed analysis of the FOC optical, ultraviolet and polarization data.

5. Conclusions

The study of the optical counterparts to radio jets with the Faint Object Camera on board the Hubble Space Telescope has already produced new and unexpected results.

The jet in PKS 0521-36, which is the most distant, has been fully resolved. Because of its length, magnetic field configuration and optical morphology, it seems to require reacceleration sites for the optical electrons. These could well be provided by shocks at the site of the brighter knots observed.

For both 3C 66B and M87, we have observed, for the first time at optical wavelengths, a filamentary structure which is similar to the radio data. In this case, we conclude that the emission comes from a boundary layer where the filaments and the strong magnetic field are located.

One of the most puzzling, but fundamental, results that must be explained by models of particle acceleration is why within an object, over more than five decades of frequency and a large variation of physical conditions, the old and young electrons have similar spatial distributions, although large differences are observed from object to object.

Our FOC observations of PKS 0521-36, 3C 66B and M87 show that, even with its present aberration problems, the HST is a uniquely important instrument for studying synchrotron jets. Future observations with the HST of other extragalactic jets should provide fundamental information about the nature of collimated activity in galaxy nuclei. **Acknowledgements.** The FOC/IDT members are: Rudolf Albrecht, Cesare Barbieri, David Baxter, J. Chris Blades, Alec Boksenberg, Phil Crane, Jean Michel Deharveng, Michael J. Disney, Perry Greenfield, Peter Jakobsen, Robert Jedrzejewski, Theo M. Kamperman, Ivan R. King, F. Duccio Macchetto, Craig D. Mackay, Antonella Nota, Francesco Paresce, William Sparks, and Gerd Weigelt.

We wish to thank Drs. F. Owen and J. Biretta for making available to us the VLA radio data for M87, and Dr. Keel for the PKS 0521-36 data.

References

Angel, J.R.P., & Stockman, H.S. 1980, *Ann. Rev. Astron. Astrophys.*, **18**, 321.
Baade, W. & Minkowski, R. 1954, *Astrophys. J.*, **119**, 221.
Bahcall, J.N., Jannuzzi, B.T., Schneider, D.P., Hartig, G.F., Bohlin, R.C. & Junkkarinen, V.J. 1991, *Astrophys. J. (Letters)*, **377**, L 5.
Bailey, J., Hough, J.H., & Axon, D.J., 1983, *Monthly Notices Roy. Astron. Soc.*, **203**, 339.
Beaver, E.A., Burbidge, E.M., Cohen, R.D., Junkkarinen, V.T., Lyons, R.W., Rosenblatt, E.I., Hartig, G.F., Margon, B., & Davidsen, A.F. 1991, *Astrophys. J. (Letters)*, **377**, L 1.
Biretta, J.A., Stern, C.P. & Harris, D.E. 1991, *Astron. J.*, **101**, 1632.
Blandford, R.D., & Eichler, D. 1987, *Phys. Report,* **154**, 1.
Blandford, R.D., & Rees, M.J. 1974, *Monthly Notices Roy. Astron. Soc.*, **169**, 395.
Bodo, G., Ferrari, A., Massaglia, S., Rossi, P., Shibata, K., & Uchida, Y. 1991, *Astron. Astrophys.*, in press.
Boisson, C., Cayatte, V., & Sol, H. 1989, *Astron. Astrophys.*, **211**, 275.
Bridle, C., Hough, J.H. Bailey, J., Axon, D.J., & Hyland, A.R. 1986, *Monthly Notices Roy. Astron. Soc.*, **221**, 739.
Burrows, C.J., Holtzman, J.A., Faber, S.M., Bely, P.Y., Hasan, H., Lynds, C.R. & Schroeder, D. 1991, *Astrophys. J. (Letters)*, **369**, L 21.
Butcher, H.R., van Breugel, W.J.M., & Miley, G.K. 1980, *Astrophys. J.*, **235**, 749.
Caganoff, S., Antonucci, R.R.J., Ford, H.C., Kriss, G.A., Hartig, G., Armus, L., Evans, I.N., Rosenblatt, E., Bohlin, R.C., & Kinney, A.L. 1991, *Astrophys. J. (Letters)*, **377**, L 9.
Cayatte, V., & Sol, H. 1987, *Astron. Astrophys.*, **171**, 25.
Curtis, H.D. 1918, Lick Observatory Publ, **13**, 11.
Danziger, I.J., Ekers, R.D., Goss, W.M., & Shaver, P.A. 1983, in *Astrophysical Jets*, ed. A. Ferrari & A.G. Pacholczyk (Dordrecht:Reidel), 131.
Danziger, I.J., Fosbury, R.A.E., Goss, W.M. & Ekers, R.D. 1979, *Monthly Notices Roy. Astron. Soc.*, **188**, 415.
de Vaucouleurs, G., & Nieto, J.L. 1979, *Astrophys. J.*, **231**, 364.
Drury, L. O. C. 1983, *Rep. Progr. Phys.*, 46, 973.
Evans, I.N., Ford, H.C., Kinney, A.L., Antonucci, R.R.J., Armus, L., & Caganoff, S. 1991, *Astrophys. J. (Letters)*, **369**, L 27.
Fraix-Burnet, D., Golombek, D., Macchetto, F., Nieto, J.L., Lelievre, G., Perryman, M.A.C., & di Sergo Alighieri, S. 1991, *Astron. J.*, **101**, (1), 88.
Fraix-Burnet, D., Nieto, J.L., Lelievre, G., Macchetto, F., Perryman, M.A.C., & di Serego Alighieri, S. 1989b, *Astrophys. J.*, **336**, 121.
Fraix-Burnet, D., Le Borgne, J.F., & Nieto, J.L., 1989, *Astron. Astrophys.*, **224**, 17.
Fraix-Burnet 1991, Proceedings 7th IAP Meeting Extragalactic Radio Sources: from Beams to Jets, Paris, France, 2-5 July, 1991.
Keel, W.C. 1986, *Astrophys. J.*, **302**, 296.

Keel, W.C. 1988, *Astrophys. J.*, **329**, 532.
Königl, A., & Choudhuri, A.R. 1985, *Astrophys. J.*, **289**, 173.
Kriss, G.A., Hartig, G.F., Armus, L., Blair, W.P., Caganoff, S., & Dressel, L. 1991, *Astrophys. J. (Letters)*, **377**, L 13.
Leahy, J.P., Jägers, W., & Pooley, G.G. 1986, *Astron. Astrophys.*, **156**, 251.
Lucy, L.B. 1974, *Astron. J.*, **79**, 745.
Lynds, R., Faber, S.M., Groth, E.J., Holtzman, J.A., Light, R.M., O'Neil, E.J., Baum, W., Currie, D.G., Ewald, S.P., Hester, J.J., Kristian, J.A., Seidelmann, P.K., Shaya, E.J., & Westphal, J.A. 1991, *Astrophys. J. (Letters)*, **369**, L 31.
Macchetto, F. 1981, in *Proc. of ESO/ESA Workshop Optical Jets in Galaxies*, ed. F. Macchetto, G. Miley, & M. Tarenghi (Noordwijk:ESA), 15.
Macchetto, F., Albrecht, R., Barbieri, C., Blades, J.C., Boksenberg, A., Crane, P., Deharveng, J.M., Disney, M.J., Jakobsen, P., Kamperman, T.M., King, I.R., Mackay, C.D., Paresce, F., Weigelt, G., Baxter, D., Greenfield, P., Jedrzejewski, R., Nota, A., & Sparks, L.B. 1991a, *Astrophys. J. (Letters)*, **369**, L 55.
Macchetto, F., Albrecht, R., Barbieri, C., Blades, J.C., Boksenberg, A., Crane, P., Deharveng, J.M., Disney, M.J., Jakobsen, P., Kamperman, T.M., King, I.R., Mackay, C.D., Paresce, F., Weigelt, G., Baxter, D., Greenfield, P., Jedrzejewski, R., Nota, A., Sparks, W.B., & Miley, G.M. 1991b, *Astrophys. J. (Letters)*, **373**, L 55.
Matthews, T.A., Morgan, W.W., & Schmidt, M. 1964, *Astrophys. J.*, **140**, 35.
Miley, G.K. 1981, in *Proc. of ESO/ESA Workshop Optical Jets in Galaxies*, ed. F. Macchetto, G. Miley, & M. Tarenghi (Noordwijk:ESA), 9.
Miley, G.K., 1980, *Ann. Rev. Astron. Astrophys.*, **18**, 165.
Miley, G.K., & van der Laan, H. 1973, *Astron. Astrophys.*, **28**, 359.
Miley, G.K., Perola, G.C., van der Kruit, P.C., & van der Laan, H. 1972, *Nature*, **237**, 269.
Morris, S.L., Weymann, R.J., Savage, B.D. & Gilliland, R.L., 1991 *Astrophys. J. (Letters)*, **377**, L 21.
Northover, K.J.E.D. 1973, *Monthly Notices Roy. Astron. Soc.*, **165**, 369.
Owen, F.N., Hardee, P.E., & Cornwell, T.J. 1989, *Astrophys. J.*, **340**, 698.
Paresce, F. 1990, FOC Instrument Handbook, Space Telescope Science Institute.
Pelletier, G., & Roland, J. 1988, *Astron. Astrophys.*, **196**, 71.
Schlötelburg, M., Meisenheimer, K., & Röser, H.J. 1988, *Astron. Astrophys. (Letters)*, **202**, L 23.
Schreier, E.J., Gorenstein, P., & Feigelson, E.D. 1982, *Astrophys. J.*, **261**, 42.
Sparks, W., Miley, G., & Macchetto, F. 1990, *Astrophys. J. (Letters)*, **361**, L 41.
van Breugel, W.J.M., & Jägers, W. 1982, *Astron. Astrophys. Suppl.*, **49**, 529.

High Sensitivity Radio Observations of NGC 4151

A. Pedlar[1], *M. Kukula*[1], *D. Longley*[1], *D. Axon*[1] *S. Baum*[2], *and C. O'Dea*[2]

[1] Jodrell Bank, U. of Manchester, Macclesfield, Cheshire. UK
[2] STScI, Baltimore, USA

1. Introduction

The relationship between the radio and optical emission from the 'Narrow Line Region' (NLR) of Seyfert nuclei has been the subject of intense study over the past decade. Conventional optical observations, unfortunately, have angular resolutions comparable to, or larger than, the size of the NLR and are typically an order of magnitude worse than radio observations. Hence meaningful comparisons between optical and radio images have only been possible in a few cases (e.g., NGC 5929: Whittle *et al.* 1986, Mkn 78 : Pedlar *et al.* 1989). However rapidly improving ground based techniques, together with a number of HST observations, are giving angular resolutions in the optical comparable to the radio. This will enable tests of models relating the optical and radio emission in the NLR (e.g., Pedlar *et al.* 1985, Taylor *et al.* 1989) to be made. However many of the earlier radio studies of Seyfert nuclei were carried before the recent MERLIN upgrade (resulting in higher sensitivity and angular resolution) or in snapshot mode with the incomplete VLA. Hence it is clear that many of these radio images can be significantly improved , which is essential if they are to be compared in detail with HST images. In this paper we report on high sensitivity 5 and 8GHz VLA observations of the archetypal Seyfert nucleus in NGC 4151, which represent the first part of a combined MERLIN/VLA radio study of this object at 1.5, 5, 8 and 15 GHz.

2. The Observations

The observations were carried out on the VLA (Thompson *et al.* 1980) at A and B configuration in 1990. Integration times of 18 hrs at 8.4GHz and 9 hrs at 5GHz were obtained. The A and B array databases were combined and selfcalibrated. In Fig. 1 we show the resulting 8GHz naturally weighted image which has been restored with a 0.4 arcsec beam and has an rms noise of 7 μJy/beam. Uniform weighting of the same UV database resulted in an angular resolution of 0.25 arcsec and sensitivity of 13 μJy/beam (Fig 2.). Less sensitive 5GHz images were also obtained with comparable angular resolution.

The present observations are consistent with the earlier work showing a 4" source elongated along PA $\sim 77^o$. A comparison of Fig. 2 with the 1.6Ghz observations of Booler *et al.* (1982) shows the eastern component of the 0.4 arcsec central double to have a flatter spectral index ($\alpha \sim -0.4$).

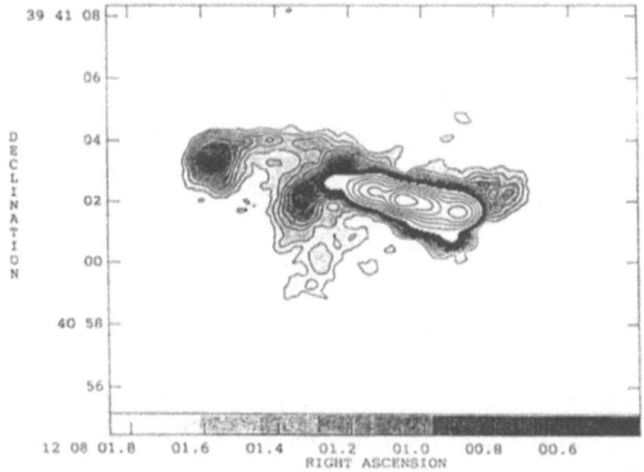

Fig. 1. Contour map of the natural weighted 8GHz image of the nucleus of NGC 4151. The angular resolution is 0.4 arcsec and the lower greyscale contour intervals are 30, 50, 70..230, 250 μJy/beam. Above 0.5mJy the contour levels are 0.5, 1, 2, 4, 8 & 16mJy/beam

It is clear that there is significant low brightness emission outside the 4 arcsec 'jet', which has been only marginally detected in earlier measurements. The emission is particularly extended to the east, terminating in a component 4.5 arcsec from the nucleus with a flux density of ~1mJy at 8GHz. In addition there are extended components at both ends of the 4 arcsec 'jet' extending in a NW/SE direction. The SE component bends to the south and is at least 4 arcsec in extent. There is also a weak component to the south of the western part of the jet. Comparison of 8 and 5GHz images indicates that almost all the extended emission has a non-thermal spectrum.

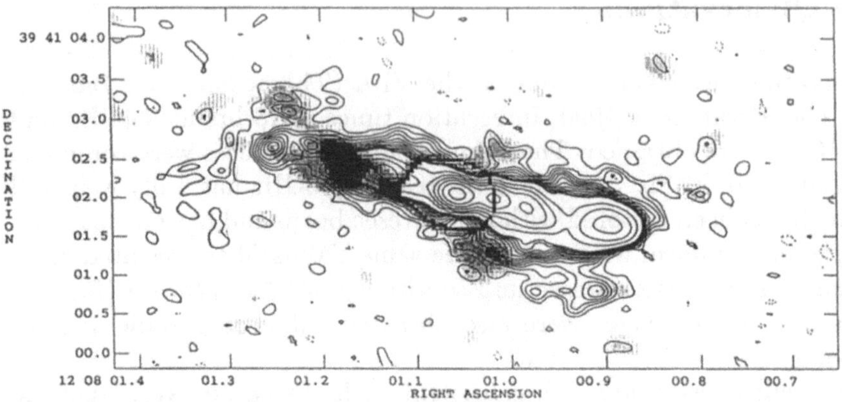

Fig. 2. A contour map of the uniform weighted 8GHz image of the nucleus of NGC 4151. The contour levels are the same as in Fig.1 and the angular resolution is 0.25 arcsec. Regions of possible thermal emission (see section 3.4) are shown as greyscale (note that regions affected by the 'flat spectrum' emission from the compact central component have been blanked out for clarity).

3. Discussion

3.1 Evidence for collimated ejection.

The present observations support the conclusions of earlier work (Booler *et al.*1982, Johnson *et al.* 1982), based on the elongated structure within 150pc of the nucleus, that collimated ejection is taking place along PA $\sim 77^o$ and 257^0 giving rise to a two sided radio jet. The detection of a component 4.5 arcsec (300pc) to the east in PA $\sim$ 77^0 reinforces this picture. The flatter spectral index (~ -0.4) of the eastern component of the central 0.4 arcsec double (see also Carral *et al.* 1990) is consistent with it containing a synchrotron self-absorbed core and hence being the component associated with the Seyfert nucleus.

3.2 'Bending' of the jet.

Although the components in the stronger parts of the jet show small deviations from a constant ejection axis, there is little evidence to support a systematic ram pressure bending of the jet caused by rotating nuclear gas as suggested by Wilson & Ulvestad (1982). In fact part of the weaker extended structure shows an 'S' shaped symmetry (Fig.1) in the opposite sense to that required for ram pressure bending by rotating nuclear gas. This is because the trailing arms observed in the optical and H1 indicate that the galactic gas is rotating clockwise.

We consider the most likely explanation for the 'S' shaped symmetry is in terms of buoyancy in which the relativistic particles associated with the

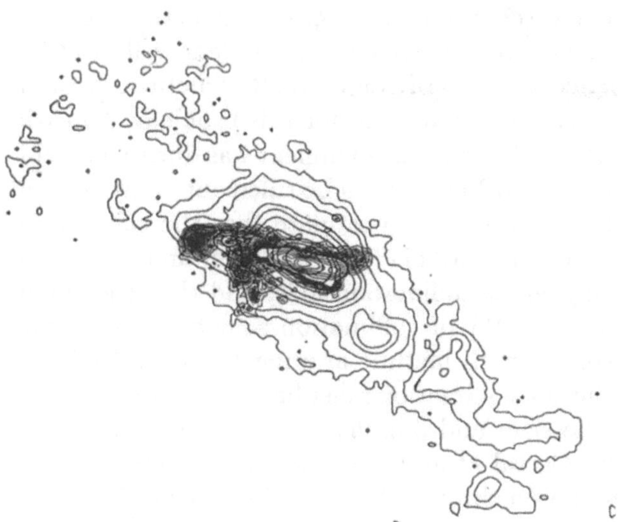

Fig. 3. The 8Ghz radio emission (Fig.1) superimposed on the [OIII] contour map of Perez et al.

jet lose energy and drift away from the disk of the galaxy. As the rotation axis in the inner galaxy is in PA 24^0 (eg Bosma *et al.* 1977) the alignment of the weak radio extensions along this PA are consistent with this picture, as they may represent flows approximately perpendicular to the plane of the galaxy.

3.3 Comparison with optical emission

NGC 4151 is one of the few Seyferts in which the NLR can be resolved by ground based optical techniques. The NLR appears to be extended over ~ 4 arcsec in PA $\sim 45^0$, and is clearly not aligned with the radio collimation axis in PA $\sim 77^0$. Schulz (1990) noted a 'protuberance' in [OIII] to the east which is aligned with the radio jet. A superposition of our 8GHz radio image and the [OIII] image Perez *et al.* (1989) of the Extended Narrow Line Region (ENLR) clearly shows the radio emission to be coincident with this protuberance (Fig.3). In addition there is evidence for an association between the southern radio extension (see section 3.2) and [OIII] emission. The predominant elongation of the ENLR , however, is in PA $\sim 228^0(48^0)$ from the nucleus and like the NLR is not aligned with the apparent collimation axis of the radio jet. There is a weak radio component to the south of the western radio jet (Fig. 2), which is in PA $234^0(54^0)$ from the nucleus, which also aligns approximately with milli-arcsecond structure in the nucleus (Harrison *et al.* 1986). There is, however, no convincing evidence from the radio images that collimated ejection is occurring in a PA other than $\sim 77^0$.

The misalignment between the radio and optical structure seen in NGC 4151 appears to be uncommon, as many (albeit less well studied) Seyferts appear to have aligned radio, NLR and ENLR structure. A number of possibilities can be considered to account for this problem. An early suggestion by Booler *et al.* (1982) that the misalignment was due to a combination of rotation and ejection of NLR clouds can be ruled out on dynamical grounds (see Unger *et al.* 1988). One could postulate that the collimation axis has recently precessed from PA$257^0(77^0)$ to PA 228^0 and hence the ENLR defines the PA of the present collimation axis, and that the PA of the radio structure represents the PA of collimation sometime in the the past. This is difficult to reconcile with lifetime arguments as the ENLR is ~2kpc in extent and hence the UV collimation must have been in PA $\sim 228^0$ for at least 6×10^3 years. The sizes and flux densities of the 15Ghz knots in the radio jet measured by Carral *et al.* (1990) imply magnetic fields of $\sim 5 \times 10^{-4}$ Gauss and hence the synchrotron half-life of the relativistic electrons producing the 15Ghz radio emission is $\sim 10^4$ years. The knots have a spectral index of $\alpha \sim -0.7$ and hence do not show an unusually steep spectral index as might be expected if significant synchrotron cooling had occurred.

It is possible that the apparent misalignment is due the combination of a wide UV cone and density bounding of the ENLR. If the active nucleus in embedded in a neutral disk, then if the collimation axis of the UV cone is not perpendicular to the disk, then simple geometrical arguments can account for the apparent misalignment between the collimation axis and the ionised gas in the ENLR. (B. Vila - private communication)

Optical observations by Terlevich *et al.* (1991) have shown evidence for a red bar elongated in PA $\sim 150^0$ which crosses the nuclear region. Terlevich *et al.* consider the possibility that the bar could be the signature of scattering in a torus 1 arcsec in diameter - which is also responsible for the collimation of the UV into the ENLR. Thus one possible scenario is that whereas the radio is collimated on scales of a few parsecs along PA $\sim 77^0$, the UV collimation is determined by this torus over the central ~ 50 pc. Thus it may be that the radio collimation axis reflects an intrinsic property of the central engine, whereas the ENLR axis is partially a consequence of the immediate nuclear environment.

It may be a coincidence that one of the most accessible and best studied Seyferts shows this misalignment between the radio and ENLR axes. Hence it is clearly important that further studies of other Seyferts, in which the radio and ENLR axes are thought to be aligned, are carried out to establish how well this alignment persists after definitive optical and radio studies.

3.4 Evidence for thermal free-free emission from the NLR.

It is clear from optical studies that large quantities of ionised gas are present in the NLR which should emit radio via free - free emission. The main problem in detecting radio free-free emission from Seyfert nuclei is the presence of the non-thermal synchrotron emission which dominates the structure particularly at low frequencies. We had hoped that the radio/optical misalignment in NGC 4151 discussed above would have resulted in regions of free-free emission being spatially offset from the non-thermal emission. However when the 8GHz image was smoothed to the 5GHz resolution (0.4 arcsec), and a spectral index image formed, there were no significant areas of flat spectral index. However spectral index of the eastern part of the jet was significantly less steep than the western part. In order to accentuate the flatter spectrum parts of the image the 5GHz image was scaled to 8GHz using the average spectral index obtained from the western part of the jet and subtracted from the 8Ghz image. The residual image showed three main components :- a central component with a peak brightness of 3.3 mJy/beam which corresponds to the flat spectrum core, together with two weaker components straddling this core with brightness temperatures of 100-200 μJy/beam. These components form a structure elongated in PA $\sim 50^0$ (see Fig. 2) and show strong similarities with the high resolution [OIII] images of the NLR of NGC 4151 shown by Ebstein *et al.* (1989) and Terlevich *et al.* (1991). The brightness temperatures of the knots implies emission measures of $\sim 5 \times 10^5 pc\ cm^{-6}$ (assuming $T_e = 10^4 K$), which given that the linear size of the knots is $< 30pc$, implies an rms electron density $> 10^2 cm^{-3}$ - which is consistent with optical measurements.

Hence there is good circumstantial evidence that the knots shown in Fig.2 could be free-free emission from the NLR gas. Observations at 1.5GHz using MERLIN and 15GHz using the VLA are required to establish the thermal spectrum of this emission and to to rule out the possibility that the knots represent regions of less steep spectral index.

References

Booler, R.V., Pedlar, A. & Davies, R.D.: 1982,*Monthly Notices Roy. Astron. Soc.*, **199**, 229

Bosma, A., Ekers, R.D. & Lequeux, J.: 1977, *Astron. Astrophys.*, **57**, 97

Carral, P., Turner, J.L. & Ho, P.T.P.: 1990, *Astrophys. J.*, **362**, 434

Ebstein, S.M, Carleton, N.P., & Papaliolios, C.: 1989, *Astrophys. J.*, **336**, 103

Harrison, B., Pedlar, A., Unger, S.W., Burgess, P., Graham, D.A. & Preuss, E.: 1986, *Monthly Notices Roy. Astron. Soc.*, **218**, 775

Johnson, K.J., Elvis, M., Kjer, D. & Shen: 1982, *Astron. J.*, **262**, 61

Pedlar, A., Dyson, J.E. & Unger, S.W.: 1985, *Monthly Notices Roy. Astron. Soc.*, **214**, 463

Pedlar, A., Meaburn, J., Axon, D.J., Unger, S.W., Whittle, D.M, Meurs, E.J.A. & Ward, M.J.: 1984, *Monthly Notices Roy. Astron. Soc.*, **238**, 863
Perez, E., Gonzalez-Delgado, R.,Tadhunter, C. & Tsvetanov, Z.: 1989, *Monthly Notices Roy. Astron. Soc.*, **241**, 31p
Schulz, H.: 1990, *Astron. J.*, **99**, 1442.
Taylor, D., Dyson, J.E., Axon, D.J. & Pedlar A.: 1989, *Monthly Notices Roy. Astron. Soc.*, **240**, 487
Terlevich, R., Portal, M.S., Diaz, A.I. & Terlevich, E.: 1991, *Monthly Notices Roy. Astron. Soc.*, **249**, 36
Thompson, A.R., Clark, B.G., Wade, C.M. & Napier, P.J.: 1980, *Astron. Astrophys. Suppl.*, **44**, 151
Ulrich, M.-H.: 1973, *Astron. J.*, **227**, L55
Unger S.W, Taylor, K., Pedlar, A.: 1988, *in "Active Galactic Nuclei"* IAU Symp. 134, Kluwer, p.331
Whittle, M.,Haniff, C.,Ward, M., Meurs, E., Pedlar, A., Unger, S.W., Axon, D.J, Harrison, B.: 1986, *Monthly Notices Roy. Astron. Soc.*, **222**, 189.

Discussion

Wilson: Could you please tell us the spectral index between 5 and 8GHz for the regions where you think thermal emission is detected?. One would, in any case, expect enhanced non-thermal emission from the optical filaments because the magnetic field is expected to be correlated with the gas density.

Pedlar: At 8Ghz the radio emission at the position of the possible thermal components is dominated by the non-thermal component and hence the measured spectral index is negative :- typically -0.4 to -0.6. Observations at more frequencies are required so that we can solve for the fraction of thermal emission present on a point by point basis. At this stage it seems simpler to me to interpret a less negative spectral index in terms of free-free emission from the NLR gas (which should be present at about the level measured), rather than to attempt to model variations in the magnetic field strengths across the source.

Tsvetanov: How much of the radio emission goes into PA 45^0 compared to PA 77°?

Pedlar: Although there is a radio component in PA $\sim 234^0(54^0)$ from the nucleus I'm not convinced that it represents evidence for collimation in this PA. Its flux is less than a mJy and hence at most it is only a few percent of the radio emission associated with the radio 'jet' in PA 77^0.

The Central Engine of Active Galactic Nuclei: Starbursts Versus Black Holes

Alexei V. Filippenko

Department of Astronomy, University of California, Berkeley, CA 94720 USA

1. Introduction

In this review, I will consider the intriguing hypothesis that the "activity" seen in a majority of active galactic nuclei (AGNs) is produced by a centrally concentrated burst of star formation, rather than by a supermassive black hole (SMBH). This point of view has been discussed by a number of investigators during the past few decades. The most recent and thorough treatment, however, is that of Roberto Terlevich and collaborators; his general scenario will be the main focus of my discussion.

Related and interesting topics that are outside the scope of this particular review include (1) the possibility that a SMBH is indeed present, but the broad-line clouds are actually the atmospheres of stars, (2) the nature of the ultraluminous far-infrared galaxies, and (3) evidence for enhanced star formation in the host galaxies of AGNs. These issues are addressed in the *Proceedings* (A. V. Filippenko, editor) of the Taipei Astrophysics Workshop (25–30 March 1991), to be published in early 1992 as part of the Astronomical Society of the Pacific Conference Series. Good discussions of these topics, as well as those covered in my review, can also be found in the excellent summary by Heckman (1991). The criticisms raised by Heckman (1991) are addressed in the recent article by Terlevich (1992). Here I will try to examine both points of view in as fair a manner as possible.

I begin by providing an overview of Terlevich's starburst hypothesis. Some potential problems with the model, as well as future tests, are presented in Section 3. Section 4 discusses a few of the possible successes of the scenario. My conclusions are given in Section 5.

2. Overview of the Starburst Hypothesis

In one form or another, the possibility that a starburst in the nucleus of a galaxy can give rise to the extreme forms of "activity" that are seen in QSOs and luminous Seyfert nuclei has been considered many times (e.g., Shklovskii 1960; Field 1964; Pronik 1973; Adams and Weedman 1975; Harwit and Pacini 1975; McCrea 1976; Koski 1978; Osterbrock 1978; Condon *et al.* 1982; Weedman 1983; Perry and Dyson 1985). The most recent, systematic, and long-term study of this hypothesis was begun by Terlevich and Melnick (1985), in a very thought-provoking paper that raised a number of important issues and challenged the commonly accepted view that all AGNs harbor a single compact "monster" (generally thought to be a SMBH). This work concentrated on narrow-lined objects such as type 2 Seyfert nuclei and low-ionization nuclear emission-line regions (LINERs; Heckman 1980). Various apparent weaknesses of the model were addressed and/or eliminated in subsequent papers (Terlevich, Melnick, and Moles 1987; Terlevich and Melnick 1987, 1988; Terlevich 1989; Terlevich 1990a,b). Moreover, these ideas were extended to include broad-line AGNs, such as type 1 Seyferts, by considering the supernovae (SNe) and supernova remnants (SNRs) produced by the massive stars; see Tenorio-Tagle *et al.* (1992) and Terlevich *et al.* (1992) for the most recent and thorough discussions of this possibility. Terlevich (1992) has also argued that QSOs could be the observational manifestation of core formation in elliptical galaxies.

Let me briefly summarize the most salient features of Terlevich's starburst scenario. It is assumed that the nuclei of galaxies, especially those of early Hubble type, can accumulate large quantities of gas that undergo vigorous bursts of star formation. This gas is probably *metal-rich* (solar abundances or slightly higher), as observed in the nuclear regions of early-type galaxies (Pagel and Edmunds 1981). In any given galaxy there can be several successive starbursts, but here for simplicity we will consider the effects of a single starburst.

During *Phase 1*, which lasts from 0 to 3 Myr, gas is photoionized by the continuum of hot main-sequence stars (Terlevich and Melnick 1985). Thus, we expect to see the spectrum of an H II region, although in many cases it might not be readily visible due to copious production of dust by very massive stars (e.g., η Carinae). The excitation of the H II region should be relatively low if the metallicity is high. Explicit calculations of the ionizing continua and the resulting emission-line spectra are given by García Vargas and Díaz (1992).

Phase 2, from 3 to 4 Myr, occurs when the most massive stars ($M_i \gtrsim 40$–$60\ M_\odot$, where M_i is the initial mass) become extreme WC or WO Wolf-Rayet stars, which Terlevich and Melnick (1985) call "Warmers." A key point is that stellar mass loss increases with increasing metal abundance in massive stars, so that very massive stars formed in high-metallicity environments can end their lives as bare He–C–O or C–O cores with surface temperatures comparable to those of the hottest known planetary nebulae [$T = (1-2) \times 10^5$ K], but with considerably higher luminosities ($L \approx 10^5\ L_\odot$). It is almost certain that such hot Warmers have been observed by Davidson and Kinman (1982), Barlow and Hummer (1982), Dopita *et al.* (1990), Pakull (1991), and others.

Calculations show that after 3 Myr, the ionizing continuum of a cluster created during a large burst of star formation (initial mass function slope = 3.0; Lequeux 1979) is nearly a power law of index $\alpha \approx -1.5$ (where $f_\nu \propto \nu^\alpha$), with a cutoff at 20–30 Ryd. Thus, the resulting emission-line spectrum from the surrounding clouds of gas having roughly solar composition is quite similar to that of Seyfert 2 galaxies. It may fit even better than the results of pure power-law photoionization models, which require rather low abundances (~ 0.3 solar). Moreover, the weak "featureless continuum" observed in some type 2 Seyferts may be the reddened spectrum of the ionizing cluster (Cid Fernandes and Terlevich 1992). Intriguing support for this interpretation comes from the discovery (Terlevich, Díaz, and Terlevich 1990) that the Ca II near-IR triplet absorption lines in Seyfert 2 nuclei are as strong as, or stronger than, those in normal galactic nuclei, as would be expected from a star cluster containing many red supergiants. As the cluster ages, the emission-line spectrum evolves into that of a LINER with a blue continuum, because of the relative dearth of very high-energy photons. A smaller starburst, with few extremely massive stars, gives rise to a lower ionization parameter; hence, there would be an absence of high-ionization emission lines in the spectrum, and a general deficit of blue photons, making it appear like a LINER with a red continuum. Most LINERs in early-type galaxies are of this type, probably because early-type galaxies have little gas from which to form a central starburst.

In *Phase 3* (Terlevich 1989), at $t = 4-8$ Myr, the initially most massive stars ($M_i \gtrsim 40\ M_\odot$) become SNe. Since their progenitors are compact stars, lacking hydrogen (and sometimes even helium) envelopes, these SNe consume large amounts of energy through adiabatic expansion; hence, they appear optically dim. Such reasoning has been invoked to explain the apparently low brightness of Cas A (Chevalier 1976). Terlevich (1990b, 1992) believes these are SNe Ib, some of which are known to be reasonably strong radio emitters. Studies of nearby young star clusters and giant H II regions (Terlevich 1987) show that the most massive stars preferentially populate the inner core, as a consequence of dynamical friction or initial conditions. Due to stellar winds and SNe, a hot "superbubble" forms in the interstellar medium (ISM) around the core of massive stars, and it subsequently expands along the steepest density gradient (e.g., the poles of a disk). Heckman, Armus, and Miley (1990; see also Filippenko and Sargent 1992) discuss galaxies in which this process may be happening: dramatic "superwinds" of material flow along the axis and produce elongated, partially collimated radio structures as well as optical filaments of shock-heated and photoionized gas. The overall emitted spectrum at optical wavelengths should resemble those of LINERs and type 2 Seyferts, since the main excitation sources are shock heating and photoionization by OB stars, Warmers, and SNe. Ionization cones, such as those seen by Tadhunter and Tsvetanov (1989) and Pogge (1988a,b), should be common, but optical variability is expected to be low because the SNe are dim. X-ray emission is produced by high-mass binaries and SNRs.

The final stage in this scenario, *Phase 4*, occurs from 8–60 Myr. At this time SNe II are produced by quite massive stars ($M_i \approx 8 - 40\ M_\odot$), presumably red supergiants. In some cases the metal-rich progenitors should form dense ($n \approx 10^7$ cm^{-3}) circumstellar shells or winds; alternatively, these objects might have been formed in a dense environment. The remnants of these SNe are expected to be very luminous at optical, UV, and X-ray wavelengths because their kinetic energy is rapidly and efficiently thermalized by the dense circumstellar gas, in marked contrast to the SNe of Phase 3. In essence, the SNRs sweep up gas and become radiative very quickly, bypassing the Sedov phase. They deposit most of their kinetic energy on short time scales, thereby reaching high luminosities. A majority of the energy is radiated in the extreme UV and X-ray regions of the spectrum because the shock velocities are so large (e.g., Shull 1980; Wheeler, Mazurek, and Sivaramakrishnan 1980; Tenorio-Tagle *et al.* 1992; Terlevich *et al.* 1992). The optical and UV spectra of these objects should resemble those of type 1 Seyfert nuclei.

Observational evidence for the existence of such SNe II is presented by Filippenko (1989a), who discusses SNe 1987F and 1988I. At optical wavelengths SN 1987F was almost indistinguishable from Seyfert 1 nuclei; even its absolute blue magnitude of −19 (near maximum) was comparable to that of moderate-luminosity AGNs. Similarly, the optical spectrum of SN 1988I was characterized by broad hydrogen Balmer lines and an otherwise nearly featureless continuum. Another SN whose early-time spectrum resembled that of a Seyfert 1 nucleus is SN 1987B, illustrated in Figure 1.

I have recently identified many additional objects of this kind, including SN 1988Z, SN 1991ae, and SN 1991av. In the case of SN 1988Z there is *direct* evidence for a dense circumstellar wind (Filippenko 1991b): narrow [O III] emission lines are visible in the early-time spectrum (Fig. 2), and the intensity ratio ([O III] $\lambda\lambda 4959 + 5007$)/([O III] $\lambda 4363$) ≈ 3 shows that $n_e \approx 10^{6.8}$ cm^{-3} if $T_e \approx 16{,}000$ K. As can be seen from Figure 11 of Filippenko and Halpern (1984), this [O III] intensity ratio yields a lower limit of $n_e \gtrsim 10^{6.3}$ cm^{-3}, irrespective of T_e. These emission lines were produced in gas photoionized by the luminous flash of ultraviolet radiation that was emitted when the supernova shock wave broke through the stellar surface. According to Tenorio-Tagle *et al.* (1992) and Terlevich (1992), such SNe II should be common in galactic nuclei. The clouds in the broad-line region

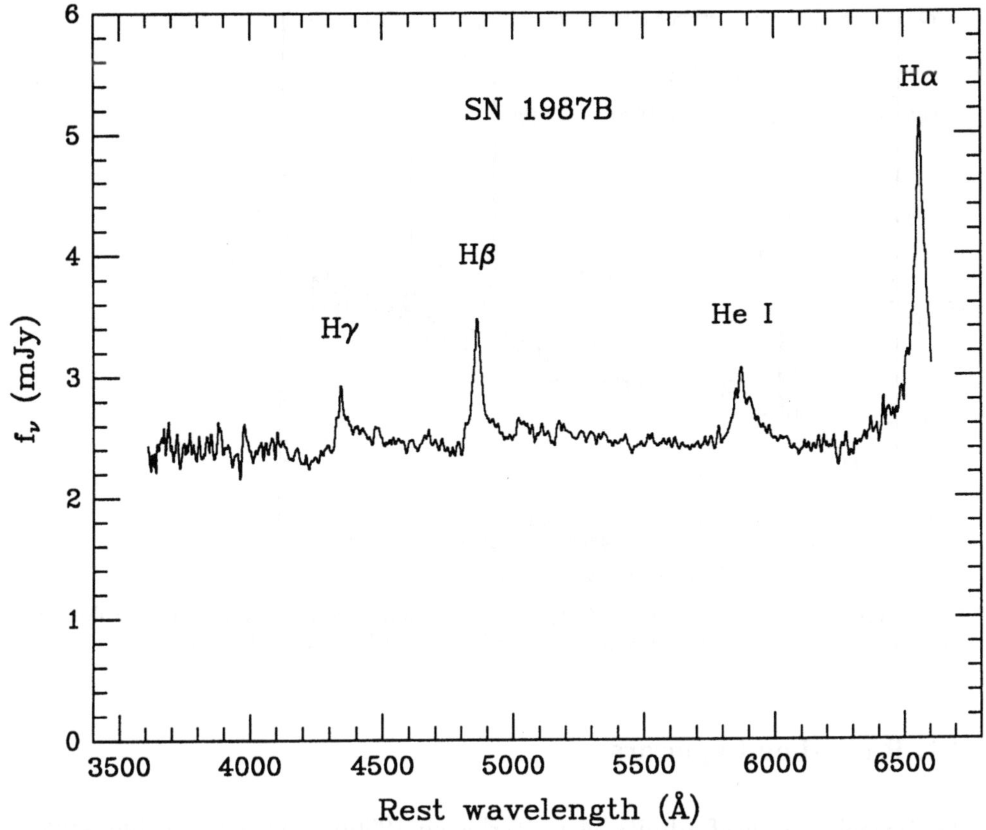

Fig. 1. Optical spectrum of the peculiar type II SN 1987B, probably obtained shortly after maximum brightness. Note the strong, broad hydrogen Balmer and He I emission lines.

are the SNe and SNRs, with variability produced by SN flashes, cooling instabilities in SNRs, and related processes.

3. Possible Problems and Future Tests

I will now summarize some possible problems associated with the starburst hypothesis, concentrating on general difficulties that are relatively model-independent, not on details that might easily be fixed in future versions of the model. Several of these issues have already been discussed by Heckman (1991), but interesting counter-arguments are given by Terlevich (1992).

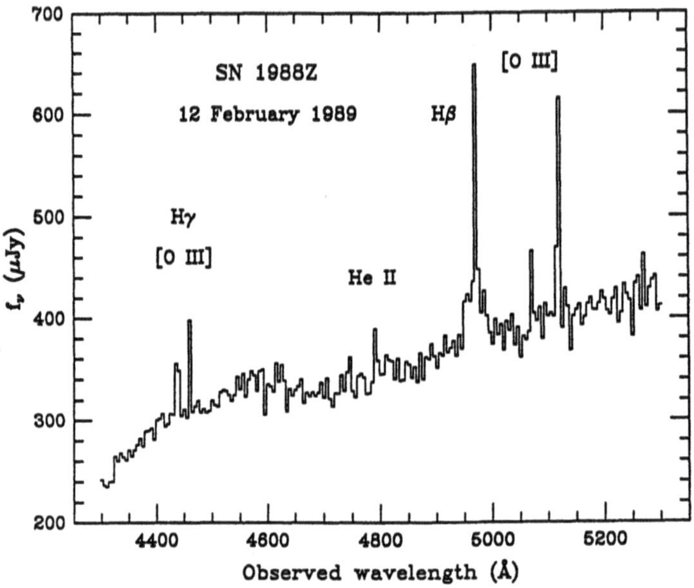

Fig. 2. Optical spectrum of SN 1988Z, obtained about two months after the explosion. The narrow emission lines are produced by photoionized circumstellar gas.

3.1. Radio-Loud Quasars

As Terlevich himself admits (e.g., Terlevich 1990b), his scenario has great difficulty accounting for the existence of radio-loud quasars and phenomena associated with them. In particular, the existence of superluminal motion, narrow jets extending tens of Mpc in a constant direction, and extremely compact (milli-arcsec) cores cannot yet be explained by a collection of SNe. These observations constitute perhaps the best indirect evidence for the existence of a spinning, stable SMBH. BL Lac objects and optically violent variables (OVVs), which fit quite well in unified schemes of AGNs (e.g., Antonucci 1989), also find no natural explanation in the starburst picture.

Terlevich points out that these radio-loud objects comprise only a small fraction ($\lesssim$ 10%) of all AGNs, and that their internal physical properties may indeed differ from those of radio-quiet AGNs for which his starburst model was conceived. Thus, he accepts this shortcoming of his model, and postulates different mechanisms for these two classes of AGNs. To me, however, it seems quite likely that radio-loud and radio-quiet AGNs are fundamentally the *same* phenomenon, with somewhat different manifestations due to secondary effects. These are not yet well identified, but they could involve the ISM, the mass of the SMBH, or the spin of the SMBH (e.g., Blandford 1990). Finding one common explanation for the entire range of phenomena is the "holy grail" of AGN research, and I don't think that the starburst model can provide it. Nevertheless, Terlevich may be correct; it is possible that two or more essentially distinct mechanisms are at work, even in high-luminosity AGNs.

3.2. Rapid X-Ray Variability

It has long been recognized that some AGNs exhibit significant X-ray variability; Barr (1991) has a list of over 30 AGNs that have shown $\Delta L_X/L_X \gtrsim 0.3$ in less than 1 day. On still shorter time scales ($\Delta t \approx 10^2 - 10^3$ s), variations of $\Delta L_X \approx 0.01 - 0.1\,L_{\rm Bol}$ are known to occur. Terlevich argues that strong variability has been documented in relatively few objects, but actually there are at least a dozen good cases, and the list is steadily growing. Among these are many *radio-quiet* type 1 Seyferts, which are to be included in Terlevich's hypothesis. Although the exact explanation is not yet understood, I find the variability to be fundamentally *plausible*, on physical grounds, in the standard model involving a SMBH. The light crossing time over 10 Schwarzschild radii is given by $\Delta t \approx 10^3(M_{\rm BH}/10^7\ M_\odot)$ s. It is intriguing that reasonable agreement is found between $M_{\rm BH}$ derived from Δt and from dynamical arguments (Wandel and Mushotzky 1986). Moreover, these values of $M_{\rm BH}$ are consistent with the masses required to sustain the large luminosities emitted by AGNs over time scales of $10^7 - 10^8$ years.

To be fair, however, I should point out that the starburst model is still quite primitive, with many unexplored possibilities. Perhaps hydrodynamical instabilities in the SN ejecta will be able to produce X-ray variability of the type observed in AGNs, although in my opinion this will be very difficult. A crucial observation would be *periodic* or *quasi-periodic* X-ray variability on short time scales; it seems almost impossible for a starburst model to explain this. Such a phenomenon may have already been found in NGC 6814 (Mittaz and Branduardi-Raymont 1989); it was also emphasized by Abramowicz (1992; see also Fiore, Massaro, and Barone 1992) at this conference. A concentrated search for more cases of (quasi)periodic X-ray variability should be made in the future.

3.3. High-Energy X-Rays and Gamma Rays

A supernova that rapidly thermalizes the kinetic energy of its ejecta (observed to be at most $\sim 3\times10^{51}$ ergs) leads to a temperature of $3\times10^8 - 10^9$ K (i.e., $kT \approx 25 - 80$ keV). Moreover, decay of $\rm Ni^{56}$, $\rm Co^{56}$, and other radioactive nuclides produces $\lesssim 10^{48} - 10^{49}$ ergs in gamma rays; most of the decay energy is converted to UV, optical, and IR photons. Thus, as pointed out by Heckman (1991), we expect to see a clear break around 50–100 keV in the spectra of AGNs driven by starbursts. Certainly a high flux density near 1 MeV would be quite alarming in Terlevich's model. (Note, however, that gamma rays are expected to be present at some non-negligible level, due to the existence of plerionic supernova remnants and accreting neutron stars.)

The existing data, though sparse, seem to be at odds with the expectations of the starburst hypothesis. Rothschild *et al.* (1983) observed roughly 10 AGNs (mostly type 1 Seyferts) and found that $\nu f_\nu \propto \nu^{0.3}$ at 100 keV. Bignami and Mereghetti (1989) observed NGC 4151, and found that a significant fraction of the bolometric luminosity comes out in the region 1–10 MeV. In MCG 8-11-11 the contribution in this range may even be dominant. Furthermore, as discussed at this conference by Ballet (1992), the SIGMA and ART-P observations between 35 keV and 1.3 MeV show that Cen A and 3C 273 have $\nu f_\nu \propto \nu^{0.6}$. In NGC 4151 a break is seen near 70 keV, but the flux at higher energies is nevertheless quite large. We clearly need additional data (e.g., from *GRO*) to settle the question, although initial results seem to indicate that many AGNs emit substantial quantities of gamma rays. On the other hand, observations of the diffuse high-energy background radiation demonstrate that *most* AGNs cannot emit gamma rays as efficiently as does 3C 273.

Again, like with rapid X-ray variability, the exact physical explanation for the high-energy X-rays and gamma rays is not obvious in the standard model involving a SMBH, but I think the production of these photons is at least plausible. After all, there is a very deep potential well with an accretion disk extending down to 10 Schwarzschild radii or less. By contrast, a high observed gamma-ray luminosity is very difficult to explain in the starburst model.

3.4. Progenitor Starbursts of QSOs

A QSO having a bolometric luminosity $L = 10^{46}$ erg s^{-1} emits $\sim 3 \times 10^{53}$ erg yr^{-1}. If most of the energy comes from SNe, the kinetic energy of *at least* 100 SNe must be thermalized each year. This estimate is a lower limit for two reasons: most supernovae probably have kinetic energies closer to 10^{51} erg rather than 3×10^{51} erg (e.g., Branch, Nomoto, and Filippenko 1991), and one does not expect 100% of the kinetic energy to be thermalized even if the circumstellar gas density is high. If a QSO lasts 8×10^6 to 6×10^7 years, as in Phase 4 of Terlevich's hypothesis, $10^9 - 10^{10}$ SNe are needed, and this requires the formation of just as many O and B stars. If the average bolometric luminosity of each star is $\sim 2 \times 10^4\ L_\odot$, the total luminosity associated with the OB stars is $\sim 10^{47} - 10^{48}$ erg s^{-1}. Unless the initial mass function is highly skewed towards massive stars, many lower-mass stars will also contribute to the "non-QSO" luminosity of the starburst. Thus, the starburst itself should appear 10–100 times more luminous than the QSO itself! Where are all the progenitor H II regions and high-luminosity Seyfert 2 galaxies that eventually give rise to QSOs?

As discussed by Heckman (1991), one possibility is that these ultraluminous starburst and Seyfert 2 nuclei are shrouded by dust which is eventually destroyed, leaving QSOs relatively free of dust. Terlevich (1992) points out that large quantities of dust are produced by massive stars during their η-Carinae phase, near the end of core hydrogen burning and during core helium burning. Extremely dusty and luminous galaxies do in fact exist; see the recent review by Sanders (1992), and references therein. These ultraluminous far-IR galaxies were first identified from observations made with *IRAS*, and they outshine QSOs at low redshifts roughly by a factor of 5 (Soifer *et al.* 1987). It is not known whether their enormous luminosities are primarily due to vigorous starbursts, dust-enshrouded AGNs, or a combination of both. However, Heckman (1991) calculates that the sub-millimeter background (0.1–0.5 mm) produced by these galaxies should have been detected by *COBE*, if such galaxies were sufficiently numerous to explain QSOs. Moreover, the amount of metals produced by the progenitors of QSOs might be uncomfortably high compared with estimates (based on number counts of faint blue galaxies) of the metal content of the universe (Cowie 1988).

3.5. Dead QSOs: Common and Massive

In one form or another, the "efficiency argument" for QSOs has been stated many times (e.g., Soltan 1982; Blandford 1987; Padovani, Burg, and Edelson 1990; Heckman 1991). Out to $z \approx 2$, QSOs produce about $10^6\,h^{-3}\,M_\odot c^2$ erg of radiated energy per luminous galaxy ($L \gtrsim 0.5\,L^*$; Schechter 1976), where $h = H_0/(100$ km s^{-1} Mpc$^{-1})$. If this energy comes from gravitational accretion onto a black hole, with a conversion efficiency of ~ 0.1, the center of each luminous galaxy should harbor a compact object with $M \approx 10^7\,M_\odot$. As noted by Blandford (1987), this is already comparable to measurements of (or upper limits to) the central mass in the Milky Way Galaxy and in nearby galaxies such as M31 and M32; see Dressler (1989) for a review.

By contrast, a SN II (whose progenitor has $M_i = 8 - 40\,M_\odot$) produces $\sim 10^{-3}\,M_\odot c^2$ of kinetic energy, leaving a compact remnant (neutron star or black hole) of $\sim 1\,M_\odot$; most of the mass is available for reprocessing into future generations of stars. Therefore, even if we assume that the SN kinetic energy is *entirely* thermalized, the conversion efficiency is only 10^{-3}. Combining this with the above result, we conclude that the nuclei of all luminous galaxies should contain central masses of $\gtrsim 10^9\,M_\odot$. The low-mass stars that form along with the cluster of OB stars increase this estimate by a factor of 10–100, yielding a central mass of $10^{10} - 10^{11}\,M_\odot$ per galaxy. If only 1% of $L \gtrsim 0.5\,L^*$ galaxies were QSOs in the past, the required mass per dead QSO would be $10^{12} - 10^{13}\,M_\odot$. Such high central masses should easily be detected, yet none has yet been seen, despite the existence of high-quality spectra of several thousand galaxies. If, instead, 10% of luminous galaxies were former QSOs, the necessary mass drops to $10^{11} - 10^{12}\ M_\odot$ — yet only one (rather controversial) candidate with such a mass has been proposed (Bland-Hawthorne, Wilson, and Tully 1991). Clearly, to decrease the mass restrictions and salvage the starburst hypothesis, we must assume that *all* luminous galaxies were QSOs at one time, and that their central regions now contain $10^{10} - 10^{11}\ M_\odot$ of stars and stellar remnants.

Suppose the QSO starburst is confined to an effective radius of 40 pc in the central region of a galaxy, and assume that this region has a mass of $3 \times 10^{10}\ M_\odot$. For a galaxy following a de Vaucouleurs (1962) $r^{1/4}$ law, the mass M ($M_\odot$) is related to the effective radius R (pc) and the one-dimensional velocity dispersion σ (km s^{-1}) by the relation $M = 2075\,R\sigma^2$ (Young 1976). Thus, the expected velocity dispersion within a diameter of 80 pc, corresponding to 1″ at the distance of the Virgo cluster ($d \approx 16$ Mpc), is 600 km s^{-1}. This is considerably higher than velocity dispersions of Virgo galaxies measured in good conditions (seeing disk $\sim$ 1″). An obvious solution, of course, is to postulate that the starburst is spatially extended over a larger radius. Even then, the observed velocity dispersions in the central regions of galaxies may be lower than expected, unless the starburst encompasses the entire core of the galaxy (Terlevich 1992). After it is fixed, the *Hubble Space Telescope* (*HST*) should be able to provide high-resolution (0.1″) measurements of the velocity dispersion in sufficiently many galactic nuclei to settle this question definitively.

3.6. The Angular Sizes of QSOs

As discussed above, the existing measurements of stellar velocity dispersions indicate that the starbursts giving rise to QSOs must be spatially extended over a reasonably large radius. Indeed, Terlevich (1992) associates the QSO phase with the formation of the cores of elliptical galaxies. He calculates that the young core of a present-day L^* galaxy would have had an effective radius of 250 pc when it was a QSO, and its effective angular diameter would be 0.05″ at $z \approx 2$. Reasonably nearby QSOs, on the other hand, would have angular sizes of several tenths of an arc second, easily resolvable with *HST* even in its current (aberrated) form. Even at $z \approx 2$, the most luminous QSOs ($M_B \lesssim -28.5$ mag) should be $\sim$ 1″ in diameter. To my knowledge, none of the QSOs thus far imaged with *HST* has been resolved at the 0.1″ level. This includes a large number of QSOs in the "snapshot survey" for gravitational lenses, the first results of which are described by Bahcall *et al.* (1992). Admittedly, most of the QSOs in this survey have rather high redshifts ($z \gtrsim 1$), but the result is nevertheless striking. If, in the future, we find that *no* QSOs at $z \lesssim 0.5$ are resolved within a central radius of 100–1000 pc, the starburst model will have to be abandoned. This is perhaps the easiest and most conclusive observation to be made during the next few years.

Another simple test that can now be done with *HST* is to look for temporal variations in the *position* of the nuclei of type 1 Seyfert galaxies. Depending on their luminosity, these objects should be powered by only 0.1–10 SNe per year. If a different SN dominates the luminosity each year, the nucleus should move around on scales of a few tenths of an arc second (i.e., 10–30 pc for an object whose distance is 20 Mpc).

I should note that a very stringent upper limit to the size of at least one QSO already exists. A probable microlensing event having a rise time of ~ 1 month, detected in one of the distinct components of the "Einstein cross" Q2237+0305, shows that the size of the QSO is only $\lesssim 10^{16}$ cm (Irwin *et al.* 1989; Corrigan *et al.* 1991). This is a luminous, *radio-quiet* QSO that cannot be ignored by proponents of the starburst scenario.

3.7. Supernovae in the Starburst Model

Recall that in Phase 3 of Terlevich's evolution of a central star cluster, which occurs in the range $t = 4 - 8$ Myr, the most massive stars become SNe. Terlevich (1992, and references therein) states that these SNe are of type Ib. However, most of the SNe Ib thus far observed in nearby galaxies have probably come from *relatively* massive stars ($M_i \approx 10 - 25\ M_\odot$) in close binary systems, where the mass is lost through transfer onto a companion star, rather than from *extremely* massive stars ($M_i \gtrsim 40\ M_\odot$) that lose their envelopes through winds; see Filippenko (1991a), as well as Branch, Nomoto, and Filippenko (1991). Moreover, SNe Ib are *observed* to be about 1.5 mag dimmer than SNe Ia at optical wavelengths, and comparable in brightness to typical SNe II. Thus, SNe Ib are not sufficiently dim to be useful in Terlevich's model.

I agree with Terlevich that Wolf-Rayet stars should produce reasonably faint explosions (e.g., Chevalier 1976), if they explode at all. On the other hand, currently we have at most one observed example (Cas A) of such an event; maybe *most* extremely massive stars simply fail to produce successful explosions, instead collapsing directly to black holes (e.g., Woosley and Weaver 1986). If so, Terlevich's calculations of the expected properties for Phase 3 will have to be modified substantially.

Also questionable is Terlevich's (1990b, 1992) assumption that the SNe Ib in Phase 3 will produce high radio luminosities. A few SNe Ib have been detected at radio wavelengths (see Sramek and Weiler 1990 for a review), but they are considerably less luminous than SNe II. Moreover, these observations suggest that the SNe Ib do *not* come from the most massive stars. If interpreted in the context of the Chevalier (1984) model, the observations of SN Ib 1983N give a value of 5×10^{-7} ($M_\odot$ yr^{-1})/(km s^{-1}) for $\dot{M}/v_w$, the mass-loss rate divided by the wind velocity. If the wind velocity is 2000 km s^{-1}, typical of Wolf-Rayet stars, the implied mass-loss rate is $\dot{M} \approx 10^{-3}\ M_\odot$ yr^{-1}, which greatly exceeds the observed rates in Wolf-Rayet stars. A real Wolf-Rayet star is surrounded by a cavity of low-density, high-velocity material which should produce little radio emission when (and if) the star finally does explode.

It is possible that Wolf-Rayet stars will behave in a different manner if located in the high-density environment of a galactic nucleus. Hence, this criticism is not as robust as some of the other, more general, arguments given previously in this section. For example, I have already mentioned (Sec. 2) the existence of galaxies having energetic "superwinds" of material flowing along their axis of rotation. These do indeed have some of the properties described by Terlevich (1992). Thus, although some of the details of Terlevich's present model might be erroneous, the general picture of Phase 3 could in fact be substantially correct. Here I chose to mention the characteristics of Wolf-Rayet stars only to emphasize that *known* individual stars of this type do not necessarily have the desired properties.

3.8. Other Arguments

There are a number of additional criticisms of the starburst hypothesis, but in many cases their validity is not yet clear. One frequently mentioned result is the "reverberation mapping" of AGNs such as NGC 5548 (Clavel *et al.* 1991; Peterson *et al.* 1991). It is found that the emission lines respond to changes in the continuum luminosity, with a lag (delay) that is quite small (several days to several weeks or months, depending on the ionization stage or the critical density). The lag may vary from one outburst to another (Netzer *et al.* 1990), although this result is still controversial (Peterson *et al.* 1992). In my opinion, the currently available data are not obviously incompatible with the starburst model. NGC 5548, for example, has an estimated bolometric luminosity of $\sim 3 \times 10^{44}$ erg s^{-1}. This requires an average of about 3 SNe per year if produced by a starburst. Substantial variations around this number are bound to occur, and these could be associated with the occasional outbursts and minima in the UV and optical light curves. Each supernova is a compact object that may be surrounded by a dense circumstellar shell. Much of the initial UV flash is produced when the shock front of the exploding star reaches the stellar surface, whereas the emission lines are formed later. Thus, a lag is expected, and it need not be exactly the same during each outburst. Reverberation mapping gives several additional results of interest (Peterson 1992), but they do not conclusively eliminate the starburst hypothesis. One clear prediction is that luminous QSOs should exhibit only low-amplitude variability, and the preliminary results seem reasonably consistent with this (Terlevich 1990b). Furthermore, we expect small or nonexistent lags between the continuum and the emission lines, because we see the effects of many SNe simultaneously. This should, in principle, be relatively easy to test.

4. Possible Successes of the Starburst Model

Terlevich's starburst hypothesis has several observational strong points. At the present time most of these are, in my opinion, confined to low-luminosity objects, but some of them show potential even for luminous type 1 Seyferts. A complete discussion can be found in Terlevich (1992) and other papers in the series. Here I concentrate on a few of the most important phenomena.

4.1. LINERs

As suggested by Terlevich and Melnick (1985), and shown in more detail by Cid Fernandes *et al.* (1992; see also Binette, Robinson, and Courvoisier 1988), some LINERs may be powered by ionizing radiation from hot stars ("Warmers"). The most obvious examples are the "blue LINERs," in which there is a sufficient number of hot stars to produce a blue continuum. The nucleus of the elliptical galaxy NGC 404, whose optical spectrum is shown in Figure 3, may be a good example. [See also Filippenko and Sargent (1986), who illustrate a high-resolution spectrum around Hα; emission severely contaminates the absorption line.] This object is clearly a LINER; [O II] λ3727, [O I] λ6300, and [S II] $\lambda\lambda$6716, 6731 are very strong. Unlike the case in most LINERs, however, deep hydrogen Balmer absorption lines are visible at blue wavelengths, revealing the presence of a substantial population of late-A and early-F stars. If some Warmers are also present, they could easily account for the observed emission lines. Despite being a "blue LINER," the nucleus of NGC 404 is substantially reddened, possibly due to dust created by the central starburst.

According to Terlevich and Melnick (1985), the more common "red LINERs" are similar to blue LINERs, except that the central star cluster probably has fewer young, high-mass stars because it is either older or less massive. In this case, diminished ionizing radiation leads to weaker emission lines, and the stellar continuum is generally redder (if we ignore the dust that may be present in blue LINERs such as NGC 404).

Another interesting point is that many LINERs in early-type galaxies have very weak [O I] λ6300; see, for example, the spectra of NGC 2681, NGC 2841, NGC 4501, NGC 4594, NGC 5194, and NGC 7217 in Filippenko and Sargent (1985), and compare them with the spectra of NGC 404, NGC 1052, NGC 1275, NGC 3031, NGC 3884, NGC 3998, NGC 4278, NGC 4579, and NGC 6500. Galaxies in the first set are classified as LINERs because of their high [N II] λ6583/Hα intensity ratio; they do not appear to satisfy Heckman's (1980) original condition that [O I] λ6300 $\gtrsim$ (1/3) [O III] λ5007. However, as shown by Filippenko and Terlevich (1992; see also Filippenko 1992), such line intensity ratios could be produced by high-metallicity gas photoionized by radiation from O-type stars, especially if

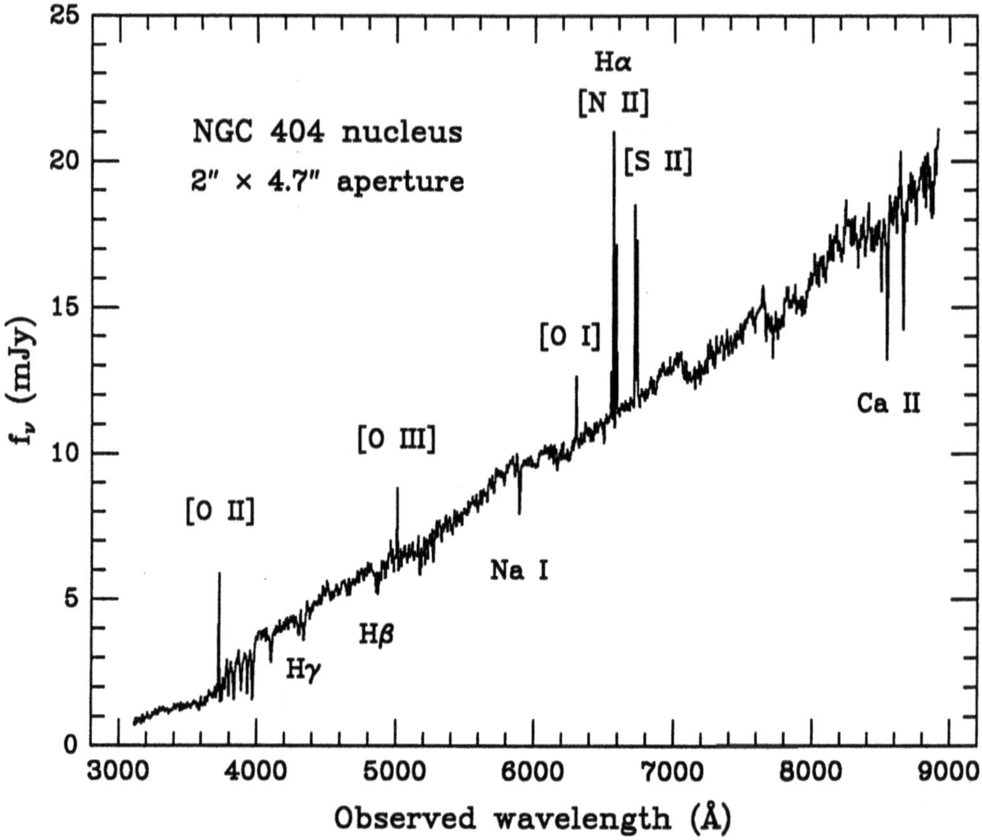

Fig. 3. Optical spectrum of NGC 404, showing prominent Balmer absorption lines.

nitrogen is selectively overabundant; one *need not* even have any Warmers. The emission lines in many of the galaxies have very low luminosities; hence, these "high metallicity H II regions" could be powered by only a few young, massive stars in the central regions. In addition, Terlevich, Melnick, and Moles (1987) point out that starbursts in low-metallicity galaxies (generally late-type galaxies) should look like normal H II regions, and their velocity dispersions are expected to be lower than in high-metallicity early-type galaxies. This explains the dearth of LINERs in late-type galaxies, as well as the relatively large widths of emission lines in LINERs.

Finally, the line-emitting clouds in some LINERs are almost certainly powered by shocks in "superwinds" emanating from the central regions. Heckman, Armus, and Miley (1990) discuss these objects as a class, while Filippenko and Sargent (1992) examine in detail the extended gas in NGC 3079. Galaxies having such properties are expected to occur in the starburst model, specifically in Phase 3 (or perhaps even Phase 1, in the case of dust-enshrouded progenitor starbursts of QSOs).

4.2. Type 2 Seyferts

An important result from Terlevich and Melnick (1985) is that some Seyfert 2 nuclei may actually be powered by star clusters in the post-starburst phase. Due to the presence of Warmers, the continuum of a massive cluster 3–5 Myr after the starburst (assumed to be coeval) resembles a power law of index 1.5, and the resulting emission lines from nearby clouds of gas have line intensity ratios that agree with those observed in many type 2 Seyferts. This possibility is examined in detail by Cid Fernandes *et al.* (1992). They use a numerical code to compute the continuum of a star cluster given its age and initial mass function. From the ionizing spectrum, nebular emission-line spectra for a grid of cloud parameters (density, density gradient, ionization parameter, and metallicity) are calculated with a photoionization code. Two-dimensional diagnostic diagrams (Baldwin, Phillips, and Terlevich 1981; Veilleux and Osterbrock 1987) are used to follow the evolution of the system. Their results agree very well with those of Terlevich and Melnick (1985), with metallicity in the range 0.5–2 times solar. Low-metallicity galaxies won't form many Warmers, and in any case the emission-line spectrum will resemble that of an H II region (e.g., Johansson, Westerlund, and Azzopardi 1990); thus, type 2 (and type 1) Seyferts should be less frequent in late-type galaxies, as observed. [An interesting exception is NGC 4395, as discussed by Filippenko and Sargent (1989).] Furthermore, by using the tables in Maeder and Meynet (1988) and Maeder (1990), Cid Fernandes *et al.* (1992) compute the progressive chemical enrichment of the interstellar medium in galactic nuclei due to stellar mass loss. The possible nitrogen overabundance of AGNs, especially LINERs (Binette 1985; Storchi-Bergmann and Pastoriza 1989, 1991), seems to arise naturally in their model.

Another significant point is that the "featureless blue continuum" observed in many Seyfert 2 nuclei may be produced by hot stars and red supergiants, rather than by nonstellar processes. This is graphically shown by Cid Fernandes and Terlevich (1992), who subtract a suitably scaled bulge spectrum of an average early-type spiral (an old stellar population) from the bulge spectrum of a late-type spiral (a mixed stellar population), and find that the continuum of the remaining spectrum closely resembles a power law. Of course, this is not surprising; the spectra of young star clusters are characterized by blue, featureless continua (e.g., Kinman and Davidson 1981). It does, however, illustrate the possibility that the nonstellar power-law component claimed to exist in many type 2 Seyferts may, in some cases, be an artifact of template mismatch. The observed absorption-line strengths in Seyfert 2 nuclei are also explained. The basic point is that a young stellar population dilutes the strength of Mg I λ5180, the G band ($\sim$ 4300 Å), and other metal features in the blue through red spectral regions, but not the Ca II near-IR triplet around 8550 Å (Díaz 1992). Indeed, the near-IR triplet is *enhanced* by the presence of red supergiants associated with the young stellar population, since they have low surface gravity. According to Terlevich, Díaz, and Terlevich (1990), the Ca II near-IR triplet is abnormally strong in many type 2 Seyferts compared with normal galaxies, especially in view of the weak absorption lines in the blue-green part of the spectrum. This was first pointed out for NGC 1068 by Malkan and Filippenko (1983).

Cid Fernandes and Terlevich (1992) also argue that the observed strength of the featureless blue continuum in some Seyfert 2 nuclei is inconsistent with the absence of broad Balmer emission lines in the spectrum, if the "hidden Seyfert 1" model (Antonucci and Miller 1985; Miller and Goodrich 1990; Miller, Goodrich, and Mathews 1991) is adopted for *all* type 2 Seyferts. The basic point is that hydrogen Balmer photons from the broad-line region should be scattered into our line of sight just as easily as continuum photons, yet the broad lines are not seen. One might argue that the broad lines are difficult to detect when superposed on a strong stellar continuum from the galactic bulge, but the quantitative simulations of Cid Fernandes and Terlevich (1992) show that this effect is not sufficiently large; the broad lines should, in fact, be visible in those type 2 Seyferts having a prominent blue continuum. Emission lines from the narrow-line region will also contaminate the spectrum, but broad wings are nevertheless expected to be discernable. Furthermore, the blue continuum should be highly polarized if it comes from scattered light, yet the observed polarizations are usually quite small (Miller 1989). In the starburst hypothesis, by contrast, the continuum of type 2 Seyferts comes from an extended region containing hot stars; little polarization is expected, and there need not be any broad Balmer emission prior to Phase 4 (see Section 2). Hot stars over an extended region were suggested by Goodrich and Miller (1989) to explain the nearly unpolarized, spatially resolved, blue continuum in Cyg A.

Another point in favor of Terlevich's general scenario is that Seyfert 2 nuclei, which are supposedly in Phase 3 of the starburst, should exhibit biconical outflows, ionization cones, and partially collimated radio emission. That is indeed the case, as reviewed at this conference by Wilson (1992). For example, striking ionization cones have been found by Tadhunter and Tsvetanov (1989) and Pogge (1988a,b). Moreover, the existence of radio emission collimated with the extended narrow-line region is documented by Haniff, Wilson, and Ward (1988), and in some cases there is a spatial correspondence between radio structures and distinct velocity components of extended [O III] emission (Whittle *et al.* 1986, 1988).

4.3. Type 1 Seyferts

Despite the criticisms presented in Section 3, some objects classified as type 1 Seyfert nuclei may actually be powered by SNe, rather than by accretion of matter onto a SMBH. In my opinion, this is unlikely to be the case for objects known to exhibit rapid X-ray variability of moderate am-

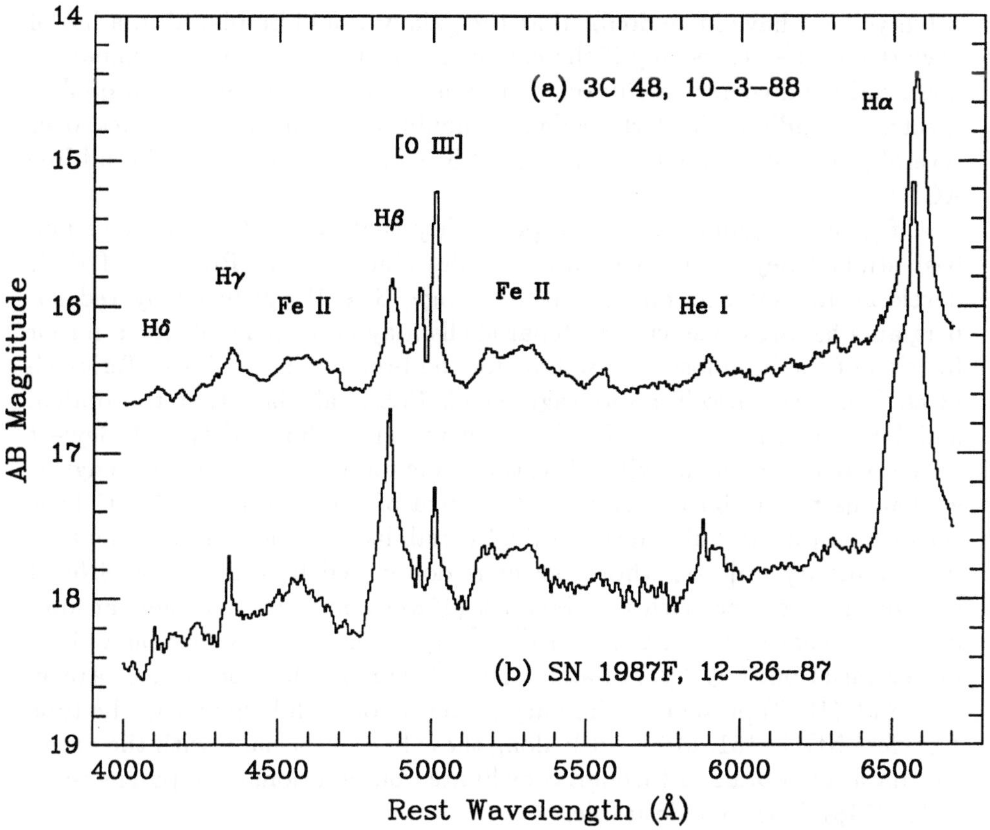

Fig. 4. Optical spectrum of the quasar 3C 48, compared with that of SN 1987F roughly 9 months after discovery. Despite some small differences, the broad-line similarities between the two objects are striking. AB magnitude $= -2.5 \log f_\nu - 48.6$, where the units of f_ν are erg s^{-1} cm^{-2} Hz^{-1}.

plitude, especially if it is periodic (as may be true in NGC 6814: Mittaz and Branduardi-Raymont 1989). Although the list of X-ray variable type 1 Seyferts is growing, and now consists of several dozen objects (Barr 1991), quite a few type 1 Seyferts do not show substantial X-ray variability. These should be considered seriously in the context of the starburst hypothesis.

A trivial case is that of mistaken identity. Given the existence of peculiar SNe II having the optical spectroscopic characteristics of Seyfert 1 nuclei and QSOs, it is possible that at least *a few* catalogued type 1 Seyferts are actually normal galaxies whose nuclei just happened to contain such a SN at the time their optical spectra were procured. As demonstrated in Figure 4, the spectrum of SN 1987F was quite similar to that of 3C 48. If an object such as SN 1987F or SN 1987B (Fig. 1) were present in a galactic nucleus, and if the nucleus were observed only once (or a few times over the course of

~ 0.5 year), I have little doubt that the galaxy would be placed in a list of Seyfert nuclei — especially if the classification spectrum had low signal-to-noise ratio and coarse resolution, as in many redshift surveys. High-quality spectra of candidate Seyfert 1 galaxies should be obtained many times, over several years, to eliminate such cases of SNe II masquerading as broad-line AGNs.

Of greater significance is the possibility that some of the nearest and best studied Seyfert 1 nuclei are actually starbursts in Phase 4 of their development (see Section 2). At this stage, SNe II produced by red supergiants having dense circumstellar shells may be responsible for a major fraction of the luminosity, as well as for the broad emission lines. Terlevich *et al.* (1992; see also Tenorio-Tagle *et al.* 1992) calculate that the optical and UV spectra of such SNe II should resemble those of type 1 Seyfert nuclei. Good agreement with observations is found at optical wavelengths, as demonstrated above, but no UV spectra of these peculiar SNe II have yet been obtained. Substantial optical variability is expected in reasonably low-luminosity objects, whose power is dominated by only a few SNe at any given time. Terlevich and Melnick (1987) point out that the variability should consist of flares of $\sim 10^{49-50}$ erg lasting a few weeks, as well as longer variations ($\gtrsim 2$ yr) of about 10^{51-52} erg. At this conference, Aretxaga *et al.* (1992) present preliminary evidence for such behavior in the type 1 Seyfert NGC 4151. More work along these lines, especially with the excellent data sets obtained through coordinated observational campaigns (e.g., NGC 5548), is clearly desirable.

5. Conclusions

My overall impression is that the starburst hypothesis is unlikely to be the correct explanation for all types of AGNs, but it might work for some classes or subclasses. The arguments *against* Terlevich's hypothesis are most compelling for extremely luminous QSOs, BL Lacs and OVVs, AGNs that show substantial and rapid X-ray variability, AGNs that emit copious quantities of gamma rays, and radio-loud quasars — especially those with well-collimated jets. Accretion onto a SMBH appears to be the best bet for these objects. Given the nearly identical properties (except at radio wavelengths) of most radio-quiet QSOs and radio-loud quasars, and the severe difficulties Terlevich encounters explaining the latter objects, to me it seems unlikely that the radio-quiet QSOs are powered by starbursts. Continuity is also often invoked to argue that low-luminosity AGNs must be powered by accretion onto a compact object, probably a black hole. Such an approach is persuasive and reasonable, but not convincing or foolproof. For example, if used without a solid physical model, it would lead one to incorrectly conclude that brown dwarfs having $M \lesssim 0.08\ M_\odot$ derive their energy from nuclear fusion of protons, as do normal stars.

Terlevich's hypothesis could indeed be the *best* explanation for several types of low-luminosity AGNs. Heckman (1987) and Filippenko (1989b) emphasize that LINERs appear to constitute a very heterogeneous class; as summarized in Section 4, there are reasons to believe that some of them are simply the result of moderate or low-level starbursts. Similarly, the starburst model of type 2 Seyferts in many cases provides good agreement with observations of emission-line intensity ratios and the strengths of various stellar absorption lines. It is not yet clear that all Seyfert 2 galaxies have hidden Seyfert 1 nuclei; perhaps "pure" type 2 Seyferts are powered by starbursts, while those with a hidden Seyfert 1 nucleus usually harbor a SMBH. Type 1 Seyferts themselves may come in two or more physically distinct varieties, with central engines consisting of either SNe or black holes. Finally, it is possible that some ultraluminous far-infrared galaxies are the starburst "progenitors" (Phase 1) of QSOs or luminous type 1 Seyferts, and galaxies with outflowing winds may represent more advanced stages (Phase 3) of starburst-powered AGNs.

Fortunately, Terlevich's starburst model has certain consequences that can be tested during the next few years. Perhaps the simplest check is to obtain high-resolution *HST* images of relatively nearby, but luminous, QSOs. I mentioned in Section 3 that the results of Bahcall *et al.* (1992) already challenge the starburst hypothesis, but in most cases the target QSOs in this study had such a high redshift that the expected angular size of the bulge's core radius is too small to be resolved with confidence. Similarly, *HST* images of nearby type 1 Seyfert nuclei should show that the center of "activity" is not stationary; individual SNe are expected to occur at different places within the starburst. *GRO* and *ROSAT* data will provide useful constraints from the high-energy end of the electromagnetic spectrum. Rapid X-ray variability of high amplitude ($\Delta L/L \approx 1$) in a significant fraction of AGNs would probably be a serious blow to the starburst hypothesis as a *general* explanation of AGNs, especially if convincing periodicities were found in the variations. Strong gamma-ray emission, at energies above 1 MeV (or even 100 keV), would also be very difficult to explain in the context of starbursts. Although well-understood mechanisms for these high-energy phenomena do not yet exist, I find it more plausible on simple physical grounds that accretion of matter into a deep gravitational potential, rather than SNe from a starburst, is ultimately responsible.

I thank Roberto Terlevich for many spirited and stimulating discussions about the physical nature of AGNs, and Joseph Shields for useful comments on the manuscript. Some of the spectra shown here were provided by R. P. Harkness (Fig. 1) and by E. M. Sadler and R. A. Stathakis (Fig. 2). This research is supported by NSF grants AST–8957063 and AST–9003829. The financial assistance of the conference organizers, as well as their patience while waiting for the written version of my review, are greatly appreciated.

References

Abramowicz, M. 1992, these *Proceedings.*

Adams, T. F., and Weedman, D. W. 1975, *Ap. J.*, **199**, 19.

Antonucci, R. 1989, in *Annals of the NY Academy of Sciences*, **571**, 180.

Antonucci, R. R., and Miller, J. S. 1985, *Ap. J.*, **297**, 621.

Aretxaga, I., Terlevich, R., Díaz, A. I., and Terlevich, E. 1992, these *Proceedings.*

Bahcall, J. N., Maoz, D., Doxsey, R., Schneider, D. P., Bahcall, N. A., Lahav, O., and Yanny, B. 1992, *Ap. J.*, in press.

Baldwin, J. A., Phillips, M. M., and Terlevich, R. 1981, *Pub. A.S.P.*, **93**, 5.

Ballet, J. 1992, these *Proceedings.*

Barlow, M. J., and Hummer, D. G. 1982, in *Wolf-Rayet Stars: Observations, Physics, Evolution*, ed. C. W. H. de Loore and A. J. Willis (Dordrecht: Reidel), p. 387.

Barr, P. 1991, private communication.

Bignami, G., and Mereghetti, S. 1989, in *Gamma-Ray Observatory Science Workshop*, ed. W. N. Johnson (Washington, DC: NRL).

Binette, L. 1985, *Astr. Ap.*, **143**, 334.

Binette, L., Robinson, A., and Courvoisier, T. J. L. 1988, *Astr. Ap.*, **130**, 279.

Blandford, R. D. 1987, in *Superluminal Radio Sources*, ed. J. A. Zensus and T. J. Pearson (Cambridge: Cambridge Univ. Press), p. 310.

Blandford, R. D. 1990, in *Active Galactic Nuclei*, ed. R. D. Blandford, H. Netzer, and L. Woltjer (Berlin: Springer-Verlag), p. 161.

Bland-Hawthorne, J., Wilson, A. S., and Tully, R. B. 1991, *Ap. J. (Letters)*, **371**, L19.

Branch, D., Nomoto, K., and Filippenko, A. V. 1991, *Comments Ap.*, **15**, 221.

Chevalier, R. A. 1976, *Ap. J.*, **208**, 826.

Chevalier, R. A. 1984, *Ap. J. (Letters)*, **285**, L63.

Cid Fernandes, R., Jr., Dottori, H. A., Gruenwald, R. B., and Viegas, S. M. 1992, *M.N.R.A.S.*, in press.

Cid Fernandes, R., Jr., and Terlevich, R. 1992, in *Relationships Between Active Galactic Nuclei and Starburst Galaxies*, ed. A. V. Filippenko (San Francisco: A.S.P.), in press.

Clavel, J., *et al.* 1991, *Ap. J.*, **366**, 64.

Condon, J. J., Condon, M. A., Gisler, G., and Puschell, J. J. 1982, *Ap. J.*, **252**, 102.

Corrigan, R. T., *et al.* 1991, *Astr. J.*, **102**, 34.

Cowie, L. 1988, in *The Post-Recombination Universe*, ed. N. Kaiser and A. N. Lasenby (Dordrecht: Kluwer), p. 1.

Davidson, K., and Kinman, T. D. 1982, *Pub. A.S.P.*, **94**, 634.

de Vaucouleurs, G. 1962, in *Classification of Galaxies by Form, Luminosity, and Color*, ed. J. R. Shakeshaft (Dordrecht: Reidel), p. 1.

Díaz, A. I. 1992, in *Relationships Between Active Galactic Nuclei and Starburst Galaxies*, ed. A. V. Filippenko (San Francisco: A.S.P.), in press.

Dopita, M. A., Lozinskaya, T. A., McGregor, P. J., and Rawlings, S. J. 1990, *Ap. J.*, **351**, 563.

Dressler, A. 1989, in *Active Galactic Nuclei*, ed. D. E. Osterbrock and J. S. Miller (Dordrecht: Kluwer), p. 217.

Field, G. B. 1964, *Ap. J.*, **140**, 1434.

Filippenko, A. V. 1989a, *Astr. J.*, **97**, 726.

Filippenko, A. V. 1989b, in *Active Galactic Nuclei*, ed. D. E. Osterbrock and J. S. Miller (Dordrecht: Kluwer), p. 495.

Filippenko, A. V. 1991a, in *Wolf-Rayet Stars and Interrelations with Other Massive Stars in Galaxies*, ed. K. van der Hucht and B. Hidayat (Dordrecht: Kluwer), p. 529.

Filippenko, A. V. 1991b, in *SN 1987A and other Supernovae*, ed. I. J. Danziger and K. Kjär (Garching: ESO), p. 343.

Filippenko, A. V. 1992, in *Relationships Between Active Galactic Nuclei and Starburst Galaxies*, ed. A. V. Filippenko (San Francisco: A.S.P.), in press.

Filippenko, A. V., and Halpern, J. P. 1984, *Ap. J.*, **285**, 458.

Filippenko, A. V., and Sargent, W. L. W. 1985, *Ap. J. Suppl.*, **57**, 503.

Filippenko, A. V., and Sargent, W. L. W. 1986, in *Structure and Evolution of Active Galactic Nuclei*, ed. G. Giuricin *et al.* (Dordrecht: Reidel), p. 21.

Filippenko, A. V., and Sargent, W. L. W. 1989, *Ap. J. (Letters)*, **342**, L11.

Filippenko, A. V., and Sargent, W. L. W. 1992, *Astr. J.*, in press.

Filippenko, A. V., and Terlevich, R. 1992, in preparation.

Fiore, F., Massaro, E., and Barone, P. 1992, these *Proceedings*.

García Vargas, M. L., and Díaz, A. I. 1992, in *Relationships Between Active Galactic Nuclei and Starburst Galaxies*, ed. A. V. Filippenko (San Francisco: A.S.P.), in press.

Goodrich, R. W., and Miller, J. S. 1989, *Ap. J. (Letters)*, **346**, L21.

Haniff, C. A., Wilson, A. S., and Ward, M. J. 1988, *Ap. J.*, **334**, 104.

Harwit, M., and Pacini, F. 1975, *Ap. J. (Letters)*, **200**, L127.

Heckman, T. M. 1980, *Astr. Ap.*, **87**, 152.

Heckman, T. M. 1987, in *Observational Evidence for Activity in Galaxies*, ed. E. Ye. Khachikian, K. J. Fricke, and J. Melnick (Dordrecht: Reidel), p. 421.

Heckman, T. M. 1991, in *Massive Stars in Starbursts*, ed. N. Walborn and C. Leitherer (Cambridge: Cambridge Univ. Press), p. 289.

Heckman, T. M., Armus, L., and Miley, G. K. 1990, *Ap. J. Suppl.*, **74**, 833.

Irwin, M. J., Webster, R. L., Hewett, P. C., Corrigan, R. T., and Jedrzejewski, R. I. 1989, *Astr. J.*, **98**, 1989.

Johansson, L., Westerlund, B. E., and Azzopardi, M. 1990, *Astr. Ap.*, **229**, 83.

Kinman, T. D., and Davidson, K. 1981, *Ap. J.*, **243**, 127.

Koski, A. T. 1978, *Ap. J.*, **223**, 56.

Lequeux, J. 1979, *Astr. Ap.*, **80**, 35.

Maeder, A. 1990, *Astr. Ap. Suppl.*, **84**, 139.

Maeder, A., and Meynet, G. 1988, *Astr. Ap. Suppl.*, **76**, 411.

Malkan, M. A., and Filippenko, A. V. 1983, *Ap. J.*, **275**, 477.

McCrea, W. H. 1976, in *The Galaxy and the Local Group*, ed. R. J. Dickens and J. E. Perry (RGO Bulletin 182).

Miller, J. S. 1989, in *Active Galactic Nuclei*, ed. D. E. Osterbrock and J. S. Miller (Dordrecht: Kluwer), p. 273.

Miller, J. S., and Goodrich, R. W. 1990, *Ap. J.*, **355**, 456.

Miller, J. S., Goodrich, R. W., and Mathews, W. G. 1991, *Ap. J.*, **378**, 47.

Mittaz, J. P. D., and Branduardi-Raymont, G. 1989, *M.N.R.A.S.*, **238**, 1029.

Netzer, H., *et al.* 1990, *Ap. J.*, **353**, 108.

Osterbrock, D. E. 1978, *Phys. Scripta*, **17**, 285.

Padovani, P., Burg, R., and Edelson, R. 1990, *Ap. J.*, **353**, 438.

Pagel, B. E. J., and Edmunds, M. 1981, *Ann. Rev. Astr. Ap.*, **19**, 77.

Pakull, M. K. 1991, in *Wolf-Rayet Stars and Interrelations with Other Massive Stars in Galaxies*, ed. K. van der Hucht and B. Hidayat (Dordrecht: Kluwer), p. 391.

Perry, J. J., and Dyson, J. E. 1985, *M.N.R.A.S.*, **213**, 665.

Peterson, B. M. 1992, in *Relationships Between Active Galactic Nuclei and Starburst Galaxies*, ed. A. V. Filippenko (San Francisco: A.S.P.), in press.

Peterson, B. M., *et al.* 1991, *Ap. J.*, **368**, 119.

Peterson, B. M., *et al.* 1992, *Ap. J.*, in press.

Pogge, R. W. 1988a, *Ap. J.*, **328**, 519.

Pogge, R. W. 1988b, *Ap. J.*, **332**, 702.

Pronik, I. I. 1973, *Sov. Astr.*, **16**, 628.

Rothschild, R., Mushotzky, R., Baity, W., Gruber, D., Matteson, J., and Peterson, L. 1983, *Ap. J.*, **269**, 423.

Sanders, D. 1992, in *Relationships Between Active Galactic Nuclei and Starburst Galaxies*, ed. A. V. Filippenko (San Francisco: A.S.P.), in press.

Schechter, P. L. 1976, *Ap. J.*, **203**, 297.

Shklovskii, I. S. 1960, *Sov. Astr.*, **4**, 885.
Shull, J. M. 1980, *Ap. J.*, **237**, 769.
Soifer, B. T., Sanders, D., Madore, B., Neugebauer, G., Danielson, G., Elias, J., Lonsdale, C., and Rice, W. 1987, *Ap. J.*, **320**, 238.
Soltan, A. 1982, *M.N.R.A.S.*, **200**, 115.
Storchi-Bergmann, T., and Pastoriza, M. G. 1989, *Ap. J.*, **347**, 195.
Storchi-Bergmann, T., and Pastoriza, M. G. 1991, *Pub. A.S.P.*, **102**, 1359.
Sramek, R. A., and Weiler, K. W. 1990, in *Supernovae*, ed. A. G. Petschek (New York: Springer-Verlag), p. 76.
Tadhunter, C., and Tsvetanov, Z. 1989, *Nature*, **341**, 422.
Tenorio-Tagle, G., Terlevich, R., Franco, J., and Melnick, J. 1992, in *Relationships Between Active Galactic Nuclei and Starburst Galaxies*, ed. A. V. Filippenko (San Francisco: A.S.P.), in press.
Terlevich, E., Díaz, A. I., and Terlevich, R. 1990, *M.N.R.A.S.*, **242**, 271.
Terlevich, R. 1987. *M.N.R.A.S.*, **224**, 193.
Terlevich, R. 1989, in *Evolutionary Phenomena in Galaxies*, ed. J. E. Beckman and B. E. J. Pagel (Cambridge: Cambridge Univ. Press), p. 149.
Terlevich, R. 1990a, in *Windows on Galaxies*, ed. G. Fabbiano, J. Gallagher, and B. E. J. Pagel (Dordrecht: Kluwer), p. 87.
Terlevich, R. 1990b, in *Structure and Dynamics of the Interstellar Medium*, ed. G. Tenorio-Tagle, M. Moles, and J. Melnick (Berlin: Springer-Verlag), p. 343.
Terlevich, R. 1992, in *Relationships Between Active Galactic Nuclei and Starburst Galaxies*, ed. A. V. Filippenko (San Francisco: A.S.P.), in press.
Terlevich, R., and Melnick, J. 1985, *M.N.R.A.S.*, **224**, 193.
Terlevich, R., and Melnick, J. 1987, in *Starbursts and Galaxy Evolution*, ed. T. X. Thuan, T. Montmerle, and J. Tran Thanh Van (Gif sur Yvette, France: Éditions Frontières), p. 393.
Terlevich, R., and Melnick, J. 1988, *Nature*, **333**, 239.
Terlevich, R., Melnick, J., and Moles, M. 1987, in *Observational Evidence for Activity in Galaxies*, ed. E. Ye. Khachikian, K. J. Fricke, and J. Melnick (Dordrecht: Reidel), p. 499.
Terlevich, R., Tenorio-Tagle, G., Franco, J., and Melnick, J. 1992, *M.N.R.A.S.*, submitted.
Veilleux, S., and Osterbrock, D. E. 1987, *Ap. J. Suppl.*, **63**, 295.
Wandel, A., and Mushotzky, R. F. 1986, *Ap. J. (Letters)*, **306**, L61.
Weedman, D. W. 1983, *Ap. J.*, **266**, 479.
Wheeler, J. C., Mazurek, T. J., and Sivaramakrishnan, A. 1980, *Ap. J.*, **237**, 781.
Whittle, M., *et al.* 1986, *M.N.R.A.S.*, **222**, 189.
Whittle, M., Pedlar, A., Meurs, E. J. A., Unger, S. W., Axon, D. J., and Ward, M. J. 1988, *Ap. J.*, **326**, 125.
Wilson, A. S. 1992, these *Proceedings*.
Woosley, S. E., and Weaver, T. A. 1986, *Ann. Rev. Astr. Ap.*, **24**, 205.
Young, P. J. 1976, *Astr. J.*, **81**, 807.

Discussion

Wilson: If I understand correctly, you are assuming that most of the mass in the nuclear region in the post-starburst phase will be made up of the neutron stars and black holes left after the supernova explosions. You then use this mass to calculate a velocity dispersion which is found to be significantly larger than indicated by observations, and hence argue against the starburst hypothesis. However, we know from pulsar proper motions that neutron stars receive a high-velocity "kick" ($\gtrsim 100$ km s^{-1}) in supernova explosions. Thus, they may not stick around in the nucleus to have their masses measured later.

Filippenko: It is true that some of the mass in the post-starburst phase is associated with compact objects, which might escape from the nucleus, but most of it actually resides in low-mass and intermediate-mass stars. I am assuming that the initial mass function during the starburst phase is not far flatter than normal. The effect you mention therefore does not fundamentally change my argument.

Winkler: The Terlevich model predicts that there are many more type 1 Seyferts than type 2s, whereas the opposite is true. Is this a major problem?

Filippenko: I don't think this is a major problem right now. Spectropolarimetric studies of Seyfert 2 nuclei, notably by J. S. Miller and colleagues (e.g., Miller and Goodrich 1990), show that a significant fraction of type 2 Seyfert galaxies actually harbor a hidden Seyfert 1 nucleus. This nucleus is blocked along our line of sight by a thick, dusty torus. Moreover, optical spectra having high resolution and excellent signal-to-noise ratios (e.g., Filippenko and Sargent 1985) show that the Hα line in many narrow-line AGNs has weak, broad wings. Hence, type 1 Seyferts may actually be the most common AGNs.

Kazanas: I wonder why you do not have optical and UV variability among your major objections to the starburst model. I doubt that the results of NGC 5548 could be explained by this model. I would think that it would have particular problems accounting for large-amplitude variability of high-luminosity objects.

Filippenko: At the present time, I feel that optical and UV variability of AGNs (ignoring BL Lacs and OVVs) is not a very convincing objection to the starburst model, compared with some of the other issues in this review. The largest and most rapid variability has been seen in relatively low-luminosity objects such as NGC 4151 and NGC 5548. In the starburst hypothesis, the bolometric luminosities of these nuclei can be explained by just a few SNe per year. With such a low SN rate, small-number statistics will lead to rapid and high-amplitude variability. Indeed, Terlevich (1990b) argues that in most AGNs the documented variability is consistent with the starburst hypothesis; large-amplitude variations are generally not seen in high-luminosity objects. It may also be possible to explain the observed differences in the lags shown by different emission lines (Clavel *et al.* 1991; Peterson *et al.* 1991). For example, my spectra of peculiar type II SNe such as SN 1991ae and SN 1991av show that the broad He I λ5876 line disappeared sooner than the broad hydrogen Balmer lines.

A Cluster of He I Emission Line Stars in the Galactic Center

R. Genzel, A. Krabbe, S. Drapatz, and V. Rotaciuc

Max-Planck-Institut für extraterrestrische Physik, Germany

Abstract. We report 1.8″/300 km s^{-1} resolution imaging spectroscopy of HeI n= $2^1P \rightarrow n=2^1S$ 2.06μm line emission in the central 45″ of the Galactic Center. There are two types of HeI emission line sources. Extended, relatively narrow emission comes from nebular knots in the Sgr A West ionized gas streamers. Second there is a group of about a dozen, broad line "HeI emission line stars". IRS16 and its "broad line region" probably is a central concentration of this 1 pc diameter cluster of HeI stars. There is no evidence for a special HeI feature related to Sgr A*, the compact radio source that lies very close to the geometric center of the HeI star cluster. We conclude that the compact, broad line HeI sources represent a cluster of late type, blue supergiants or Wolf-Rayet stars with heavy mass loss. The HeI emission comes from the extended atmospheres of these stars that have mass loss rates of a few 10^{-5} to 10^{-4} $M_\odot$/yr and outflow velocities of about 1000 km s^{-1}. The total mass outflow rate created by the central cluster is about 3x10^{-4} $M_\odot yr^{-1}$. The high ratio of HeI/HI(Brα,Brγ) fluxes suggests an overabundance of helium in the atmospheres of these stars. The HeI stars are members of a cluster of massive stars that can account for the total luminosity, the flux of Lyman continuum photons and the fairly low excitation of the central few parsecs. Our data show very clearly that massive star formation has taken place in the Galactic center within the last few 10^6 years.

1. Introduction

The luminosity of the central few parsecs of our Galaxy is about 10^7 $L_\odot$ most of which emerges at $\lambda \geq 20\mu m$ (Becklin et al. 1982, R_o=8 kpc). The majority of that far-infrared luminosity originates from a source or sources with an effective radiation temperature $\sim$ 35,000 K (Lacy et al. 1982). The nature of the source(s) has been a matter of intense debate during the past decade and two lines of interpretation have developed. One holds that there has been a episode of star formation a few million years ago, when a few tens of massive stars formed (Rieke and Lebofsky 1982, Lacy et al. 1982, Rieke et al. 1989, Allen 1987, Allen et al. 1990). In this "star formation" hypothesis the remaining late type O stars left over from the past star formation activity presently power the Galactic center. The other, "central engine", interpretation is based mainly on the presence of the compact radio source Sgr A* (Lo 1989) and on the evidence for a central concentration of mass and luminosity (Genzel and Townes 1987, Gatley 1987). It proposes the existence of a single luminosity source, perhaps an accretion disk around a massive central black hole (Lacy et al. 1982, Serabyn and Lacy 1985, Gatley 1987).

2. Observations

The measurements were taken in September 1990 with the MPE 1→5μm Fabry-Perot imaging spectrometer FAST (Krabbe et al. 1991) on the 3.9m Anglo-Australian telescope. FAST uses a 58x62 pixel InSb array detector manufactured by Santa Barbara Research Corporation behind a a Queensgate scanning Fabry-Perot spectrometer resulting in a resolving power of 103 in the K-band. For the AAT observations an image scale of 0.95″ per pixel was chosen and the FWHM spatial resolution of our measurements is 1.8″.

3. Results and discussion

3.1 A Cluster of HeI Emission Line Stars

The HeI line maps at 5 different velocities, along with a 2.06μm continuum map are shown in Fig.1. There are two types of 2μm HeI emission line sources. The first is relatively faint, extended emission (3 to 10″) seen only at v_{LSR}=–300 and 0 km s^{-1}. This component of HeI emission shows an excellent correlation with the spatial distribution and velocity field of the ionized gas streamers in the Sgr A West HII "mini-spiral", as traced, for example, by the 5 GHz radio continuum (Lo and Claussen 1983). We identify this component of HeI emission with nebular material in the central parsec. A second component of HeI emission is more intense, is spatially compact or unresolved and has broad line widths (Δv_{FWZP} ≈1300→2000 km s^{-1}).

We have identified at least 12 broad line sources in the central 20″ (0.8 pc). Their centroid is near Sgr A*, and at least 10 of its members are positionally coincident with compact 2μm continuum sources within ~1″. In the following we thus will refer to all compact broad HeI emission line sources as "HeI (emission line) stars". The brightest HeI star is located about 12″ southwest of IRS16C. It was discovered and previously discussed by Allen et al.(1990) and will henceforth be referred to as the AHH star.

The two types of HeI emission sources are also easily distinguished by their very different HeI/HI(Brα,γ) or HeI/S(5GHz) flux ratios. Following Geballe et al.(1984) for the analysis of the 2.06μm HeI line flux we find

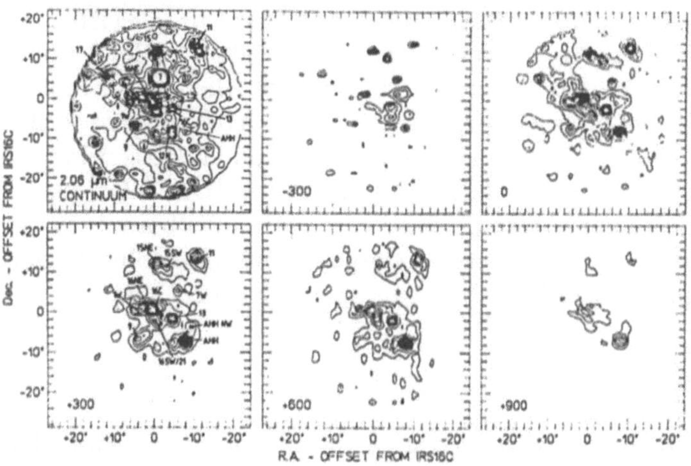

Fig. 1. Contour maps of 2.06μm continuum (upper left) and HeI line emission in five selected velocity channels in the central 40″ of the Galaxy. The center velocity is indicated for each line map, the FWHM spectral resolution is 300 km s^{-1}. The FWHM spatial resolution is 1.8″. Contour units are 2mJy per pixel for the continuum and 1.8×10^{-5} erg s^{-1} cm^{-2} per pixel for the line maps.

that observed line ratios in the narrow line, nebular knots thus indicate gas with only partial ionization of He ($He^+/H^+ \leq 0.04$). In contrast the ratio of HeI to HI(Brα/γ) or radio continuum in the HeI stars is an order of magnitude greater ($F_{HeI}/F_{HI(Br\alpha)} \geq 2$) than in the nebular knots. This implies that helium is fully ionized toward these stars and probably that helium is overabundant ($He^+/H^+ \geq 0.3$).

3.2 IRS16 as a Central Condensation of the HeI Star Cluster

The characteristics of the central IRS16 complex and its HeI "broad line region" (Geballe et al. 1984, 1991) are remarkably similar to those of the isolated HeI stars. The only significant differences are the unique position of the IRS16 complex at the center of the cluster and within a few arcsec of Sgr A* and the fact that its HeI broad line emission is extended over more than 3″. At high resolution the near-infrared continuum of IRS16 splits up into half a dozen or so very compact stellar objects (Simon et al. 1990). We thus propose that the IRS16 complex consists of several HeI stars and is a central concentration in the 0.8pc diameter HeI cluster (see also Allen et al. 1990).

3.3 Nature of the HeI Stars

The HeI stars must be hot in order to produce He^+ and they must be losing mass because of the $\geq 10^3$ km s^{-1}linewidths. The average $\approx$ 100 redshift of the HeI line centroids is very likely the result of a P-Cygni profile (Allen et al. 1990). On the basis of the strength of HeI 2.058μm relative to HeI 2.12μm and HI(Brγ) and the lack of HeII recombination lines Allen et al. (1990) conclude that the AHH object is a Ofpe/WN9 star. The AHH object and the other HeI stars in the Galactic center have similar line and line to continuum ratios as stars of this type in our Galaxy and in the LMC (McGregor et al. 1988a,b). The evolutionary state of these rare stars (Walborn 1977) is uncertain and they may be more closely related to luminous blue/S Dor variables than to classical Wolf-Rayet stars (Appenzeller, priv.comm.).

3.4 Mass Loss Rate

A simple estimate of the mass loss rates can be obtained if the 2.058μm HeI line is optically thin and if case B recombination theory is applicable. In that case we derive mass loss rates of $\dot{M}$ $\approx$3x10^{-5} to 10^{-4} M$_{\odot}$yr^{-1} for the HeI stars. Another estimate comes from the strength of HI(Brα) and the theory of optically thick line emission in dense winds (Simon et al. 1983). For the brightest HeI star (AHH) we then find a mass loss rate of 4x10^{-5} M$_{\odot}$yr^{-1}. Finally, an upper limit to the mass loss rate can be obtained from the radio and 2.2μm continuum flux densities. Assuming that the infrared and radio continuum is free-free emission from the wind a mass loss can be obtained by application of equation (20) of Wright and Barlow (1975). From the upper limits to the 5GHz emission and the 2.2μm flux densities of the HeI stars we find upper limits of a few 10^{-4} M$_{\odot}$yr^{-1} if the emission originates in a region where the flow already has reached the terminal velocity. In summary then it is fairly clear that the HeI stars in the Galactic center must be loosing mass at a high rate ($\dot{M}$ $\approx$ a few 10^{-5} to 10^{-4} M$_{\odot}$yr^{-1}). The combined mass loss rate of the entire HeI cluster amounts to about 3x10^{-4} M$_{\odot}$yr^{-1}.

3.5 Effective Temperature

A rough estimate of the effective temperature of the radiation field of the HeI stars is obtained by comparing the number of ionizing photons required for explanation of the total HeI flux and of the radio continuum flux density in the same region under the assumption of an ionization bounded, photoionized nebula. Based on the same assumptions about the 2μm line formation as before, the number of photons in the He^{+}–continuum required to explain the total dereddened HeI flux found in Fig.1 is about 4x10^{48} s^{-1}. For comparison, the number of Lyman continuum photons in the same region estimated from the radio continuum flux density is about 7x10^{49} s^{-1}. Taking the recent non-LTE, line-blanketed model atmospheres of Kudritzki et al. (1991) that include the effect of the stellar winds, such a ratio of helium to hydrogen ionizing photons is obtained for main sequence stars of Teff$\approx$35,000 K and for supergiants of Teff$\approx$31,000 K.

The good agreement of this estimate with that obtained from the excitation of the Sgr A West HII region in various fine structure lines suggests that the HeI stars are closely connected to the production of hydrogen and helium ionizing photons in Sgr A West.

3.6 Luminosities

The 2.2μm continuum flux densities constrain the luminosities of the HeI stars. First, assuming blackbody emission with a temperature near 31,000 K the dereddened 2μm flux densities of the HeI stars correspond to luminosities between 4 and 50x10^5 $L_\odot$. This is an upper limit as all additional effects (2.2μm emission from the wind, extended stellar atmosphere etc.) raise the 2.2μm flux relative to the UV/visible flux. Second, the data of McGregor et al. (1988a,b) and Williams et al. (1987) on the near-infrared fluxes and luminosities of late WN/Ofpe stars and other blue supergiants in our Galaxy and the LMC suggest an empirical ratio of luminosity to 2.2μm flux density of about 7x10^5 $L_\odot$/Jy. Despite the fairly wide range in $L/S_{2.2}$ ratios found in different stars the mean value is actually close to the simple blackbody estimate. With that ratio the HeI stars in the Galactic center would be supergiants with luminosities between 1.4x10^5 and 2x10^6 $L_\odot$. The HeI cluster then has a total luminosity of 1.2x10^7 $L_\odot$, sufficient to account for a significant fraction or even the majority of the infrared luminosity of the central few parsecs, corrected for photons escaping this central region (Becklin et al. 1982).

3.7 Massive Star Formation in the Central Parsec

Our data show very clearly that massive star formation has taken place in the central parsec within the last few 10^6 years, in agreement with the original proposal of Rieke and Lebofsky (1982). The blue supergiant/Wolf-Rayet phase of massive (15 to 80 $M_\odot$) stars lasts less than about 10^6 years (Chiosi and Maeder 1986). Luminous blue variables may even have a lifetime significantly less than 10^6 years (Hillier, priv.comm.). One interpretation of our data, therefore, is that massive star formation has been proceeding continuously in the Galactic center. Then there should be between 5 and 30 times as many main sequence O stars as post main-sequence blue supergiants/Wolf-Rayet stars (Humphreys and McElroy 1984, Chiosi and Maeder 1986). The present data allow the presence of additional main-sequence O stars in the central parsec as long they are cooler than 35,000 K and thus later than O9 ($\sim$24 $M_\odot$). Late O stars are almost an order of magnitude less luminous and fainter at 2μm than Ofpe/WN9 stars and don't emit HeI lines. 150 O9 stars would contribute about the same total luminosity as that estimated for the HeI cluster. The O stars may then actually contribute a large fraction of the $\lambda \leq$504Å radiation in the central parsec as a significant fraction of the He-ionizing continuum in the HeI stars is probably absorbed out by the dense winds (Hillier, priv.comm.) The question also arises whether the environment of a metal rich (high mass loss) and dense (many binaries) stellar cluster could lead to an increased abundance of blue supergiants with heavy mass loss. IRS7 and the other possible red supergiants in the central parsec (IRS11,19,12) could be members of the same massive star cluster as the HeI stars but are presently in their red supergiant phase. The fairly large ratio of blue to red supergiants (3 to 10) is probably consistent with this scenario (Chiosi and Maeder 1986).

An alternate explanation of the presence of such a large number of rare HeI emission line stars is to postulate that massive stars formed in a brief episode of active star formation not much more than 10^6 years ago.

Acknowledgements: We are grateful to I.Appenzeller and D.J.Hillier for very valuable comments and discussions.

References

Allen, D.A., 1987: in *The Galactic Center*, ed. D.C. Backer, AIP **155**, 1.
Allen, D.A., Hyland, A.R., Hillier, D.J., 1990: *Monthly Notices Roy. Astron. Soc.*, **244**, 706.
Becklin, E.E., Gatley, I., Werner, M.W., 1982: *Astrophys. J.*, **258**, 135.
Chiosi, C., Maeder, A., 1986:, *Ann. Rev. Astron. Astrophys.*, **24**, 329.
Gatley, I., 1987: in *The Galactic Center*, ed. D.C. Backer, AIP **155**, 8.
Geballe, T.R., Kriuscinas, K., Lee, T.J., Gatley, I., Wade, R., Duncan, W.D., Garden, R., Becklin, E.E., 1984: *Astrophys. J.*, **284**, 118.
Geballe, T.R., Krisciunas, K., Bailey, J.A., Wade, R., 1991: *Astrophys. J. (Letters)*, **370**, L73.
Genzel, R., Townes, C.H., 1987: *Ann. Rev. Astron. Astrophys.*, **25**, 377.
Humphreys, R.M., McElroy, D.B., 1984: *Astrophys. J.*, **284**, 565.
Krabbe, A., Rotaciuc, V., Hofmann, R., Drapatz, S., Genzel, R., 1991: in preparation.
Kudritzki, R.P., Gabler, R., Kunze, D., Pauldrach, A.W.A., Puls, J., 1991: in *Proceedings STScI Workshop on Starburst Galaxies*, ed. C. Leitherer, in press
Lacy, J.H., Townes, C.H., Hollenbach, D.J., 1982: *Astrophys. J.*, **262**, 120.
Lo, K.Y., Claussen, M.J., 1983: *Nature* **306**, 647.
Lo, K.Y., 1989: in *The Center of The Galaxy*, ed. M. Morris, Kluwer (Dordrecht), 527.
McGregor, P.J., Hyland, A.R., Hillier, D.J., 1988a: *Astrophys. J.*, **324**, 1071.
McGregor, P.J., Hillier, D.J., Hyland, A.R., 1988b: *Astrophys. J.*, **334**, 639.
Rieke, G.H., Lebofsky, M.J., 1982: in *The Galactic Center*, eds. G.R. Riegler and R.D.Blandford, AIP **83**, 194.
Rieke, G.H., Rieke, M.J., Paul, A.E., 1989: *Astrophys. J.*, **336**, 752.
Serabyn, E., Lacy, J.H., 1985: *Astrophys. J.*, **293**, 445.
Simon, M., Felli, M., Cassar, L., Fischer, J., Massi, M., 1983: *Astrophys. J.*, **266**, 623.
Simon, M., Chen, W.P., Forrest, W.J., Garnett, J.D., Longmore, A.J., Gauer, T., Dixon, R.I., 1990: *Astrophys. J.*, **360**, 95.
Walborn, N.R., 1977: *Astrophys. J.*, **215**, 53.
Williams, P.M., van der Hucht, K.A., The, P.S., 1987: *Astron. Astrophys.*, **182**, 91.
Wright, A.E., Barlow, M.J., 1975: *Monthly Notices Roy. Astron. Soc.*, **170**, 41.

Near–infrared Imaging of Hard X–ray Selected Active Galaxies

Jari Kotilainen

Institute of Astronomy, Madingley Road, Cambridge CB3 0HA

Abstract. We present near–IR observations of a hard X–ray selected sample of AGN, mainly Seyfert 1s. We decompose the luminosity profiles into AGN, bulge and disk components and subtract the stellar contribution to the nuclear light. For the nonstellar emission, we find a good correlation between hard X–rays and near–IR. We assume an underlying nonthermal powerlaw with modification by dust and find that both nonthermal and thermal components are present, but dust emission is dominant in most cases, especially in low–luminosity sources.

1. Introduction

There are generally thought to be two different kinds of AGN; those dominated by thermal emission (radio–quiet AGN) and nonthermal emission (BL Lacs, radio–loud AGN) (Bregman 1990). Further subdivision is caused by luminosity (e.g. Sy1 – QSO) and orientation (e.g. Sy1 – Sy2) effects (Lawrence 1987). Continuum emission from AGN covers the whole observable spectrum, and for a typical radio–quiet object the energy output νF_ν is roughly constant from IR to X–rays with two broad bumps, the larger in UV/soft X–ray region, due to thermal emission from an accretion disk, and the other in the IR from rearadiation by dust at different temperatures. The origin of the near–IR emission in AGN is not yet clearly understood. Rapid variability indicates a nonthermal origin for blazars. Steep spectral shapes and extended 10 μm emission argue for thermal origin for Seyfert 2s. It is therefore plausible that both components exist in QSOs and Seyfert 1s.

To study this, we use a complete, flux–limited hard X–ray selected sample of Piccinotti *et al.* (1982), which includes 34 AGN, mainly Seyfert 1s. Importantly, it is not biased with respect to dust. Previous studies of near–IR continuum have been hampered by the uncertainty about the degree by which the host galaxy dilutes the nuclear light. Using near–IR imaging data, we decompose luminosity profiles into AGN and galaxy components and derive stellar and nonstellar fractions for the nuclear aperture, to correct for the host galaxy. Further details are given in Kotilainen *et al.* (1991a,b).

2. Observations and data reduction

We have obtained near-IR broad band images at J, H and K of the sample during the period of 1987 to 1989 using telescopes at Cerro Tololo, Kitt Peak and Mauna Kea using 62x58 px InSb arrays with pixel scale varying from 0.3 to 0.6 px/arcsec. Usually multiple exposures were taken for each galaxy to avoid saturating the nucleus and to increase S/N away from nucleus. The data reduction comprised of bias and dark subtraction, sky subtraction using sky frames taken immediately before and after the galaxy frame, removal of bad pixels, flatfielding using H sky frames, removal of any residual sky, and finally calibration against standard stars. We estimate that the internal statistical error of the photometry is about 0.02 mag and the error due to the estimation of the sky level about 0.05 mag.

3. Decomposition

Surface brightness profiles of spiral galaxies can be decomposed into bulge and disk components (Kent 1985):

$I_b(r) = I_{ob} \exp_{10} [-3.33 ((r/r_b)^{1/4} - 1)]$

$I_d(r) = I_{od} \exp (-r/r_d)$.

We extend this approach to the near–IR wavelengths and add a point source at the center of the galaxy, representing the AGN component:

$I_n(r) = F_T \delta(r)$.

These theoretical profiles are convolved with a two–gaussian seeing point spread function:

$PSF(r) = \frac{I_o}{1.05}[exp(-\frac{r^2}{2\sigma^2} + aexp(-\frac{r^2}{9.68\sigma^2})]$.

To determine the best fit, we use a method, where starting from initial values, we iteratively minimize the least–squares fit to the observed profile with five free parameters, I_{on}, I_{ob}, I_{od}, r_b and r_d. The main source of error in the decomposition is the seeing accuracy. There are no stars in the small field of IR array, so we had to use the standard stars before and after the galaxy. We studied the effect of seeing changes and estimate an accuracy of about 10 % for the derived fractions.

We find, that the fraction of the nonstellar component increases towards longer wavelengths. This is expected, since the stellar flux distribution peaks at about 1.5 μm. The change is very significant from H to K, and most of the sample galaxies are dominated by nonstellar light in K. There are a few uncertain decompositions for heavily reddened sources (e.g. 3A 0557, IC 4329A). We also find, that low–luminosity sources have a larger fraction

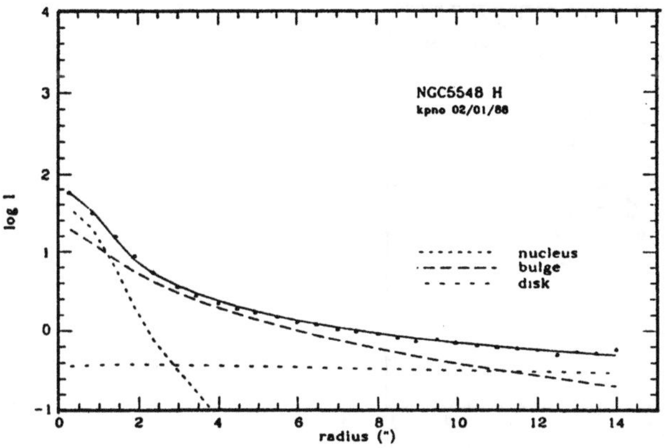

Fig. 1. The luminosity profile fit of NGC 5548 in H band. Triangles represent the observed profile, short dash line the nucleus, long dash line the bulge and dot–dash line the disk.

of the nuclear light arising from the host galaxy. An example of a profile fit for NGC 5548 in H band is given in Fig. 1.

Very little previous decomposition analysis exists in the near–IR, because of the recent advent of imaging instruments. Previously two methods were used: a) to use a red optical CCD image, fit the profile with the three components, and extrapolate the stellar contribution into IR assuming normal stellar colours (Malkan & Filippenko 1983, Ward *et al.* 1987) or b): to compare growth curve from multiaperture IR photometry directly with a stellar growth curve (Malkan & Filippenko 1983). The two main previous studies in near–IR are by Ward *et al.* (1987) and McAlary & Rieke (1988). Our nonstellar fractions are generally consistent with theirs (or slightly larger, probably due to an aperture difference).

4. Results of the photometry

Several of the galaxies were observed more than once. We find flux variations of more than 30 % within 1 – 2 months only for 3C 120 and 3C 273. A few more galaxies had similar variation within about one year. These variations ar compatible with a thermal origin and the variability does not affect the conclusions we draw.

In most cases, the JHK colours become redder with decreasing aperture size. Most galaxies have about constant J–H gradient, while there is a great range in H–K gradient. Other galaxies exhibit comparable J–H and H–K gradients. While some of these gradients can be explained by obscuration and extinction by dust, others may have a contribution from very hot dust. Most of the nuclear JHK colours (fig. 2) lie in the region between normal

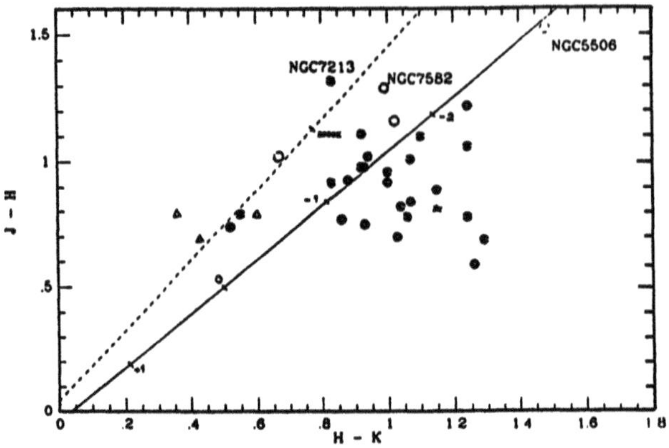

Fig. 2. J–H vs H–K diagram at 3 arcsec aperture. Oval regions represent the areas populated by normal galaxies and quasars. The solid line is a powerlaw locus for different spectral indices. The dashed line is a blackbody locus for different temperatures. Reddening vector of $A_V = 2.0^m$ and the direction of hot dust are indicated. In all figures filled circles represent Sy 1 nuclei, open circles Sy 2 nuclei, filled triangle the quasar 3C 273 and open triangles BL Lac objects.

galaxies and quasars, with reddening and hot dust explaining the discrepant cases.

5. Discussion

Several authors have noted a close link between X–ray and near–IR luminosities suggesting a nonthermal contribution to the NIR emission (e.g Edelson & Malkan 1986, Carleton *et al.* 1987). We can reproduce this good correlation (Fig. 3). The correlation does not get better after the stellar subtraction, but this could be due to the uncertainty of the subtraction. On the other hand, the correlation slope approaches a universal value of unity after subtraction.

Respectively, several authors have stressed the importance of thermal emission in the NIR (e.g. McAlary & Rieke 1988, Sanders *et al.* 1989). In a sample of PG quasars, Sanders *et al.* found only a weak correlation between X–rays and IR, a universal minimum in the energy output at around 1 μm corresponding to the minimum contribution from thermal emission from both the accretion disk and reradiating dust, and steep NIR spectral shapes. They argue that dust dominates the NIR emission and developed a warped disk model, with IR emission at different wavelength coming from different parts of the disk. Another strong evidence for thermal emission is the variability study of Fairall 9 by Clavel *et al.* (1989), who found that the NIR flux responds to UV/X variations with a lag of 1 year, in agreement with the distance of the hottest dust grains, and with the time delay

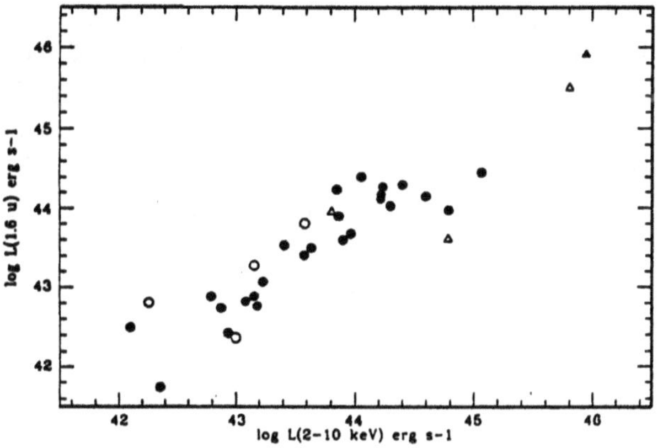

Fig. 3. Hard X-ray luminosity vs H band luminosity at 3 arcsec aperture, corrected for the galactic contamination.

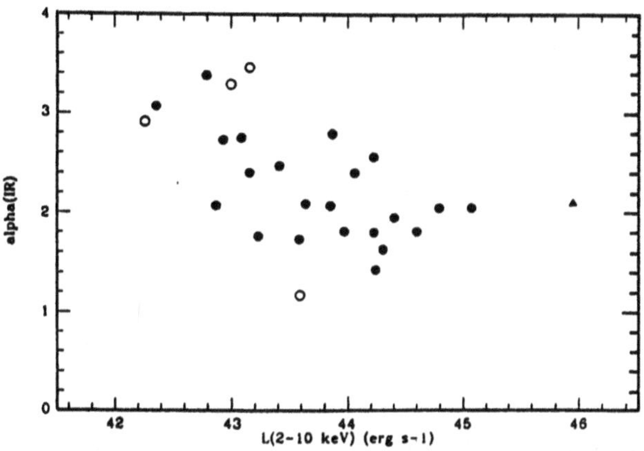

Fig. 4. Near-infrared spectral index vs the hard X-ray luminosity.

increasing with wavelength. Finally, Done *et al.* (1990) found no IR variability in NGC 4051 while X-rays flickered by a factor of 2, indicating that the extended thermal emission overwhelms any possible nonthermal connection between the two regions.

Is there an underlying powerlaw connecting the NIR to the X-rays? An upper limit to the nonthermal contribution to the near-IR continuum can be obtained assuming an underlying powerlaw of α = -1 from X-rays to J band (which we therefore assume to be totally nonthermal) and all modifications from it in the NIR as due to dust (Carleton *et al.* (1987). We find an average upper limit for the Sy 1s to be $\sim$ 50 % suggesting that thermal emission is dominant in most galaxies.

If the steep infrared slope is due to dust emission, it should correlate with reddening indicators.

However, we do not find a correlation between the near–IR slope and a)Balmer decrement, b) Paschen line ratios and c) the ratio of hard X–ray to Hα luminosity. Possible reasons for this are discussed in Kotilainen *et al.* (1991b). A slight correlation is found with the axial ratio of the galaxy, and a good anticorrelation with the hard X–ray luminosity (Fig. 4), explainable by decreasing dust content with increasing luminosity, Lawrence & Elvis 1982).

6. Conclusions

Based on near–IR imaging and profile decomposition, we find that emission from the underlying galaxy is a significant component to the nuclear light in especially low-luminosity Seyfert galaxies. We find a good correlation between hard X–ray and near–IR emission. Although this correlation does not get better after the stellar subtraction, the slope of the correlation gets closer to a universal ratio of 1, suggesting an underlying nonthermal powerlaw between the two regions. The steep NIR spectral slopes, on the other hand, suggest a contribution from dust. We suggest that both thermal and nonthermal component is present, with emission from dust probably dominant in most cases.

References

Bregman, J.N., 1990: *Astron. Astrophys. Rev.*, **2**, 125.
Carleton, N.P., *et al.*, 1987: *Astrophys. J.*, **318**, 595.
Clavel, J., Wamsteker, W., Glass, I.S., 1989: *Astrophys. J.*, **337**, 236.
Done, C., *et al.*, 1990: *Monthly Notices Roy. Astron. Soc.*, **243**, 713.
Edelson, R.A., Malkan, M.A., 1986: *Astrophys. J.*, **308**, 59.
Kent, S.M., 1985: *Astrophys. J. Suppl.*, **59**, 115.
Kotilainen, J.K., *et al.*, 1991a,b: *Monthly Notices Roy. Astron. Soc.*, submitted.
Lawrence, A., 1987: *Publ. Astron. Soc. Pacific*, **99**, 309.
Lawrence, A., Elvis, M., 1982: *Astrophys. J.*, **256**, 410.
Malkan, M.A., Filippenko, A.V., 1983: *Astrophys. J.*, **275**, 477.
McAlary, C.W., Rieke, G.H., 1988: *Astrophys. J.*, **333**, 1.
Piccinotti, G., *et al.*, 1982: *Astrophys. J.*, **253**, 485.
Sanders, D.B., *et al.*, 1989: *Astrophys. J.*, **347**, 29.
Ward, M.J., *et al.*, 1987: *Astrophys. J.*, **315**, 74.

Discussion

Wilson: I think that perhaps the correlation between near-infrared and X-ray emission is not surprising even in thermal models. The infrared would be acting as a "calorimeter", simply reflecting the UV/X-ray dust heating source.

Kotilainen: I agree that there may be a general luminosity correlation, in the sense that sources luminous in X-rays would also be luminous in the infrared. But I don't think it would produce this tight a correlation!

Mushotzky: Does the detected variability of 3C 120 constrain the emission mechanism?

Kotilainen: No. The variability is consistent with the thermal models.

Winkler: What do your results tell us about the colours of the underlying galaxies?

Kotilainen: Our study of the host galaxies is in a preliminary stage, but indicates that the nuclear stellar colours are somewhat redder than in normal galaxies.

Models of Anisotropic Ionizing Radiation in Seyfert Galaxies

J.A. Acosta-Pulido[1], *I. Pérez-Fournon*[1], *and A.S. Wilson*[2]

[1] Instituto de Astrofísica de Canarias, Tenerife, Spain.
[2] Astronomy Program, University of Maryland, USA.

We study the photoionization of the ENLR by anisotropic nuclear radiation fields. Several explanations have been proposed to explain the concentration of the line emission along the radio axis (Unger *et al.* 1987):

1– A large obscuring torus located between the BLR and the NLR (Antonucci and Miller 1985).

2– A geometrically thick accretion disk (Acosta–Pulido *et al.* 1990).

3– A blazar beam produced by a relativistic jet (Morganti *et al.* 1991).

We calculate the variation of the intensity[1] of several lines with the inclination angle w.r.t. the symmetry axis in the three cases above. We assume that the gas at the ENLR is distributed uniformly in a disk and the angular variation is calculated at constant radius. The value of U_t at $\theta = 0°$ is determined from the best fit with the observed line ratios and then its value is varied with θ according to each model prescriptions (Fig. 1). A detailed description of the calculations, as well as the observational implications for each model, will be published elsewhere (Acosta–Pulido, Wilson and Pérez–Fournon 1991). We have found that the angular variation of the intensity of certain lines depends on the nature of the ionizing nuclear emission. The actual angular variation of the different emission lines should be firmly established observationally.

[1] We use the photoionization code MAPPINGS (Binette, Dopita and Tuohy 1985). We assume a plane parallel geometry and each cloud to be radiation bounded.

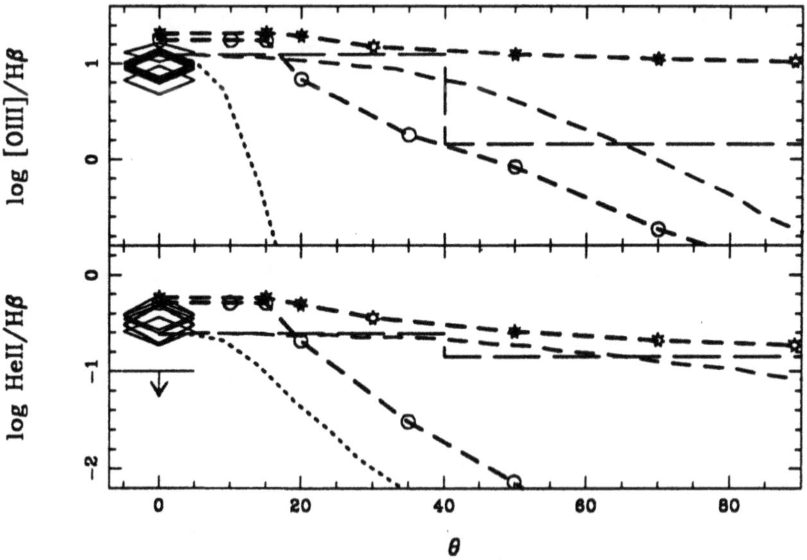

Fig. 1. Predicted line intensities relative to H recombination lines [$U_t(\theta = 0°) \sim 10^{-2}$]. The observed data along the radio axis correspond to the line intensities at $\theta \sim 0°$ (NGC 3516: Pogge 1989, Ulrich and Péquignot 1980; NGC 4151: Penston *et al.* 1990; NGC 4388: Pogge 1988). The relativistic jet models are represented by the dotted ("fast" jet) and short-dashed ("slow" jet) curves. The long-dashed curve represents the molecular torus model: a ratio of direct to scatter light ~ 30 and $\theta_{open} \sim 40°$ (Sokolowski, Bland-Hawthorn and Cecil 1991). The spectral shape is $F_\nu \propto \nu^{-1.5}$, in the above cases. The points joined by dotted lines represent accretion disk models (Madau 1988) for $M = 10^6\ M_\odot$, $r_{out} = 1000\ r_g$ (stars) and for $M = 10^8\ M_\odot$, $r_{out} = 1000\ r_g$ (empty circles).

References

Acosta-Pulido, J.A., *et al.* , 1990: *Astrophys. J.*, **365**, 119.
Acosta-Pulido, J.A., Wilson, A.S., Pérez-Fournon, I., 1991: in preparation.
Antonucci, R.R.J., Miller, J.S., 1985: *Astrophys. J.*, **297**, 621.
Binette, L., Dopita, M.A., Tuohy, I.R., 1985: *Astrophys. J.*, **297**, 476.
Madau, P., 1988: *Astrophys. J.*, **327**, 116.
Morganti, R., *et al.* , 1991: *Monthly Notices Roy. Astron. Soc.*, **249**, 91.
Penston, M.V., *et al.* , 1990: *Astron. Astrophys.*, **236**, 53.
Pogge, R.W., 1988: *Astrophys. J.*, **332**, 702.
Pogge, R.W., 1989: *Astron. J.*, **98**, 124.
Sokolowski, J., Bland-Hawthorn, J., Cecil, G., 1991: *Astrophys. J.* **375**.
Ulrich, M.-H., Péquignot, D., 1980: *Astrophys. J.*, **238**, 45.
Unger, S.W., *et al.* , 1987: *Monthly Notices Roy. Astron. Soc.*, **228**, 671.

Double and Extended Emission Line System in the Radio Galaxy 0114+074S

C.E. Akujor[1,2] *and N. Jackson*[1,3]

[1] NRAL, Jodrell Bank, Macclesfield, Cheshire SK11 9DL
[2] On leave from the University of Nigeria, Nsukka, Nigeria
[3] Current address: Sterrewacht, Postbus 9513, 2300 RA Leiden, The Netherlands

0114+074S is a redshift 0.344 radio galaxy 30″ from a redshift 0.861 quasar, 0114+074N (Akujor 1989). We report a detection of complex split emission lines in 0114+074S, and a continuum knot coincident with a radio hotspot

Radio continuum observations of 0114+074S were made with the VLA in A-configuration at 327MHz, 1446 and 4866 MHz (fig 1). The depolarisation ratios (Garrington et al. 1988) of the west and east sides are 0.21 and 0.04. Although both sides of the source exhibit strong depolarisation at 1.4 GHz, there is more depolarisation on the east side, and in the hotspots.

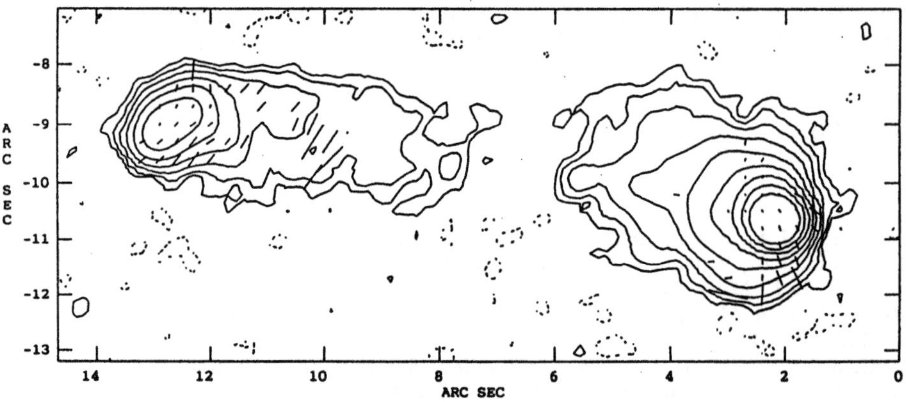

Fig. 1. VLA 0.5″resolution radio map of 0114+074S at 5 GHz. Contours are plotted at 0.0464 mJy × (1,2,4,8,16,32,64,128,298). A 0.1″vector represents 13% polarization.

The optical continuum emission 3″ west of the nucleus (fig 2) is almost equal in brightness to the total continuum emission from the central object. It has approximately the same spectral slope as the nuclear continuum and coincides exactly with the radio hotspot. One of the only other clear examples of this phenomenon in radio galaxies is 3C 33 (Meisenheimer & Heavens 1986) in which the optical polarization properties of the extended emission indicate a synchrotron origin. A synchrotron model is not impossible here,

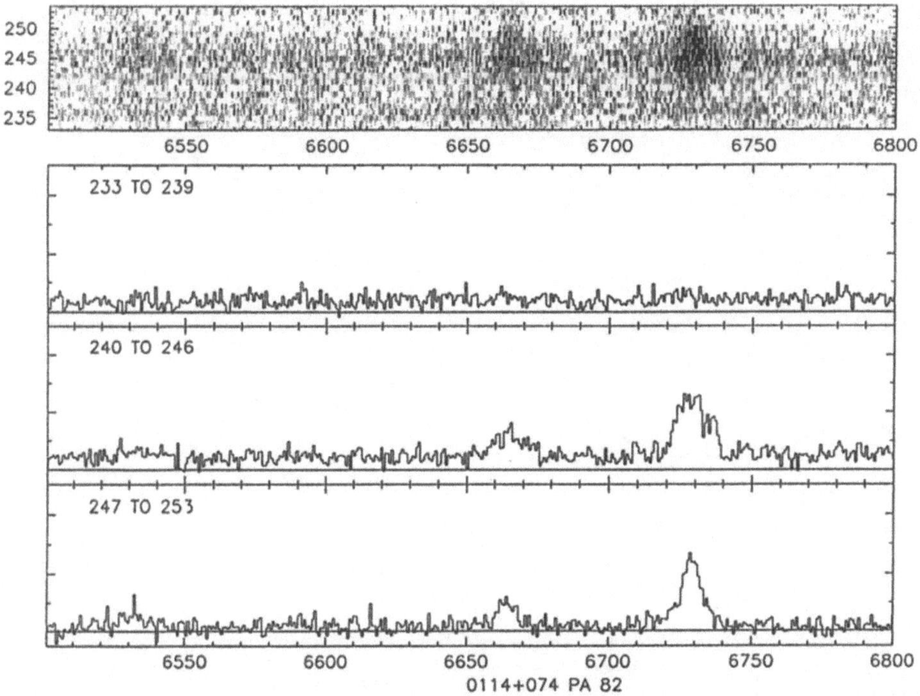

Fig. 2. (Top) Greyscale of the CCD frame containing the spectrum of 0114+074S at a position angle of 82° . The spectral direction runs from left to right: east is at the top of the plot. (Lower panels) Cuts across the spectrum at row numbers corresponding to the labelling on the greyscale plot. Note the continuum region 3″(10 pixels) to the west of the nucleus. Note also the narrowing of the [OIII] profile between the cut across the nucleus (middle spectrum) and the easterly extension. This narrow component may obscure the split–line effect of figure 3.

but would require a rather flat ($\nu^{-0.6}$) overall spectral index between the optical and radio.

The alternative explanations for the continuum emission include light scattered from an obscured nucleus and star formation associated with a radio jet. The redness of the extended continuum in 0114+074S suggests that it is not scattered nuclear light as seen in 2152-69 (Tadhunter et al. 1987). Also, the coincidence with the radio hotspot suggests that interaction is taking place, but optical polarimetry is needed to rule out scattering or synchrotron models.

Complex emission line morphologies are quite common in 3C radio galaxies (Baum et al. 1990), but only one definite split [OIII] line is seen, in 3C 159 (Tytler & Browne 1985). Our results for 0114+074S (fig 2–3) suggest that the lines are split along spatial directions perpendicular and parallel to the radio axis. There are probably at least two components, as the spectrum parallel to the radio axis contains a narrow extended [OIII] component as well as the double component.

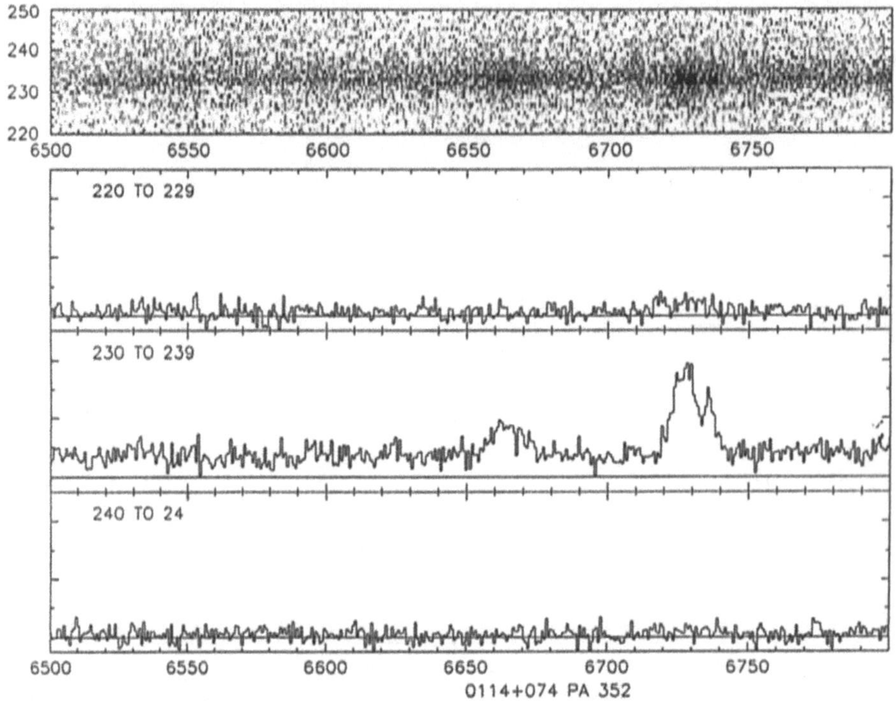

Fig. 3. Greyscale of the CCD frame containing the spectrum of 0114+074S at a position angle of 352°, together with spectral cuts across it. Note the lack of extended emission along this direction, and the clear splitting of the emission line (middle spectrum).

One plausible interpretation (Pedlar et al. 1985) is that the radio jet shocks the ambient gas, and forms a cocoon of excited gas around itself which is then photoionized by the central source. Because of the velocity field of this expanding cylinder of gas, we see split narrow emission lines as the gas cools.

References

Akujor, C.E., 1987: *Astron. J.*, **94**, 867.
Akujor, C.E., 1989: *Astron. J.*, **89**, 1226.
Baum, S.A., Heckman, T., van Breugel, W., 1990: *Astrophys. J. Suppl.*, **74**, 389.
Garrington, S.T., Leahy, J.P., Conway, R.G., Laing, R.A., 1988: *Nature*, **331**, 147.
Meisenheimer, K., Heavens, A.F., 1986: *Nature*, **323**, 419.
Pedlar, A., Dyson, J.E., Unger, S.W., 1985: *Monthly Notices Roy. Astron. Soc.*, **214**, 463.
Tadhunter, C.N., Fosbury, R.A.E., Binette, L., Danziger, I.J., Robinson, A., 1987: *Nature*, **325**, 504.
Tytler, D. Browne, I.W.A., 1985: *Astron. J.*, **90**, 1676.

Emission line maps of the Seyfert-2 galaxies Mkn 3 and Mkn 78

Klaus Anton and Stefan J. Wagner

Landessternwarte, Königstuhl, D-6900 Heidelberg, Germany

1. Observations and Reduction

Narrow band imaging in Hβ, [O III]λ5007 and the adjacent continuum of the Seyfert-2 galaxies Mkn 3 and Mkn 78 have been carried out during two observing campaigns in December 1989 and February 1990 at the Calar Alto observatory. We used the 2.2m telescope and an RCA-CCD mounted at the f/8 Cassegrain focus. The spacial scale was 0.35" per pixel. Reduction procedures included debiasing and dark-subtraction, a procedure to correct for a low-amplitude, non-linear off-set of all columns in the images and flat-fielding.

2. Mkn 3

Mkn 3 is one of the most luminous Seyfert 2 galaxies in the local surroundings. Like NGC 1068 it has fairly broad "Narrow Lines" (FWHM [O III] λ 5007 = 1100 km s^{-1}). Goodrich and Miller found the spectropolarimetric properties of Mkn 3 to be similar to those of NGC 1068. In polarized light the Balmer lines show broad wings. A weak radio jet emerges perpendicular to the major axis of the stellar component. The luminous ENLR has already been described by several authors. In earlier work we found the gas of the ENLR to show clear rotation (Wagner, 1987, Astron. Astrophys. **185**, 77), which indicated that the ENLR is actually a disk of gas rotating about the projected major axis of the galaxy.

Our deep images trace the ENLR gas over a dynamic range of 30,000. We find the same exponential decrease of intensity along the major axis and along the minor axis of the ENLR feature. H β is significantly weaker. The [O III]/Hβ ratio varies only by a small amount from about 11 in the nucleus to 8 for the least luminous parts. We find no significant radial variation of the [O III]/Hβ ratio along the major or minor axis outside the nucleus. Inside the central seeing disk (The FWHM of the seeing was 0.8" in all filters) the [O III]/Hβ ratio is slightly enhanced. In the central 4" the ENLR and the radio jet have identical position angles. Further out the ENLR bends slightly. A similar change in position angle is indicated in the radio-maps of Pedlar et al., 1985, M.N. **238**, 863.

The constancy of the [O III]/Hβ ratio and the spatial extension of the line-emitting gas perpendicular to the long axis poses some problems for the standard model of radiation cones as an explanations of ENLR. Since the gas along the minor axis of the ENLR has the same range and distributions of gas temperature and densities and constant [O III]/Hβ, it is most likely that this gas is ionized by the same continuum radiation. The different spatial extent would then reflect matter-boundedness of the ENLR rather than ionization-boundedness. We therefore interpret the velocity field as rotation rather then bipolar outflow. A likely model for Mkn 3 would therefore be an early-type galaxy, which was essentially gas-free, eventually experienced gas-infall with the gas settling in a disk perpendicular to the main plane and at low inclination towards the line-of sight. This gas is now fully ionized by the nuclear continuum source. Additional dust and remnants of the obscuring torus might still play an additional role.

3. Mkn 78

The host galaxy of this Seyfert-2 galaxy is probably early-type with symmetric isophotes orientated along PA $\approx$ 80°. The peak of the line emission coincides with the center of the galaxy defined by the peak of the continuum emission.

The ENLR seen in [O III] and Hβ is elongated and extends over about 11 kpc with a major axis PA $\approx$ 270°. Two faint filaments of ionized gas seen in [O III] extend out to the west.

As has been reported by previous work (e.g. Sargent, 1972, Ap.J.**173**, 7), the [O III] and the Hβ map of Mkn 78 show a point-like source of ionized gas west of the nucleus which is kinematically (by about +800 km s^{-1}) and spatially (distance about 2.5" $\widehat{=}$ 2.8 kpc) separated from the center of the host galaxy (De Robertis, 1987, Ap.J.**316**, 597). The secondary component is not only visible in our [O III] and Hβ images but also in the [O III]/Hβ -map indicating that the ionization state in non-uniform.

The morphology of the ENLR is very similar to that of the radio emission. The radio continuum emission is elongated in the same direction as the ionized gas with a second component west of the nucleus separated by $\approx$ 2.5". The radio nucleus coincides probably with the center of the host galaxy and the western knot with the corresponding emission line maximum. The secondary component is elongated towards the stronger nuclear source in both maps. Because of the coincidence of the PA and the location of individual point sources seen in the emission line maps and the radio continuum, the ionization of the ENLR and the formation of the linear radio structure seems to have the same origin.

Forbidden High Ionization Lines in Quasars

Immo Appenzeller and Stefan J. Wagner

Landessternwarte, Königstuhl, 6900 Heidelberg, Germany

Abstract. We observed [Fe VII] and [Ne V] lines in a sample of QSOs. As observed in lower-luminosity AGNs, the forbidden high-ionization lines of QSOs tend to be blueshifted and broader than the lower-ionization NLR features. The widths of the forbidden high-ionization lines of AGNs appear to be correlated with the widths of the permitted BLR line components. The observations are consistent with the assumption that the forbidden high-ionization lines are emitted by photoionized gas located somewhere between the BLRs and NLRs.

1. Observations

Between March 12 and March 15, 1989 we used the B&C spectrograph at the 2.2-m MPI telescope of the European Southern Observatory at La Silla, Chile to obtain spectrograms of 10 QSOs. Depending on the redshifts, for each object we observed a wavelength range containing either the [Ne V] λ3426 or the [Fe VII] λ6086 lines (which correspond to very similar ionization potentials) and a comparison region containing lower-ionization forbidden lines. For all frames we used gratings giving a reciprocal dispersion of 59Åmm^{-1}, corresponding to spectral resolutions $R = \lambda/\Delta\lambda = c/\Delta v$ ranging between 2500 in the blue and about 4000 in the red spectral region. The exposure times were chosen to reach S/N values allowing the detection and measurement of emission lines down to an equivalent width of about 0.5 Å.

2. Line Strengths

In the spectra of four of the observed ten objects we found only (3σ) upper limits (ranging between 0.3 Å and 0.7 Å) for the forbidden high-ionization lines ("FHILs") mentioned above. For the remaining six QSOs we measured [Ne V] equivalent width values ranging between 0.4 and 4.7 Å and [Fe VII] equivalent width values ranging between 0.9 Å and 2.5 Å. These values are quite similar to the corresponding equivalent width values observed by us earlier in a sample of lower-luminosity Seyfert nuclei (cf. Appenzeller and Östreicher 1988). On the other hand, our QSO sample showed considerably weaker lower-ionization lines (such as the [O III] lines). Hence in the sample of QSOs observed by us the FHILs appear enhanced relative to the lower-ionization forbidden lines. In one object (PG 1211+143) we find prominent [Fe VII] λ6086 and [Fe X] λ6374 emission but no trace of [O I] $\lambda\lambda$6300 and 6364. The observed [Fe VII]/[Fe X] ratio is found to be in the range predicted for photoionization models by Korista and Ferland (1989).

3. Line Profiles

All observed QSO FHILs showed moderately broad profiles with FWHM values ranging between 562 km s^{-1} and 1189 km s^{-1}. Again following the trend known from lower-luminosity Seyfert nuclei, in all cases where a direct comparison was possible, the FHIL lines were broader (in two cases *much* broader) than the lower-ionization [O III] lines. Moreover, we found the FHILs on average blueshifted (by about 220 km s^{-1}) relative to the low-ionization forbidden lines.

By combining our QSO FHIL line width data with corresponding data for lower-luminosity Seyfert I nuclei (which tend to have less broad permitted lines than QSOs), we find a linear relation between the FWHM values of the FHILs and the FWHM of the broad permitted lines. No such relation is indicated for the lower-ionization [O III] lines. The correlation between the FHIL and the (broad) permitted line widths seems to argue for a common broadening mechanism. If the observed line profiles are caused by orbital motions of the emitting gas in the gravitational potential of the AGN's central object, the observed line widths may indicate that the FHILs are formed at distances from the AGN center intermediate between the NLR and the BLR. The observed blueshift of the FHILs can be explained assuming that the hot gas producing these lines forms a wind where the radiation of the receding (redshifted) part of the flow is weakened by opaque regions of a circumnuclear accretion disk.

References

1. Appenzeller, I., Östreicher, R., 1988: *Astron. J.*, **95**, 45.
2. Korista, K.T., Ferland, G.J., 1989: *Astrophys. J.*, **343**, 678.

4.7 μm Spectroscopy of NGC 1068

Kazuya Ayani[1], *Takao Nakagawa*[2], *and Tetsuya Nagata*[3]

[1] National Astronomical Observatory, Japan
[2] Institute of Space and Astronautical Science, Japan
[3] Kyoto University, Japan

Abstract. The nucleus of NGC 1068 was observed around 4.7 μm in order to detect absorption features by gaseous CO in the molecular torus against the nuclear continuum. The features were not detected. The torus, therefore, must be optically so thick that the true nucleus cannot be directly seen even at 4.7 μm.

The unified scheme of Seyfert galaxies assumes that a geometrically thick molecular torus surrounds the active nucleus (Antonucci and Miller 1985; Krolik and Begelman 1988). In this scheme, the nucleus of a Seyfert 2 galaxy, NGC 1068, is hidden by such a torus in our line of sight. Although the torus must be quite opaque in the optical, Chelli et al. (1987) argues that the nuclear continuum is seen *through* the torus around 4 μm. Such a torus would produce detectable CO (v = 1–0) absorption lines around 4.7 μm. We, therefore, observed the nucleus of NGC 1068 at 4.6 – 4.8 μm. If each absorption line is resolved, direct measurement of the column density and the temperature of the molecular torus can be made.

The observation was carried out with the Cooled-Grating Array Spectrometer (CGAS) mounted on the NASA IRTF 3-m telescope with a spectral resolution of $\lambda/\Delta\lambda \simeq 1300$ and with a $2''\!.7$ aperture. The telluric absorption was corrected by taking ratios of the spectra of NGC 1068 to those of γ Cet (type A2V).

The ratio spectrum of NGC 1068 to γ Cet around 4.7 μm is shown in figure 1. The absorption lines of gaseous CO cannot be recognized in our spectra. This means that the torus, if exist, must be optically too thin or too thick. Assuming that the torus is thin, we derived the upper limits of the column density of the gaseous CO. They are $3 \times 10^{17}\,\mathrm{cm}^{-2}$, $5 \times 10^{17}\,\mathrm{cm}^{-2}$, and $3 \times 10^{18}\,\mathrm{cm}^{-2}$ for the rotational temperature of 20 K, 100 K, and 1000 K, respectively, if the intrinsic line FWHM is $300\,\mathrm{km\,s^{-1}}$. Smaller line FWHM give smaller upper limits. If we adopt $n(\mathrm{CO})/n(\mathrm{H_2}) = 1 \times 10^{-4}$, our results give the upper limits of H_2 column density, $N(\mathrm{H_2}) \lesssim 10^{22}\,\mathrm{cm}^{-2}$ for most cases. This is inconsistent with the argument that the torus is opaque in the visible. Hence we conclude that the torus is optically so thick that most of the observed continuum at 4.7 μm does *not* pass through the torus. This

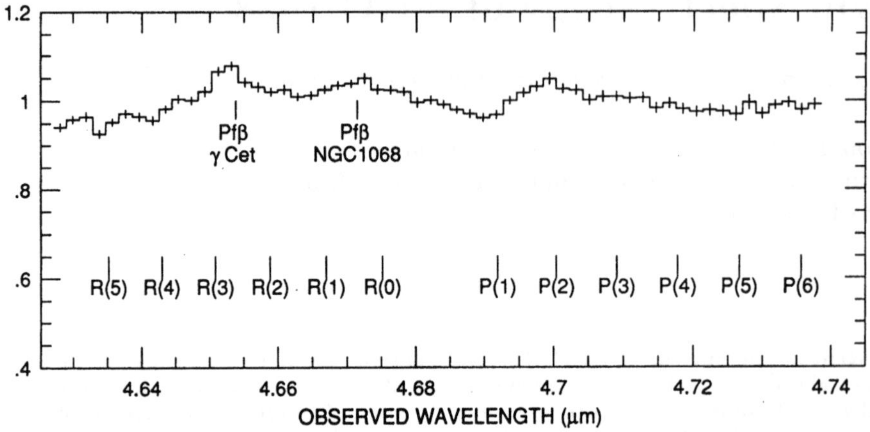

Fig. 1. The ratio spectrum of NGC 1068 to γ Cet. The absorption lines of gaseous CO cannot be recognized. The positions where the gaseous CO absorption lines (both P-branch and R-branch) would appear are indicated in this figure. The emission-like feature at 4.653 μm must be caused by dividing the spectrum of γ Cet in which the Pfβ line (4.6538 μm) should appear in absorption. Another weak emission-like feature at 4.672 μm is probably a Pfβ emission line of NGC 1068. The feature at 4.68 – 4.70 μm might be due to a broad absorption feature of solid CO. The systemic velocity of NGC 1068 is assumed to be 1137 km s^{-1}(Baldwin et al. 1987).

is consistent with the argument that the torus is also optically thick in the X-ray, i.e., the hydrogen column density exceeds 10^{24} cm^{-2} (Krolik and Begelman 1988). The continuum at 4.7 μm must be dominated by thermal emission from the hot dust in the circumnuclear region or the torus of NGC 1068.

References

Antonucci, R.R.J., Miller, J.S., 1985: *Astrophys. J.*, **297**, 621.
Baldwin, J., Wilson, A.E., Whittle, M., 1987: *Astrophys. J.*, **319**, 84.
Chelli, A., Perrier, C., Cruz-González, I., Carrasco, L. 1987, *Astron. Astrophys.*, **177**, 51.
Krolik, J.H., Begelman, M.C., 1988: *Astrophys. J.*, **329**, 702.

Essential Properties of the cD Galaxy in Abell 1795

F.W. Baier, T. Kirchner, and H. Tiersch

Zentralinstitut für Astrophysik, Sternwarte Babelsberg, D-O-1597 Potsdam, Rosa-Luxemburg-Str. 17a

The most promising scenarios for the formation of cD galaxies are galaxy cannibalism and growing by cooling flow. So the ratio of the contributions of both processes is of great interest. Because there seem to be different kinds of cD galaxies (Tonry 1987) we discuss the properties of the central galaxies in connection with the global cluster properties.

The cD galaxy of A 1795 (Zw162010) is one of the most extreme examples with respect to X–ray emission, optical luminosity, colour, radio emission and peculiar velocity. We estimated the luminosity profile in B and V with an excess luminosity relatively to de Vaucouleurs law for the outer region (Baier et al. 1991). The characteristic parameter for this is r_{exc} (i.e. from here the excess luminosity is brighter than the de Vaucouleur's law). For Zw162010 we found $r_{exc} \cong 27$ kpc which is typical for a poor cluster. The positional angle of the central galaxy in both colours (Fig.1) is $10^o < \varphi < 17^o$ which is in agreement with Binggeli's value. The axial ratio reaches from $b/a = 0.7$ at the centre of the galaxy to $b/a = 0.5$ at a radius value of $r = 60$ arcsec. As N-body simulations (Binney 1977) have shown dynamical friction increases the anisotropy of a massive galaxy in a cluster if the initial whole cluster is aspherical because of an anisotropic velocity dispersion. This finding is the basis of our intention to search for a connection between large scale structures of galaxy clusters, the shape of single galaxies, and the structure of the innermost parts of the central galaxy.

From our $B - V$ profiles for the cD galaxy in A 1795 (Fig. 2) we find very blue values in the inner region. This result is in accordance with former observations.

An influence of the cooling flow $400 M_{\odot} yr^{-1}$ (Stewart et al. 1984) on the formation of the cD galaxy in A 1795 is assumed. van Breugel et al. (1984) derived a cooling time $t_{cool} = 2 \times 10^7$ yrs of this gas. During this time a cooler optical emission line gas ($T = 10^4 - 10^5$K) can be formed from the cooling halo gas in the inner parts of Zw162010.

The tangential motion of Zw162010 relative to the cluster from SE to NW can be seen from IPC– and HRI measurements (Morganti et al. 1988)

which show a small difference between the position of the cD galaxy and the centre of the X–ray distribution, i.e. a shift of the galaxy to the NW.

We derived a central antisegregation in luminosity for A 1795 which is considered as a clear evidence for a deficit of bright galaxies in the cluster centre and so a strong cannibalism in the past is favoured. During this process the interstellar material from companion galaxies is thrown out by ram pressure and so the density of the intergalactic gas is increased. Therefore, it promotes the accretion by a cooling flow.

We compared optical and radio properties. The distribution of radio emission in the cD galaxy (Zw162010=4C 26.42) was observed by Thompson et al. (1980) (Fig. 3). The measurement shows a position angle of about 20° east of north which agrees with the orientation of the whole cluster and that of the cD galaxy. The Z–shaped structure of 4C 26.42 was explained by van Breugel et al. as due to a "wobbling" jet ejection axis caused by tidal interactions. Mergers possibly change the orientation of the radio source. It seems to be promising to assume the following history of the radio source: The existence of the pairs P_1, P_1' and P_2, P_2' of regions with polarized radio emission speaks in favour of two early successive plasma ejections in directions 1 and 2. The lines $\overline{P_1 P_1'}$ and $\overline{P_2 P_2'}$ (which corresponds to the direction of all other cluster populations as general galaxy distribution, cD orientation, cooling flow) show a position angle of 20^0 east of north and are parallel. Possibly the latter circumstance corresponds to a simple translation of the galaxy. After this the jet ejection axis changed (perhaps because of a merger) to directions 3 and 4 connected with a simultaneous motion of the galaxy along its way.

The indentation northerly of the region P_3 is presumably caused by the resistance of the interstellar medium against the upstreaming jet material. van Breugel et al. discussed a bright emission line region (ELR) westerly of this small indentation in the northern radio lobe of 4C 26.42 and explained the origin of this ELR as due to a distortion of the radio structure upstream by an existing interstellar material in this region.

These arguments favour an earlier ejection of the jets in a direction of roughly 20^0 east of north and the following change of the jet axis. From the existence of hot spots in different directions it is suggested that the radio engine would have to be intermittent.

Alternatively to the intermittent ejection scenario, the radio morphology of 4C 26.42 can be explained in the frame of the "standard" jet model (Blandford & Rees 1974): According to this model a continuous outflow of hot relativistic plasma from the core of the galaxy takes place.

Sumi & Smarr (1984) found, that jets, emerging through cooling inflow atmospheres, rapidly evolve through parameter space, crossing over into the unstable region, i.e. from an initially stably propagating supersonic jet (edge-brightened, Fanarofff & Riley class II) to a disruptive transonic one (edge-darkened, FR I).

The radio structure of 4C 26.42 can be interpreted as a transition at the inner hot spots (at about 1kpc) from supersonic stable FR II type jets to subsonic unstable FR I type structures (plumes) in an inversion symmetric configuration, which is possibly due to galactic cannibalism (merger). The deviations from a S-shaped configuration seem to stem from a translational motion of the cD galaxy relatively to the intracluster gas.

Therefore, the coincidence of orientation of the radio structure, the cD galaxy and the cluster in the N–S direction was obviously disturbed by an merged object.

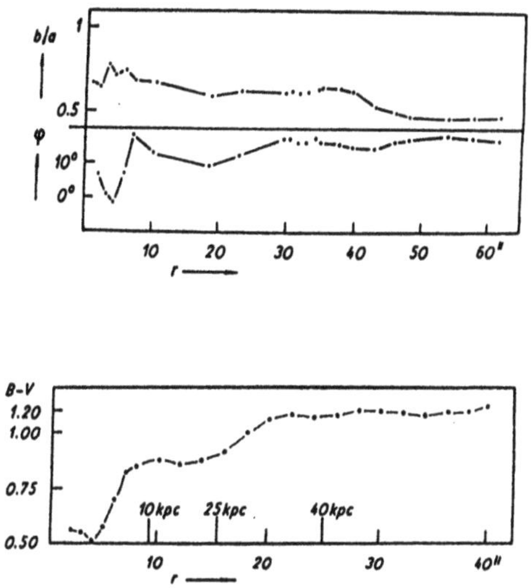

Fig. 1. The axial ratio b/a and the positional angle φ of the cD galaxy. **Fig. 2.**(bottom) Radial colour distribution $B - V$ for the cD galaxy

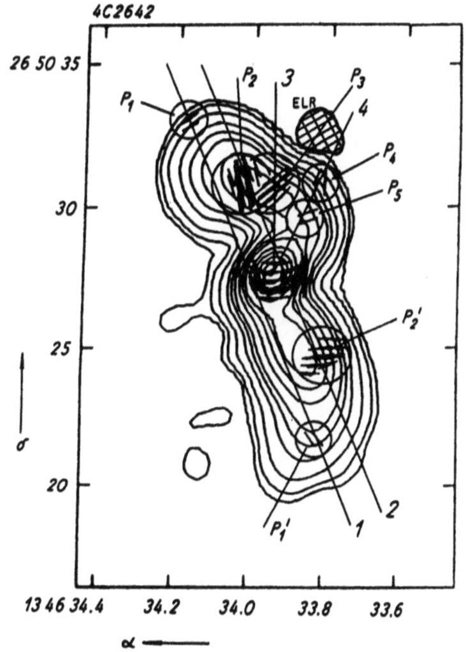

Fig. 3. Contour map of 4C 26.42 at 6 cm with 1."3 × 1."3 resolution with superimposed polarization vectors and emission line region (van Breugel et al. 1984)

References

Baier, F.W., Fritze, K., Tiersch, H., and Ziener, R., 1991: *Astron. Nachr.*, **312**, 149.
Binney, J., 1977: *Monthly Notices Roy. Astron. Soc.*, **181**, 735.
Blandford, R.D., Rees, M.J., 1974: *Monthly Notices Roy. Astron. Soc.*, **169**, 395.
Breugel, W. van, Heckman, T., Miley, G., 1984: *Astrophys. J.*, **276**, 79.
Morganti, R., Fanti, R., Gioia, J.M., Harris, D.E., Parma, P., 1988: *Astron. Astrophys.*, **189**, 11.

Stewart, G., Canizanes, C., Fabian, A., Nulson, P., 1984: *Astrophys. J.*, **278**, 536.
Sumi, D.M., Smarr, L.L., 1984: in: *Physics of Energy Transport in Extragalactic Radio Sources*, eds. A.H. Bridle, J.A. Eilek, NRAO Workshop No.9, p. 168.
Thompson, A.R., Clark, B.G., Wade, C.M., Napier, P.J., 1980: *Astrophys. J. Suppl.*, **44**, 151.
Tonry, J. L., 1987: in: *Structures and Dynamics of Elliptical Galaxies,* ed.: T. de Zeeuw, IAU-Symposium No. **127**, D. Reidel, 1987, p. 89.

'Composite' Ionization Mechanisms in Active Galaxies

B. Boer

Astronomical Institute, Ruhr-University, Postfach 10 21 48, 4630 Bochum, Germany

1. Introduction

The classification of emission-line galaxies as active galactic nucleus (AGN), LINER or starburst galaxy, is generally based on line ratios measured from spatially unresolved spectra. Although in most cases the classification is straightforward, a number of galaxies shows intermediate line ratios, suggesting that their nuclei are subject to a mixture of ionization mechanisms. We present long-slit spectra of three 'composite galactic nuclei' (CGN). All three show spatial variations of the line ratios. In two cases, NGC 2782 and NGC 4102, the existence of two spatially and physically distinct ionization mechanisms could indeed be proven.

2. NGC 2782

From our spatially resolved spectra we find that the intermediate line ratios observed in NGC 2782 are due to an extended high-velocity high-ionization 'cloud' at 4"-8"($\sim$500-1000 h^{-1} pc) south of the center. In this region the spatial distribution of the [OIII]λ5007 line has a second maximum, with approximately the same strength as the central maximum. The intensity distributions of the Hβ, Hα and [NII] lines are not symmetric, but have extended southern wings. Since this intensity enhancement is seen only along the minor axis (PA 180°) and not in spectra at PAs 45°, 90° and 135°, we refer to it as a cloud. The [OIII]λ5007/Hβ ratio reaches values up to $\sim$ 5 in the region of the cloud, whereas it is $\sim$ 1 elsewhere in the nuclear region. The [NII]λ6583/Hα and [SII]$\lambda\lambda$6716+31/Hα ratios are also significantly larger in the south than at other positions. The line ratios of the nuclear region are those of a starburst galaxy, while the level of excitation in the cloud is consistent with shock ionization. Along the major axis (PA 90°), and along PAs 45° and 135° the velocity field is consistent with rotational motions in the plane of the galaxy. Along the minor axis (PA 180°) the [OIII]-lines are blueshifted by up to 100 km s^{-1} in the region from 4"-8" south. In the same region the FWHMs of [OIII] and, to a lesser extent of [NII], increase drastically to ~500 km s^{-1}, from ~150 km s^{-1} in the north. The kinematical behaviour of the cloud is consistent with outflow along the minor axis. Radio maps (Wilson&Willis,1980, ApJ240,429) show, in addition to the central peak, a southern peak coincident with the optical cloud as well as a peak at 5" north, suggesting a bipolar outflow. The receding northern part of the outflow is probably obscured in the optical by the large amount of dust in the plane of the galaxy. Our observations are in good agreement with theoretical models by Tomisaka and Ikeuchi (1988, ApJ330,695) for outflow from a nuclear starburst, which forms a thin dense shell of swept-up material. The shell gradually elongates in the direction of the minor axis, until it finally breaks at the top, and becomes a cone-like 'chimney' through which the hot gas expands freely. At the radial distance from the center where the line-of-sight is tangential to the shell, a large amount of emitting material as well as a large range of velocities is encountered. This explains both the observed extranuclear line intensity maximum at r=6" south, and the large FWHMs at cloud positions. In the free expansion phase, however, the emission comes mainly from gas in the walls of the cone, and the intensity distributions are not likely to have an extranuclear maximum. Since NGC 2782 has an inclination angle of 43°, and the line profiles at r≥3" north do not show high-velocity wings, the expansion within the plane of the galaxy is less than half the expansion perpendicular to the plane of the galaxy. The dynamical time scale for the cloud is t_{cl} ~3 10^6 h^{-1} yr. The shell-geomtery of the TI-models A and B for t=3 10^6 yr is indeed elongated and closed.

3. NGC 4102

A blueshifted highly excited component is also present in NGC 4102. Along the minor axis (PA 135°) the Hα and [NII] lines are blueshifted by ~80 km s^{-1} and [OIII] by ~180 km s^{-1}, relative to the systemic velocity derived from CO and HI data and from symmetrizing the major axis velocity curve. Along the major axis (PA 45°), PA 90° and PA 135°, the Hα- and [NII]-velocities are consistent with rotational motion, but the [OIII] lines are again blueshifted relative to Hα and [NII]. The FWHMs of the [OIII] lines are larger than the FWHMs of Hα at nearly all positions. The gas in the N-NW-W quadrant seems to be highly excited ([OIII]/Hβ ratio ~3), whereas the [OIII]/Hβ ratio is constant at ~1 elsewhere. The [NII]/Hα ([SII]/Hα) ratio is also larger in the same region: ~1.5 (~0.9) compared to ~0.9 (~0.4) elsewhere. Qualitatively the observations can again be explained by a starburst nucleus with outflow along the minor axis causing extranuclear high-excitation.

4. NGC 6764

Since the line emission of NGC 6764 is confined to the central few arcseconds, the spatial variation of the line ratios is difficult to determine. Nonetheless, spatial variation is indicated by the different line ratios derived from spatially integrated spectra at various position angles. E.g., the [NII]/Hα and [SII]/Hα ratios are 0.68($\pm$0.02) and 0.39($\pm$0.01) for PA 73° (major axis), compared to 1.01($\pm$0.01) and 0.71($\pm$0.01) for PA 208°. However, this does not necessarily imply the presence of two ionization mechanisms, since the radiation field of the central source may not be the same in different regions, due to e.g. geometrical dilution or dust. The kinematics give no indication of a high-velocity component in addition to the rotational component.

Spatially Resolved, Kinematic Evidence of Mass Inflow to the Circumnuclear Region (<100 pc) in NGC 1068

Gerald Cecil[1], *Jonathan Bland-Hawthorn*[2], *Steve Lubow*[3]

[1] Dept. of Physics and Astronomy, Univ. of North Carolina
[2] Dept. of Space Physics and Astronomy, Rice University
[3] Space Telescope Sci. Inst. and Johns Hopkins University

NGC 1068 remains one of the most comprehensively studied of the Seyfert galaxy class, not least because of its position on the celestial equator which allows the object to be observed from both hemispheres. However, little is known of the large-scale gas motions in this nearby galaxy and, in particular, their connection with the nuclear activity. Currently, we are using CFHT and CTIO Fabry-Perot observations of NGC 1068 in Hα, [NII]λ6583 and [OIII]λ5007 to produce the most detailed picture at any wavelength of the gas kinematics throughout the inner disk. These same CFHT observations led to the discovery of the extended, diffuse ionized medium within NGC 1068 (Bland-Hawthorn, Sokolowski and Cecil 1991).

Deep broadband visual images of NGC 1068 show an outer θ-shaped ring with inclination, $i \approx 35°$, and position angle, $\alpha \approx 90°$, and a bright central disk with diameter ~20 kpc oriented perpendicular to the outer ring. In Fig. 1, we present a preliminary velocity field for the ionized gas over the inner disk. The velocity field is shown for a field of view of roughly 200″ (1″ $\approx$ 75 pc) and reveals a series of concentric "spiral waves" that appear superimposed on ordered, large-scale rotation. The large-scale disk appears to undergo flat rotation beyond a radius of 1.5 kpc with a kinematic major axis that is consistent with the photometric position angle of the outer ring. Across the inner 30″, the Z-symmetric velocity contours are the hallmark of highly radial orbits along a bar at ~45° for the inferred disc orientation. Thus, the observed elliptic streaming is consistent with the presence of the stellar bar detected at near infrared wavelengths (Scoville *et al.* 1988; Telesco & Decher 1988).

By subtracting the simple model of elliptic streaming overlayed on flat rotation, we emphasize bisymmetric spiral streamers (Fig. 2) in both the ionized line flux and kinematics that penetrate from ~1 kpc to within 100 pc of the AGN. We find evidence for this effect in both the CO line flux and kinematics, albeit at lower spatial resolution (Kaneko *et al.* 1989). By implication, it is likely that the observed non-circular motions are confined to the galactic plane and arise from bar-driven density waves. Recently, in collaboration with Harvey Liszt (National Radio Astronomical Observatory), we find tentative evidence that, while the λ21cm and CO kinematics are largely consistent, the Hα kinematics show substantial deviations from the

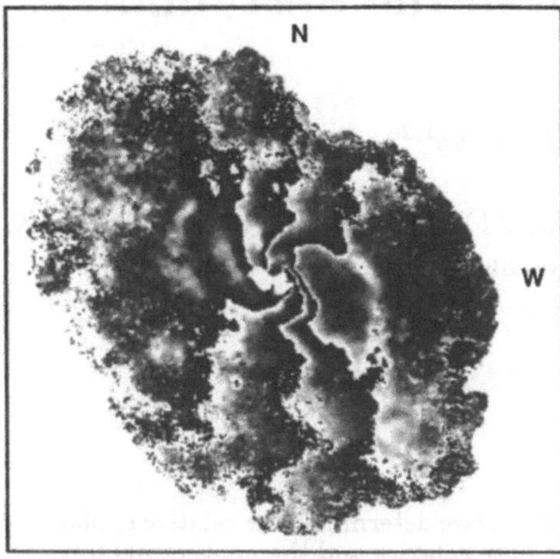

Fig. 1. The ionized gas velocity field in NGC 1068 shown for a ~200″ field of view at spatial and kinematic resolutions of (<1″, ~50 km s^{-1} FWHM) respectively.

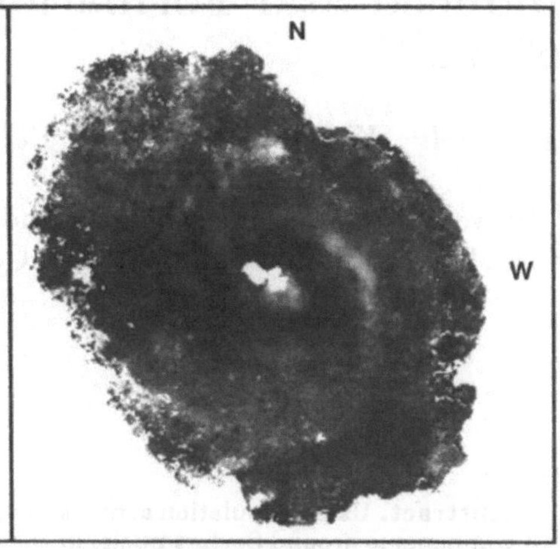

Fig. 2. The residual velocity field in NGC 1068 after subtracting a simple model of flat rotation + elliptic streaming from Fig. 1. The residual motions can reach 50 km s^{-1} in projection with respect to the systemic velocity.

cold phase at many positions within the inner disk away from the spiral streamers.

In the context of standard spiral density waves, the expected large-scale shocks lead to a phase shift between the gas response and driving stellar bar. The resulting mass inflow may dominate standard viscous accretion. On much smaller scales, it is unlikely that such shocks can be sustained. Therefore, the shock mechanism is likely to be important at intermediate scales (0.1 to a few kpc). The mass inflow rate can then be computed in terms of the gas density, galactic rotation parameters and shock strength. Before we publish the outcome of this study, we shall attempt a self-consistent fit to these parameters, particularly the shock strength, which will require gas dynamic modeling. Ultimately, we would like to address the implications of these observations for fueling nuclear activity.

References

Bland-Hawthorn, J., Sokolowski, J., and Cecil, G., 1991: *Astrophys. J.*, **375**,,78.
Kaneko, N., *et al.*, 1989: *Astrophys. J.*, **337**,,691.
Scoville, N.Z., *et al.*, 1988: *Astrophys. J. (Letters)*, **327**,,61.
Telesco, C.M., and Decher, R., 1988: *Astrophys. J.*, **334**,,573.

Induced Star Formation in the Surroundings of AGN

A. Goerdt[1], *K.J. Fricke*[1], *and W. Kollatschny*[1,2]

[1] Universitätssternwarte, Geismarlandstr. 11, D-3400 Göttingen, Germany
[2] Dr. Remeis-Sternwarte, Bamberg, Germany

Abstract. Using population synthesis methods we determined the relative stellar components around Seyfert nuclei in their host galaxies and the nonthermal flux contribution from the nuclei; both are correlated.

We took deep optical 2D-spectra with high spectral and spatial resolution of five nearby face-on Seyfert/Starburst galaxies (Table 1). The spectral energy distributions were synthesised as a function of the distance to the nucleus using a special Simplex algorithm developed by us. The presence of a possible nonthermal component and reddening by dust were taken into account. The stellar library we used for the synthesis contains O5 to M8 main sequence and giant stars.

An observed spectrum of the nucleus of the Seyfert 1 galaxy NGC 1365 together with its synthesised spectrum is shown in Fig.1. We found by our syntheses that the spatial extent of a nonthermal component in Seyfert galaxies peaks at the nuclei. Fig.2 shows the relative light contributions of the stellar OBA and KM populations relative to the FG-stars in NGC 1365 after subtraction of the central nonthermal component. Generally, strong starburst components were found within 2kpc of the AGN superimposed on an old stellar component.

Of our five program galaxies only NGC 1097 shows no evidence for a nonthermal component thus proving its starburst nature. The strength of the circumnuclear starburst is found to be clearly correlated with the strength of the nonthermal continuum flux coming from the AGN (Fig.3) indicating the induced nature of the circumnuclear starburst.

This work has partly been supported by BMFT grant Verbundforschung Astronomie FKZ 50 9004 5

Table 1. Observed galaxies

Name	Hubble-type	AGN-type	z
NGC 1365	SBb	Seyfert 1	0.006
NGC 1566	Sc	Seyfert 1	0.005
NGC 1808	Sbc pec	Seyfert 2	0.003
NGC 3081	SBa	Seyfert 2	0.007
NGC 1097	RSBbc	Starburst/Liner	0.004

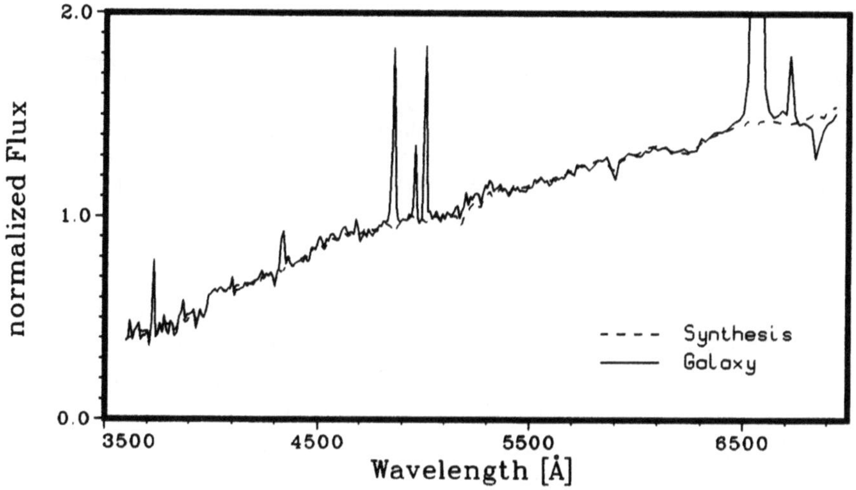

Fig. 1. Observed and synthesised spectra of the nucleus of NGC 1365

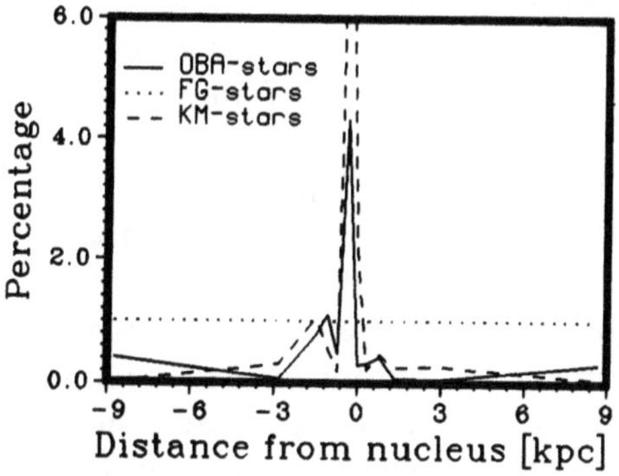

Fig. 2. NGC 1365: Fraction of spectral types relative to FG-stars

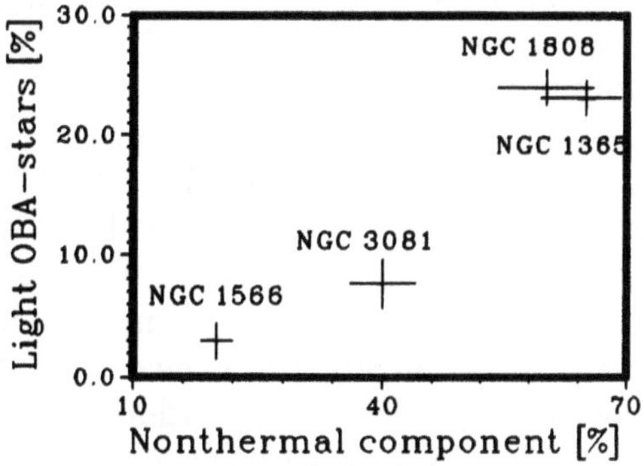

Fig. 3. Starburst-luminosity versus non-thermal nuclear flux

Optical Emission Associated with the Radio Jet in B2 1243+26

J.I. González-Serrano[1] *and I. Pérez-Fournon*[2]

[1] Departamento de Física Moderna, Universidad de Cantabria, Santander, Spain
[2] Instituto de Astrofísica de Canarias, La Laguna, Tenerife, Spain

Abstract. We present deep CCD photometry of the low luminosity radio galaxy B2 1243+26 which contains a two-sided, well defined radio jet at the kpc scale. The galaxy is located in a region of high galaxy density and has a nearby companion of similar luminosity. Close to the nucleus of the radio galaxy, we detect elongated optical emission parallel to the radio jet, presumably associated with it. Also, a resolved source is found at the position of the bright knot at the end of the jet.

We are investigating a sample of low-luminosity radio galaxies with the aim of detecting the optical counterpart of the jets. In this paper we analyse broad-band CCD observations of B2 1243+26 which is located in the Abell cluster A1609 at a distance of 356.4 Mpc (z = 0.0891; H_o=75 km s^{-1} Mpc^{-1}) (Fanti et al. 1987). This galaxy presents a double jet which extends 75 kpc to the NE and SW of the nucleus.

B2 1243+26 was observed using a CCD GEC camera attached to the prime focus of the 2.5m INT at the Observatorio del Roque de los Muchachos (Spain). Several exposures in the V, R, and I filters were obtained. Surface photometry was performed using the procedure described in González-Serrano and Pérez-Fournon (1991). Smooth models of the galaxy light distribution were constructed and subtracted to the original frames in order to detect underlying structures.

Our images show that B2 1243+26 is located in a high galaxy density region. The second brightest galaxy in the field lies at a projected distance of 122 kpc from the radio galaxy (at P.A. 201°) and both have similar luminosity. These galaxies seem to be interacting. Figure 1a shows a contour map of the central part of the original I–image. A contour map of the image obtained by adding the residual frames in V, R, and I is presented in Fig. 1b. Three faint structures, labelled with letters A to C on the Figure, are detected along the direction of the inner radio jet. The resolved source labelled D coincides with the bright knot located at the end of the southern radio jet.

The inner emission (A to C) is elongated and parallel to the radio jet, indicating some physical association. Two different interpretations may explain the optical emission: (1) emission line regions originated by the interaction of the jet with the interstellar medium; (2) stripped matter resulting from the interaction between B2 1243+26 and its companion. Faint extended emission is in fact detected between these two galaxies (Fig. 1a).

Spectral indices from the radio to the visible for source D are α(1.4GHz, I)=0.41, α(I, R)=1.3, and α(R-V)=2.0. These values are similar to those measured in the final knots of the M87 jet (Pérez-Fournon et al. 1988), and imply a cutoff frequency in the range 3.6 – 4.6$\times$ 10^{14} Hz. Although source D could be alternatively a cluster member in view of the high galaxy density in the field, we note that even if the spectral index between radio and optical was $\sim$ 0.6, the knot should be detected in our images.

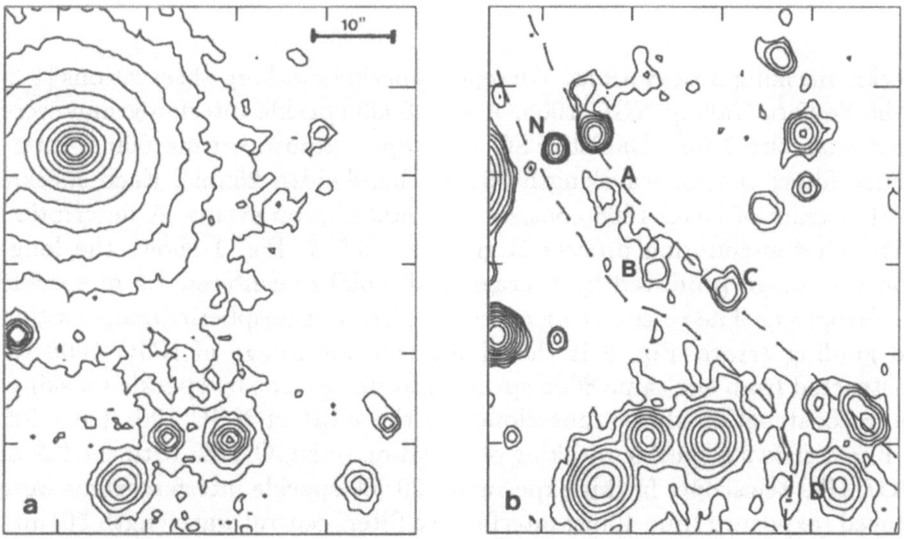

Fig. 1. (a) Contour map of B2 1243+26 in the I band. (b) Contour map of the total (smoothed) V+R+I residual image. The nucleus is indicated with N and the detected features with A, B, C, and D. The dashed line represents the faintest contour at 1.4 GHz (Fanti et al. 1987).

References

Fanti, C., Fanti, R., de Ruiter, H.R., Parma, P., 1987: *Astron. Astrophys. Suppl.*, **69**, 57.
González-Serrano, J.I. and Pérez-Fournon, I., 1991: *Astron. Astrophys.*, in press.
Pérez-Fournon, I., Colina, L., González-Serrano, J.I., and Biermann, P.L., 1988:, *Astrophys. J. (Letters)*, **329**, L81.

High-Resolution Speckle Masking Observations of the Seyfert Galaxy NGC 1068

K.-H. Hofmann, W. Mauder, and G. Weigelt

Max-Planck-Institut für Radioastronomie, Auf dem Hügel 69, 5300 Bonn 1, Germany

Abstract. We show high-resolution speckle masking observations of NGC 1068 in the O III line. The photon-counting speckle interferograms were taken with the 1.5-m Danish/ESO telescope and the 2.2-m ESO/MPG telescope. Several high-resolution images were reconstructed from ~ 28 000 speckle interferograms. The reconstructed images show five individual clouds in the Narrow Line Region of NGC 1068.

Speckle masking observations. We report speckle masking observations [1-3] of the Seyfert Galaxy NGC 1068. About 8000 speckle interferograms were taken with the 1.5-m Danish/ESO telescope (exposure time 60ms; interference filter: central wavelength 501nm, bandwidth 25nm). Each speckle interferogram of NGC 1068 consists of ~ 400 photon events. A description of the photon-counting process is given in Ref. 4. Fig. 1 shows the long-exposure image produced by averaging all 8000 re-centered 1.5-m speckle interferograms The re-centering overcomes both atmospheric image motion and guiding errors. Fig. 2 is the high-resolution image of NGC 1068 reconstructed from the same 8000 speckle interferograms by speckle masking. The reconstruction shows four clouds in the NLR of NGC 1068 (see Ref. 6). Fig. 4 shows a speckle masking observation of NGC 1068 with the 2.2-m ESO/MPG telescope. In this experiment 20 000 speckle interferograms were reduced (exposure time 40ms; interference filter: central wavelength 501nm, bandwidth 15nm). Each speckle interferogram consists of ~ 120 photon events. The long-exposure image is shown in fig. 3. The reconstructed image of NGC 1068 (fig. 4) is very similar to the HST image recorded with the WFPC. The speckle masking reconstruction shows five clouds within a region of 1" diameter (cone opening angle ~ 50° to 60°).

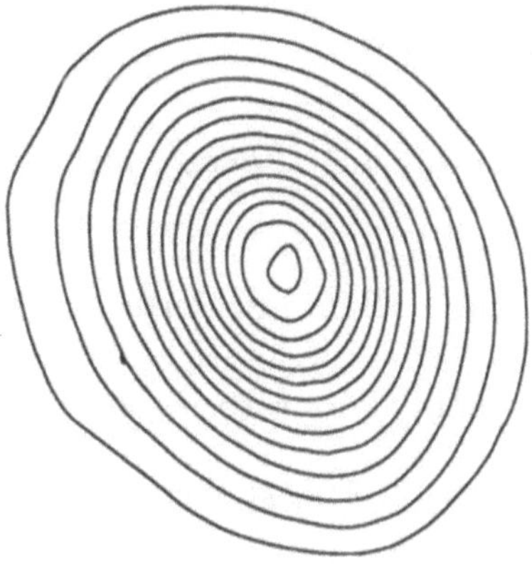

Fig. 1.

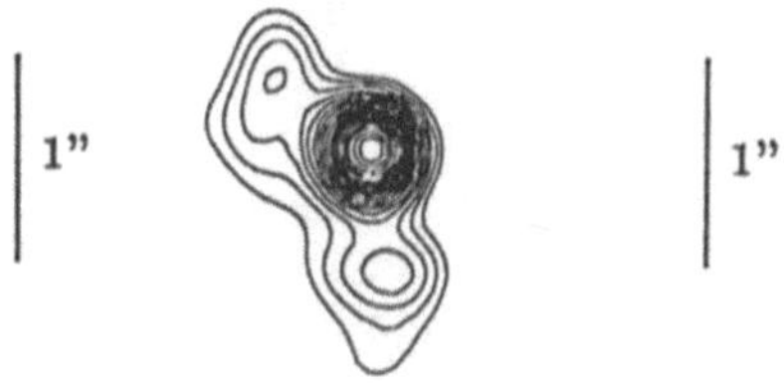

Fig. 2.

Fig. 3.

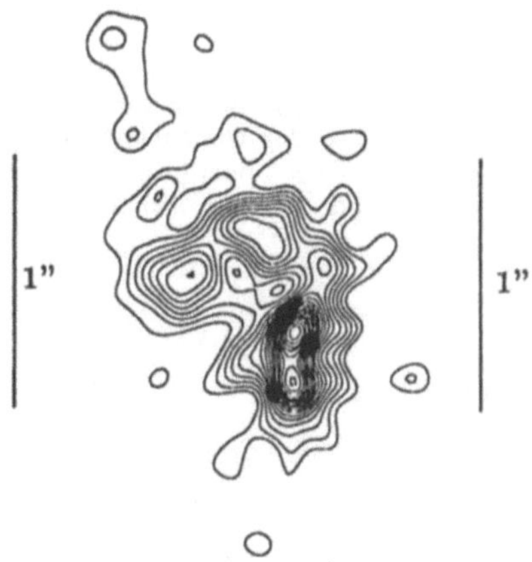

Fig. 4.

References

1 Weigelt, G., 1977: *Opt. Commun.* **21**, 55.
2 Weigelt, G., and Wirnitzer, B., 1983: *Opt. Lett.* **8**, 389.
3 Lohmann, A., Weigelt, G., Wirnitzer, B., 1983: *Appl. Opt.* **22**, 4028.
4 Hofmann, K.-H., Mauder, W., and Weigelt, G., 1990: in: *Photon-Counting Speckle Masking*: Proc. ICO 15, 1990, SPIE Vol. **1319** 442.
5 Hofmann, K.-H., and Weigelt, G.: *Astron. Astrophys.*, **167**, L15.
6 Hofmann, K.-H., Mauder, W., Weigelt, G.: in: *Proc. of the ESO Workshop on Extranuclear Activity in Galaxies*, Garching, 16-18 May 1989, p. 35.

Evidence for an Obscured BLR in the Early-Type Radio Galaxy IC5063

M.D. Inglis[1], *C. Brindle*[1], *J.H. Hough*[1], *D.J. Axon*[2], *J.A. Bailey*[3], *and M.J. Ward*[4]

[1] Division of Physical Sciences, Hatfield Polytechnic, Hatfield, Herts AL10 9AB

[2] University of Manchester, Nuffield Radio Astronomy Laboratories, Jodrell Bank, Macclesfield, Cheshire SK11 9DL

[3] Joint Astronomy Centre, 665 Komohana Street, Hilo, Hawaii 96720, USA

[4] University of Oxford, Astrophysics, Nuclear Physics Laboratory, Keble Road, Oxford OX1 3RD

Abstract. We report the detection of a strong, broad Hα emission line seen in polarized flux for the early-type radio galaxy IC5063. This suggests that the galaxy has a BLR obscured from our direct view with the broad line radiation being scattered into our line of sight by scatterers outside the obscuring regions.

1. Introduction

In this paper we describe new observational results on the active galaxy IC5063 = PKS 2048-572, usually classified as an SO type galaxy, z = 0.0110, with relatively narrow (350 km s^{-1}) emission lines, similar to those of a Seyfert 2 type galaxy. The possibility of a Seyfert 1 type nucleus has been suggested from the detection of a faint broad Hα emission line component (Bergeron, Durret and Boksenberg 1983, Colina et al. 1990) along with an off-centre Narrow Line Region (NLR) (Wagner and Appenzeller 1990).

2. Observations and Reductions

Polarization observations were obtained on the 12th August 1990, using the new waveplate modulator built by Hatfield Polytechnic. This was used with the RGO spectrograph, the 25 cm camera and the Thomson CCD on the 3.9m Anglo-Australian Telescope. A dekker with 2 apertures, each of size 1.3 by 2.7 arc secs, separated by 23 arc seconds was used. The object was switched between the 2 apertures every 16 minutes and the total on chip exposure time was 80 minutes.

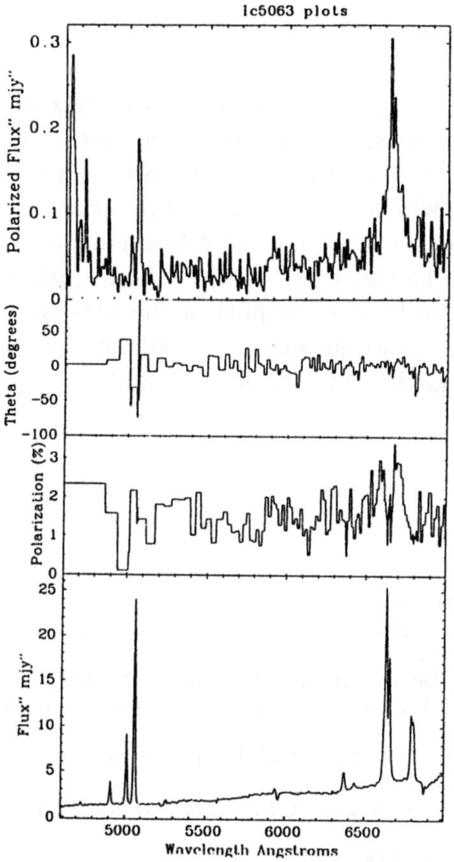

Fig. 1. Spectra of total and polarized flux of IC 5063

3. Discussion

Fig. 1 shows the spectrum of IC5063. The plot shows the polarized flux, the position angle, the percentage polarization (data binned to 0.4 % error per bin) and the total flux. It is immediately apparent that in the polarized flux the Hα line is broadened considerably, with FWZI of at least 9000 km s^{-1}. this is the largest line width detected by such means and is typical of that for Seyfert 1 galaxies. We note that the polarization of the Hα narrow line ($\sim$ 1 %), is less than that of the continuum as found in a number of Seyfert 2 galaxies, e.g. NGC 1068 (Bailey et al. 1988). If the narrow lines are intrinsically unpolarized, then the observed degree of polarization should correspond to interstellar polarization produced by aligned grains in IC5063. Brindle et al. 1990, and Hough et al. 1987 calculate a value of $P_{max} = 1.5$ % and $\lambda_{max} = 0.5 l\mu m$ for the interstellar polarization, consistent with the observed narrow line polarization.

4. Conclusion

Line Region (BLR) in the early-type radio galaxy IC5063. This clearly indicates that scattering is important, at least at optical wavelengths. As discussed by Antonucci and Barvainis (1990), IC5063 is probably similar to the Broad Line radio galaxy 3C 234, which also shows high polarization with the p.a. perpendicular to the radio axis. Antonucci and Barvainis have also discovered other narrow-line radio galaxies with high K-band polarization perpendicular to the radio axis, in further support of the idea that narrow-line radio galaxies are obscured radio quasars or broad line radio galaxies with radio axes viewed at a large angle to the line of sight.

References

Antonucci, R.R.J., and Barvainis, K., 1990, *Astrophys. J.*, **363**, 117
Bailey, J., Axon, D.J., Hough, J.H., Ward, M.J., McLean, I., and Heathcote, S.R., 1988, *Monthly Notices Roy. Astron. Soc.*, **234**, 899
Bergeron, J., Durret, F., and Boksenberg, A., 1983, *Astron. Astrophys.*, **127**, 322
Brindle, C., Hough, J.H., Bailey, J.A., Axon, D.J., and Sparks, W.B., 1990, *Monthly Notices Roy. Astron. Soc.*, **247**, 327
Colina, L., Sparks, W.B., and Macchetto, F.D., 1990, Proc. Cof. Paired and Interacting Galaxies, Ed: J.W. Sulentic, W. Keel, in press
Hough, J.J., Brindle, C., Axon, D.J., Bailey, J., and Sparks, W.B., 1987, *Monthly Notices Roy. Astron. Soc.*, **224**, 1013
Wagner, S. and Appenzeller, I., 1990, *Astron. J.*, **225**, L13.

Measuring the BL Lac Beam in Centaurus A

R.Morganti[1], *A.Robinson*[2], *Z.Tsvetanov*[3], *T.Oosterloo*[4], *and*
R.A.E. Fosbury[5]

[1] Istituto di Radioastronomia, via Irnerio 46, Bologna, Italy
[2] Institute of Astronomy, Madingley Road, Cambridge, UK.
[3] Astronomy Program, University of Maryland, USA
[4] European Southern Observatory, Garching bei München, Germany
[5] Space Telescope–European Coordinating Facility, ESO, Germany

Centaurus A, the closest Fanaroff-Riley Class I radio galaxy, has a system of emission line filaments associated with its NE radio lobe which extends from 7 to 20 kpc from the nucleus (assuming a distance of 3.7Mpc for Cen A). In a recent study, Morganti *etal.* (1991) showed that the filaments are photoionized by a continuum source with spectral properties characteristic of AGN and further argued that the source is located in the heavily obscured nucleus and is probably relativistically beamed towards the filaments. The inferred beam power is comparable to the luminosities of BL Lac objects, providing strong support, if correct, for unified schemes in which FRI radio galaxies form the parent population of BL Lac objects where the relativistic beaming axis is randomly oriented with respect to our line of sight (Browne 1983, Ulrich 1989, Padovani & Urry 1990, Urry & Padovani 1991).

Here we present striking new evidence for a "blazar beam" in Centaurus A in the form of narrow band images in [OIII]λ5007 and Hα+[NII] obtained with the ESO 3.5m NTT. The [OIII]/Hα+[NII] ratio map (Fig.1) displays ionization structure which is clearly related to the direction of the incoming photon stream: the [OIII]/Hα+[NII] ratio decreasing sharply across the filament from the edge facing the nucleus ([OIII]/Hα+[NII] $\sim$ 2.8) to the opposite side ([OIII]/Hα+[NII] $\sim$ 0.4). This gradient is naturally explained by the ionization stratification which would result if the filaments are illuminated from the direction of the active nucleus.

A lower limit to the ionizing photon luminosity received by the filaments may be obtained by balancing the photon budget. The observed Hα flux then implies $q \geq 10^{53}$ photons s^{-1} sr^{-1}, in good agreement with the luminosity (q$\geq$ $10^{54\pm1}$ photons s^{-1} sr^{-1}) which Morganti *etal.* found to be necessary to account for the ionization state of the gas. The greatest source of uncertainty in their analysis arises from the rather loose constraints on the gas density. In order to obtain a better estimate of n_e, new high dispersion spectra were taken with EMMI at the ESO 3.5m NTT in the region of [SII]$\lambda\lambda$6717,6731 ($\sim$ 1Å) and [OII]$\lambda\lambda$3726,3729 ($\sim$ 0.3Å). The mean values

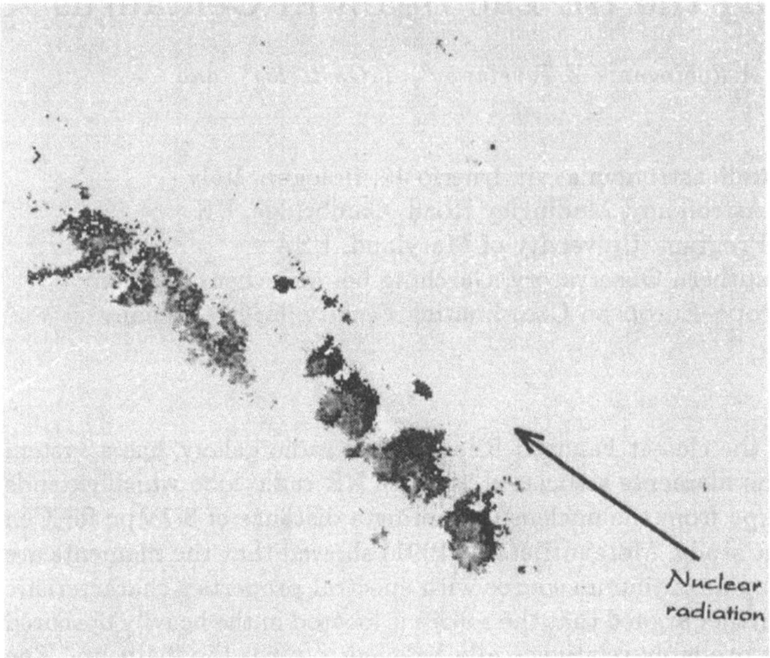

Fig. 1. Ionization map pf part of the inner optical filament derived from [OIII]λ5007 and Hα+[NII] images. The ratio varies from ~2.8 (black) to ~0.4 (light grey). The direction to the centre of Cen A is marked.

obtained for these ratios (1.53 ± 0.14 and 1.47 ± 0.18 respectively) confirm that the gas density in the filament is, on average, < 100 cm^{-3}.

These new results thus provide confirmation that the ionizing radiation field in Centaurus A is strongly beamed in the direction of the filaments with a luminosity similar to that of BL Lac.

References

Browne, I.W.A., 1983. *Monthly Notices Roy. Astron. Soc.*, **204**, 23.
Morganti, R., Robinson, A., Fosbury, R.A.E., di Serego Alighieri, S., Tadhunter, C.N., Malin, D.F., 1991. *Monthly Notices Roy. Astron. Soc.*, **249**, 91.
Padovani, P. & Urry, C.M., 1990. *Astrophys. J.*, **356**, 75.
Ulrich, M.-H. 1989. In *"BL Lac Objects"* (Eds. Maraschi, L., Maccacaro, T. & Ulrich, M.-H.) p.45.
Urry, C.M. & Padovani, P. 1991. In *"Variability of Active Galactic Nuclei"* (eds. Miller, H.R. and Wiita, P.J.) in press.

Spectroscopy of the Nucleus in 3 C 120

T.W.B. Muxlow[1], *D.J.Axon*[1], *A. Pedlar*[1], *M.J. Ward*[2], *and S. W. Unger*[3]

[1] Nuffield Radio Astronomy Laboratories, Jodrell Bank, University of Manchester.
[2] Institute of Astronomy, University of Cambridge.
[3] Royal Greenwich Observatory, Cambridge.

1. Introduction

Superluminal motion, indicative of highly relativistic bulk flows, is now well-established in the nuclear jets in many active galaxies and quasars (see e.g. Porcas, 1987). Such motion is seen in the inner regions of the one–sided arcsecond jet with measured velocities up to 2.5 mas yr^{-1} corresponding to $\sim 4h^{-1}c$ (Benson et al., 1988, H_0=100 h km s^{-1} Mpc^{-1}). The interim results presented here are part of a study at several epochs using the IDS at the f /15 Cassegrain focus of the 2.5 m Isaac Newton Telescope using the IPCS detector covering the spectral lines of Hβ, λ4861, and [OIII], $\lambda\lambda$4959, 5007. Earlier epochs, already published (Axon et al., 1989) have found split [OIII] emission associated with the one–sided radio jet which they interpret as evidence for strong shocks from a jet–interstellar medium interaction.

2. The Observations

The results presented here are partially processed scans from observations in November 1990. Four position-angles have been selected and are shown in Figure 1 together with L-Band radio images of the outer structure from the VLA with an angular resolution of 12 arcseconds and the inner jet region from MERLIN with an angular resolution of 0.35 arcseconds. Regions where strong split–line [OIII] emission is seen are marked with a thick line between tick marks. Within these regions the cross-sections have been integrated together to form the spectra displayed in Figure 2.

3. Discussion

Although [OIII] emission is found elsewhere in the nuclear region of 3C 120, we find no evidence for **strong split–velocity systems** except associated with the one–sided radio jet and South East quadrant within a few arcseconds of the core. The spectra have a cross–section separation of about 0.6 arcseconds. The [OIII] line is broadened in the nucleus itself although no separate

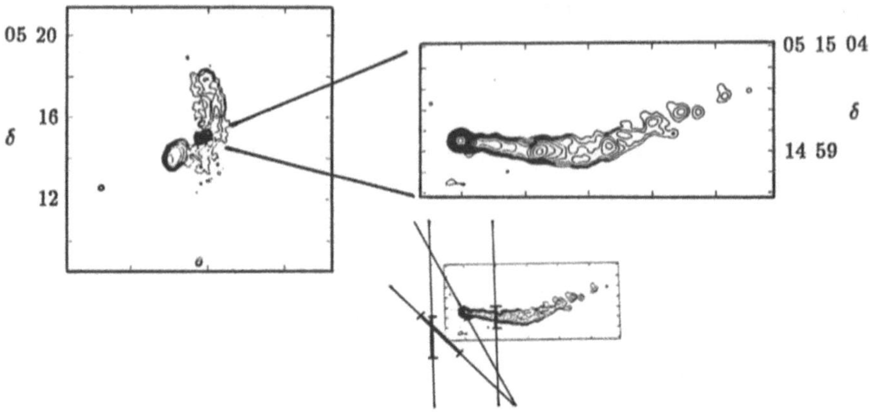

Fig. 1. Example position angle cuts overlayed on the radio structure of 3C 120.

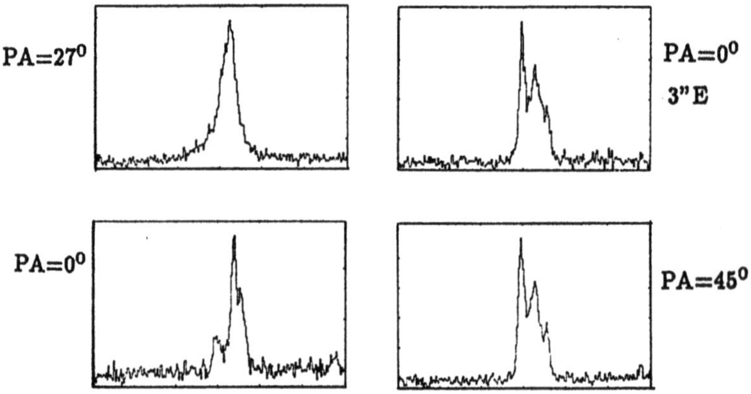

Fig. 2. Integrated spectra for those position angles shown in Figure 1.

components can be resolved. The scan in PA 0 3" West shows the well established split velocity system as it crosses the radio jet. The scan in PA 0 3" East again shows strong velocity splitting and appears to show an almost complete velocity ring indicating that we may be detecting shocked gas blasted radially away from an interaction with an invisible counter–jet. The scan in PA 45 3" South–East shows very strong multiple velocity emission together with an underlying systemic rotation. The velocity distribution along this scan together with an example fit to the multiple velocity profile, (cross–section 64) are shown in Figure 3.

4. Conclusions

It is clear that strong split–line [OIII] emission is a valuable diagnostic of an interaction between a radio jet and the ambient medium. Our [OIII] detection on the counter–jet side in the position angle joining the nucleus to

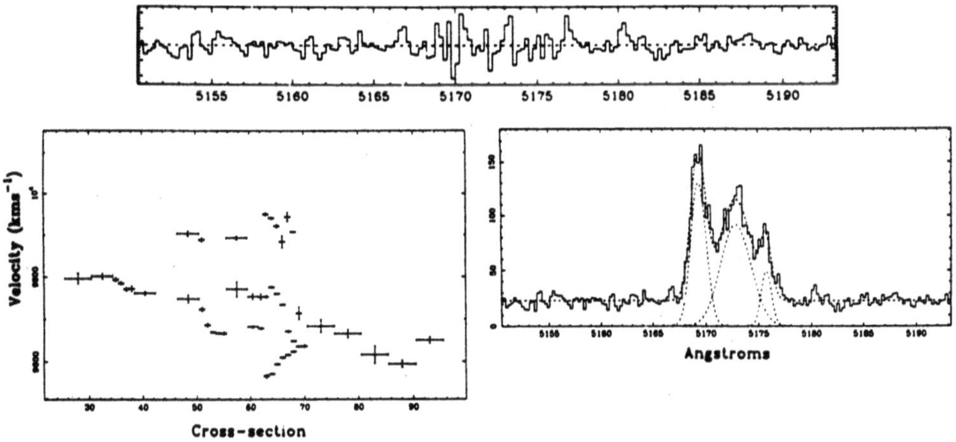

Fig. 3. Velocity distribution and fit for scan in position angle 45 degrees 3" south–east of the core.

the outer radio structure indicates that some form of interaction is occurring. Whether this is due to the blockage and disruption of a previous counter–jet, or whether this is evidence for the passage of an invisible jet is unclear. The sharp bending seen in the radio structure together with the observed super–luminal motion indicates rapid motion with alignment close to the line of sight. If the counter–jet was blocked in this region we might well expect to find some extended region of radio emission in this quadrant. Thus far there is no evidence for this in the high angular resolution images made of the nucleus of 3C 120.

References

Axon, D.J., Unger, S.W., Pedlar, A., Meurs, E.J.A., Whittle, D.M., and Ward, M.J., 1989, *Nature*, **341**, 631

Porcas, R.W. 1987, in *Superluminal Radio Sources,* ed. J.A. Zensus and T.J. Pearson, Cambridge University Press, p. 12.

Benson, J.M., Walker, R.C., Unwin, S.C., Muxlow, T.W.B., Wilkinson, P.N., Booth, R.S., Pilbratt, G., and Simon, R.S. 1988, *Astrophys. J.*, **334**, 560

Walker, R.C., Walker, M.A., and Benson, J.M. 1988, *Astrophys. J.*, **335**, 668

Muxlow, T.W.B., and Wilkinson, P.N., 1991, *Monthly Notices Roy. Astron. Soc.*, in press..

Galaxy Clustering around B2 Quasars

I. Pérez–Fournon[1], *M. Serra-Ricart*[1], *J.I. González-Serrano*[2], *L. Gregorini*[3], *L. Padrielli*[3] *and H. de Ruiter*[3]

[1] Instituto de Astrofísica de Canarias, Tenerife, Spain.
[2] Dpto. de Física Moderna, Universidad de Cantabria, Santander, Spain.
[3] Istituto di Radioastronomia, Bologna, Italy.

We report on CCD imaging and low dispersion spectroscopy at the Nordic Optical Telescope (NOT) and at the William Herschel Telescope (WHT) of a complete sample of radio loud quasars (and quasar candidates) selected from the Bologna (B2) catalogue of radio sources. The B2 quasar sample consists of 72 objects with UV excess, complete down to $m_{ph} = 21$ and $S_{408MHz} = 0.25$ Jy. It was extensively observed with the WSRT at 6 and 21 cm and with the VLA at 6 cm (Rogora et al. 1986, 1987, and references therein). The radio structures have shown that a relation exists between the largest angular size and the off-axis angle of the core with respect to the line connecting the hot spots of the lobes, the more distorted sources having smaller angular size. Actually, Barthel et al. (1988) noted that high redshift quasars seem to be more distorted than quasars at low redshifts, which suggests that the radio structure is epoch-dependent.

Our preliminary results from the high-resolution CCD images obtained with the NOT and WHT reveal an enhancement of the galaxy density near some of the B2 quasars (Fig. 1) which can be interpreted as clusters of galaxies associated with the quasars or as foreground clusters acting as gravitational lenses of the intrinsically fainter quasars. The first interpretation supports the presence of the rich intracluster medium needed to explain the distorted radio structures. On the other hand, in a few fields there is evidence of gravitational interaction of the quasars with nearby companions or optical emission associated with extended radio structures (Fig. 1). This result is similar to the alignment between optical and radio structures found in high redshift radio galaxies. Possible interpretations include: a) star formation triggered by the interaction of the radio jet with a dense intergalactic medium, and b) scattered nuclear anisotropic radiation.

The details of the analysis of the radio/optical data will be published elsewhere (Serra-Ricart et al. 1992) and will also address the following questions: a) cosmological evolution of radio structure, b) dependence of radio structure on luminosity, and c) dependence of radio structure on parent object.

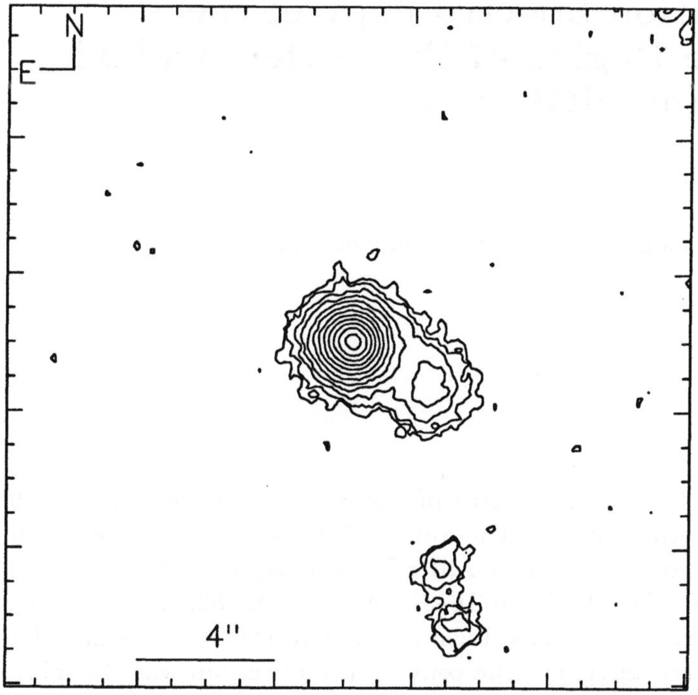

Fig. 1. Contour map of CCD image (r filter) of the field containing B2 1213+30B obtained with the WHT. Three faint, resolved objects are detected at less than 10" from the quasar position. The object ~ 3" SW of the quasar is spatially associated with the extended radio emission found by Rogora et al. (1987).

References

Barthel, P.D., et al., 1988: *Astron. Astrophys. Suppl.*, **73**, 515.
Rogora, A., et al., 1986: *Astron. Astrophys. Suppl.*, **64**, 557.
Rogora, A., et al., 1987: *Astron. Astrophys. Suppl.*, **67**, 267.
Serra-Ricart, M., et al., 1992: in preparation.

High Resolution Spectroscopy of the Narrow Line Region of the Seyfert Galaxies NGC 7469 and NGC 1566

P. Rafanelli and P. Belloni

Department of Astronomy, University of Padova, Italy

1. Introduction

High resolution (0.45-0.60 Å) spectra of the Seyfert galaxies NGC 7469 and NGC 1566 covering the spectral range 3500-8000 Å were taken at the ESO 3.6 m telescope using the CASPEC spectrograph. The profiles of the emission lines Hα, Hβ, [SII]$\lambda\lambda$6716–31, [NII]$\lambda\lambda$6548–83, [FeVII]λ6087, [OIII]$\lambda\lambda$4959–5007 were decomposed using gaussian profiles by means of a nonlinear least-squares algoritm. The convergence criterium was the minimization of the reduced χ^2.

The kinematical parameters (FWHM,V–V_{sys}) of the single gaussian components isolated in this way were then used to investigate the structure of the nuclear and neighboring regions of both galaxies and to verify if they were correlated with their critical density N_c and ionization potential I.P..

The valuable diagnostic line ratios [OIII]λ5007/Hβ, [SII]$\lambda\lambda$6716-31/Hα and [NII]λ6583/Hα show that in the two galaxies both components of the narrow lines are ionized by a nonthermal source.

2. The results of the analysis

In NGC 7469 (a Seyfert 1.5 galaxy) the contour maps of the emission lines [NII]$\lambda\lambda$6548–83 and [SII]$\lambda\lambda$6716–31 and of the narrow components of Hα and Hβ show two peaks having an angular separation of 0.8 arcsec (as measured in Hα and [NII]) and a difference in velocity $\Delta v \sim 160\ km\ sec^{-1}$. The structure of the broad Balmer emission is better investigated through the profile of Hβ which is far from other lines, Fig. 1. In addition to the the double narrow component, the asymmetry towards the red of the broad underlying profile is well reproduced with two gaussian profiles with a larger FWHM. The contour maps of the [OIII]$\lambda\lambda$4959–5007 and [FeVII]λ6086 lines show an evident asymmetry towards the blue but no sign of double peaks. Their profile can be fitted assuming the existence of two narrow components superposed to a blueshifted semibroad one.

In NGC 7469 the single gaussian components of the narrow lines have FWHM and (V-V_{sys}) well correlated with the I.P and with the N_c (FWHM increasing with I.P. and N_c, V-V_{sys} decreasing for increasing values of I.P. and N_c).

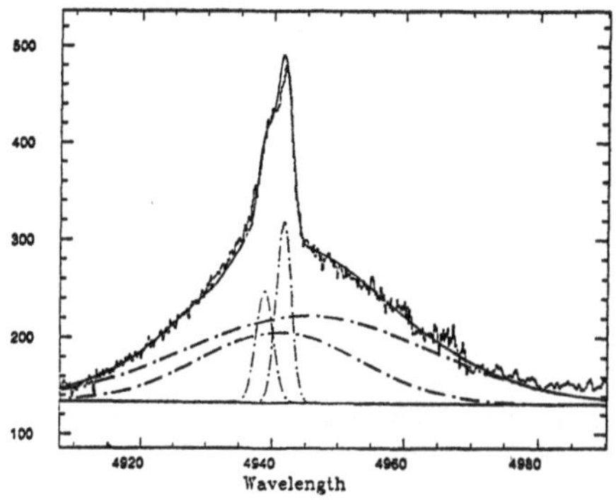

Fig. 1.

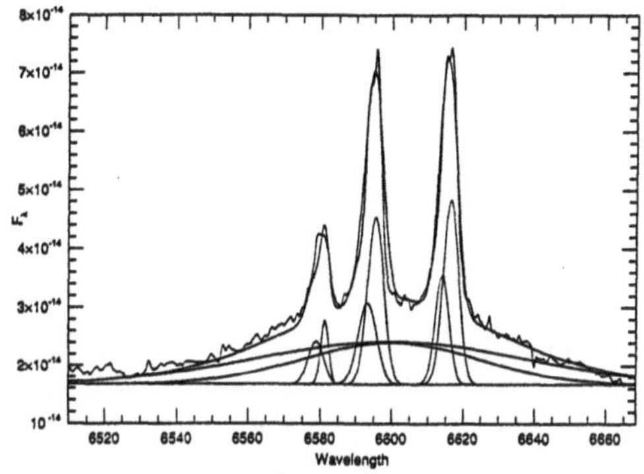

Fig. 2.

NGC 1566 is a Seyfert galaxy of type changing from S 1.9 to S 1.2. (Alloin D. *et al.*,1985); at the time of our observations, Aug 1984, the galaxy was of type S 1.9, namely without any sign of broad component under Hβ.

The contour maps of [OIII]$\lambda\lambda$4959–5007,[NII]$\lambda\lambda$6548–83 and of the narrow components of Hα show the presence of two knots whose velocity separation is $\Delta v \sim 80$ $km\ sec^{-1}$ in Hα and [NII] and $\Delta v \sim 130$ $km\ sec^{-1}$ in [OIII]. In the [OIII] lines it is also evident an asymmetry towards the blue. The deconvolution of the profiles of the Hα and [NII] lines is shown in Fig. 2. The broad component under the Hα profile is fitted by two gaussians the narrower of which is probably due to the presence of a semibroad component of Hα. The asymmetry of the [OIII]$\lambda\lambda$4959, 5007 flat topped profiles is reproduced also by two narrow components and by a semibroad one.

In NGC 1566 the two components of the low ionization lines do not show any sign of correlation between line widths, V–V_{sys}, N_c and I.P. This peculiar behavior has been even observed in NGC 3081 and in NGC 5506 (Appenzeller and Östreicher, 1988), both Seyfert–2 galaxies namely very similar to the class S 1.9 of NGC 1566.

References

Alloin D. *et al.* : 1985, *Astrophys. J.*, **288**, 205
Appenzeller I., and Östreicher R.: 1988, *Astron. J.*, **95**, 45.

IRAS Observations of UCM galaxies

M. Rego[1], *J. Zamorano*[1], *and M. Cordero-Gracia*[2]

[1] Departamento de Astrofísica, Universidad Complutense, Madrid
[2] ETSI Aeronáuticos, Universidad Politécnica, Madrid

Abstract. The far-infrared propierties of a sample of 97 galaxies selected from an Hα objetive-prism survey have been analyzed using IRAS data. The intensity of the radiation heating the dust grains have been estimated and the infrared propierties studied. A three component model was applied in order to stablish the cirrus, starburst and power-law relative contribution.

We have begun a preliminary study of the properties of a sample of UCM galaxies with IRAS counterparts with the goal of discriminating among the different classes of galaxies and understanding the physical processes responsible of the infrared emission. The sample selected for study was identified from Schmidt plates taken with the Schmidt telescope at Calar Alto Observatory (Rego et al. 1989).

As has been suggested by several investigators (Salzer (1989)), the galaxies selected by its relevant IR luminosity frequently present starburst nuclei. In fact one might consider the IRAS survey as equivalent to an objetive-prism survey sensitive to Hα emission.

Models including the different contributions to the observed fluxes have been performed by different authors (Sekiguchi (1987), Rowan-Robinson et al. (1989)). The specific model developped here overlaps with that performed by the last authors. However there are several differences. We considered the IRAS fluxes as the result of the contribution of three components, cool, warm and power-law. The cool component is heated by UV and blue photons , the last are originated by old stars. The density of both colors photons beeing the same, the radiative transfer equation of dust provided a temperature of T=35 K. The warm component is heated by photons supplied by the starburst nucleus and has been parametrized as a black body with temperatures ranging from 70 to 120 K. The non-thermal component is represented by an α=0.7 power-law continuum. To obtain the relative contributions, the flux uncertainties was taken into account.

To evaluate this model, a sample has been selected taken characteristic objets from different samples, excluding those showing higher IRAS flux uncertainties. Only the last criterion has been used to pick up UCM galaxies. As an example, the component contribution fractions for a warm temperature of 120 K is given in the Table.

References

IRAS Point Catalog (V2), 1986, Joint IRAS Science Working Group (Washington: US Gov. Printing Office)
IRAS Catalog and Atlases Explanatory Suppl., 1985, ed. Beichman, C.A., Neugebauer, G. (Washington: US Gov. Printing Office)
Rego, M., Zamorano, J., Gonzalez-Riestra, R., 1989: *Astron. Astrophys. Suppl.*, **79**, 443. Astrophys. Suppl. 79, 443
Rowan-Robinson, M., Crawford, J., 1989: *Monthly Notices Roy. Astron. Soc.*, **238**, 523.
Salzer, J.J., McAlpine, G.M., 1989: *Astrophys. J. Suppl.*, **70**, 447.
Sekiguchi, K., 1987: *Astrophys. J.*, **316**, 145.

A Study of Sérsic-Pastoriza Galaxies

D.J. Saikia[1], *A. Pedlar*[2], *G.J. Yates*[2], *P. Collison*[2], *D.J. Axon*[2], *and S.W. Unger*[3]

[1] Tata Institute of Fundamental Research, National Centre for Radio Astrophysics, Poona University Campus, Post Bag No. 3, Ganeshkhind, Pune 411007, India
[2] University of Manchester, Nuffield Radio Astronomy Laboratories, Jodrell Bank, Macclesfield, Cheshire SK11 9DL, UK
[3] Royal Greenwich Observatory, Madingley Road, Cambridge CB3 0EZ, UK

The Sérsic-Pastoriza or S-P galaxies have morphologically peculiar nuclei and include the well-known hot-spot systems. These galaxies sometimes exhibit both starburst and Seyfert activity which makes them particularly interesting for studying their kinematics, evolution and possible relationships between the two forms of activity. We have made a high-resolution radio survey of a large number of these galaxies which reveal compact components likely to be SNRs, HII regions, galactic nuclear components, or perhaps intense regions of emission which may be completely resolved out in observations of significantly higher angular resolution.

The observations reveal a wide variety of structures which range from a reasonably collinear structure with a compact, central component and S-shaped structures to those with diffuse extended emission with compact components superimposed on it. For a subsample of these sources for which we have made scaled array observations with the VLA A and B arrays at λ20 and 6cm, the extended emission appears to have a steep radio spectrum ($\alpha >$ 0.5, where $S \propto \nu^{-\alpha}$). The extended non-thermal emission is likely to be the result of a burst of star formation during which period relativistic electrons are produced and diffuse out into the interstellar medium of the galaxy. These electrons interact with the interstellar magnetic field, producing the observed radio emission.

The total energy in the compact components is similar to those found in Seyfert galaxies by Unger et al. (1986) and in a sample of Sbc galaxies by Vila et al. (1990). The median value of the total energy is about 10^{52} ergs. The radio properties show signs of activity in these galaxies which are similar to those of Seyferts and LINERs. In a few of these galaxies we have found evidence of compact components which are likely to be supernova remnanats. One of the best examples is NGC 1808 which is an archetypal starburst galaxy in the southern hemisphere (Saikia et al. 1990). In the next page we present some of the results of our radio observations.

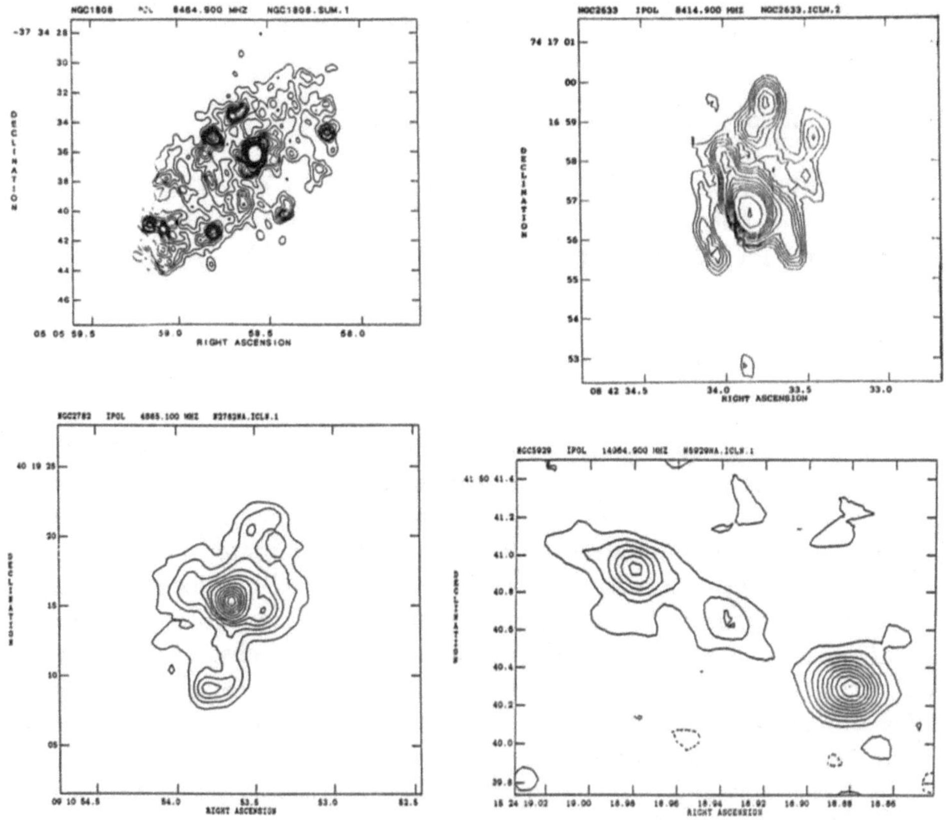

Fig. 1. High-resolution radio images of NGC 1808 and NGC 2633 at 8 GHz, NGC 2782 at 5 GHz and the Seyfert galaxy NGC 5929 at 15 GHz.

References

Saikia, D.J., Unger, S.W., Pedlar, A., Yates, G.J., Axon, D.J., Wolstencroft, R.D., Taylor, K. and Gyldenkerne, K., 1990: *Monthly Notices Roy. Astron. Soc.*, **245**, 397.

Unger, S.W., Pedlar, A., Booler, R.V. & Harrison, B.A., 1986: *Monthly Notices Roy. Astron. Soc.*, **219**, 387.

Vila, M.B., Pedlar, A., Davies, R.D., Hummel, E. & Axon, D.J., 1990: *Monthly Notices Roy. Astron. Soc.*, **242**, 379.

Quasar Jets and Off-Nuclear Emission-Line Gas

P. Shastri[1], *L. Brown*[2], *D. Wills*[1], *and B.J. Wills*[1]

[1] McDonald Observatory, University of Texas, Austin TX, USA
[2] Laboratory of Astronomy & Solar Physics, NASA/Goddard Space Flight Center, Greenbelt MD, USA

We present preliminary results on the connection between kpc-scale radio jets and the extended emission-line region in quasars, using narrow-band optical imaging of the off-nuclear [OIII] emission and high-fidelity radio imaging. That core-dominated (CDQ) and lobe- dominated (LDQ) radio-loud quasars differ only with respect to orientation (Blandford & Konigl, 1979) has been disputed by the finding that the two classes differ in their off-nuclear emission-line properties (Boroson et al. 1985; Stockton & MacKenty, 1987 ApJ 316, 584: SM). Unlike these studies, our final sample consists of CDQs and LDQs matched in their lobe luminosities, which will allow us to meaningfully confront the above hypothesis.

The [OIII] imaging is being done on the 2.7m telescope at McDonald Observatory, using the Goddard Astronomical Fabry-Perot Imaging Camera (Oliversen et al. 175th AAS Meeting, Washington, 1990) that has a tunable etalon that covers a wide wavelength range, and sets wavelengths accurate to within an Å. (bandpass used: ~25Å). The 2 images were obtained under good seeing conditions with exposures of ~1 hr.

1028+313 (Fig. 1): We have discovered an emission line region of $\sim 12''$ extent, implying a projected linear size of ~47 kpc (H_o=50 km s^{-1} Mpc^{-1}, q=0.5). It is far from isotropic and strongly reminiscent of spiral arms, but is not seen in the nearby-continuum image. Overlaid is the 1.4GHz radio image of the quasar (Gower & Hutchings, 1984, AJ 89, 1658). The radio jet is well-aligned with the direction of separation of the [OIII] extensions from the nucleus. It is interesting that the radio jet is on the side of the *fainter* [OIII] lobe, because if the surface-brightness asymmetry of the radio jets is not due to Doppler enhancement of the approaching jet, then it is expected that the more dissipative, i.e., brighter jet would be on the side of higher gas density.

1100+772 (Fig. 2): Our 8GHz VLA radio image is shown superimposed on the [OIII] image. The [OIII] emission structure again shows extensions (also cf. SM). The direction of the radio axis is within $\sim 30°$ of that of the [OIII] emission. SM had suggested that the [OIII] structures could be tidal

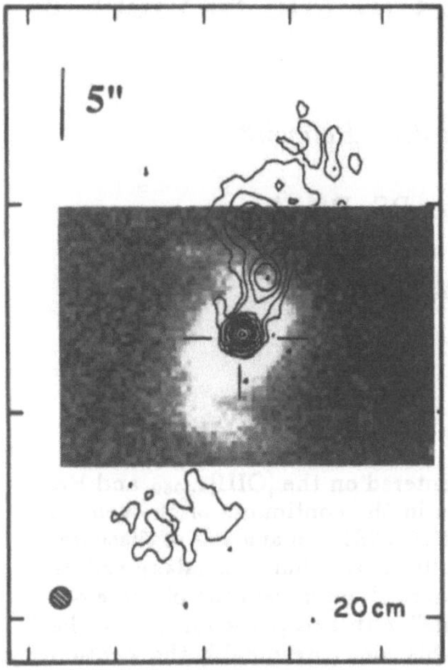

Fig. 1. 1028+313: CDQ

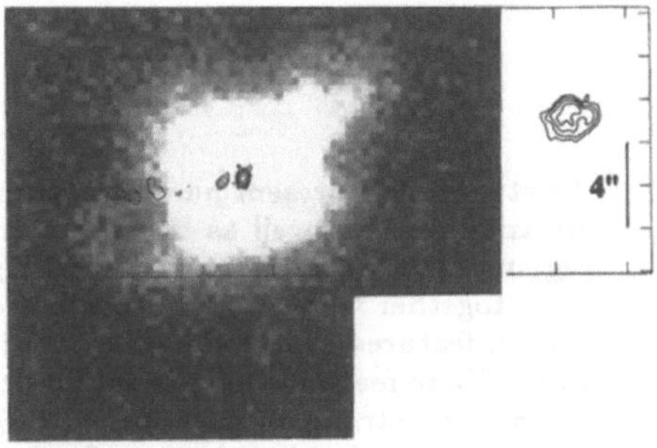

Fig. 2. 1100+772: LDQ

tails associated with interacting/merging galaxies. It is interesting though that the faint curved south-eastern [OIII] tail is parallel to the adjacent curved radio jet.

This work was supported by NSF grant AST-8714937. The National Radio Astronomy Observatory is operated by Associated Universities Inc. under contract with the National Science Foundation.

High Latitude Excited Gas in the Seyfert 2 Galaxy NGC 3281

T. Storchi-Bergmann[1,2], *A.S. Wilson*[2] *and J.A. Baldwin*[3]

[1] Instituto de Fisica, UFRGS, Porto Alegre, RS, Brasil
[2] Astronomy Program, University of Maryland, MD, USA
[3] Cerro Tololo Inter-American Observatory, Chile

Abstract. We present narrow band images centered on the [OIII]$_{\lambda4959}$ and Hα emission lines as well as broader band images in the continuum of the central region of the highly inclined Seyfert 2 galaxy NGC 3281. An analysis of these images, together with long-slit spectroscopic results, shows that this galaxy exibits several features expected in the "obscured Seyfert 1" interpretation of these objects. These features include an "ionization cone" with axis perpendicular to the stellar disk, strong obscuration at the apex of this cone (presumably the location of the true nucleus) and an outflowing wind within the cone.

1. Introduction

A currently popular model of Seyfert 2 galaxies involves a nuclear ionizing source and broad line region hidden behind an obscuring torus (Antonucci and Miller, 1985). This torus shadows the escaping ionizing radiation into a conical morphology. Krolik and Begelman (1986) have proposed a model in which the nuclear continuum drives a wind off the inner edge of the torus. Conical emission-line morphologies have been found in several Seyfert 2's (Haniff, Ward and Wilson, 1991 and references therein). Here we present images of the Seyfert 2 galaxy NGC 3281, which seems to fit the above model, showing an ionization cone, a dusty torus and a wind. The cone is seen in the pure [OIII] image, being similar to the ones in NGC 1365 and NGC 7582 (Storchi-Bergmann and Bonatto, 1991).

2. Discussion

CCD images and long-slit spectroscopy were obtained with the 4m and 1.5m telescopes of the Cerro Tololo Inter-American Observatory and were processed in the usual way.

We show in figure 1a the continuum subtracted Hα + $[NII]\lambda\lambda$6548,6583 image superposed on the green continuum (λ5461) image. Figure 1b presents the pure [OIII]λ4959 image (continuum subtracted) superposed on an "obscuration map". This obscuration map was constructed as follows: from long-slit spectroscopy, it has been possible to derive the characteristics of the stellar population in the central parts of NGC 3281 (Storchi-Bergmann, Wilson and Baldwin, 1991; SWB, from now on). We find that, on the basis of the stellar spectral synthesis, the intrinsic flux ratio between the red and green continuum images should be essentially equal to unity. The ratio of the red and green continuum images may then be used as a measure of the continuum obscuration. We also show in the figures, as a reference, the peak of the green continuum light.

The first thing that can readily be seen in the images is the conical morphology of the [OIII] emission to the NE of the nucleus. This morphology can also be seen, but somewhat less clearly, in $H\alpha + [NII]$. The NE side of the stellar disk is the far one, so this cone, with axis roughly perpendicular to the disk, suffers little obscuration. The conical shape is confirmed by the detection of double peaks in the [OIII] and $H\beta$ emission lines in spectroscopic observations of the cone region (SWB).

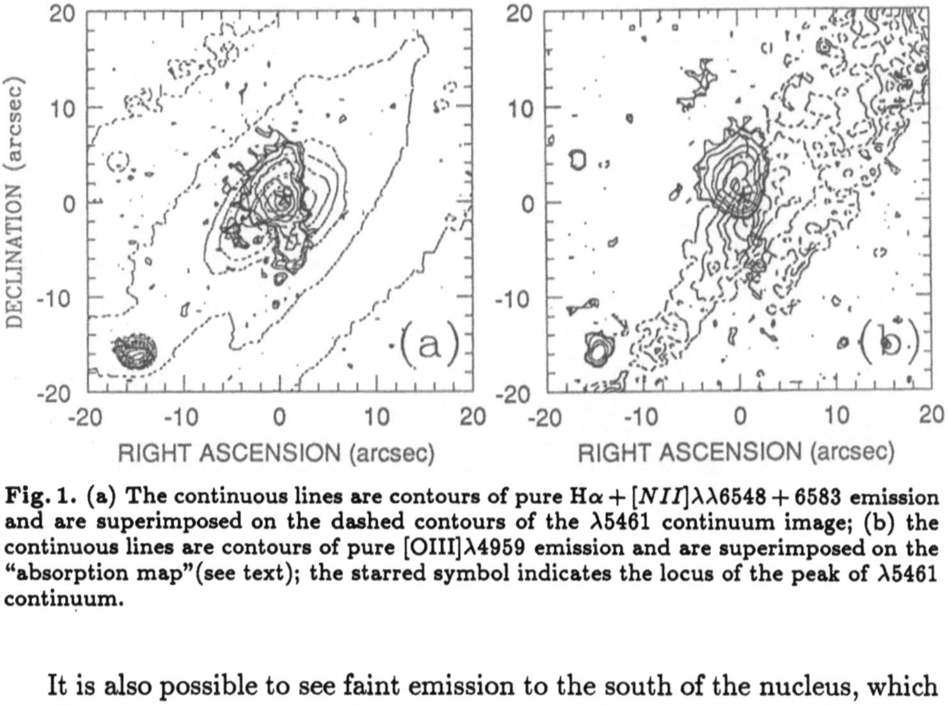

Fig. 1. (a) The continuous lines are contours of pure $H\alpha + [NII]\lambda\lambda 6548 + 6583$ emission and are superimposed on the dashed contours of the $\lambda 5461$ continuum image; (b) the continuous lines are contours of pure [OIII]$\lambda 4959$ emission and are superimposed on the "absorption map"(see text); the starred symbol indicates the locus of the peak of $\lambda 5461$ continuum.

It is also possible to see faint emission to the south of the nucleus, which could originate in a cone on the opposite side of the nucleus, partially hidden by obscuration in the disk. When we compare the [OIII] emission with the obscuration map an interesting result is found: the apex of the cone coincides with the locus of highest obscuration! This is a strong piece of evidence that the true nucleus of the galaxy does not coincide with the peak of continuum light, but is instead hidden behind the obscuration in the galaxy disk (perhaps in the form of a torus).

We have used the long-slit spectroscopy of the [OIII]–$H\beta$ region to study the kinematics of the gas. After subtraction of a model for the rotation of the galaxy (SWB), we find that most of the gas to the NE is blueshifted, with velocities of about 100 km/s, and characterizing an outflow from the nuclear region.

References

Antonucci, R.R.J., Miller, J.S., 1985: *Astrophys. J.*, **297**, 621.
Haniff, C.A., Wilson, A.S., Ward, M., 1991: *Astrophys. J.*, **368**, 168.
Krolik, J.H., Begelman, M.C., 1986: *Astrophys. J. (Letters)*, **308**, L55.
Storchi-Bergmann, T., Bonatto, C.J., 1991: *Monthly Notices Roy. Astron. Soc.*, **250**, 138.
Storchi-Bergmann, T., Wilson, A.S., Baldwin, J., 1991: in preparation. (SWB)

Physical Conditions in the Extended Narrow Line Region of Mkn 573

Zlatan Tsvetanov

Astronomy Program, University of Maryland

1. Introduction

Mkn 573 has received much attention in recent years. Narrow band Hα and [OIII] λ5007 images of the object (Haniff, Wilson & Ward 1988) reveal an Extended Narrow Line Region (ENLR) elongated approximately 13 arcsec in the direction of the radio axis at PA 125° (Ulvestad & Wilson 1984). The line ratio $\Re$ = [OIII] λ5007/(Hα+[NII]) map published by Haniff, Wilson & Ward (1991 – HWW91) suggests the presense of two oppositely directed cones of ionizing radiation, which is in general agreement with our results (Tsvetanov et al. 1989) showing that the maximum of the line ratio $\Re$ is shifted ~4 arcsec to the SE of the nucleus.

Mkn 573 was observed as part of a program to study the ionizing radiation field in nearby Seyferts. Detailed analysis of the entire observational material is reported elsewhere (Tsvetanov & Walsh 1991). Here I discuss some of the most important results concerning the excitation gradients and angular dependence of the ionizing radiation.

2. Observations and results

The ENLR of Mkn 573 was mapped spectroscopicaly by a grid of five 1.5 arcsec wide slit positions covering the wavelength range of 5000–7000 Å with a scale of 3.8 Å px^{-1} along the dispersion and 0.675 arcsec px^{-1} along the slit. The estimated seeing was 1.3 arcsec. Emission line intensities were measured from the Gaussians fitted to the line profiles at each spatial increment along the slit. The absolute flux calibration is accurate to $\leq$20% while the accuracy of the relative line intensities varies with the strength of the lines, and is less than 10% for the brightest lines and much worse for the weakest cases.

Several important diagnostic lines are measured throughout the ENLR. Line ratios [NII] $\lambda\lambda$6548,83/Hα, [SII] $\lambda\lambda$6716,31/Hα, [OI] λ6300/Hα and to lesser extent [OI] λ6300/[OIII] λ5007 show similar behaviour with distance, both radially and along the axis at PA 125°. In general, these ratios increase with distance up to ~2 arcsec and then decrease further out. The general trend for the [OIII] λ5007/Hα ratio is to increase outwards and there is a strong peak at ~4 arcsec SE of the nucleus, which coincides with the peak in the excitation map derived from the images. The density, as estimated from the [SII] line ratio, decreases outwards especially beyond 2 arcsec from the nucleus.

3. Discussion

Line ratios throughout the ENLR of Mkn 573 are typical of gas ionized by a hard nonthermal spectrum (eg., Veilleux & Osterbrock 1987). To get an estimate of the ionization parameter U (the ratio of ionizing photon number density over the hydrogen number density) as a function of the distance from the source we have compared the observed line ratios with photoionization models computed through the multipurpose code MAPPINGS (cf. Binette, Robinson & Courvoisier 1988). The distribution of the points in the line diagnostic diagrams is well explained by excitation trends. Density gradients, though undoubtly present, may account for some of the spread of the points, but can not produce a fit alone. The fact that we see basically an U sequence rules out any significant contribution to the ionization by young hot stars.

Fig. 1 shows the radial dependence of the ionization parameter U as determined from our best fit to the line ratio diagnostic diagrams. A clear tendency of U rising with distance is seen in this figure. It is important to note that similar behaviour has been found in other objects as well (Colina, Sparks & Macchetto 1991; HWW91; Pogge 1989). The observed rising outwards excitation is best explained by a density gradient. The spread of the points in the density dependence is rather high, but the distribution is consistent with a profile of an isothermal, hydrostatic atmospher – $N(r) = N(r_0)[1+(r/r_0)^2]^{-3\beta/2}$, where r_0 is the core radius and β is a parameter – which naturally leads to an increasing ionizing parameter at larger distances and we favour this interpretation.

Fig. 2 presents a plot of the number of ionizing photons, Q_{ion}, *vs* angle θ from the axis at PA 125°. Q_{ion} is determined from the estimate of the ionizing parameter U combined with the [SII] doublet density measurements. There is a clear tendency of the ionizing photon luminosity being higher along the axis than in the perpendicular direction. Given the accuracy and spatial resolution of the data we can only estimate that the "polar diagram" has a general triangular shape with FWHM~80°, which compares well with the cone opening angle as determined from the analysis of the velocity pattern (Tsvetanov & Walsh 1991).

The picture emerging from both the observations and from the modelling can be explained in terms of the Unified Schemes, which as applied to Seyfert galaxies requires that ionizing radiation is emitted from the nucleus into two wide oppositely directed cones. The overall elongated shape of the ENLR

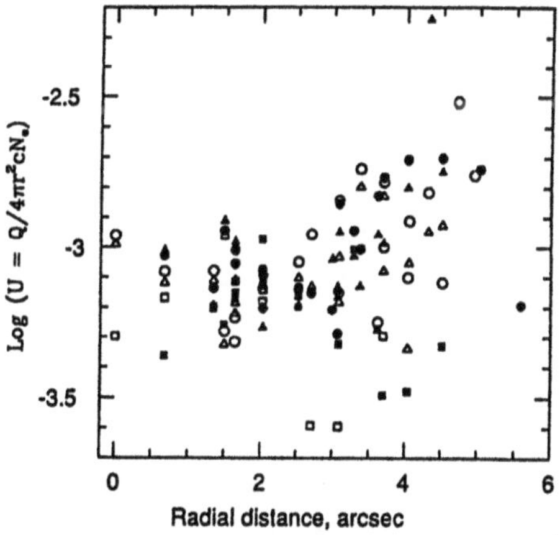

Fig. 1. Ionization parameter U *vs.* radial distance. Circles, squares and triangles correspond to U obtained through [NII] $\lambda\lambda$6548,83/Hα, [OI] λ6300/Hα and [SII] $\lambda\lambda$6716,31/Hα line ratios, resp.

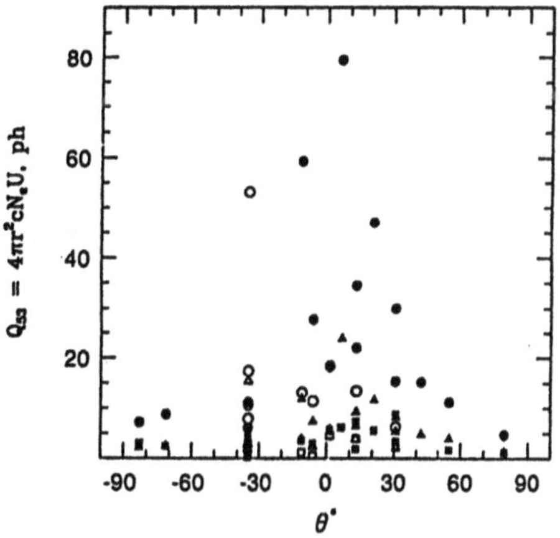

Fig. 2. The number of ionizing photons Q in units of 10^{53} s^{-1} *vs.* angle θ from the ENLR/radio axis at PA 125°. All symbols have the same meaning as in Fig. 1.

of Mkn 573 is consistent with a biconical ionizing fieldfff illuminating the ISM of the host galaxy with the line of sight close to the edge, but within the cone opening angle. This is in accordance with the presence of the high ionization lines in the nuclear spectrum as well as with the the small photon budget deficit (if any) found by Kinney et al. (1991).

References

Binette, L., Robinson, A., & Courvoisier, T.J.-L. 1988, *Astron. Astrophys.*, **194**, 65

Colina, L., Sparks, W.B, & Macchetto, F. 1991, *Astrophys. J.*, , in press.

Haniff, C.A., Wilson, A.S., & Ward, M.J. 1988, *Astrophys. J.*, **334**, 104

Haniff, C.A., Wilson, A.S., & Ward, M.J. 1991, *Astrophys. J.*, **368**, 168

Kinney, A.L., Antonnucci, R.R.J., Ward, M.J., Wilson, A.S., & Whittle, M., 1991, *Astrophys. J.*, , in press.

Pogge, R. 1989, *Astrophys. J.*, **345**, 730

Tsvetanov, Z., Tadhunter, C., Pereź, E., & González-Delgado, R. 1989, in *Extranuclear Activity in Galaxies*, ed. E.J.A. Meurs and R.A.E. Fosbury, (Garching: ESO), p.19.

Tsvetanov, Z.I., & Walsh, J.R 1991, *Astrophys. J.*, , submitted.

Ulvestad, J.S. & Wilson, A.S. 1984, *Astrophys. J.*, **285**, 439

Veilleux, S., & Osterbrock, D.E. 1987, *Astrophys. J. Suppl.*, **63**, 295.

Gas and Dust in NGC 1275

Balthasar Vila-Vilaró[1], *Ismael Pérez-Fournon*[1], *and José I. González-Serrano*[2]

[1] Instituto de Astrofísica de Canarias, Tenerife, Spain.
[2] Dpto. de Física Moderna, Universidad de Cantabria, Santander, Spain.

NGC 1275, the cD galaxy at the center of the Perseus cluster, shows strong extranuclear emission line systems at two discrete redshifts (Minkowski 1957). The Low Velocity System (LVS, at 5300 km/s) has been interpreted as condensations from the cooling flow. The High Velocity System (HVS, at 8200 km/s) is considered a late type infalling galaxy lying in front of the cD (Rubin *et al.* 1978). Some issues to be solved for this object are if the LVS and HVS are interacting, if this interaction could explain the high Hα luminosity of the LVS (Hu *et al.* 1983) and if there is star formation associated with the cooling flow.

Observations: high-resolution Hα and [OIII] $\lambda\lambda$5007, 4959 CCD images for both LVS and HVS and continuum images were taken with the 4.2m WHT + TAURUS at La Palma. Long-slit spectra at four P.A. through the nucleus were taken with the 2.5m INT (La Palma).

Results: the apparent avoidance of the line emitting regions of both systems could be the result of high extinction of the LVS by the foreground HVS galaxy, and does not imply any interaction. However, there are some hints on this in our velocity curves; in the NW along P.A. 293° those positions with higher recession velocities also show hig her velocity dispersions (see Figs. 1 and 2) and are located in local enhancements of the line flux. The dip in the LVS velocity close to the nucleus in the NW (also present in P.A. 311°) could be due to interaction, off-center ordered rotation or just a radial motion (Vila-Vilaró *et al.* 1991). Our broad-band colour maps show some regions of recent star formation: a large region (kpc scale) with a typical spectrum of A stars SE of the nucleus (Rubin *et al.* 1978) and the region of massive star formation detected by Shields and Filippenko (1990) in the NE. Our spectra of the first zone indicate an age for the region of $\geq 10^7$yr and a mass of $\leq 10^9$ M_o (Vila-Vilaró *et al.* 1991).

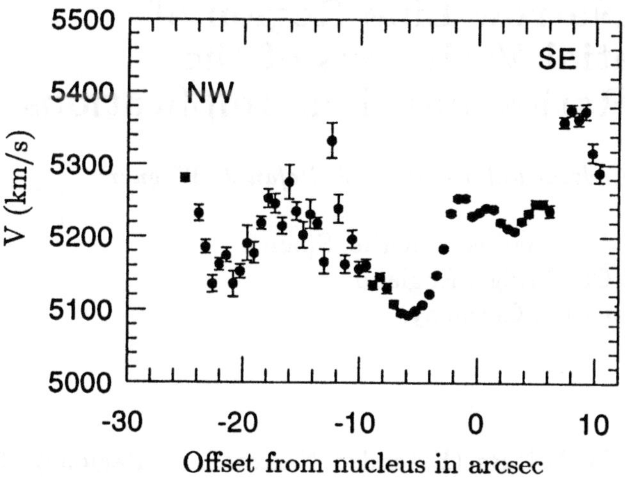

Fig. 1. Velocity curve of the LVS in P.A. 293° (the HVS is in the NW).

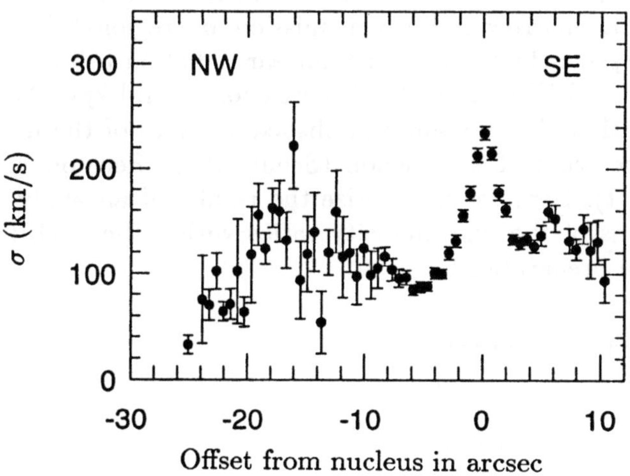

Fig. 2. Velocity dispersion of the LVS in P.A. 293°.

References

Hu, E.M., *et al.*, 1983: *Astrophys. J. (Letters)*, **275**, LL27.
Minkowski, R., 1957: in *IAU Symp No.* 104, p. 107.
Rubin, V.C., *et al.*, 1978: *Astrophys. J. Suppl.*, **37**, 325.
Shields, J.C., and Filippenko, A.V., 1990: *Astrophys. J. (Letters)*, **353**, L7.
Vila-Vilaró, B., Pérez-Fournon, I., and González-Serrano, J.I. 1991, in preparation.

The Extended Narrow Line Region of NGC 4151: Spatial Variations of the Emission Line Ratios and their Implications

Balthasar Vila-Vilaró[1], *Andrew Robinson*[2], *and Stefan J. Wagner*[3]

[1] Instituto de Astrofísica de Canarias, Tenerife, Spain.
[2] Institute of Astronomy, Cambridge, England.
[3] Landessternwarte, Königstuhl, Germany.

The physical picture of the ENLRs (Extended Narrow Line Regions) of Seyfert galaxies is that they are kinematically undisturbed gas in the disk of the host galaxy photoionized by ionizing UV radiation from the active nucleus (Unger *et al.* 1987). Since they are spatially resolved (and easier to model than the "classical" narrow and broad emission line regions) they can be used to trace the "polar diagram" of the nuclear radiation field.

The brightest part of the ENLR in NGC 4151 extends several kpc SW of the nucleus. Several authors have already published analyses of the integrated emission line spectrum of this region (Schulz 1988, Penston *et al.* 1990, hereafter **Paper I**). Here we summarize the results of an analysis of the *spatial* variations of the line intensity ratios within the ENLR (Vila-Vilaró *et al.* 1991, in preparation).

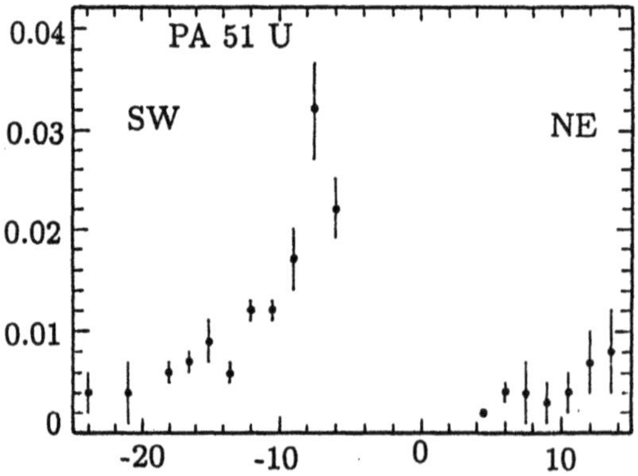

Fig. 1. **U for PA 51, horizontal units in arcsec from center**

Observations: Several spectra were obtained at La Palma Observatory in several slit positions parallel and perpendicular to the major axis of the ENLR. These include spectra at PA 51 along the ENLR through the nucleus and at PA 141, through a bright knot in the ENLR 11" SW of the nucleus, obtained with the 2.5m INT using the IDS and IPCS at a resolution $\delta\lambda$ =1Å, together with spectra at PA135, 5" SW of the nucleus, obtained with the 4.2m WHT using ISIS and an EEV CCD.

Results: The emission line spectrum of the ENLR is characteristic of photoionization by an AGN continuum; for example, a power-law ionizing continuum ($F_\nu \propto \nu^{-\alpha}$) with spectral index $\alpha = 1.6 \pm 0.1$ is required to produce the observed HeII/Hβ ratio. An examination of the classical density diagnostics ([OII] and, in particular, [SII]) suggests that the density of the emitting gas is roughly constant at ~ 200 cm^{-3} along the radial direction (PA51) in the SW. The value of the ionization parameter in the ENLR was inferred from the [OII]λ3727/[OIII]λ5007 ratio using the empyrical formula (**Paper I**) $\log U = -2.74 - \log(I_{[OII]}/I_{[OIII]})$ (Figs.1,2). We note that along PA141 U decreases steeply outwards from the main ENLR axis while it also decreases systematically with radial distance from the nucleus (along PA51). In fact, in the radial direction, the decrease in U to the SW appears to follow an r^{-2} law, consistent with photoionization by a geometrically diluted nuclear continuum source.

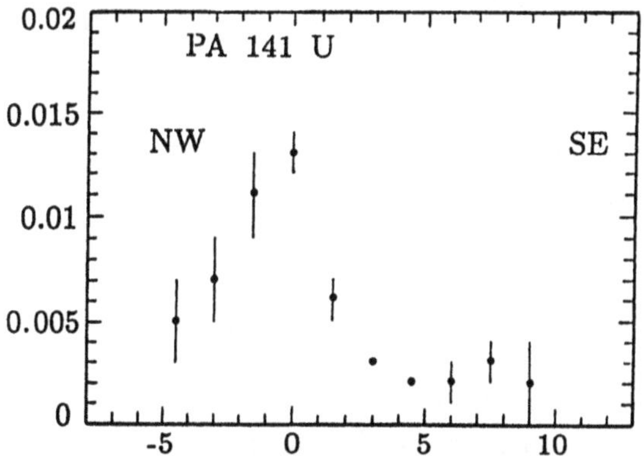

Fig. 2. U for PA 141, horizontal units in arcsec from center

References

Penston, M.V. *et al.*, 1989: *Astron. Astrophys.*, **236**, 53. (Paper I)
Schulz, H., 1988: *Astron. Astrophys.*, **203**, 233.
Unger, S.W., Pedlar, A., Axon, D.J., Whittle, M., Meurs, E.J.A., and Ward, M.J., 1987: *Monthly Notices Roy. Astron. Soc.*, **228**, 671.

Mapping the ^{12}CO J=2–1 Emission from the Dust Lane in Centaurus A

W. Wild[1], *M. Cameron*[2], *A. Eckart*[2], *R. Genzel*[2], *H. Rothermel*[2], *and G. Rydbeck*[3]

[1] European Southern Observatory, Chile
[2] Max-Planck-Institut für extraterrestrische Physik, Garching, Germany
[3] Onsala Space Observatory, Sweden

1. Introduction

We present a fully sampled map at a resolution of 22″ of the ^{12}CO J=2–1 emission from the dust lane of Centaurus A recently obtained with the SEST. Centaurus A (NGC 5128), our closest radio galaxy (distance 3 Mpc), morphologically resembles an elliptical light distribution sliced by a severely warped dust lane. In two previous papers we presented maps of the ^{12}CO J=1–0 and far-IR emission from the dust lane (Eckart *et al.* 1990a), and measurements of molecular absorption lines towards the nuclear continuum source (Eckart *et al.* 1990b). From a comparison with the ^{12}CO J=2–1 CSO measurements of Phillips *et al.* (1987) at a few positions in the disk, we concluded that the molecular gas in the disk must be cold (<10K) or subthermal. However, our recently completed first fully extensive map in ^{12}CO J=2–1 indicates that the disk emission is significantly stronger than suggested by Phillips *et al.* (1987). We used the Swedish-ESO-Submillimeter-Telescope in La Silla, Chile, in December 1989, March 1991 and May 1991 to map 220 positions in the dust lane over an area of approximately 224″ by 56″, mostly sampling at 8″ spacing. Each position was typically measured on four separate occasions. The beamwidth (FWHM) of the SEST at 230 GHz is 22″. Velocity resolution was 0.9 km s^{-1} over 650 km s^{-1}.

2. First Results

A preliminary map of the ^{12}CO J=2–1 velocity integrated emission is presented in Figure 1. The contour levels are 25, 30 , 35 ... K km s^{-1} and the peak intensity is 53 K km s^{-1}. The most striking discovery is that the J=2–1 emission is strong throughout the disk at a level of up to T_{mb} = 0.6 K, comparable to the strength in the center. The J=2–1/J=1–0 ratio in parts of the disk is close to unity and thus higher than previously thought. This implies that the molecular gas in certain regions of the disk may be warmer than the previously estimated ≤10 K (Eckart *et al.* 1990a) – a conclusion based on the J=2–1 spectra of Phillips *et al.* (1987). Our J=2–1 map, if

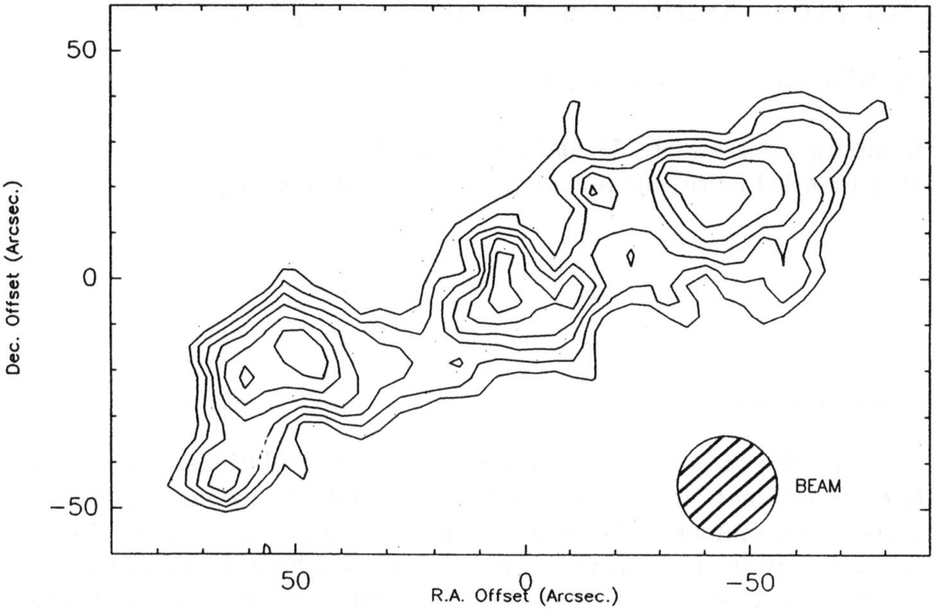

Fig. 1. Preliminary map of the ^{12}CO J=2–1 emission from Cen A

convolved to the 44″ resolution of our CO J=1–0 data, is morphologically similar to our previously published CO J=1–0 map (Eckart *et al.* 1990a) but, in addition, reveals considerable structure at the resolution of 22″.

A comparison of the CO J=2–1 and CO J=1–0, when convolved to similar resolutions, indicates variation of the molecular excitation through the disk.

References

Eckart A., Cameron M., Rothermel H., Wild W., Zinnecker H., Rydbec G., Olberg M., & Wiklind T., 1990a, *Astrophys. J.*, **363**, 451

Eckart A., Cameron M., Jackson J. M., Rothermel H., Stutzki J., Rydbeck G. & Wiklind T., 1990b, *Astrophys. J.*, **365**, 522

Phillips T. G. *et al.*, 1987, *Astrophys. J. (Letters)*, **322**, L73.

The Host Galaxies of Nearby QSOs

Lutz Wisotzki[1], *Hans Hagen*[1], *Ulrich Hopp*[2], *and Dieter Reimers*[1]

[1] Hamburger Sternwarte, D-2050 Hamburg 80
[2] Max-Planck-Institut für Astronomie, D-6900 Heidelberg

1. Introduction

The properties of quasar host galaxies and their evolutionary link to the AGN phenomenon are still not well understood. Direct investigations of intrinsically luminous QSOs that are close enough to allow good observational access to their hosts are required. We present here observations of two recently discovered nearby QSOs, both not known as strong radio sources.

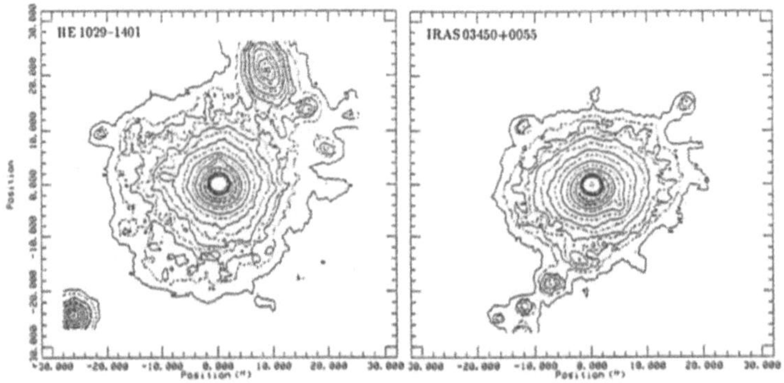

Fig. 1. Contour plots of *R* band images. The contours correspond to logarithmic surface brightness and are spaced at 0.5 mag intervals, starting with 26.5 mag/□$''$. A box-shaped smoothing was applied to the outer contours; the faintest contours were most heavily smoothed. Linear scales at QSO distance are 2.4 kpc/$''$ (HE 1029–1401) and 0.89 kpc/$''$ (IRAS 03450+0055).

2. Observations

We have used the ESO faint object spectrograph and camera (EFOSC1) with the ESO 3.6 m telescope to obtain deep *R* band images of HE 1029–1401 and IRAS 03450+0055. A coronographic mask was employed to shield off the central part of the QSO seeing disk (Fig. 1).

HE 1029–1401: Discovered only a few months ago (Wisotzki et al. 1991), HE 1029–1401 is now the third brightest QSO in the sky ($V = 13.86$), and with $z = 0.086$ it is among the few QSOs in a distance of less than 500 Mpc

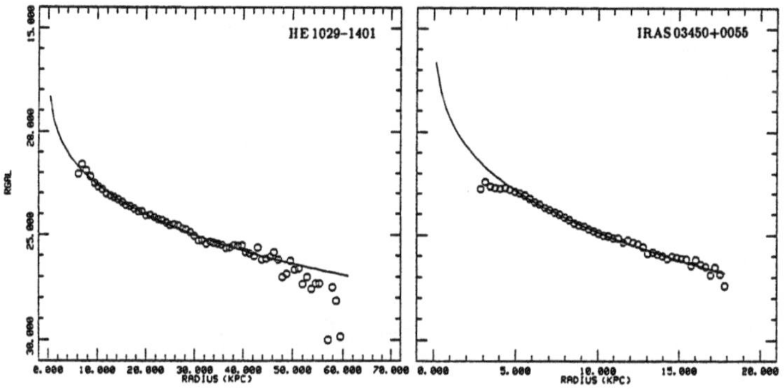

Fig. 2. Azimuthally averaged brightness profiles of the host galaxies, with the scaled point spread functions subtracted (open circles). The solid lines show fits to spheroidal models.

($H_0 = 50\,\mathrm{km\,s^{-1}Mpc^{-1}}$). Model fits to the brightness profile (for $r < 45\,\mathrm{kpc}$) clearly favours an $r^{\frac{1}{4}}$ law (Fig. 2). Resulting fit parameters are the central surface brightness $R_0 = 14.^m9$ and the effective radius $r_e = 7.0\,\mathrm{kpc}$. The total galaxy magnitude derived from the fit is $R_G = 15.2$, corresponding to an absolute brightness of $M_R = -23.5$ ($E(B-V) = 0.05$ assumed), quite typical for giant elliptical galaxies. Recent investigations of a few other nearby radio-quiet QSOs have yielded similar results (Véron-Cetty & Woltjer 1990). Certainly the dichotomy: radio-loud QSOs are hosted by ellipticals, while radio-quiet QSOs reside in spirals, is no longer tenable, at least not in this simple version.

IRAS 03450+0055: One of the few QSOs discovered by the IRAS satellite (Low et al. 1988), it was independently identified by us on HQS objective prism plates and subsequently spectroscopically confirmed ($z = 0.031$, $V = 14.5$). Most remarkable in our deep image is the chain of little "clouds" that seems to emerge out of the QSO galaxy. It will be difficult to assess wether this is a chance projection of unrelated faint galaxies or a physically real phenomenon, e.g. a fragmented jet or a tidal arm. A spheroidal fit to the profile is again to be preferred. The resulting central surface brightness is $R_0 = 12.2$, the effective radius 1.0 kpc. The total magnitude derived is $R_G = 14.5$, corresponding to an absolute magnitude of $M_R = -22.2$ (with a galactic reddening of $E(B-V) = 0.1$).

References

Véron-Cetty, M.P., Woltjer, L., 1990: *Astron. Astrophys.*, **236**, 69.
Low, F.J., Huchra, J.P., Kleinmann, S.G., Cutri, R.M.: *Astrophys. J.*, **327**, L41.
Wisotzki, L., Wamsteker, W., Reimers, D., 1991: *Astron. Astrophys.*, in press.

High-Resolution Speckle Masking Observations of the Seyfert Galaxy NGC 1386

P. Zeidler[1], *I. Appenzeller*[2], *K.-H. Hofmann*[1], *W. Mauder*[1], *S. Wagner*[2], *and G. Weigelt*[1]

[1] Max-Planck Institut für Radioastronomie, Auf dem Hügel 69, 5300 Bonn 1, Germany
[2] Landessternwarte Königsstuhl, 6900 Heidelberg, Germany

Abstract. We show the first speckle masking observations of the Seyfert galaxy NGC 1386. Two sets of speckle interferograms were taken with the 2.2-m ESO/MPG telescope. The first data set consists of about 12000 speckle interferograms and was taken with an Hα –filter (filter bandwidth 60nm). The second data set consists of about 8000 speckle interferograms and was taken with the edge filter RG 630. The two high-resolution images reconstructed from the two data sets show four dominant clouds within a region of 2.3" diameter. All four clouds lie on a line in position angle 20 degrees.

1. Introduction

NGC 1386 is one of the nearest Seyfert galaxies, which makes this object particularly interesting for high resolution work. An extensive study in the radio wavelength region was published by Weaver et al. (1991), who realized that the ENLR gas is highly ionized and found indications of a two-component velocity field. We carried out speckle masking observations (Weigelt, 1977, 1983, Lohmann et al., 1983 and Hofmann and Weigelt) of the ENLR of NGC 1386 with an Hα–filter and a broad-band continuum filter (RG 630).

2. Speckle masking observations

About 20000 speckle interferograms were recorded in both filters. The Hα speckle interferograms consist of $\sim$ 130 photon events, the RG 630 speckle interferograms of $\sim$ 200 photon events. A description of the photon-counting process is given in Hofmann et al., 1990. Fig. 1 is the long-exposure image obtained by averaging all 12000 re-centered Hα speckle interferograms. Fig. 2 shows the high-resolution $H\alpha$ image of NGC 1386 reconstructed by speckle masking from the same 12000 speckle interferograms. The image shows four dominant clouds within a region of 2.3" diameter. All four clouds lie on a line in position angle 20 degrees. Fig. 3 is the long-exposure RG 630 image obtained by averaging all 8000 re-centered RG 630 speckle interferograms. Fig. 4 shows the broad-band continuum image (RG 630 filter) of NGC 1386 reconstructed by speckle masking.

The reconstructions (Figures 2 and 4) show a dominant knot, emitting the bulk of the $H\alpha$ luminosity, and several clouds to the north and south.

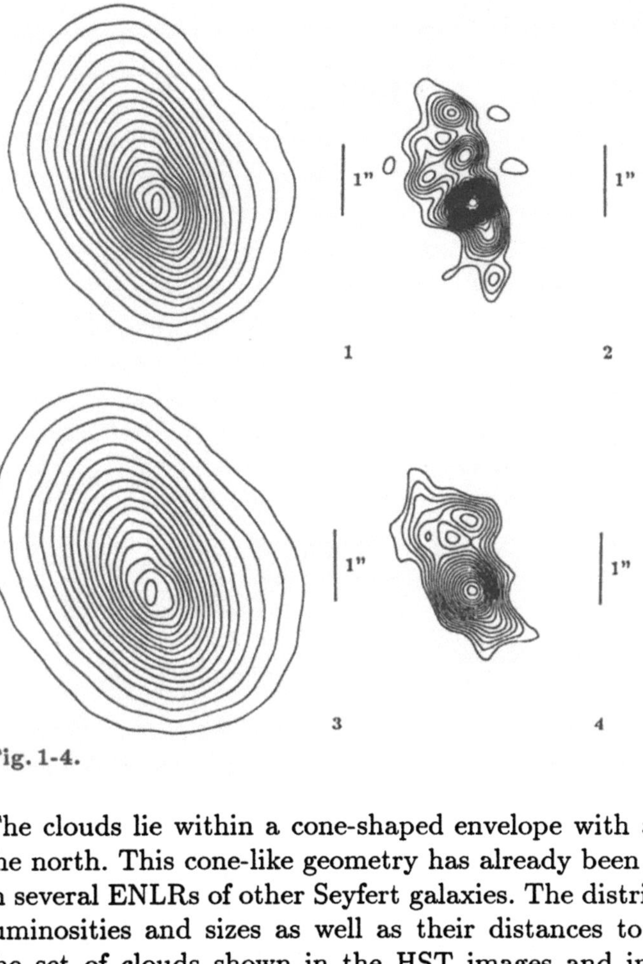

Fig. 1-4.

The clouds lie within a cone-shaped envelope with a slight asymmetry to the north. This cone-like geometry has already been found on larger scales in several ENLRs of other Seyfert galaxies. The distribution of clouds, their luminosities and sizes as well as their distances to the nucleus resemble the set of clouds shown in the HST images and in the speckle masking reconstructions of NGC 1068 (Hofmann, Mauder and Weigelt 1991, in this conference procedings and in Hofman et al., 1989. The axis of the cone is nearly parallel to that of the elongated radio feature reported by Weaver et al. (1991). As in several other Seyfert galaxies the bulk of the radio emission and the most luminous line emitting clouds lie on opposite sides of the centre.

References

Weaver, K. et al.: 1991, *Astrophys. J.*, **366**, 50
G. Weigelt: Opt. Commun. **21** (1977) 55
G. Weigelt and B. Wirnitzer: Opt. Lett. **8** (1983) 389
A.W. Lohmann, G. Weigelt and B. Wirnitzer: Appl. Opt. **22** (1983) 4028
K.-H. Hofmann and G. Weigelt: AUA167,L 15
K.-H. Hofmann, W. Mauder and G. Weigelt: Photon-Counting Speckle Masking: Proc. ICO 15,1990, SPIE Vol. **1319**, p., 442
K.-H. Hofmann, W. Mauder and G. Weigelt: Speckle Masking Observations of NGC 1068 and NGC 7469: Proc. of the ESO Workshop on Extranuclear Activity in Galaxies, Garching, 16-18 May 1989, p., 35

4 – Magnetic Fields and Jet Formation

MHD Effects in AGN Accretion Disks

Jean Heyvaerts

Laboratoire d'Astrophysique Extragalactique et de Cosmologie, Université Paris 7 et Observatoire de Paris, 92195 Meudon, France

1. AGNs and Magnetic Fields

In this review, I shall take the view that AGNs are fed by accretion disks orbiting massive black holes. The physics involved is similar to that in other accretion disks, but for two important aspects. First, the dynamics very close to the central object should be general-relativistic. Second, the central black hole would not be a source of magnetism by itself, so that any field present should be advected with the accreted fluid or generated in it by dynamo effects.

Are magnetic fields really present and dynamically significant? Observations certainly indicate their presence. For example, optical (Fraix Burnet et al. 1991) and radio polarization measurements of objects like the M87 jet are evidence for an organized magnetic field. Radio and infrared polarization measurements in the vicinity of the center of our own galaxy also reveal quite large, possibly dynamically significant, fields (Aitken 1989, Aitken et al. 1991, Tsuboi et al. 1986). Dynamo action is expectable in such accretion disks. MHD is then likely to be relevant to them.

2. Angular Momentum Transport

2.1 Various Possibilities

A piece of material endowed with angular momentum can only accrete onto a compact object by releasing its specific angular momentum. Classical views consider this to be effected by some kind of viscosity associated with turbulence (Shakura and Sunyaev 1973). This "effective" kinematic viscosity is often written as:

$$v = \alpha h c_s \tag{1}$$

Where c_s is the sound speed, h the disk thickness and α a coefficient of order unity. Some physical arguments make this formula plausible, and, indeed, it can be recovered under the assumption that the necessary rate of viscous heating implied by the accretion results from the dissipation of a 3D Kolmogoroff turbulent cascade, with an injection scale of the order of the disk thickness. However, up to now, the origin of the assumed turbulence has not been described in detail. Moreover, the idea that turbulence acts as an effective viscosity, without any other qualitatively different effect, is a disputable assumption. An anisotropic turbulence lacking invariance by mirror reflection would give rise to a component of the Reynolds stresses proportional to the velocity, not to the velocity gradients. This so-called "AKA effect" (Alpha Kinetic Anisotropic, Frisch et al. 1987) is the kinetic equivalent to the magnetic α-effect, well known in turbulent dynamo theory. Finally, angular momentum could be transported by a number of other processes, such as

- transport by wave fields,
- effective viscosity from MHD turbulence,
- torques transmitted by magnetic connection through a corona,
- angular momentum release to an MHD wind.

2.2 Purely Hydrodynamical Aspects

A major difficulty with the source of turbulence is that the keplerian differential rotation is stable to incompressible hydrodynamical motions in the sense of the Rayleigh criterion: its specific angular momentum, $r^2\Omega$, increases outwards. Similar stability results have recently been obtained for compressible fluids (Lerner and Knobloch 1988). However, a turbulent state could conceivably result from global, rather than local, instabilities, or from finite amplitude perturbations, or else from instabilities involving some other cause than just shear (convection or MHD effects).

Shear-associated global instabilities as considered by Papaloizou and Pringle (1987) depend on boundary conditions, as shown by Blaes (1987), and seem to lead to weak and localized dissipation (Kaisig, 1989).

Finite amplitude perturbations to plane Couette flow, a simple analogue to circular shear flow, have been considered by Dubrulle and Zahn (1990), who found

instability for large enough perturbations. The threshold is proportional to the inverse of the Reynolds number, however, and could be very low indeed. They found that unstable secondary fluctuations do not survive in 2-dimensional flows, but could do so in 3-dimensional ones. A plausible source of turbulence, related to shear only, is perhaps to be found in these finite amplitude instabilities.

By contrast, Vishniac and Diamond (1989) discussed the role of stable, but permanently emitted, waves in angular momentum transport. They have shown that internal gravity waves with m = 1 may transport negative angular momentum from the disk boundary inwards and deposit it at all radii by nonlinear dissipation. This non-local transport, though not of a viscous nature, is estimated to have an efficiency comparable to Shakura and Sunyaev's turbulent viscosity.

2.3 MHD Aspects

Tagger et al. (1990) have shown that ionized thin disks embedded in a locally vertical field are subject to global spiral instabilities of low azimuthal m-number, driven by the differential rotation. The unstable waves are similar to fast magnetosonic waves, but are strongly modified by the small vertical extent of the massive medium and by differential rotation. The dispersion relation resembles that of spiral density waves, the magnetic contribution playing a role similar to a negative self-gravity. The growth time is a few tens of rotation periods for the optimal amplification conditions, which however require reflection of waves at some inner boundary.

Recently, Balbus and Hawley (1991) have described a local and quite strong instability of differentially rotating magnetized disks, which exists even when the Rayleigh criterion indicates stability of purely hydrodynamical modes. The basic state is an axisymmetric shear flow permeated by a magnetic field with no radial component. Axisymmetric perturbations, locally represented as plane waves varying as $expi(k_r r + k_z z - \omega t)$ are considered. The fluid motion is regarded as quasi-incompressible. A dispersion relation is obtained, which reduces to that of internal waves, driven by the stable radial stratification of specific angular momentum, in the absence of magnetic field and to that of Alfvèn waves in the absence of rotation and buoyancy. Neglecting buoyancy but retaining the effect of differential rotation and magnetic field, perturbations with a given k_z are found to be stable, whichever k_r, if

$$k_z^2 v_{Az}^2 + \frac{d\Omega^2}{dlnr} > 0 \tag{2}$$

Here v_{Az} is the Alfvèn velocity of the z-component of the magnetic field. The Θ-component is irrelevant. In keplerian disks, $d\Omega^2/dr$ is negative, and instability is found for k_z^2, smaller than some critical value, k_{zc}^2:

$$k_{zc}^2 = -\frac{1}{v_{Az}^2}\frac{d\Omega^2}{dlnr} \tag{3}$$

The larger B_z, the smaller k_{zc}. When the critical wavelength $2\pi/\ k_{zc}$ becomes larger than the disk thickness, the local approximation breaks down and instability is expected to disappear, lacking a physically acceptable range of k_z. This happens for large B_z fields. At very small B_z the instability growth rate becomes smaller than viscous damping rate, and the instability is quenched. The physics of the instability can be understood as follows. Consider a cylindrical shell of the disk at radius r_0. Then perturb it by displacing radially its fluid elements by $\xi(t) \exp ik_z z$. An infinitesimal fluid element now suffers a magnetic tension force caused by the sinusoidal shape of the displaced field lines, which is proportional to $-(k_z^2 v_{Az}^2 \xi)$. It also suffers the resultant of gravity and centrifugal forces, which balance each other in the unperturbed state. This net force is proportional to $(r_0 + \omega^2(r_0 + \xi) - g(r_0 + \xi))$, where $\omega(r_0 + \xi)$ is the perturbed rotation frequency and $g(r_0 + \xi)$ the gravitational pull at the displaced position. If the fluid sheet were not connected by the magnetic field, the lagrangean changes of angula r velocity would obey a law of conservation of specific angular momentum for each fluid element. But this is not so here, since the fluid elements are tethered to each other by magnetic field lines. Suppose that the perturbations evolve slower than the Alfvèn transit time. If so, it is the total angular momentum of the sheet as a whole which is conserved, not the one of each fluid element separately. At the linear approximation, this means that the angular velocity is conserved: $\omega(r_0 + \xi) = \Omega(r_0)$. This finally gives, for the radial displacement $\xi(t)$, the evolution equation:

$$\frac{d^2\xi}{dt^2} + \left(k_z^2 v_{Az}^2 + \frac{d\Omega^2(r_0)}{dlnr_0}\right)\xi = 0 \tag{4}$$

and we find instability when the spring constant is negative, that is, when $d\Omega^2/dr_0$ is negative and when at the same time $k_z^2 v_{Az}^2 < -d\Omega^2/dlnr_0$. We recover the previous result. Note, however, that this picture holds only near the threshold and does not incorporate the azimuthal dynamics which tends to propagate ordinary torsional Alfvèn waves. Hawley and Balbus (1991) illustrate their point by 2-dimensional numerical calculations including compressibility and azimuthal components of vector fields. Periodic boundary conditions in z and reflecting boundary conditions on the radial walls of the computational box are used. The predicted instability is indeed observed in these computations. Saturation occurs by reconnection of the distorted fields (there is some numerical resistivity). The unstable development of the perturbations has also been observed in geometries grossly different from the one assumed in the analytical study, like poloidal loops.

Some aspects of Balbus and Hawley's work would deserve some more elaboration in the future. For example, their modes are axisymmetric: it would be important to study the fate of non-axisymmetric perturbations. Also, their analysis is local. Since the azimuthal dynamics tends to induce propagating perturbations, it would be important to study the effect of the boundary conditions to see whether transmission or reflection of the Alfvèn waves allow the buildup of a locally increasing perturbation inside the disk. Note that their numerical work uses reflecting boundary conditions.

Both aspects have been recently examined by Tagger and collaborators, who find that m=0 modes are unstable too, but the instability conditions are different, in particular their modes remain unstable at large magnetic fields. They also found some dependence of the instability conditions on the assumed boundary properties.

3. Field Transport and Dynamo Action

3.1 Passive Field Transport

Several authors (Königl 1989, Khanna and Camenzind 1991) addressed the question of the field structure generated by dissipative transport in stationary accretion flows. Königl assumes self-similarity, while Khanna and Camenzind found numerical solutions. No such solution would be realistic, however, without allowance for the α-effect and also for buoyancy (Stella and Rosner 1984). Therefore these results should be regarded as a first step towards a more complete solution.

3.2 Dynamo Action

The concept of turbulent dynamo action is certainly relevant to accretion disks. The role of turbulent fluctuations is easily understood. Suppose that we neglect molecular dissipation, and split the velocity field into an average part, v and a fluctuation, ω, and similarly the magnetic field into $B + b$. Taking an average, over the possible realizations of the turbulent fields, of the induction equation gives for B the evolution equation:

$$\frac{\partial B}{\partial t} = rot(v \times B) + < rot(\omega \times b) > \tag{5}$$

The effect of turbulence is contained in the averaged second term, with suitable approximations (Moffatt 1978), can be brought to the form:

$$< rot\ (\omega \times b) > \ = \ rot\ \alpha_D B - rot\ \eta\ rot\ B \tag{6}$$

The second term represents effective dissipation. The first one represents the so-called α-effect. The coefficient α_D is proportional to the kinetic helicity, $< \omega \cdot rot\ \omega >$. It vanishes if the properties of the turbulent velocity field are invariant by mirror symmetry. Any dynamo theory should explain the origin of the kinetic helicity. Pudritz (1981) suggests that a vertical density gradient and differential rotation would generate helical internal waves. Vishniac, Jin and Diamond (1990) show that an inward travelling flux of such waves indeed gives rise to kinetic helicity, because of the strong asymmetry of the wave distribution. Most studies of the dynamo effect, up to now, have been kinematic, that is v, α_D and η are regarded as fixed, known, functions of space. Such an approach, which cannot describe the eventual dynamo saturation (i.e. the amplitude of the generated field), reduces equations (5-6) to a linear eigenvalue problem. The latter depends on a dimensionless control parameter, Q, the dynamo number, which is the geometrical mean of the Reynolds numbers that can be formed with representative values of α_D and the rotation velocity:

$$Q = \sqrt{R_\alpha R_\omega} \tag{7}$$

where

$$R_\alpha = \frac{\alpha_D l}{\eta} \quad , \quad R_\omega = \frac{\Omega l^2}{\eta}$$

In these expressions, l is some characteristic scale of the disk. Growing modes are found when Q exceeds some critical value $Q_\star$. The linear analysis gives the spatial structure of these growing magnetic configurations. In his analytic study, Pudritz (1981) found that the critical dynamo number $Q_\star$, defined with some appropriate scale l, is of order unity, while he estimated the actual dynamo number to be approximately equal to the inverse of the Mach number of the turbulence. This means that the turbulence needs to be sufficiently subsonic for dynamo action to occur. Any classical α-disk is then on the verge of dynamo action. Whether or not field generation will actually occur should then depend on such details as how strong the turbulence exactly is and how it is distributed in the disk! It seems that no firm conclusion can be reached.

Recently, Stepinsky and Levy (1990) took a numerical approach to the problem. In their model, the most easily excited mode is one with a quadrupolar type of symmetry, in agreement with earlier results. This is because fields with dipolar symmetry cross the disk vertically and suffer extremely weak stretching by differential rotation inside the disk. Above threshold, they find that the field is generated in patches of larger and larger number of arcades, each of a size slightly large than the disk thickness. This can be understood because the local dynamo number (the one defined as in eq (7), with α_D, Ω and l representative of the local values) is not uniform. When the global dynamo number grows, larger and larger fractions of the disk area become subject to the dynamo instability. This leads us to expect that above threshold the dynamo acts locally, and probably incoherent between different parts of the disk. Accretion disks, if they exhibit dynamo action a all, probably have "activity centers". Camenzind and Lesch (1991) also find that quadrupole fields are favoured near threshold. They find oscillatory solutions at higher dynamo numbers, adding a time-dependent character to the actual solution, which may stabilize when non-linear feedback effects are included.

Buoyancy and non-linear effects, like feedback on the large-scale flow or on the turbulence level, remain a crucial issue, and possibly also the turbulent behaviour of the medium external to the disk itself.

3.3 Buoyancy

Buoyancy is a process by which magnetic fields escape out of the fluid where they are generated. Basically, since the sum of gas and magnetic pressure is almost equal inside and outside a flux tube, the gas pressure, and then the gas density, tends to be lower in more strongly magnetized regions. Flux tubes are then like emptier ropes immersed in a denser fluid, and they tend to float to the surface. Stella and Rosner (1984) have shown that buoyancy ejection times are quite short, since the velocity may reach the Alfvèn speed.

When the field comes in the form of a continuum, rather than in the form of individualized flux tubes, the buoyancy manifests itself as an instability of the magnetic structure. For example, a layer of field parallel to the disk's plane tends to kink, and to develop parts which emerge out of the field source region. These parts, when reaching larger heights, lose their material which drains down along field lines under the effect of gravity. This causes the emerged parts to yield even more to the destabilizing Lorentz forces. The non-linear development of this Parker instability has been studied numerically in two dimensions by Matsumoto et al. (1990). In accretion disks, the differential rotation generates an azimuthal component of the magnetic field that cannot be ignored. Shibata et al. (1990) considered this situation in the shear-layer model, performing a 2 1/2 dimensional computation, i.e. $\partial/\partial y$ is taken to be zero but the y-components of vector fields are not. They find that shear limits the buoyancy escape speed, and that in the presence of a strong shear a strongly magnetized (low-β) disk medium can be formed even against buoyancy. Dense blobs of condensed gas form amidst this structure, apparently suspended against gravity by magnetic tension forces in a way similar to solar prominences. In the absence of shear, the scenario is entirely different, with a developing Parker instability and the maintenance of a high-β disk. Their conclusion is that accretion disks may exist in two forms: high-β disks surrounded by a corona of ejected magnetic fields, and low-β disks, which cannot eject field as fast as it grows, and develop blobby structures where vertical mechanical balance is between gravity and magnetic tension.

4. Disks with Coronae

4.1 Origin of the Disk's Corona

In the case of high-β disks (low magnetization inside the disk), a corona is likely to form around the disk due to the buoyant escape of dynamo-generated fields. Stresses are exerted on this outer region by small-scale perturbations created by turbulence in the disk, and by large-scale shearing, much in the same way as the solar corona is acted upon by photospheric motions. Learning from solar physics we expect this outer region to be heated and to exhibit flaring activity. The heating agents are the electric currents which flow in this outer medium as a response to the drive exerted by the denser disk medium. These currents would dissipate by processes similar to those studied in the solar context (for recent reviews, see Priest and Krishan (1990) and be an efficient source of heating, thus creating a hot corona as first pointed out by Galeev et al. (1979), and studied since then by many authors (Kuperus and Ionson (1985), Burm (1986), Zuccarello et al. (1987).

Coronal heating processes are believed to involve turbulent reconnection. Reconnection is a finite resistivity effect which, in a laminar regime, occurs in restricted regions of the plasma. In these regions, electric fields aligned to the magnetic field may appear and field lines may change topology, allowing the global magnetic structure to find a state of lesser stress. When the flow which supports reconnection becomes turbulent, the resistive regions tend to form a very complex pattern. This complex process however is still ruled by simple properties. It has been known for long (Woltjer, 1958) that in each flux tube of a closed, perfect MHD system, a quantity, K, called the magnetic helicity, is conserved, whichever the motions of the system. K is defined by:

$$K = \iiint_{Tube} A \cdot B \, d^3r \tag{8}$$

where A is a vector potential for B. This is, strictly speaking, not so for reconnecting systems, which are dissipative, but Taylor (1974) conjectured that the total magnetic helicity of a high magnetic Reynolds number system would approximately be conserved (see also Taylor (1986)). This idea has gained support and is understood in modern turbulence theory as a manifestation of an inverse cascade of magnetic helicity in 3-dimensional MHD turbulence. One might then expect that a turbulent dissipative episode would result in a magnetic configuration of minimum energy for its given total magnetic helicity. The energy released is the difference between the energy content in the initial state and the energy of this relaxed post-turbulent "Taylor state". For a low- *beta* plasma, the relaxed state is a linear force free field.

4.2 Coupled Disk-Corona Model

An actual corona is not a closed system, nor can it reach a completely relaxed Taylor state, even though it is turbulently reconnecting all the time, because the driver medium always recreates the stresses which induce reconnection. Nevertheless it might be a fair enough approximation to the magnetic coronal structure to represent it by such a relaxed Taylor state. Heyvaerts and Priest (1989) used such a modelization of the disk's corona to investigate the effects of coupling between them. Given this ansatz, it becomes possible to calculate the distribution of rotation in the disk, which is slightly non-keplerian because of the Lorentz forces exerted by the currents which close their coronal circuits through the disk. These forces may be regarded as an interaction at distance between magnetically conjugated regions of the disk. The Poynting flux from the disk to the corona can also be calculated, giving the coronal heating rate. This coupled disk-corona model turns out to depend on a magnetization parameter, μ. R being the disk radius, $M_\star$ the mass of the central black hole and $\dot{M}$ the accretion rate, this parameter is:

$$\mu = \left(4\pi R^3 \, \frac{B_0^2}{\mu_0}\sqrt{\frac{GM_s}{R^3}}\right) \Big/ \left(\frac{GM_\star \dot{M}}{R}\right) \tag{9}$$

When μ becomes of order unity, the deviations from the Kepler law due to magnetic stresses transmitted through the corona become substantial and the rate of coronal heating is found to become comparable to the accretion luminosity.

5. Coupled Disks and Jets

This review would not be complete if wind-driven accretion were not briefly mentioned. Other speakers in this meeting review this subject. As stressed earlier, a rotating magnetized wind may carry away enough angular momentum that this loss alone allows accretion to proceed at an interesting rate. An important progress here has been the selfsimilar disk model derived by Königl (1989) which matches to a self-similar wind model derived by Blandford and Payne (1982). In the Königl model the radial momentum equation is treated at the keplerian approximation. The field-aligned dynamics is also not solved for. Only the angular momentum and the induction equations are considered, the other equations being assumed to be compatible with self-similarity. Also, the field evolution inside the disk has been modeled ignoring the α-dynamo effect, which may be an important restriction. The fact that turbulent viscosity is not demanded in this model to transport angular momentum by no means imply that the disk should be laminar. Königl's calculation still shows that self-consistent solutions connecting a disk to a rotating magnetized wind which carries the lost angular momentum to infinity together with a significant part of the accreted energy are quite plausible.

The particular case of AGNs would require a relativistic modeling of their relativistic speed efflux and of the effects of the proximity of the black hole. Camenzind (1991) reviewed such winds. Lovelace et al. (1987) produced a model of these central regions of an AGN which consists of a resistive disk with a stationary magnetic configuration of the odd-B_z symmetry, matched to a force-free magnetosphere (total electric force density $\rho E + j \times B = 0$), the structure of which has been calculated assuming a simple distribution of poloidal currents. The odd-B_z symmetry adopted resulted in a small extraction of magnetic energy by the wind. This is because the azimuthal field component, B_Θ, is not large at the disk surface in this model, resulting in a weak torque. The smallness of B_Θ is due to the fact that in the limit of zero disk thickness the tangential components of the magnetic field are continuous across the disk's plane in this symmetry, meaning that the thin disk then carries not net surface currents. The situation is of course completely different in even-B_z symmetry (Wang et al. (1990)). However, kinematic dynamo models seem to prefer odd-B_z solutions!

6. Conclusions

The origin of turbulence in accretion disks remains an open but not desperate question. The role of the Balbus-Hawley instability should be precised, and the non-axisymmetric modes considered. Non linear hydrodynamical instabilities also constitute an interesting possibility. Dynamo action in α-disks are not certain, and could depend on details of turbulence strength and distribution. It is desirable to improve on kinematical dynamo models. Buoyancy is an important piece of physics to consider. The exotic possibility of chumpy low-β disks deserves attention. More classical high-β disks certainly have active coronae, which are likely to give rise to important observational features, like flaring activity. Coupled disk-corona models and disk-wind ones are attractive concepts which one would like to model even more self-consistently than has been hitherto possible.

References

Aitken, D. K., 1989, in "The Center of the Galaxy", IAU Symp. **136**, M. Morris ed., Kluwer Acad. Publ., p. 457-464

Aitken, D.K., Gezari, D., Smith, C.H., McCaughrean, M., Roche, P.F., 1991, preprint

Balbus, S.A., Hawley, J.F., 1991, *Astrophys. J.*, **376**, 214

Blaes, O., 1987, *Monthly Notices Roy. Astron. Soc.*, **227**, 975

Blandford, R.D., Payne, D.G., 1982, *Monthly Notices Roy. Astron. Soc.*, **199**, 883

Burm, H., 1986, *Astron. Astrophys.*, **165**, 120

Camenzind, M., 1991, in "Variability of Active Galaxies", Lecture Notes in Physics 377, Springer-Verlag, W. Duschl et al. ed., p. 201

Camenzind, M., Lesch, H., 1991, in preparation

Dubrulle, B., Zahn, J.P., 1990, in press in J. Fluid Mech.

Fraix Burnet, D., Le Borgne, J.F., Nieto, J.-L., 1991, *Astron. Astrophys.*, , in press

Frisch, U., She, Z.S., Sulem, P.L., 1987, Physica 28D, 382

Galeev, A.A., Rosner, R., Vaiana, G.S., 1979, *Astrophys. J.*, **229**, 318

Hawley, J.F., Balbus, S.A., 1991, preprint, *Astrophys. J.*, in press.

Heyvaerts, J., Priest, E.R., 1989, *Astron. Astrophys.*, **216**, 230

Kaisig, N., 1989, *Astron. Astrophys.*, **218**, 89

Khanna, R., Camenzind, M., 1991, this meeting

Königl, A., 1989 *Astrophys. J.*, **342**, 208

Kuperus, M., Ionson, J.A., 1985, *Astron. Astrophys.*, **148**, 309

Lerner, J., Knobloch, E., 1988, J. Fluid Mech. 189, 117

Lovelace, R.V.E., Wang, J.C.L., Sulkanen, M.E., 1987, *Astrophys. J.*, **315**, 504

Matsumoto, R., Horiuchi, T., Hanawa, T., Shibata, K., 1990, *Astrophys. J.*, **356**, 259

Moffat, H.K., 1978, "Magnetic field generation in electrically conducted fluids", Cambridge Univ. Press

Papaloisou, J.C., Pringle, J.E., 1987, *Monthly Notices Roy. Astron. Soc.*, **225**, 267

Priest, E.R., Krishan, V., 1990, Proc. IAU Symp. **142** "Basic Plasma processes on the sun"

Pudritz, R., 1981, *Monthly Notices Roy. Astron. Soc.*, **195**, 881

Shakura, N.I., Sunyaev, R.A., 1973, *Astron. Astrophys.*, **24**, 337

Shibata, K., Tajima, T., Matsumoto, R., 1990, *Astrophys. J.*, **350**, 295

Stella, L., Rosner, R., 1984, *Astrophys. J.*, **277**, 312

Stepinski, T.F., Levy, E.H., 1990, *Astrophys. J.*, **362**, 318

Tagger, M., Henriksen, R.N., Sygnet, J.F., Pellat, R., 1990, *Astrophys. J.*, **353**, 654

Taylor, J.B., 1974, Phys. Rev. Letters **33**, 1139

Taylor, J.B., 1986, Ref. Mod. Phys. **58**, 741

Tsuboi, M., Inoue, M., Handa, T., Tabara, H., Kato, T., Sofue, Y., Kaifu, N., 1986, *Astron. J.*, **92**, 818

Vishniac, E.T., Diamond, P., 1989, *Astrophys. J.*, **347**, 435

Vishniac, E.T., Jin, L., Diamond, P., 1990, *Astron. J.*, **365**, 648

Wang, J.C.L., Sulkanen, M.E., Lovelace, R.V.E., 1990, *Astrophys. J.*, **355**, 38

Woltjer, L., 1958, Bull. Astron. Netherlands 14, 39

Zuccarello, F., Burm, H., Kuperus, M., Raadu, M., Spicer, D.S., 1987, *Astron. Astrophys.*, **180**, 218.

Discussion

Rees: The magnetic energy dissipated in a corona can be in two forms. Either it goes into Alfvèn waves, and then into thermal energy of the plasma, or it is dissipated in small localized regions of reconnection (going into relativistic particles). Can anything be said about the relative importance of these different processes?

Heyvaerts: It is difficult to say how much of the energy dissipated in the corona ultimately goes into thermal or suprathermal particles form. The answer would result from a rather detailed description of the damping of large scale perturbations evolving into a form of small scale turbulence which is intermediate between classical "collisional" MHD and kinetic (Vlassov) turbulence in an inhomogeneous medium. To my knowledge, there is no field of astrophysics (i.e. solar flare theory) nor geophysics (i.e. auroral physics) where such a problem has yet been given an answer that would be any quantitative. What is known, however, is that a significant fraction of the energy release may go into accelerated particles, i.e. in solar flares (see for example the review by Benz, 1987, Solar Phys. **111**, 1-18). This acceleration may, for a small part, result from a direct DC electric field acceleration, the beams subsequently coupling to the plasma by plasma waves (as described for example in the poster by H. Lesch at this conference) or, more indirectly, from the eventual coupling of turbulent MHD flows to small, kinetic, scales. Incidently, MHD waves, just as reconnection flows, may follow this route to dissipation too (see for example Roux, 1985, ESA SP **235**, 151). The calculation of this "branching ratio" is in fact a major unsolved problem.

Courvoisier: When a significant part of the energy is dissipated in the corona, the disk structure must be strongly influenced. Do you take this into account when you calculate the negative coupling between the disk and the corona?

Heyvaerts: Yes, we calculate the effect of the corona-disk coupling on the rotation law of the disk, which is self-consistently (slightly) non-keplerian. Certainly, the deposition of a fairly large amount of power on the superficial layers of the disk rather than inside should change its vertical structure, but we did not have to solve this problem to calculate the total rate of coronal heating in this model, since we treated the disk as thin.

Camenzind: The work of Hawley and Balbus does not include magnetic diffusivity and helicity. What happens to the instability if turbulence is excited?

Heyvaerts: I do not think that the work of Balbus and Hawley intended to give a description of the fate of a field line sitting in a turbulent disk (hence endowed with diffusivity and helicity). Its usefulness, as I understood it, resides rather in the fact that it demonstrates the laminar disk to be impossible, which had up to now never been shown beyond doubt, and then that turbulence should be the rule. Note, however, that their work does not rule out the possibility of other routes to turbulence, for example by the development of hydrodynamic non-linear instabilities (Dubrulle and Zahn). If, however, the Balbus Hawley process is the dominant source of turbulence, this would probably imply that magnetization is the rule too.

Preuss: Is it generally true that every magnetized disk develops a flaring corona? and that it is (so far) only uncertain which fraction of the energy dissipated/converted will be radiated away?

Heyvaerts: Yes, I think that the inevitability of buoyancy or Parker instability indeed imply that any magnetized disk should be surrounded by a corona (with the proviso of low beta disks described by Japanese colleagues, which are, so to speak, "coronae without disks"). Since the footpoints of coronal magnetic field lines are acted upon by a turbulent velocity field in the disk, the latter is also very likely to be permanently flaring. How much of the total energy delivered to the corona returns by irradiation, conduction or high energy particle impact to the denser disk, how this deposition reacts on the coronal physical conditions, and how much is eventually radiated directly away has up to now not been calculated, but it should be possible to do so, even in some detail, under some assumption concerning which fraction of the total energy dumped in the corona goes into the various possible forms (thermal, high energy particles, of which characteristic energy, etc. .... see answer to M. Rees).

Blandford: Do you think that the dynamo cycle can ever be fast enough to dominate the dynamical instabilities operating on the shearing timescale $\sim \Omega_K^{-1}$?

Heyvaerts: The turnover time of a turbulent eddy is, in the Shakura-Sunyaev hypothesis, of the order of a dynamical time or longer. Since it takes at least an eddy turnover time scale to operate the "dynamo alpha-effect", because, physically, this is associated with the restauration of some poloidal field from the toroidal one by such eddies, it results that the keplerian time definitely is a lower bound to the dynamo time scale. Whether the dynamo effect can survive instabilities operating on a dynamical time scale or not depends on which instabilities one has in mind: in fact the dynamo process grows on a state of turbulence that should itself originate in some instability of some supposedly laminar flow. Such instabilities should then be regarded as feeding the dynamo rather than killing it. So is probably the Balbus Hawley one (see answer to Camenzind). By contrast some instability of the axisymmetric *mean* field and flow (with underlying small scale turbulence) which would disrupt in such a short time this large scale state, supposedly operating the alpha-omega dynamo, would certainly be fatal to it. Or at least would it be fatal to the picture of field generation that is presently favoured by most of us.

Field: In your last viewgraph you emphasized the flow of energy from a magnetic corona. Do the same parameters allow you to explain the angular momentum flow required to release the gravitational energy of the accreting flow?

Heyvaerts: The model for a coupled disk-corona system which I briefly presented here assumes the coronal field structure "plugged" on the disk to be closed. The B_Φ Bz stresses exerted by the disk on the corona (or conversely) transfer angular momentum from one towards the other. Since, however, our model corona is force-free, it cannot support stresses, but only serves at transferring them from one point in the disk to another one, magnetically connected to it (this is shown in detail in our paper). So, a closed force-free corona is no sink for the disk's angular momentum, but only a redistributor of it. In this sense, it plays a role similar to magnetic tethering inside the disk itself, with the important difference that the partially emerged field lines in the corona are no more subject to buoyant ejection. A realistic corona will however have closed and open field lines attached to it, and it is on the latter ones that angular-momentum-carrying winds will be blown. Just the same angular momentum flux would in fact be transmitted along these open lines as we calculated for the closed ones, if only the azimuthal field at the disk-corona interface would be the same in both cases, all other physical parameters being equal. But there is no reason for such an equality, even in an approximate sense, and in fact we should expect the open structure to give rise to a smaller azimuthal B_Φ component. How much smaller cannot be precised without a modeling of the disk-field system, which, in fact, is just the point of papers like the one by Königl, for example.

Hydromagnetic Disk Winds in AGN: The YSO Connection

Ralph E. Pudritz

1. Introduction

Many different types of mechanisms have been proposed to account for the formation and collimation of jets in active galactic nuclei (AGNs) (eg Rees 1984, Blandford 1990). Of this host, one mechanism is able to efficiently drive an outflow without the necessity of an intense radiation field or special disk geometry. This drive is essentially a centrifuge; gas can be flung out if it can be forced to co-rotate with a fast rotating central object. The mechanism that enforces the co-rotation of accelerating gas with the rotor is readily provided by strong magnetic fields that thread it. Of course such an outflow must also be collimated into jets. Here again, hydromagnetic winds are an attractive possibility because they have the desirable ability to self-collimate due to the strong $\mathbf{J} \times \mathbf{B}$ pinching forces exerted by the strong toroidal field of the jet.

In the context of black hole/accretion disk models for AGNs, there are two obvious candidates for the role of the rotor, namely a rapidly spinning black hole (Blandford and Znajek 1977) or the surrounding accretion disk (Blandford and Payne 1982, henceforth BP). A rapidly spinning black hole can have its rotational energy drawn off by threading an external field through the exterior event horizon of the hole. Whether or not a hole spins sufficiently quickly in an accretion disk, the surrounding Keplerian accretion disk always locally spins at near "break-up" speed, and should constitute an ideal source for the flow, as BP first showed. The speed of rotation of the central regions of the disk depends only upon the depth of the central gravitational well of the central object and as such, suggests a simpler physical model. Is there some additional consideration that would compel us to favour the disk versus spinning hole hypothesis?

In order to answer this question, it is useful to note that there is a striking parallel with the driving mechanism of bipolar outflows in the vicinity of young stellar objects (YSOs). The observed momentum associated with optical jets and CO outflows around young stars is too large by several orders of magnitude to be accounted for by radiative drives (eg Lada 1985, Cabrit 1989). The large measured thrust is found for both low and high mass stars and appears to be a universal attribute of forming star/disk systems. A hydromagnetic drive is therefore implicated and again the possibilities are a fast rotating central young star (eg. Hartmann and McGregor 1982), or its surrounding accretion disk (Pudritz and Norman 1983, Pelletier and Pudritz 1991, henceforth PP).

The YSO problem has the distinct advantage that many essential properties of the flow, and sometimes of the underlying star as well can be directly measured. In particular, a significant number of T-Tauri stars have measured projected rotational velocities $vsin(i)$. While the typical rotation speed of a solar mass speed at break-up is of order 300 km s^{-1}, the average spin speed measured in these stars is only about 1/10 of this value. This is insufficient to account for the high speed jets and outflows near YSOs and a hydromagnetic disk wind becomes an increasingly attractive possibility. While there is abundant evidence for accretion disks around such stars, there seems to be little correlation between the spin rate of stars, and the infrared-excess of the star. The IR excess is usually interpreted as arising by the reprocessing of stellar or accretion disk UV photons by the surrounding dusty disk. If one supposes that the spin down of stars is due to a stellar hydromagnetic wind, then younger stars should rotate systematically faster than stars nearer the main sequence. Thus, if one plots the spin rate of stars on an infrared colour-colour diagram, one would expect to see that the youngest stars (and hence the most deeply embedded) are the fastest rotators. One would by the same reasoning, anticipate that stars without infrared excess are the slowest rotators. In fact, the observations summarized in Figure 1 (from Gomez de Castro and Pudritz 1991) show that there is no correlation of stellar spin and infrared excess. For any infrared colour, one can find objects that are slow or fast rotators. Apparently, there is another torque in the problem that is or has efficiently counter acted the spin-up accretion torque. In the end, the star is left so slowly spinning that it is not a good centrifugal drive for the observed jets.

Thus, by with the theory of disk winds in YSOs, we might find that spinning magnetized black-holes also rotate at relatively low angular velocities. The disk wind idea is attractive by virtue of its simplicity. We do not need to trouble ourselves with the complicated physics of boundary layers or exotic stellar properties (YSO case) in order to predict the properties of the wind. The depth of the central gravitational well is all that matters.

In this contribution, I will show that many of the detailed physical properties of hydromagnetic winds developed for understanding jets in YSOs very naturally extend into the AGN regime. One idea in particular, the production of strong hydromagnetic shocks at magnetic focal points in such outflows, may have a great deal to do with the production of hard X-ray sources in AGNs.

2. General Properties of 2-D, Hydromagnetic Disk Winds

A review of hydromagnetic disk wind theory including recent work by PP can be found in Pudritz, Pelletier, and Gomez de Castro (1991). Here, I confine myself to a brief digest and refer the reader to the these papers. Imagine that we thread a Keplerian accretion disk by magnetic field lines

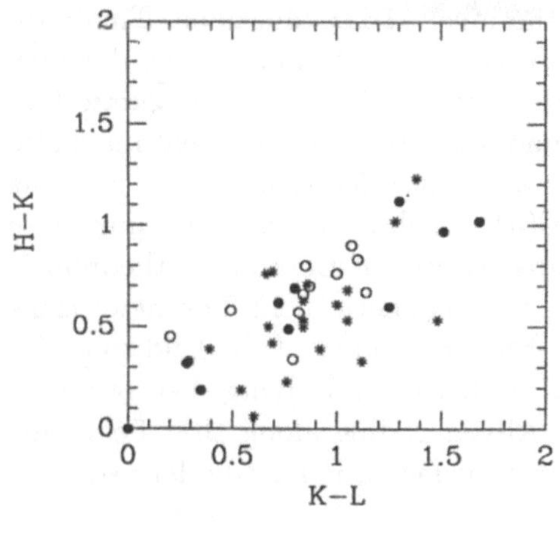

Fig. 1. Measured rotation speeds of T-Tauri stars, plotted in an infrared colour-colour diagram; (i) filled circles, $v sini >$ 20 km s^{-1}, (ii) open circles $20 > v sini >$ 15 km s^{-1}, and (iii) asterisks, $v sini <$ 15 km s^{-1} (from Gomez de Castro and Pudritz 1991).

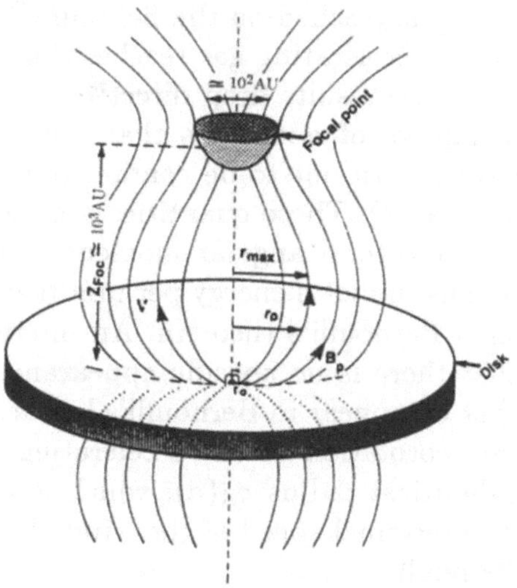

Fig. 2. Shocks at magnetic focal points above and below an AGN disk (adapted from Gomez de Castro and Pudritz 1991).

so that the resulting magnetic geometry resembles an hour-glass with the disk being on the plane that constitutes the narrow waist of the structure. A general feature of rotating hydromagnetic flow is that toroidal magnetic field is produced. Thus, it is convenient to divide all vectors into toroidal and poloidal parts $\mathbf{B} = \mathbf{B}_\phi + \mathbf{B}_\mathbf{p}$. For axisymmetric and time independent flow, one can from the divergence condition simply show that the poloidal field components can be derived from one scalar function, the magnetic flux a; ie $\mathbf{B}_\mathbf{p} = (\nabla a \times \hat{\phi}/r)$ where r is the cylindrical radius.

The acceleration of the wind from an accretion disk can be accomplished by centrifugal forces. Imagine that a magnetic field line makes an angle θ_o with the disk surface at the radius r_o on in the disk. A parcel of gas attached to the field line undergoes a centrifugal force as it is forced by the line to co-rotate with the disk. The parcel is also gravitationally attracted to the central star if the disk is relatively massless. Thus, the component of the gravitational attraction along the field line, is balanced somewhat by the field aligned component of the centrifugal force. Clearly as one moves out along the field line, there comes a point of zero effective gravity, which is known as the slow magnetosonic point. Below this point, one has a hydrostatic corona. Beyond this, centrifugal force, dominates, and the gas parcel is accelerated away. Acceleration effectively ceases when the Alfvén point

$r_A(r_o)$ is reached on the field line, that is where the kinetic energy density in the accelerating gas reaches that of the poloidal field.

These results drop directly out of the ideal MHD equations. There are a number of quantities that are conserved along streamlines of the flow (which turn out to be contained within surfaces of constant magnetic flux $a = const$). These quantities are the mass flux, the poloidal field magnetic flux, the total angular momentum per unit mass $l(a)$ (units cm^2 s^{-1}), as well as the total energy per unit mass $E(a)$, which is of course nothing more than Bernoulli's theorem. An interesting feature of Bernoulli's theorem is that there is no specific appearance of the magnetic field The new terms that do appear in Bernoulli's law arise from the ability of the field to maintain corotation of the accelerating gas with the underlying disk out to a cylindrical radius $r_A(a)$, which is the Alfvén radius along that field line. The terminal speed of the wind along any surface a is readily derived from Bernoulli,

$$v_\infty = 2^{1/2}\Omega_K(a)r_A(a) \tag{2.1}$$

This result makes the simple claim that the terminal speed of the flow depends upon the rotational frequency of the rotor times the lever arm of the flow. In order to make a sensible model for the distribution of magnetic flux across the accretion disk, it is simplest to assume a power law, ie

$$a(r_o) \propto \Omega_K(r_o)^{-1/\alpha} = r_o^{3/2\alpha} \tag{2.2}$$

where the magnetic field line on the surface $a(r_o)$ penetrates the disk at the radius r_o. The self-similar disk wind theory of BP has the particular value $\alpha = 2$ but in general, non self-similar models of any type can be arranged with other flux distributions α.

The *collimation* of hydromagnetic winds is a technically far more difficult problem, but the physics is again essentially simple. Consider the forces that act perpendicularly to the magnetic flux surface. The component of the centrifugal force on the accelerating gas that is perpendicular to the field line clearly is decollimating; the field lines are pushed open. Any magnetic field configuration has an associated magnetic pressure $B^2/8\pi$, and this must also act to push open the magnetic geometry and decollimate the flow. The collimating ingredient is the magnetic tension associated with the toroidal field. The accelerating flow achieves terminal speed slightly beyond the Alfvén surface. It is in this beyond this region that the field becomes ever more toroidal and flow collimation sets in.

There is a unique structure of the toroidal field in the jet that one can deduce from the simplest consideration of jet stability. The current carried by the flow is, from Maxwell $\mathbf{J_p} = (c/4\pi)\nabla \times \mathbf{B}_\phi$. The current intensity up the z-axis is just found by integrating J_z over a cross-sectional area of the flow. A line of algebra reveals that $I_z = (c/2)rB_\phi$. This is important because it shows that the current intensity diverges at either the rotation axis (as in the BP model), or at infinity *unless* $B_\phi \propto r^{-1}$. One can show (see PP) that there is then only one unique flux distribution across the accretion disk that is compatible with this scaling of the current, ie $\alpha = 3$. Thus we see that the minimum energy configuration of the magnetic field in a jet-disk system involves a non-self similar outflow.

The *collimation* of rotating hydromagnetic winds comes about because such winds cannot help but generate a large toroidal magnetic field B_ϕ. This means that pinching forces of the type $\mathbf{J_p} \times \mathbf{B}_\phi$ become important in squeezing the flow towards the axis. The actual equation that describes the force balance is found by merely taking the component of the equations of motion that is perpendicular to the magnetic flux surfaces $a = const.$ It is known as the Grad-Shafranov equation and it has been studied by extensively in plasma physics as well as by jet theorists (eg Camenzind 1990). We find that it can be written in a simplified form in the region beyond the Alfvén surface;

$$\Delta a + \nabla\chi(a).\nabla a = \eta(a)\lambda\frac{a}{r^2} \tag{2.3}$$

(from Pelletier and Pudritz 1991). Here, χ and η are slowly varying functions of a. The function λ measures the amount of angular momentum in the flow,

$$\lambda = \frac{(\Omega_K r_A)^2}{V_A^2}|(\nabla r_A)_{r=r_A}|^2 \tag{2.4}$$

We see from the Bernoulli equation that the terminal velocity scales as the Alfvén speed, so that we might well expect that $\lambda \simeq const$ over the flow. In this case, we see that the GS equation is essentially linear. Further study shows that its solutions are nearly Bessel functions and that the flow surfaces can possibly converge. These conclusions are verified by very detailed calculations (PP) and in fact, can also be seen in the asymptotic analysis of BP. We can find a large class of analytical solutions if we in fact fix $\lambda = const.$ In this case, equation (2.3) becomes a differential equation for the Alfvénic surface that can be solved once we know about the wind density or mass loss rate.

3. Angular Momentum Extraction From Accretion Disks

The wind torque upon the disk can readily be incorporated into the disk angular momentum equation. This equates the wind torque with the flow of angular momentum in the disk,

$$\dot{M}_a\frac{d(\Omega r_o^2)}{dr_o} = 2\frac{d\dot{M}_w}{dr_o}.\Omega r_A^2 \tag{3.1}$$

This equation is the statement that in order to drive a steady state mass accretion rate $\dot{M}_a$ through each annulus of the accretion disk, that a particular mass loss rate per annulus must be arranged, if we are given the lever arm ($= r_A$) of the torque. The wind torque can be shown to dominate any viscous torque in the disk by a factor that is the ratio of their lever arms, that is, $r_A/\delta z_o$ where $\delta \leq 1$ and z_o is the disk half thickness. We see that the result (3.1) allows us to solve the wind mass loss rate, and hence the wind density as a function of r_A. Knowing the wind density, we can find the Alfvén speed of the wind in equation (2.4), and assuming that $\lambda = const$, one can solve for the shape of the Alfvén surface.

Let us note in passing that the disk angular momentum equation has in this analysis been imposed as a general boundary condition on the disk wind. We see that it removes any ambiguity we might have had in modelling the wind mass loss rate. The physics says that independent of the details of how the mass loss is arranged, steady state accretion disk theory ***prescribes*** what the mass loss rate must be in order that the required angular momentum in a Keplerian disk be transported away to enable accretion.

Finally, we have found that a generic feature of our disk wind solutions is that they not only collimate, but for fast enough flows, actually recollimate towards the axis forming *magnetic focal points*. It is convenient to measure the speed of the flow $v_{p,\infty}$ in units of the total magnetic field in the flow, which is the fast magnetosonic (or FM) Mach number n. We find that if $n \geq n_T$, where $n_T = [(2^{3/2}\lambda^{1/2}) - 3]^{1/2} > 1$ is some modest critical value, then flows recollimate towards the axis. It is then easy to show that the focal length of the flow under the most general conditions is

$$z_{foc} = 2[3(n^2+1)]^{1/2} r_{max} \tag{3.2}$$

where r_{max} is the outer maximum outer radius of the outflow.

4. Solutions

By assuming steady state, axisymmetric flow in the wind and disk, and solving equations (3.1) and (2.4), and employing equation (2.1) we can make a detailed set of predictions about the flows. Only two parameters need to be set (this is the same as the more restrictive BP self-similar theory). One of these parameters is λ, and the second measures the magnetic field that threads the inner part of the accretion disk in the source. For AGN conditions it is,

$$\varepsilon_i = 0.34 \frac{(\dot{M}_a/10^{-1} yr^{-1})(M/10^8 M_\odot)^{1/2})}{(B_i/4\times 10^3 G)^2 (r_i/10 r_G)^{5/2}} \tag{4.1}$$

As in Blandford and Payne, we take the inner radius of the flow to be 10 Schwarzschild radii (r_g) since inside this scale, relativistic effects become important and the theory would have to be accordingly modified. We adopt the mass and mass accretion rates onto the central black hole of $10^8 M_\odot$ and $10^{-1} M_\odot yr^{-1}$.

I shall only give our results for the various physical variables for the minimum energy jet ($\alpha = 3$); the interested reader can find the general formulae in PP. For this unique choice of the magnetic flux distribution upon the accretion disk, the magnetic flux scales as $a(r_o) \propto r_o^{1/2}$, and the magnetic field as $B_o \propto r_o^{3/2}$. We set $\lambda^{1/2} = 5$ which, together with the value for ε_i, fixes the two adjustable parameters in the theory. The lever arm of the flow (the Alfvénic surface) scales as

$$\frac{r_A(r_o)}{r_o} = 4.9 \left(\frac{r_i}{r_o}\right)^{1/2} \tag{4.2}$$

where the scale factor in front is derived from the values of our two parameters. From this formula, we find that the Alfvén surface of material leaving the inner part of the disk lies at $49r_G$. An interesting feature of non-self similar models for the flow is that the outflow width now has a characteristic scale. While angular momentum and energy can be removed from all disk radii, most of the outflow stems from the region whose outer border is that radius for which $r_A(r_e) \simeq r_e$. In our context, this lies at about $r_e \simeq 87r_G$. The terminal velocity follows from equations (4.2) and (2.1) so that $u_\infty \propto r_o^{-1}$ Thus, flow from the inner part of the disk reaches speeds that are a factor $(r_e/r_i) = 8.7$ greater than from the outer parts. The mass loss rate in the disk wind scales as $\dot{M}_w(r_o) \propto \dot{M}_a[(r_o/r_i) - 1]$. Integrating over the entire surface of the accretion disk we obtain $\dot{M}_w(r_e) = 0.09\dot{M}_a$. We see that the wind mechanism is indeed efficient; only one disk particle in every ten is required to carry off the angular momentum of material in the disk in the disk wind. The corresponding disk density for material in the wind is, for material that left the disk at radius r_o, and is presently at a radius r in the wind

$$n(r) = 2 \times 10^8 \frac{1}{\mu\chi} \left(\frac{r_o}{r}\right)^2 cm^{-3} \tag{4.3}$$

where μ is the reduced mass and $\chi = (n^2/2) + 1$. A typical outflow speed of material leaving the inner portions of the disk is $< v_\infty/c >= 0.38$ so that the wind is of high velocity but still subrelativistic.

Finally, if one sums up the wind luminosity from each part of the disk $L_w = 0.5\dot{M}_w v_\infty^2$, one finds that it is just half the total gravitational binding energy released by material accreting through the disk, $L_w = 0.5(G\dot{M}_a M_{compact}/r_i)$. This confirms that the mechanical energy of the wind is nothing other than the gravitational energy released during accretion. The process of energy extraction cannot be 100 % efficient, and the residual energy goes into disk heating. The point is that *most* of the energy release could be going out in the wind.

It is somewhat remarkable that the parameters that give a reasonable fit to the data are very similar to those for YSO outflows. If one demands that the observed thrust measured in the CO bipolar outflow surrounding YSOs derives from a disk wind, and uses accretion rates estimated from accretion disk theory, then YSO winds and jets fit nicely with the parameters (PP)

$$\epsilon_i \simeq 0.24 \quad \lambda^{1/2} \simeq 3.2 \quad \text{(YSOs)}$$

This correspondence implies that the physics of disk/jet systems may be universal, relying more upon the depth of gravitational potential well created by the central object than any other property.

5. Recollimation of Flows and Magnetic Focal Points in AGNs

One of the properties of the BP self-similar theory is that flows at high enough FM number, refocus towards the axis. PP have found that this behaviour is generic to most models, including the minimum energy solution discussed above. The critical Mach number for which refocussing occurs is $n_T = 3.3$. Inserting this in our focal length formula (where $r_{max} \simeq r_e$), and using our fiducial outflow radius, one finds the focal length is of order

$$z_{foc} \simeq 1.1 \times 10^3 r_G \tag{5.1}$$

which for a $10^7 - 10^8 M_\odot$ black hole is at $10^{15} - 10^{16}$ cm.

The existence of such focal points is important because shocks are likely to occur there (eg. Achterberg et al. 1983, Gomez de Castro and Pudritz 1991). For the conditions discussed above, the mechanical luminosity of the winds is at least $L_w = 0.8 \times 10^{45} (\dot{M}_a / 10^{-1} M_\odot yr^{-1})(M_8)(r_i/10 r_g)^{-1}$ erg s^{-1}. In the shock, a reasonable fraction of this energy is available to produce the hard X-ray emission that is detected in the ROSAT experiment. The velocity of the gas being about .4c implies that shock emission is indeed possible. One has at the magnetic focal points therefore, an MHD shock that converts a reasonable fraction of the wind energy into hard X-rays, which illuminate the disk. Figure 2 shows the likely geometry. Moreover, since the shock stands off the disk at a distance of $10^{15} - 10^{16}$ cm, one expects that $10^5 - 10^6$ sec variability is possible, as the current data show. Hard X-rays from the MHD shocks at the focal points illuminate the upper and lower surfaces of the AGN disk and will be absorbed and re-emitted by that material if it has sufficiently high column depths.

It appears that reprocessing of hard X-ray radiation from the magnetic focal points is a self-consistent mechanism that explains the recent X-ray observations of AGNs that have been highlighted at this meeting.

I have reviewed work carried out in the main with my collaborators Guy-Pelletier and Ana Gomez de Castro and I thank them for their many stimulating conversations and insights. I also wish to thank the organizers of this interesting meeting for the opportunity to speak. This work was supported by NSERC of Canada.

References

Achterberg, B., Blandford, R.D., Goldreich, P., 1983: *Nature* **304**, 607.

Blandford, R.D.: in: *Active Galactic Nuclei*, SAAS FEE Advanced Course 20, R.D. Blandford, H. Netzer, and L. Woltjer (Springer-Verlag; New York), p.161.

Blandford, R.D., Payne, D.G., 1982: *Monthly Notices Roy. Astron. Soc.*, **199**, 883.

Blandford, R.D., Znajek, R.L., 1977: *Monthly Notices Roy. Astron. Soc.*, **179**, 433.

Cabrit, S., 1989: in; *ESO Workshop on Low Mass Star Formation and Pre-Main Sequence Evolution*, B. Reiputh ed.

Camenzind, M., 1990: *Rev Modern Astr*, **3**.

Gomez de Castro, A.I., Pudritz, R.E., 1991: *Astrophys. J.*, submitted.

Hartmann, L., McGregor, K.B., 1982: *Astrophys. J.*, **259**, 180.

Lada, C.J., 1985: *Ann. Rev. Astron. Astrophys.*, **23**, 267.

Pelletier, G., Pudritz, R.E., 1991: *Astrophys. J.*, submitted.

Pudritz, R.E., Norman, C.A., 1983: *Astrophys. J.*, **274**, 677.

Pudritz, R.E, Pelletier, G., Gomez de Castro, A.I. 1991, in: *The Physics of Star Formation and Early Stellar Evolution*, C.J. Lada and N.D. Kylafis eds. (Kluwer; Dordrecht), p.539.

Rees, M.J., 1984: *Ann. Rev. Astron. Astrophys.*, **22**, 471.

Discussion

Blandford: I would like to comment that the focussing work (with Achterberg and Goldreich) was an act of hydromagnetic desperation perpetrated to account for the Sco X-1 radio source - a source that is now believed to be extragalactic and correlated to the galactic X-ray source! I have a question. What boundary conditions do you (effectively) impose at large cylindrical radius in the wind solution to relax self-similarity?

Pudritz: If one imposes the condition of divergence-free current flow in the jet, there is a unique distribution of magnetic flux across the disk.

Collin: I would like to comment on the fact that the link between the outflowing wind and the accretion flow may give an explanation to an old standing problem: the fine-tuning between the rate of accretion and outflowing mass rate which is observed through narrow lines and broad lines - the outflowing mass rate being of the same order as the inflowing one.

Pudritz: With this highly efficient wind mechanism, only a small fraction of the accretion rate M_a is actually involved in "dissipation" and disk heating. Most of the gravitational binding energy goes out in the wind so that these disks could be "colder" than the standard viscous models people have been using.

Kundt: I am glad you have mentioned the strong resemblance between the extragalactic radio sources and the bipolar flows driven by PTT-stars. In both cases, I find it difficult to extract the dominant fraction of the power directly from the accretion disk; [c.f. Astrophys. Space Sci. **148**, 343 (1988)].

Pudritz: I have not discussed the efficiency of the extraction mechanism, but it can be high in principle. One should investigate antipolar diffusion or the Balbus-Hawley instability in some detail in order to check this (as in Königl, ApJ, 1989).

Camenzind: The global topology of the magnetic fields depends very much on the form of the current in the Grad-Shafranov equation. Did you estimate the influence of non-linearities for the asymptotic solutions?

Pudritz: We self-consistently showed (Pelletier and Pudritz 1991) that asymptotically, nearly linear equations result in the combined Grad-Shafranov and Bernoulli coupled equations. This result appears to conform with similar equations in the Blandford and Payne solution.

Field: The analogy between YSOs and AGNs is fascinating. But in the YSO problem we are given a poloidal field by flux freezing in the ISM collapsing to form the YSO. In the AGN case we can form predominantly a toroidal field from differential rotation. How do we generate the poloidal field you require in AGN?

Pudritz: In AGNs one can always rely upon dynamo action to generate fields within the disk (e.g. Pudritz, 1981, MNRAS). The dominant modes in the existing calculations tend to be even (closed loops) in nature. However, a slight change in boundary conditions could produce predominantly odd (open field) geometry. Even without this, some activity in the disk corona would be able to open up the field lines.

How do Jets Establish Themselves ?

Hélène Sol[1] *and Guy Pelletier*[2]

[1] Observatoire de Paris–Meudon, 92195 - Meudon - France
[2] Université Joseph Fourier 38041 - Grenoble - France

Abstract. Jets from quasars and active galactic nuclei are usually thought of as a mixture of synchrotron radiating relativistic particles, thermal material and magnetic fields, propagating through the ambient interstellar or intergalactic gas, with somewhat well defined boundaries between the jet material and the ambient medium. To reach this situation it is necessary that close to the nucleus some interaction between the 'jet' particles and the ambient medium takes place to generate the jet cavity. Before such an interaction occurs, there is not really a jet but rather a beam of energetic particles streaming through the particles of the ambient medium. From the study of the stability properties of such beams, we deduce that electron-positron beams with bulk Lorentz factors smaller than 43 can be responsible for VLBI features. Then different types of VLBI structure would be expected according to the relative values of parallel magnetic field and background plasma density inside the 'jet'.

1. Introduction

Roughly speaking, galactic nuclei can be described as an energetic medium of protons, electrons, and possibly positrons, surrounded by some ambient medium (interstellar gas, intercloud medium, wind from accretion disk or from star cluster). To generate a jet, it seems that one needs to add rotation and to get some magnetic structure and acceleration zone in the vicinity of the central engine. Relativistic protons, electrons and positrons can thus be extracted from the central part. They stream like beams of particles through the ambient medium without noticing it as long as they have no interaction with it. Once an interaction between the two media takes place, set up of a jet with specific boundary and proper cavity can occur. Two major questions arise. Which type of interaction will develop first? (usual coulombian collisions appear quite negligible). How far from the active nucleus this will happen compare to the VLBI scale ?

The importance of streaming relativistic particles in AGN has been emphasized by several authors (Woltjer, 1966; Camenzind, 1984; Coleman, 1984). A few works studied the radiative properties of particle beams (Rose et al., 1984 and 1987; Baker et al., 1988) and their stability (Achterberg, 1981; Coleman, 1986; Pelletier et al., 1988; Sol et al., 1989; Achatz et al.,

1990). We have recently investigated the case of one-dimensional relativistic beams of electrons-positrons and of electrons-protons streaming through a thermal plasma of electrons-protons (Pelletier, Sol, 1991). The magnetic field is assumed longitudinal. This is a typical beam-plasma configuration where the inversion of population due to the beam can be a source of energy and of instability. There are two main types of self-generated instabilities that we discuss below, one described by the electrostatic branch of the beam-plasma instability, corresponding to the excitation of Langmuir waves, the other described by the electromagnetic branch, corresponding to the excitation of alfvénic waves.

2. Electrostatic beam-plasma instability

This instability seems a priori important since it has the highest linear growth rate and therefore is a good candidate to be the fastest interaction to develop between the beam and the ambient medium.

2.1 Quenching by density fluctuations

It is known from the study of type III solar bursts that density fluctuations can quench the electrostatic beam-plasma instability. We showed that, for different physical reasons, such a result also applies to the case of one-dimensional relativistic cold beams, that is to say with no pitch angle scattering and no velocity spread in the beam (works motivated by solar physics describe three-dimensional scenarios and hot beams). The beam-plasma system is described by two sets of equations, the continuity and momentum equations for the beam and for the ambient plasma, coupled by the Poisson equation for the electric field generated by the instability, $E = E_p + E_b$, E_p being due to the ambient plasma and E_b to the beam. Assuming that the density fluctuations of the ambient medium δn_p are not highly variable and that the electric field admits a slowly varying envelop $\mathcal{E}_{p(b)}$ such that $E_{p(b)} = \frac{1}{2}(\mathcal{E}_{p(b)} e^{-i\omega_p t} + c.c.)$, the equations reduce to :

$$\begin{cases} \left(-2i\omega_p \frac{\partial}{\partial t} - 3v_t^2 \frac{\partial^2}{\partial x^2} - 2i\omega_p \hat{\Gamma}_L\right) \mathcal{E}_p = -\omega_p^2 \mathcal{E}_b - \omega_p^2 \frac{\delta n_p}{n_p} \mathcal{E}_p \\ \left(-i\omega_p + \frac{\partial}{\partial t} + v_b \frac{\partial}{\partial x}\right)^2 \mathcal{E}_b + \frac{n_b}{n_p} \frac{\omega_p^2}{\gamma^3} (\mathcal{E}_p + \mathcal{E}_b) = 0 \end{cases}$$

Here t is the time, x the spatial coordinate along the beam, ω_p the ambient plasma frequency, v_t the thermal velocity of the ambient plasma, $\hat{\Gamma}_L$ the Landau operator, n_p the ambient plasma density, v_b the beam velocity,

n_b the beam density and γ the beam Lorentz factor. After a Fourier-Laplace transformation and using the Bourret approximation to describe the effect of random density fluctuations (but the same result can be obtained for coherent density variation), the dispersion relation is deduced :

$$L_0(k,\omega) - \frac{\lambda^2}{2}\left[L_0^{-1}(k-\bar{k},\omega-\bar{\omega}) + L_0^{-1}(k+\bar{k},\omega+\bar{\omega})\right] = 0$$

$L_0(k,\omega) = 0$ is the usual dispersion relation without density fluctuations, where

$$L_0 = \omega - \frac{3}{2}k^2 + i\nu_L(k) - \frac{g^3}{[\omega - \delta(k)]^2}$$

expressed with reduced variables, the g^3- term describing the beam effect, $g = \left(\frac{1}{2\gamma^3}\frac{n_b}{n_p}\right)^{1/3}$. The λ^2-term is the modification induced by the density fluctuation, $\lambda^2 = \frac{1}{4}\frac{<\delta n_p^2>}{n_p^2} << 1$, $\bar{k}$ and $\bar{\omega}$ being the typical wavenumber and pulsation of the density fluctuation spectrum. For a negligible Landau effect, we found that for density fluctuation with small enough wavelengths (namely $|\bar{k}| >> gk_0$ where k_0 is the resonant wavenumber) this general dispersion relation admits three real roots when $\lambda^2 \geq \frac{3}{2^{2/3}}g^2$. Thus sufficiently large density fluctuations are able to stabilize the beam-plasma system. Fluctuations broaden the sharp resonance of relativistic beams and stability is achieved by a detuning effect.

2.2 Generation of density fluctuations in strong magnetic field zones

We have shown in Pelletier et al., 1988, in the context of strong Langmuir turbulence, that relativistic beams induce lattices of stable Langmuir wavepackets in the ambient medium in highly magnetized regions with $\omega_c > \omega_p$ (ω_c is the electron cyclotron frequency), namely when the longitudinal magnetic field is larger than a critical value $B_c = 3.2 \times 10^{-3}\, n_p^{1/2}$. As the ponderomotive force pushes the particles out of the wavepackets, this results into self-generated density fluctuations which can detune the beam. This appears to be a self-consistent mechanism to quench the electrostatic beam-plasma instability in presence of a strong magnetic field.

3. Electromagnetic beam-plasma instability

As emphasized by Achatz et al., 1990, this instability can have an important effect and rapidly destroy beams with high Lorentz factor $\gamma \simeq 10^3$. It is described by the following dispersion relation :

$$c^2 k^2 - \omega^2 + \omega_{p_i}^2 \frac{\omega}{\omega \pm \omega_{c_i}} + \omega_p^2 \frac{\omega}{\omega \pm \omega_c}$$

$$+ \frac{\omega_{p_a+}}{\gamma} \frac{\omega - kc}{\omega - kc \pm \frac{\omega_{c_a+}}{\gamma}} + \frac{\omega_{p_a-}^2}{\gamma} \frac{\omega - kc}{\omega - kc \pm \frac{\omega_{c_a-}}{\gamma}} = 0$$

Index i is used for the ions (protons) of the ambient plasma, a^+ and a^- for the beam constituents (e^+e^- or p^+e^-). We studied the excitation of alfvénic waves, perturbations of Alfvén waves with pulsation ω close to $k\,v_a$, assuming that the Alfvén velocity is small compare to the light velocity, $v_a << c$ (this is true as long as $\omega_c/\omega_p \ll 40$). For e^+e^- beams, we found that the instability window remains narrow and contains no low-frequency waves of alfvénic type with $|\omega| < \omega_{c_p}$ as long as $\gamma < \left(\frac{m_p}{m_e}\right)^{1/2} \left(\frac{\omega_c}{\omega_p}\right)$, where m_p and m_e are the proton and electron mass. Beams of electron-positron are stable relatively to the excitation of alfvénic waves if $\gamma < 43\,(\omega_c/\omega_p)$ and they can propagate in strong magnetic field zone ($B > B_c$, $\omega_c > \omega_p$) if their Lorentz factor γ is smaller than 43. For p^+e^- beams, it is impossible to avoid the instability window and the high linear growth rate deduced from the dispersion relation suggests that beams of electron-proton are rapidly destroyed, on scales several orders below the parsec-scale, unless important non linearities develop and modify the picture.

4. Models for VLBI structures

From the above study we conclude that beam-plasma systems have no destructive instabilities in strong magnetic field zones, for relativistic $e^+\,e^-$ beams with $\gamma \leq 40$. Other beam dissipation mechanisms such as radiative synchrotron losses, relativistic bremsstrahlung, ionization losses, $e^+\,e^-$ annihilation, inverse-Compton losses (far enough from the central engine) have been discussed in Sol et al., 1989. They can have negligible effects on global dynamics of the beams. Therefore we reach the following scenario. Starting from the most general case of relativistic $p^+\,e^+\,e^-$ beams extracted from the AGN, a first interaction zone arises, below the VLBI scale, where

$e^- p^+$ and too energetic $e^- e^+$ beams are destroyed by excitation of alfvénic waves. Only $e^+ e^-$ beams with $\gamma \leq 40$ can survive on the VLBI scale if the magnetic field parallel to the VLBI "jet" is high enough, $B > B_c$. Such beams can be responsible for VLBI features. This is coherent with the extreme superluminal motion observed in the blazar AO 0235+164 which has a Lorentz factor $\gamma \geq 25$ (O'Dell et al., 1988). In the frame of the standard model for superluminal motion one consequence of our proposal is that pair creation should occur in all superluminal sources.

According to the values of the magnetic field parallel to the VLBI structure and of the density of the thermal ambient plasma two main different regimes for the beam can occur, (i) propagation for high parallel magnetic field $B > B_c$, (ii) dissolution for low parallel magnetic field $B < B_c$. This suggests that different types of VLBI structures should be observed. For $B > B_c$ close to the AGN, $e^+ e^-$ beams propagate until they reach a region with $B \simeq B_c$ where they start to be destroyed by excitation of Langmuir waves. As we described in Sol et al., 1989, this beam relaxation zone at $B = B_c$ corresponds to heating of the ambient medium, with subsequent radio emission enhancement, and to entrainement of the ambient medium (mildly relativistic asymptotic speed can be reached). Realistic values of physical parameters in jets suggest that these critical zones are located somewhere between 10^{-3} and 10^3 pc from the central engines, likely observable at VLBI scale in several sources. For the radio galaxy $3C$ 120 , a good example of high magnetic field sources ($B > B_c$), data by Walker et al., 1987, allow to estimate the location of the critical zone at 1.4 kpc from the nucleus, close to the peculiar 4"-knot. In low magnetic field sources with $B < B_c$ close to the AGN, beams are immediately in a regime where they lose energy and momentum by excitation of Langmuir waves. The situation is then quite propitious for the occurrence of intermittent events during the process of beam dissolution which could be related to the knotty structure of VLBI sources and to radio variability. We showed that the length required for total beam destruction is mainly dependent on the bulk Lorentz factor. Thus estimates of Lorentz factors could be deduced from the extent of VLBI sources.

Several problems such as confinement, magnetic field generation and energy-momentum requirements to feed the large scale radio sources, still remain in such a scenario for VLBI structures. A way to solve them is to assume a two-flow model for jets as we proposed earlier in Sol et al., 1989. The ambient plasma pervaded by the relativistic beam is then a collimated wind from the accretion disk (or from a star cluster). This offers natural explanation for differences between VLBI and large scale properties of jets together with a continuity in the jet structure. This is especially interesting for the velocity which seems to be highly relativistic at VLBI scale and likely much slower, classical or mildly relativistic, at larger scales. Another outcome of two-flow models for jets is the open possibility to describe different types of extragalactic radio sources by different relative importance of the small scale beam to the large scale wind in terms of carried energy and momentum.

References

Achterberg, A., 1981, *Astron. Astrophys.*, **98**, 161

Achatz, U., Lesch, H., Schlickeiser, R., 1990, *Astron. Astrophys.*, **233**, 391

Baker, D.N., Borovsky, J.E., Benford, G., Eilek, J.A., 1988, *Astrophys. J.*, **326**, 110

Camenzind, M., 1984, *Astron. Astrophys. (Letters)*, **130**, L9

Coleman, C.S., 1984 : "X-ray and UV emission from AGN, MPE report 184," Brinkmann, W., and Trümper, J., eds, p. 203 ; and 1986 : "Structure and evolution of AGN", G. Giuricin et al., eds, Reidel publishing company, p. 521

O'Dell, S.L., et al., 1988, *Astrophys. J.*, **326**, 668

Pelletier, G., Sol, H., 1991, submitted

Pelletier, G., Sol, H., Asseo, E., 1988, *Phys. Rev.*A, **38**, 2552

Rose, W.K., Beall, J.H., Guillory, J., Kainer, S., 1987, *Astrophys. J.*, **314**, 95

Rose, W.K., Guillory, J., Beall, J.H., Kainer, S., 1984, *Astrophys. J.*, **280**, 550

Sol, H., Pelletier, G., Asseo, E., 1989, *Monthly Notices Roy. Astron. Soc.*, **237**, 411

Walker, R.C., Benson, J.M., Unwin, S.C., 1987, *Astrophys. J.*, **316**, 546

Woltjer, L., 1966, *Astrophys. J. (Letters)*, **146**, L 597.

Discussion

Lesch: How does the stability of the beam depend on the ratio of background density to beam density ?

Sol: Our results apply to beam to background density ratio between 10^{-4} and 10^{-1}, typically of the order of 10^{-2}.

Burbidge: What is the efficiency of the process, i.e. what is the ratio between the energy in the mildly relativistic jet and the energy initially put in by the "black box" ?

Sol: Beams are totally destroyed after some relaxation length and transmit all their energy and momentum to the ambient plasma. In that sense, the efficiency is 100 %. The energy available to build the large scale radio sources is the sum of the energy put in the relativistic beam by the "black box" and of the energy carried by the ambient plasma if it is a wind from the galactic nucleus.

Kundt: For a few AGN sources (including 3C 273 and our own galaxy), Lorentz factors above 10^4 have been 'observed'. Such high Lorentz factors are probably required for the bulk motion of the pair plasma, among others to explain the one-sided optical jets (cf my contribution to "Hot spots in extragalactic radio sources", Lecture notes in Physics, 327, Meisenheimer and Röser, eds., Springer, 1989).

Sol: One possibility would be to assume that the beam propagates in a vacuum with no background plasma close to the central engine. In such a case, the interactions described here would occur only in a mixing layer around the beam. As interaction likely leads to radio emission enhancement this picture could be used to explain the edge-brightening observed in some jets.

The Stability of Current-Carrying Jets

Stefan Appl and Max Camenzind

Landessternwarte Königstuhl, W-6900 Heidelberg, Germany

1. Introduction

A global electric circuit model for radio galaxies was developed by Benford [1]. The magnetosphere of the central engine is connected to the hot spots by large scale currents that flow in the jet. Any jet that emanates from a rotating magnetosphere inevitably produces an electric current. Estimates typically yield currents of the order of 10^{18} Ampères. In order not to charge up the nucleus a return current of equal size has to close the circuit. Further important features are that the plasma flow goes through the three critical MHD-points terminating at supermagnetosonic velocities, and that the jets eventually collimate asymptotically to a cylindrical shape. As is known from fusion research magnetically confined plasmas are violently unstable, a major concern have been the instabilities of current-carrying jets, in particular w.r.t. nonaxisymmetric perturbations.

2. The Equilibria

As the equilibria cannot be derived from observations one has to rely on theoretical arguments and simulations. As the jet propagates it will evolve into a state of minimum energy. Relaxation theory [2, 3] provides a useful tool in finding such equilibria. It states that the minimum energy configuration is given by a special force-free field $\nabla \wedge \mathbf{B} = \alpha \mathbf{B}$, where α is a constant independent of the flux surface. Solutions for cylindrical geometry are given by the Besselfunctions $B_\phi = B_0 \, J_1(\alpha r)$ and $B_z = B_0 \, J_0(\alpha r)$. Making use of the condition that no net current flows and assuming that current and return-current flow within the jet we can fix α to the value of the first zero of J_1, $\alpha R = 3.83$, where R is the jet radius. Such an equilibrium is independent of the initial conditions at the footpoint of the jet and of the detailed processes that lead to it. Configurations of this kind are known in fusion research as *reversed field pinches* (RFP), because B_z changes sign within the plasma column. We later call it also relaxed jet.

The second current-carrying equilibrium that we are considering applies to a jet on the kpc-scale. Axisymmetric MHD-simulations with a helical field [4] help to define an equilibrium for this domain. Typically the jet possesses a beam-cocoon structure, i.e. the beam propagates outwards from the central engine and after being shocked in a Mach disk at the head of the jet it is flowing back along the beam forming a cocoon. During propagation the current diffuses out of the beam. It flows in the beam-cocoon interface and returns in the interface between cocoon and external medium. The cocoon is dominated by a toroidal magnetic field, while in the beam it is mainly longitudinal. We model this equilibrium by a jet with a homogeneous longitudinal field surrounded by a static zone with a purely toroidal field that extends out to $r = 3R$. This equilibrium is referred to as cocoon-jet.

For both equilibria the jet is described as an infinite plasma column with a homogeneous velocity profile. This is the simplest assumption that is still reasonable and at the same time it allows to compare the results with those of hydrodynamic jets and equilibria with a homogeneous longitudinal magnetic field. For the latter two exist analytic dispersion relations. The jet is in pressure equilibrium with an external thermal plasma without magnetic fields. Internal and external densities are taken to be equal, the adiabatic indices are $\gamma = \frac{4}{3}$ and $\gamma = \frac{5}{3}$, respectively.

As parameters we use the fast magnetosonic Mach number $M = v/\sqrt{v_a^2 + c_s^2}$ on the axis (v_a and c_s are the Alfvén and sound velocities, respectively) and the plasma-β, $\beta = 8\pi p/B^2$. While the first gives the ratio of the kinetic and the sum of magnetic and thermal energy, the latter measures the ratio of thermal and magnetic energy. The fast magnetosonic Mach number is in fact the relevant one. For low-β jets it becomes the Alfvén Mach number and for hydrodynamic jet it reduces to the familiar sonic Mach number.

3. Stability

The stability of the above equilibria was investigated using a linear stability analysis for an infinite cylindrical ideal compressible MHD-equilibrium and compared to the corresponding hydrodynamic configuration and one with a purely longitudinal field [5]. A normal mode expansion of the perturbed quantities $\propto \exp\{i(m\phi + kz - \omega t)\}$ leads to an eigenvalue problem for the two ordinary differential equations

$$\frac{1}{r}\frac{d}{dr}(r\xi_r) = C_1\xi_r - C_2P'$$
$$\frac{d}{dr}P' = C_3\xi_r - C_1P'$$

for the radial displacements ξ_r of the fluid elements and the total perturbed pressure $P' = (p+B^2/8\pi)'$. The above system, supplemented by Sommerfeld boundary condition, was solved numerically for complex ω and real k. As for supersonic motions spatial growth lengths are more meaningful than temporal growth rates, the inverse growth length k_g was computed by means of the group velocity $v_g = \partial\Re\omega/\partial k$ as $k_g = \Im\omega/v_g$ [6].

Jets suffer from Kelvin-Helmholtz-instabilities (KH). For current-free jets they are classified [7] into fundamental and reflection modes. The first are essentially backwards traveling waves in the jet. They exist for all wavelengths and Mach numbers. In their nonlinear development they may lead to the disruption of the jet, in particular the kink mode. The reflection modes in turn exist only for supersonic velocities and above a critical wavenumber. They consist of waves that are growing in amplitude by resonant reflection at the boundary of the jet. They apparently form weak oblique shocks and do not destroy the coherence of the jet. Magnetized jets with a purely longitudinal magnetic field show the same mode structure, but with somewhat reduced growth rates. The *static* RFP is known to be liable to current-driven instabilities, in particular to the free boundary kink mode.

In the case of current-carrying jets one has to deal with the combined effects of KH and current-driven instabilities. The entire spectrum will therefore be very complex. For supermagnetosonic velocities, however, the kinetic energy, which is the driver of the KH instability exceeds the sum of thermal and magnetic energy. Therefore, the KH instability will dominate, and the effect of the electric current will consist in a modification of the instabilities of hydrodynamic jets. The most dangerous instability is the fundamental kink mode.

4. Results

The results are displayed as inverse growth lengths k_g as a function of the longitudinal wavenumber k, both quantities normalized w.r.t. the inverse jet radius. The three equilibria involving magnetic fields have a plasma-β of zero.

Figure 1 shows the nonaxisymmetric fundamental modes for the two current-carrying and two current-free equilibria for Mach numbers of 3 and 10. In the latter case (Fig. 1b) the differences are most evident, but what follows holds also for lower Mach numbers. The most unstable configuration

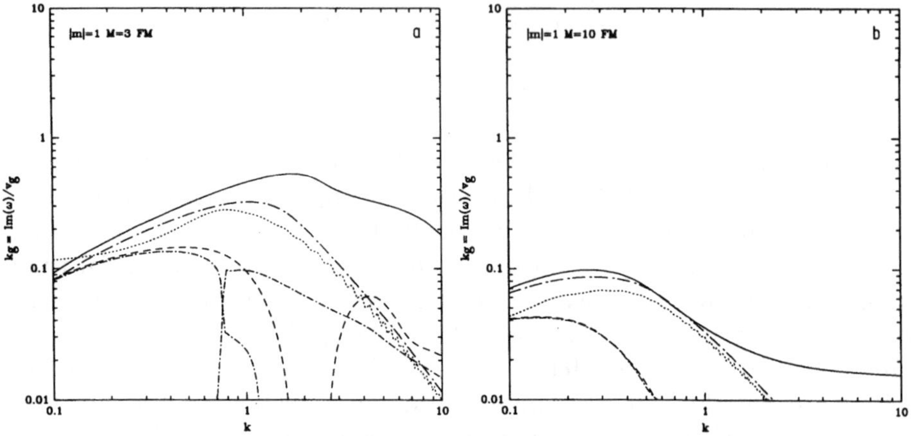

Fig. 1. Inverse growth length as a function of wavenumber of the $|m| = 1$ fundamental mode for four different equilibria at a Mach number a) $M = 3$, b) $M = 10$. The curves represent the hydrodynamic (solid), longitudinal field (long dash-dotted), cocoon (dotted), and RFP ($m = 1$ short dashed; $m = -1$ short dash-dotted) equilibrium

is the hydrodynamic jet, followed by the flow with longitudinal field. The cocoon-jet, which is identical to the previous apart from the toroidal field wrapped around, behaves similarly, with however slightly reduced growth rates, mainly at long wavelengths. The most stable configuration corresponds to the relaxed jet. For all three magnetically dominated equilibria the growth rates at short wavelenghts are drastically reduced. The behaviour at smaller Mach numbers (Fig. 1a) is generally similar though somewhat complicated by the growing importance of the details of the magnetic structure.

Figure 2 shows for the same equilibria the first axisymmetric reflection mode. The stability properties here are essentially the same. More clearly than for the fundamental modes, the reflection modes show that, in particular for the relaxed jet, the maximum of the instability is shifted towards longer wavelengths and the maxima are broader.

Finally, Fig. 3 shows for the nonaxisymmetric fundamental mode of the relaxed jet the dependence on Mach number. Due to the convective nature of the instability high Mach numbers are stabilizing just as is the case for hydrodynamic jets. In addition however, the magnetic structure has a significant influence depending on the Mach number, e.g. for high values the $m = 1$ fundamental mode (Fig. 3a) splits up into two unstable branches, separated by an interval where the mode is entirely stabilized. The $m = -1$ mode (Fig. 3b), in turn, divides into two overlapping branches for lower Mach numbers. The details depend on the specific current profile chosen. The influence of the plasma-β is discussed elsewhere [8].

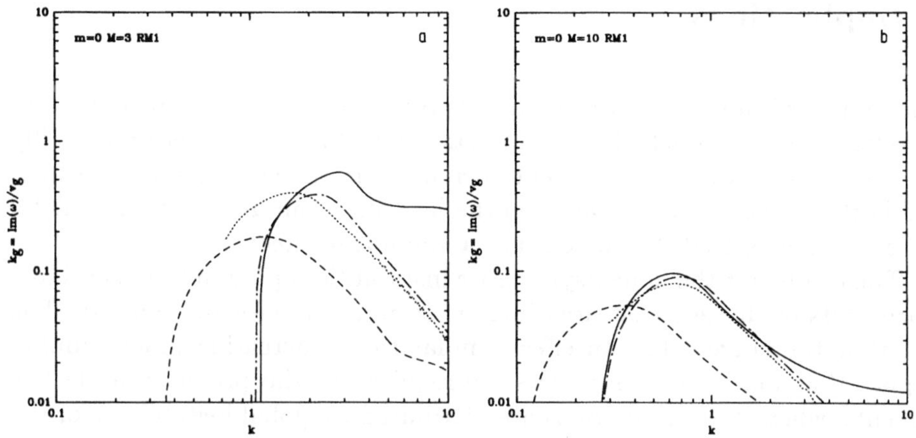

Fig. 2. Inverse growth length as a function of wavenumber of the $m = 0$ first reflection mode for four different equilibria at a Mach number a) $M = 3$, b) $M = 10$. For the meaning of the different lines see Fig. 1

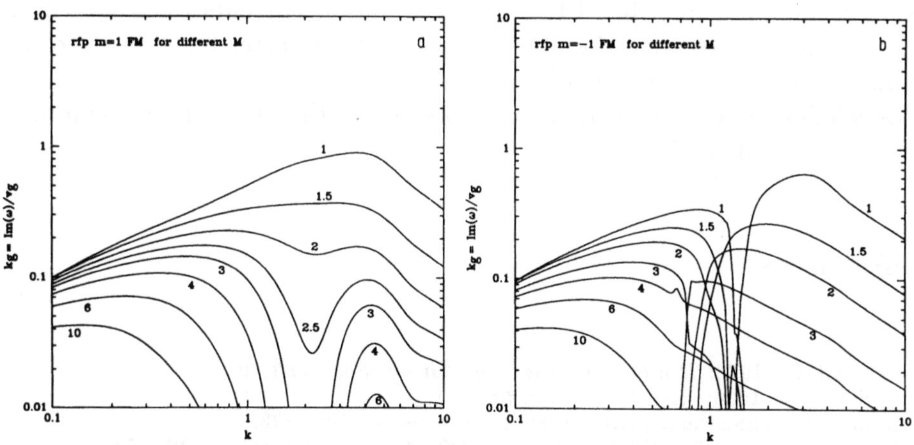

Fig. 3. Inverse growth length as a function of wavenumber of the nonaxisymmetric fundamental mode for the relaxed jet with $\beta = 0$ for various Mach numbers. a) $m = 1$, b) $m = -1$

Summarizing, at *all wavelengths and Mach numbers* the current-carrying jets are more stable than their current-free counterparts. Low plasma-β and high Mach numbers effectively stabilize the beam. The currents tend to shift the instability towards longer wavelengths and act therefore similar as an increased Mach number.

5. Implications

The enhanced stability gives good support to jets carrying currents, and therefore also for models that create jets from rotating magnetospheres [9]. The fact that two widely different current-carrying equilibria were tested, and both are more stable than even the jet with a purely longitudinal field, indicates that indeed the current has a stabilizing effect.

The results for the nonaxisymmetric instabilities apply to the lateral displacements of the jets that are observed in radiogalaxies and quasars. The fact that the current has an effect similar to an increasing Mach number implies that lower Mach numbers are required in the presence of electric currents, when these oscillations are attributed to global instabilities of the flow. Furthermore, low plasma-β jets appear to reduce significantly small wavelength instabilities. According to general belief those produce turbulence and dissipate kinetic energy. Therefore, magnetically dominated jets can propagate more effectively and will produce less particle acceleration resulting in a low surface brightness. Consequently, looking at the morphology, it seems plausible that FR 2 jets have larger currents than the FR 1 sources. This agrees with the observation that the latter have much lower estimates for the magnetic field.

Acknowledgement. This work was supported by the Deutsche Forschungsgemeinschaft (SFB 328).

References

1. Benford, G., 1978: *Monthly Notices Roy. Astron. Soc.*, **183**, 29.
2. Taylor, J.B., 1986: *Rev. Mod. Phys.* **58**, 741.
3. Königl, A., Choudhouri, A.R., 1985: *Astrophys. J.*, **289**, 173.
4. Kössl, D., Müller, E., Hillebrandt, W., 1990: *Astron. Astrophys.*, **229**, 378.
5. Appl, S., Camenzind, M., 1991: *Astron. Astrophys.*, submitted.
6. Birkinshaw, M., 1984: *Monthly Notices Roy. Astron. Soc.*, **208**, 887
7. Ferrari, A., Trussoni, E., Zaninetti, L., 1981: *Monthly Notices Roy. Astron. Soc.*, **196**, 1051
8. Appl, S., Camenzind, M., 1991: in *Extragalactic Radio sources – From Beams to Jets*, Eds. G. Pelletier, J. Roland, H. Sol, Cambridge University Press, Cambridge
9. Camenzind, M., 1990: *Rev. Mod. Astron.* **3**, 234.

Discussion

Wilson: I'd just like to draw the attention of the hydromagneticists to the jet in the galaxy NGC 4258. This is a jet in a spiral galaxy and appears to lie more or less in the galaxy disk. The jet is seen in *both* radio synchrotron and optical emission lines. Cecil, Tully and myself have recently mapped it both spatially and kinematically in an optical emission line (Ap.J. submitted). We find that the inner jet is actually comprised of two strands which wrap around each other in two helical structures. These helices are seen in both space and velocity mapping over about three helix wavelengths. It looks very much like two magnetic flux tubes being wrapped through connection to a nuclear rotating disk. This disk cannot be a standard (small) accretion disk around a black hole, because the observationally inferred rotation period is much longer than the disk's dynamical rotation period. Instabilities in current-carrying jets are unattractive explanations in view of the presence of two helices and the clear periodic structure over three wavelengths.

Fields: In your force-free field solution the magnetic field does not match at the boundary. Therefore you have to have a surface current which separates the jet from the rest of the world. I wonder how you handle the energy associated with this current sheet.

Appl: The transition between the jet to the surrounding medium is approximated by step functions in velocity and magnetic field. This is done in order to compare with previous stability results for jets and not to complicate the model. Phenomena due to and within the boundary, such as current sheets and turbulent layers, can therefore not be addressed by this model. Using ideal MHD, in particular, the surface current does not dissipate energy. It shields the jet electromagnetically. Magnetosonic waves, however, which are most relevant for instability, can propagate across the boundary.

Observations of Compact Active Galactic Nuclei - High Frequency VLBI and Intraday Variability

Arno Witzel

Max-Planck-Institut für Radioastronomie, Bonn, Germany

Für Peter Höcker

1. Introduction

The observations of variability in extragalactic radio sources and the observational method of VLBI are closely related. The discovery of variations in the intensity of the radio emission of extragalactic objects by Sholomitsky (1965) and Dent (1965) and subsequent work has not only shown that incoherent synchrotron radiation is the dominant emission process for the radio radiation in such objects, but in addition has allowed first derivations of the sizes of the emitting regions. Causality arguments have allowed source sizes as small as lightweeks to be deduced from the variability timescales known at that time. The postulate of such compact objects at large distances with angular sizes well below arcseconds was the prime impetus for developing the technique of radio interferometry with very long baselines (VLBI) and the VLBI observations in turn proved nicely the existence of the postulated compact regions by means of direct observations.

The discoveries then made by VLBI are manifold and have played a major role in our present day understanding of the physical processes in Active Galactic Nuclei (AGN). In particular, the development of the relativistic jet model would not have been possible without direct observations with resolutions on the milliarcsecond scale, which corresponds to pc-dimensions in distant quasars and similar objects. Numerous books have been written on the subject and two recent ones may be mentioned here, since they summarize the advances and the present state of research in a splendid way: "Superluminal radio sources" (ed.: Zensus and Pearson, 1987) and "Parsec scale radio jets" (ed.: Zensus and Pearson, 1990). As a result of steady improvements on the technical side, it is nowadays justified in numerous cases to talk about images of radio sources with mas-resolution obtained by VLBI rather than maps or models. Leaving beside the intriguing results from polarization VLBI observations which show marked differences between quasars and BL Lac-objects (e.g. Roberts *et al.*, 1990), I want

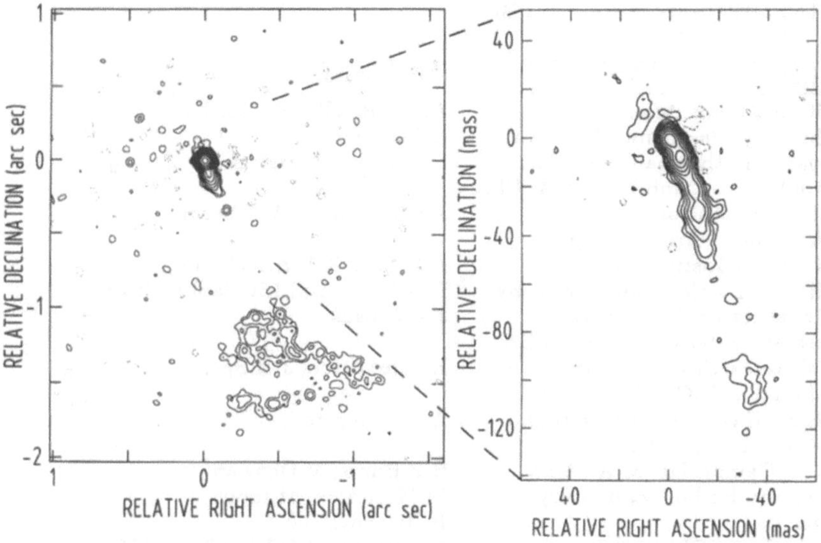

Fig. 1. **The combined MERLIN-VLBI maps of the quasar 0836+71 (z=2.16) as observed at 18 cm in 1988.97 (Hummel *et al.*, 1992).**

to mention here the recent results of the statistical investigations of complete samples of radio sources, which prove that bulk relativistic motion is present in the majority - if not essentially all - compact flat spectrum radio sources (Witzel *et al.*, 1988; Pearson and Readhead, 1988; Schalinski *et al.*, this volume), and in a significant and constantly increasing fraction of lobe-dominated quasars as well (Zensus and Porcas, 1987, Hough and Readhead, 1989). Also noteworthy is the high degree of complexity in some mas-jets, in which stationary and apparently superluminal moving components coexist. As an example of such objects, the quasar 0836+710 is shown in fig. 1. This high redshift quasar (z=2.16) shows one-sided emission even at a dynamic range of 1:1500. The figure presents the results of so called "world array" observations at 18 cm wavelength, i.e. observations with the Merlin Interferometer and an intercontinental VLBI-array taken simultaneously and reduced together, thus giving a continuous range of resolutions from 200 " to 3 mas (Hummel *et al.*, 1992).

Obviously, still unresolved structures are present in the images obtained from earthbound cm VLBI observations, thus showing that still better resolving power is required. Therefore the evolution of the VLBI technique has to go into two directions:

1) Space-VLBI, which will enlarge the baselines and thus the resolution and under certain conditions the quality of the imaging.
2) mm-VLBI, which can also increase the resolution and, in addition, allows the observation of the sub-parsec scale nuclear regions, which are optically thick at longer wavelengths.

Table 1. The participants of 7 mm-VLBI

Institution	Collaborators
California Institute of Technology, Pasadena, CA, USA	A.C.S.Readhead, C.R.Lawrence, R.Vermeulen
Harvard-Smithsonian Center for Astrophysics, Cambridge, MA, USA	I.I.Shapiro, N.Bartel
Instituto de Astrofisica de Andalucia, Granada, Spain	J.M.Marcaide, A.Alberdi
Max-Planck-Institut für Radioastronomie, Bonn, Germany	A.Witzel, T.P.Krichbaum, D.A.Graham, I.I.K.Pauliny-Toth
National Radio Astronomy Observatory, Socorro, N.M., USA	J.A.Zensus
Naval Research Laboratory, Washington, D.C., USA	K.J.Johnston, J.H.Spencer
NEROC Haystack Observatory, Westford, M.A., USA	A.E.E.Rogers
M.I.T., Cambridge, M.A., USA	B.F.Burke, V.Dhawan[1]
Nobeyama Radio Observatory, Japan	M.Morimoto, M.Inoue, H.Hirabayashi[2]
Onsala Space Observatory, Sweden	R.S.Booth, B.O.Rönnäng, A.Kus[3], F.Colomer

[1] now at Raman Research Institute, Bangalore, India.
[2] now at Institute of Space and Astronautical Science, Tokyo, Japan.
[3] now at Torun Radio Astronomy Observatory, Poland.

2. Observations with mm-VLBI

At present unfortunately no results are available from dedicated space-VLBI experiments except for the TDRSS missions, which have demonstrated the technical feasibility of orbiting VLBI (e.g. Levy *et al.*, 1989). Future space missions ('Radioastron' and 'VSOP') are expected to provide results during this millenium still.

mm-VLBI, on the other hand, has been successfully tested and used at several frequencies up to 100 GHz (e.g. Baath *et al.*, 1991). At 7 mm wavelength, the 'breakthrough' towards almost normal 'production runs' has been achieved recently. Table 1 lists the institutions and scientists involved in the work with the 7 mm-VLBI-array, which was used for the observations described below. Due to the still limited sensitivity the number of scientists involved is at present larger then the number of objects which have been investigated in detail. Table 2 indicates these sources in the upper section, whereas the lower part lists those sources for which fringes have been detected in snapshot 'fringe-finding' experiments.

The sub-parsec structure of the sources 3C 84, 3C 273, and 3C 345 will be presented in the following in the context of their images at longer wavelengths. In particular, the kinematic evolution of the individual components and the phenomenon of curved trajectories in jets will be addressed.

Table 2. The sources detected so far with VLBI at 43 GHz

Source	1985.16[1]	1986.35[2]	1987.44[3]	1988.49[4]	1989.19[5]	1990.48[6]
3C 84	S	I	I	I		I
4C 39.25						I
3C 273	S	S		I	I	
3C 345	S	S	M	I		I
1803+78	S	S	M	I	I	M
BL Lac						M
3C 454.3		S	M			
0133+47	S	S				
0212+73						S
0234+28	S					
NRAO 150	S	S				
0716+71			S	M		
0804+49						S
0917+62						S
OJ 287	S	S				
3C 279			S			
1308+32	S	S				
NRAO 512			S			
1928+73			S			
2007+77			S			S

[1] see also Marcaide *et al.*, 1985.
[2] see also Bartel *et al.*, 1988.
[3] see also Dhawan *et al.*, 1990; Krichbaum, 1990.
[4] see also Krichbaum and Witzel, 1991(a,b); Zensus *et al.*, 1992.
[5] see also Krichbaum, 1990; Krichbaum *et al.*, 1990.
[6] data analysis not yet complete.

I image, map and Gaussian modelfit obtained
M amount of data allows only modelfits
S snapshot observations yielding detections

3C 273

The quasar 3C 273 (z=0.158; 1 mas corresponds to 1.78 pc at $H_0 = 100$ km s^{-1} Mpc^{-1}, $q_0 = 0.5$) is one of the archetypes of a superluminally expanding radio source and has been observed intensively with VLBI at several wavelengths. Owing to its high brightness 3C 273 is one of the prime candidates for 7 mm-VLBI observations. In the beginning of 1988 the source underwent a large and rapid flux density outburst, first observed at optical wavelengths and subsequently in the radio regime (Courvoisier *et al.*, 1990). Fig. 2 shows a composite picture of the flux density variations at mm-wavelengths and the 7 mm-VLBI maps taken four and thirteen months after the time of the (optical) outburst. The identification of the components as shown in fig. 2 leads to an angular expansion of 0.8 mas/yr. This corresponds to an apparent superluminal motion with $\beta = 5.5$ for an assumed $H_0 =$ 100 km s^{-1} Mpc^{-1}, $q_0 = 0.5$ and with the assumption of a cosmological explanation of the redshift. The velocity derived from this observation is in good agreement with the values obtained at longer wavelengths. (A detailed discussion is given in e.g. Krichbaum *et al.*, 1990; Krichbaum and Witzel, 1991(a,b)).

It has been convincingly demonstrated that the apparent superluminal motion in 3C 273 follows a bent trajectory (e.g. Biretta *et al.*, 1986, Zensus *et al.*, 1988). The degree of curvature seems to increase with decreasing separation from the core (see Krichbaum *et al.*, 1990 and Baath *et al.*, 1991). The observations available at present are consistent with one common trajectory for the different components in contrast to the situation in the 3C 345 nucleus (see below). At 3 mm wavelength 3C 273 shows at the epoch 1988.3 an elongated core-jet structure with several components oriented towards the west, in good agreement with the results described above (Baath *et al.*, 1991).

3C 345

The superluminal quasar 3C 345 (z=0.595; 1 mas corresponds to 3.78 pc) has a well-documented history of its nuclear kinematics, too. VLBI observations over a wide range of frequencies have recently been discussed by Zensus, 1990 (see also references therein). The source structure at 7 mm wavelength at the epochs 1988.5 and 1990.5 is shown in the accompanying contribution by Krichbaum *et al.*, this workshop. It is in excellent agreement with observations at e.g. 1.3 cm wavelength, but obviously shows more details due to the higher angular resolution.

Fig. 3 shows the curved trajectories of the individual components identified in 3C 345 since the early eighties. The data for components C4 and C5 are taken mainly from observations at 22 GHz (Zensus *et al.*, 1990) and 10.7 GHz (Biretta *et al.*, 1986). Close to the core, measurements at 7 mm revealed the two new components C6 and C7 which are located significantly further north than C4 and C5. Future measurements will have to show, whether their trajectories differ grossly from the two well- established paths of C4 and C5. In contrast to 3C 273, a common bent trajectory for the individual components is excluded by the observational data for 3C 345.

3C 84

Due to its complicated radio structure the peculiar galaxy NGC 1275, also named 3C 84, is also "technically" a great challenge to mm-VLBI observations with the present arrays. Long term monitoring of this relatively nearby object (z=0.018; 0.1 mas corresponds to 30 light days) has shown subluminal motion of components embedded in an extended lobe-like structure extending about 10 mas from the core towards the south. Romney *et al.*, 1984 find a velocity of 0.25 c based on 10.7 GHz data, Marr *et al.*, 1989 give 0.5 c based on 22 GHz observations. The 7 mm-VLBI data taken between 1986.4 and 1988.5 show structural variability of the northern core

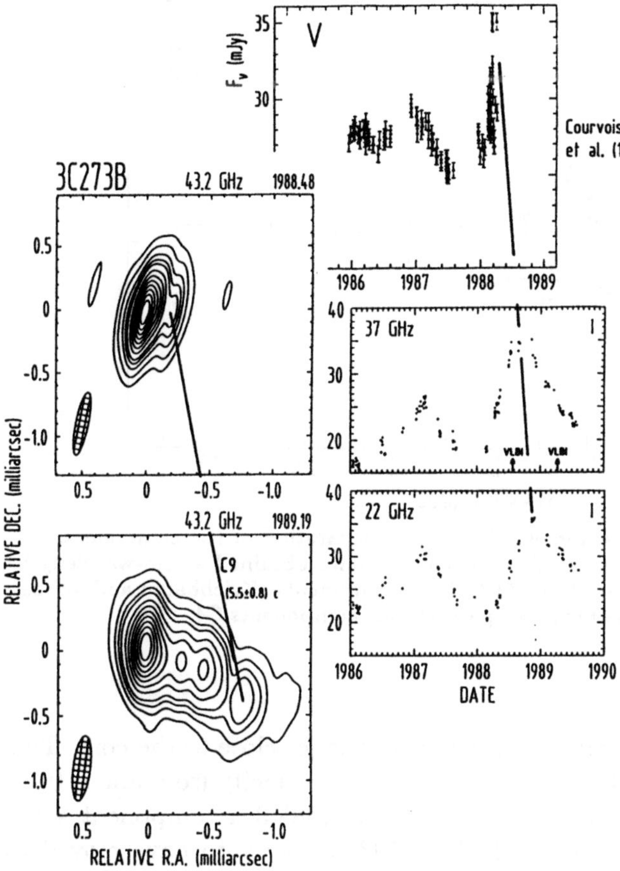

Fig. 2. **43 GHz-maps of 3C 273 at 1988.48 and 1989.9 (Krichbaum *et al.*, 1990). The straight line indicates the motion of component C9. Optical (V-band) and radio lightcurves (22 GHz and 37 GHz) of the pronounced flux density outburst in 1988 are shown on the right (Courvoisier *et al.*, 1990).**

region (fig. 4). Additional southern structures found at 7 mm up to a core distance of 7 mas are consistent with southern structures seen at 22 GHz (see the figure in Krichbaum *et al.*, this volume, and Readhead *et al.*, 1990; Readhead *et al.*, *in preparation*).

Fig. 4 presents models of the jet-like structure seen (north) in the nucleus. (A diffuse surrounding halo-like component has been omitted in the plots). In addition, cuts through the brightness distribution along the mean jet axis are presented. The obvious changes in the structure are most readily explained by an outward motion of the components labeled B, C and D with velocities of 0.1 to 0.2 c. Component A is proposed to be the core on the basis of its inverted spectrum and its variability. N is a new component visible since 1988.5. The position angle of this "northern mas-jet" is about 220° and thus significantly deviates from the preferential direction in the

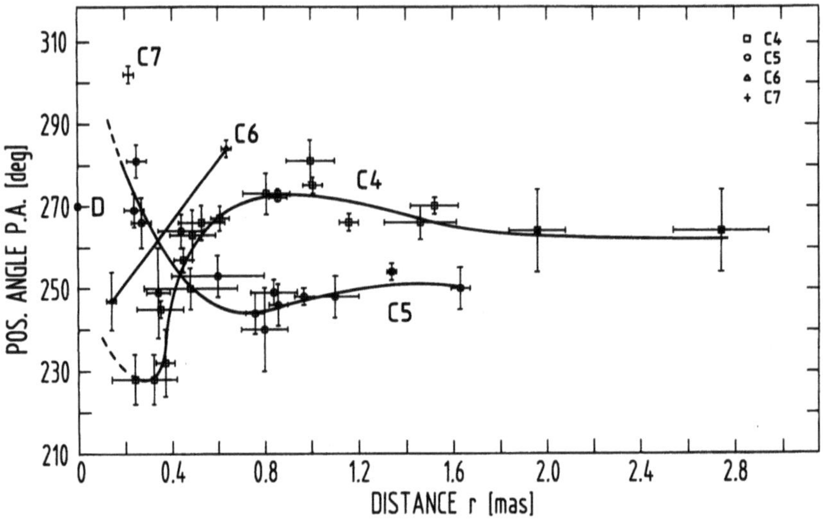

Fig. 3. A plot of position angle versus relative core separation of the components C4, C5, C6 and C7 in 3C 345. Open symbols denote for data obtained at cm-wavelengths (Zensus, 1990), filled symbols are from 43 GHz measurements (Krichbaum and Witzel, 1991 b). The lines sketch a possible mean path for the components C4 – C6.

southern lobe region indicating a curvature of the jet close to the core. This curvature is accompanied by a change in apparent velocity from about 0.1 c to 0.5 c. The 3 mm-VLBI measurements show 1988.3 a component separated by 0.1 mas from the core, which in 1989.3 seems to have moved to 0.25 mas distance (Baath, 1991). This corresponds to a velocity of 0.1 c in good agreement with the value deduced from the 7 mm-VLBI observations.

In all three sources discussed above the jets are significantly bent and the degree of curvature seems to increase with decreasing distance from the core. Also in other, less well observed, sources (e.g. 1803+78 (Krichbaum, 1990); BL Lac, (Krichbaum, *unpublished data*)) enhanced jet curvature in the proximity of the core is found. In 3C 273, 3C 345 and 1803+78 quasi-sinusoidal jet bendings (i.e. bends followed by back-bends) are observed (Krichbaum and Witzel, 1991 a,b). In 3C 345 individual components seem to follow their own trajectories, whereas in 3C 273 at present a common trajectory for all knots seems adequate. It seems tempting to identify these paths with projections of three-dimensionally curved (helical) trajectories. The existence of helical jet structures is theoretically expected and discussed in various magneto-hydrodynamic jet models (Königl and Choudhuri, 1985; Lovelace et al., 1987; Hardee and Norman, 1989). In particular, helical jet structures could be produced by a magnetic field anchored in a rotating accretion disk (e.g. Camenzind 1990).

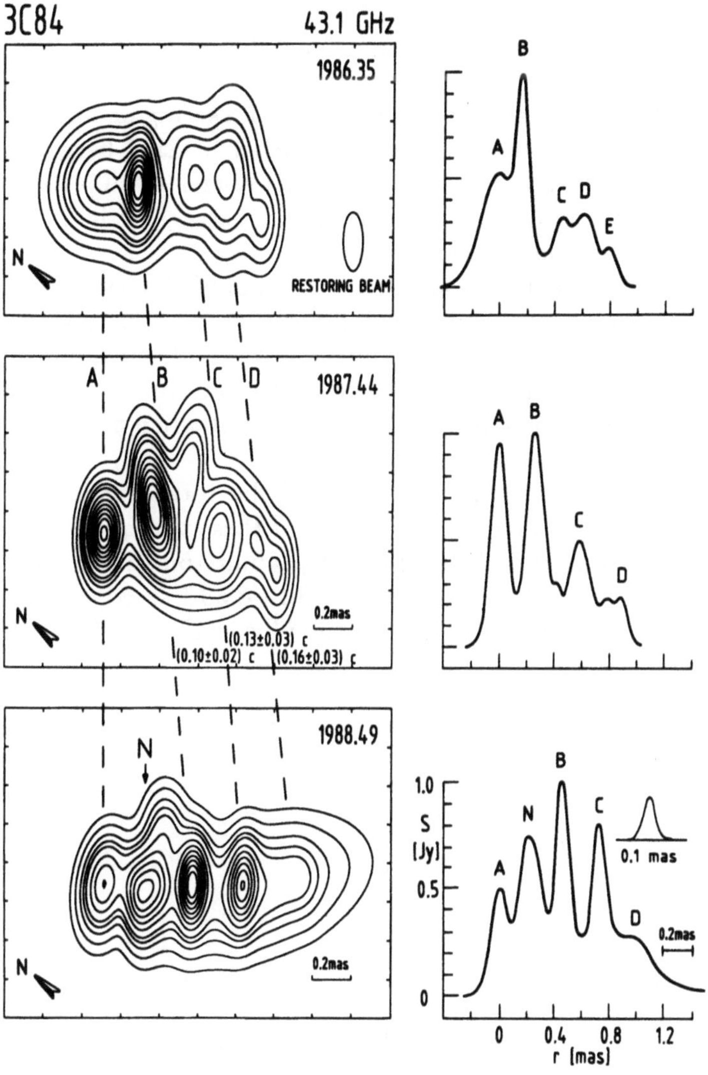

Fig. 4. Left: 43 GHz-models of the northern core region in 3C 84 at 1986.4 (top), 1987.4 (middle), and 1988.5 (bottom) (Krichbaum and Witzel, 1991 b; Krichbaum *et al.*, 1992). The models have been rotated by +50° , a common surrounding halo has been omitted. Straight lines indicate the motion of the components B, C, and D with respect to the stationary assumed core A. Right: To peak flux density normalized profiles of the jet's brightness distribution are shown. Labels identify components.

3. Intraday Variability of compact radio sources

In 1984 Heeschen reported on the observation of low amplitude variability on timescales of weeks and longer in many, especially compact, extragalactic radio sources. This phenomenon called "Flickering" was subsequently explained by refractive scattering effects in the interstellar medium (Heeschen and Rickett, 1987). – The importance of the ISM for the explanation of low frequency variability of radio sources has been proposed as early as 1978 by Shapirovskaya. –

During an attempt to investigate the occurrence of flickering on shorter timescales with the 100 m radiotelescope of the MPIfR in Effelsberg the effect of intraday variability (IDV) was discovered (Witzel et al., 1986; Heeschen *et al.*, 1987): Rapid intensity variations in extragalactic radio sources on timescales shorter than or comparable to a day.

It is readily obvious that in case of an intrinsic origin of the variability one would be able to study structures well beyond the reach of any present VLBI observations. – It is also obvious that intraday variability can affect aperture synthesis observations, if the flux density variations during the observation are not carefully recorded and taken into account when analyzing the data]

In the following I will discuss the variations of the total and of the polarized radio flux densities as well as correlations of these two quantities. The statistical occurrence of IDV and the optical variability on similar time scales and its relation to the radio observations will be addressed, too. A short description of present attempts to explain the physical causes for IDV will then conclude this section.

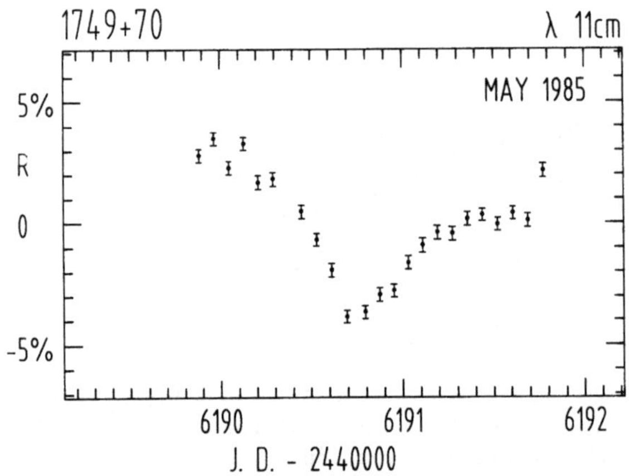

Fig. 5. The intensity variations of the BL Lac 1749+70 observed at 11 cm wavelength in May 1985, $R := 100 \cdot \Delta S / < S > \%$ (see e.g. Heeschen *et al.*, 1987).

Fig.5 shows the variations of the flux density of the BL Lac object 1749+70 as observed in May 1985 at 11 cm wavelength. This observation initiated our investigations of IDV, because the timescale of the phenomenon well below canonical values for intensity changes in combination with the smooth shape of the variation recalling 'lightcurves' were totally unexpected (see Witzel, 1990). Subsequently other sources exhibiting IDV were found

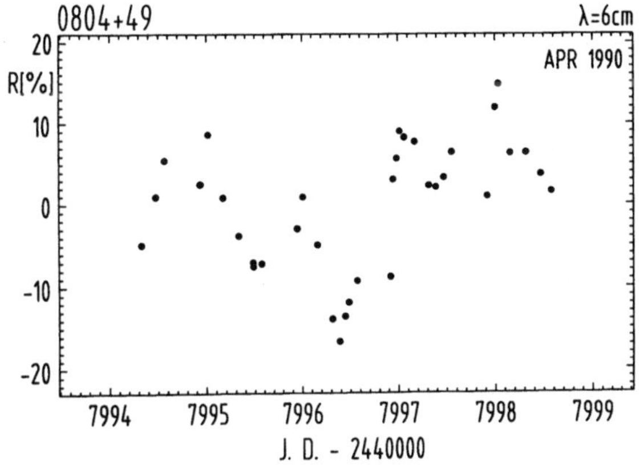

Fig. 6. **The intensity variations of the QSO 0804+49 observed at 6 cm wavelength in April 1990 (see e.g. Witzel and Quirrenbach, 1991)**

with amplitudes of up to 35 % (see fig. 6). But also relatively small variations (amplitudes of about 2 % or less, see e.g. Krichbaum *et al.*, 1991 c) could be reliably determined on the basis of a thorough calibration with the aid of numerous extended radiosources showing steep radio spectra. These objects were assumed to be non variable on short timescales - an assumption which was proved by the observations. The accuracy of the flux density determinations at 11 cm wavelength with the 100 m telescope varied from 0.25% to 0.5% depending mainly on the integration times on the source (e.g. Krichbaum *et al.*, 1991 c). The reliability of the observations was proven by a comparison of data taken simultaneously with the 100 m telescope and the VLA. Fig. 7 shows the variability profile of the quasar 0917+62 at 6 cm wavelength in May 1989 measured with the Very Large Array-interferometer in New Mexico and several thousands of kilometers away under very different atmospheric conditions with the single dish 100 m-telescope at Effelsberg in the Eifel.

A defining property of the sources exhibiting IDV is the fact that their structure function exhibits a clear maximum indicative of the presence of a well defined time scale of appropriate length. Other objects, whose variability is more random or monotonic (e.g. part of a longer timescale event), show structure functions without well defined extrema (fig. 8). The frequency dependence of the variability in the radio regime has been studied for 0917+62, 0716+71, and 0954+65 in May 1989. Fig. 1 of the accompanying paper by Wegner *et al.*, this volume, show that for 0716+71 the amplitudes at the higher frequencies increase, whereas for 0917+62 the variability-amplitudes peak near 11 cm wavelength (e.g. Qian etal., 1991). For 0954+65 not much change in the variability amplitudes has been found at that epoch, with a tendency of decreasing amplitudes at higher radio frequencies (see fig. 9).

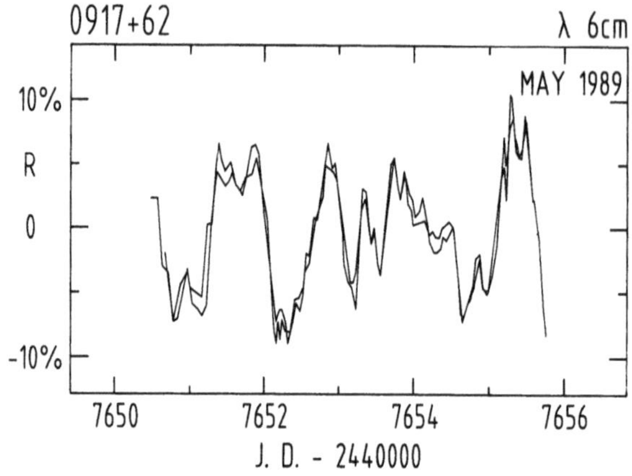

Fig. 7. The intensity variations of the QSO 0917+62 observed at 6 cm wavelength in May 1989 with the 100 m telescope and the VLA (see e.g. Witzel, 1990).

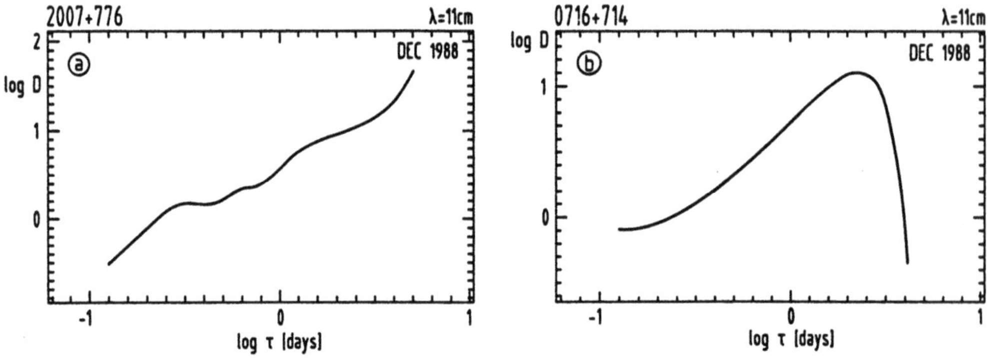

Fig. 8. Typical structure functions of an object showing IDV: 0716+71 and an object varying on longer timescales: 2007+77 (see e.g. Witzel, 1990).

The number of sources investigated for intraday variability still is rather limited. 56 objects with flat radio spectra in the cm regime have so far been observed at least during one session of at least about one week duration. All objects are selected from the 1 Jy-catalogue (Kühr *et al.*, 1981). The objects at declinations greater than 35 degrees form a flux density limited complete sample of flat spectrum sources. 15 additional sources were included in the observations, since they seemed to be most promising candidates (BL Lacs, OVVs, etc.). The sources at declinations greater than 60 degrees have been observed with first priority and repeatedly since a lot of additional information has been accumulated for those (circumpolar and therefore easily observable) objects (e.g. Witzel *et al.*, 1988; Pearson and Readhead, 1988). From these observations the following conclusions can be drawn:

i) IDV occurs predominantly in objects which contain more than 80 % of their flux density in an unresolved core even on mas-scales. For sources

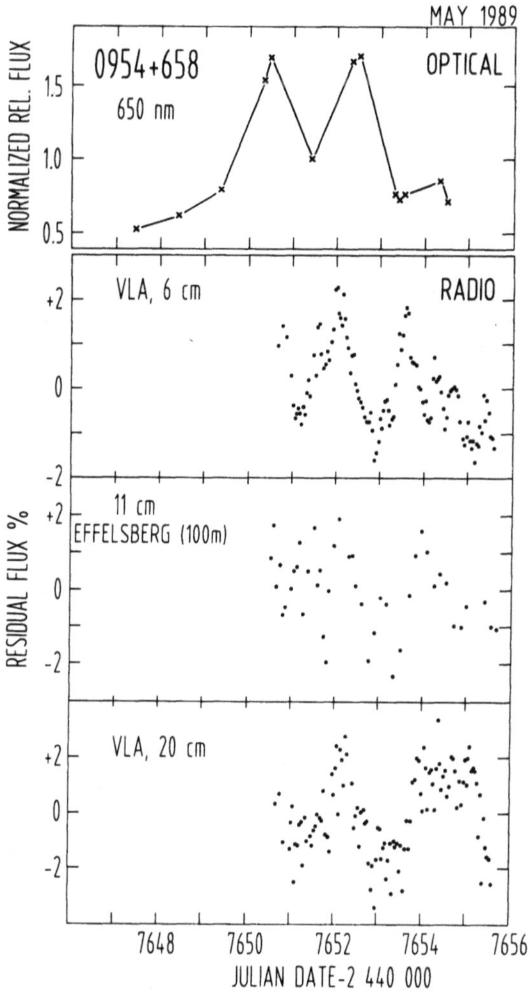

Fig. 9. The variations of the BL Lac 0954+65 observed in May 1989 at radio and optical wavelengths.

with prominent VLBI jets, like 0836+71, no variations on such short timescales have been found yet.

ii) If a source has been shown to vary at one epoch, it usually varied at all epochs observed so far. The amplitude of variability can be different.

iii) At any given epoch roughly 1/3 to 1/2 of all flat spectrum radio sources display IDV with amplitudes greater than 2 to 3 %.

Fig. 10 shows the histograms of variability amplitudes for steep and flat spectrum sources. The shaded area in fig. 10 b represents sources with prominent mas-jets. Omitting those, one can state that at any given epoch all highly compact sources show intensity variability on short timescales

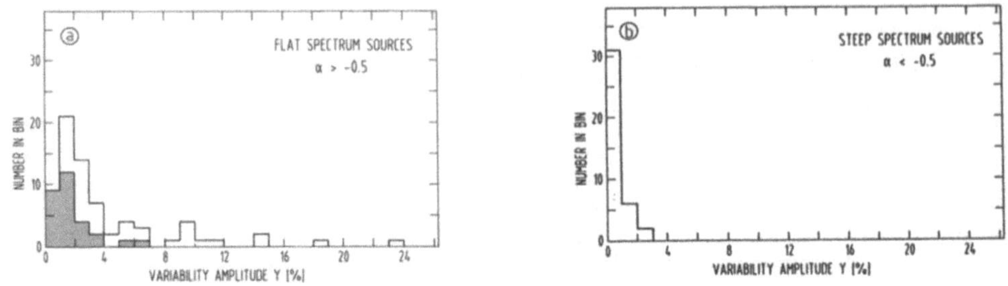

Fig. 10. Distribution of the variability amplitudes Y of (a) flat spectrum and (b) steep spectrum sources at 6 cm wavelengths. The sources with a prominent VLBI jet or double structure on the VLBI scale are shaded in (a).

(this includes variations whose structure functions do not show a preferential short timescale) (Witzel and Quirrenbach, 1991).

Starting in June 1988 we measured the polarized emission in addition to the total flux densities of the sources. Significant ($\Delta S_{pol} \geq 20$ %) simultaneous variations have been found in 6 out of 13 objects observed. The BL Lac object 0716+71 and the quasar 0917+62 show distinct differences: whereas the variations in 0917+62 seem to be anticorrelated in total and polarized flux density (fig. 11) at 3 epochs in 1988 and 1989, in 0716+71 a direct correlation between the two quantities was found in may 1989 during an event on somewhat longer timescale (see Wegner et al., this volume). In addition, the position angle in 0917+62 did show a swing in December 1988 at the time of a pronounced minimum of the polarized flux density (Qian *et al.*, 1991). Further drastic changes in the polarization position angle coincident with minima of the polarized emission have been found for this source at more recent epochs.

In a direct attempt to decide upon the intrinsic or extrinsic nature of IDV we took optical intensity measurements in parallel to the radio observations, since it is generally agreed that the observation of correlated radio/optical IDV would rule out an explanation by effects of the interstellar medium. In 1990 Wagner et al. showed that optical variability on timescales comparable to the ones observed at radio wavelengths occurred in 3 out of 6 objects observed in May 1989. The amplitudes of the variations in the optical range were significantly greater than in the radio data. Significant optical variability of AGN on short timescales ('microvariability') has been reported independently by Miller *et al.*, 1989 (see also this volume).

A correlation of the radio and optical variability was found by Quirrenbach *et al.*, 1991 in the BL Lac object 0716+71. Fig.12 presents the results from parallel observations in February 1990 with the VLA at 6 cm wavelength and with the 71 cm optical telescope of the Landessternwarte in Heidelberg as well as the 2.2 m telescope on Calar Alto at 650 nm. The apparent change in the mode of variability around J.D. 2447930, when the preferential timescale changed from about 1 day to about 7 days, took

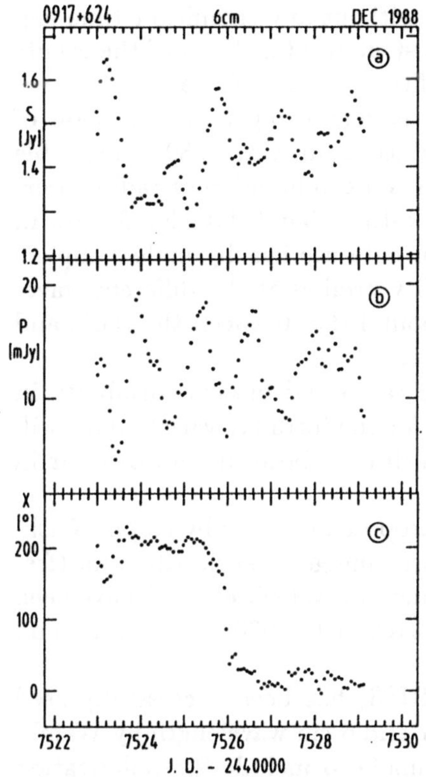

Fig. 11. The variations of the total fluxdensity (12 a), the polarized flux density (12 b), and the polarization position angle (12 c) of 0917+62 at 6 cm wavelength observed in December 1988 (Quirrenbach *et al.*, 1989 b).

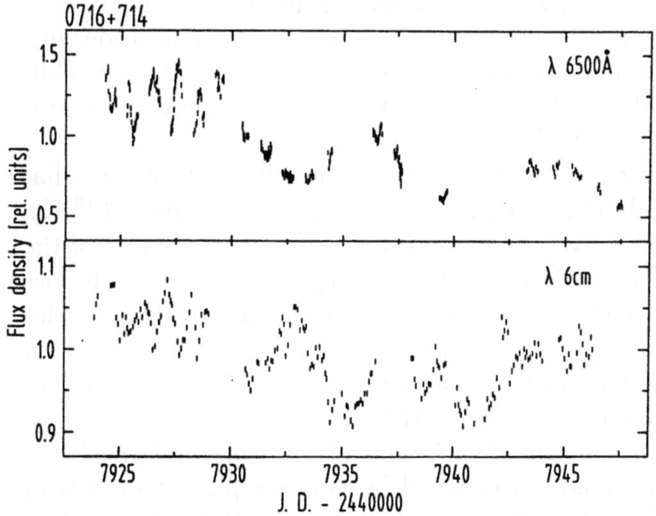

Fig. 12. The variations of the 6 cm flux density and the R-band emission of 0716+71 observed in February 1991 (see Quirrenbach *et al.*, 1991).

place in the radio and the optical regime without any significant time lag. An analysis of the power spectra and the structure functions of the events (e.g. Wagner, 1991) demonstrates the reality of the correlation.

Some evidence, although by far not as convincing as in the case of 0716+71, for correlated radio/optical variations in May 1989 is available for the BL Lac object 0954+65. Fig.9 shows a combination of radio observations at 6, 11 and 20 cm wavelengths with R band data for 6 days in May 1989. Although especially the optical data are heavily undersampled, it seems tempting to identify the variability profiles at the different wavelengths with each other. A time lag of about 1 day between the radio and the optical events would then be suggested.

Although the correlation between the radio and optical variability in 0716+71 strongly favours intrinsic causes for the intraday variability, I will briefly summarize all the basic effects which have been discussed so far in the literature.

i) Gravitational microlensing has been proposed recently by Gopal-Krishna and Subramanian, 1991. The shortcomings of such attempts (frequency dependence, high apparent transverse velocities, etc.) have been listed by Quirrenbach, 1989 a,b; Quirrenbach, 1990; and Witzel and Quirrenbach, 1991.

ii) Refractive Interstellar Scintillation (RISS) has been successfully used to model lightcurves of 0917+62 at 11 and 6 cm wavelength by Wambsganss *et al.*, 1989. No attempt was made to include the polarization properties in that paper, which were unknown at that time. It seems plausible, however, that with two-component models the variability of the polarized flux densities can be explained by RISS, too. A model based on RISS to explain the swing in the polarization angle of 0917+62 has recently been proposed by Simonetti (1991). The wavelength dependence of the amplitude of IDV in e.g. 0716+71, however, seems to contradict simple RISS models, which predict a decreasing amplitude with increasing frequency (Rickett, 1990). Especially the correlated variability of the optical and the radio intensities of 0716+71 seems to contradict any explanation by RISS.

iii) An intrinsic explanation in the context of incoherent synchrotron radiation leads to extremely high brightness temperatures (of up to 10^{19} K), clearly in excess of the inverse Compton limit of 10^{12} K because of the short timescales of the variability. However, Qian *et al.*, 1991, discuss the model dependence of the transformation of the observed timescales into the rest frame of the source. In the case of cylindrical geometries an additional correction factor of γ^{-2} can be derived, which would reduce the brightness temperatures to about 10^{12} K.

The simultaneous and anticorrelated variations of total and polarized flux densities observed can be readily explained by two component models of the type proposed by Qian *et al.*, 1991 and Marscher, 1991 (and Marscher, this volume). The correlated variability in the radio and the optical regime seems at present to pose the most severe difficulties for such explanations: the rapid rise of the light curves in the optical seems to be too fast for stochastic or diffusive shock wave acceleration models.

In view of the difficulties mentioned above, also models of coherent emission processes as proposed by Benford, 1984, and Weatherall and Bendford, 1991, will have to be thoroughly investigated (see also Lesch, this volume, and Benford, this volume).

Finally, even if an intrinsic cause for the IDV is accepted, the extreme compactness of the sources has to lead to scattering effects in the interstellar medium (see e.g. Witzel and Quirrenbach, 1991). Further investigations are necessary (and worthwhile) to disentangle the various effects.

Irrespective of the physical explanation, sources which show IDV are ideal candidates for future space VLBI experiments, too, by virtue of their extreme compactness.

Acknowledgements: The work described in this paper was a truly collaborative effort of numerous scientists and I gratefully acknowledge the cooperation of T. Krichbaum, A. Quirrenbach and S. Wagner who made the results possible and the work pleasant. Also the contributions of R. Wegner, A. Zensus, C. Hummel, C. Schalinski, K.-H. Wüllner, K. Standke, D. Graham, A. Alberdi, M. Rioja, M. Ott, K. Anton and U. Erkens were of invaluable help. Thanks are due to I. Pauliny-Toth and T. Pearson for critically reading the manuscript.

References

Baath, *et al.*, 1991: *Astron. Astrophys.*, **241**, L1.
Baath, L., 1991: in: *Variability of Blazars*, ed. E. Valtaoja and M. Valtonen (Cambridge University Press), in press.
Bartel, N., *et al.*, 1988: *Nature*, **334**, 131.
Benford, G., 1984: in: *Physics of Energy Transport in AGNs*, NRAO proceedings, 1984.
Biretta, J.A., *et al.*, 1986: *Astrophys. J.*, **308**, 93.
Camenzind, M., 1990: in: *Reviews in Modern Astronomy 3, Accretion and Winds*, ed. G. Klare, Springer (Berlin), p. 3.
Courvoisier, T.J.-L., *et al.*, 1990: *Astron. Astrophys.*, **234**, 73.
Dent, W.A., 1965: *Science*, **148**, 1458.
Dhawan, V., *et al.*, 1990: *Astrophys. J. (Letters)*, **360**, L43.
Gopal-Krishna, and Subramanian, K., 1991: *Nature*, **349**, 766.
Hardee, P.E., and Norman, M.L., 1989: *Astrophys. J.*, **342**, 680.
Heeschen, D.S., 1984: *Astron. J.*, **89**, 1111.
Heeschen, D.S., *et al.*, 1987: *Astron. J.*, **94**, 1493.
Heeschen, D.S., and Rickett, B.J., 1987: *Astron. J.*, **93**, 589.
Hough, D.H., and Readhead, A.C.S., 1989: *Astron. J.*, **98**, 1208.
Hummel, C.A., *et al.*, 1992: in preparation.
Königl, A., and Choudhuri, A.R., 1985: *Astrophys. J.*, **289**, 188.

Krichbaum, T.P., 1990: in: *Parsec-scale radio jets*, ed. J. A. Zensus and T. J. Pearson (Cambridge University Press), p. 83.
Krichbaum, T.P., *et al.*, 1990: *Astron. Astrophys.*, **237**, 3.
Krichbaum, T.P., and Witzel, A., 1991a: in: *Frontiers of VLBI*, ed. H. Hirabayashi, M. Inoue and K. Kobayashi (Universal Academy Press, Tokyo), in press.
Krichbaum, T.P., and Witzel, A., 1991b: in: *Variability of Blazars*, ed. E. Valtaoja and M. Valtonen (Cambridge University Press), in press.
Krichbaum, T.P., Quirrenbach, A., and Witzel, A., 1991c: in: *Variability of Blazars*, ed. E. Valtaoja and M. Valtonen (Cambridge University Press), in press.
Krichbaum, T.P., *et al.*, 1992: in preparation.
Kühr, H., Witzel, A., Pauliny-Toth, I.I.K., Nauber, U., 1981: *Astron. Astrophys. Suppl.*, **45**, 367.
Levy, G.S., *et al.*, 1989: *Astrophys. J.*, **336**, 1098.
Lovelace, R.V.E., *et al.*, 1987: *Astrophys. J.*, , 315,504.
Marcaide, J.M., *et al.*, 1985: in: *Proc. IRAM-ESO-Onsala Workshop on (Sub-) Millimeter Astronomy*, ed. Shaver,P.A. and Kjär,K., p. 157.
Marr, J.M., *et al.*, 1989: *Astrophys. J.*, **337**, 671.
Marscher, A.P., 1991: in: *Variability of Blazars*, ed. E. Valtaoja and M. Valtonen (Cambridge University Press), in press.
Miller, H.R. *et al.*, 1989: *Nature*, **337**, 627
Pearson, T.J and Readhead, A.C.S., 1988: *Astrophys. J.*, **328**, 114
Qian, S.J., *et al.*, 1991: *Astron. Astrophys.*, **241**, 15.
Quirrenbach, A., *et al.*, 1989a: *Nature*, **337**, 442.
Quirrenbach, A., *et al.*, 1989b: *Astron. Astrophys. (Letters)*, **226**, L1.
Quirrenbach, A., 1990: Ph.D. thesis, University of Bonn.
Quirrenbach, A., *et al.*, 1991: *Astrophys. J.*, **372**, L71
Readhead, A.C.S., *et al.*, 1990: in: *Parsec-scale radio jets*, ed. J. A. Zensus and T. J. Pearson (Cambridge University Press), p. 71.
Rickett, B.J., 1990: *Annu. Rev. Astron. Astrophys.* **28**, 561.
Roberts, D.H., *et al.*, 1990: in: *Parsec-scale radio jets*, ed. J. A. Zensus and T. J. Pearson (Cambridge University Press), p. 110.
Romney, J.D., *et al.*, 1984: in: *IAU Symposium 110, VLBI and Compact Radio Sources*, ed. R. Fanti, K. Kellermann, and G. Setti (Dordrecht: Reidel), p. 137.
Schalinski, C.J., *et al.*, 1991: in: *Variability of Blazars*, ed. E. Valtaoja and M. Valtonen (Cambridge University Press), in press.
Shapirovskaya, N.Y., 1978: *Astron. Zh.*, **55**, 953.
Sholomitskii, G.B., 1965: *Astron. Zh.*, **42**, 673.
Simonetti, J.H., 1991: *Astron. Astrophys.*, submitted.
Wagner, S., *et al.*, 1990: *Astron. Astrophys.*, **235**, L1.
Wagner, S., 1991: in: *Variability of Blazars*, ed. E. Valtaoja and M. Valtonen (Cambridge University Press), in press.
Wambsganss, J., *et al.*, 1989: *Astron. Astrophys. (Letters)*, **224**, L9.
Weatherall, J.C., and Benford,G., 1991: *Astrophys. Journal*, submitted.
Witzel, A., *et al.*, 1986: *Mitt. Astron. Ges.*, **65**, 239.
Witzel, A., *et al.*, 1988: *Astron. Astrophys.*, **206**, 245.
Witzel, A., 1990: in: *Parsec-scale radio jets*, ed. J. A. Zensus and T. J. Pearson (Cambridge University Press), p. 206.
Witzel, A., and Quirrenbach, A., 1991: in: *Proceedings of the conference "Propagation Effects in Space VLBI" held in Leningrad, June 1990*, ed. N. Kardashev and L. Gurvits, in press.
Zensus, J.A., and Porcas, R.W., 1987: in: *Superluminal Radio Sources*, ed. J. A. Zensus and T. J. Pearson (Cambridge University Press), p. 126.
Zensus, J.A., and Pearson, T.J., (editors), 1987: *"Superluminal Radio Sources"*, Cambridge University Press.
Zensus, J.A., *et al.*, 1988: *Nature*, **334**, 410.
Zensus, J.A., and Pearson, T.J., (editors), 1990: *"Parsec-scale Radio Jets"*, Cambridge University Press.

Zensus, J.A., 1990: in: *Parsec-scale radio jets*, ed. J. A. Zensus and T. J. Pearson (Cambridge University Press), p. 28.
Zensus, J.A., *et al.*, 1990: *Astron. J.*, **100**, 1777.
Zensus, J.A., *et al.*, 1992: in preparation.

Discussion

Burbidge: Your results are remarkable. I just want to stress that there is by now a great deal of evidence from associations between QSOs and bright galaxies that the QSO redshifts are not measures of distance so superluminal motions are not real but an artifact of the distance assumed. The only normal galaxy in your list NGC 1275 shows subluminal motions!

Field: The radio is polarized, hence synchrotron. Is the optical polarized, hence synchrotron also?

Witzel: Yes

Kovalev: You tell about non-linear motion of components. Have you the strong evidence of this effect or, may be, this is a result of a superposition of several linear motions? If curved motion is real, how many sources show it?

Witzel: Curved trajectories seem to be a common phenomenon in the core (mas) jets of AGN. In the S5-sample more than a third of the sources (i. e. most of the well observed cases) show evidence for motion on curved paths (but e. g. 0153+744 seems to be stationary!). The best documented cases like e. g. 1928+738, 0836+710, and 3C 345 are most easily explained by motions along curved trajectories.

Varshalovich: 1) What is the shortest time of variability?
2) What is early optic or radio, and what is the delay?

Witzel: ad 1) We have seen events with amplitudes of 5% in 0917+62 whose duration was of order 1 hour.
ad 2) In 0716+71 we don't know since the variations at both regimes are quasiperiodic and thus determinations of the time lags are difficult. If you believe in a correlation in 0954+65 in May 1989 the optical came first (see fig. 9).

Radiative Acceleration of Blobs in AGN

Eduardo Trussoni, Gabrielle Ghisellini, and Gianluigi Bodo

Osservatorio Astronomico di Torino, Pino Torinese, Italy

Abstract. We investigate the motion of a magnetized blob in the radiation field emitted by a source in the core of AGN. The interaction between the relativistic electrons and the photons is the synchrotron absorption, with cross section that can be much larger than the Thomson cross section. This allows the blob to be accelerated to highly relativistic velocities, as required by the models for superluminal motion in AGN.

1. Introduction

Processes of plasma acceleration and collimation are present in almost all AGN. In particular, the VLBI detection of superluminal motion of radio emitting knots is likely to be related to the outwards motion of clouds moving at relativistic velocities (bulk Lorentz factors $\Gamma \sim 5 - 10$).

The large amount of luminosity present in the inner region of the AGN suggested that radiation could be an effective mechanism of plasma acceleration. However, detailed studies have shown that this process allows only mildly relativistic velocities if the interaction is through the Thomson scattering. This is due to the strong drag effect of aberrated photons, when the speed of the flow approaches the speed of light (see e.g. Abramowicz and Piran, 1980; Ferrari *et al.*, 1985).

The cross section σ_s of the interaction between photons and relativistic electrons via the synchrotron absorption can be much larger (by several orders of magnitude) than the Thomson cross section σ_T (Ghisellini and Svensson, 1990). Furthermore, as σ_s depends from the photons frequency, the drag effect is reduced. This suggests that a cloud of relativistic electrons in the inner core of an AGN can be accelerated much more efficiently than in the case of Thomson scattering. The acceleration of an optically thick cloud has been already analyzed by Ghisellini *et al.* (1990), who have shown that it can reach relativistic velocities even with source luminosities much less than the Eddington luminosity (L_E). Besides the reduced drag, this is possible due to the fact that, with respect to the Thomson interaction, a very light blob can be optically thick to synchrotron absorption.

In this paper it will be shown that also a cloud optically thin to synchrotron absorption can be accelerated up to relativistic velocities.

2. The Model

The central source has a total luminosity $L \lesssim L_E$, with nonthermal self-absorbed spectrum $I(\nu) \propto \nu^{-\alpha_1}$ for $\nu > \nu_t$, and $I(\nu) \propto \nu^{\alpha_0}$ for $\nu < \nu_t$, where ν_t is the peak frequency. Close to the central source a blob is present with a component of relativistic electrons and a randomly distributed magnetic field. The blob interacts only with the photons coming from the underlying source, therefore no external medium and/or large scale magnetic field is assumed to affect the blob dynamics. In this conditions the motion of the blob in the radiation field is given by (neglecting gravity) $\mathrm{d}\Gamma/\mathrm{d}r = F'_{\|}(\Gamma, r, \sigma_s, \nu)$ where $F'_{\|}$ is the radiative force in the cloud frame of reference and r is the radial coordinate.

For an optically thick gas, σ_s is the cross sectional area of the blob. Conversely, in the optically thin case the synchrotron cross section depends on several parameters, in particular from the photon frequency ν, the electron energy γ and the magnetic field B. Neglecting the geometrical details of the interaction, from Ghisellini and Svensson (1991) σ_s is given by ($\nu_c = \frac{3}{2}\gamma^2\nu_B$, where ν_B is the gyrofrequency and θ the pitch angle):

$$\sigma_s(\nu,\gamma,\theta) = \frac{16\pi^2}{3\sqrt{3}}\frac{e}{\gamma^5 B\sin\theta}K_{5/3}\left(\frac{\nu}{\nu_c}\right), \qquad \nu_c = \frac{3}{2}\gamma^2\,\nu_B\sin\theta, \qquad (1)$$

where e is the electron charge, B the magnetic field and $K_{5/3}$ the modified Bessel function of order 5/3. It is easy to see that for $\nu \gg \nu_c$, $\sigma_s \approx 0$ while, for $\nu \ll \nu_c$, $\sigma_s \sim e/(B\sin\theta\,\gamma^5)(\nu_c/\nu)^{5/3}$ (in particular, for $\nu \to \nu_B/\gamma$, $\sigma_s \sim r_L r_e/\gamma$ with r_L and r_e the Larmor and electron radii, respectively). As σ_s depends on γ, we considered two cases: a) $\gamma = $ const (i.e. there is an *ad hoc* heating mechanism that works on time scales much faster that the synchrotron time scale), and b) γ is set by the balance between the synchrotron losses and the energy gains due to the radiating source. In the latter case a second equation must be provided for γ.

From the definition of σ_s (Eq. 1), by expressing in the laboratory frame of reference the force $F'_{\|}$ acting on the blob and the total energy balance for the absorbing electron of energy γ, some algebra leads to the following equations for γ and Γ (for simplicity it has been set $\theta = 1$):

$$\frac{\mathrm{d}\gamma}{\mathrm{d}x} = \frac{Ar_o}{m_e c^3\beta}\Gamma\int_{\mu_c}^{1}(1-\beta\mu)^2 F(\nu_t,\mu,\Gamma,\gamma)\mathrm{d}\mu - \frac{\sigma_T r_o}{4\pi m_e c^2}\frac{B_{eq}^2}{\Gamma\beta}(\gamma_r^2-1),$$

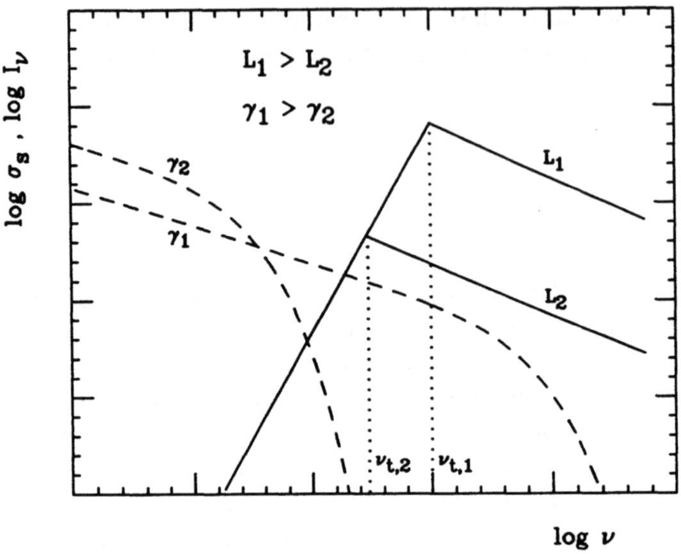

Fig. 1. Plot (in arbitrary units) of the source radiation intensity I_ν (full lines) and synchrotron cross sections σ_s (dashed lines). For σ_s a fixed magnetic field B is assumed.

$$\frac{\mathrm{d}\Gamma}{\mathrm{d}x} = \frac{Ar_o}{m_p c^3}\Gamma^2 \int_{\mu_c}^{1} (\mu - \beta)(1 - \beta\mu) F(\nu_t, \mu, \Gamma, \gamma) \mathrm{d}\mu \tag{2}$$

where μ is the cosine of the angle between the incident photon and the cloud, $\mu_c = \sqrt{(1 - 1/x^2)}$, with $x = r/r_o$ (r_o is the initial position of the blob), and A, ξ and F are given by:

$$F(\nu_t, \mu, \Gamma, \gamma_r) = \int_{\nu_{min}}^{\nu_t} \left(\frac{\nu}{\nu_t}\right)^{\alpha_o} K_{5/3}(\xi)\,\mathrm{d}\left(\frac{\nu}{\nu_t}\right)$$

$$+ \int_{\nu_t}^{\nu_{max}} \left(\frac{\nu}{\nu_t}\right)^{-\alpha_1} K_{5/3}(\xi)\,\mathrm{d}\left(\frac{\nu}{\nu_t}\right),$$

$$\xi = \frac{\nu}{\nu_c'}\Gamma(1 - \beta\mu), \qquad A = C x_g \frac{e}{B\gamma^5}\frac{m_p c^3}{\sigma_T}\frac{L}{L_E} \tag{2'}$$

where $x_g = r_g/r_o$ with r_g the gravitational radius ($r_g \approx 3 \times 10^{13} M_8$ cm, with M_8 the central mass in units of $10^8 M_\odot$), $\nu_{min} = 3/2\nu_B$ and C is a constant $\mathcal{O}(1)$.

The set of Eqs. 2 is a system of two integro–differential equations. However, by means of the asymptotic expansions for $K_{5/3}$, the two integrals on ν and μ can be solved analytically, and the system reduces to two ordinary differential equations.

3. Results

The several parameters appearing in Eqs. 2 have been chosen has follows.
a) The radius of the source is likely to be $\gtrsim$ the size of the central body, while the cloud is assumed to begin its motion at the edge of the source. We have assumed $x_g = 0.2$.
b) The thin spectral index is assumed to be $\alpha_1 = 0.7$. For $\nu < \nu_t$ the value of α_o cannot be deduced from observational data. If the electron population at low energies is a Maxwellian, then $I(\nu) \propto \Theta\nu^2$, with $\Theta = kT/(m_e c^2)$ (T is the brightness temperature); while if it is a power law, we have $I(\nu) \propto \sqrt{(\nu^5/B_S)}$, where B_S is the magnetic field inside the source. Therefore we discuss separately the two cases $\alpha_o = 2$ and $\alpha_o = 2.5$, with $5 < \Theta < 50$, and $10^3 < B_S < 10^4$ Gauss, respectively.
c) The source luminosity (assumed $\leq L_E$) is not independent of the value of ν_t. In fact, by comparing the fraction of the luminosity coming from the optically thick spectrum ($\nu \leq \nu_t$) with the radiation intensity in the same range of frequencies, we get $\nu_t \propto (L/\Theta)^{1/3}$ (for $\alpha_o = 2$) and $\nu_t \propto (L^2 B_S)^{1/7}$ (for $\alpha_o = 2.5$). In both cases ν_t increases with L (for fixed Θ or B_S) so that the number of photons in the thick spectrum is unchanged. This implies that, if the cross section cut–off lies in the self-absorbed part of the spectrum (i.e. $\nu_c < \nu_t$), the acceleration process is not expected to depend on L.
d) The cloud has been assumed to have physical characteristics not very different from the source. Therefore is has been assumed $10^3 < B < 10^4$, while the electrons at the injection have Lorentz factors $5 < \gamma < 10^2$.

The numerical solutions of Eqs. (2) are discussed for the two cases $d\gamma/dx = 0$ and $\neq 0$.

Acceleration with $\gamma =$ const.

In this case only the equation for Γ is integrated, with γ assumed as a parameter. In Fig. 2 the asymptotic value Γ_∞ is plotted vs γ for the two cases $\alpha_o = 2$ and $\alpha_o = 2.5$: in both cases the blob can attain relativistic velocities ($\Gamma_\infty \approx 10$) for $\gamma \sim 10 - 30$, while lower velocities are found for larger or smaller values of γ. This behaviour is understood recalling that (for fixed B) $\nu_c \propto \gamma^2$ and, for $\nu \ll \nu_c$, $\sigma_s \propto \gamma^{-10/3}$ (see Fig. 1). If $\nu_c > \nu_t$, an increase of γ means a sharp decrease of σ_s at low frequencies and an increase of ν_c, that however falls in a spectral range poor of photons. Therefore in such a case we expect a decrease of Γ_∞ for larger γ. When conversely $\nu_c < \nu_t$, σ_s increases by lowering γ, but also the cut off frequency ν_c is shifted to lower values strongly reducing the number of photons useful for acceleration, so that Γ_∞ is reduced. The final consequence is that there is a value of γ for which a maximum Γ_∞ is found, as seen in Fig. 2.

Similar arguments explain the behaviour of Γ_∞ with L (plotted in Fig. 3). For high values of γ, $\nu_c > \nu_t$, and

an increase of L means also an increase of the number of photons useful for acceleration. For lower values of γ, such that $\nu_c < \nu_t$, the number

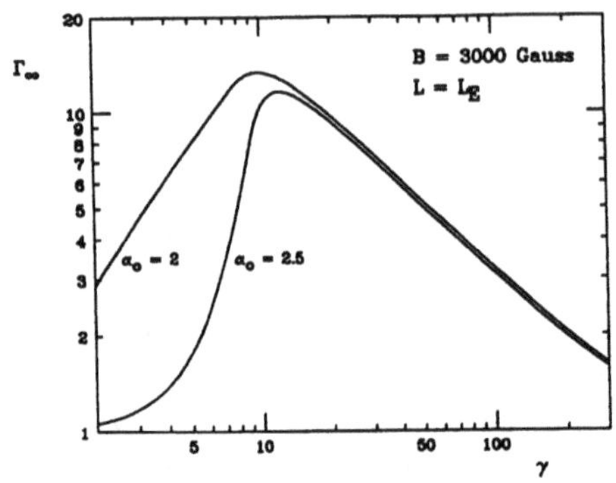

Fig. 2. Plot of Γ_∞ vs γ ($\gamma =$ const during the acceleration phase), for $\alpha_o = 2.$ (with $\Theta = 20$), and $\alpha_o = 2.5$, (with $B_S = 1000$ Gauss).

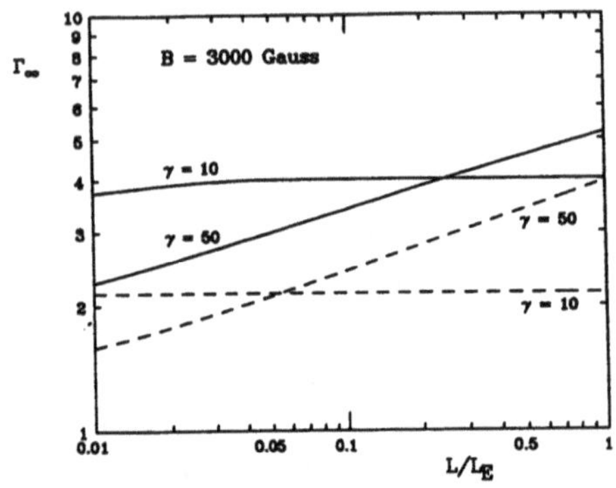

Fig. 3. Plot of Γ_∞ vs L/L_E ($\gamma =$ const during the acceleration phase) for $\alpha_o = 2.$ (full line, with $\gamma = 10$, $\Theta = 5$, and $\gamma = 50$, $\Theta = 20$), and $\alpha_o = 2.5$ (dashed line, with $B_S = 10^4$ Gauss).

of photons interacting with the electrons is unchanged by varying the luminosity, as discussed above, so that Γ_∞ is almost independent from L. The important consequence is that, with suitable parameters, the blob can be accelerated to relativistic velocities even though the source has a very sub-Eddington luminosity, $L \ll L_E$.

Acceleration with $\gamma \neq$ const.

The numerical integration of the two coupled Eqs. (2) shows that, whichever is its initial value, γ reaches almost immediately an equilibrium value (for which $\mathrm{d}\gamma / \mathrm{d}x = 0$). During all the acceleration phase the electron energy is always very close to this value (and $\gamma \to 1$ for $x \gg 1$, as the electron heating $\propto L \propto 1/x^2$). The possibility to attain this equilibrium however depends on the parameters: in fact for $\alpha_o = 2.5$ and $\nu_c < \nu_t$ there is no equilibrium values for γ. In such a case, on a very short distance $\gamma \to 1$ and $\nu_c \to \nu_B$, strongly reducing the number of photons interacting with the electron.

The plots $\Gamma(x)$ and $\gamma(x)$ (see Fig. 4) show that in general also in this case relativistic bulk velocities can be attained by the blob within distances $\lesssim 10^2$ the initial position. In particular, for $\alpha_o = 2.5$, we see that at $x \gtrsim 3$ γ drops rapidly to one, and the bulk acceleration stops. Therefore, these are values of the parameters for which the blob cannot be accelerated at all from the central source.

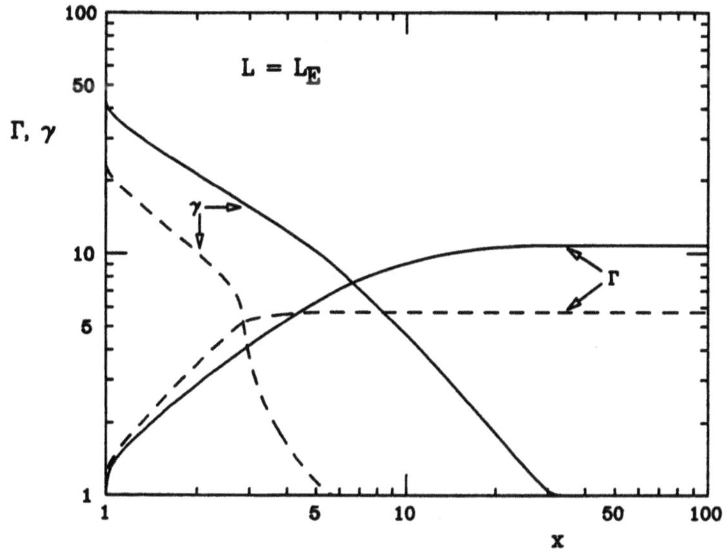

Fig. 4. Plot of Γ and γ vs x for $\alpha_o = 2$ (full line, with $\Theta = 40$ and $B = 3000$ Gauss) and $\alpha_o = 2.5$ (dashed line, with $B_S = 1000$ Gauss and $B = 10^4$ Gauss).

4. Conclusions

The main results of our analysis are the following:
1) The large value of σ_s allows asymptotic ultrarelativistic velocities for the clouds ($\Gamma_\infty \sim 5 - 10$), assuming reasonable values for the parameters, typical for the standard models of AGN. The very important result is that these velocities are obtained also for luminosities $\ll L_E$.
2) The asymptotic bulk Lorentz factor is strongly dependent from the value of the spectral index of the thick part of the spectrum. An increase of α_o means a much less number of photons at low frequencies, where σ_s is larger, with a less efficient acceleration.
3) If the energy of the absorbing electrons is not kept constant by some extra mechanism, it decreases during the acceleration phase. In any case, apart from some peculiar values of the parameters, the source photons can heat them long enough to allow the blob to move relativistically, before it asymptotically cools ($\gamma \to 1$).

From the astrophysical point of view, we can say that radiative interaction can be an effective mechanism to get relativistic moving clouds of synchrotron emitting electrons. This model therefore could be easily related to the superluminal motion in AGN, provided that a reacceleration of electrons occurs on the VLBI scale (~ 1 pc). It is not discussed here which kind of process could heat electrons in the cloud. In fact the problem of the synchrotron losses from a blob of electrons arises anyway, independently from the kind of the bulk acceleration process for the cloud.

References

Abramowicz, M. A., & Piran, T., 1980, *Astron. Astrophys. (Letters)*, **241**, L7
Ferrari, A., Trussoni, E., Rosner, R. & Tsinganos K., 1985, *Astrophys. J.*, **294**, 397
Ghisellini, G., Bodo, G. & Trussoni, E., 1990, *Astron. Astrophys. (Letters)*, **362**, L1.
Ghisellini, G. & Svensson, R. 1991, *Monthly Notices Roy. Astron. Soc.*, , in press.

Discussion

Field: Is the blob "weighted" by non relativistic matter (e.g. protons) ? If so, the acceleration will be reduced. If, on the other hand, there are no non-relativistic ions (for example only positrons), the blob will expand at $\sim c$, vitiating the assumption that the blob remains coherent.

Trussoni: The dynamics of the cloud is still an open problem. Obviously, if too much thermal matter is present, the asymptotic velocity is lowered. We have just assumed a proton thermal gas enough to have a neutral gas, that also provides part of the cloud inertia. A further component that contributes to keep coherent the blob during the acceleration is the random magnetic field. In fact it must also be noticed that, in the assumed conditions, the cloud can expand, but at a velocity of the order of the Alfvén velocity ($\ll c$).

Salvati: Does Γ_∞ depend on the details of the assumed geometry? In particular on the initial location of the blob with respect to the photon source?

Trussoni: The geometry of the source of photons can be quite complicate. However, in order that the acceleration be effective, it is just enough that at the beginning the could "feels" a non-zero photon flux. If this condition is fulfilled, the value of Γ_∞ should not be critically dependent from the details of the inner region.

Wilson: Do not you still have a problem with photon drag, limiting the bulk Γ_∞ attained, if there is "any" ambient population of low frequency photons to which the cloud is optically thick in synchrotron radiation ?

Trussoni: In the inner core a possible population of diffused low frequency photons is overwhelmed by the source luminosity. Farther away their effect could be larger but, as the cloud is relativistic, such photons are blueshifted and their drag effect strongly reduced.

Blandford: How important are the magnetic (or surface pressure) stresses acting on the cloud, relative to the radiation pressure stress ?

Trussoni: We made the quite restrictive assumption that the cloud interacts only with the source photons, neglecting the possible strong effects of an external gas or magnetic field. On the other hand our condition could be met if the acceleration region is inside a jet (with a much lower density), that swept a sort of channel through which the cloud can move with small interaction with the external ambient.

Emission Models for Compact Jets

Alan P. Marscher[1,2]

[1] Department of Astronomy, Boston University, 725 Commonwealth Ave., Boston, MA 02215, USA
[2] Harvard-Smithsonian Center for Astrophysics, 60 Garden St., Cambridge, MA 02138, USA

Abstract. Models for the nonthermal emission from compact jets in active galactic nuclei are reviewed and discussed. The observed characteristics of the emission support incoherent synchrotron radiation as the emission mechanism. The "standard working model" of a relativistic jet with stationary and moving shocks reproduces most of the observational data quite well. Nevertheless, recent observations of $\sim$ 1 day timescale variations at centimeter wavelengths are only barely consistent with the model. I review the limits on variability timescales and make predictions that must be fulfilled to avoid substantial revision to the standard model.

1. Introduction

The continuum spectra of radio-loud active galactic nuclei (AGN) are dominated by nonthermal emission over a wide frequency range. In many of these objects, as in radio-quiet AGN, thermal emission is important at optical-ultraviolet frequencies, and perhaps over some portions of the infrared and X-ray domains. At centimeter wavelengths, where VLBI can resolve the emission regions at the milliarcsecond scale, many of the radio-loud AGN exhibit compact, one-sided, knotty jets (cf. Pearson & Readhead 1988). The compact jet is often connected to, or at least pointing toward, a larger jet and/or extended radio lobe. The narrow end of the jet terminates at a "core," which is usually the most compact feature, and the knots tend to be larger the farther they are from this core. Figure 1 shows a typical core-jet source. The spectrum of any given component peaks at a frequency that is generally higher the closer that component is to the core. The overall spectrum of the source is a superposition of the spectra of the individual components, and is typically quite flat up to a frequency where the core spectrum turns over. The nonthermal spectrum is rather steep with a negative slope toward frequencies higher than this, and usually extends to infrared and often optical frequencies (see, e.g., Brown et al. 1989).

The emission mechanism within compact jets is generally accepted to be incoherent synchrotron radiation with random electron pitch angles. This fits the continuum spectra of the individual components as well as the po-

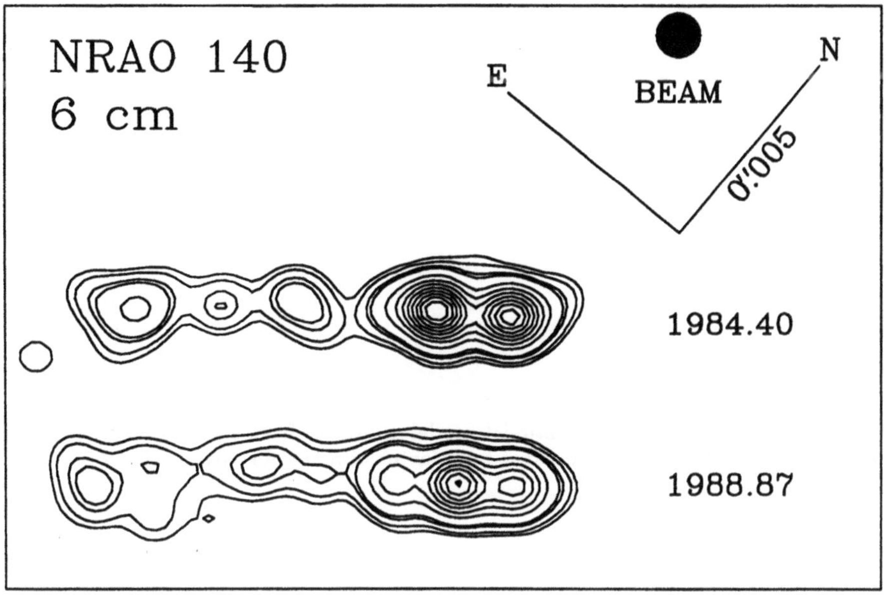

Fig. 1. VLBI image of the core-jet quasar NRAO 140 at a wavelength of 6 cm at two epochs. Contours are 2, 3, 5, 10, 20, 30, 40, 50, 60, 70, 80, and 90 percent of the 1984.40 peak of 3.2×10^{10} K (= 0.65 Jy beam^{-1}). Motion of the knots relative to the core (at the right-hand end of the jet), corresponding to an apparent speed of $\sim 6.5c$ ($H_0 = 100$, $q_0 = 0.5$), can be seen. From observations obtained by the author.

larization properties (very low circular polarization, substantial linear polarization) observed in compact jets. It also explains why the observed (as opposed to inferred) brightness temperatures of the individual components tend to obey the "inverse Compton catastrophe" limit of $\sim 10^{12}$ K (Jones et al. 1974; Burbidge et al. 1974). Coherent processes, on the other hand, tend to be narrow-band and in most cases cause high circular polarizations.

The literature concerning the nonthermal emission in radio-loud AGN is vast and thus cannot be reviewed adequately in the space available here. I will therefore review what I consider to be the work most relevant to the current problems associated with nonthermal emission from compact jets. These relate mainly to the variability in flux density, polarization, and structure observed in these sources. I will briefly outline the "standard working model" (SWM) used by most workers in the field to interpret the observations, show how the simplest versions of that model can be adjusted to accommodate many of the observational details, and demonstrate that the SWM is stretched to its limit in explaining observations of rapid centimeter-wave variability.

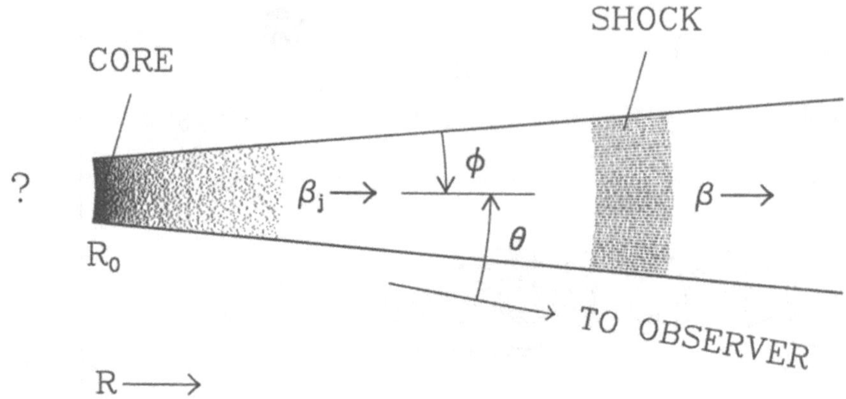

Fig. 2. Basic geometry of the relativistic jet model as viewed from a direction transverse to the axis. The vertical scale is expanded by a factor of a few for visual clarity.

2. The Relativistic Jet Model

The SWM of compact jets is the relativistic jet model, the basic outline of which was first presented by Blandford & Rees (1978) and elaborated upon by Blandford & Königl (1979) and Marscher (1980). In the simplest version of the model, the jet is assumed to be first visible at some point R_0, beyond which it flows at a constant Lorentz factor Γ_j (speed $\beta_j c$), confined to a cone of constant opening half-angle ϕ (see Fig. 2). This simple geometry conforms quite well to the available VLBI observations, although apparent accelerations and curved trajectories of knots are sometimes seen (Biretta et al. 1986). If one further assumes that the jet is free (confined solely by its own inertia), it resembles a conic section of a spherical wind. The density therefore falls off as R^{-2}, while the component of the magnetic field parallel to the jet $B_{\|} \propto R^{-2}$ and the component transverse to the jet $B_{\perp} \propto R^{-1}$. The region near $R = R_0$ contains the highest density and magnetic field strength of the conic portion of the jet, and therefore appears as a stationary, bright feature even though plasma streams through it at a relativistic speed. In the model, this region is identified with the VLBI core. Since it is not clear whether the portion of the jet at radii $R < R_0$ is directly observable at any waveband, one can at present only speculate as to the origin and source of collimation of the jet. It is possible, for example, that a strong standing shock at R_0 converts bulk flow energy into internal energy, causing the jet to become suddenly visible beyond this point.

The relativistic jet model relieves several major difficulties (apparent superluminal motion, brightness variability on timescales shorter than the light-travel time across the source, and excessive energy requirements) associated with radio-loud AGN. Relativistic beaming of the emission toward the observer causes one to miscalculate the luminosity and photon density of the source. In order to derive these quantities, one usually multiplies the observed integrated flux F by $4\pi d_\ell^2$ to obtain the luminosity at the source, where d_ℓ is the luminosity distance. However, if relativistic beaming occurs, the radiation is not emitted isotropically and the timescales are contracted in the observer's frame. Therefore, one must further multiply F by a factor δ^{-4} for a variable emission region moving within the jet, where $\delta \equiv [\Gamma(1-\beta\cos\theta)]^{-1}$ and θ is the angle between the jet axis and the line of sight. The Doppler factor δ ranges from $(1+\beta)\Gamma$ for $\theta = 0$, to Γ for $\theta = \sin^{-1}(1/\Gamma)$, to Γ^{-1} for $\theta = \pi/2$. Any given flux-limited survey tends to be dominated by beamed sources, so one has no problem with statistics as long as there is a sufficient supply of "parent" objects that have low apparent luminosities because most of their radiation is beamed away from us. Since the actual luminosities of sources with jets beamed toward us are much lower than previously inferred, the energy requirements and "Compton catastrophe" are greatly diminished.

Since we have yet to understand how a relativistic jet is focussed and accelerated, we cannot obtain a firm theoretical upper limit to the Lorentz factor of the jet. However, if the jet is accelerated near an accretion disk, Compton drag should limit the Lorentz factor to $\Gamma_{\rm j} \lesssim$ 10-100 (Phinney 1987; Melia & Königl). There are other arguments that lead to a similar constraint (see Phinney 1987). One can use the measured angular size, turnover frequency, and flux density to derive the magnetic field and energy density in relativistic electrons of a source. When this is done for a typical compact component in a blazar without correcting for relativistic beaming, one finds that the energetics are dominated by the relativistic electrons (see, e.g., Burbidge et al. 1974; Marscher & Broderick 1985). If one assumes that the component is moving relativistically with a Doppler factor δ, this imbalance is lessened and the total energy requirement is correspondingly diminished. But once the Lorentz factor exceeds about 10, the magnetic field begins to dominate and the energy requirements become extremely high. The other consequence of very high Lorentz factors is that the more highly beamed blazars are, the greater the required number of unfavorably beamed parent objects is. This also restricts the *typical* value of Γ to be $\lesssim 10$, although *a few* extraordinary objects could have higher values. The "critical" value $\Gamma \sim 10$ is a rough one, but the arguments are general.

The magnetic field can be derived, and the combined magnetic and relativistic electron pressure estimated, by measuring the spectra and angular sizes of individual components in the compact radio jet using VLBI observations (see Marscher 1987 for a review). One can then use these results to estimate the values of these physical parameters at the core. Marscher & Gear (1985; hereafter M&G), who worked out a model for the compact jet of 3C 273, obtained $B_{\rm core} \sim 0.1$ to 1 gauss and $P_{\rm core} \sim 10^{-1\pm1}$ dynes. This pressure is of the same order as that of the broad emission-line clouds

($\sim 10^{-2}$ dynes for number densities $n_e \sim 10^{10}$ cm^{-3}). This, plus the similar size scales of the VLBI core and the broad emission-line region (BLR), leads to the expectation that there may be a physical relationship between the jet and BLR. Perhaps the same hot, thermal medium ($T \sim 10^8$ K, $n_e \sim 10^6$ cm^{-3}) thought to confine the clouds also confines the inner jet.

3. Shock-in-Jet Models for Nonthermal Flares

Blandford & Königl (1979) proposed that disturbances in the jet flow would cause shocks to form and propagate down the jet. They associated these shocks with the knots observed to move at apparent superluminal speeds in VLBI images of compact extragalactic jets. Jones (1982), M&G, and Hughes et al. (1985) subsequently showed that such shocks could reproduce quite well many of the detailed characteristics of the structural, spectral, and polarization variability of blazars. Figure 2 shows the basic geometry of the model.

Shocks are created when a significant disturbance occurs in the otherwise steady jet flow. (See Marscher 1990 for a definition of "significant disturbance.") As in the case of prolonged disturbances in the solar wind, the basic shock structure (pressure, density, or velocity vs. distance R along the jet) can be a "shock pair," with a forward-facing shock sweeping up the previously undisturbed jet medium and a reverse shock (moving backward relative to the forward shock but moving forward relative to the observer) that decelerates the enhanced "driver gas" flow propagating from inner radii. A contact discontinuity separating the shocked ambient jet plasma from the enhanced flow is situated between the forward and reverse shocks. For more impulsive disturbances, the reverse shock will not form.

One important result (M&G) is that, at high frequencies where the radiative lifetimes of the shock-accelerated electrons are short, the emission region can be confined to a very thin layer behind the shock front. This can shorten the timescale of variability relative to the often assumed limit $\lesssim R/c$. Furthermore, the emission region is larger at lower frequencies, so that the shortest timescales of variability occur at infrared-optical frequencies, as observed (e.g., Bregman et al. 1990). Marscher et al. (1992; hereafter MGT) illustrate this effect by considering the variations caused by a shock moving through a hydromagnetically turbulent jet. Rapid fluctuations in both total flux density and polarization are seen in the numerical simulations. Of particular interest is the common occurrence of apparent polarization position angle rotations (by as much as ~ 1 cycle) that are nothing more than the superposition of random fluctuations in the ambient magnetic field in the jet (see also Jones 1988).

A shock-induced flare begins with a sharp rise in the spectrum, with the turnover frequency ν_m decreasing slightly as the flux density rises simultaneously at all optically thin frequencies. This abrupt rise corresponds to either the initial formation of the shock or the progressive weakening of Compton losses as the shock expands. Synchrotron losses soon dominate, and the entire spectrum shifts toward lower frequencies while the flux density at the turnover remains roughly constant. Once the adiabatic-loss phase begins, the flare subsides such that the low-frequency side ($\nu < \nu_m$) "slides" downward in the sense that the flux density remains roughly constant until the turnover passes by. The entire evolution is given in Fig. 3. Quantitatively, the evolution proceeds according to the formulas of M&G, Marscher (1990; note that the minus sign in the exponent of expression (8a) of that paper should be deleted), and MGT.

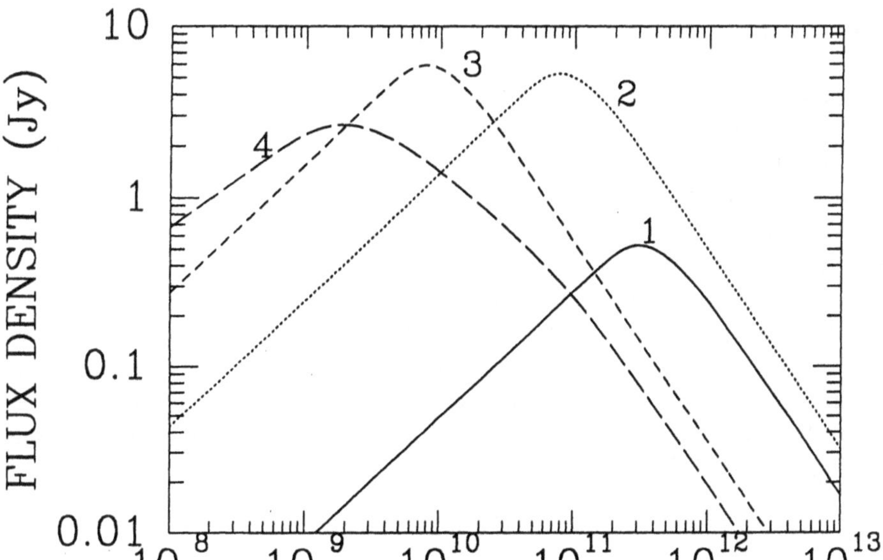

Fig. 3. Schematic spectral evolution of a flare arising from a shock propagating down a jet. The outburst progresses from the Compton-loss stage (1 to 2) to the synchrotron-loss stage (2 to 3) to the adiabatic-loss stage (3 to 4).

VLBI and VLA observations of jets show that they often bend considerably. For relativistic jets, the small angle of the jet axis to the line-of-sight amplifies the bending angle. Even so, bends of at least several degrees are required by the observations. Although it is not necessarily true that the jet axis itself bends, this is the most straightforward interpretation of the observations. Hardee (1990) has shown that macroscopic fluid instabilities can induce a jet to twist by several degrees. Curvature in a relativistic jet causes variations in Doppler boosting and observed motion transverse to the line of sight. These effects lead to systematic changes in the observed brightness and apparent superluminal motion of a compact source. The observational consequences of variations in the orientation angle of the jet θ are described by Marscher (1990) and MGT, who give the spectral variations expected from a shock in a curved jet. The flux density can rise and fall simultaneously at all frequencies, with little change in the self-absorption turnover frequency, as observed in some flares.

4. Variability Timescales

4.1 General Principles

Variations in flux density, spectrum, and structure can provide excellent diagnostics for determining fundamental properties of a source. Unfortunately, the interpretation is highly model dependent. Still, one can exclude certain models on the grounds that they cannot reproduce the nature of the variations. For example, the so-called "Christmas tree" model for apparent superluminal motion can be excluded on the grounds that, in all well-documented cases, the motion is systematic (away from the core).

The most important, and often misleading, parameter that can be derived from observations of variability is the size of the varying emission region. In the most straightforward case, a source should not be able to vary simultaneously throughout its volume any more rapidly than over a "variability timescale"

$$t_{\rm var} < (D/2c)[(1+z)/\delta], \tag{1}$$

where D is the size of the largest dimension, z is the redshift, and δ is the Doppler factor if the emitting region is moving relativistically. The most conservative estimate of the observed value of $t_{\rm var}$ is that given by Burbidge et al. (1974), who take into account that only a fraction of the flux density F_ν varies:

$$t_{\rm var} \approx \frac{t_2 - t_1}{|\ln(F_{\nu,1}/F_{\nu,2})|}. \tag{2}$$

In interpreting their data, many observers replace the dimension D above with the distance R from the central engine to the emission region. Hence the conclusion by some authors (who will go unnamed) that they have derived a limit to the distance of the emission region from a putative black hole, and therefore a limit to the Schwarzschild radius, and thus mass, of the black hole. In reality, they have only estimated the extent of the plasma responsible for the variations, and the emitting region could be much more distant from the central engine. [These same authors often also set both the Doppler factor and the denominator of eq. (2) equal to unity, thereby making their conclusions as extreme as possible.]

It is also possible for the timescale of variability to violate expression (1) in the context of certain geometries that allow pattern speeds ("phase velocities") faster than the speed of light. "Light echoes" (Lynden-Bell 1977) are the simplest version of such a phase effect. Within the context of the SWM, the intersection of a shock with a density or magnetic field enhancement can provide the sort of signal-screen interaction that leads to rapid variability. The fastest phase effect results when all the interactions take place along the line of sight, as would occur, for example, in the collision of two planar regions oriented transverse to the line of sight and moving directly toward the observer. A more realistic geometry is one in which a radial "signal" (e.g., a shock) strikes a feature that subtends an angle ψ within the jet. The minimum timescale of variability is the time difference (in the observer's frame) between when the signal first strikes the feature and when it stops striking the feature. If the internal variations (i.e., how long the feature stays lit up) are shorter than this, the timescale of variability is

$$\begin{aligned} t_{\mathrm{phase}} &= [\frac{r}{\beta c}\{[1-\beta\cos(\theta+\psi)]-[1-\beta\cos(\theta-\psi)]\} \\ &\quad + \frac{r}{c}\sin\frac{\psi}{2}\sin\frac{\psi}{4}](1+z) \\ &= \frac{r}{c}[(2\sin\theta\sin\psi)+\sin\frac{\psi}{2}\sin\frac{\psi}{4}](1+z) \\ &\approx \frac{D\theta}{c}(1+z) \qquad\qquad (\psi\lesssim\theta\lesssim 1) \\ &\approx \frac{D\psi}{8c}(1+z) \qquad\qquad (\theta\lesssim\psi\ll 1). \qquad (3) \end{aligned}$$

Here, $D = r\sin\psi \approx r\psi$. The above expression is strictly valid only for a feature that is centered on the jet axis. For off-axis features, the limits of the last expression should be changed such that the velocity vector of the signal lies within an angle ψ of the line of sight at the point where it interacts with the screen.

The phase effect is washed out if the internal variations (in the observer's frame) occur over a timescale longer than that given by eq. (3). Another limit is the timescale required for the signal to pass the feature. If the feature's dimension (or, if greater, the signal's dimension) parallel to the line-of-sight is x, this timescale is

$$t_{\mathrm{pass}} = (x/\delta\Gamma c)(1+z) \approx (x/2\Gamma^2 c)(1+z) \qquad (\theta \ll 1/\Gamma). \qquad (4)$$

The timescale of the variations decreases as the fraction of the jet covered by the perturbation ψ/ϕ is made smaller, but this comes at the expense of a diminished amplitude of variation, which falls off as $(\psi/\phi)^2$. Hence, the timescale of variability as defined by expression (2) in fact increases for smaller values of (ψ/ϕ). In comparing theoretical timescales with the observations, the longest theoretical value must be adopted.

4.2 Rapid Variability at Centimeter Wavelengths

Quirrenbach et al. (1989) report rapid (timescales $\sim$ 1 day) brightness variations at centimeter wavelengths for certain high declination sources. Subsequent observations of the source 0917+624 (Qian et al. 1991) show the variations to be broad-band (from about 20 to 2 cm) and to include polarization variations. In the source 0716+714 two quasi-periodic episodes are observed at both radio and optical wavelengths, with the variations in the two wavebands very similar although not in phase (Quirrenbach et al. 1991). The latter observations (which need to be confirmed through further monitoring) exclude any known interstellar scattering mechanism as the underlying cause of the variability.

As is discussed in the analysis of 0917+624 by Qian et al. (1991), intraday variability at the longer wavelengths implies such high brightness temperatures ($\sim 10^{18}$ K) as to create an "inverse Compton catastrophe" if the basic radiation mechanism is incoherent synchrotron emission with random pitch angles. Yet the polarization characteristics of the source (moderate linear polarization, very low circular polarization) are precisely those expected for the standard incoherent synchrotron process. It therefore makes sense to determine whether it is possible to explain the rapid variations in the context of the SWM. This has been discussed by Qian et al. (1991) in the context of a specific model, which assumes that the timescale of eq. (4) governs the observed variability. This requires that the timescale derived in expression (3) be less than the value of $t_{\mathbf{pass}}$ of eq. (4). Here I wish to generalize and also to provide the reader with a sense of the severe constraints on such models as well as how the models can be tested.

As outlined in the previous section, the key ingredients of a successful model are (1) a favorable orientation such that phase effects occur and (2) a sufficiently high Doppler factor that the crossing time of the "signal" passing the feature [eq. (4)] is shorter than the minimum timescale allowed by the phase effect [eq. (3)]. If, in the context of the jet model, we are favorably oriented such that $\theta < \phi$, the timescale of variability is then governed by the last expression of eq. (3). For the variations of 0917+624 at 11 cm (which give the most severe difficulties), the timescale is 1.3 days at an amplitude of 0.27 Jy. If we then use eq. (3) to derive the diameter D and assume a cosmological distance with $H_0 = 100h$ km s^{-1} Mpc^{-1} and $q_0 = 0.5$, the variations imply a brightness temperature $T_B \approx 1 \times 10^{13} h^{-2} (\psi/0.01 \text{ rad})^2$.

By comparing this with the inverse Compton brightness temperature limit, $\sim 10^{12}[\delta/(1+z)]^{6/5}$ K, we find that for this source $\psi \lesssim 0.0025h\Gamma^{3/5}$ radians in order for the phase effect to be successful. For $\Gamma \sim 10h^{-1}$, this corresponds to $\psi \sim 0^\circ\!.6h^{2/5}$. The opening half-angle of the jet is then constrained to be $\phi \lesssim 1^\circ$ in order to maintain the observed amplitude of variability (see above). This is consistent with the small value of ϕ inferred for the 3C 273 jet by M&G and Unwin et al. (1985). The orientation angle must therefore also be $\theta \lesssim 1^\circ$, so that we are looking straight down the jet. The low probability of such a favorable viewing angle (even considering beaming effects, which have a broad – FWHM $= 2/\Gamma$ radians – peak centered on $\theta = 0$) implies that 0917+624 must represent an extreme case of rapid centimeter-wave variability. Therefore, the implied brightness temperature of similar sources should, with few exceptions, be less than the 0917+624 value of $\sim 10^{18}$ K.

The quasi-periodic behavior of the brightness fluctuations in 0716+714 could prove to be very interesting if borne out by further observation. In the model in which a shock moves through a turbulent jet (MGT), false periodicities can appear for several cycles, but they are not in general repeated. If confirmed, the periodicities would suggest oscillations in the jet or some type of coherent emission process.

A limit on the extent x of the "signal" along the jet axis is given by eq. (4). For the variations from 0917+624, the limit $t_{\rm pass} \lesssim 1.3$ days gives the limit $x \lesssim 0.09$ pc. The distance along the jet [from eq. (3) 9where the variations occur is $r \sim 4$ pc, so that the ratio $x/r \sim 0.02$. This is considerably thinner than that expected for a shock ($x/r \gtrsim 0.15$) based on nonrelativistic gas dynamics, although relativistic effects may diminish the ratio. Radiative energy losses of the electrons are unlikely to be sufficiently important to cause the emission region behind the shock to be thin, as may occur at high frequencies (M&G).

4.3 Non-Standard Models

In the event that future observations push the SWM beyond its reasonable limits, there are a number of high-brightness temperature models that are potentially capable of explaining the rapid variability. Most such processes, however, produce the radiation close to the central engine (e.g., Lesch & Pohl 1992). It is very difficult to see how this radiation can be transported from deep within the source, given the intervening plasma, highly opaque at centimeter wavelengths, that is known to exist. Many coherent processes, such as stimulated inverse Compton emission, are too narrow-band to explain variability over a decade of frequencies (Blandford, private communication). Most high-brightness temperature mechanisms that can operate out in the jet result in substantial amounts of circular polarization that must be rendered somehow unobservable, for example by appealing to almost exact symmetry between left- and right-handed emission (see Jones et al. 1974), which is possible for coherent synchrotron emission by an electron-positron plasma (e.g., Sanders & DaCosta 1978).

A long-standing trivial solution to the timescale problem is that the distances are not as great as derived from a standard cosmological interpretation of the redshifts (e.g., Burbidge et al. 1974). This increases the angular size of a source of given physical size, and therefore decreases the implied brightness temperature below the 10^{12} K limit. It does not, however, solve the overprediction of self-Compton X-rays in some sources whose angular sizes have been measured with VLBI, which requires either relativistic motion or clusters of high brightness-temperature emission regions (e.g., Marscher & Broderick 1981).

5. High-Energy Emission

As discussed above, a compact nonthermal jet should produce X-ray and γ-ray emission through the self-Compton process. Indeed, these objects are brighter X-ray sources (for a given optical magnitude) than are objects with weak or extended nonthermal emission (e.g., Zamorani et al. 1981). One therefore expects a direct correlation between variability at radio to infrared and X-ray to γ-ray frequencies. Although there is still insufficient monitoring at all but radio frequencies to test this hypothesis in detail, the observation in 3C 279 of a major X-ray flare (Makino et al. 1989) and, recently, a hard γ-ray flare (Fichtel et al. 1992) during millimeter–submillimeter flares (Robson, private communication) certainly supports this. Hard ($\gtrsim$ 100 MeV) γ-rays produced near the central engine are expected to pair produce off the intense X-ray photon field there, and thus are unlikely to escape (McBreen 1979). It is therefore probable that the hard γ-rays arise from inverse Compton scattering in the compact jet.

In general, since the self-Compton flux density is proportional to the synchrotron photon density as well as the density of relativistic electrons, both of which increase during an outburst, the high-energy flare should be more pronounced than the synchrotron flare, after subtraction of the "quiescent" emission. However, the self-Compton X-ray emission from a component moving in a bending jet should vary simultaneously, and with about the same fractional amplitude, as the optically thin radio to infrared synchrotron emission from an isolated component, since the Doppler factor δ boosts both types of emission identically.

If the jet is turbulent, essentially simultaneous flickering of the optically thin synchrotron and inverse Compton emission should occur as the shock encounters inhomogeneities. However, those synchrotron variations that are primarily caused by changes in magnetic field *direction* will correspond to minor, if any, high-energy self-Compton fluctuations.

A significant amount of high-energy emission could arise from the inner portion of the jet between the central engine and the observed VLBI jet. If the inner jet is tapered (Marscher 1980; Ghisellini et al. 1985), the high-energy emission could be either synchrotron or self-Compton radiation from the innermost regions (Celotti et al. 1991). Alternatively, rapidly streaming electrons deep in the jet can up-scatter optical-uv photons from an accretion disk to X-ray and γ-ray energies, a process which also tends to slow down the jet to a Lorentz factor $\Gamma \sim 10$ (Melia & Königl 1989; Begelman & Sikora 1987).

6. Summary

The standard relativistic jet model is attractive because it explains simultaneously the extraordinary array of exotic observational properties of compact, nonthermal extragalactic sources. The rapid centimeter wave variability detected recently by the German group cited above stretches the model to its limits. It is extremely important to pursue and extend these observations so that we can either confirm the ability of the model to allow such extreme variability or adopt a different approach that could alter significantly our concepts of the physics of AGN.

The work discussed in this paper has been supported in part by U.S. National Science Foundation grant AST-8815848 and NASA grant NAGW-1068.

References

Begelman, M.C., Sikora, M. 1987, ApJ, 322, 650
Biretta, J.A., Moore, R.L., Cohen, M.H. 1986, ApJ, 308, 93
Blandford, R.D., Königl, A. 1979, ApJ, 232, 34
Blandford, R.D., Rees, M.J. 1978, in BL Lac Objects, ed. A.M. Wolfe (Univ. of Pittsburgh), p. 328
Bregman, J.D., et al. 1990, ApJ, 352, 574
Brown, L.M.J., et al. 1989, ApJ, 340, 129
Burbidge, G.R., Jones, T.W., O'Dell, S.L. 1974, ApJ, 193, 43
Celotti, A., Maraschi, L., Treves, A. 1991, ApJ, in press
Fichtel, C., et al. 1992, in Proceedings of the 2nd Gamma Ray Observatory Workshop (NASA), in press
Fiedler, R.L., Pauls, T., Johnston, K.J., Dennison, B. 1991, ApJ, submitted
Ghisellini, G., Maraschi, L., Treves, A. 1985, A&A, 146, 204

Hardee, P.E. 1990, in Parsec-Scale Radio Jets, ed. J.A. Zensus & T.J. Pearson (Cambridge Univ. Press), p. 266
Hughes, P.A., Aller, H.D., Aller, M.F. 1985, ApJ 298, 301
Jones, T.W. 1982, BAAS, 14, 963 (Abstract)
Jones, T.W. 1988, ApJ, 332, 678
Jones, T.W., O'Dell, S.L., Stein, W.A. 1974, ApJ, 188, 353
Lesch, H., Pohl, M. 1992, A&A, in press
Lynden-Bell, D. 1977, Nature, 270, 396
Makino, F., et al. 1989, ApJ, 347, L9
Marscher, A.P. 1980, ApJ, 235, 386
Marscher, A.P. 1987, in Superluminal Radio Sources, ed. J.A. Zensus & T.J. Pearson (Cambridge Univ. Press), p. 280
Marscher, A.P. 1990, in Parsec-Scale Radio Jets, ed. J.A. Zensus & T.J. Pearson (Cambridge Univ. Press), p. 236
Marscher, A.P., Broderick, J.J. 1981, ApJ, 247, L49
Marscher, A.P., Broderick, J.J. 1985, ApJ, 290, 735
Marscher, A.P., Gear, W.K. 1985, ApJ, 298, 114 (M&G)
Marscher, A.P., Gear, W.K., Travis, J.P. 1992, in Variability of Blazars, ed. E. Valtaoja (Cambridge Univ. Press), in press (MGT)
McBreen, B. 1979, A&A, 71, L19
Melia, F., Königl, A. 1989, ApJ, 340, 162
Pearson, T.J., Readhead, A.C.S. 1988, ApJ, 328, 114
Phinney, E.S. 1987, in Superluminal Radio Sources, ed. J.A. Zensus & T.J. Pearson (Cambridge Univ. Press), p. 301
Qian, S.J., Quirrenbach, A., Witzel, A., Krichbaum, T.P., Hummel, C.A., Zensus, J.A. 1991, A&A, 241, 15
Quirrenbach, A., Witzel, A., Krichbaum, T., Hummel, C.A., Alberdi, A., Schalinski, C. 1989, Nature, 337, 442
Quirrenbach, A., et al. 1991, ApJ, 372, L71
Sanders, R., DaCosta, L. 1978, in BL Lac Objects, ed. A.M. Wolfe (Univ. of Pittsburgh), p. 375
Zamorani, G., et al. 1981, ApJ, 245, 357

Discussion

Camenzind: What are the physical processes which determine the spectral index and the density of relativistic particles in the jet model?

Marscher: For the shocks, our models assume merely that the electrons are heated only by compression, which is a very conservative assumption. The density is enhanced by the usual compression ratio. As for the underlying jet, no detailed model has been developed for the injection of electrons near the core.

Wandel: Can you estimate how much radiation would be emitted by charged particles moving at a Lorentz factor of about 10 and curving by an angle of about 0.1 radians over the timescales of variability observed?

Marscher: Since the bending of the jet is by a few degrees over a sizescale ($\sim$ parsecs) orders of magnitude larger than the gyroradius of an individual particle, the effect you are suggesting is negligible compared to normal synchrotron radiation. In addition, the frequency would be extremely low, so the waves would be re-absorbed by the particles.

Lesch: Is it possible to avoid circularly polarized emission by using beams running exactly along the magnetic field?

Marscher: Yes, but then there is no radiation unless the field lines are curved ("curvature radiation"), in which case the circular polarization is quite high. It would seem that the best way to eliminate large circular polarizations in models of this sort is to assume that the emitting plasma is composed of equal numbers of electrons and positrons.

Kundt: Isn't Phinney's conclusion that the bulk Lorentz factor is limited to ~ 10 based on the assumption that the pairs are formed at the small radii predicted by the black-hole model? Bulk Lorentz factors $\gg 10$ are implied by other constraints (such as the existence of one-sided optical jets) and consistent with my SMD model.

Marscher: Yes, that is quite correct – if the acceleration of the jet occurs outside the high-radiation environment of the putative accretion disk, the limit is greatly relaxed.

Kovalev: You have shown the results of spectral evolution calculations in a jet model. Did you calculate these spectra as if the magnetic field of the emission region is moving in the frame of the observer or in the co-moving frame of the jet?

Marscher: I calculate the emission and absorption coefficients in the co-moving fluid frame, then transform these to the observer's frame to perform the radiative transfer calculations.

Lesch: How do you avoid that the distribution function of the relativistic electrons becomes more and more anisotropic in pitch angle, when it enters diverging magnetic field lines (conservation of magnetic moment)?

Marscher: I assume that the pitch angles are continuously re-isotropized, as apparently occurs in Galactic cosmic rays. This presumes that plasma processes operate inside the jets much as they do in our Galaxy.

The Spectra of Extended Radio Jets and Hot Spots

Klaus Meisenheimer

Max-Planck-Institut für Astronomie, Königstuhl 17, D-6900 Heidelberg 1, Germany

Abstract. The state of the art in determining the overall spectra of extended (> 1 kpc) radio jets and hot spots is reviewed. Comparing the observations with model spectra allows to draw conclusions about the relevant particle acceleration processes and the physical conditions in the emission regions. New results will be presented for the hot spots of 3C 303 and Cygnus A, and the radio jets of PKS 0521-36 and M87. The perfect agreement of the brightness distribution and polarization of the M87 jet at radio and optical frequencies seems to require continuous particle acceleration all along the jet.

1. Introduction

A classical, extended radio source can be divided into three main constituents: (1) the core region which powers the radio source and other AGN phenomena, (2) a compact radio jet on parsec and subparsec scales, and (3) the extended radio source which consists of huge lobes and a long radio jet observed in some sources. Bright hot spots may be embedded in the lobes of the most powerful sources. Most of this conference has been devoted to the physical processes in the core region. The compact jets have been discussed by Alan Marscher, Arno Witzel, Hélène Sol and others. This talk focuses on the outer regions of radio emission – in particular the kiloparsec jets and the hot spots which are some tens or hundreds of kpc away from the core. I will discuss the topic of synchrotron emitting plasma only, while the problems concerning the interactions between the radio source with its surroundings are omitted.

The basic observables of extended radio sources are the two-dimensional brightness distribution $B_\nu(\alpha, \delta)$ and the (linear) polarization characterized by its degree $\Pi_\nu(\alpha, \delta)$ and orientation angle $\chi_\nu(\alpha, \delta)$. Since the spectral characteristic of synchrotron radiation is very smooth, the brightness and polarization of a pure synchrotron source vary only gradually with observed frequency. It is this lack of pronounced spectral features together with additional complications due to propagation and projection effects which prohibit one from deriving a unique physical model even from the most detailed observations. Usually one has to proceed the opposite way: One builds a set of plausible models which are characterized by a few parameters only and compares them with the observations.

The spectral dependence of brightness and polarization potentially provides the most valuable information about the physical conditions in the source – but the smoothness of synchrotron spectra usually requires a very large range in observed frequency in order to see significant spectral features and to study their spatial variation across the source. Thus observations beyond the classical radio band ($\nu < 30$ GHz) are of great importance. In this talk I give an overview of the state of the art in determining the synchrotron spectra of extended radio sources between 10^9 and 10^{15} Hz. The central part of this contribution is focussed on the comparison of the data with simple source models. This allows us to attack some of the basic questions, that is magnetic field strengths and configurations, particle acceleration processes and their efficiencies and time scales. Unfortunately, it will become clear that very little can be learned about such basic parameters as jet speed, density and composition from the investigation of the synchrotron emission alone.

2. Observational work

2.1 Some stepping stones

Instead of giving a complete account of the observations of jets and hot spots beyond 30 GHz ($\lambda < 1$ cm), I will rather list a few stepping stones towards establishing their overall synchrotron spectra. Concerning hot spots, a more complete review of the literature up until 1988 may be found in Röser (1989).

The first path to be followed led to the detection of those eight or nine extended radio sources which definitely emit optical synchrotron radiation. The following list contains five prominent sources, the reference for their first optical detection and the reference for establishing the synchrotron nature of the optical emission by polarimetry:

Source	Optical detection	Optical polarization found
M87 jet	Curtis 1918	Baade 1956
3C 273 jet	M. Schmidt 1963	Röser & Meisenheimer 1986
3C 33 hot spot	Simkin 1979	Meisenheimer & Röser 1986
Pic A hot spot	Röser & Meisenheimer 1987	therein
PKS 0521-36 jet	Danziger *et al.* 1979	Sparks *et al.* 1990

As a recent example of the type of work which can be done nowadays with groundbased optical telescopes Fig. 1 displays an image and polarization map of the radio jet in PKS 0521-36. This should be compared with

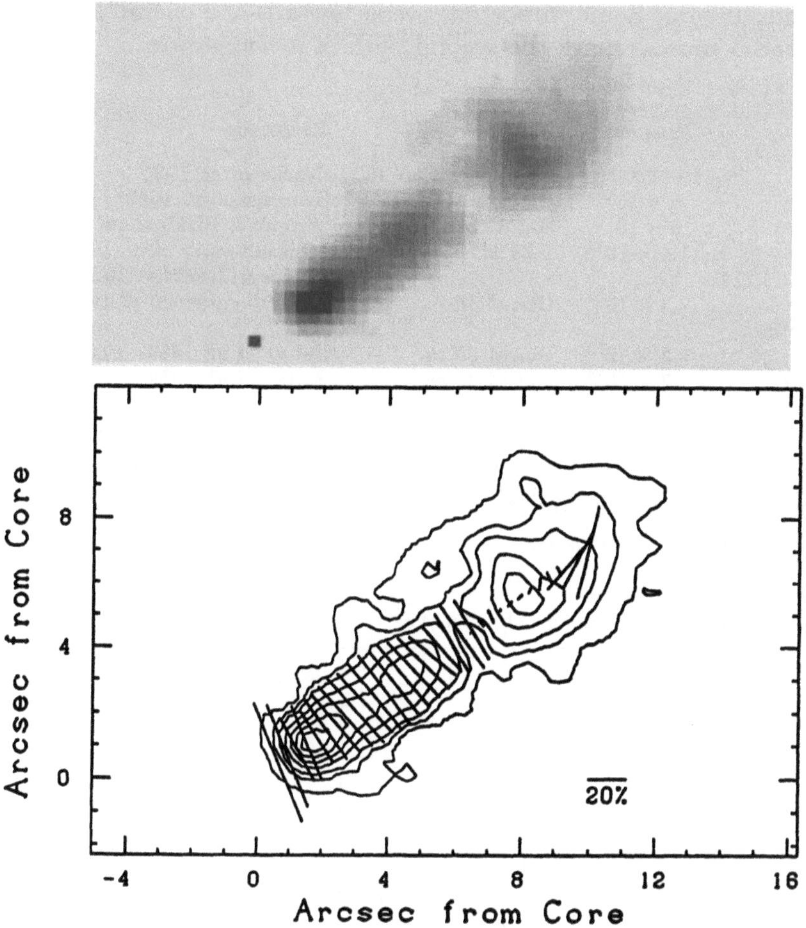

Fig. 1. **Red CCD–image and polarization map of the jet in PKS 0521-36 (from Schlötelburg *et al.* 1991). The light of the galaxy has been subtracted. *Effective* resolution is 1.″3 FWHM.**

the HST results of Macchetto *et al.* (this conference and Macchetto *et al.* 1991). One can see that the high signal to noise obtainable within a few hours with a groundbased 2.2 m telescope compensates well for the higher resolution that is in principle available from deconvolved HST images. Optical polarization maps of similar quality have also been obtained with the Durham polarimeter (e.g. Scarrott & Rolph 1989).

When the synchrotron nature of the radiation is established in at least one high frequency band (typically in the optical) one may safely assume that synchrotron radiation dominates the entire (continuum) spectrum between the radio (10^9 Hz) and optical regime (10^{15} Hz). Thus flux measurements at intermediate frequencies (mm, near-infrared) can be used directly as a constraint for the overall synchrotron spectrum in the entire range between 10^9 and 10^{15} Hz. Some important steps towards the overall spectra of extended radio sources mark the second path of investigation:

Source	Frequency	Instrument	Reference
M87 jet	$1-2\times10^{14}$	IR photom. on 90"	Stocke *et al.* 1981
M87 jet	5×10^{17}	EINSTEIN HRI	Schreier *et al.* 1982
Cyg A hot spots	9×10^{10}	Hat Creek Interfer.	Wright & Birkinshaw 1984
Hot spots in 3C 20, 3C 33, 3C 111	1.4×10^{14}	UKIRT photometer	Meisenheimer *et al.* 1989 Crane & Stockton 1989
Hot spots in 3C 20, 3C 33 etc.	2.3×10^{11}	IRAM 30 m	Meisenheimer *et al.* 1989
M87 jet	2.3×10^{11}	IRAM 30 m	Salter *et al.* 1989
Cyg A hot spots	2.3×10^{11}	IRAM 30 m	Salter *et al.* 1989

Examples for the overall spectra which result from these observations are given by Stocke *et al.* (1981), Meisenheimer *et al.* (1989), Crane & Stockton (1989) and in section 2.3, below.

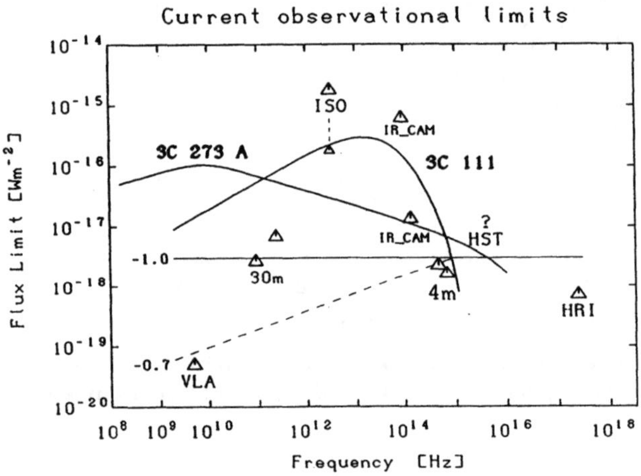

Fig. 2. Observational state of the art. The flux limit νS_ν for various instruments is compared with two power law spectra $S_\nu \sim \nu^\alpha$, $\alpha = -1.0, -0.7$ and the observed spectra of the hotspots in 3C 273 A and 3C 111.

2.2 The state of the art

From the previous paragraph it is obvious that the fraction of extended radio sources which have been detected at frequencies $\nu >> 10^{10}$ Hz is small. This reflects the limitations of present day observational feasibility. Fig. 2 summarizes the state of the art as reached from the ground or as it will be reached from space in the near future (ROSAT X–ray satellite, ISO: infrared space observatory). The figure demonstrates that a radio source easily detectable at the VLA (assumed limit: 1 mJy at 5 GHz) could only be detected at optical frequencies if its spectrum is flatter than $\nu^{-0.7}$ between 10^{10} and 10^{15} Hz (this may be true for some radio jets, see Fig. 5). If the source spectrum is represented by the – more typical – slope $\alpha = d\log S_\nu / d\log\nu = -1$ one needs a radio flux $S_\nu \gtrsim 100$ mJy, well above the VLA limit and a not too extended source to allow an optical detection. Presently, observations at intermediate wavelengths (1.3 mm, 100 μm) are only feasible if the source is either much brighter than this or exhibits an infrared-peaked spectrum (in terms of νS_ν e.g. 3C 111 hot spot, see below). In fact, we find that the spectra of some hot spots cut off dramatically at frequencies $\nu > 10^{14}$ Hz. Spectral cutoffs at slightly lower frequency $\nu_c \leq 5 \cdot 10^{13}$ Hz would render the source undetectable at optical wavelengths or prohibit any high frequency polarimetry to verify the synchrotron nature of a possible counterpart near the detection limit. This is the most likely explanation for the fact that even among the brightest radio hot spots ($S_\nu > 200$ mJy at 5 GHz) only a fraction < 20% has been detected (Röser 1989).

Although the EINSTEIN satellite detected X–rays from the vicinity of several extended radio sources (M87 jet, Cen A, 3C 273 jet ?) the origin of this radiation is unclear. We hope that the superior resolution of the ROSAT HRI and some information about the X–ray spectra will allow us to decide whether the exciting possibility of synchrotron X–rays occurs in some of these sources.

2.3 Synchrotron spectra

Typical hot spot spectra are relatively flat, $\alpha \cong -0.5$ at radio frequencies $\nu \lesssim 10^{10}$ Hz, but steep or very steep, $\alpha \lesssim -2$ in the near-IR–optical regime (fig. 3, more examples are given in Meisenheimer *et al.* 1989, hereafter referenced as MEA'89). A smooth synchrotron spectrum through the measured flux values in the radio (1 GHz$< \nu <$ 15 GHz), at 1.3 mm (230 GHz), at 2.2μm ($1.3 \cdot 10^{14}$ Hz), and in the optical ($\nu > 3 \cdot 10^{14}$ Hz) requires one to invoke a spectral cutoff at some frequency $\nu_c \geq 5 \cdot 10^{13}$ Hz which most naturally is explained by the ocurrence of a high energy cutoff in the underlying energy distribution of relativistic electrons. In fact, model synchrotron spectra based on theoretical considerations about the energy spectrum produced by diffusive shock acceleration (first order Fermi acceleration, e.g. Drury 1983, Webb *et al.* 1984, Heavens 1984) provide an excellent fit to the observations (fig. 3 and 5, details are given in section 3.). There are, however, a few exceptions: The western hot spot of Pictor A may show a straight radio-to-optical power law spectrum $S_\nu \sim \nu^{-0.9}$ which may even reach out to 10^{17} Hz (see fig. 6 in MEA'89). Another example is given in Fig. 4a which displays the radio-to-optical spectrum of the western hot spot in 3C 303 (Leliévre & Wlerick 1975, Kronberg 1976)

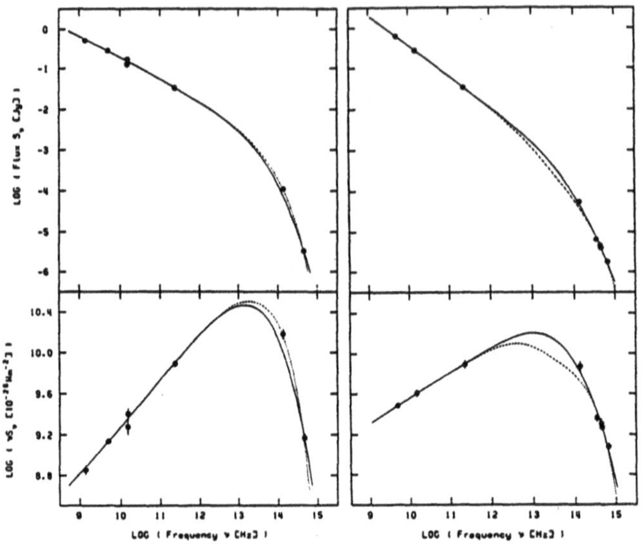

Fig. 3. Spectra of th hotspots of 3C 33 S *(left)* and 3C 111 E *(right)*. The *upper* panels give log S_ν *vs.* log ν, the *lower* panels show log νS_ν *vs.* log ν (from MEA'89).

At intermediate frequencies, $10^{10} < \nu < 10^{14}$ Hz, the behaviour of the spectra can only be studied if both millimeter and near-infrared data are available. They seem to form two distinct groups: (i) spectra in which the (flat) radio power law continues up to the high frequency cutoff (examples: 3C 20, 3C 33, 3C 111, see fig. 3 and MEA'89), and (ii) spectra exhibiting a low frequency 'break' of $\Delta\alpha \cong -0.5$ somewhere around $\nu_b \cong 10$ GHz (e.g. hot spot 3C 273 A, Meisenheimer & Heavens 1986, or Pic A, MEA'89). Very prominent examples of the second type are the three dominant hot spots in Cygnus A (A, B, C in Hargrave and Ryle 1974). Although there is definitely no detection down to an optical flux limit of about 0.4μJy the overall spectra of the Cygnus A hot spots are now well defined by the flux measurements up to $2.3 \cdot 10^{11}$ Hz and the upper limits at $\nu > 10^{14}$ Hz (Fig. 4b). I will come back to the physical interpretation of the observed cutoff and break frequencies in the following section.

Constructing the overall spectra of extended radio jets is even more ambitions than in the case of hot spots. If the spectrum varies significantly along the jet it is necessary to combine resolution matched observations in order to get a reliable result. Fortunately, the spectral variations along the jet of M87 are smooth (Biretta *et al.* 1991 and Fig. 9a, below), so the overall spectrum of the brightest knots A+B+C ($\sim$ 9" to 19" from the core, contributing 83% to the total jet flux) can be derived from data with rather different resolution. The combined results of the efforts of about a dozen groups are shown in Fig. 5. The overall spectrum is dominated by a straight powerlaw $S_\nu \sim \nu^{-0.65}$ between $3 \cdot 10^{10}$ Hz and 10^{14} Hz, flattening slightly at low frequencies to a radio spectral index $\alpha_R = -0.51$. Although

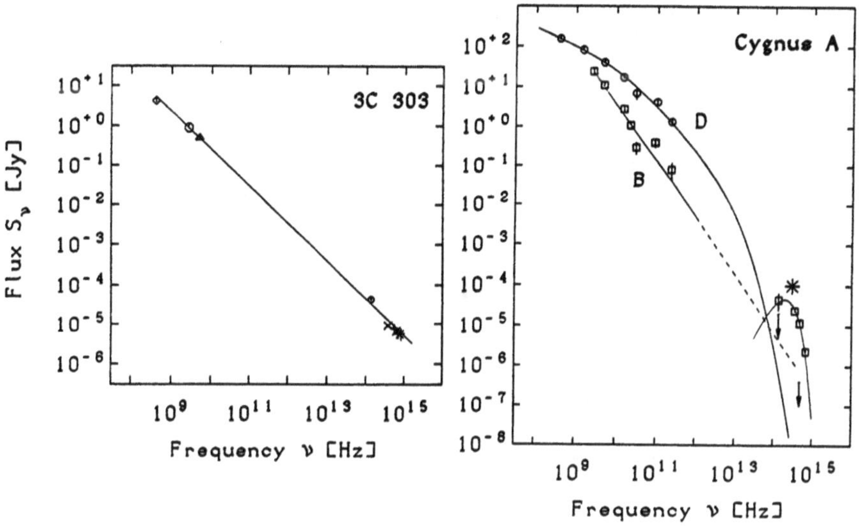

Fig. 4. Overall spectra of the hot spots in 3C 303 *(left)* and Cygnus A *(right)*. Radio data for 3C 303 are from Lonsdale *et al.* 1983 (◇, Laing 1981 (○), and Pooley & Henbest 1974 (△). The K-band flux is from Yates *et al.* 1991. Optical data are from Leliévre & Wlerick 1975 (+) and Keel 1988 (×).
Radio data for Cygnus A are from Muxlow *et al.* (1988) and Alexander *et al.* 1984. The 3mm fluxes are from Wright & Birkinshaw (1984), the 1.3mm fluxes are from Salter *et al.* (1989). Optical and near-IR data are from Yates *et al.* 1991. Note that at the position of hot spot B there is an M-star (*) with a thermal spectrum. Only upper limits could be derived for the high frequency flux of hot spot D. Together with the mm-data, they constrain the cutoff frequency to $\nu_c = 6^{+6}_{-3} \times 10^{12}$ Hz.

the nature of the particle acceleration process in this jet is unknown (see discussion in sect. 5, below) the pronounced steepening beyond $\nu = 10^{14}$ Hz is perfectly described by the synchrotron spectrum of an electron energy distribution with sharp high energy cutoff (continuous line in Fig. 5b). Note that most measurements in the IR – optical – UV region agree amazingly well with each other with only two exceptions: the JHK measurements of the Australian group (Smith *et al.* 1983, Killeen *et al.* 1984), and the RI photometry of Biretta *et al.* 1991. If the adopted cutoff spectrum is correct, a different spectral component is needed to account for the X-ray flux. With the large uncertainties in the present UV data and the lack of spectral information at X-ray frequencies, however, it cannot be excluded that some (theoretically unjustified) $\Delta\alpha = -0.8$ break smoothly connects the radio-optical spectrum to the X-ray point.

In a joint effort of radio and optical astronomers we are currently analysing multifrequency data of the jet of 3C 273 between 408 MHz and 10^{15} Hz with a common resolution of 1.3 arcsec. Preliminary results show that the main body of the optical jet (12″ to 20″ from the QSO) emits a radio–IR spectrum $\nu^{-0.83}$ with a high frequency cutoff which moves to lower

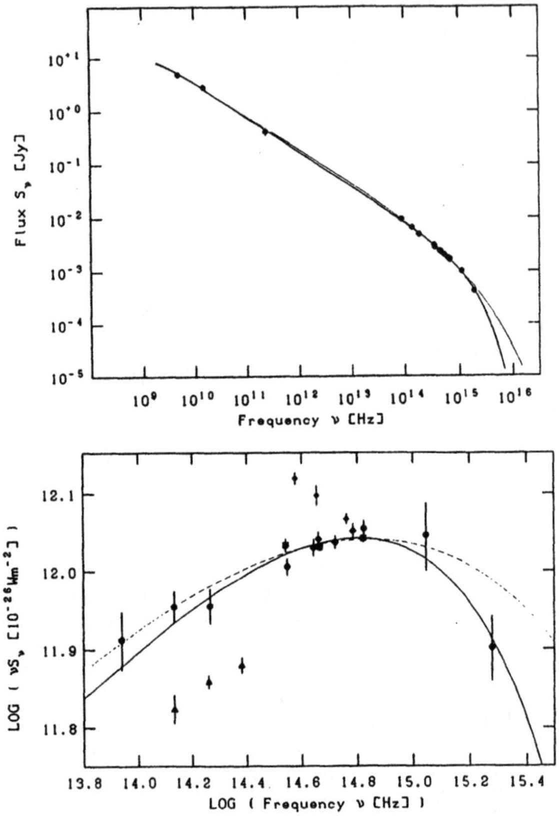

Fig. 5. Spectrum of the brightest part in the jet of M87 (knots A+B+C). **(Top)** Spectrum between 10^9 and 10^{16} Hz. Radio data are from Owen (priv. comm.). The 1.3 mm flux has been derived from the map of Salter *et al.* (1989) by fitting a radio image which has been convolved to the IRAM 30 m beam to the data.
(Bottom) The near-IR to optical part of the spectrum. Near-infrared measurements are from Stocke *et al.* 1981 (•) and an Australian group (△, Smith *et al.* 1983, Killeen *et al.* 1984). Optical data are from Keel 1988 (•), Perez–Fournon *et al.* 1988 (◇), Biretta *et al.* 1991 (◇) and our work (□). The UV points are from Perola & Tarenghi (1980). Note that a sharp cutoff in the electron energy distribution (continuous line in bottom panel) provides a better fit to the data.

and lower frequencies as one goes outwards along the jet. At variance with earlier claims (Röser & Meisenheimer 1986, Keel 1988 and Fraix-Burnett & Nieto 1988) the optical spectrum of the dominant radio hot spot is not exceedingly steep, but deviates only a slightly from a straight powerlaw continuation of the $S_\nu \sim \nu^{-1.2}$ spectrum between 15 GHz and the optical (Röser & Meisenheimer 1991).

Little is known about the overall spectra of other radio jets. For PKS 0521-36 we recently obtained accurate optical photometry (Schlötelburg *et al.* 1991) but the published radio data are insufficient to derive a reliable spectrum. The fourth good candidate for an optical synchrotron jet,

3C 66B (Butcher *et al.* 1980), is of such low contrast that no useful optical polarimetry or photometry could be obtained so far.

3. Physical conditions in hot spots

Since the coverage of the frequency range 10^{10} to 10^{14} Hz is rather sparse one has to compare the observations with theoretically predicted model spectra in order to extract information about the physical conditions.

3.1 Hot spot model spectra

The simplest model for the synchrotron emission of radio hot spots assumes that the hot spot represents the strong terminal shock (Mach disk) of a highly supersonic jet. In the shock's frame the jet plasma streams in with some upstream velocity $u_- = v_-/c$ and is compressed by the ratio r, i.e. $u_+ = u_-/r$ (Fig. 6). The most likely location for strong particle acceleration is in the immediate vicinity of the shock discontinuity, no matter which kind of acceleration process is at work. For a strong planar shock the flow pattern should remain essentially one-dimensional for some distance downstream (say to $x \lesssim D$).

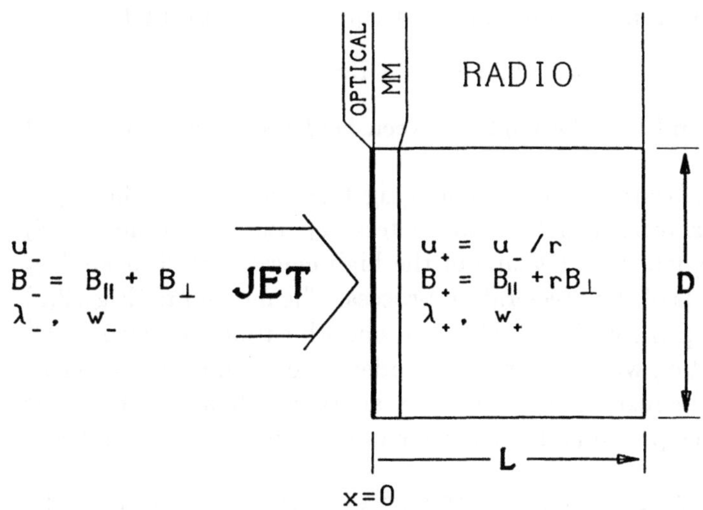

Fig. 6. Schematic sketch of the hot spot model.

In general, the overall spectrum of an unresolved hot spot does not originate from the close vicinity of the accelerating shock alone but will contain also radiation which is emitted from the region downstream. Thus the observed spectrum does *not* represent the spectrum of the freshly accelerated electrons.

If one assumes diffusive shock (first order Fermi-)acceleration and a constant downstream magnetic field B_+ an analytic expression can be given for the downstream evolution of the electron spectrum and the integral spectrum (Heavens & Meisenheimer 1987): At the shock ($x = 0$) the balance between acceleration gains and synchrotron losses will limit the maximum energy to some value $\gamma_c = E_{max}/m_e c^2$. Further downstream, $x > 0$, the cutoff energy γ_c moves to lower and lower energies

$$\gamma_c(z) = \frac{4}{3} \frac{r-1}{rw_- + w_+(1+z)} \tag{1}$$

with the 'specific loss rate' $w_\pm = 1.65 \cdot 10^{-11} (B/nT)^2 (\lambda_\pm/pc) u_\pm^{-2}$ where $\lambda_\pm$ is the electron mean free path and $z = \frac{3u_+}{\lambda_+} x$ is a dimensionless downstream coordinate.

For a hotspot emission region of finite length L the integral electron spectrum follows the initial powerlaw $N(\gamma) \sim \gamma^{-q}$ (with $q = \frac{r+2}{r-1}$ for 1st order Fermi) only up to a break energy $\gamma_b = \gamma_c(Z_{max})$ with $Z_{max} = 3u_+ L/\lambda_+$. Between γ_b and the maximum energy $\gamma_c = \gamma_c(0)$ the spectrum is steeper: $N(\gamma) \sim \gamma^{-q-1}$.

The resulting synchrotron spectrum is relatively flat $\alpha_0 = -\frac{q-1}{2}$ (i.e. $\alpha_0 = -0.5$ for $r = 4$ or $\alpha_0 = -0.25$ for $r = 7$) at frequencies below the break frequency ν_b where it steepens to $\alpha_1 = \alpha_0 - 0.5$. The characteristic energies γ_b and γ_c are converted to emitted frequencies according to

$$\nu_{c,b} = 42 B_+ \sin\theta \, \gamma_{c,b}^2 \text{ Hz} \tag{2}$$

where B_+ is in nT and θ is the angle between field vector and the line of sight.

That is, the synchrotron spectra emerging from this simple hot spot model are fully characterized by *four* parameters: α_0, ν_b, ν_c and the flux S_0 at some fiducial frequency. The shape of the high energy cutoff around γ_c depends on the details of the acceleration process. The model fits (Figs. 3,5) assume two extreme possibilities: (i) the very smooth cutoff produced by 1st order Fermi acceleration with a momentum-independent diffusion coefficient (Heavens & Meisenheimer 1987) and (ii) a sharp cutoff at γ_c. Realistic acceleration processes presumably produce cutoff spectra in between these extrema.

The examples in Figs. 3–5 and in MEA'89 show that a good fit can be obtained in most cases. In principal it should be possible to determine even the shape of the high energy cutoff if one measures the spectrum between 10^{14} and $2 \cdot 10^{15}$ Hz with a relative accuracy of a few percent (see Fig. 5).

3.2 Magnetic field strength

For sufficiently long hot spot emission regions, $L \gg \lambda_+/(3u_+) \cong 10pc$, the break frequency ν_b depends only on the downstream field B_+ and the time $t_L = L/cu_+$ it takes an electron to be advected through the entire emission region of length L:

$$\nu_b = 3.1 \cdot 10^{16}(r-1)^2\ u_+^2\ (L/kpc)^{-2}\ B_+^{-3}\ \mathrm{Hz}.$$

If one is able to determine the length L on radio maps (at $\nu < \nu_b$) and one observes a spectral break at ν_b an estimate of the downstream field is (for $r = 4$):

$$B_+ = 250\ u_-^{2/3}(L/kpc)^{-2/3}(\nu_b/\mathrm{GHz})^{-1/3}\ \mathrm{nT} \qquad (3)$$

which is completely independent from the widely used but poorly established minimum energy argument (see Leahy 1991).

Assuming $u_+ = u_-/r = 0.3/r \cong 0.075$ the field estimate (3) should be correct within a factor of 2 for mildly relativistic jet speeds $0.1 < u_- < 0.8$. The comparison between the field estimate (3) and the minimum energy estimate [1] gives a comforting agreement to within a factor of 2 (see details in MEA'89). It is remarkable that all the field estimates cluster around $\langle B_{HS} \rangle = 35\,\mathrm{nT}$ although the hot spots span a factor of > 100 in luminosity !

3.3 Maximum energy, acceleration time and efficiency

With the field strengths B_{HS} the observed cutoff frequencies (Fig. 7) can directly be converted into maximum energies reached by the acceleration process. One finds a wide range $3 \cdot 10^4 < \gamma_c < 3 \cdot 10^6$. The distribution of ν_c (or γ_c) is now much broader than in MEA'89 due to the inclusion of two new values (3C 303, Cyg A) and the confirmation of the 'minimum cutoff' alternative for 3C 123 (see MEA'89: tab. 4). As presumed in MEA'89 the narrow range found in small samples is due to an observational selection effect (see Fig. 2). So models which try to predict a common $\nu_c \cong 3 \cdot 10^{14}$ Hz (Biermann & Strittmatter 1987) seem obsolete.

On the contrary, the wide variety in maximum energies reached in almost identical fields B_{HS} indicates that the acceleration time scales $\tau_{acc} = \frac{E}{dE/d\gamma}$ can be very different ($100 < \tau_{acc} < 10000$) years from source to source. Since the total time to bring an electron to γ_c should be only a few times longer, the stability of the shock front should be no severe problem for maintaining

1 In the hot spot model the ratio κ between energy stored in protons and in electrons depends on z. Therefore the downstream evolution of a common initial κ_0 should be taken into account when estimating the minimum energy field (see Meisenheimer 1989).

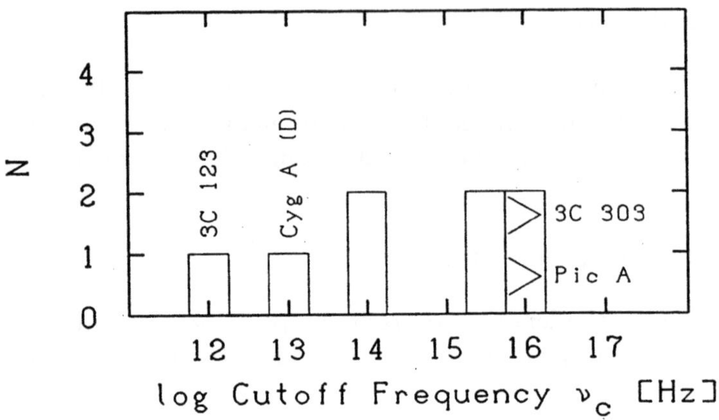

Fig. 7. Cutoff frequencies

the acceleration process (typical dynamical timescale is $1kpc/0.1c = 3 \cdot 10^4$ yrs).

So far the conclusions about the physical conditions have been independent of the specific acceleration process involved. In order to test whether a self-consistent diffusive shock acceleration model is able to explain the observed spectra one has to carry out an efficiency analysis: From the requirement that the jet's ram pressure $n_p m_p c^2 u_-^2$ should balance at least the magnetic field and relativistic particle pressure in the hot spot, I find that 0.2% to 8% of the electrons have to be accelerated to relativistic energies. In the non-relativistic model ($u_{jet} \cong 0.3$) the conversion from kinetic energy into accelerated particle energy must be rather efficient, $0.2 < \epsilon_{acc} < 0.5$ (MEA'89). Altogether, first order Fermi acceleration at a strong non-relativistic shock seems to be (just) able to account for the observed synchrotron spectra of radio hot spots. However, the observed spectra could equally well be explained by 1st order Fermi acceleration at a strong *relativistic* shock (e.g. Kirk & Schneider 1987, Ballard & Heavens 1991). In this case a lower acceleration efficiency would be sufficient.

4. Magnetic field configuration

4.1 Field configuration and polarization

The typical polarization values, $\Pi \cong 20\%$, observed in extended radio jets and hot spots are much lower than the value expected for well ordered magnetic fields, $\Pi \cong 70\%$. This led to the general idea that tangled field geometries (with essentially random field vectors) might play an important role in extended radio sources. It has been shown by several authors that higher polarization values, $\Pi \gtrsim 50\%$, occasionally observed could be explained by compressing or shearing a tangled field (e.g. Laing 1980, Matthews & Scheuer 1990). Here I would like to emphasize the opposite possibility that perfectly ordered fields could lead to relatively low average polarization.

I will demonstrate this by using one specific field geometry as an example – a helical field in which the poloidal component B_z and the toroidal component B_φ are equal in the jet (pitch angle $\alpha_B = 45°$, see fig. 8). Strictly speaking the following calculations apply only for one 'onion shell' of a cylindrical jet. In a realistic MHD jet (Kössl *et al.* 1990, Appl, these proceedings) many such onion shells must be superimposed. At some point (z_s in Fig. 8) the jet is disturbed by an axial compression, as is expected for a hot spot or a strong knot.

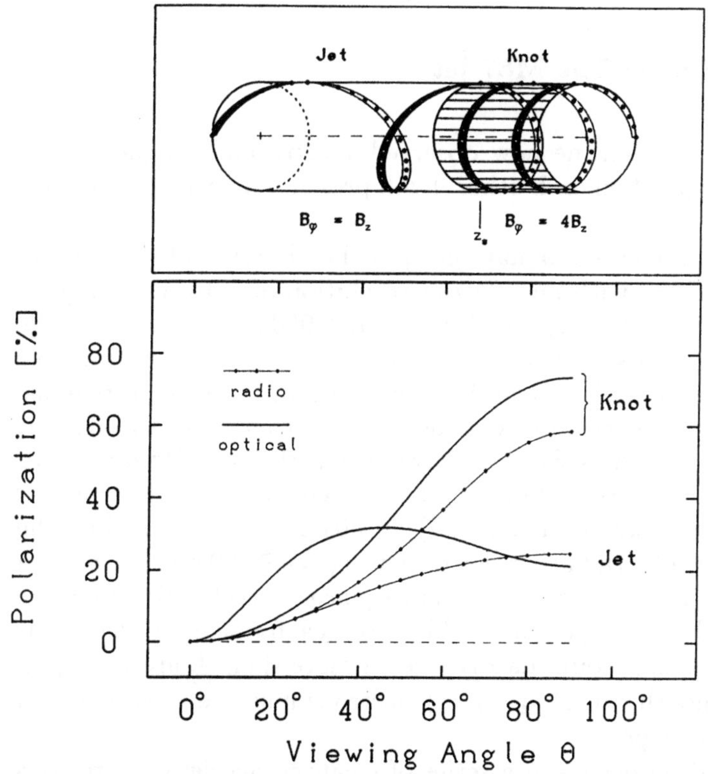

Fig. 8. Jet with helical magnetic field (pitch angle: 45°). The *upper* panel shows the field configuration by following one single field line. At $z = z_s$ a strong shock occurs. The *lower* panel gives the degree of polarization Π observed from the jet and from the knot as a function of the viewing angle between the jet axis and the line of sight.

The lower panel of Fig. 8 shows the polarization which would be observed if the jet is *not resolved* perpendicular to its axis. Note that for *all* viewing angles (between the jet axis and the line of sight) the mean B-vector of the polarization is B *parallel* axis in the jet and B *perpendicular* axis in the knot.

In fact the '*radio*' polarization (calculated for a powerlaw $S_\nu \sim \nu^{-1}$) remains low $\Pi < 25\%$ in the jet for all viewing angles but the knot polarization can become very high if viewed from the side. Note that for steep 'optical' spectra (near $\alpha = -2.5$ in a cutoff spectrum) the jet's polarization reaches a broad maximum $\Pi \cong 25\%$ at about 45°. Regarding the resolved morphology and polarization structure of a hollow jet with $\alpha_B \cong 45°$ one finds that in measuring the total flux the rim-brightening (expected for a random field) is strongly suppressed when the jet is seen from the side, while the polarized flux is extremely edge brightened. Such a behaviour is observed in the inner 10 arcsec of the M87 jet (Owen *et al.* 1989). If one looks into the jet under viewing angles $\lesssim 40°$ perpendicular B vectors are seen near the axis of the jet (perhaps observed in NGC 6251, Perley *et al.* 1984).

4.2 Spectra and Fields: The M87 jet

This paragraph tries to outline how observed spectra and magnetic field configuration may depend on each other. I will present a heuristic model for the jet in M87.

Recently we have been able to map the optical and radio-optical spectral indices in the jet of M87 with an *effective* resolution of 1.3 arcsec. Fig. 9a displays the run of the optical spectral index (400–900 nm) along the ridge line of the jet. From knots D to G the spectral index shows a remarkable steadiness with only small differences between knot and inter-knot regions. The impression that the spectrum remains essentially constant throughout most of the jet is confirmed by Fig. 9b, showing that the cutoff frequency ν_c is constant within a factor of < 3 out to knot G. This is a very astonishing result since for an equipartition field $\langle B \rangle \cong 50\,\mathrm{nT}$ one derives that the loss time of an optically radiating electron is $\tau(\gamma_c) \cong 90\,\mathrm{yrs}$. So even a freely moving electron could not move from its place of acceleration by more than $\frac{1}{3}$ arcsec before becoming unobservable at optical frequencies. Electrons gyrating around field lines could hardly stream more than $\frac{1}{10}$ arcsec (8 pc). To reproduce the smooth run of ν_c (or γ_c) electrons must be re-accelerated also in the inter-knot regions.

If the re-acceleration occurs in a series of localized acceleration regions (e.g. shocks) one needs at least two accelerators per arcsec to reproduce the smooth variation of ν_c. The typical distance between the knots is 3 arcsec. From the perfect agreement between the radio and optical polarization (Meisenheimer *et al.* 1989) we conclude that the agreement between the brightness distribution at radio and optical frequencies should be perfect down to a scale of $\lesssim 0\farcs5$. This is impressively confirmed by the deconvolved HST images (Macchetto, this conference) which are almost identical with the radio images at 15 GHz ($0\farcs1$ resolution, Owen *et al.* 1989). Thus ν_c seems to be smooth down to spatial resolutions of about $0\farcs2$. Obviously,

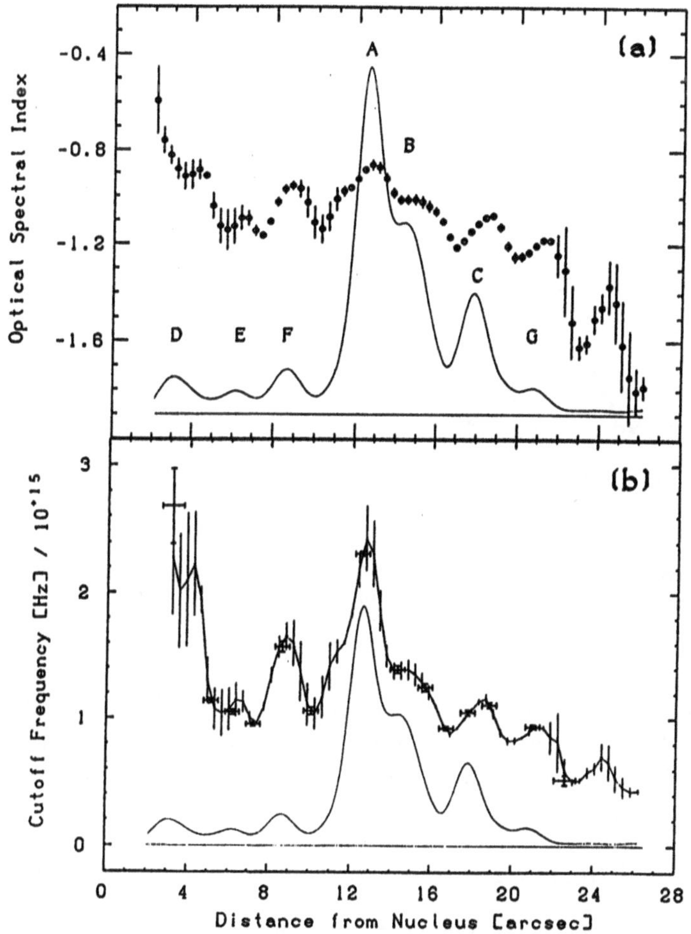

Fig. 9. **(a)** Optical spectral index α_{opt} along the M87 jet at 1.''3 *effective* resolution. A thin line shows the corresponding intensity profile.
(b) Cutoff frequency ν_c as derived from a fit of the radio–to–optical spectrum (*crosses*) or from the optical spectral index α_{opt}.

the relativistic electrons are trapped within some 100 milliarcsecs of their place of acceleration and a continuous acceleration process is required.

Since γ_c is constant to within a factor of 2 it is natural to assume it is exactly constant or at least a very smooth function of z along the jet axis (in the following model I will assume $\gamma_c \sim R_{jet}^{-1}$ where R_{jet} is the jet radius on the 15 GHz map). Than the apparent variations of $\nu_c(z)$ are due to variations of $|B| \sin\theta$ (eq. 2). Ignoring the $\sin\theta$, Fig. 9b can directly be read as a plot of $|B|$ with

$$|B(z)| = 30 \frac{\nu_c}{10^{15} Hz} nT$$

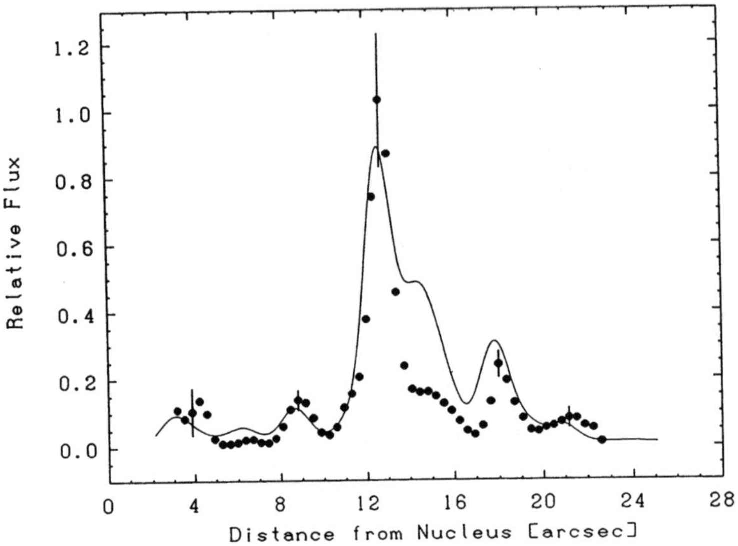

Fig. 10. Model of the brightness distribution along the jet of M 87. The observed distribution (thin line) is compared with the prediction of eq. (5). Typical error values are shown.

where the conversion factor comes from an equipartition analysis of knot A: $B_{knot\ A} = 70\,\mathrm{nT}$. Assuming a helical field in a hollow jet $|B| = (B_z^2 + B_\varphi^2)^{-1/2}$ (as suggested by the radio polarization, see 4.1) which undergoes only axial compressions by some ratio $r(z)$ we find

$$| B(z) |= B_0\sqrt{1 + r^2(z)}. \tag{4}$$

Choosing $r(knotA) = 4$ in agreement with the observed polarization pattern, one gets $B_0 = 17\,\mathrm{nT}$. Now $r(z)$ can directly be derived from the observed $\nu_c(z)$, and we are able to predict the (optical) synchrotron emissivity:

$$\epsilon_{opt} \sim n_{rel}(z) \cdot (B(z))^{1-\alpha}$$

where $n_{rel}(z)$ is the local density of relativistic electrons. Since $n_{rel}(z) \sim r(z)$ and $\alpha \cong -1.0$ we predict that the flux distribution along the jet is

$$S_\nu(z) \sim r(z)[1 + r^2(z)]^{0.95}. \tag{5}$$

Fig. 10 shows that the agreement between the predicted $S_\nu(z)$ and the observed flux profile along the jet is rather good. Allowing for small differences caused by our neglection of $\sin\theta$ the only severe discrepancies occur between knots B and C. Here the apparent distortions of the magnetic field may have caused higher losses and/or a lower efficiency of the acceleration process.

I do not claim that this model is the ultimate description of the M87 jet. My aims have been only (i) to convince you that γ_c = constant provides a viable description even for the small scale variations of ν_c and (ii) to demonstrate in some detail how variations of $B \sin\theta$ alone can account for a knotty appearance and spectral variations.

I admit openly that I am not in the position to propose any acceleration process which could produce $\gamma_c = constant$ everywhere in the jet. Therein lies a large playing ground for the theoreticians. But the detailed high frequency studies of the brightest and nearest classical radio jet certainly rule out the idea that 1st order Fermi acceleration does everything you need. There are two models which try to side step the problem but do not convince me: Assuming that the M87 jet is highly relativistic ($\gamma_{jet} > 10$, Kirk & Begelman, priv. comm.) weakens the loss time argument. But where is the hot spot? The lossfree axial pipeline for highly relativistic electrons proposed by Owen *et al.* (1989) would most probably be destroyed quickly if one allows the electrons to interact with the magnetized jet mantle. But they have to if they are to be responsible for the synchrotron light.

5. Summary

The main results of current studies of the overall spectra and polarization characteristics of extended radio jets and hot spots are:

- The determination of the overall spectrum ($10^9 < \nu < 10^{15}$ Hz) is required to reach firm conclusions about the physics.
- Fermi acceleration models are (just) able to account for the observed synchrotron spectra of radio hot spots.
- In hot spots the magnetic field has about the equipartition value ($\cong$ 35 nT). Typical maximum energies are $E_c \gtrsim 10^5 m_e c^2$, implying acceleration time scales on the order of 1000 years.
- Optically detected jets require a continuous re-acceleration along the entire jet. Shock acceleration seems very unlikely.
- In some jets the field pattern alone may explain the knotty structure. In this case the apparent proper motion would represent a pattern speed.

References

Alexander, P., Brown, M.T. and Scott, P.F. 1984, *Mon. Not. R. astr. Soc.* **209**, 851
Appl, S. 1991, *these proceedings.*
Baade, W. 1956, *Astrophys. J.*, **123 550**
Ballard, K.R. and Heavens, A.F. 1991, *Mon. Not. R. astr. Soc.* **251**, 438
Biermann, P.L. and Strittmatter, P.A. 1987, *Astrophys. J.*, **322 643**
Biretta, J.A., Stern, C.P. and Harris, D.E. 1991, *Astron. J.*, **101 1632**
Butcher, H.R., van Breugel, W., and Miley, G.K. 1980, *Astrophys. J.*, **235 749**

Crane, P. and Stockton, A. 1989, in K. Meisenheimer and H.-J. Röser (eds.): *Hot Spots in Extragalactic Radio Sources*. Lecture Notes in Physics Vol. **327**. Springer Heidelberg, p. 115.
Curtis, H.D. 1918, *Lick Obs. Publ.* **13**, 11
Danziger, I.J., Fosbury, R.A.E., Goss, W.M., and Ekers, R.D. 1979, *Mon. Not. R. astr. Soc.* **188**, 415
Drury, L. O'C. 1983, *Rep. Prog. Phys.* **46**, 973
Fraix-Burnet and J., Nieto, J.-L. 1988, *Astron. Astrophys.* **198**, 87
Fraix–Burnet, D., Nieto, J.–L. and Poulain, P. 1989, *Astron. Astrophys. (Lett.)* **221**, L1
Hargrave, P.J. and Ryle, M. 1974, *Mon. Not. R. astr. Soc.* **166**, 305
Heavens A.F. 1984, *Mon. Not. R. astr. Soc.* **210**, 813
Heavens A.F. and Drury, L.O'C. 1988, *Mon. Not. R. astr. Soc.* **235**, 997
Heavens, A.F., Meisenheimer, K. 1987, *Mon. Not. R. astr. Soc.* **225**, 335
Keel, W.C. 1988, *Astrophys. J.*, **329 532**
Killeen, N.E.B., Bicknell, G.V., Hyland, A.R. and Jones, T.J. 1984, *Astrophys. J.*, **280 126**
Kirk, J.G. and Schneider, P. 1987a, *Astrophys. J.*, **315 425**
Kössl, D., Müller, E. and Hillebrandt, W. 1990, 229 378,397,401
Kronberg, P.P. 1976, *Astrophys. J. (Letters)*, **203 47**
Laing, L R.A. 1980, *Mon. Not. R. astr. Soc.* **193**, 439
Laing, R.A. 1981, *Mon. Not. R. astr. Soc.* **195**, 261
Leahy, P. 1991, in P.A.Hughes (ed.): *Beams and Jets in Astrophysics*, Cambridge Univ. Press, p. 100.
Lelièvre, G. and Wlérick, G. 1975, *Astron. Astrophys.* **42**, 293
Lonsdale, C.J., Hartley-Davies, R. and Morison, I. 1983, *Mon. Not. R. astr. Soc.* **202**, 1p
Macchetto, F. 1991, *these proceedings*
Macchetto, F. *et al.* 1991, *Astrophys. J. (Letters)*, **369 55**
Matthews, L A.P. and Scheuer P.A.G. 1990a, *Mon. Not. R. astr. Soc.* **242**, 616
Matthews, A.P. and Scheuer P.A.G. 1990b, *Mon. Not. R. astr. Soc.* **242**, 623
Meisenheimer, K. 1989, in K. Meisenheimer and H.-J. Röser (eds.): *Hot Spots in Extragalactic Radio Sources*. Lecture Notes in Physics Vol. **327**. Springer Heidelberg, p. 253
Meisenheimer, K. and Heavens, A.F. 1986, *Nature*, **323 419**
Meisenheimer, K. and Röser, H.–J. 1984b, *Astrophys. J. (Letters)*, **279 39**
Meisenheimer, L K. and Röser, H.-J. 1986,*Nature*, **319 459**
Meisenheimer, K., Röser, H-J., Hiltner, P., Yates, M.G., Longair, M.S., Chini, R. and Perley, R.A. 1989a, *Astron. Astrophys.* **219**, 63
Meisenheimer, K., Röser, H.-J. and Schlötelburg, M. 1989, in E. Meurs and R. Fosbury (eds.): *Extranuclear Activity in Galaxies*, ESO Proceedings Garching.
Muxlow, T.W.B., Pelletier, G., Roland, J. 1988, *Astron. Astrophys.* **206**, 237
Owen, F.N., Hardee, P.E. and Cornwell, T.J. 1989, *Astrophys. J.*, **340 698**
Perez-Fournon, I., Colina, L., Gonzalez-Serrano, J.I. and Biermann, P.L. 1988, *Astrophys. J. (Letters)*, **329 81**
Perley, L R.A., Bridle, A.H. and Willis, A.G. 1984, *Astrophys. J. Suppl.*, **54 291**
Perola, G.C. and Tarenghi, M. 1980, *Astrophys. J.*, **240 447**
Pooley, G.G. and Henbest, S.N. 1974, *Mon. Not. R. astr. Soc.* **169**, 477
Röser, H-J. 1989, in K. Meisenheimer and H.-J. Röser (eds.): *Hot Spots in Extragalactic Radio Sources*. Lecture Notes in Physics Vol. **327**. Springer Heidelberg, p. 91.
Röser, H.-J. and Meisenheimer, K. 1986, *Astron. Astrophys.* **154**, 15
Röser, H.-J. and Meisenheimer, K. 1987, *Astrophys. J.*, **314 70**
Röser, H.-J. and Meisenheimer, K. 1991, *Astron. Astrophys.* ., in press
Salter, C.J., Chini, R., Haslam, C.G.T., Junor, W., Kreysa, E., Mezger, P.G., Spencer, R.E., Wink, J.E. and Zylka, R. 1989, *Astron. Astrophys.* **220**, 42
Scarrott, S.M. and Rolph, C.D. 1989, *Mon. Not. R. astr. Soc.* **238**, 349
Schlötelburg, M., Röser, H.-J. and Meisenheimer, K. 1991, *in preparation*
Schmidt, M. 1963, *Nature*, **197 1040**
Schreier, E.J., Gorenstein, P. and Feigelson, E.D. 1982, *Astrophys. J.*, **261 42**
Simkin, S.M. 1978, *Astrophys. J. (Letters)*, **222 55**

Smith, L R. M., Bicknell, G.V., Hyland, A.R. and Jones, T.J. 1983, *Astrophys. J.*, **266** **69**
Sparks, W.B., Miley, G.K. and Macchetto, F. 1991, *Astrophys. J. (Letters)*, **361** **41**
Stocke, L J.T., Rieke, G.H. and Lebofsky, M.J. 1981, *Nature*, **294** **319**
Webb, G.M., Drury, L.O'C. and Biermann, P. 1984, *Astron. Astrophys.* **137**, 185
Wright, M.C.H. and Birkinshaw, M. 1984, *Astrophys. J.*, **281**, 135
Yates, M.G., Meisenheimer, K. and Röser, H.-J. 1991, *in preparation.*

Discussion

Preuss: Have you looked at both hot spots of the FR type II radio galaxies in your sample (e.g. 3C 111)? If so, did you find any systematic difference w.r.t. the one-sidedness typical for this type of object ?

Meisenheimer: We have looked at only two cases up to now. We did not detect the eastern hot spot of 3C 20 but found a patch of emission near Pictor A (east). Both observations are consistent with symmetrical hot spot spectra.

Rees: There is an argument, due I think to Peter Scheuer, that the jets supplying the lobes of (e.g.) Cygnus A must consist of electrons and positrons – if the electrons were accompanied by protons (which carry more energy and momentum), the jets wouldn't carry enough electrons to yield the synchrotron emission from the entire radio lobes. Do you believe this argument ?

Meisenheimer: Presently, I can't tell since I did not go through the calculation. But I feel that Peter Scheuer might be right.

Wandel: I am puzzled by the high efficiency of the acceleration required in the hot spot. Typically the efficiency of the 1st Fermi process (e.g. in shock acceleration of cosmic rays) is much lower (of order $(1-p) - ln(E/E_0)$, when p is the escape probability).

Meisenheimer: The efficiency in terms of the *fraction* of accelerated electrons is less than a few percent. In terms of *energy* (conversion of kinetic energy into relativistic particles) efficiencies of 10 to 50% are required in the mildly relativistic models.

Pudritz: Are there any systematic variations in field direction in pre– compared to post–knot regimes? If so, do these persist from knot to knot?

Meisenheimer: For the M87 jet, the answer is definitely *no*.

Depolarization in Compact Radio Sources

C-I. Björnsson

Stockholm Observatory, S-133 36 Saltsjöbaden, Sweden

Abstract. Assuming that the flat spectrum of compact radio sources (CRS) is due to optical depth effects in an inhomogeneous source, possible origins of the low polarization are discussed. It is argued that the depolarization is caused neither by synchrotron self-absorption nor a turbulent magnetic field. Instead it is suggested that the magnetic field is coherent and that the low polarization is due to large Faraday depths in a medium which is internal to the source and whose density scales with radius in a way similar to that of the synchrotron emitting electrons.

1. Depolarization caused by synchrotron self-absorption

Consider a source model characterized by (i) a power-law electron distribution: $N(E) = N_o E^{-s}$ and (ii) spherical symmetry, where the magnetic field (B) and density are given by; $r < r_o : B = B_o = const.$ and $N_o = N_{oo} = const.$; $r > r_o : B = B_o(r/r_o)^{-m}$ and $N_o = N_{oo}(r/r_o)^{-n}$. The fiducial radius r_o causes the source to become optically thin at a frequency $\nu \equiv \nu_o$. For $\nu < \nu_o$, the region $r < r_o$ is unimportant and one can show, in the case of negligible Faraday rotation, that the linear polarization is given by $P = P_o(u_o - \beta)/(1 - u_o\beta)$, where $\beta = [(1 + k_o)^{1-\delta} - (1 - k_o)^{1-\delta}]/[(1 + k_o)^{1-\delta} + (1 - k_o)^{1-\delta}]$, $\delta = (4 + m)/(2(n - 1) + m(s + 2)) \equiv (2\alpha + 5)/(s + 4), u_o = (s + 1)/(s + 7/3)$, and $k_o = (s + 2)/(s + 10/3)$. P_o takes into account depolarization due to the magnetic field geometry.

Synchrotron self-absorption has been invoked to explain $\approx 90^o$ swings of position angle (PA) in CRS both as a function of frequency in unresolved sources and as a function of position in spatially resolved sources (Cotton et al. 1984). Likewise, the low value of P in the unresolved VLBI-core has been explained as resulting from optical depth effects (Gabuzda, Wardle, and Roberts 1989). As can be deduced from the expression above for P, the major factor determining the depolarization is the spectral index α. Only for sources with α close to $-5/2$ can the observed characteristics be reproduced. Hence, if the flat or slightly inverted spectra of CRS are due to an inhomogeneous source-structure neither the $\approx 90^o$ flip in PA nor the low value of P in the VLBI-core is likely to be due to synchrotron self-absorption. Rather, changes in PA are most straightforwardly explained as a direct result of a varying magnetic field.

2. Depolarization caused by a turbulent magnetic field

Observations show that the circular polarization for an individual CRS has a preferred sign both as a function of time (Komesaroff et al. 1984) and frequency (Weiler and de Pater 1983). Independent of its cause, either due to transfer effects (Björnsson 1990) or from the emission process directly, this indicates a large scale ordering of the magnetic field. This inference pertains to the optically thick core, since this is the most likely place of origin for the circular polarization. Further arguments against a high level of turbulence are provided by the observed large and smooth variations in PA as well as the rapid time variations in the mm/radio flux. These can be given simple explanations by invoking a large scale magnetic field structure.

3. Depolarization caused by Faraday rotation

It has often been argued that the lack of observed frequency dependent polarization is evidence against the presence of significant amount of Faraday rotation. However, this argument is not necessarily valid, at least not for a spatially unresolved, optically thick source; if the density of both the Faraday rotating medium and the synchrotron emitting electrons scale with radius in the same way and $n = m + 1$, giving $\alpha = 2(1 - m)/m$, the polarization is independent of frequency for $\nu < \nu_o$, since particles radiating at different frequencies all have the same energy. Thus, the low polarization of the unresolved VLBI-core is conceivably due to Faraday depolarization. In resolved sources one expects the polarized flux to be more extended than the total flux (cf. 3C 345, Cotton et al. 1984).

References

Björnsson, C-I., 1990 : *Monthly Notices Roy. Astron. Soc.*, **242**, 158.

Cotton, W.D., Geldzahler, B.J., Marcaide, J.M., Shapiro, I.I., Sanroma, M., Rius, A., 1984: *Astrophys. J.*, **286**, 503.

Gabuzda, D.C., Wardle, J.F.C., Roberts, D.H., 1989: *Astrophys. J.*, **338**, 743.

Komesaroff, M.M., Roberts, J.A., Milne, D.K., Rayner, P.T., Cooke, D.J., 1984: *Monthly Notices Roy. Astron. Soc.*, **208**, 409.

Weiler, K.W., de Pater, I., 1983: *Astrophys. J. Suppl.*, **52**, 293.

Diamagnetic Effects in Synchrotron Sources

Gianluigi Bodo, Gabrielle Ghisellini, and Eduardo Trussoni

Osservatorio Astronomico di Torino, Pino Torinese, Italy

The standard synchrotron theory provides the electromagnetic energy emitted by a cloud of relativistic electrons moving in a constant magnetic field. However the loop of current from the moving electrons creates an additional and opposite magnetic field (i.e. the cloud is diamagnetic), whose effects are usually neglected (Ginzburg and Syrovatskii, 1969; Singal, 1986). In order to evaluate these effects, assume that in a region embedded in a magnetic field H created by external *batteries*, there is a cloud of relativistic electrons, with number density N and energy density W_K. Each electron orbit can be considered as a loop of current, but inside the region the net current vanishes, due to the opposite contributions of adjacent loops. Only at the border of the cloud the currents are not balanced and this induces a counterfield, which partially screens the external magnetic field. At the boundary, the particles feel a non uniform magnetic field, and are subject to a drift in a direction perpendicular both to the field and to the field gradient (Fig. 1). This drift originates a new current, which must also be taken into account when computing the internal field (the drift current was *not* considered by Singal, 1986). From the Ampere law one finally obtains the equilibrium field B through

$$\frac{1}{3}W_K = \frac{1}{8\pi}(H^2 - B^2). \tag{1}$$

This equation provides the value of B–field which leads to equilibrium between the magnetic ($W_B = B^2/8\pi$) plus electron ($P_k = W_k/3$) pressure of the internal zone and the magnetic pressure ($W_H = H^2/8\pi$) of the surrounding region. The equilibrium curve in the $B - W_K$ plane (Fig. 2) results to be a parabola, with $B = 0$ for $W_K = 3W_H$, so that no equilibrium is possible for $W_K > 3W_H$. In fact, if an *arbitrary* large W_K is initially injected into the region, there will be a transient phase, while the counterfield is establishing, with a large dB/dt. The changing B induces a large electric field which *decelerates* the particles, *decreasing* their energy density. At the end of the transient phase, the particle energy density, no matter how large its initial value, cannot exceed $3W_H$ (systems with $W_K > 3W_H$ are in non–equilibrium). If the time scale of variation of the field is longer than the dynamical time scale, then the transient phase is described by the *adiabatic* invariant curve W_K^2/B = const (see Fig. 2).

After equilibrium is established, the system slowly (compared to the dynamical timescale) evolves, due to synchrotron losses. By loosing energy, the electrons produce less counterfield, and the equilibrium B field inside the region increases. The changing B induces an electric field E, which opposes to the variation: *i.e.* tends to reaccelerate the particles. The system evolves through successive equilibrium states on a time scale $\gtrsim$ the synchrotron cooling time. By using the pressure equilibrium condition and the constancy of the magnetic flux, we can derive the total energy ($\int Pdt$) radiated by one particle from the initial time (when $B = B_0$ and $N = N_0$) to the final time (when $B = H$, the particles having radiated all their energy)

$$\int Pdt = \gamma mc^2 + \frac{3}{2N_0}(W_H - W_B) . \tag{2}$$

The total radiated energy exceeds the particle energy. The extra energy involved is supplied by the accelerating electric field induced by the changing magnetic field, and, ultimately, by the batteries maintaining the field H.

During the transient phase and the during the evolution along the equilibrium curve the luminosity of the cloud is varying. Therefore it can be expected that some observed radio variabilities could be related to the diamagnetic behaviour of the regions of radiating electrons.

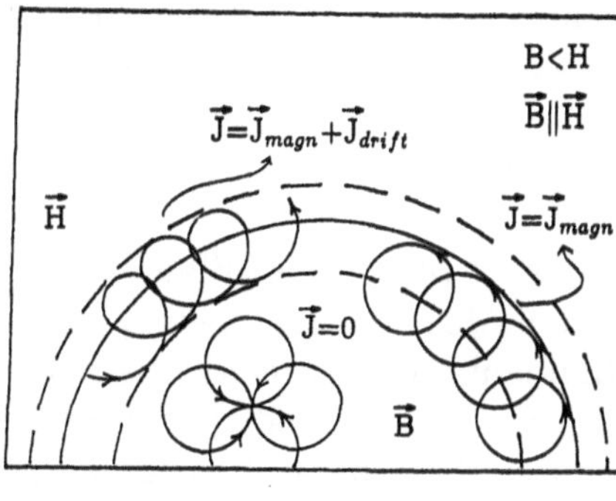

Fig. 1. Inside a region of magnetic field H, a cloud of gyrating particles produces a counterfield, with an inside equilibrium field $B < H$. The value of B depends only from total current produced by the electrons on the external edge of the cloud.

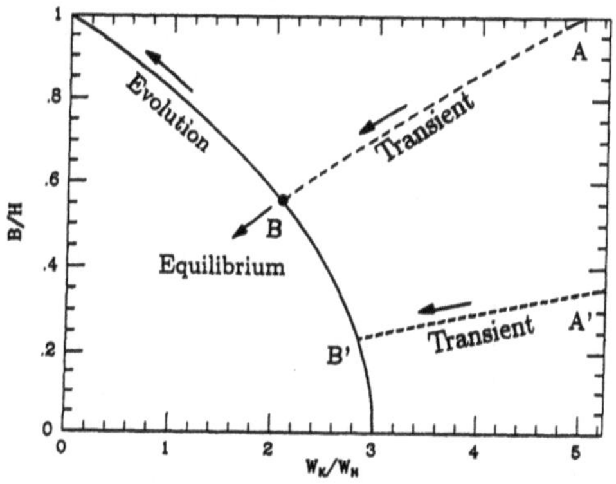

Fig. 2. Equilibrium curve of B/H vs the electron energy density W_K/W_H (full line), and transient phase (dashed lines with $W_K^2\ /B =$ const, path A, A' → B, B'). The arrows show the direction of the evolution during the transient phase and after the equilibrium is attained.

References

Ginzburg, V.L., Syrovatskii, S.I., 1969: *Ann. Rev. Astron. Astrophys.*, **7**, 375.
Singal, A.K., 1986: *Astron. Astrophys.*, **155**, 242.

Non-linear Filaments Evolution in Radio Sources

G. Bodo[1], *S. Massaglia*[2], *P. Rossi*[3], *A. Ferrari*[1]

[1] Osservatorio Astronomico di Torino, Pino Torinese, Italy
[2] Istituto di Fisica Generale dell'Universitá, Torino, Italy
[3] Harvard-Smithsonian Center for Astrophysics, Cambridge, U.S.A.

Recent high resolution optical and radio observations of extragalactic sources show in their jets and lobes regions of enhanced brightness that appear in form of elongated inhomogeneities of the emitting plasma (filaments). This kind of features have been observed, at radio frequencies, in the lobes of Cygnus A (Dreher, Carilli and Perley, 1987), in the jet and lobes of M87 (Owen, Hardee and Cornwell, 1989; Hines, Owen and Eilek, 1989). Polarization maps show that, in general, such filaments are aligned to the magnetic field. In the lobes of Cygnus A the filaments have typical transverse size of ~ 1 kpc and a typical length of ~ 20 kpc and are about twice as bright as the background in the 6 cm band. In the case of M87, the filaments in the radio lobes are scaled down by a factor of about 10 in size with respect to Cygnus A and their emissivity contrasts to the background, at 6 cm, is typically ~ 10.

At optical frequencies, HST observations show presence of filaments in the jet of 3C 66B (Macchetto *et al.*, 1991) and of M87 (see Macchetto, these proceedings). In the case of 3C 66B, the filaments are not evident in the radio maps (Leahy, Jägers and Pooley, 1986), while in optical they appear longitudinal with respect to the jet with emissivity enhanced by a factor $\gtrsim 4$ to the background. In the case of M87, the optical filaments follow the pattern shown in radio VLA maps and appear wrapped around the jet, giving rise to the well known knotty structure.

Bodo *et al.* (1990) proposed that such structures can be formed by thermal instabilities driven by synchrotron radiation in a relativistic plasma. They have performed a linear stability analysis of a infinite homogeneous magnetized medium affected by synchrotron losses under the following assumption: i) the pressure is provided by relativistic electrons while the inertia is given by thermal protons and ii) thermal and relativistic components are strictly coupled. They found that the system is, in general, unstable to stationary condensation modes and slow MHD waves provided the particle pressure dominates over the magnetic one (i.e. non equipartition conditions), and argued that filaments are non–linear transient structures that increase

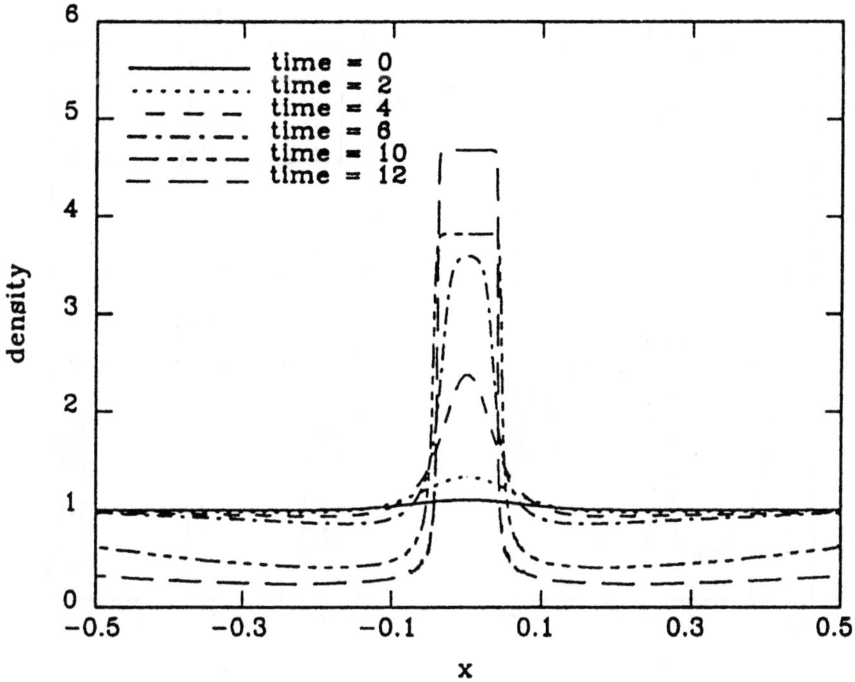

Fig. 1. Temporal evolution of the perturbation for $\beta_0 = 8\pi$ in case of free boundary conditions.

their emissivity on timescales of the order of some synchrotron cooling times and become eventually dark.

We present here some results of the non–linear evolution of the condensation mode by means of numerical integration of the full set of 1-D MHD equations. We study the problem under the above assumptions and limit ourselves to consider a Maxwellian (or monochromatic) distribution for the relativistic electrons; in this case the results are approximatively valid also for steep ($\Gamma > 3$) power–law distribution functions. The system is perturbed with a Gaussian–like perturbation applied to density and magnetic field; the evolution time is measured in units of the synchrotron cooling time defined as $t_{sync} = 3P/\mathcal{L}$ (P = pressure, $\mathcal{L}$ = emissivity) and the main parameter in the initial plasma–$\beta_0 = 8\pi P_0/B_0^2$. The results found indeed show that condensations form along the magnetic field and their emissivity increases over times of the order of some t_{sync}; moreover these filaments should appear limb–brightened to observations. The optical filaments observed in the jets of 3C 66B have spectral index $\alpha \simeq 1.1$ (as in the radio emission) and are visible between $\gtrsim 1.5$ to 3 kpc from the nucleus; since the resulting index for the electron distribution is $\Gamma \sim 3.2$, we can apply our results and interpret the initial gap of 1.5 kpc, in the present scheme, as the distance travelled by the jet's plasma in some t_{sync}, at radio frequencies, at $V_j \sim 10^8$ cm s^{-1} before the instability fully develops.

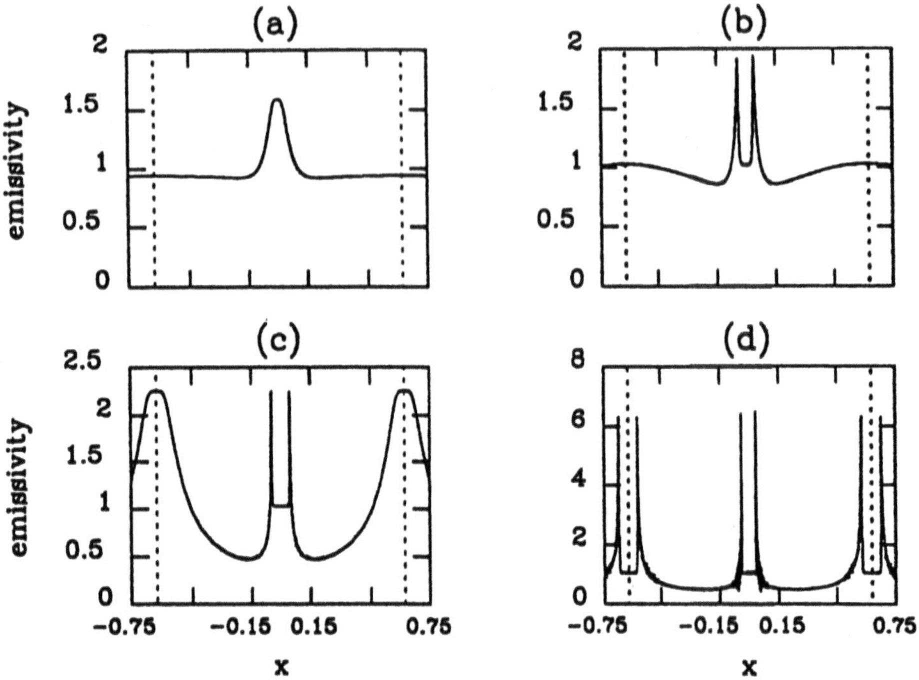

Fig. 2. Temporal evolution of radiative losses for $\beta_0 = 8\pi$ and periodic boundary conditions. Panel (a) at $t = 4t_{sync}$, (b) at $t = 8t_{sync}$, (c) at $t = 12t_{sync}$, and (d) at $t = 16t_{sync}$.

References

Bodo, G., Ferrari, A., Massaglia, S., Trussoni, E., 1990: *Monthly Notices Roy. Astron. Soc.*, **244**, 530.

Hines, D.C., Owen, F.N., Eilek, J.A., 1989: *Astrophys. J.*, **347**, 713.

Leahy, J.P, Jägers, W.,, Pooley, G.G., 1986: *Astron. Astrophys.*, **156**, 173.

Dreher, J.W., Carilli, C.L.,, Perley, R.A., 1987: *Astrophys. J.*, **316**, 611.

Macchetto, et al., 1991: *Astrophys. J. (Letters)*, submitted.

Owen, F.N., Hardee, P.E., Cornwell, T.J., 1989: *Astrophys. J.*, **340**, 698.

The Quasi–Periodicity and the Origin of Knots in Quasars

Max Camenzind and Martin Krockenberger

Landessternwarte Königstuhl, D-6900 Heidelberg, Germany

The question of how jets are formed in quasars, BL Lac objects and radiogalaxies is fundamental for the understanding of the emission from these objects. VLBI observations in the mm–wavelength region have shown that these jets are already well collimated within a distance of less than a light year with quite high jet radii. This rules out formation of jets by the accretion funnel of geometrically thick disks.

Relativistic jets are formed under quite unique conditions. Relativistic speeds cannot be attained in pure hydrodynamic acceleration, magnetic fields are a necessary element for jet formation [1, 2, 3]. Dynamo action in the inner disk around a supermassive black hole creates a rotating magnetosphere which is filled with disk plasma or with a relativistic electron–positron plasma. The plasma is accelerated along open flux surfaces Ψ and collimated outside the light cylinder $R_L = c/\Omega_K(R_F)$ for foot point radii R_F near the inner last stable orbit around a rapidly rotating Kerr black hole

$$R_L \simeq 10\, GM_H/c^2 \simeq 1.5 \times 10^{15}\,\mathrm{cm}\, M_H/10^9 M_\odot\,. \tag{1}$$

The total energy E and the total angular momentum L are conserved along the axisymmetric magnetic flux surfaces. This determines the Lorentz factor γ and the specific angular momentum $j = R^2\Omega$ for the plasma

$$\gamma(R,\Psi) = \frac{E}{\mu}\,\frac{1 - \Omega^F L/E - M^2(R,\Psi)}{1 - M^2(R,\Psi) - (R/R_L)^2}\,, \tag{2}$$

$$j(R,\Psi) = \frac{R^2\Omega^F\,(1 - \Omega^F L/E) - M^2(R,\Psi)c^2 L/E}{1 - M^2(R,\Psi) - \Omega^F L/E}\,. \tag{3}$$

In a jet with non–vanishing opening angle Θ_0, the Mach–number M is not constant, but the specific angular momentum $j = R^2\Omega$ stays constant, $j_\infty = R_A^2\,\Omega^F \simeq R_L^2\,\Omega^F$ for $M^2 \gg 1$.

Non–axisymmetric bubbles of relativistic particles and fields are dragged along the magnetic surface of the jet and describe a helical trajectory [4]

$$\boldsymbol{x}_B(t) = \left(R(t)\,\cos\Phi(t), R(t)\,\sin\Phi(t), \int_0^t v_z(t')\,dt'\right)\,. \tag{4}$$

The spectral flux S_ν observed from such an unresolved knot is strongly amplified by beaming and is in general time–dependent, since the Doppler factor $D(t)$ is a function of t. Due to the rotation of the bubble, the beaming factor can strongly vary and lead to quasi-periodic flaring as the bubble moves along a magnetic surface of constant opening angle [4]. A long coherence in these pulsations is only visible, if the jet is perfectly collimated with opening angle smaller than 0.1 degree and for large Lorentz factors $\gamma_{\text{jet}} \geq 5$. The lightcurves of the optical flares in 3C 273 [5] (Fig. 1) and of some BL Lac objects can be very well reproduced [4]. The angular velocity of the bubble decreases with time because of the conservation of angular momentum in the opening cone. The radius of the jets in quasars is of the order of 0.1 lyr, the intrinsic rotation period is of the order of years. This period is, however, drastically shortened by the Doppler effect.

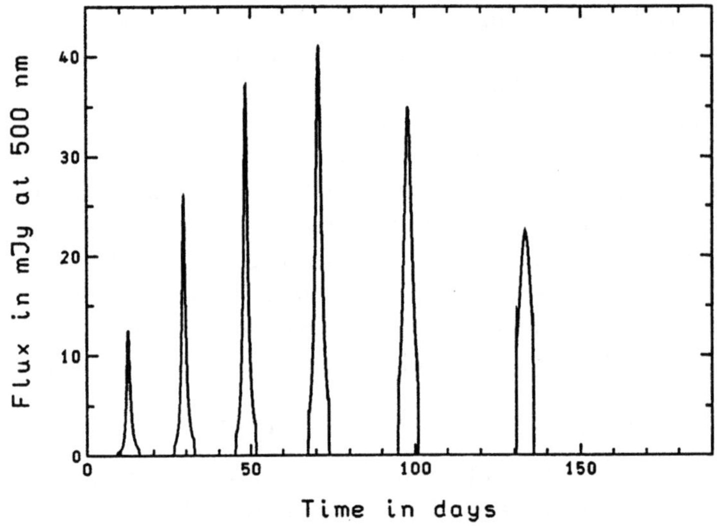

Fig. 1. Quasi-periodic flaring for the superluminal quasar 3C 273. $u_p = 15c$, $\Theta_0 = 0.1°$ and $i = 3°$, $\alpha = 0.2$, $M_H = 7 \times 10^9\, M_\odot$, $a_H = 0.9 M_H$. The knot is injected at $R_0 \simeq 10 R_L \simeq 1 \times 10^{17}$ cm and rotates with the period $P_0 \simeq 660$ days. These knots are only visible in the optical, if beaming is sufficiently strong to compete with the quiescent emission from the disk and the hot dust in this quasar (27 mJy).

References

1. Camenzind, M., 1989: in *Accretion Disks and Magnetic Fields in Astrophysics*, ed. G. Belvedere, Kluwer (Dordrecht), p. 129.
2. Camenzind, M., 1990: *Rev.Mod.Astronomy* **3**, 234.
3. Camenzind, M., 1991: in **Variability of Active Galaxies**, Lecture Notes in Physics **377**, Springer (Heidelberg), p. 201.
4. Camenzind, M., Krockenberger, M., 1991: *Astron. Astrophys.*, submitted.
5. Courvoisier, T.J.-L., et al., 1988: *Nature* **335**, 330.

The Microvariability of PKS 0735+178

M. Carini[1], *R.H. Miller*[2], *A. Sadun*[2], *J. Noble*[2]

[1] CSC/NASA Goddard Space Flight Center
[2] Georgia State University

The radio source 0735+178 was first identified with a blue stellar object by Blake (1970) and Moseley *et al* (1970). Its polarization was shown to be variable by Carswell *et al* (1974). Variability has also been observed at radio wavelengths by Andrew *et al* (1971) and in the infrared by Rieke (1972). Burbidge and Strittmatter (1972) found the spectrum to be featureless, while Carswell *et al* (1974) determined a z of 0.424 from absorption lines observed in the spectrum.

The long term variability of this object was investigated by Pollock (1975), who found a total range of 2.0 magnitudes for the observed variations. Other studies of the long term variability of this object have been carried out by Webb *et al* (1988) and Pollock (1975). Variations with timescales of days to months have been observed by Pollock (1975), Wing (1973), Carswell (1974) and Bertaud (1974). Veron and Veron investigated 0735+178 for the presence of variations with timescales of less than one day, but were unsuccessful in their attempt to detect such variations.

Beginning in 1987, we have undertaken a program to investigate the nature of the microvariability for PKS 0735+178. No evidence for the presence of microvariability was detected for a number of nights; specifically on the nights of 11 and 14 November 1987, no significant variability was observed. However, from the 11th to the 14th, a decline in the object's brightness of 0.45 magnitudes was detected. PKS 0735+178 was next observed on the night of 17 March 1990, and once again no significant variability was detected. From 19 March through 24 March 1989, the object was intensively monitored to search for microvariability variations. From 19 March to 20 March, the source rose in brightness by 0.09 magnitudes. From 20 March to 21 March, the source declined in brightness by 0.14 magnitudes. The source continued this decline from 21 March to 24 March, fading an additional 0.1 magnitudes over those 3 days. On the night of 19 March, the source was observed to be 1.08 magnitudes fainter than on 17 March. In addition, from UT 0430 to 0530, a decline of 0.25 magnitudes was observed.

These variations were subjected to a power spectrum analysis. No periodicities were found in the data and the power spectrum showed no evidence of a $1/f$ or flicker noise process responsible for the variations. The most rapid timescale event observed for 0735+178 was the 0.25 magnitude decline in one hour observed on the night of 19 March 1989. This leads to a mass of the supermassive black hole of $M \leq 9.33 \times 10^8 M_\odot$. This corresponds to an Eddington luminosity of $L_E = 1.22 \times 10^{47} ergs/sec$. When this event was observed, the object was at a brightness of V = 16.02. This yields a luminosity of $L = 2.05 \times 10^{42} ergs/sec$. Thus, relativistic beaming is not required to explain the observed variations.

References

Andrew, B.H., Medd, W.J., Harvey, G.A., Locke, J.L., 1972: *Nature* **236**, 445.
Bertaud, Ch., Dumortier, B., Pallas, C., 1974: *IAU Circ.* **2678**.
Blake, G.M., 1970: *Astrophys. Letters*, **6**, 201.
Burbidge, E.M., Strittmatter, P.A., 1972: *Astrophys. J. (Letters)*, **174**, L57.
Carswell, R.F., Strittmatter, P.A., Williams, R.E., Kinman, T.D., Serkowski, K., 1974: *Astrophys. J. (Letters)*, **190**, L101.
Moseley, G.F., Brooks, C.C., Douglas, J.N., 1970: *Astron. J.*, **75**, 1015.
Pollock, J.T., 1975: *Astrophys. J. (Letters)*, **198**, L53.
Veron, P., Veron, M.P., 1975, *Astron. Astrophys.*, **39**, 289.
Webb, J. R., Smith, A.G., Leacock, R.J., Fitzgibbons, G.L., Gombola, P.P., Shepard, D.W., 1988: *Astron. J.*, **95**, 374.
Wing, R.F., 1973: *Astron. J.*, **78**, 684.

Limits on the Magnetic Field in the Compact Regions of Active Galactic Nuclei

A. Celotti[1,2], *G. Ghisellini*[3], *and A.C. Fabian*[1]

[1] Institute of Astronomy, Madingley Road, Cambridge CB3 0HA
[2] International School for Advanced Studies, Trieste, Italy
[3] Osservatorio Astronomico di Torino, Pino Torinese, Italy

We estimate upper limits on the magnetic energy density in the compact X–ray emission region of five active galactic nuclei using synchrotron self-Compton theory (SSC). We assume that the X–rays are EUV/soft X–ray bump photons Compton–scattered by a non–thermal distribution of relativistic electrons and that the size R of the region is given by the X-ray variability timescale. The minimum electron density required to produce the observed X-ray flux is obtained by maximizing the temperature of the EUV/soft X-ray blackbody consistent with the same size region and the observed UV and soft X-ray fluxes. The maximum magnetic field B in the X-ray emission region is then determined by requiring that the the SSC radiation does does not exceed the observed flux at any frequency.

The limiting value of B depends on the energy extent of the electron distribution since the value of the maximum Lorentz factor γ_{max} determines which process dominates at a given frequency. We are therefore able to define a forbidden area in the $B - \gamma_{max}$ plane in which the predicted flux exceeds the observed flux at some frequency.

The physical parameters for the five sources that we have considered are given in Table 1 (assuming $H_0 = 100$ km/s/Mpc, $q_0 = 0$); the $B - \gamma_{max}$ constraints obtained for NGC 6814 are shown in Fig. 1.

We find that equipartition between the magnetic and X-ray energy densities is only marginally possible and in several cases is strongly ruled out. This poses problems for many theories of the acceleration and radiation of particles in the X-ray emission region which rely on the magnetic field to accelerate the particles. The most promising alternative picture, suggested by the work of Rees (1987), paradoxically requires the magnetic field to be strong. It can then confine many small, dense, clouds (the observed soft X-ray flux limits the total column density to $N_H < 10^{21}$ cm^{-2}) which, through free–free absorption, cause the optical and UV photosphere to lie beyond the X-ray emission region. Most of the limits in Table 1 do not then apply and primary radiation is SSC emission. We are considering, in collaboration with M.J. Rees, the model in more detail.

For further references and details see Celotti, Ghisellini & Fabian (1991).

Table 1. Rest frame parameters.

Source	z	R (cm)	T_{bb} (K)	U_x	U_B
NGC 4051	0.0023	9.0×10^{12}	3.6×10^4	5.9×10^4	2.5×10^4
NGC 6814	0.0053	1.5×10^{12}	2.4×10^4	6.4×10^7	4.8×10^6
3C 273	0.158	1.1×10^{15}	8.5×10^4	2.1×10^6	2.5×10^3
3C 279	0.538	2.6×10^{14}	3.1×10^5	2.8×10^7	$> 4 \times 10^8$
H1821+64	0.297	2.0×10^{16}	1.6×10^4	2.4×10^5	1.8×10^3

z is the redshift, R the dimension of the X–ray region, T_{bb} the temperature of the maximum blackbody. U_x is the radiation energy density, estimated by integrating the X–ray spectrum up to the maximum observed frequency; U_B is the upper limit to the magnetic energy density, corresponding to the minimum γ_{max} needed to emit this same frequency by scattering EUV photons.

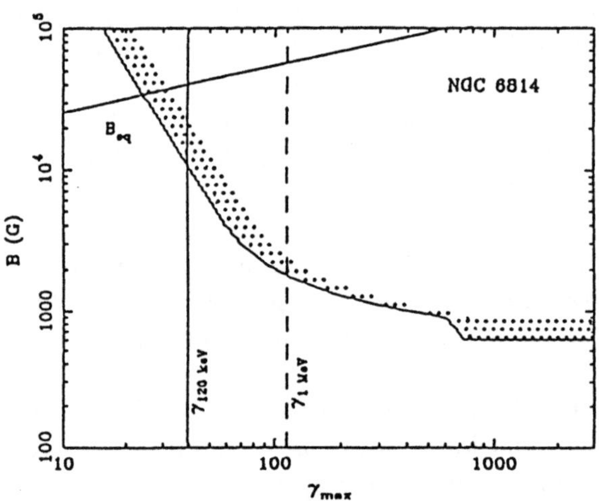

Fig. 1. The maximum allowed magnetic field versus γ_{max} for NGC 6814. The dashed region on the right side of the oblique line is not allowed. The vertical continuous line represents the minimum value $\gamma_{120\ keV}$: the permitted region therefore extends to the right of this line. The dashed vertical line represents the value of γ_{max} required to emit a reference frequency of 1 MeV. The line labelled B_{eq} shows the magnetic field in equipartition with the radiation energy density, obtained by integrating the X–ray flux up to the frequency $(4/3)\nu_{bb}\gamma^2_{max}$, neglecting the contribution from the (unobserved) EUV radiation field.

References

Celotti, A., Ghisellini, G., Fabian, A.C., 1991: *Monthly Notices Roy. Astron. Soc.*, in press.

Rees, M.J., 1987: *Monthly Notices Roy. Astron. Soc.*, **228**, 47P.

Pair Cascades in Ultrarelativistic Outflows from AGN

P.S. Coppi, A. Königl, J.F. Kartje, and D.Q. Lamb

Department of Astronomy and Astrophysics, University of Chicago, 5640 S. Ellis Avenue, Chicago, IL 60637, USA

Abstract. We investigate the interaction of an initially ultrarelativistic jet with the radiation field of an AGN accretion disk. The jet is slowed to a typical Lorentz factor $\gamma \sim 10$, and large numbers of e^+e^- pairs are produced for sufficiently high initial Lorentz factors. The radiation Compton upscattered by the jet may be relevant to the spectra of BL Lacs and gamma-ray bright AGN.

A way of producing jets with characteristic Lorentz factors $\gamma \lesssim 10$ is to accelerate particles near the central black hole to much higher initial energies and then let them decelerate through their interactions with the ambient radiation field (e.g., Compton drag – see Phinney 1987). The mechanism is robust and gives similar terminal jet Lorentz factors for a wide range of initial jet particle energy distributions. A detailed calculation by Melia & Königl (1989) showed that the upscattered spectrum resulting from the deceleration of a jet in the thermal radiation field of an AGN accretion disk can also account for the hard X-ray spectral component inferred in several BL Lac objects. In light of growing evidence for a moderately beamed X-ray component in AGN showing radio jets (e.g., a correlation between X-ray luminosity and spectral index with jet orientation – see Wilkes, this conference proceeding), it is worth pursuing further the possible contributions of a jet to AGN X- and gamma-ray emission. They may be particularly relevant to hard or gamma-ray bright sources such as 3C 273. Current emission models based on isotropic pair plasmas have difficulty reproducing such spectra without requiring a fine-tuning of the input parameters, overproducing the gamma-ray background, or introducing unwanted spectral features through the accumulation of cooled e^+e^- pairs.

The work discussed here extends the study by Melia & Königl. (1989). Because of computational difficulties, that study treated only the case of initial jet Lorentz factors $\gamma_i < \gamma_{KN} \equiv m_e c^2/kT_{disk}$, for which Thomson scattering dominates the jet-radiation interaction.

The maximum initial Lorentz factor is unknown, however, and mechanisms have been proposed (e.g., Fermi acceleration in a shock) that could accelerate protons to Lorentz factors as high as $\sim 10^8$. For $\gamma_i > \gamma_{KN} \approx 10^4$, several processes (including Compton scattering in the Klein-Nishina regime, photon-photon and triplet pair production, and photomeson production) lead to the rapid development of a photon-e^+e^- cascade. The cascade acts both as a "pair multiplier" and as a "gamma-ray filter," producing, for each injected particle, $\sim (\gamma_i/\gamma_{KN})$ e^+e^- pairs with Lorentz factors $\sim \gamma_{KN}$, and eliminating photons with energies above $\sim \gamma_{KN} m_e c^2$. The created pairs are further decelerated by Thomson scattering to the terminal Lorentz factors γ_∞. The resulting scenario is similar to that discussed in Burns and Lovelace (1982), although we consider additional jet-radiation interactions and do not specify the initial jet production mechanism.

We study the structure of such cascades with an accurate numerical code, allowing for the possibility that production of pairs may render the beam opaque to the ambient radiation field. At present, only the case of an electron beam injected mono-energetically into an isotropic, uniform radiation field has been completed. Future versions of the code will allow for proton injection and a non-uniform, moderately anisotropic background radiation field. Because of the high degree of beaming, interactions between high energy particles (Compton upscattered photons, pairs) are neglected. The cascade kinetic equations (containing the *exact* interaction terms) are then discretized and solved implicitly for the steady state photon and pair distributions as a function of distance down the jet. Using this pair distribution, the emergent X- and gamma-ray spectrum as a function of viewing angle is computed via a Monte Carlo code. The code takes into account the multiple scatterings of escaping photons off pairs. These scatterings are critical in determining the spectrum seen by an off-axis observer.

Support for this work under NASA grants NAGW-830,-1301 is gratefully acknowledged.

References

Burns, M.L., Lovelace, R.V.E., 1982: *Astrophys. J.*, **262**, 87.
Melia, F., Königl, A., 1989: *Astrophys. J.*, **340**, 162.
Phinney, E.S., 1987: in *Superluminal Radio Sources*, ed. J. A. Zensus and T. J. Pearson (Cambridge: Cambridge University Press), p. 301.

Brightness Temperatures in Excess of 10^{12} K from monoenergetic electrons

André R. Crusius-Wätzel

Max-Planck-Institut für Astrophysik, Karl–Schwarzschild–Straße 1, W-8046 Garching, Germany

Recent observations of quasars and BL Lac objects with space VLBI have revealed radiation brightness temperatures of $T_b = 1-4 \times 10^{12}$ K at 2.3 GHz (Linfield et al., 1989). These *direct* results clearly violate the inverse Compton limit of $10^{11}-10^{12}$ K for a power–law distribution of relativistic electrons (Kellermann and Pauliny–Toth, 1969). The physical reason for this restriction is that the ratio of the intensity of the Compton scattered photons to that in the synchrotron photons grows as $\propto T_b^5$. Considering subsequent scatterings leads to an instability, when the power in high frequency radiation becomes comparable to the power in synchrotron radiation. But even if the first order scattered photons leave the emission volume without further scattering, the timescale on which the electrons loose their energy gets very short compared to the synchrotron loss time, if the brightness temperature is larger than $10^{11} - 10^{12}$ K. Exceeding this value by a factor of 4 already reduces the lifetime of the electrons by three orders of magnitude.

These limitations can be somewhat relaxed, if the radiation is produced by a peaked electron energy distribution (or a power–law with a low–energy cutoff). Then the bulk of the radiation energy density can be optically thin, and therefore lower as in the opposite case. The existence of quasi–monoenergetic electron spectra is suggested by several observations, as e.g., the radio continuum measurements of the galactic center region including the compact source Sgr A* (Lesch, Schlickeiser and Crusius, 1988). Such distribution functions can be generated in accretion flows in active galactic nuclei via diffusive shock acceleration in combined action with spontaneous synchrotron losses. The particles pile up where the timescales for both processes are equal (Schlickeiser, 1984). The synchrotron spectrum from a δ–distribution in energy is $\propto \nu^2$ in the self–absorbed part and $\propto \nu^{1/3}$ in the optically thin region, until it cuts off exponentially at the frequency ν_c. The brightness temperature in the optically thick region is given by

$$T_b = 4.45 \times 10^9 \gamma_0 \, \text{K},$$

where γ_0 is the Lorentz factor of the relativistic electrons (Crusius–Wätzel, 1991). The crucial point now is to assume that the radiation spectrum is self–absorbed only up to $\nu_s = \xi\nu_c$, with $\xi < 1$. To avoid the inverse Compton catastrophe, it is required that the energy density in the radiation, $W_{rad} = (4\pi/c)\int d\nu\, I_\nu$, is less than the energy density in the magnetic field, $W_{mag} = B^2/8\pi$. This leads to the condition that the Lorentz factor be limited to

$$\gamma_{0,max} \approx 1.4 \times 10^2 B^{-1/7} \xi^{-5/21},$$

which also limits T_b. The brightness temperature at the frequency ν_s, at which the optical depth is one, can then be written as

$$T_b(\nu_s) = 3.0 \times 10^{12} (\nu_s/\mathrm{GHz})^{-5/11} (B/\mathrm{G})^{2/11}\,\mathrm{K}.$$

Calculations with a power–law distribution that has a cutoff at low energies gave similar results.

To conclude, it is possible to have brightness temperatures exceeding the 10^{12} K limit by a factor of a few from a relativistic electron distribution with a low–energy cutoff in general (e.g., also a relativistic Maxwellian will work). This happens when the emission is not self–absorbed up to the typical synchrotron frequency for the lowest energy electrons. Of course, brightness temperatures of the order of 10^{18} K, as derived from variability arguments in some compact sources (see Witzel, this volume), are not within the scope of these considerations and require different physics.

References

Crusius-Wätzel, A.R., 1991: *Astrophys. J.*, **368**, 390.
Kellermann, K.I., Pauliny-Toth, I.I.K., 1969: *Astrophys. J. (Letters)*, **155**, L71.
Lesch, H., Schlickeiser, R., Crusius, A., 1988: *Astron. Astrophys. (Letters)*, **200**, L9.
Linfield, R.P. et al., 1989: *Astrophys. J.*, **336**, 1105.
Schlickeiser, R., 1984: *Astron. Astrophys.*, **136**, 227.

VLBI Observations of 5 Radio Galaxies

G. Giovannini[1,2], *G. Comoretto*[3], *L. Feretti*[1,2], *J. Marcaide*[4], *T. Venturi*[2], *R. Vermeulen*[5], *and A.E. Wehrle*[6]

[1] Dipartimento di Astronomia - Bologna
[2] Istituto di Radioastronomia - Bologna
[3] Osservatorio Astrofisico di Arcetri
[4] Istituto de Astrofisica de Andalucia
[5] Owens Valley Radio Observatory - Caltech
[6] Jet Propulsion Laboratory - Los Angeles

A complete sample of 27 radio galaxies, having an arcsec core flux density $>$ 100 mJy at 6 cm has been selected from the B2 and 3CR galaxy samples in order to study the properties of radio galaxies on the milli-arcsec (mas) scale (Giovannini et al., 1990, Ap. J. 358, 159). On the basis of snapshot observations, we determined that all these galaxies, but one, have radio emission on the mas scale.

Our aim is now to obtain VLBI maps of the galaxies of the complete sample to study in detail their mas structure. The study of the physical properties of a complete sample of radio galaxies from the mas to the arcsec and arcmin scale is critically needed to analyse the validity of the "unified scheme" and to test models of origin and evolution of radio sources.

We present the maps of 5 radio galaxies, obtained so far (Fig. 1). The observations were carried out with the global VLBI array, with the MkII and/or MkIII recording systems, and each source was tracked for $\sim$7 hours. In all sources, the flux density in the parsec radio structure accounts for all the flux density of the arcsec core. Therefore we are confident to have mapped the whole radio structure.

NGC 315 and NGC 2484 shows asymmetric structures, with the jet aligned with the large scale structure. 3C 109 shows a one-sided jet on the same side of the kpc jet, but with a different position angle. A precession model could explain this difference. 3C 338 shows a remarkably symmetric structure on both the kpc and pc scale. The arcsec core is variable. B2 1144+35 is characterized by two strong resolved components, separated by 20 mas. This source is variable at the arcsec scale.

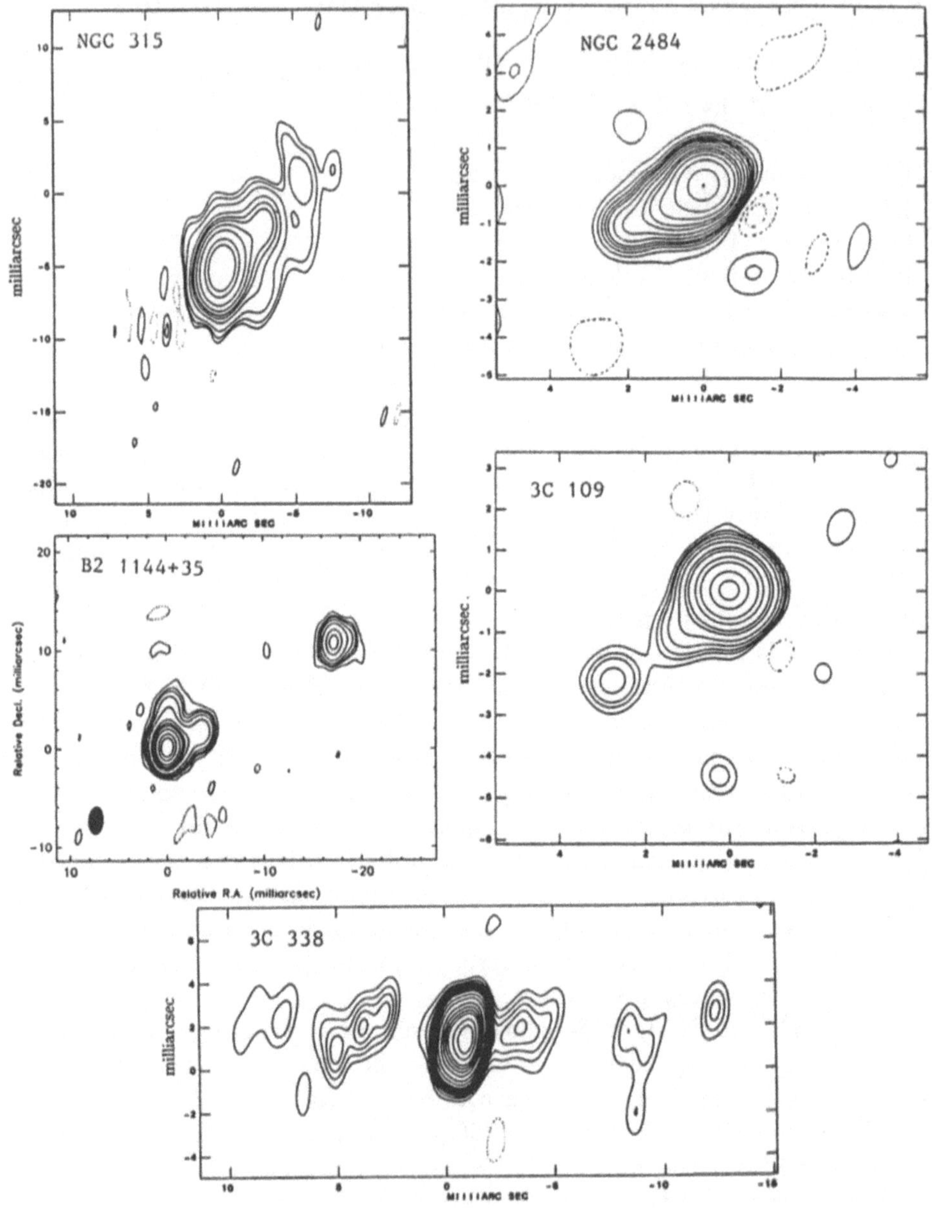

Fig. 1. VLBI contour maps of the sources. The first contour is always ~2 r.m.s. of the noise.

Synchrotron Emission from Bent Shocked Relativistic Jets

J.L. Gómez, A. Alberdi, and J.M. Marcaide

Instituto de Astrofísica de Andalucía, Apd. 3004, Granada 18008. Spain

The parsec scale radio structures of compact sources can be explained in terms of shock waves propagating along bent jets. These jets consist of plasma at bulk relativistic velocities with assumed equation of state and emitting synchrotron radiation. We have developed a numerical code to compute the transfer equations and obtain the total and polarized flux emission in a variety of scenarios (Gómez et *al.* 1991).

Our code divides the jet in cells, calculating for each one all the physical parameters required to solve the transfer equation for the synchrotron radiation. To save computation we restrict ourselves to the plane defined by the jet ridge line and the line to the observer. For each direction we have taken all the jet cells in a column, solving the synchrotron radiation transfer equation for each cell, considering as the incident radiation the radioemission of the previous one. When solving the transfer equation, we consider the differential Doppler boosting between consecutive cells, and for the last cell -the closest one to the observer- the boosting with respect to the observer. Since we know the values for the Doppler factors and the observer frequency, we can compute the frequency at which we must integrate the transfer equation for each cell through the expression $\omega_{obs} = \omega_{int} \Pi \delta_{i,i+1}$, where i varies between 1 and n, being n the total number of cells of a column to the observer and Π the product of the δ of the cells.

The existence of the superluminal VLBI components can be explained in terms of shock waves travelling along the jets. The shock waves will increase the energy and local density of particles passing through the shock, producing an enhancement of the opacity and radio emission. The component of the magnetic field parallel to the shock front is enhanced, inducing a rotation of the electric field vector and of the polarization angle (See Figure 1). These enhancement factors can be expressed, in the ultrarelativistic case, in terms of the velocity of the quiescent flow in the shock frame (Hughes et *al.* 1989).

We can explain the existence of stationary VLBI components through bends in the jet. The quasar 4C39.25 is a good example of this behaviour

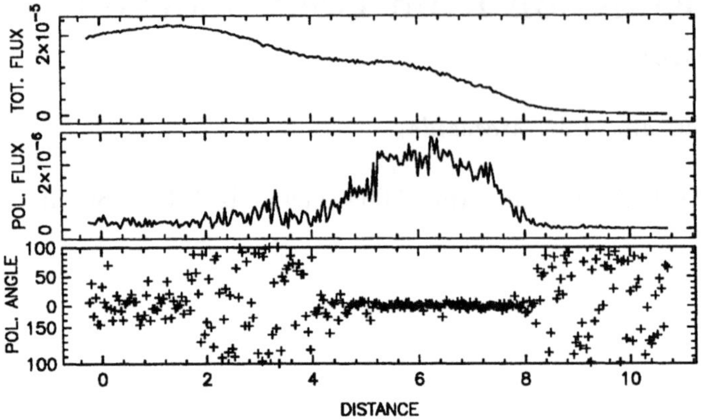

Fig. 1. Flux densities (in arbitrary units) and polarization angle for a rectilinear jet with a shock wave.

(Alberdi et *al.* 1991). A curvature in a relativistic jet leads to variations in the Doppler boosting of the flux density by a factor of 2 or so even if the jet bends by only a few degrees, showing themselves as components in the parsec scale radio jet (See Figure 2). In the case of a shock wave propagating through the jet, these bends can produce a noticeable change in the flux density, in the spectral index (Marscher et *al.* 1991), etc. of both the individual components and the source as a whole (see Figure 2). They can also produce a change in the apparent superluminal speed of the VLBI component associated to the moving shock wave. However, a bend does not produce neither a rotation in the polarization angle, nor a significant change in the polarized flux.

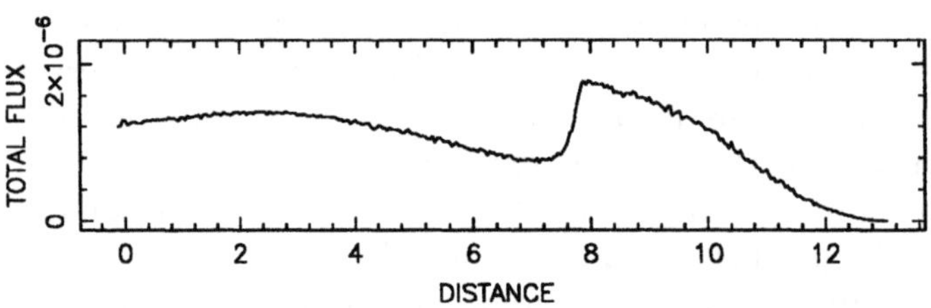

Fig. 2. Total flux (in arbitrary units) for a jet with a bend of 5 degrees towards the observer.

References

Alberdi, A., Marcaide, J.M., Marscher, A.P., Zhang, Y.F., Elósegui, P., Gómez, J.L. and Shaffer, D.B., 1991: *Astron. Astrophys.*, submitted.

Gómez, J.L., Alberdi, A., Marcaide, J.M., 1991: in preparation.

Hughes, P.A., Aller, H.D., and Aller, M.F., 1989: *Astrophys. J.*, **341**, 54.

Marscher, A.P., Gear, W.K., and Travis, J.P., 1991: to appear in *Variability of Blazars*, ed. E. Valtaoja (Cambridge U. Press).

Collimation of Jets in a Kerr Magnetosphere

Martin Haehnelt[1] *and Max Camenzind*[2]

[1] Institute of Astronomy, Madingley Road, Cambridge CB3 0HA
[2] Landessternwarte Königstuhl, W-6900 Heidelberg 1, Germany

1. Introduction

The collimation of jets is one of the least understood phenomenon in the physics of AGN. Recent observations have shown that it takes place on very small scales (< 1pc). An magnetized accretion disc around a rotating black hole, which builds up a magnetosphere and creates a magnetized wind and jet along the rotation axis is the most promising model to explain the collimation of jets (Camenzind 1990). We therefore investigated the innermost region (a few Schwarzschild radii) of the magnetosphere of a Kerr black hole.

2. Model

Assuming stationarity, axisymmetry, force-freeness, ideal MHD and using the 3+1 split of space-time the magnetic flux distribution $\Psi(r, \Theta)$ is determined by the relativistic Grad-Schlüter-Shafranov equation (Macdonald &Thorne 1982)

$$\nabla \cdot \left\{ \frac{\alpha}{\tilde{\omega}^2} \left[1 - \left(\frac{(\Omega_F - \omega)\tilde{\omega}}{\alpha} \right)^2 \right] \nabla \Psi \right\} + \frac{8\pi^2}{\alpha \tilde{\omega}^2} \frac{dI^2}{d\Psi} = 0.$$

Crucial for the understanding of the possible topology of a magnetosphere is the Neumann boundary condition that has to be fulfilled on the light cylinder. Free parameter (functions) are the specific angular momentum of the black hole a, the integrated poloidal current $I(\Psi)$, the angular velocity of the magnetic field lines $\Omega_F(\Psi)$ and the distribution of flux in the equatorial plane. α (lapse function), ω (shift function), and $\tilde{\omega}$ are parameters of the Kerr metric.

3. Solutions

The figure on the next page shows a meridional cut through the magnetosphere outside the inner light cylinder with a quadrupolar distribution of flux in the disc. Plotted are lines of Ψ= const. which are identical with the magnetic field lines. The topology of the field lines shows a feature that is typical for the solutions we obtained. The magnetosphere is strongly deformed by the Neumann boundary condition that has to be fulfilled on the light cylinder. The magnetic field lines are bent upward and collimated along the rotational axis. We were able to obtain convergence for our solutions for current strengths up to 10^{18} A. The detailed topology depends strongly on the boundary condition on the disc.

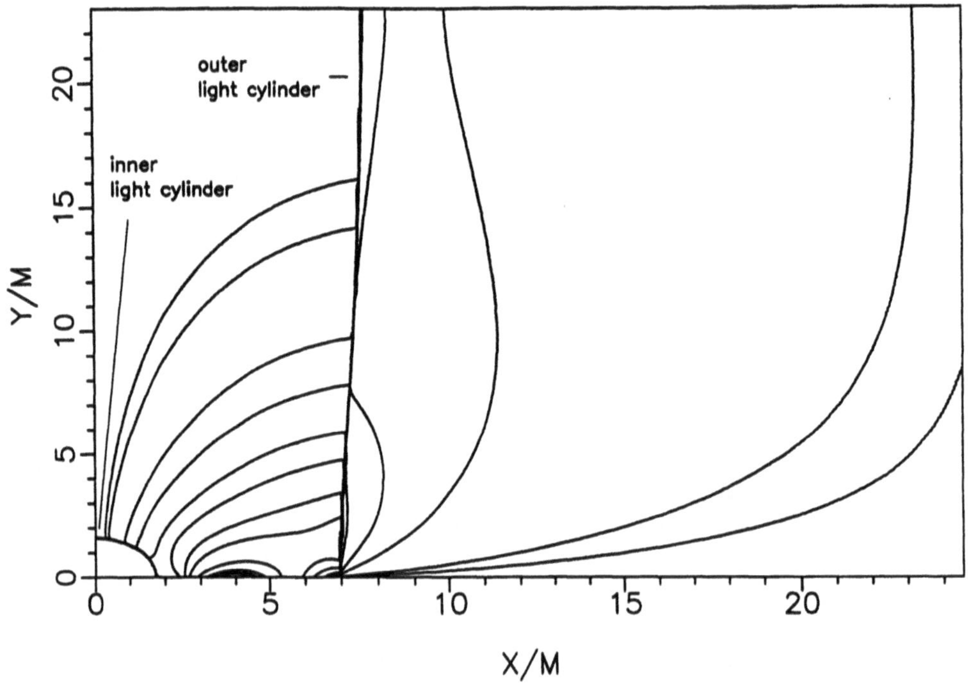

Fig. 1. Poloidal magnetic field outside the inner light cylinder of a rotating black hole ($a = 0.8M$), assuming a quadrupole distribution of flux in the equatorial plane.

4. Conclusion

The topology of the magnetosphere of a Kerr black hole allows the collimation of jets. An electron-positron plasma jet seems possible as well as a proton plasma jet. The current strengths obtained are sufficient for a magnetic confinement of the jet. The strong curvature of the field lines that is found at the light cylinder may allow particle acceleration and suggests this region as a possible X-ray source. The corresponding timescale of $10^3 - 10^4$ for a $10^8\,M_\odot$ black hole is well in agreement with observations of X-ray variability. We suggest that the different species of AGN especially the difference between radio-loud and radio-quiet objects might be explained by different topologies of their magnetosphere on small scales.

References

Camenzind, M., 1990: *Review in Modern Astronomy*, **3**, 234.
Haehnelt, M., Camenzind, M., 1991: *Monthly Notices Roy. Astron. Soc.*, submitted.
MacDonald, D.A., Thorne, K.S., 1982: *Monthly Notices Roy. Astron. Soc.*, **198**, 345.

Nonlinear Evolution of the Kelvin-Helmholtz Instability in 2-D Supersonic Slab Jets

Michał Hanasz

Institute of Astronomy, Nicolaus Copernicus University, ul. Chopina 12/18, PL-87-100 Toruń, Poland

The K-H instability is considered as a basic physical mechanism responsible for the existence of a number of morphological features like wiggles and knots observed in extragalactic jets (see Zaninetti, Van Horn, 1988; Hardee, 1987 and references therein). If we want to know whether an example of wiggling jet can be explained by the K-H model, we should check *if the nonlinear stabilization suppress the K-H instability modes to evolve up to the amplitude high enough to be observed.* The aim of this work is to obtain some insight into *nonlinear stabilization properties* for the K-H instability in astrophysical jets.

For the K-H instability of the interface of two fluids *the method of multiple scales (MMS)* has been applied (see Nayfeh, 1973; Weissman, 1979 and references therein). Some formal generalization of the standard approach is introduced in this work to consider a case of linearly unstable modes. In this method dependent variables as well as independent variables are expanded in power series of a small parameter ε related to the magnitude of perturbation. The linear perturbation is taken in the form: $f_1(x,y,t) = A\,F_1(x)\,exp(i(k\,y - \omega\,t)) + c.c.$ where A is a constant in linear approximation, x and y are respectively coordinates transversal and parallel to the jet axis.

The change of frequency can be regarded by allowing the amplitude A to depend on slow time scales $t_1 = \varepsilon\,t$, $t_2 = \varepsilon^2 t$, $y_1 = \varepsilon\,y$ and $y_2 = \varepsilon^2 t$, The basic idea of MMS is making the solution free of secular terms (something like $t\,cos(k\,y - \omega\,t)$). The requirement that the third order solution of equations of hydrodynamics is non-secular leads to the *nonlinear Schrödinger equation* for the evolution of A in slow time and space scales:

$$i\frac{\partial A}{\partial t_2} - \frac{1}{2}P\frac{\partial^2 A}{\partial y_1^2} = A\,|\,A\,|^2\,Q, \qquad (1)$$

where $P = d^2 w/dk^2$, and Q is the Landau constant. The simplest solution of (1) is $A(t_2, y_1) = a\,exp(-i\,\Omega\,t_2) = const(y_1)$, where a is real and Ω is complex. The $\Omega_i = Im(\Omega) = a^2 Im(Q)$ is the nonlinear correction to the growth rate. A given mode is nonlinearly stabilized if $\Omega_i < 0$, and for this case the stabilization amplitude can be determined as: $A_s = \sqrt{-\omega_i/\Omega_i}$.

The exemplary numerical results for the stabilization properties in nonlinear range are presented in the Figs. 1a and b together with the solutions of the linear dispersion relation for a Mach number $M = 5$ and density contrast $\nu = 1$. The following properties are common for different values of parameters M and ν:

- *for the waves short compared to the jet radius, both the fundamental (FM) and reflection (RM) modes are strongly stabilized,*
- *near the resonance RMs are nonlinearly destabilized, whereas the FM can be stabilized or destabilized,*
- *the stabilization amplitude decreases while k increases.*

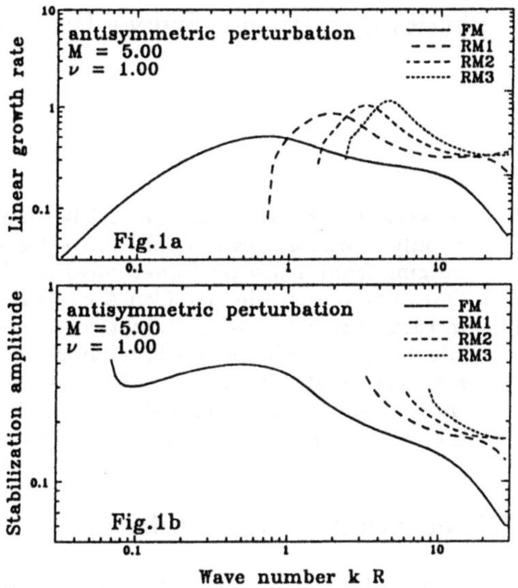

The promising feature of the applied formalism is its ability to treat more general cases of *cylindrical, magnetohydrodynamic and relativistic jets*, however the way of generalization of the method, allowing to treat nonlinear stability properties out of the points of marginal stability, deserves a deeper understanding.

The author is grateful to the Robert Bosch Foundation for the financial support to the participation at the conference.

References

Zaninetti, L., Van Horn, H.M., 1988: *Astron. Astrophys.*, **189**, 45.
Hardee, P.E., 1987: *Astrophys. J.*, **318**, 78.
Nayfeh, A.H., *Perturbation Methods* (Wiley,1973).
Weissman, M.A., 1979; *Philos.Trans.R.Soc.London Ser.A* **290**, 639.

The Large-Scale Radio Structure of Centaurus A

Norbert Junkes[1], *Raymond F. Haynes*[2], *and Julienne I. Harnett*[3]

[1] Institut für Theoretische Physik und Sternwarte, Kiel, Germany
[2] Australia Telescope National Facility, Sydney, Australia
[3] Physics Department, Sydney University, Australia

Abstract. Centaurus A (Cen A) is the nearest active radio galaxy. Its distance of only 3 Mpc allows a detailed investigation of different structural components, ranging from a few pc (inner core) to almost one Mpc (outer radio structure). Our radio observations of Cen A cover the complete area of the outer radio lobes in total intensity and polarization. A comparison of different radio frequencies allows us to evaluate the Faraday rotation and the intrinsic orientation of the magnetic field. The outer radio lobes show a distinct precession structure (S-shape) which can be reconstructed with a precession of the central engine. The radio results show Cen A as the closest example of an early-type active galaxy.

Observations and Results

We present radio maps of Cen A which are, by far, more sensitive than previous ones and contain polarization data at high frequencies (first complete polarization map of Cen A at 6 cm wavelength, first maps at 3.5 cm wavelength).

The 6 cm maps (Fig. 1a,b) show two remarkable details: The first is the S-shaped structure of the outer radio lobes which implies the impact of precession, the second is the obvious difference of polarization features in the northern and southern lobe of Cen A. While total and polarized emission in the northern lobe appear quite similar, the polarization south of $-44°30'$ splits in several blobs and the polarized intensity decreases. The depolarization in the southern lobe indicates its backward direction while the northern lobe is inclined towards the observer. This is supported by the detection of HI absorption south of the central position (van Gorkom et al., 1990) and the existence of a one-sided jet in the direction of the northern lobe.

Comparison of the results at the two wavelengths allows us to derive the Faraday rotation and the intrinsic direction of the magnetic field in Cen A. The rotation measure of the northern lobe is about $-70\ rad/m^2$, hence polarization vectors of the 3.5 cm map are close to the intrinsic value (less than 5° deviation).

The polarization of Cen A shows different directions for the magnetic field in the inner and outer lobes. The field is radial in the northern part of

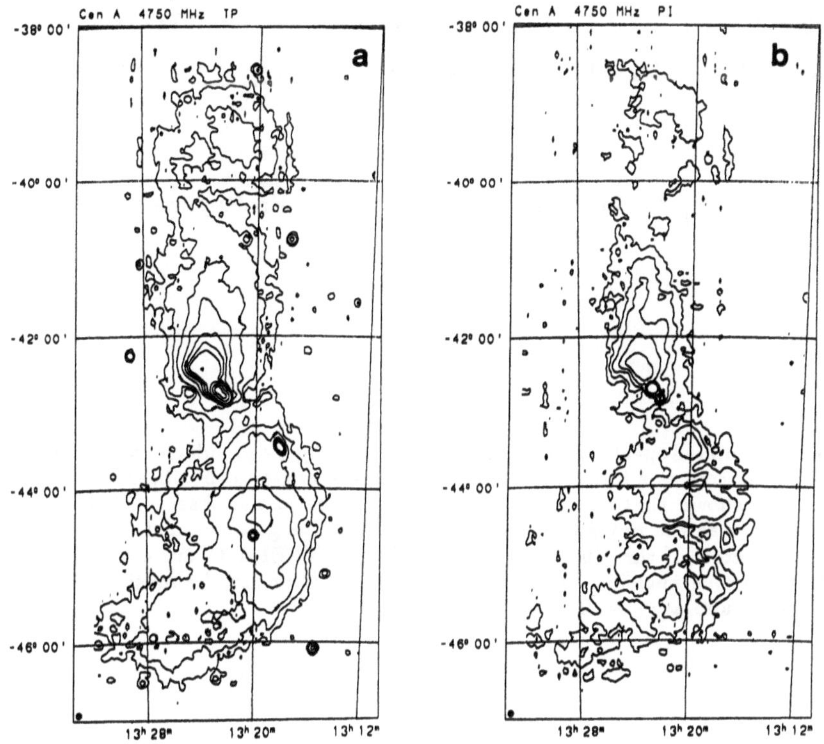

Fig. 1a,b. **6cm radio contour map of the complete area of Cen A (5°·9°) observed with the Parkes-64m radio telescope. a) total emission, b) linear polarized emission.**

the inner lobe, while it is tangential in the northern middle lobe. This holds for a part of the southern lobe as well but the field direction changes in the transition region south of −44°30′.

The S-shape structure of the outer radio lobes of Cen A has its counterpart in the warped dust lane of the galaxy. Model computations for a merger between an elliptical and a spiral galaxy after $5 \cdot 10^8$ years reproduce this warp convincingly (Tubbs, 1980).

The radio features of Cen A show a striking similarity to other active radio galaxies with double lobe structure. The reputation of Cen A as a peculiar object may be caused by its proximity only.

References

Tubbs, A.D., 1980: *Astrophys. J.*, **241**, 969.
van Gorkom, J.H., van der Hulst, J.M., Haschik, A.D., Tubbs, A.D., 1990: *Astron. J.*, **99**, 1781.

Physical Characteristics from Observations of Active Galactic Nuclei in a Jet Model

Yu.A. Kovalev

Astro Space Center, Lebedev Physical Institute, Profsoyuznaya 84/32, 117810 Moscow, Russia

Abstract. Calculated spectra evolution of jets in the hedgehog model for variable extragalactic sources as a function of the time and the angle to the observer are analyzed. It is shown that the fitting the model curves to the observed curves allows to extract as the physical parameters of the active galactic nuclei as the physical parameters of the Universe, if the additional VLBI data are available. The radius of active nuclei is estimated as $r_0 \leq 6 \cdot 10^{18} cm$ for the most of the variable objects and as $r_0 \sim 2 \cdot 10^{18} cm$ for BL Lac.

The main purposes of the present work are (i) to discuss some results of spectra evolution for active galactic nucleus, obtained from calculation of jet emission in the hedgehog model and (ii) to demonstrate what type of physical parameters and also how they may be extracted from the fitting the model results and results of observations.

The earlier work (Kovalev and Mikhailutsa, 1980) provide all data concerning the evolution of synchrotron emission for quasi point active source in that model. On the basis of that data one can calculate the flux density F of a narrow jet as a function of four parameters only: time T, frequency f, angle ξ between directions of the jet motion and to observer, a parameter γ of energetic distribution of emitted relativistic electrons, if f, F and T will be understood as normalized values, accordingly, to frequency f_{mo}, (at which intensity has maximum value in initial moment), to flux F_{mo} in initial moment at this frequency, multiplied by a geometry parameter G, and to the time r_0/c (where c is the light velocity, r_0 is active nucleus radius – if only the jet is started from "the nucleus surface"). So, one can have the following integral equation in the hedgehog jet model:

$$F/(G \cdot F_{mo}) = f(T, f/f_{mo}, \xi, \gamma),$$

where $T = t \cot c/r_0$.

Giving steps on f/f_{mo} and T one can calculate the evolution of spectrum of jet emission in the time T under several fixed ξ and γ. The results of that procedure give two stages of evolution: (i) spectra and spectral width are varied if $T \leq \cos\xi$, (ii) they are not varied if $T > \cos\xi$. So, the width of jet spectra is a function of time and the angle to the observer.

It can be shown that the model fitting allows to extract the following parameters (from the observations of flux spectra evolution only): GF_{mo}, f_{mo}, r_0, ξ and, as a result, $(\Delta\psi \cdot r_0^2/R^2H^{1/2})$, where $\Delta\psi$ is an angle of width of the jet, H is the magnetic field, R is the distance to the observer.

If we will additionally know $\Delta\psi$ and time evolution of the jet length from the VLBI observations, that distance R (like Kovalev, 1980) and magnetic field H can be extracted (from radio observations only). Next steps allow to calculate Hubble constant H_0 and deceleration parameter q_0 of the Universe, if the red shift Z and the known cosmological equation $f(Z, H_0, q_0, R) = 0$ are used and it is exist a lot of the sources.

This model was successfully fitted to many years monitoring observations of BL LAC, 3C 345 and 3C 454.3 in the range (1.3 - 31) cm (Berlin et al., 1991).

It can be supposed on the basis of observations of spectra variability and calculated curves that $t < 3years$ and $T \gtrsim 0.5$ for the most of active galaxies. Then $r_0 \lesssim 6 \cdot 10^{18} cm$. The detailed fitting of the model and observations for one of the flares in BL Lac gives $t \sim 2weeks$, $T \sim 0.02$ and $r_0 \sim 2 \cdot 10^{18} cm$.

References

Berlin, A.B., Kovalev, Yu.A., Kovalev, Yu.Yu., Larionov, G.M., Nidgelski, N.A. and Soglasnov, V.A., 1991: in: *Proceedings of the Conference 'Variability of Blazars'*, Turku

Kovalev, Yu.A., Mikhailutsa, V.P., 1980: *Sov.Astronomy* 57, 696.

Kovalev, Yu.A., 1980: *Sov.Astronomy* 57, 22.

The Sub-parsec Structure of AGN

T.P. Krichbaum[1], *A. Witzel*[1], *D.A. Graham*[1], *and J.A. Zensus*[2]

[1] Max-Planck-Institut für Radioastronomie, Bonn, Germany
[2] National Radio Astronomy Observatory, Socorro, N.M., USA

Global VLBI observations at $\nu = 43$ GHz ($\lambda = 7$ mm) now yield radio-images with angular resolutions of up to 100μas (corresponding to 0.43 pc at a redshift of z=1 for $H_0 = 100$ km s^{-1} Mpc^{-1}, $q_0 = 0.5$). High observing frequency and small angular resolution facilitate detailed investigations of sub-parsec scale regions in jets and active galactic nuclei (AGN), being self-absorbed at lower observing frequencies. Technical and scientific aspects of 7 mm-VLBI are given in more detail in e.g. Krichbaum, 1990, Krichbaum and Witzel 1991a,b.

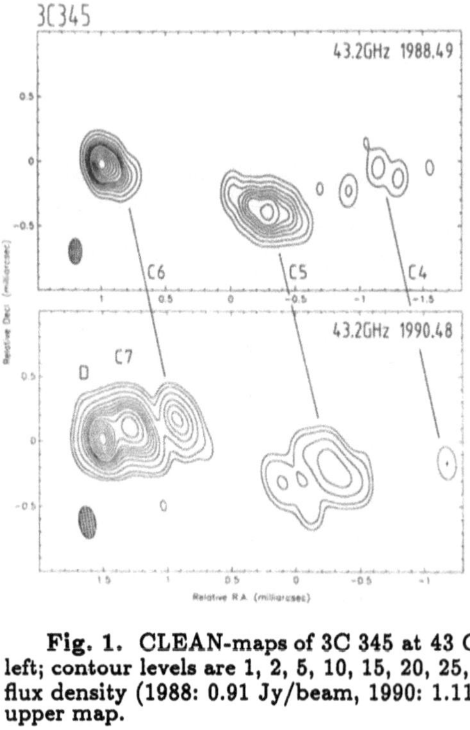

Fig. 1. **CLEAN-maps of 3C 345 at 43 GHz. The beam size is indicated in the lower left; contour levels are 1, 2, 5, 10, 15, 20, 25, 30, 40, 50, 60, 70, 80, and 90 % of the peak flux density (1988: 0.91 Jy/beam, 1990: 1.11 Jy/beam). The 1 % level is omitted in the upper map.**

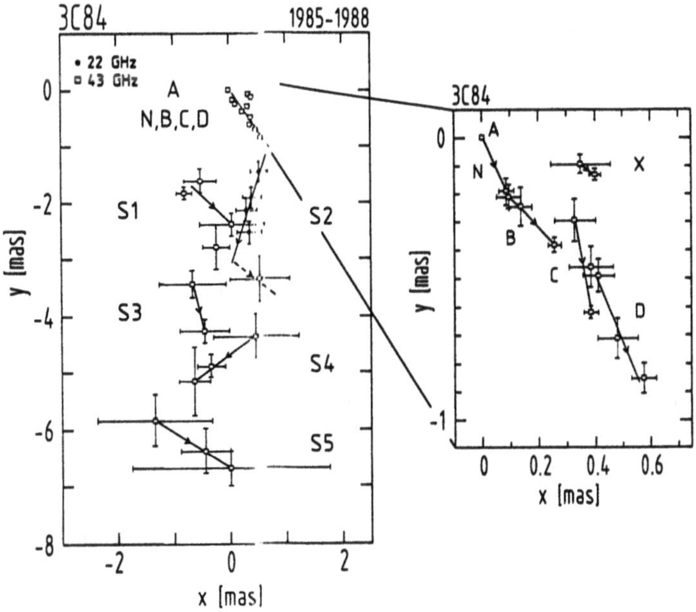

Fig. 2. **Motion of VLBI-components seen at 43 GHz in 3C 84 (epochs: 1986.4, 1987.4, 1988.5). 22 GHz-data for S2 from Readhead *et al.* are indicated by filled symbols. Lines with arrows connect adjacent positions.**

Our monitoring of sources with 7 mm-VLBI (e.g. 3C 273: Krichbaum *et al.*, 1990; 3C 84: Dhawan *et al.*, 1990, Krichbaum *et al.*, 1992, 3C 345: Krichbaum and Witzel, 1991, Zensus *et al.*, 1992) reveals apparent superluminal motion of jet components along paths indicating enhanced jet-curvature close to the (unresolved) 'core-region'. This effect is best illustrated in 3C 345. CLEAN-maps obtained from 43 GHz-VLBI observations in 1988.5 and 1990.5 are shown in figure 1. The comparison with 22 GHz VLBI-maps (see Zensus, 1990) leads to the component identification, indicated by labels in fig.1. The components C4 and C5 – both seen at 22 and 43 GHz – continue their separation from the core D along bent trajectories, which are clearly different from each other (see the plot of component positions for 3C 345 in the review of A. Witzel, this workshop). In 1988 a new component C6 appeared close to D. The velocity of C6 between 1988 and 1990 is $\beta_{app} = (4.3 \pm 0.3)$, in good agreement with velocities of other jet components. If our identification is correct, C6 moves to the north-west along a path similar to the path of C4, however rotated in position angle by $\Delta(P.A.)_{C6-C4} \sim +15°$ to the north (see the figure in A. Witzel's review). This and the fact that the new component C7 is located further north than any other earlier component suggest that in 3C 345 new jet components follow trajectories, which are different from the paths of their precursors. Further monitoring is required to follow the motion of C6 and C7.

VLBI-monitoring of the Seyfert-galaxy 3C 84 (= NGC 1275) at 43 GHz since 1986 revealed *subluminal* motion from 3 epochs of observation (Krichbaum *et al.*, 1992, Krichbaum and Witzel, 1991a,b). In a northern 'mas-jet' the components N, B, C, and D separate from the core A with β_{app} = 0.1...0.2 c (see the figure in A. Witzel's review). Five components S1-S5 located south of A (within a region of extended emission seen at lower frequencies) seem to move with faster velocities (typically (0.4...0.7 c). A major kink seen in the source structure at $r \sim 1$ mas suggests a geometrical interpretation of the observed apparent acceleration. The motion of the components is sketched in fig.2; corresponding positions from our 3 observations are connected by straight lines. The northern 'mas-jet' is displayed enlarged (right). A component S2 ($\beta_{app} \simeq 0.5$) has been seen earlier also at 22 GHz (Readhead *et al.*, 1990) and possibly moves along a bent path. Back-extrapolation of its motion in time yields strong evidence that S2 was ejected during a large 90 GHz flux density outburst beginning in 1979. Still large uncertainties for the positions of S1-S5 do not yet allow to determine their path of motion unambiguously, however a complex (turbulent ?, quasi-helical ?) motion pattern is suggested.

The participation of more antennas in future 7 mm-VLBI experiments (first fringes to the VLBA-antenna at Pie Town have been seen in June 1991; a first 43 GHz VLBI-observation with the 30 m mm-telescope at Pico Veleta is scheduled for autumn 1991) will further enhance the imaging capabilities at 43 GHz, allowing for more detailed investigations of the sub-parsec scale structure of radio-sources.

References

Dhawan, V., *et al.*, 1990: *Astrophys. J. (Letters)*, **360**, L43.

Krichbaum, T.P., 1990: in: *Parsec-scale radio jets*, ed. J. A. Zensus and T. J. Pearson (Cambridge University Press), p. 83.

Krichbaum, T.P., *et al.*, 1990: *Astron. Astrophys.*, **237**, 3.

Krichbaum, T.P., and Witzel, A., 1991a: in: *Frontiers of VLBI*, ed. H. Hirabayashi, M. Inoue and K. Kobayashi (Universal Academy Press, Tokyo), in press.

Krichbaum, T.P., and Witzel, A., 1991b: in: *Variability of Blazars*, ed. E. Valtaoja and M. Valtonen (Cambridge University Press), in press.

Krichbaum, T.P., *et al.*, 1992: *in preparation.*

Readhead, A.C.S., *et al.*, 1990: in: *Parsec-scale radio jets*, ed. J. A. Zensus and T. J. Pearson (Cambridge University Press), p. 71.

Zensus, J.A., 1990: in: *Parsec-scale radio jets*, ed. J. A. Zensus and T. J. Pearson (Cambridge University Press), p. 28.

Zensus, J.A., *et al.*, 1992: *in preparation.*

Variability Timescales of Quasars and BL Lac Objects

Markku Lainela and Esko Valtaoja

Tuorla Observatory, University of Turku, Finland

A program of monitoring compact extragalactic radio sources with the 14-m telescope in Metsähovi, Finland, was begun in 1980. Flux measurements have been done mainly at 22 and 37 GHz. The 100 and 230 GHz observations with the new 15-m SEST telescope in Chile were initiated in 1987. The total number of flux measurements of over 50 sources exceeds 12000. The results up to 1986 have been published in Salonen et al. (1987) and Teräsranta et al. (1987), where details of the observing programs can be found.

Our sample consist of the 26 best-observed sources from our monitoring program. This is the same sample as in Valtaoja et al. (1988, 1991). We have analyzed the average variability characteristics of the sources using parameter

$$T(n) = \frac{1}{N}\sum_{i<j} \frac{|S_i - S_j|}{\sqrt{\sigma_i^2 + \sigma_j^2}},$$

(Wardle et al. 1981) where S_i is the flux density measured at time t_i; S_j at time t_j; and σ_i, σ_j are their associated errors. The data S_i, S_j are separated by a time interval $50(n-1) < |t_i - t_j| < 50$n days, n=1 to 20; N is the number of pairs of data falling in the nth time interval.

In most sources the average variability exceeds the observational error (between 3-10 % depending on the source strength) first on timescales in excess of weeks. An exception is OJ 287, with variability down to daily timescales. In our sample the variability behaviour can be divided into three different main classes: In the first class the values of T increase first rapidly but shortly after that the values become steady. In the second class the values of T increase until time scale is very long (several years). In the third class sources show evidence for preferred timescales of 600-1400 days. Figure 1 shows examples of these three classes. It seems that T does not correlate with classification or spectral shape.

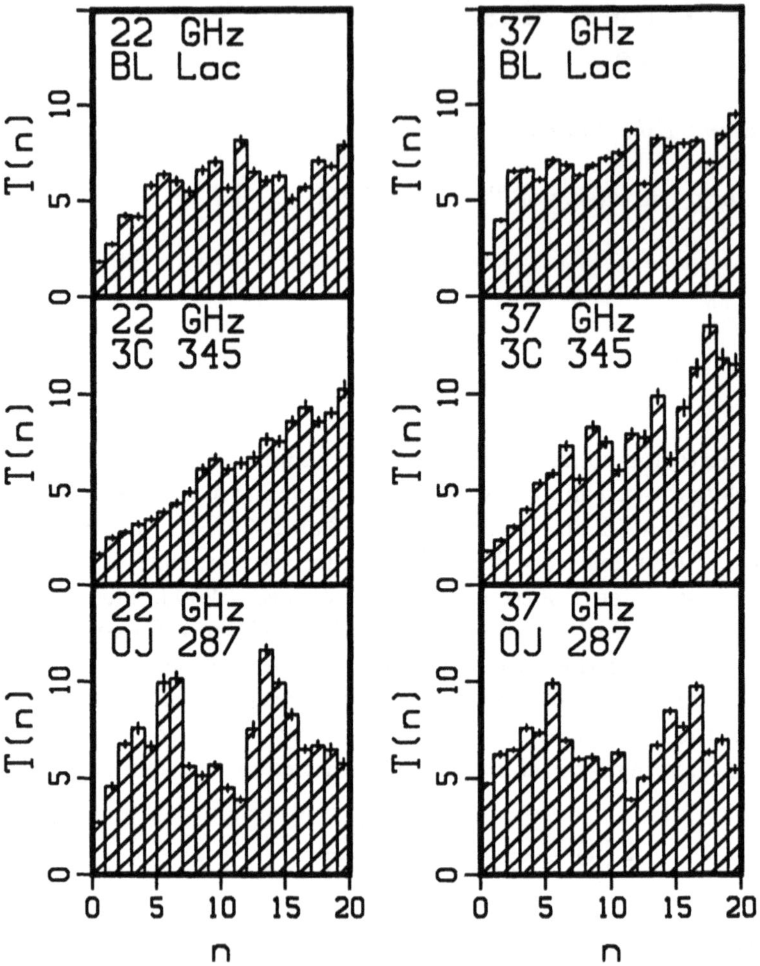

Fig. 1. The average variability of BL Lac, 3C 345 and OJ 287. Time intervals are linear from T(1)=0-50 days to T(20)=950-1000 days.

References

Salonen, E. et al.: 1987, *Astron. Astrophys. Suppl.*, **70**, 409
Teräsranta, H. et al.: 1987, *Astron. Astrophys. Suppl.*, **71**, 125
Valtaoja, E. et al.: 1991, *Astron. Astrophys.*, submitted.
Valtaoja, E. et al.: 1988, *Astron. Astrophys.*, **203**, 1
Wardle,J.F.C. et al.: 1981, *Astron. J.*, **86**, 848.

Coherent Radiation and Intraday Variability: Magnetic Reconnection in AGN

Harald Lesch

Max-Planck Institut für Radioastronomie, 5300 Bonn, Germany

A model for rapid variability (less than one day) in active galactic nuclei is developed. At small disk radii $\simeq 5 \cdot 10^{14}$ cm, (which correspond to 10 Schwarzschild radii for a black hole with $10^8\, M_\odot$) accretion disks contain strong turbulent magnetic fields of the order of 100 Gauss. Caused by buoyancy forces magnetic flux tubes escape from the disk and encounter the magnetic flux in the disk corona. If field lines with different polarity approach each other the magnetic energy is partly dissipated by magnetic reconnection. Thereby strong electric fields are built up which mainly accelerate electrons. Dependent on the plasma turbulence which is excited by the current in the reconnection sheet (caused by the strong magnetic field gradient on both sides of the layer) the electrons can reach relativistic energies. Since all the accelerated electrons gain almost the same amount of energy, their energy distribution function is more or less monoenergetic, that is relativistic electron beams (REB) emerge from the reconnection regions. Such beams are unstable and excite Langmuir waves with a frequency $\omega_{pe} \simeq 5.6 \cdot 10^4 \sqrt{n}$ which depends only on the electron number density n. The finite lengths of different beams emerging from different reconnection sheets allows that the waves (excited form a first beam) are upscattered by inverse Compton scattering by a following REB to frequencies $\gamma^2 \omega_{pe}$ (Baker et al., 1988). For Lorentz factors $\gamma \simeq 10^3$ and densities typical for disk coronae (derived from iron line observations) $n \simeq 10^6\, \mathrm{cm}^{-3}$ one easily reaches the optical frequencies.

The length of the outburst is related to the stability of the electron beams. Stability considerations, including weak and strong turbulence effects (Lesch and Schlickeiser, 1987), are taken to estimate the relaxation length of a REB. With the above parameters and a temperature for the coronal plasma of about 10^8 K the relaxation length corresponds to about one light day.

A slight enhancement of the magnetic field strength dissipated in the layers leads to an increase of the final electron energies. The resulting inverse

Compton scattered emission then appears in the X-ray range (for details see Lesch and Pohl, 1991).

When considering possible astrophysical applications of inverse Compton scattering of Langmuir waves, the following should be kept in mind. Apart from factors of the order of unity, the power radiated by an electron due to synchrotron radiation, inverse Compton scattering and scattering of Langmuir waves are in a ratio of the energy densities in the magnetic field, the unscattered photons and the Langmuir waves, respectively. It requires exceptional circumstances for the energy density in the Langmuir waves to exceed globally the energy density in the magnetic field in the whole system. It is implausible that this inequality could be satisfied other than in localized regions (like reconnection sheets) of an extended source. The mechanism of inverse Compton scattering of Langmuir waves is suitable to explain the variability but not the continuum emission.

Nevertheless, we think that in any kind of AGN coherent radiation processes could account for at least a part of the variability, since reconnection and the accompanying electron acceleration of electrons to relativistic energies should appear in any plasma where the magnetic field is turbulently mixed.

The coexistence of both coherent and incoherent emission processes is probably responsible for the variable emission characteristics of AGN.

References

Baker, D.N., Borovsky, J.E., Benford, G., Eilek, J.E.: 1988, *Astrophys. J.*, **326**, 110
Lesch, H., Schlickeiser, R.: 1987, *Astron. Astrophys.*, **179**, 93.
Lesch, H. Pohl, M.: 1991, *Astron. Astrophys.*, in press.

A Novel Mechanism for Nonthermal X-ray Emission

Karl Mannheim

Max-Planck-Institut für Radioastronomie, 5300 Bonn 1, Germany

A successful interpretation of nonthermal continuum emission from knots and hot spots in jets of radio-loud AGN pictures these regions as shocks in a bulk flow of fully ionized hydrogen plasma with a superimposed ordered magnetic field structure. The theory of diffusive shock-acceleration then explains the emission as synchrotron radiation of relativistic electrons (Meisenheimer et al., 1989; Bregman, 1990). In a recent effort to understand the radiative properties of the plasma caused by relativistic nuclei which are almost inevitably co-accelerated with electrons (Biermann and Strittmatter, 1987) it was found that they can cause a significant contribution to X- and γ-ray emission exceeding synchrotron-self-Compton emission (Mannheim and Biermann, 1989; Mannheim et al. 1991).

Biermann (1991) has shown that the escaping proton flux from hot spots in the jets of FR II galaxies or similar objects can indeed explain the highest energy component of cosmic rays observed at earth.

The novel mechanism paraphrased proton-initiated-cascade emission (PIC) works very simple: High energy protons (the most abundant hadrons) immersed in radio to optical photons produce pions decaying into pairs and γ-rays which in turn cool catastrophically building up a shower of electromagnetic energy – similar to the interaction of cosmic rays with the earth's atmosphere, yet with a photon "atmosphere".

The *novelty* of the idea is that it has not been recognized before that shock-accelerated nuclei inevitably suffer rapid energy-losses due to the presence of synchrotron-photons from the accelerated electrons.

Here I present a possible *empirical test* of the hypothesis that a flat spectrum proton distribution exists in hot spots of FR II galaxies.

- Inferring physical properties of hot spots at the working end of jets from radio to optical data (Meisenheimer et al., 1989) leads to the *prediction* of PIC X-rays with $\alpha_{\mathbf{keV}} = 0.7^{+0.2}_{-0.1}$ (Fig. 1).
- X-rays from radio-selected BL Lacs can be explained by PIC (Fig. 2). The spectral index $\alpha_X \approx 1$ agrees with the observed mean X-ray index of BL Lacs.
- PIC-radiation is anisotropic close to the accretion disc, because of the predominant thermal UV-photons coming from the bottom of the jet.

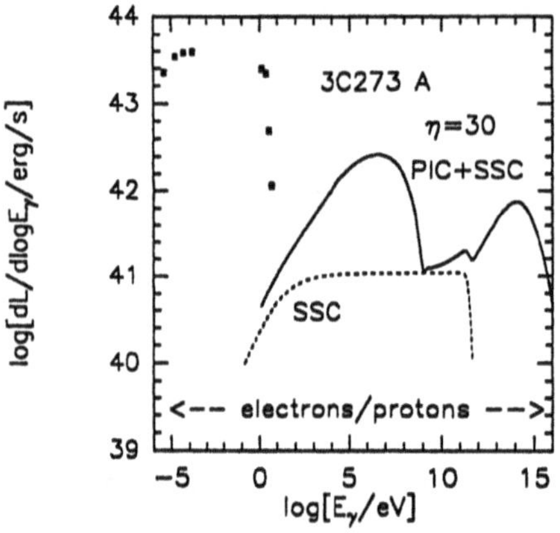

Fig. 1.

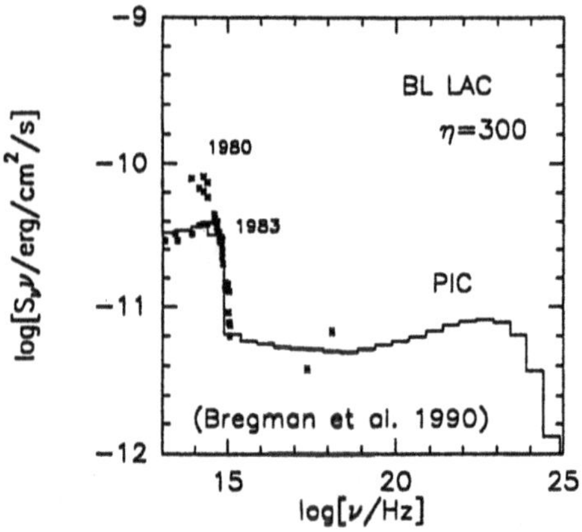

Fig. 2.

Hence proton-initiated "Compton-reflection" models are recovered explaining nuclear X-ray emission (cf. Lightman and Zdziarski, 1987).

References

Biermann, P.L., Strittmatter, P.A.: 1987, *Astrophys. J.*, **322**, 643

Biermann, P.L.: 1991, in *Frontiers in Astrophysics*, eds. Silberberg, R., Fazio, G., Rees, M., Cambridge University Press

Bregman, J.N.: 1990, *Astron. Astrophys. Rev.* **2**, 125

Lightman, A.P., Zdziarski, A.A.: 1987, *Astrophys. J.*, **319**, 643

Mannheim, K., Biermann, P.L.: 1989, *Astron. Astrophys.*, **221**, 211

Mannheim, K., Krülls, W.M., Biermann, P.L.: 1991, *Astron. Astrophys.*, in press.

Meisenheimer, K., Röser, H.-J., Hiltner, P.R., Yates, M.G., Chini, R., Perley, R.A.: 1989, *Astron. Astrophys.*, **219**, 63.

The Optical Microvariability of 3C 371

R.H. Miller[1], *J. Noble*[1], *and P. Wiita*[1], *and M. Carini*[2]

[1] Georgia State University
[2] CSC/NASA Goddard Space Flight Center

Microvariability has now been conclusively demonstrated to exist for a number of blazars (Miller, Carini, and Goodrich, 1989, Carini *et al* (1990, 1991), Carini and Miller (1991) and Wagner (1991)). As part of a long term program to investigate the nature of the microvariability for a large sample of blazars, we selected 3C 371 for intensive monitoring during a five night period from 27 June to 01 July 1989. The results of this campaign are discussed in the following paragraphs.

During the first three nights of this campaign, the source steadily faded approximately 0.5 magnitudes followed by a modest brightening during the fourth night and a continuation of the decline on the last night. During each of these nights, substantial variability was detected on the timescales of hours or less. The most dramatic variation which we detected was observed on 01 July 1991 when we observed an approximate 0.35 magnitude decline over a five hour period.

Assuming such events arise from global rather than local changes in the source, one may use such events to calculate a characteristic timescale, or doubling–time, which, in turn, can be used to estimate the size of the emitting region from $R \leq ct$, assuming no relativistic bulk motion of the material is present. If one further assumes that the radiation is generated in the vicinity of a supermassive black hole, *e.g.* $R = 100R_s$ where R_s is the Schwarzschild radius, then an upper limit on the mass of the black hole (which we define as the *variability mass* (Carini *et al* 1990)) can be determined along with a corresponding Eddington luminosity.

The rapid variations which we detected on 01 July 1989 lead to an estimate of the mass of the supermassive black hole of $M \leq 3.6 \times 10^7 M_\odot$. This corresponds to an Eddington luminosity of $L_E = 4.7 \times 10^{45} ergs/sec$. Following Carini (1990), we find a bolometric luminosity of 3C 371 at the time of these observations to be $L = 1.5 \times 10^{45} ergs/sec$. Thus, relativistic beaming is not required to explain the observed variations. No periodicities were found in these variations, and an analysis of the PDS of the variations yields a slope for the PDS of approximately 2.0.

References

Carini, M.T.: 1990, Ph.D Thesis, Georgia State University.
Carini, M.T. and Miller, H.R.: 1991, *Astrophys. J.*, in press.
Carini, M.T., Miller, H.R. and Goodrich, B.D.: 1990, *Astron. J.*, **100**, 347.
Carini, M.T., Miller, H.R., Noble, J. C., and Sadun, A.C.: 1991, *Astron. J.*, **101**, 1196.
Miller, H.R., Carini, M.T., and Goodrich, B.D.: 1989, *Nature*, **337**, 627.
Wagner, S.J. 1991, in *Active Galactic Nuclei* ed. W.J. Duschl, S. J. Wagner and M. Camenzind, p. 163.

Jet Formation due to Accretion Disk Instability

Mikio Nagasawa[1] *and Takuya Matsuda*[2]

[1] Institute of Computational Fluid Dynamics, 1-22-3 Haramachi, Meguro, Tokyo, 152
[2] Dept.of Aeronautical Eng., Kyoto University, Kyoto 606

Abstract. We present the three-dimensional simulations of Roche-lobe overflow in a close binary system by smoothed particle hydrodynamics. The nonaxisymmetric structure which is manifest in the outer region causes an effective torque and transports the angular momentum outward. However, it is not sufficient for the infalling gas to reach the compact star. For the adiabatic gas, the hot torus forms at the inner edge of the disk. After a few tens of revolutions, the instability of hot torus triggers the jet ejection. The hot torus could be an energy reservoir for the catastrophic accretion. This is a new jet mechanism from the non-magnetic accretion disk.

1. Roche Overflow and Disk Formation

As one of the standard initial and boundary conditions for the accretion problem, we start the simulation as the Roche problem in the semi-detached binary. Our SPH is carried out with the particles number of 10^5. The structure of a mass-losing star is solved by placing particles in it, while the compact object is represented by a massive particle with the radius $R_2=0.01R_1$. The model has the mass ratio $M_1/M_2=1$ and the initial separation $A=2.0R_1$, which is smaller than the critical value for the tidal disruption.

The tidal effect is so strong that the gravitational torque initiates the accretion flows. It can amplify the large oscillation of the giant, then the overflow goes through the wide channel near the Lagrange L1 point. Without the specific inner boundary condition such as a strong vacuum, the conservation of angular momentum halts the mass infall at a certain distance from the compact star. Then, the overflowed gas makes the accretion disk around the compact star. At the inner edge, this torus becomes fat and fat since we treat the adiabatic gas of $\gamma=1.2$ modeling the cooling effect.

2. Disk Instability and Jet Formation

For the continuation of accretion flow, the tidal viscosity is dominant only in the region of $r \gtrsim 0.33A$ (ref.1), and the turbulent viscosity becomes manifest near the compact star. The process of effective torque is needed in the intermediate region. In our model, a nonaxisymmetric disk instability can play the role. After the revolutions of $\sim 10 t_{\mathrm{Kep}}(r_{\mathrm{in}})$, the catastrophic disruption of hot torus occurs and the jet triggers in the polar directions.

The 2-D models of thin disk approximation are unable to calculate the instability of hot torus (ref. 2). From the linear analysis of the normal mode, ref. 3 showed that the azimuthal wavenumber m=1 (ring disruption) is the most unstable. This is in agreement with our result. The instability occurs from the inner edge of the torus, which is good for the jet collimation. The ejected gas is very hot, but the mass ratio to the accreted gas is very small as ~0.01.

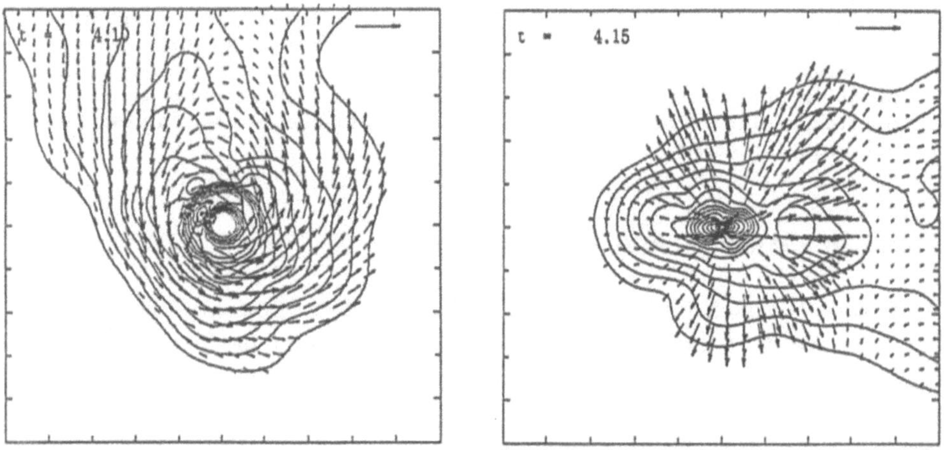

Fig. 1. Hot torus structure of the 3-D accretion disk. The appearance of m=1 mode disk instability (left). Vertical cross section of the disk and the jet (right).

References

1. Paczynski, B., 1977: *Acta Astr.* **28**, 91
2. Sawada, K., Matsuda, T., Hachisu, I., 1986: *Monthly Notices Roy. Astron. Soc.*, **219**, 75
3. Kojima, Y., 1986: *Prog. Theor. Phys.* **75**, 251

VLBI Observations of Quasars 3C 345 and CTA102 at 18 cm Wavelength

Fredrik T. Rantakyrö[1] *and Lars B. Bååth*[2]

[1] Department of Astronomy/Astrophysics, Chalmers S-412 96 Göteborg, Sweden
[2] Onsala Space Observatory, S-439 00 Onsala, Sweden

1. Introduction

We present results from very high dynamic range VLBI observations of the quasar 3C 345 and two preliminary maps of the quasar CTA102 at 18 cm wavelength. Observations made at several epochs and frequencies show that the components seem to move outwards along a curved jet (Biretta *et.al.* 1986, *ApJ*, **308**, 93).

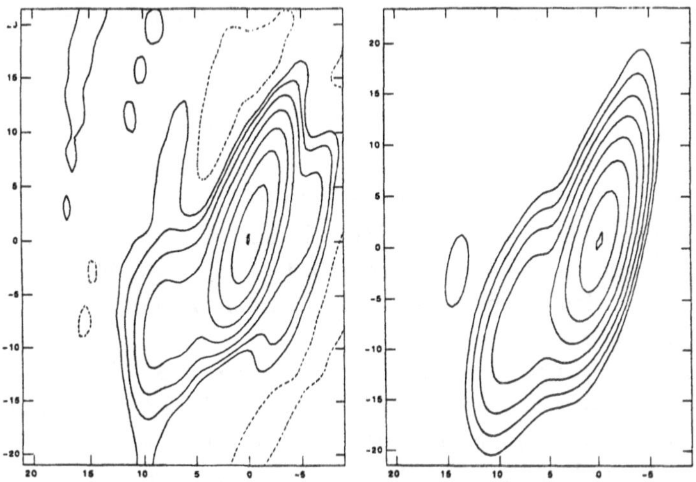

Fig. 1. Maps are from left to right CTA102 May 1988 and June 1989.

2. Viewing angle of the jet of 3C 345

There is a power law decrease(-2.33±0.16) in the overall luminosity of the jet with distance from the core(Figure 2). We suggest two models and make a χ^2-fit to the observed power law for the flux. A *constant comoving flux model* requires a constant ultrarelativistic velocity for the plasma flow and

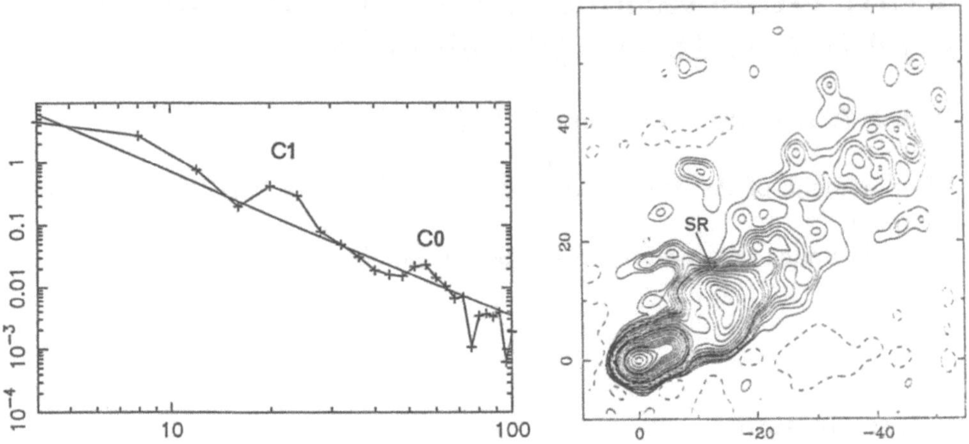

Fig. 2. Figures are *left* Integrated flux of jet with distance from the core(mas) and *right* Map of 3C 345 April 1985.27.

that the viewing angle increases with increasing distance from the core. An adiabatically expanding jet model requires that the viewing angle will suddenly jump as the plasma starts to expand in a cone. The viewing angle will at the start of the expansion be close to the line of sight and then slowly increase with distance from the core.

3. C1 – a stationary shock ?

The observed positions for component C1b (r_0=16.4±0.9 (1981.11), r_0=17.5 ±1.5 (1985.27 (Rantakyrö *et.al.* 1991)), C1(r_0=16.5±1.4 (1981.78, Waak *et.al.* 1988,*AJ*,**95**,1023) and r_0=15.5±1.9 (1981.89, Cohen *et.al.* 1983, *ApJ*, **272**, 383)) are consistent with C1 being an extended component at a constant distance from the core.

A distance of 1 mas in our map corresponds to a projected distance of 10^{19} cm (4 pc). With a viewing angle of $\theta \sim 7°$(Unwin *et.al.* 1983, *ApJ*, **271**, 536) of the jet close to the core, the apparently shocked region at r≥12 mas corresponds to a deprojected distance of 10^{21} cm from the core.

This distance is of the same order as expected for the narrow line regions(10^{21}-10^{22} cm) in active galactic nuclei (Netzer 1987, in: *Astrophysical Jets and their Engines*, eds Kundt, W., 103-124). The beginning of the shocked region is consistent with being the inner edge of the stationary component C1, elongated perpendicular to the jet axis. We suggest this to be the stationary shock formed when the relativistic plasma passes through the narrow line region.

The Radio Jets of the S5-Quasar 1928+738

C.J. Schalinski[1,2], *A. Witzel*[1], *C.A. Hummel*[1,3], *T.P. Krichbaum*[1], *A. Quirrenbach*[1,3], *K.J. Johnston*[4]

[1] Max-Planck-Institut für Radioastronomie, Bonn, Germany
[2] Geodätisches Institut der Universität Bonn, Germany
[3] U.S.N.O./N.R.L. Interferometer Project, Washington D.C., U.S.A.
[4] Center for Advanced Space Sensing, N.R.L., Washington D.C., U.S.A.

1. Introduction

Bulk relativistic motion (b.r.m.) is now found to be a common phenomenon among quasars and BL Lac objects: 12 out of 13 sources comprising a statistically complete sample of flat spectrum radio sources selected from the S5-survey show evidence for superluminal motion of the core-jets (Witzel *et al.*, 1988; Schalinski, 1990). B.r.m. is independently confirmed by the lack of measured inverse Compton X-rays as well as variability of both the total and core component flux densities applying light travel time arguments. At least 1/3 of the sources also displays "stationary" features or different component velocities (Schalinski *et al.*, 1991). Motion along curved trajectories indicates that helical geometries have to be included in relativistic jet models. A detailed investigation requires the analysis of individual objects. S5-quasar 1928+738[1] is well suited for this, since it displays jet-like radio emission on scales of a few parsecs to kpc (e.g. Johnston *et al.*, 1987), and structural variability has been established on the basis of multiple components in VLBI-maps at different wavelengths (Eckart *et al.*, 1985; Schalinski, 1990).

2. Structure and kinematics of 1928+738

Whereas the arcsecond structure is dominated by a compact component and a double-sided bent jet, the mas-structure may be followed out to $\sim 250mas$ from the compact "core" with no indication of a counterjet (dynamic range 200:1). At 0.25mas resolution ($\lambda 1.3cm$) the structure consists of an unresolved component and a jet with wiggles with an amplitude of $\sim 0.2mas$

---

[1] $z = 0.30$; $1mas = 2.75pc$: $H_0 = 100km/s/Mpc$, $q_0 = 0.5$

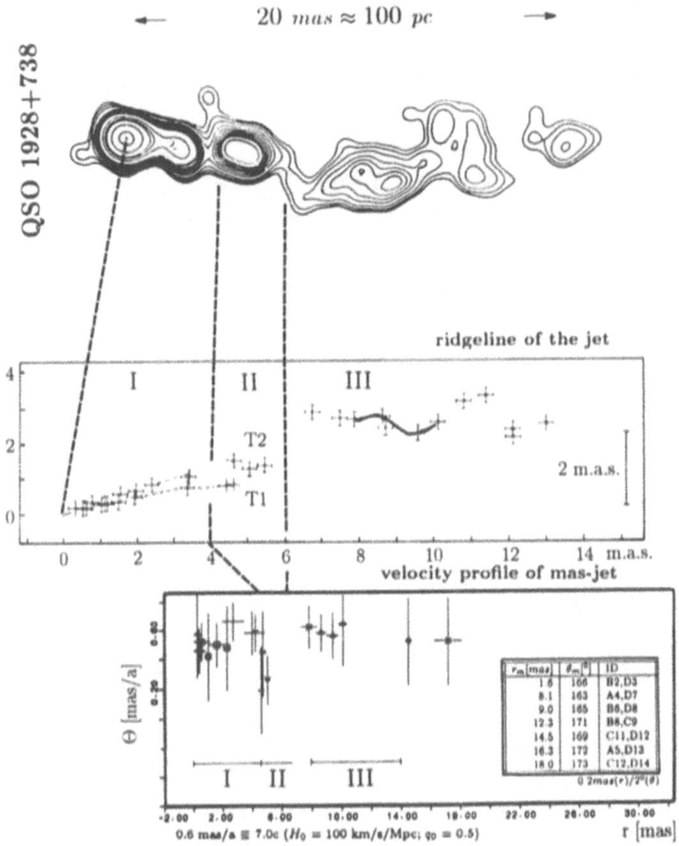

Fig. 1. Velocity profile (left), ridgeline (middle), and λ6*cm*-image (right) of the milliarcsecond structure of S5-quasar 1928+738.

(e.g. Hummel *et al.*, 1991). At 1mas resolution (λ6*cm*) the source displays a curved structure with at least 14 distinguishable features within the inner 18mas (fig.1). The amplitude of the wiggles rises to ~ 2*mas*, with two kinks at ~ 5 and ~ 10*mas* "core"-separation. The following velocity profile along the mas-jet is determined (fig.1): between ~ 6...18 *mas* (zone III) a velocity of 6.5 ± 0.5 *c* (0.56 ± 0.04 *mas/a*) is derived on the basis of five components identified in four λ6*cm*-maps. This is confirmed by the two epochs at λ18 *cm*. Thus superluminal motion is detected at 50pc projected core-distance and comparable to the value of 46pc derived for the 3C 273-jet by Zensus et al. (1990). From 6 epochs at λ1.3 *cm* Hummel *et al.*(1991) determine a mean velocity of 4.0 ± 0.6 *c* (0.34 ± 0.05 *mas/a*) for the inner part of the jet (0.25...4.5 *mas*: zone I). A "transition" zone (II), with lower velocities (0.19, 0.26 mas/a) is located at the positions of the two major kinks in the λ6*cm*-images (~ 4.5...6 *mas*). This is confirmed by a further epoch at λ3.6 *cm* (1988.89) (0.21mas/a).

Comparable velocities of components along almost identical paths can be explained applying the relativistic beaming concept by the assumption of a *uniform* Lorentz factor of the jet (here: $\gamma_{min} = 6.6$, $\theta_{LOS} = 9^0$). The occurrence of both (lower) non-uniform velocities and curvature in the "transition zone" can be attributed to directional changes of the relativistic jet towards larger viewing angles $\theta_{LOS} = 18^0 \ldots 41^0$, adopting moderate Doppler factors ($\delta \leq 2.5$) as derived from SSC-calculations and flux density variability. A unique velocity ($\beta \geq 0.988$) throughout the jet can also solve the apparent contradiction between the one-sided pc-jet and the two-sided kpc-structure, assuming $\theta_{LOS}^{kpc} \sim 64^0$ (taking the jet-counterjet-ratio R=16 for the "inner" kpc-jet into account). The alternative (extreme) solution is *constant* $\theta_{LOS} = \theta_{LOS}^{VLBI} \leq 17^0$, implying β is subrelativistic on kpc-scales ($\beta^{VLA} \sim 0.45$). A growing number of sources is now found to display increased curvature of the jet axis towards the "core" (see Witzel; Krichbaum, this vol.). One model to be tested is based on helically magnetic field structures originating in a rotating magnetosphere anchored in an accretion disk (e.g. Camenzind, 1991). The non-ballistic motions along "stationary", curved trajectories (zone III) can be qualitatively described by 2nd order reflection mode shocks (e.g. Hardee, 1990, and refs.). The "decoupling" of the observed ($\beta_{obs} = 6.5 \pm 0.5$) from the intrinsic velocity β (and δ) allows for larger θ_{LOS}: with $\delta_{SSC} = 1.7$ and $\beta = 0.8$ a *phase velocity* $\beta_{ph} = 1.11$ and $\theta_{LOS} = 36^0$. Phase velocities, however, are not required to explain the high velocity within the standard relativistic beaming model, and represent a contradiction to the non-detection of a counterjet on the mas-scale in case of larger θ_{LOS} and smaller β. The properties of 1928+738 make it an interesting candidate to test *two fluid* models (Pelletier, Roland & Sol, 1991).

References

Camenzind, M., 1991: in *Lecture Notes in Physics* **377**, W.J. Duschl, S.J. Wagner, M. Camenzind (Eds.), (Heidelberg: Springer Verlag), p. 201.

Eckart, *et al.*, 1985: *Astrophys. J. (Letters)*, **296**, L 23.

Hummel, *et al.*, 1991: *Astron. Astrophys.*, in press.

Hardee, P.E., 1990: in: *Parsec-scale radio jets*, ed. J. A. Zensus and T. J. Pearson (Cambridge University Press), p. 266.

Johnston, *et al.*, 1987: *Astrophys. J. (Letters)*, **313**, L 85.

Pelletier, R. & Sol, H., 1991: in *Lecture Notes in Physics* **377**, op. cit., p. 225.

Schalinski, C.J., 1990: PHD-thesis, university of Bonn.

Schalinski, C.J., *et al.*, 1991, in: *Variability of Blazars*, ed. E. Valtaoja and M. Valtonen (Cambridge University Press), in press.

Witzel, A., *et al.*, 1988: *Astron. Astrophys.*, **206**, 245.

Zensus, J.A., *et al.*, 1990: *Astron. J.*, **100**, 1777.

A Very Simple Model to Explain the "Simultaneous" Variations at Optical and Radio Wavelengths as Observed in Some Blazars

A. Sillanpää, K. Nilsson, and L.O. Takalo

Tuorla Observatory, SF-21500 Piikkiö, Finland

We present a very simple model to explain the mystery of the nearly simultaneous variations at different wavelenghts as observed in some blazars by the Heidelberg-group (Wagner, 1991). Our model is based on a very sharply edged jet with an opening angle about 1-2 degrees and with similar angle between the jet and line of sight. In order to get nearly simultaneous variations from the optical and radio region we have to make some assumptions:

-the jet has to be highly relativistic and to have nearly constant velocity between optical and radio regions ($v \geq 0.99c$).
-the jet has sharp edges, so that we can ignore optical depth effects.
-the angle between the jet and line of sight is very small ($\leq 3°$).
-the distance between the optical and radio region in the jet is only something like 0.1 pc.

If these assumptions are true, we have a situation where the emitting blob in the jet is moving only with a slightly smaller velocity than the information from the optical region, coming almost directly towards us. In this case we will observe the variations at optical and radio regions nearly simultaneously because the observed time delay is shortened from the true delay with the relation:

$$t_{obs} = \delta^{-1} t_{em}$$

where δ is the well known Doppler factor. In Figure 1 we show the behaviour of the Doppler factor with the different line of sight angle (Θ) and with different jet-speeds ($v = \beta c$). In Figure 2 we see the behaviour of the observed time delays between optical and radio regions with different Doppler factors and different distances between the emitting regions.

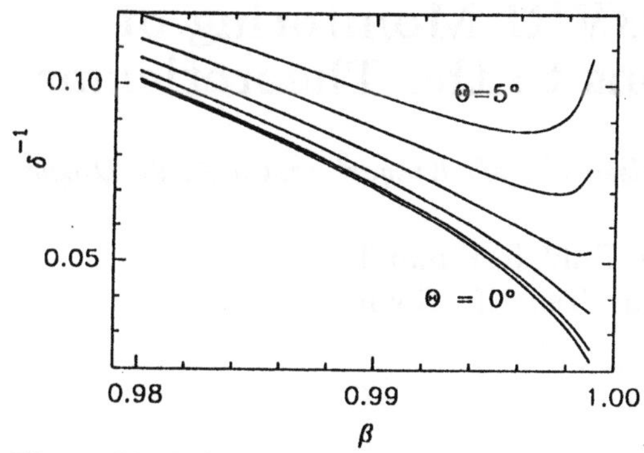

Fig. 1. The behaviour of the Doppler factor with the different Θ-values and with the different jet speeds ($v = \beta c$).

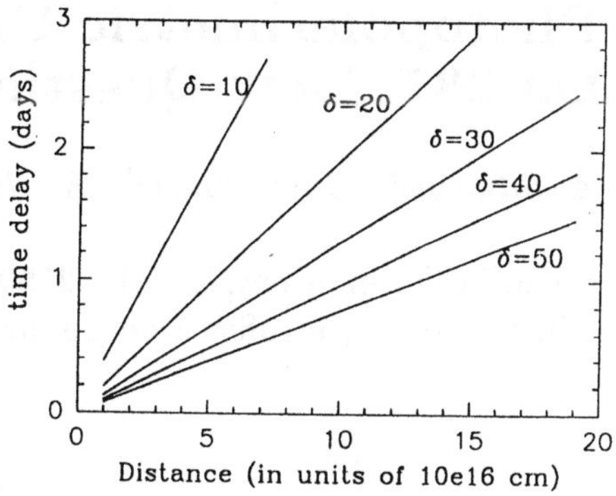

Fig. 2. The behaviour of the observed time delays between optical and radio regions with different Doppler factors and different distances between emitting regions.

References

Wagner, S.: 1991, In " Variability of Blazars", ed. E.Valtaoja and M. Valtonen, Cambridge Univ. Press (in press)

Photopolarimetric UBVRI Monitoring of OJ 287: New Questions to the Theoreticians

A. Sillanpää[1], *L.O. Takalo*[1], *K. Nilsson* [1], *M. Kidger*[2] *and J.A. de Diego*[2]

[1] Tuorla Observatory, Tuorla, 21500 Piikkiö, Finland
[2] Instituto de Astrofisica de Canarias, Tenerife, Spain

We present preliminary results of simultaneous UBVRI photopolarimetric monitoring of OJ 287 in January 1991. The observations were made at the Nordic Optical Telescope (NOT), using the Turku photopolarimeter. The observing procedure has been described in Sillanpää et al. (1991). In Figure 1 we show the observed R light curve for OJ 287. Notice the clear anticorrelation between flux and position angle; when flux increases the position angle decreases. On the other hand polarization level show no correlation with either the flux or position angle. On the second night we observed a brightening of 40% in all colour bands and a 60 degree swing in the position angle, all in five hours! This is the fastest ever observed position angle swing in OJ 287. Note also that the polarization level was low (5%) during the swing. On the next night OJ 287 was back at its "normal" state. After which it started to brighten slowly, with the position angle decreasing again at the same rate with the flux. The polarization was first increasing and then decreasing. We have considered two possible causes for the position angle swing:

1. A shock in the jet: Königl and Choudhuri (1985) proposed a model for the position angle rotation in blazars, in which the rotation is caused by shock waves illuminating successive transverse cross sections of the jet.

2. A polarized flare in the accretion disk: The position angle swing is then due to the flare rotation around the black hole. During the position angle swing also the spectrum of OJ 287 changed.

We believe that the accretion disk flare model can more easily explain the observations than the shock in the jet model. Mainly because in it the spectral changes arrive naturally from the temperature gradient in the accretion disk. But as these observations have demonstrated, there is still a lot in blazars, what we do not understand. For better understanding of these fascinating objects more accurate observations, simultaneously in several wavelenghts are needed. Only this will provide theoreticians with enough good quality data for accurate model calculations.

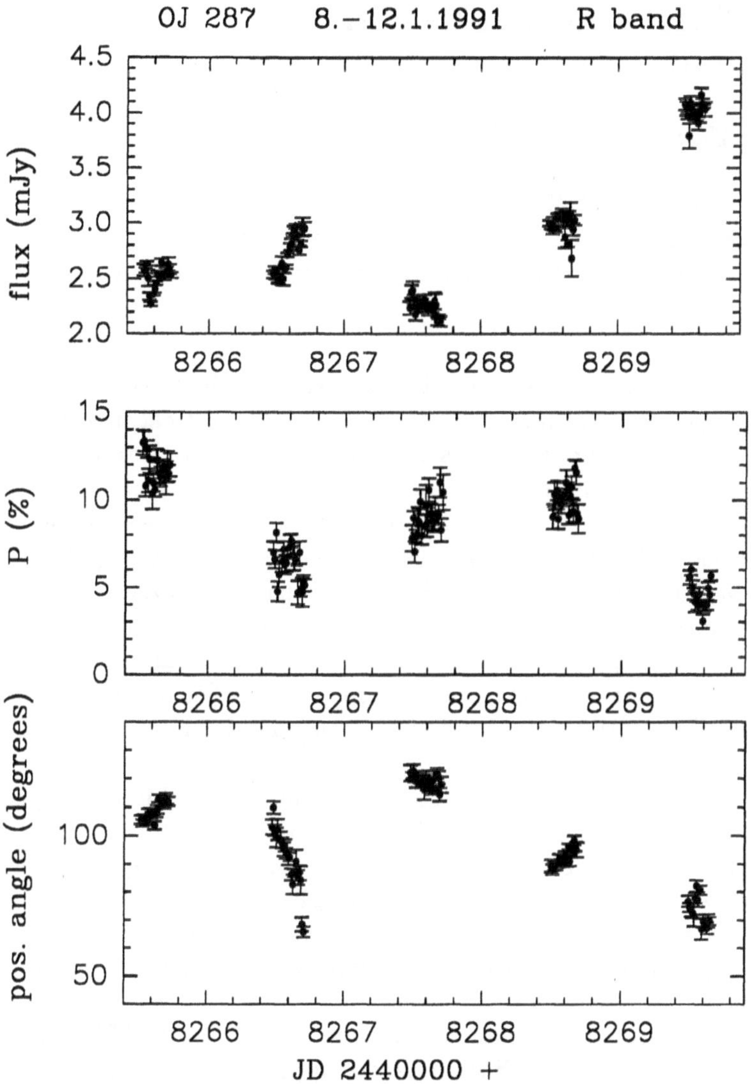

Fig. 1. The observed R band light curve for OJ 287 in January 1991. We show four point averages, giving a time resolution of 15 minutes.

References

Königl, A., and Choudhuri, A.: 1985, *Astrophys. J.*, **289**, 188

Sillanpää, A. et al. 1991, *Astron. J.*, **101**, 2017.

Radiative Unstable Modes in Astrophysical Jets

E. Trussoni[1], *S. Massaglia*[2], *G. Bodo*[1], *P. Rossi*[3], *and A. Ferrari*[1]

[1] Osservatorio Astronomico di Torino, Pino Torinese, Italy
[2] Istituto di Fisica Generale dell'Universitá, Torino, Italy
[3] Harvard-Smithsonian Center for Astrophysics, Cambridge, U.S.A.

In the last fifteen years the possible onset of the Kelvin–Helmholtz (K–H) instability in extragalactic jets has motivated several theoretical investigations, due to its relation to the dynamics and to the morphological features of jets. Even though several ingredients have been taken into account (e.g. magnetic fields, velocity shears, rotation, etc.), previous analyses followed an adiabatic approach. However, if relativistic electrons share an important fraction of the total internal energy of the jet, synchrotron losses cannot be neglected in the stability analysis, as can be seen by comparing the synchrotron time scale (t_{syn}) with the typical K–H time scale. We have performed the stability analysis for a confined magnetized jet including the synchrotron losses. The MHD equations have been linearized, with the following assumptions:
1) Synchrotron losses are the fastest cooling mechanism, and the heating is independent of the physical parameters of the plasma.
2) Relativistic particles provide the gas internal energy, and thermal protons the inertia: these two components are strictly coupled.
3) The spectrum of electrons is a maxwellian or a very steep power law.
4) The perturbed quantities have a form $f(r)\exp[\omega t - i(kz + m\theta)]$, with k real and ω complex.

The dispersion relation is similar to the one for adiabatic MHD gas (Ferrari *et al.*, 1981), with the radiation terms appearing only in the radial wavenumber. Here results are presented for a pinching perturbation ($m = 0$) with wavelength $ka = 0.5$ (a is the jet's radius), for two values of the nondimensional time $\tau = t_{syn}V_s/a$ (V_s is the sound speed) and of the ratio $\sigma = V_A/V_s$ (V_A is the Alfvén velocity). In Fig. 1 is plotted the nondimensional growth rate Re(ϕ) [$\equiv Re(\omega)a/V_s$] vs the Mach number M for $\sigma = 0.8$ and with no cooling ($\tau = \infty$) and strong losses ($\tau = 0.5$). With respect to the adiabatic case, the K–H process is modified by cooling as follows: i) the low Mach number cut–off is removed, ii) for low Mach numbers also the ordinary mode with positive phase velocity is destabilized, and iii) the slow MHD modes have much larger growth rates. In Fig. 2 the same plots are reported for $\sigma = 0.2$: in this case the cooling effects are much stronger because the onset of the thermal stationary modes (present also in a homogeneous plasma, see Bodo *et al.*, 1990) with high growth rates, independent of the M. Helical modes ($m = 1$) and perturbations with smaller wavelengths ($ka > 1$), have similar behaviour.

These results show that radiative losses are important only for weakly magnetized jets ($\sigma < 1$), while close to equipartition conditions ($\sigma \sim 1$) strong cooling effects are found for small Mach number. Therefore for jets associated with the FR II radio sources (e.d. Cyg A), which are likely to be highly supersonic, previous results obtained on the adiabatic K–H instability still hold. Conversely larger effects should be found in jets associated to FR I radio sources. For beams far from equipartition (with $\sigma < 1$) conversely thermal condensation modes can develop on short time scales. On the basis of non-linear analysis by Bodo *et al.* for a homogeneous plasma (these proceedings) we can argue that these modes could evolve (on a synchrotron time scale) into filamentary structures of high density and magnetic field. These filamentary structures can be associated with the peculiar morphologies observed very recently in few jets (e.g. M87) at radio (VLA) and optical (HST) frequencies (see Macchetto, these proceedings).

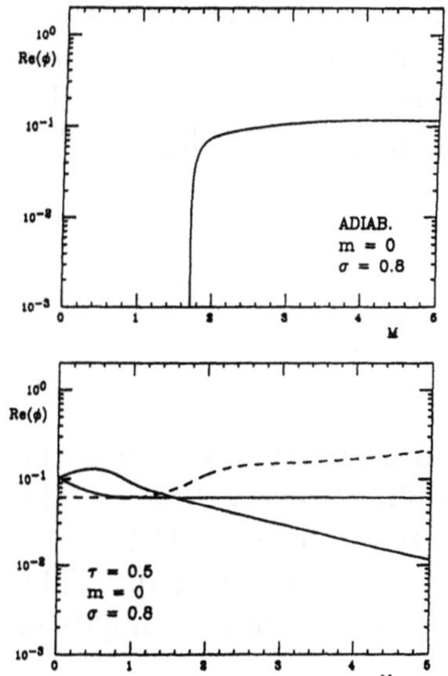

Fig. 1. Undimensional growth rate Re(ϕ) vs M for $\sigma = 0.8$. Full and dotted lines refer to the ordinary and slow MHD perturbations.

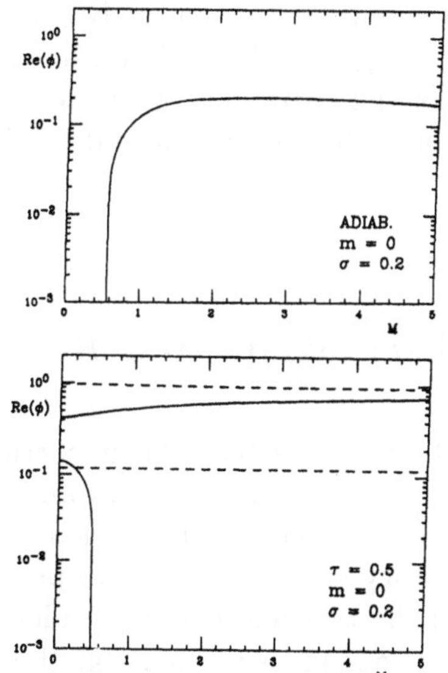

Fig. 2. The same as Fig. 1, with $\sigma = 0.2$.

References

Bodo, G., Ferrari, A., Massaglia, S. & Trussoni, E.:1990, *Monthly Notices Roy. Astron. Soc.*, **244**, 530.

Ferrari, A., Trussoni, E. & Zaninetti, L.:1981, *Monthly Notices Roy. Astron. Soc.*, **196**, 1054.

Radio- and Optical Variability in 0716+714

S.J. Wagner[1], *J. Heidt*[1], *A. Witzel*[2], *R. Wegner*[2], *A. Quirrenbach*[3]

[1] Landessternwarte Heidelberg - Königstuhl, Germany
[2] Max-Planck-Institut für Radioastronomie, Bonn, Germany
[3] U.S. Naval Observatory, Washington D.C., USA

1. Introduction

The object S5 0716+714 is a compact, flat-spectrum radio source exhibiting all properties of typical BL Lac objects. It is variable in all wavelength regimes, highly polarized and has a featureless continuum. The short time-scales of the variability both in the radio and the optical bands make this blazar an interesting candidate for investigating intrinsic intraday variability.

2. Observations and Results

The observations have been carried out in February 1990. We used the VLA to observe S5 0716+714 and several other objects every 2 hours. The optical observations were carried out at the 2.2m telescope of the Calar-Alto observatory and a 0.7 m telescope of the Landessternwarte Königstuhl. The data of the two telescopes are identical within the errors for all 8 nights of simultaneous measurements at the two stations.

Figure 1 shows the lightcurves in the optical bands (B, R and I) for the whole campaign, and a comparison of the 650 nm, 3.6 cm and 6 cm data in a 3-day period. All lightcurves are given in normalized intensities.

Both, the optical and radio measurements show quasiperiodic variations in the first week of the campaign. The time-scales of this variation are identical in the two regimes. At the same time (JD 2 447 930) the mode of the variability changed from rapid fluctuations to slower variations in both regimes. During both parts the statistical properties of the variability are very similar (e.g. Quirrenbach et al., 1991, Wagner, 1991). The lag between the optical and radio data is different during the two modes. Its sign cannot be determined unambiguously. The short lag, indicated by the nearly simultaneous change of modes, provides evidence that the volumes of the emitting and scattering particles producing the variable radio and optical emission are similar in size and not far apart from each other. The "saw-tooth" shaped lightcurve indicates an important role of geometric effects for the observed modulation.

There is no spectral variability within the optical bands at short time-scales but a slight reddening throughout the campaign. The spectral index in the radio-regime slightly varies on short time-scales. The variations at 3.6 cm lag behind the variations at 6 cm by 0.2 days. Further information on the spectral properties is given by Wegner et al., elsewhere in this volume.

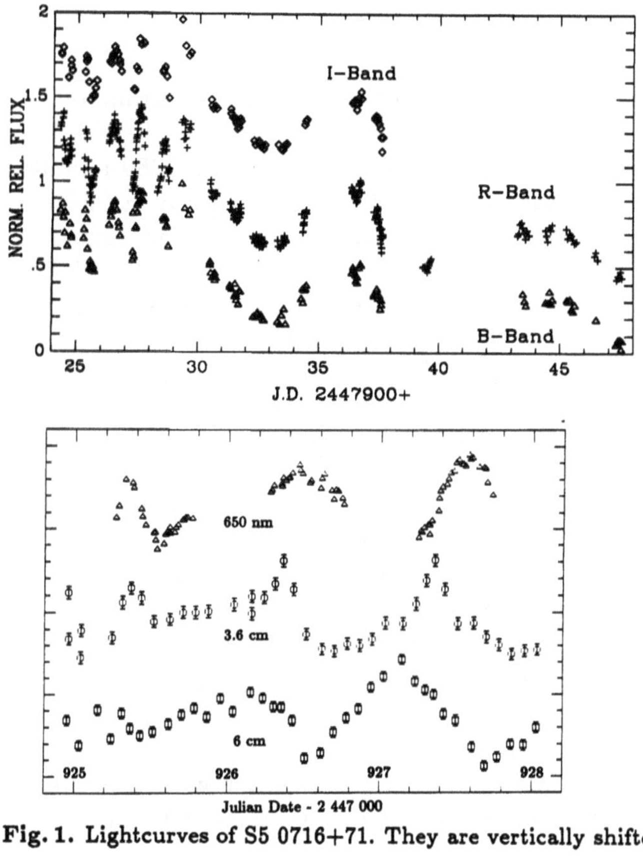

Fig. 1. Lightcurves of S5 0716+71. They are vertically shifted by an arbitrary amount.

References

1. Quirrenbach, A., *et al.*, 1991: *Astrophys. J. (Letters)*, **372**, L71.
2. Wagner, S.J., 1991: in: *Variability of Blazars*, E. Valtaoja (ed.)

Intraday Variability of Compact Radio Sources

R. Wegner[1], *A. Witzel*[1], *A. Quirrenbach*[2], *T.P. Krichbaum*[1], *and C.J. Schalinski*[1]

[1] Max-Planck-Institut für Radioastronomie, Bonn, Germany
[2] now at Universities Space Research Association, Washington,D.C., U.S.A.

We present new results of the flux density monitoring program of the intraday-variable radio sources 0716+714, 0917+624, and 0954+658. The measurements were carried out between May 3 and May 9 1989 simultaneously with the VLA in Socorro and the 100 m telescope in Effelsberg. A total of 5 wavelengths between 2 and 20 cm was covered. Between January 29 and February 23 1990 we observed these sources with the VLA at 4 wavelengths between 2 and 20 cm. These observations were accompanied by optical measurements (Wagner et al., this volume).

Fig. 1 presents the variations of the BL Lac object 0716+714 in May 1989: Fig. 1a shows the polarized flux density at 6 cm wavelength, figs. 1b to 1d show the total flux densities at 6, 11 and 20cm (bottom) wavelengths. The uncertainties of the observations have been determined from the calibration sources to be 1 % for the total flux densities and 3 % for the polarized flux densities. Two facts are clearly visible for fig. 1:

The amplitude of variability increases with frequency and the variations of the total and the polarized intensity at 6 cm wavelength are directly correlated, in contrast to the anticorrelation of these quantities in the quasar 0917+624 (Witzel, this volume), which displays a peak of variability amplitude at 11 cm wavelength (see fig. 2).

Fig. 3 shows the variations of the flux densities of the BL Lac objects 0716+714 (fig. 3a), and 0954+658 (fig. 3c) together with the quasar 0917+624 (fig. 3b) as observed in January/February 1990 at 6 cm and 20 cm wavelengths.

In 0716+714 the amplitude of variability seems to increase with frequency, a behaviour which as visible at other epochs, too (compare with fig. 1). A power spectrum analysis of all available observations for this source shows two characteristic time scales of about 1 and about 7 days. These timescales were detected in the optical emission in February 1990 as well (see Quirrenbach et al., 1991 and Wagner et al., this volume).

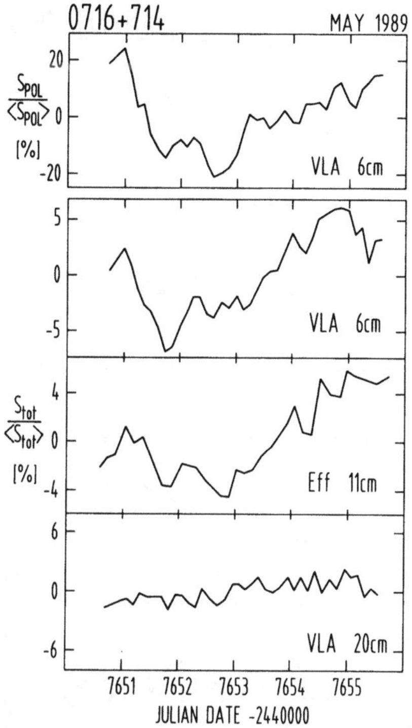

Fig. 1. Variations of the polarized (top area) and the total flux density of 0716 +714 in May 1989.

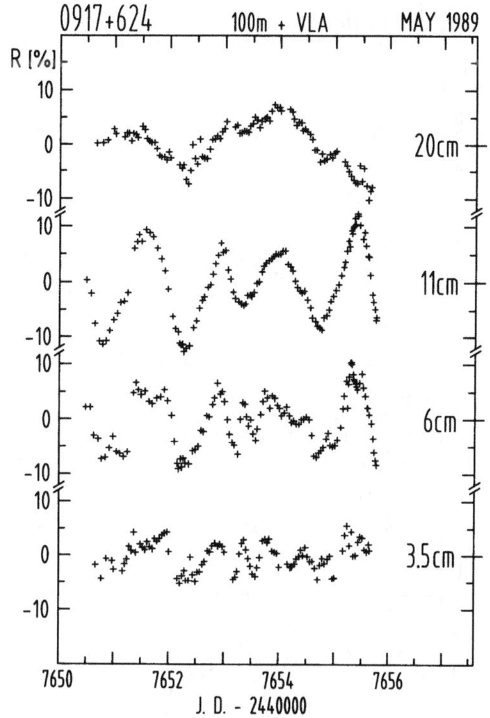

Fig. 2. Variations of the the total flux densities of 0917+624 in May 1989. The 11 cm data were taken with the 100 m telescope. The others were observed with the VLA (Quirrenbach, 1991)

References

Quirrenbach, A., *et al.*, 1991: *Astrophys. J. (Letters)*, **372**, L71

Quirrenbach, A., 1991: in: *Variability of Active Galaxies*, eds. W.J. Duschl and S.J. Wagner, Springer Heidelberg, 131 Wagner, S., *et al.*, 1990: *Astron. Astrophys. (Letters)*, **235**, L1.

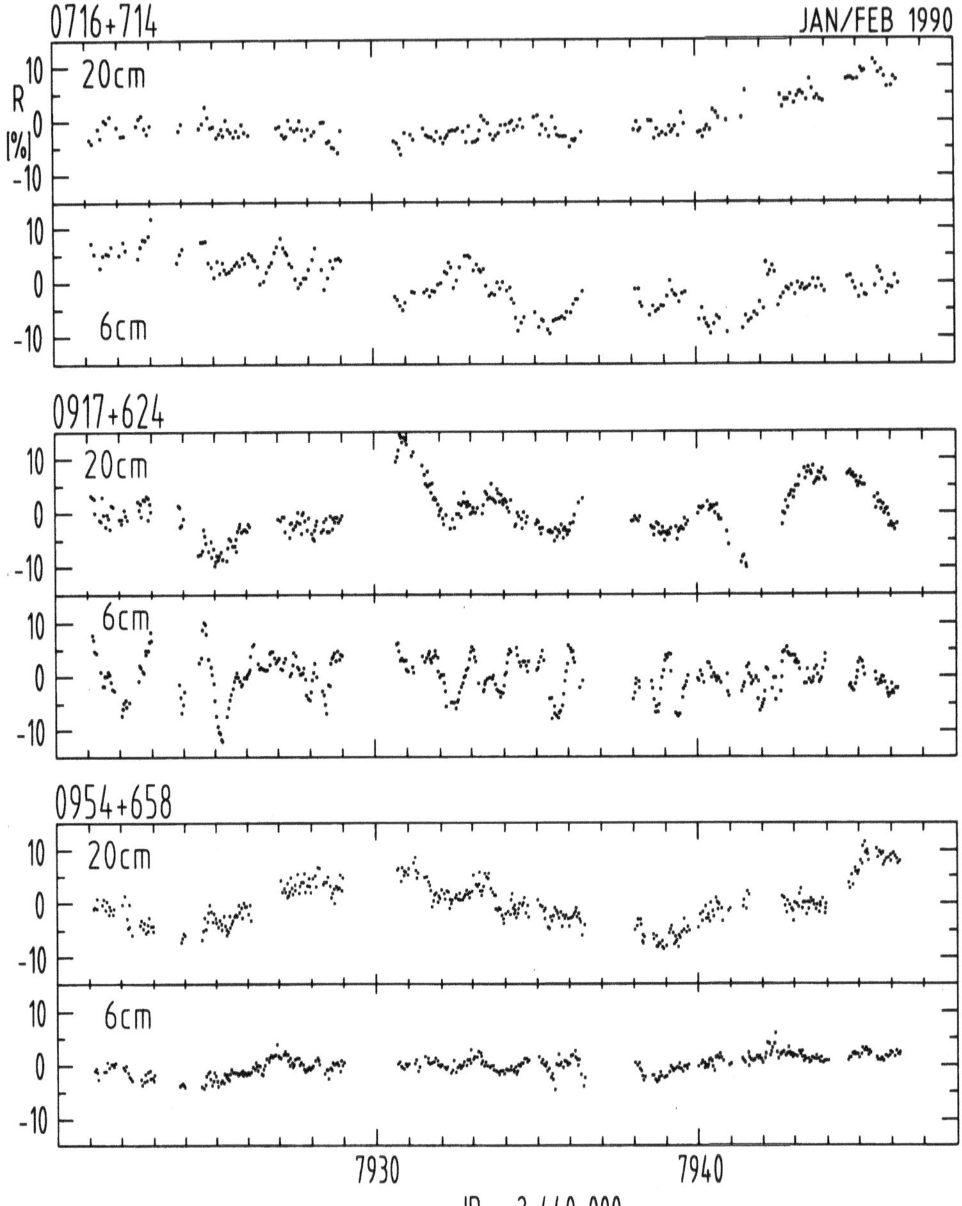

Fig. 3. Variations of and the total flux density of 0716+714, 0917+624 and 0954+658 at 20 and 6 cm wavelengths in Jan/Feb 1990.

5 – Unified Schemes and Evolution

‘Wide’ Jets and the X-ray and Radio Properties of BL Lac Objects

Laura Maraschi[1], *Annalisa Celotti*[2], *and Gabrielle Ghisellini*[3]

[1] Dipartimento di Fisica, Universitá di Milano, 16, via Celoria, 20133 Milano, Italy
[2] Scuola Internazionale Superiore di Studi Avanzati, Trieste, Italy
[3] Osservatorio Astronomico di Torino, Pino Torinese, Italy

1. Introduction

The properties of BL Lacs found from X–ray surveys (XBL hereinafter) are remarkably different from those of “classical” radio selected BL Lacs (RBL). In particular, the radio to X–ray flux ratio, as measured by the spectral index α_{RX}, is substantially smaller for XBL than for RBL, while the average X–ray luminosity of the two groups is comparable. Thus XBL can be thought of as radio weak, in comparison to radio strong RBL, with similar X–ray luminosity.

It follows that X–ray selection should be unbiased with respect to the radio properties, yet it produces only (or mostly) radio weak objects. On the basis of this argument it was suggested that, for a given X–ray luminosity, radio weak XBL represent the majority of the population, *i.e.* their space density is larger than that of radio strong objects (Maraschi *et al.* 1986). Statistics, at that time, was poor (only 13 XBL) but the above points have been confirmed by the results of the *EINSTEIN* EMSS (containing 34 BL Lacs, 28 with redshift) and by the 11 XBL discovered by EXOSAT.

A further point concerns the energy distributions: these are extremely smooth and “coherent” in the sense that objects with flat/steep α_{RX} also have flat/steep spectra in individual bands from the IR up to the UV (for recent reviews see Bregman 1990, Maraschi 1991).

A widely accepted model for the energy distribution of BL Lacs involves a relativistic jet, emitting at high frequencies near the core and at lower frequencies further out (Ghisellini, Maraschi & Treves, 1985). Based on that, we proposed that the relative numbers of XBL and RBL and the different energy distributions could be understood at the same time, if the X–ray emission was more isotropic than at lower frequencies. Observers at large and small angles would see similar X–ray fluxes but only observers at small angles would see a strong relativistically enhanced radio flux.

We discussed in detail a model in which the flow velocity of the relativistic plasma in the jet increases (accelerates) with radial distance, so that relativistic beaming is mild in the X–ray emitting region of the jet but strong in the radio emitting region (Ghisellini & Maraschi 1989). The spectral differences between XBL and RBL could indeed be explained by differences in the viewing angle θ, if a smooth acceleration of the plasma in the jet yields a bulk Lorentz factor $\gamma \sim 1$ in the inner (X–ray emitting) regions, $\gamma \sim 2-3$ in the intermediate (optical emitting) zones, and $\gamma \sim 4-5$ in the outer (radio emitting) parts. Roughly, if RBL are seen at $\theta < 1/\gamma$, their percentage in an X–ray selected sample should be the ratio of solid angles: $\cos(1/\gamma)/[1-\cos(1/\gamma)]$, which yields 1/50 for $\gamma = 5$, consistent with the absence of RBL in the EMSS sample.

Recently, complete radio and X–ray selected samples of BL Lacs have become available which allow to compute the respective luminosity functions (Stickel *et al.* 1991, Morris *et al.* 1991). Therefore it is now possible to use the absolute space densities to constrain beaming models. In a recent series of papers, Urry and Padovani tested the hypothesis that *all* BL Lacs (both XBL and RBL) are low luminosity radio galaxies (FR1 galaxies) beamed at us. They considered separately and independently the X–ray and radio luminosity functions of the parent population (FR1 galaxies) and derived the X–ray and radio luminosity functions of BL Lac objects (Padovani & Urry 1990, Urry, Padovani & Stickel 1991) following the procedure developed by Urry & Shafer (1984) and extended by Urry & Padovani (1991). They find that the scheme is successful if the X–ray emitting plasma has $\gamma \simeq 3-4$ and the radio emitting plasma has $<\gamma>= 7.4$ with a power law distribution extending up to $\gamma = 40$.

However the above models do not address the question of which physical acceleration mechanism could work on the appropriate scale, a factor $10^2 - 10^3$ at least in distance from the core. Furthermore it may be difficult to account for large and fast luminosity variations observed in the X–ray band (Treves *et al.* 1989) without invoking relativistic beaming. For these reasons we discuss here an alternative to the above models, in which the Lorentz factor γ for the bulk flow is kept constant along the jet, but the solid angle subtended by the streamlines is larger than $1/\gamma$ near the core and decreases with distance. Therefore we will consider a jet with increasing collimation rather than with increasing velocity.

In Section 2 we discuss the main features of the model and the comparison with the observed radio and X–ray luminosities of the objects. We discuss in Section 3 joint constraints from the radio and X–ray luminosity functions assuming that γ is the same in both bands but the opening angle of the jet is wider for the X–ray emitting plasma. In Section 4 we suggest that the three classes of FR1 sources, XBL and RBL can be unified by our model, but we remark the inherent remaining uncertainties.

A more detailed description of the model is in preparation.

2. Wide relativistic jets

Consider a relativistic flow with fixed bulk Lorentz factor γ but with velocity directions uniformly spread in a cone of semiaperture Θ_j, larger than the critical semiaperture angle for relativistic beaming $\Theta_c \simeq 1/\gamma$. Assume also that Θ_j decreases along the jet, possibly reaching Θ_c in the outer parts. At an angle θ with the jet axis the observed luminosity $L(\theta)$ is

$$L(\theta) \equiv \ell R(\theta) = \frac{\ell}{\Delta\Omega_j} \int_0^{\Theta_j} \sin\psi d\psi \int_0^{2\pi} \delta^p(\theta,\psi,\phi) d\phi \qquad (1)$$

where ℓ is the intrinsic jet luminosity, $R(\theta)$ is the enhancement factor due to beaming, and the beaming factor $\delta(\theta,\psi,\phi)$ is a function of the polar coordinates ψ and ϕ within the cone of half angle Θ_j. The exponent p is assumed hereafter to be $p = 3+\alpha$, where α is the spectral index $[F(\nu) \propto \nu^{-\alpha}]$. Such a value is appropriate for amplification of monochromatic luminosity from a moving blob of plasma (other more complicated cases have been discussed by Lind & Blandford 1985).

Eq. (1) has to be calculated numerically, and Fig. 1 shows the enhancement factor $R(\theta)$ for $\gamma = 10$ and for some values of Θ_j. The general behaviour of $R(\theta)$ can be understood considering that, for $\theta < \Theta_j$, the radiation reaching the observer is mainly produced in a cone of angle Θ_c around the line of sight. The number N of such small cones within the Θ_j cone is simply the ratio of the subtended solid angles: $N = (1-\cos\Theta_j)/(1-\cos\Theta_c)$. As long as $\theta < \Theta_j$, there always is a little cone pointing at the observer, and therefore the luminosity is approximately constant and given by

$$L(\theta) \simeq \frac{\ell}{N}\delta^p(0) \sim \frac{2\pi}{\Delta\Omega_j}\ell(2\gamma)^p[1-\cos(1/\gamma)], \qquad \theta < \Theta_j \qquad (2)$$

where ℓ/N is fraction of the intrinsic luminosity emitted by particles with velocity direction within Θ_c, and $\delta^p(0)$ is the relativistic correction of the flux observed at zero angle with respect to the small cone. The solid angle subtended by the jet is $\Delta\Omega_j = 2\pi(1-\cos\Theta_j)$.

Comparing with the value obtained in the case of parallel velocities, for which $L(0) = \ell\delta^p(0) \sim (2\gamma)^p\ell$, we see that the maximum observed luminosity from a wide angle jet is smaller by a factor $(1/N) \sim [1/\gamma^2(1-\cos\Theta_j)]$, but it is approximately constant for all directions within Θ_j.

Similarly to the Urry and Shafer's formalism all the above can be generalized in the assumption that only a fraction f of the unbeamed luminosity is enhanced by beaming. In this case $L = \ell + f\ell R(\theta) = \ell[1+fR(\theta)]$. Assuming also that f is constant for all sources, we can define $\bar{R}(\theta) = 1 + fR(\theta)$.

Let us now compare the X-ray and radio luminosities of BL Lacs with those of FR1 radio galaxies. In Fig. 2 we report the X-ray vs 5 GHz radio luminosity for the FR1 galaxies in the 3C sample (Fabbiano *et al.* 1984) and

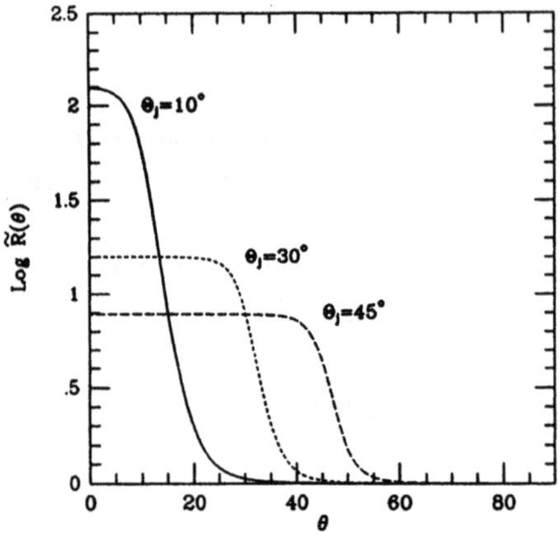

Fig. 1. The enhancement factor $\bar{R}(\theta)$ calculated for $\gamma = 10$, $p = 3$, $f = 0.1$ and different values of the semiaperture angle of the jet $\Theta_j = 10°$, $30°$ and $45°$.

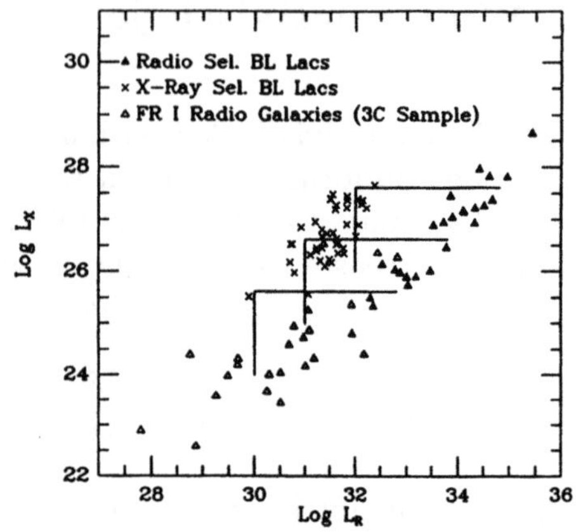

Fig. 2. 2 keV X-ray luminosity vs. 5 GHz radio luminosity for FR1 sources (circles), XBL (crosses) and RBL (triangles). The 'L' curves are calculated using $\Theta_X = 34°$, $\Theta_R = 1°$, $\gamma = 12$, $p_R = 3$, $p_X = 4$, $f_x = 0.018$ and $f_r = 0.05$.

for a sample of BL Lacs derived from the lists of Ghisellini *et al.* (1986), Maraschi *et al.* (1986), Morris *et al.* (1991), Giommi *et al.* (1991).

In Fig. 2 we also show the curves predicted by our scheme corresponding to a source whose X-ray emission comes from a jet of semiaperture $\Theta_X = 45°$ and radio emission coming from a jet with $\Theta_R = 10°$, but with the same bulk Lorentz factor $\gamma = 10$. The 'L' shape of these curves is due to the different dependence of the radio and the X-ray luminosities on the viewing angle. For large θ, but approaching Θ_X, L_x changes rapidly, while L_r remains almost constant (the unbeamed radio luminosity dominates the received flux). On the contrary, when θ approaches Θ_R, L_x remains almost constant (line of sight within the large X-ray jet), but L_r changes rapidly. According to the viewing angle, the 'L' curves connect FR1 sources with XBL and RBL, and are suggestive of the unifying picture discussed in Section 4.

3. Luminosity functions

Using the relation between the intrinsic and the observed luminosity given by Eq. (1), we can compute the luminosity function (LF) of the beamed population of objects $\Phi(L)$, from the intrinsic LF of the parent population $\Phi_u(\ell)$ (Urry & Shafer 1984)

$$\Phi(L) = -\int_{\Theta_1}^{\Theta_2} \Phi_u\left(\frac{L}{\bar{R}(\theta)}\right)\frac{\sin\theta}{\bar{R}(\theta)}d\theta \qquad (3)$$

where Θ_1, Θ_2 are the limiting angles at which objects of intrinsic luminosity $\ell_{min} < \ell < \ell_{max}$ have an observed luminosity L. In our scheme the parent population and the beamed one are distinguished by the value of $\bar{R}$, and we define *beamed* an object with $\bar{R} \geq 2$, *i.e.* an object with $L \geq 2\ell$.

Due to the complexity of the function $\bar{R}(\theta)$ an exact analytical solution of Eq. (3) is not possible. However, we can find an approximate value of the luminosity function at the particular observed luminosity L_3 corresponding to objects of *minimum* intrinsic luminosity ℓ_{min} maximally beamed. When $\theta < \Theta_j$, the observed luminosity is almost constant and equal to $L_3 = \ell_{min}\bar{R}(0)$. If the LF of the parent sources $\Phi_u(\ell)$ is a power law of slope B, $[\Phi_u(\ell) = \Phi_0 \ell^{-B}]$, then at L_3 Eq. (3) can be approximated by setting $\bar{R}(\theta) = \bar{R}(0)$, yielding

$$\Phi(L_3) \sim \Phi_0 L_3^{-B} [R(0)]^{B-1} \int_0^{\Theta_j} \sin\theta d\theta = \Phi_0 \frac{\ell_{min}^{-B}}{R(0)} (1 - \cos\Theta_j) \quad (4)$$

The ratio of the parent and the beamed LF at L_3 is $\sim (1 - \cos\Theta_j)\bar{R}(0)^{B-1}$, and can be used to put constraints on the choice of the parameters, once we identify a particular class of sources as the parent population of BL Lacs.

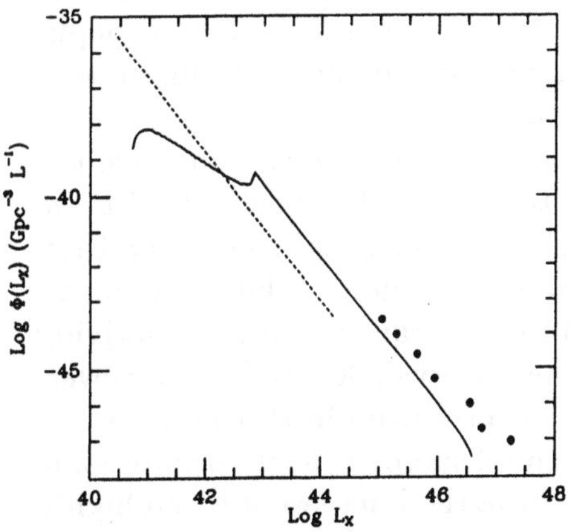

Fig. 3a. The differential X-ray luminosity function of BL Lacs calculated assuming FR1 as parent population (shown by the dashed line), for $\gamma = 11.2$, $\Theta_X = 16°.2$, $f_x = 0.03$. Also shown (dots) is the LF calculated by Morris *et al.* (1991), assuming luminosity evolution with $L = L(z = 0)(1 + z)^{-7}$.

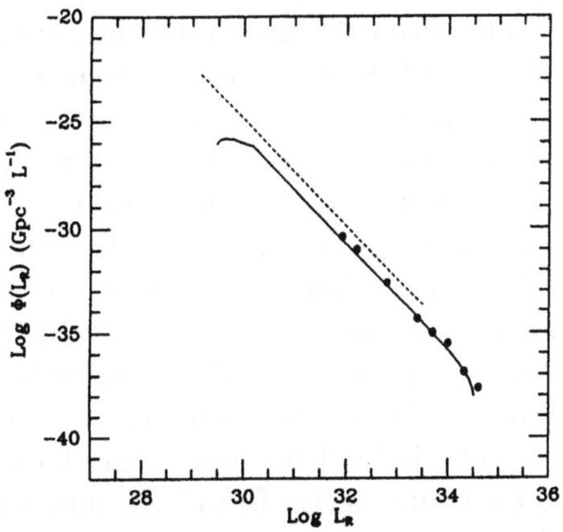

Fig. 3b. The differential radio luminosity function of BL Lacs calculated assuming FR1 as parent population (shown by the dashed line), for $\gamma = 11.2$, $\Theta_R = 5°$, $f_r = 0.023$. Also shown (dots) is the LF calculated by Stickel *et al.* (1991).

4. Unification scheme

In this section we apply the above formalism to the unification model between Radio selected, X–ray selected BL Lac objects and FR1 galaxies.

Suppose that the X–ray jet has a greater opening angle ($= \Theta_X$) than that of the radio jet ($= \Theta_R$), but that the bulk Lorentz factor is the same. For $\Theta_R < \theta < \Theta_X$, the X–ray luminosity is enhanced and approximately constant, but the radio luminosity is dimmed. These sources could be XBL. For $\theta < \Theta_R$, both the X–ray and the radio luminosity are enhanced. These sources could be RBL. Finally, for $\Theta_X < \theta < 90^0$, both the X–ray and the radio luminosity are dimmed. These could be FR1 sources. The ratio of number density of XBL and FR1 sources is simply $(1 - \cos\Theta_X)/\cos\Theta_X$, which is 0.015, 0.155 and 0.414 for $\Theta_X = 10^0$, 30^0 and 45^0, respectively.

Figs. 3a,b show the calculated LF in radio and in X–rays, and the comparison with the LF functions derived by Stickel *et al.* (1991) and by Morris *et al.* (1991). The agreement is good for $\gamma = 11.2$, $\Theta_X = 16°.2$, $\Theta_R = 5°$, $f_r = 0.023$ and $f_x = 0.03$. In our model, all the above parameters are fixed by assuming a relative space density between XBL and RBL, the ratio between the maximum luminosities of BL Lacs and the parent populations, and the ratio of the luminosity function at the particular luminosity $L_3 = \ell_{min}\bar{R}(0)$, where the beamed LF breaks.

Since we assume FR1 sources as the parent populations, and since the opening angle of the radio jet is $\sim 1/\gamma$, to fit the observed radio LF we necessarily require approximately the same radio parameters as in the Urry & Padovani scheme, even if we have assumed a unique value of γ for all sources. We then use the found value of γ for the X–ray emitting jet, deriving a suitable opening angle in order to fit the observed X–ray LF. However, we would like to mention some important uncertainties in this procedure:

i) The LF of FR1 sources are based on small samples. Furthermore some BL Lacs may derive from FR2 sources (*i.e.* have the lines swept by an highly enhanced continuum).

ii) The hypothesis that f is constant for all sources may be wrong, and some of the beamed object, instead to have a large γ, may be characterized by a large f.

iii) Only a fraction of FR1 sources could now have an active nucleus, and give rise to the BL Lac phenomenon.

In any case, if our model has some significance, there must be a connection between XBL and RBL, independently on the fraction of FR1 sources presently active. Their relative space density can be derived very simply in our model, by the ratio of the respective jet solid angles: $(1 - \cos\Theta_X)/(1 - \cos(1/\gamma)$, which yields ~ 10 for $\Theta_X = 16°.2$ and $\gamma = 11.2$. It is therefore very important to have large and unbiased complete sample of BL Lacs observed in X–rays, to find the relative ratio between XBL and RBL.

References

Bregman, J.N., 1990: *Astron. Astrophys. Rev.*, **2**, 125.
Fabbiano, G., Miller, L., Trinchieri, G., Longair, M. & Elvis, M., 1984: *Astrophys. J.*, **277**, 115.
Ghisellini, G., Maraschi, L., & Treves, A., 1985: *Astron. Astrophys.*, **146**, 204.
Ghisellini, G. & Maraschi, L., 1989: *Astrophys. J.*, **340**, 181.
Giommi, P., *et al.*, 1991: preprint
Lind. K.R., & Blandford, R.D., 1985: *Astrophys. J.*, **295**, 358.
Maraschi, L., Ghisellini, G., Tanzi, E. & Treves, A., 1986: *Astrophys. J.*, **310**, 325.
Morris, S.L., *et al.*, 1991: *Astrophys. J.*, in press.
Padovani, P. & Urry, C.M., 1990: *Astrophys. J.*, **356**, 75.
Padovani, P. & Urry, C.M., 1991: *Astrophys. J.*, **368**, 373.
Stickel, M., Padovani, P., Urry, C.M., Fried, J.W. & Kuhr, H., 1991: *Astrophys. J.*, **374**, 431.
Treves, A., *et al.*, 1989: *Astrophys. J.*, **341**, 733.
Urry, C.M. & Shafer, R.A., 1984: *Astrophys. J.*, **280**, 569.
Urry, C.M., Padovani, P. & Stickel, M., 1991: *Astrophys. J.*, in press.
Urry, C.M. & Padovani, P., 1991: *Astrophys. J.*, **371**, 60.

Spectropolarimetry of Radio Galaxies as a Constraint on Orientation

Neal Jackson

NRAL, Jodrell Bank, Macclesfield, Cheshire SK11 9DL, England.[1]

1. Introduction: spectropolarimetry of AGN

Some AGN have polarized continua. In the highly polarized objects known as blazars, the polarization is thought to be due to synchrotron processes. In other objects the polarization is generally lower and is probably due to scattering. The third possible polarizing mechanism, dichroic absorption, can occur both in the host galaxy and in our own galaxy, especially at low galactic latitude.

Antonucci & Miller (1985) discovered that the Seyfert 2 (narrow lined) galaxy NGC 1068 contained polarized broad lines. They interpreted this in terms of a model in which the broad line emitting region (BLR) cannot be seen directly because of an obscuring medium, but is scattered into our line of sight by scatterers outside the plane of the medium. These scatterers can be dust or electrons, and different mechanisms probably operate in different objects, or even in different parts of the same object (see Kinney et al. 1991 for a recent review in the case of NGC 1068).

In order to test models such as the Antonucci and Miller picture it is important to find out how common broad polarized lines are. Such a study was undertaken by Miller & Goodrich (1989) in eight Seyfert 2 galaxies, and in three cases broad polarized lines were found. In the five cases where no broad polarized lines were found, Miller & Goodrich speculated that the scatterers were not present or were themselves obscured by the same medium that hides the BLR.

Antonucci (1984) found one radio loud object – 3C 234 – with polarized broad lines and a highly polarized continuum. This object is a broad line radio galaxy, a class of object which may contain substantial amounts of dust (Rudy 1984). A further detection of a polarized broad line has been made in the narrow line radio galaxy IC 5063 (Inglis et al. 1991, these proceedings).

---

[1] **Current address: Sterrewacht, Postbus 9513, 2300 RA Leiden, The Netherlands**

Support for the "hidden nucleus" model of Antonucci & Miller has been steadily accumulating over the last few years. The discovery of a high off–nuclear continuum polarization in 2152-69 (Tadhunter et al. 1987), 3C 277.2 and 3C 368 (di Serego Alighieri et al. 1989, Tadhunter et al. 1990) suggests that radiation from a hidden bright continuum object is being scattered.

Could we therefore envisage an Antonucci & Miller model for radio loud objects? This was proposed by Barthel (1989: see also Scheuer 1987, Peacock 1987). The model is that narrow line radio galaxies are "really" quasars: they have a continuum and broad lines, but these features are hidden by an obscuring torus. Our only chance of seeing them is via a scattering screen lying outside the torus.

In order to test the hypothesis that radio galaxies and quasars can be unified by orientation effects in the same way that Antonucci & Miller proposed to unify Seyfert galaxies, we might try to look for polarized broad lines in narrow line radio galaxies exactly as AM looked for polarized broad lines in Seyfert 2 objects. Unfortunately, most of the detections of highly polarized continuum in narrow line radio galaxies have been made in high redshift objects. In these objects, broad Balmer lines of the type found by Miller and collaborators are redshifted out of the optical spectrum. However, we are fortunate to have a unique local high–powered narrow line radio galaxy: Cygnus A.

Cygnus A has been studied spectroscopically for many years. Its optical continuum contains stellar absorption features and strong narrow emission lines. Pierce and Stockton (1986) suggested that a large fraction of the continuum light seen in the near–nuclear region is scattered from a hidden nucleus. Support for this view was provided by a high quality polarization map of Tadhunter et al. (1991), which shows a "bipolar nebula" polarization structure extending several arcseconds outside the nucleus. There is significant dilution of the polarization around the emission line peak, which is offset 3 arcseconds NW from the nucleus, suggesting that the polarization seen is real and not due to galactic foreground emission. The degree of polarization is about 2%, although it should be noted that Goodrich & Miller (1990) find a smaller degree of polarization and no significant dilution at the position of the emission–line maximum. Cygnus A is therefore an obvious candidate for further spectropolarimetry, in order to attempt to detect further evidence of a hidden broad nucleus such as polarized emission lines.

2. Observations

Observations were carried out at the William Herschel Telescope (WHT) using the ISIS spectrograph with polarimetry attachment, on 1990 September 29. Owing to bad weather only 50 minutes' integration were obtained

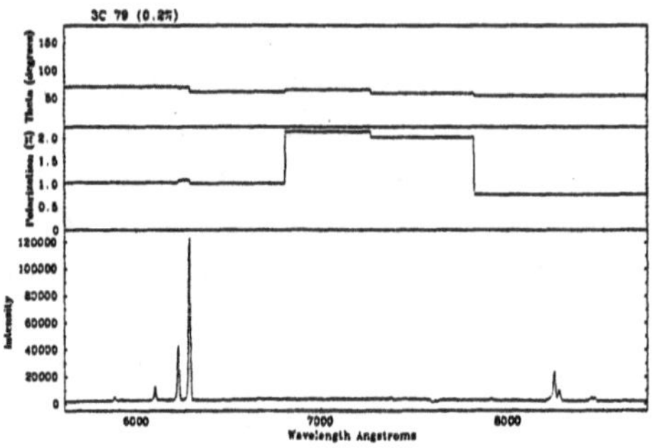

Fig. 1. Percentage polarization and polarization position angle derived for 3C 79, plotted in bins with an error of 0.2%

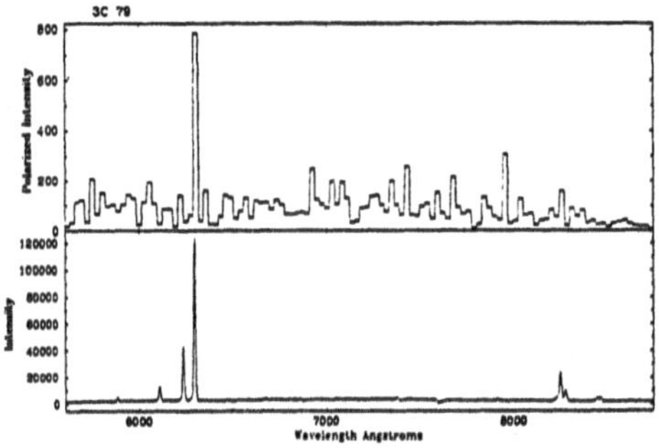

Fig. 2. Polarized flux spectrum of 3C 79

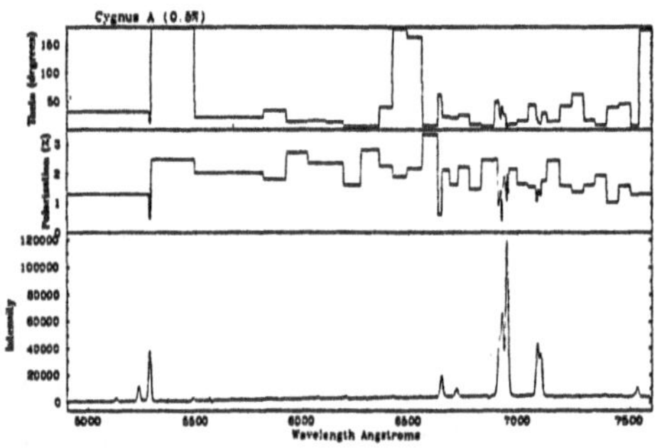

Fig. 3. Percentage polarization and polarization position angle derived for Cygnus A, plotted in bins with an error of 0.5%

on Cygnus A and 2 hours on 3C 79. Sequences of four exposures were obtained at different positions of the halfwave plate, of which two were used to calculate Stokes Q and two for Stokes U. A 1 arcsecond slit was used, and instrumental calibration was performed using the standard star HD14433. Full details of the reduction will be give in Jackson and Tadhunter (1991, in preparation).

Figure 1 shows the percentage polarization and the polarization position angle derived for 3C 79. There is a detection of polarization at the 0.8% level in the continuum, which is in good agreement with the value of Antonucci (1984). No significant broad line is visible. To emphasize this point, figure 2 shows the polarized flux spectrum ($P \times F_{\lambda}$).

Figures 3 and 4 show the percentage polarization, the polarization position angle and the polarized flux in Cygnus A. Again, no polarized broad line is visible.

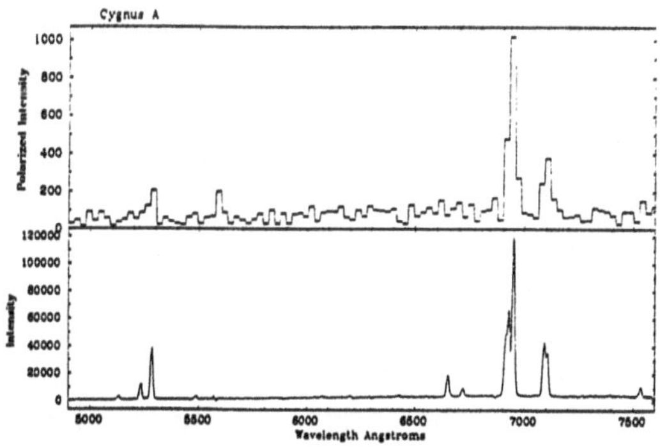

Fig. 4. Polarized flux spectrum of Cygnus A

3. Non–detections of polarized broad lines

3.1 Reasons for non–detection

If we are to believe the radio galaxy/quasar unified model, we must find an explanation for the non–detection of polarized broad lines in narrow line radio galaxies.

Various explanations for non–detections of polarized broad lines were considered and rejected by Miller & Goodrich for their Seyfert 2 sample. These included optically thick scattering, which would broaden the Balmer lines unacceptably. The simplest plausible explanation is that the scatterers are not there, or if they are, they are hidden inside the obscuring torus.

However, if we detect a polarized continuum which is not due to foreground absorption – as we do in Cygnus A – we should also detect a polarized broad line if light from a quasar nucleus is being scattered into our line of sight. In cases like 3C 79, where evidence of continuum scattering is not present, such a prediction cannot be made. Objects like 3C 79 are still worth investigating, as a serendipitous discovery of a polarized broad line would be powerful evidence in favour of the radio galaxy/quasar unified scheme.

3.2 Constraints in the case of Cygnus A

Consider first the narrow lines. Tadhunter et al. found that [OIII] is polarized, but less so than the continuum. We can marginally confirm this result (Figure 3) although we do not have sufficient signal to noise to say this definitely.

The polarization properties of the narrow line region are complicated because the narrow line region is extended and we do not know where the scatterers lie in relation to it. It has been suggested (Jackson & Browne 1990) that if the radio galaxy/quasar unified scheme is correct, at least some of the narrow line region must lie inside the obscuring medium, because radio galaxies tend to have weaker [OIII] lines than quasars of equivalent extended radio power. In this case we would expect at least some part of the narrow line region to be polarized by scattering. But let us be conservative and assume that the [OIII] polarization that we do see is due entirely to foreground effects.

Now suppose that a typical quasar continuum and broad line region is being scattered into our line of sight. Most lobe-dominated quasars have Hβ lines of equivalent width approximately 100 Å and full width at half maximum of 80 to 150 Å (Miley & Miller 1979, Oke, Shields & Korycansky 1982, Bergeron & Kunth 1984, Jackson and Browne 1991). The polarized spectrum should contain a line of similar width (unless the line is broadened, for example by large optical depth in hot electrons) and of a smaller equivalent width. This reduction in equivalent width arises because some of the polarized continuum is not from the nucleus, but is scattered or foreground polarized starlight. It can be estimated as approximately a factor of 2 by supposing that the continuum and narrow lines are polarized by the same external mechanism, presumably dichroic absorption in the Cygnus A galaxy or our own galaxy. The factor of 2 follows since the continuum is about a factor of 2 more polarized than the narrow lines, and is in agreement with the work of Osterbrock (1983) who suggested that starlight made up about half of the observed optical continuum.

We would therefore expect a polarized broad line with an equivalent width between 40Å and 80Å . At present we do not have the data to rule this out. Depending on how wide the intrinsic broad Balmer line is, we would expect to see a polarized broad line if it had an equivalent width of 100Å – 150 Å . Further observations are in progress to tighten this limit.

If we do not find a polarized broad line, can we explain this in terms of the radio galaxy/quasar unified scheme? One possible explanation (Tadhunter et al. 1990, see also Chambers & Miley 1989) is that the scattered continuum is blazar–like and does not contain lines. This is unlikely in terms of the unified radio galaxy/quasar picture. Any central blazar beam would have an opening angle of 10° or less, in order to explain the range of Lorentz factors seen in observations of superluminal motion in quasars, and it is not plausible to assume that all the scatterers lie within this beam.

Acknowledgment. The William Herschel Telescope is operated on the island of La Palma by the Royal Greenwich Observatory at the Spanish Observatorio del Roque de los Muchachos of the Instituto de Astrofisica de Canarias.

References

Antonucci, R.R.J., Miller, J.S., 1985: *Astrophys. J.*, **297**, 621.
Antonucci, R.R.J., 1984: *Astrophys. J.*, **278**, 499.
Barthel, P.D., 1989: *Astrophys. J.*, **336**, 606.
Bergeron, J., Kunth, D., 1984: *Monthly Notices Roy. Astron. Soc.*, **207**, 263.
Chambers, K.C., Miley, G.K., 1989: in: *Cooling flows in clusters and galaxies*, A. Fabian (ed), Cambridge University Press
Goodrich, R.W., Miller, J.S., 1990: *Astrophys. J. (Letters)*, **346**, L21.
Jackson, N., Browne, I.W.A., 1990: *Nature*, **343**, 43.
Jackson, N., Browne, I.W.A., 1991: *Monthly Notices Roy. Astron. Soc.*, **250**, 414.
Kinney, A., et al., 1991: *Astrophys. J.*, in press.
Miley, G.K., Miller, J.S., 1979: *Astrophys. J. (Letters)*, **228**, L55.
Miller, J.S., Goodrich, R.W., 1990: *Astrophys. J.*, **355**, 456
Oke, J.B., Shields, G.A., Korycansky, D.G., 1982: *Astrophys. J.*, **277**, 64.
Osterbrock, D.E., 1983: *Publ. Astron. Soc. Pacific*, **95**, 12.
Peacock, J., 1987: in *Astrophysical Jets and their Engines*, W. Kundt (ed), D. Reidel, Dordrecht, Holland
Pierce, M.J., Stockton, A., 1986: *Astrophys. J.*, **305**, 204
Rudy, R.J., 1984: *Astrophys. J.*, **284**, 33.
Scheuer, P.A.G., 1987: in *Superluminal Radio Sources*, R.J. Zensus and J.A. Pearson (eds), Cambridge University Press
di Serego Alighieri, S., Binette, L., Courvoisier, T.J.L, Fosbury, R.A.E., Tadhunter, C.N., 1988: *Nature*, **334**, 591.
Tadhunter, C.N., Scarrott, S.M., Rolph, C.D., 1990: *Monthly Notices Roy. Astron. Soc.*, **246**, 163.

Discussion

Wilson: Recent papers by Djorgovski et al. (*Astrophys. J. Lett.*, **372**, L67, 1991) and by Ward, Blanco, Nishida and myself (Ap. J. in press) provide strong evidence for a quasar nucleus in Cygnus A. The evidence comprises (i) a point source at the nucleus in K and L' bands, but invisible at shorter wavelengths, (ii) excellent agreement between three independent methods of determining the extinction to this point source ($50\pm$ 10 mag in V) and (iii) the presence of strong H_2 emission in the K band (1-0 S(1) and 1-0 S(3)) which are most plausibly explained through heating of a blocking molecular torus by the known nuclear X-ray source.

Unified Schemes of Radio Galaxies and Quasars

Ian W.A. Browne[1] *and Neil Jackson*[1,2]

[1] NRAL, Jodrell Bank, Macclesfield, Cheshire SK11 9DL, England
[2] Current address: Sterrewacht, Postbus 9513, 2300 RA Leiden, The Netherlands

1. Introduction

Unified schemes have become synonymous with attempts to attribute a lot of the diversity in observed properties of active galaxies to variations in a single parameter, the orientation of some axis in the object with respect to the observer. For non spherically symmetric objects there are several ways in which orientation can affect what is seen. Radiation initially emitted isotropically can be made anisotropic by propagation effects. Selective absorption or scattering by dust or electrons is one way in which the intensity can be affected; the polarization state can also be modified by either scattering or the Faraday effect. When anisotropic motion is present, profiles of emission lines change with viewing angle and if the motions are relativistic, Doppler boosting can greatly enhance the continuum emission in some directions compared to others. Radiation emitted near the radius of a black hole (e.g. from the inner part of an accretion disk) can be gravitationally focussed.

Of course orientation is not the only parameter that makes intrinsically similar objects appear different. We see objects at different stages in their life-histories and we should always be aware of the possibility that one type of object evolves to or from another.

To be of real interest unified schemes require a believable orientation indicator. The most widely used is the ratio of the radio core luminosity to that of the extended radio emission in radio loud objects. It is believed that this is a good indicator because there is convincing observational evidence from superluminal motions for highly relativistic bulk motions in the radio cores and such motions will give rise to Doppler boosting of the core emission relative to the unbeamed extended emission. Just how good an orientation indicator this ratio R is, depends on how closely the intrinsic core power is related to the lobe power (i.e. the ratio R_T). There are indications that the dispersion in this relation cannot be very large; for one thing, the known range of R is close to the range of R one would predict just from the Doppler boosting arising from the kind of Lorentz factors deduced from measurements of superluminal motion. Secondly, the correlation between extended radio luminosity and optical line luminosity indicates that the former is closely related to another measure of nuclear activity, this time the power of the ionizing radiation from the core.

Part of the reason for our interest in unified schemes is the search for other observable parameters which correlate with R. Independent orientation indicators are important because they might be useable in radio quiet objects and also the physical mechanism that gives the orientation dependence can give strong clues as to the three dimensional distribution of material within the active galaxy.

In this review we will first discuss whether or not all radio loud AGN can be unified within one scheme. We conclude that there are at least two independent types of object requiring two schemes. We present a "low luminosity" scheme which attempts to unify low luminosity radio galaxies with BL Lac objects and a "high luminosity" scheme for quasars and high luminosity (FR2) radio galaxies. Some aspects of this latter scheme are still controversial and we give a critical review for and against quasars being intrinsically similar to FR2 radio galaxies. One recent development has been the discovery that various asymmetry parameters (jet sidedness, depolarization sidedness, arm length ratio, differences in lobe spectral indices, emission line sidedness) seem to be strongly inter-related. This clearly has relevance to unified schemes and we try and summarize what we think this new evidence is telling us.

2. Disunification of schemes

Radio astronomers have long divided sources into two classes according to radio morphology and luminosity. FR1 sources (Fanaroff & Riley, 1974) have low luminosities ($\leq 10^{26.5}$W Hz^{-1} at 178MHz) and radio structures which though double have no prominent hotspots in their lobes. Their radio brightness falls off with distance from the parent galaxy. By contrast the FR2 sources have high luminosities and radio lobes dominated by bright hotspots situated at their leading edges away from the galaxy. The FR1 sources tend to have symmetric jets while in FR2s the jets are almost invariably one-sided, at least to the limits of present day observations (Bridle 1990). There is also a difference in the optical emission line properties of the two classes in the sense that the FR1s have weak lines while the FR2s, even those which are not quasars, have strong lines. Sources in the two classes also show different rates of cosmological evolution. The FR1s evolve slowly or not at all, whereas the population FR2s shows strong evolution with them being more common or more luminous in the past. It would therefore not be surprising if there were two distinct types of beamed object corresponding to FR1 and FR2 radio sources viewed beam-on.

Is there evidence for more than one type of highly beamed radio loud object? The label "blazar" has been given to all extragalactic objects which have rapid, large amplitude optical continuum variations (OVVs) or high percentage optical linear polarization (HPQs) or both, together with a compact radio core. Because of the extreme properties of all objects displaying blazar behaviour, it is widely assumed that in the majority of objects the continuum at all wavelengths is relativistically beamed non–thermal emission. But are we seeing beamed versions of just one type of object? The evidence is now quite strong that we are not. There is one group consisting of high luminosity objects which show broad emission lines in their spectra; they have all the properties of quasars. There is also a group of low luminosity objects with no detectable emission lines or very weak narrow lines; these are the BL Lac objects. In the following we will assume that there are just two classes of object displaying blazar behaviour, the OVV/HPQs which are the high luminosity broad line objects and the BL Lac objects which are the low luminosity narrow and very weak lined objects (but see also Burbidge & Hewitt 1989). We will subsequently argue that the two classes represent the beamed counterparts of quasars/FR2 radio galaxies and FR1 radio galaxies respectively. However, before we do that we will show some more of the evidence for there being two classes represented within the population of "beamed objects".

2.1 Redshift distributions of BL Lacs and OVV/HPQs in a flux limited sample

Could BL Lac objects be quasars viewed at a very small angle to the line of sight? If BL Lacs really were a subset of the HPQ/OVV quasar population we would expect the relative numbers in the two highly beamed populations to remain roughly constant with redshift and, moreover, the space density of BL Lacs should always be less than that of flat spectrum quasars. In Fig. 1 we show redshift distributions for flat spectrum quasars, OVV/HPQs and BL Lacs from the Kühr et al. (1981) complete 5GHz 1Jy survey. The BL Lac identifications are from Stickel et al. (1991). It is clear that the OVV/HPQs and the flat spectrum quasars have very similar redshift distributions with the numbers peaking at a redshift of about unity. On the other hand the BL Lac sample has a more uniform redshift distribution with a strong deficit of objects with $z > 1$. The relative numbers are not constant with redshift. This cannot be due to objects without measured redshifts since there are very few of these and there is no reasonable way to distribute them to make the histograms have a similar shape. (N.B. The objects without redshift are nearly all fainter than magnitude 18 so they are likely to lie at high redshift). The largest difference in redshift distributions is at low redshift; there is one OVV/HPQ quasar with z<0.2 but there are seven BL Lacs. Locally BL Lacs outnumber quasars which is certainly a result inconsistent with BL Lacs being a subset of quasars. An equivalent way of looking at the population statistics is in terms of luminosity functions and their evolution (Padovani & Urry, 1990). BL Lacs are relatively low luminosity objects whose population shows little cosmological evolution while the quasars are a strongly evolving high luminosity population.

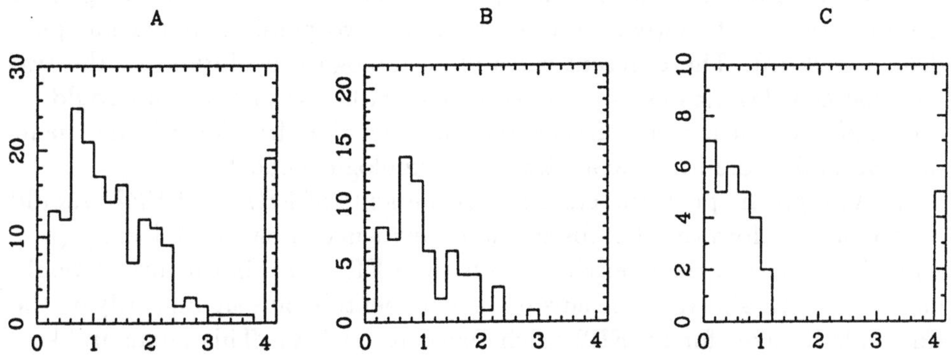

Fig. 1. Redshift distributions of (A) flat spectrum quasars, (B) OVV/HPQs and (C) BL Lac objects in the Kühr et al. 5GHz survey. Objects without redshifts are plotted in the last redshift bin (z=4.1).

2.2 X-ray spectral indices of BL Lacs and OVV/HPQs

Worrall & Wilkes (1990) have examined the 0.2-3.5 keV spectra of compact extragalactic radio sources measured with the Einstein Observatory. Their results show that, while the average spectral index for HPQs and flat radio spectrum quasars are similar at about 0.5, for BL Lacs it is about 1. This implies that BL Lacs are not simply highly beamed versions of quasars.

2.3 The case for two schemes

Since there is strong evidence for two populations, both amongst the radio lobe-dominated objects which form potential parent population(s) and amongst the core-dominated optically active objects which are the beamed counterparts of the parent population(s), it seems logical to look at two unified schemes, one for the low luminosity FR1 objects, i.e. a BL Lac scheme, and one for the high luminosity FR2 objects, i.e. a quasar/high-luminosity-radio-galaxy scheme.

2.4 Can time reunify the schemes?

Is it profitable to speculate on whether or not FR2 objects evolve into FR1 objects? We can conceive of a picture in which individual objects have one or more outbursts of activity each of which start in FR2 mode and develop with time into FR1 mode. This hypothesis leads to testable predictions:

1. In its simplest form it would predict that each object would spend a fixed fraction of its active life in each of the two possible modes independent of redshift. The different rates of cosmological evolution for the two Fanaroff & Riley classes are not consistent with this picture, but could be accommodated if it were further hypothesized that fraction of time spent in each mode were itself dependent on cosmological epoch.

2. At a given epoch the cluster environments of FR1s and FR2s should be the same. However the observational evidence shows that locally ($z < 0.2$) FR1s occur in richer environments than FR2s. To fix this up we would have to hypothesize that the cluster environments evolve significantly on the timescale it takes for an FR2 to change into an FR1. This seems unlikely to us.

3. In general the radio structures of FR1s do not look as if they have evolved from FR2s, or vice versa.

We conclude, therefore, that the above evolutionary scenario does not work.

Another less interesting picture is one in which with passage of cosmological time objects that in the past might have had FR2 outbursts at more recent epochs have FR1 outbursts. This may indeed be the truth but such a hypothesis has little predictive power. Vagnetti et al. (1991) have considered an evolutionary unified scheme of the above type for BL Lacs and quasars based on a joint analysis of radio and optical data. They construct a scheme in which the continuum emission in both types of object is composed of the sum of beamed and unbeamed components which evolve with cosmological time at different rates in the optical and radio band. Parameters for such a scheme can be found which are consistent with the observed population statistics.

2.5 Gravitational lensing as a unifying mechanism

Ostriker & Vietri (1985) have proposed that BL Lac objects are high redshift HPQ/OVV quasar nuclei gravitationally lensed (and amplified) by the galaxy in which the BL Lac image appears to lie. There has been considerable discussion of this suggestion in the literature (Gear, 1991; Gopal-Krishna & Subramanian, 1991). We favour the view that the proposed mechanism may operate in a few individual cases, but statistically it is unimportant. One of the most persuasive arguments for this view is put forward by Worrall & Wilkes based on their results on the X-ray spectral indices of BL Lacs and flat radio spectrum quasars. The fact that their spectral indices are systematically different argues strongly against BL Lacs being lensed quasars.

3. The low luminosity radio galaxy/ BL Lac unified scheme

3.1 Statistical results

In the last two years a lot of progress has been made in testing unified models of low luminosity radio galaxies and BL Lac objects. Much of this progress has been possible because of the work done to produce well-defined samples of BL Lac objects, both X-ray selected and radio selected. In the radio there are 5GHz samples from Kühr & Schmidt (1990) and Stickel et al. (1991), and in the X-ray the Einstein Observatory EMSS survey (Stocke et al. 1991) and the HEAO-1 strong source sample (Schwartz et al. 1989). Though these samples are primarily radio or X-ray selected it is important to keep in mind that a secondary selection process always takes place at optical wavelengths and this process may itself introduce subtle selection biases (see below).

The hypothesis being explored is that in the cores of FR1 radio galaxies there are relativistic jets whose nonthermal emission, when viewed at small angles of sight, can dominate the emission from the rest of the galaxy at all wavelengths, giving rise to what we recognize as a BL Lac object. This picture has been developed by Blandford and Rees (1978), Browne (1983) and Antonucci & Ulvestad (1985) amongst others, but has been most thoroughly tested recently by Urry et al. (1991) and references therein. Their approach is to construct the luminosity function for BL Lac objects and compare it with what they would predict from knowledge of the FR1 luminosity function and its modification by relativistic beaming (Urry & Shaffer, 1984; Urry & Padovani, 1991). The predicted beamed luminosity function is the result of the convolution of the parent population luminosity function with the Doppler boosting probability distribution.

Urry et al. (1991) apply this procedure to samples of radio selected BL Lac objects. A distribution of Lorentz factors is suggested by the available observations of superluminal motion in BL Lacs. They assume a power law form for this distribution and allow the index of this power law to be a free parameter. The minimum Lorentz factor is constrained by the observations to be about 5, while the maximum is another free parameter in the model. A further parameter is the maximum value of R which is constrained, but not very tightly, by observation. With these parameters an excellent fit can be obtained between the predicted luminosity function for BL Lacs and that deduced from the observations of the complete radio samples. They deduce a maximum Lorentz factor of 36 and an index of -4 for the Lorentz factor distribution. The observed number counts and the redshift distributions are also in agreement with the observations.

It appears therefore that a self-consistent scheme can be set up in which FR1 radio galaxies are the parent population of BL Lac objects. One further constraint on such a model would be to use it to predict the distribution of R values and compare these with observations which exist for the majority of the objects in the complete samples. As far as we know this has not yet been done.

The situation for X-ray selected BL Lacs is not so clear. Padovani & Urry (1990) have applied their formalism to these objects but at a time when the data on the EMSS X-ray survey were unavailable to them. Within these limitations the scheme again seems to be consistent with FR1 galaxies being the parent population, but possibly with a lower Lorentz factor. More recently, Wolter et al. (1991) and Morris et al (1991) have examined the luminosity function and redshift distributions of X-ray selected BL Lacs with the benefit of the new data (Stocke et al. 1991) on the EMSS survey objects. Morris et al conclude that the BL Lac population has strong negative cosmological evolution, while Wolter et al. find that, though the FR1 model can predict the correct number counts, it does not give a good fit to the observed redshift distribution of BL Lacs. Strong negative cosmological evolution for the X–ray selected BL Lacs does not seem consistent with the conclusion of Stickel et al. (1991) that there is evidence for slight positive evolution at the 2σ level from the radio selected complete samples. At face value these results suggest that X-ray and radio selected BL Lacs may belong to different populations.

Is a resolution for this confused situation possible? We believe that part of the problem may lie in the difficulty that arises in identifying optically faint BL Lacs in the presence of the light of an underlying bright elliptical galaxy. This difficulty is most marked for the relatively deep X-ray surveys like the EMSS. For a BL Lac object with an X-ray flux near the survey limit, it is possible to predict the expected optical flux density assuming a "typical" optical/X-ray spectral index. The result is that the BL Lac flux is likely to be substantially less than that of a giant elliptical galaxy until the galaxy redshift exceeds about 0.5. Thus we reach the almost counterintuitive conclusion that it's the nearby BL Lacs in bright galaxies that will be the most easily missed in deep surveys. We therefore expect the source counts to be artificially flattened at low flux densities and there to be a deficit of low redshift BL Lacs. This is exactly what is observed in the EMSS survey and is quite likely to be "explanation" for the difference in behaviour of the radio and X-ray selected complete samples. These conclusions are to be discussed in more detail by Browne & Marcha (in preparation).

3.2 Other relevant observations

Ulrich (1989) presents a review of the properties of BL Lac host galaxy properties and how they compare with those of FR1 radio galaxy properties. Her conclusions are that the BL Lac galaxies are giant ellipticals and that they have a distribution of absolute magnitudes which is consistent with what would be expected for a population of radio galaxies with the observed spread of extended radio luminosities. Thus this evidence supports the unified scheme. However, McHardy et al. (1991) report an interesting result on the BL Lac object PKS 1413+135. This object appears to lie in a disk galaxy which is surprising as FR1 radio galaxies all seem to lie in ellipticals. Clearly there is important work to be done in following up more of the host galaxies of BL Lac objects.

Any extended radio emission possessed by PKS 1413+135 is very weak (Perley, 1982). This raises the interesting question of just how low the luminosity of an FR1 radio galaxy can be and still support a relativistic jet producing BL Lac-type emission. Most elliptical and S0 galaxies are radio emitters at some low level, so are these capable of producing mini-BL Lacs? This is a difficult question to answer because these mini-BL Lacs would be much fainter than the galaxies in which they lie. Nevertheless it would be very interesting to look for such objects in the cores, say, of bright nearby flat spectrum radio galaxies exploiting high resolution or colour or both to discriminate against the starlight from the underlying galaxy.

Another important observation that would boost confidence in the BL Lac/ FR1 unified scheme would be the confirmation of motion with speeds of the order of that of light in the cores of FR1 radio galaxies. Several such cores have been observed with VLBI, but it is only in M87 that we are aware of a detection of motion. In this case the nuclear jet speed is found to be 0.28±0.08c (Reid et al. 1989) and that of knots in the arcsec jet to vary from 0.1c to 1c (Biretta & Owen, 1990). Though M87 may not be a typical FR1 radio galaxy this detection of apparent relativistic motions is encouraging for the unified scheme.

4. High luminosity radio galaxy/quasar unified schemes

4.1 Introduction

Objects that have extended radio luminosities above the FR2 limit are narrow line radio galaxies, all steep radio spectrum quasars and the vast majority of flat radio spectrum quasars. Can all these be incorporated into a single unified scheme? Barthel (1989) advocated such a picture, drawing on the analogy with low luminosity AGN in which there is considerable evidence for anisotropic emission to suggest the existence of a "blocking torus" surrounding the nuclear region and hiding the nuclear continuum and broad line region from some lines of sight (Lawrence & Elvis, 1982; Antonucci & Miller, 1985). We are seduced by this picture, but not totally. What we find most attractive is the idea that blocking tori and "ionization cones" (Whittle, 1989) are common to all AGN; what we worry about are some of the small details. Below we try and present the important evidence as we see it and suggest some of the minor fixes which may be necessary if the scheme is to work.

4.2 Systematic asymmetries in high luminosity radio sources

One of the most fascinating developments of the last couple of years has been the discovery that some of the symmetry properties of radio sources are very strongly related to each other in ways which are highly relevant to the interpretation of unified schemes.

4.2.1 Radio depolarization and jet-sidedness asymmetries (the Laing/Garrington effect)

When jets are detected in FR2 radio sources they are almost invariably one-sided (Bridle, 1990). Garrington et al. (1988) and Laing (1988) have measured the Faraday depolarization in the lobes of sources with jets and find that almost without exception the lobe on the side of the source opposite the jet depolarises more than the other. A simple interpretation is that the sources are axisymmetric with two relativistic jets and that, because of Doppler boosting, only the approaching jet is seen. The reason for the depolarization asymmetry is that there is a halo of depolarizing material centered on the galaxy (X-ray halo?) and the near-lobe is seen through less of this halo than the far one (Garrington & Conway, 1991). This interpretation strongly supports the unified scheme.

4.2.2 Lobe spectral index and jet-sidedness asymmetries

Garrington et al. (1991) and Liu & Pooley (1991) that the radio lobe on the side of the jet nearly always has a flatter spectral index than the other lobe. This may indicate some beamed flat spectrum emission in the approaching lobe, a possibility considered by Garrington et al., or intrinsic differences between the lobe being supplied by the jet and the other lobe. If the latter, this raises a doubt about the jet-sidedness being a purely relativistic selection effect, and in turn about the support the Laing/Garrington effect gives to unified schemes.

4.2.3 Emission line asymmetries

In high redshift radio galaxies (McCarthy et al. 1991), and probably in quasars (Heckman et al. 1991), asymmetries observed in the extended emission lines with respect to the continuum from the galaxy correlate strongly with the ratio of the separations of the radio lobes from the galaxy. In galaxies, if an emission line asymmetry is detected, the lines are always stronger on the short-arm side of the source. Given that light travel time arguments would identify the short-arm side as the far-side of the source, it is difficult to conclude anything other than that the line asymmetry result indicates intrinsic asymmetry in the radio sources. Possibly the emission lines are tracers of gas density and high density gas impedes the progress of radio lobes as they plough through it.

What can we conclude from all these results? One critical question is if, and if so how, jet-sidedness correlates with emission line sidedness. Unfortunately most of the objects with measured emission line sidedness are galaxies and these rarely have detected radio jets. However, Heckman et al. (1991) have measured 9 quasars with jets and asymmetric Lyα emission; in 8 out of 9 cases the emission lines are strongest on the jet side. Thus for quasars at least we can conclude that depolarization is weak on the jet-side and that emission lines are strong. This implies that the emission-line gas is not responsible for the depolarization. In turn this is consistent with orientation being the cause of the Laing/Garrington effect and suggests that the reason that line emission is stronger on the jet side is that the jet-side is the near-side. If true, this is a very interesting conclusion; it suggests that there is sufficient absorption in the quasar parent galaxy to hide emission-line gas extended over >10kpc.

4.3 Sizes, redshifts and luminosity functions

4.3.1 Object of the exercise

If radio galaxies and quasars are intrinsically the same type of object but are seen from a different angle to the line of sight, then any intrinsic properties should be similar between the two classes of object. Specifically, *in an unbeamed sample*, i.e. one selected by unbeamed emission,

- the redshift distributions of the two classes of object should be similar
- the radio luminosity functions should be identical (although not the optical luminosity functions, as optical emission in the model is affected by an obscuring torus)

- the radio size distributions should be consistent with the model: that is, radio galaxies should be larger than quasars over all ranges of redshift.

There is only one completely identified unbeamed sample: 3CRR (Laing, Riley & Longair 1983), which is selected at 178 MHz. At this frequency the radio structure is dominated by steep–spectrum, extended emission. There is a wide consensus that most of this emission is in fact unbeamed (Murphy 1988, Browne & Perley 1986, Macklin 1981). Only two objects get into the sample by virtue of core and, therefore probably beamed emission. These are 3C 345 and 3C 454.3; 3C 245 is a marginal case. Moreover, if a strong beamed core component were affecting the 3C sample enough to cause a bias in the relative redshift distribution of radio galaxies and quasars within it, we would expect that the R value of 3C quasars would increase with redshift, and it does not: in fact, there is a marginal correlation in the opposite sense.

4.3.2 The size problem

Radio galaxies are generally larger than quasars, consistent with their being seen with their radio axes closer to the sky plane. However, Kapahi (1990) claimed that this effect becomes smaller at low redshift and finally disappears at redshifts less than 0.5.

The small number of objects in the sample is a problem for such investigations. Hooimeyer (1991) finds no inconsistency between the two classes of object, supporting the Barthel (1989) claim. The discrepancy with Kapahi arises because the samples compared are different: in order to increase the size of the quasar sample at low redshift, Kapahi includes eight sources from the Parkes catalogue selected at 408 MHz. In addition, as Kapahi points out, whether broad line radio galaxies (BLRGs) are included with the quasars or the radio galaxies strongly affects the result. A larger sample could be valuable here, since a discrepancy in linear size distribution would be very difficult to explain without modification of the simple unification scheme.

4.3.3 Statistical redshift tests

We can use the 3CRR sample to compare the redshift distributions of radio galaxies and quasars. Excluding the BLRGs, together with 3C 345 and 3C 454.3, we obtain a difference between the two classes of object significant at the 99% level. If, however, we include the BLRGs with the quasars, which is probably the fairest thing to do, the significance of the difference falls to 90%. By removing 3C 245 we can push the significance to below 90%. The main limitation of this test is that there are only 30 quasars in 3CRR[1].

4.3.4 Implications for luminosity functions

Padovani and Urry (1991, these proceedings) have applied their beamed luminosity function analysis to quasars and narrow-line radio galaxies and have investigated if, starting from the radio galaxy luminosity function,

[1] We have excluded the compact steep spectrum (CSS) objects from these comparisons

they can correctly predict the luminosity function for both steep and flat spectrum quasars. They conclude that they can, and with reasonable values for the few free parameters in their model. This is a powerful plausibility argument in favour of the radio galaxy/quasar unified scheme. Further investigations are needed to see if the luminosity functions predict the correct redshift distributions and R distributions for samples of objects selected at different frequencies. For instance, one would predict that in the sample made up of the 3C sample and the Allington–Smith 1Jy sample (Allington–Smith 1982) together, the redshift distributions should be very similar for radio galaxies and quasars. For a sample selected at a high frequency, say 2.7 or 5GHz, where beamed emission becomes important, the log R distribution becomes a powerful diagnostic giving information about the intrinsic ratio of jet to lobe luminosity and the the range of Lorentz factors (Orr & Browne 1982). More radio maps of 2.7 GHz radio galaxies with weak cores are required.

4.4 Optically thin regions

4.4.1 Where is the optically thin region?

The hypothetical obscuring torus in the radio galaxy/quasar unification model intercepts most of the optical and UV light in certain directions. Provided that the infrared radiation arriving along the line of sight is predominantly thermal, the infrared flux densities of radio galaxies and quasars should become equal at some long infrared wavelength. Recent evidence for thermal IR emission in lower luminosity objects is summarized by Sitko and Zhu (1991). We discuss high luminosity objects below.

4.4.2 Statistics of IRAS sample

As has previously been pointed out (Scheuer 1987, Lawrence 1990, Hooimeyer 1991) it should be possible to test the radio galaxy/quasar unification model by comparing fluxes of radio galaxies and quasars in the far infrared. Unfortunately, the only instrument that has worked in that wavelength range – IRAS – was not sensitive enough to detect large numbers of radio galaxies. Figure 2 shows a plot of the detections and limits in 3C, taken from Golombek et al (1988), Neugebauer et al. (1986) and Yates and Longair (1989). Again, the interpretation depends on the nature of the BLRGs. If they are counted as quasars, there is an obvious difference between the IRAS fluxes of quasars and radio galaxies (in the sense that quasars have larger fluxes) at a given redshift. Out to the furthest redshift at which detections were made, six out of the seven quasars are detected compared to four out of 22 radio galaxies, and there is no reason to suppose the quasars were on average looked at harder than the radio galaxies. If the BLRGs are excluded, the difference is difficult to assess: the large number of limits make survival statistics difficult to apply, and comparisons treating limits as detections show no significant difference.

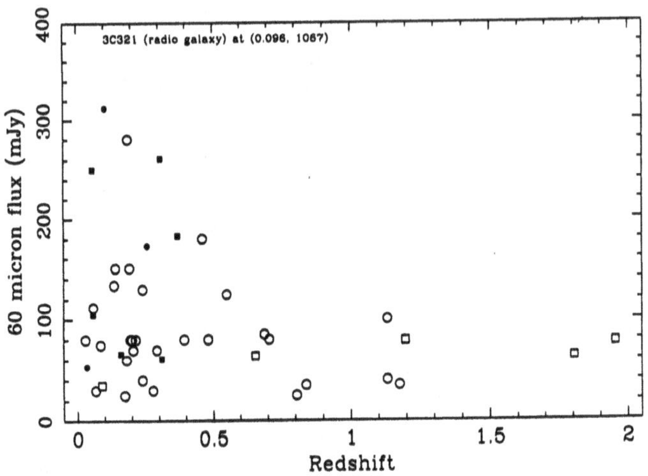

Fig. 2. IRAS detections of quasars (squares) and radio galaxies (circles) in 3CRR. Limits are represented by open symbols.

Until the advent of ISO the best possibility of resolving this question consists of observations at wavelengths accessible to ground based telescopes (e.g. $10\mu m$). Such a study is under way (Lawrence 1990).

4.5 Emission line properties

4.5.1 [OIII] properties

Rawlings and Saunders (1991) have pointed out that quasars and radio galaxies lie roughly on the same line in a plot of radio jet power against narrow line luminosity: this is evidence that both types of object contain a similar photoionizing central source and thus is an argument for the radio galaxy/quasar unified scheme. On slightly closer examination, quasars of redshift 0.8 or less tend to have larger [OIII] luminosities for a given extended radio luminosity (Jackson and Browne 1990). A slight modification of the scheme, such as locating of some of the narrow line region inside the obscuring torus, can be invoked to explain this effect.

4.5.2 [OII] properties

Figure 3 shows the luminosity of narrow lines in radio galaxies and quasars plotted against redshift for the 3CRR sample. The data are taken from McCarthy (1988), Jackson & Browne (1990), Yee & Oke (1978), Costero and Osterbrock (1977), Cohen and Osterbrock (1981) and Saunders et al. (1989). [OII] is plotted for quasars above redshift 0.5. It is obvious that, although the [OIII] fluxes are different for radio galaxies and quasars, the [OII] fluxes are much more similar. Is this a redshift or luminosity effect, or is it an effect of a different photoionizing continuum in the two type of object? The simplest explanation consistent with the unified scheme is that the [OII] region is less strongly absorbed by the obscuring torus than the

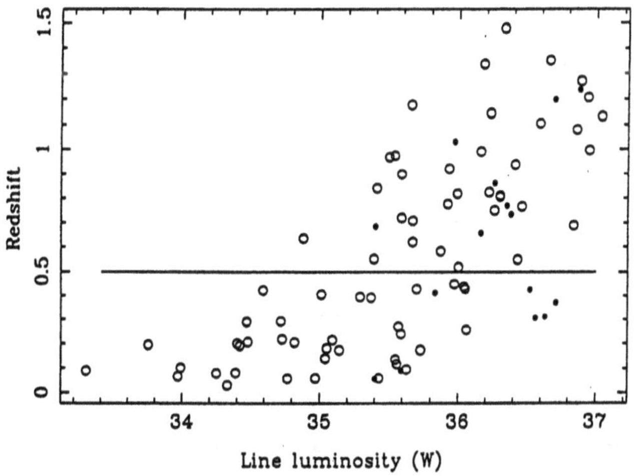

Fig. 3. Luminosities of narrow lines in quasars (filled circles) and radio galaxies (open circles) in 3CRR. [OII] is plotted for objects of redshifts greater than 0.5, and [OIII] for other objects.

[OII] region. Further observations of [OII] and [OIII] in carefully selected objects are required.

4.6 Hidden monsters: the evidence

4.6.1 Scattered continuum

An alternative approach to comparing statistical properties of radio galaxies and quasars is to attempt to infer the existence of hidden quasar nuclei in radio galaxies by their effects on other parts of the source that we can see. In particular, scattered – and hence polarized – light from the central engine might be visible although the direct line of sight to the centre is blocked by the obscuring torus.

A recent list of candidate hidden nuclei in radio–loud objects is given by Antonucci and Barvainis (1990). Some radio galaxies have highly polarized off–nuclear emission: for example, 3C 277.2 (di Serego Alighieri et al. 1988), 2152-69 (Tadhunter et al. 1989) and 3C 368 (Scarrott et al. 1990). Polarized off–nuclear emission has also been found in the prototypical radio galaxy Cygnus A (Tadhunter et al. 1990). In the latter case, the inferred central ionizing flux appears to resemble that of a typical quasar. A central obscured nucleus is also indicated in this object by the discovery of a heavily reddened infrared nucleus (Djorgovski et al. 1991).

4.6.2 Broad optical lines

If a hidden quasar continuum can be scattered outside the plane of the obscuring matter, could we also expect to see polarized broad lines? Jackson (1991, these proceedings) reports on a so far unsuccessful search in Cygnus A. Broad polarized lines have been discovered in 3C 234 (Antonucci

1984) although this object is sometimes classed as a broad line radio galaxy. Inglis et al. (these proceedings) report the detection of a polarized broad line in the narrow line radio galaxy IC 5063. Thus there is now limited direct evidence for a few radio loud AGN having hidden broad-line regions.

4.7 Summary: fixing up the scheme

We summarize the tests of the radio galaxy/quasar scheme, together with the suggested status of the test and possible explanations if the test result disagrees with the scheme.

Table 4.1. Summary of evidence for/against the radio galaxy/quasar scheme

	status	possible explanation
redshift difference	not proven	cone angle evolution
size difference	not proven	very difficult to explain
NL difference	established	[OIII] in cone
[OII]/[OIII] difference	not proven	cone angle evolution
IR difference	not proven	very difficult to explain
host galaxy difference (Hutchings 1987)	not proven	very difficult to explain
scattered continuum	established	argument for hidden nucleus but need to get energies right
scattered BLR	some evidence	strong argument for US if generally established

What is the minimum modification required to make the radio galaxy/ quasar scheme consistent with the data?

Figure 4 shows a plot of the 5 GHz extended radio luminosity in the 3CRR sample plotted against the degree of radio core dominance R. In the simplest version of the scheme, the division between radio galaxies and quasars should occur at the same R value at all luminosity. In practice the division occurs at lower values of R at higher luminosity (or redshift: the two quantities are impossible to separate). This could in principle be solved if the opening angle of the obscuring torus decreased with increasing luminosity (or redshift), so that objects would be more likely to be obscured at low luminosity. This would agree with the explanation suggested in section 4.3.1 for the [OIII] problem, as more of the narrow line region might be expected to be hidden from view at low luminosity.

What about the broad line radio galaxies? Inclusion of the BLRGs with quasars largely removes the redshift problem (section 4.1.3) but makes the

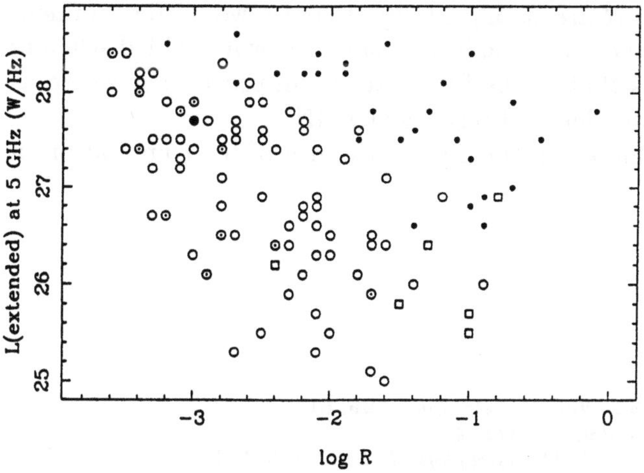

Fig. 4. 5GHz extended radio luminosity plotted against the degree of radio core dominance *R*. Quasars are represented by filled circles, radio galaxies by open circles (upper limits on log *R* contain central dots) and BLRGs by squares.

infrared problem (section 4.2) worse. Are BLRGs really low luminosity dusty quasars (Rudy 1984)? Or should many narrow line radio galaxies in fact be classified as BLRGs? Claims exist in the literature of suggestively strong, possibly broad, infrared hydrogen lines (3C 219, Fabbiano et al. 1986). If such discoveries are confirmed, this might provide powerful evidence that radio galaxies are not all that they seem, but would still require evolution of the radio galaxy/quasar population, because the excess of quasars relative to radio galaxies at high redshift would be larger than we now think.

5 Conclusions

Two unified schemes seem to be necessary if we are to account for the properties of both high and low luminosity objects. We are convinced by the basic concept of unified schemes; in general terms they work well but some detailed observations require fixes to keep everything consistent. More fixes are required for the FR2/quasar scheme but this may simply reflect the fact that BL Lac objects have very few measurable parameters (like emission line properties) compared to quasars and radio galaxies.

What of the ambition to find a convincing orientation indicator which might be used in radio quiet objects? The present state of affairs is not very encouraging. In the radio loud objects Hβ width loosely anti-correlates with *R* (Wills & Browne 1986). There are also suggestions of systematic trends of various emission line ratios with *R* (Jackson & Browne 1991). So far none of these correlations with R are strong enough that we would be confident to use the parameter as an orientation indicator on its own. Unified schemes seem to work; the hope is that in the future they will prove their worth as tools for understanding the detailed physics of AGN.

Acknowledgments. We thank J.P. Leahy for the use of his compilation of 3CRR data.

References

Allington-Smith, J., 1982: *Mon. Not. R. astr. Soc.*, **199**, 611.
Antonucci, R.R.J., 1984: *Astrophys. J.*, **278**, 499.
Antonucci, R.R.J., Barvainis, R., 1990: *Astrophys. J. Lett.*, **325**, L21.
Antonucci, R.R.J., Miller, J.S., 1985: *Astrophys. J.*, **297**, 621.
Antonucci, R.R.J., Ulvestad, J., 1985: *Astrophys. J.*, **294**, 158.
Barthel, P.D., 1989: *Astrophys. J.*, **336**, 606.
Biretta, J.A., Owen, F.N., 1990: *Parsec-Scale Jets*, ed. J.A. Zensus, T.J. Pearson, p125, Cambridge University Press
Blandford, R.D., Rees, M.J., 1978: *Mon. Not. R. astr. Soc.*, **169**, 395.
Bridle, A.H., 1990: *Parsec-Scale Jets*, ed. J.A. Zensus, T.J. Pearson, p186, Cambridge University Press.
Browne, I.W.A., 1983: *Mon. Not. R. astr. Soc.*, **204**, 23P.
Browne, I.W.A., Perley, R.A., 1986: *Mon. Not. R. astr. Soc.*, **222**, 149.
Burbidge, G., Hewitt, A., 1989.
Cohen, R.D., Osterbrock, D.E., 1981: *Astrophys. J.*, **243**, 81.
Costero, R., Osterbrock, D.E., 1977: *Astrophys. J.*, **211**, 675.
Djorgovski, S., Weir, N., Matthews, K., Graham, J.R., 1991: *Astrophys. J. Lett.*, **372**, L67.
Fabbiano, G., Willner, S.P., Carleton, N.P., Elvis, M., 1986: *Astrophys. J. Lett.*, **304**, L37.
Fanaroff, B.L., Riley, J.A., 1974: *Mon. Not. R. astr. Soc.*, **167**, 31.
Garrington, S.T., Leahy, J.P., Conway, R.G., Laing, R.A., 1988: *Nature*, **331**, 147.
Garrington, S.T., Conway, R.G., 1991: *Mon. Not. R. astr. Soc.*, **250**, 198.
Garrington, S.T., Conway, R.G., Leahy, J.P., 1991: *Mon. Not. R. astr. Soc.*, **250**, 171.
Gear, W.K., 1991: *Nature*, **249**, 676.
Golombek, D., Miley, G.K., Neugebauer, G., 1988: *Astron. J.*, **95**, 26.
Gopal-Krishna, Subramanian, K., 1991: *Nature*, **249**, 766.
Heckman, T.M., Lehnert, M.D., van Breugel, W., Miley, G.K., 1991: *Astrophys. J.*, **370**, 78.
Hooimeyer, J.A., 1991: PhD thesis, University of Leiden.
Hutchings, J.B., 1987: *Astrophys. J.*, **320**, 122.
Jackson, N., Browne, I.W.A., 1990: *Nature*, **343**, 43.
Jackson, N., Browne, I.W.A., 1991: *Mon. Not. R. astr. Soc.*, **250**, 422.
Kapahi, V.K., 1990: in *Parsec scale jets*, eds J.A. Zensus and T.J. Pearson, Cambridge University Press.
Kühr, H., Witzel, A., Pauliny-Toth, I.I.K, Nauber, U., 1981: *Astron. Astrophys. Suppl.*, **45**, 367.
Kühr, H., Schmidt, G.D., 1990: *Astron. J.*, **233**, 379.
Laing, R.A., Riley, J.M., Longair, M.S., 1983: *Mon. Not. R. astr. Soc.*, **204**, 151.
Laing, R.A., 1988: *Nature*, **331**, 149.
Lawrence, A., Elvis, M., 1982: *Astrophys. J.*, **256**, 410.
Lawrence, C., 1990: in *Parsec scale jets*, eds J.A. Zensus and T.J. Pearson, Cambridge University Press.

Liu, R., Pooley, G., 1991: *Mon. Not. R. astr. Soc.*, **249**, 343.
Macklin, J.T., 1981: *Mon. Not. R. astr. Soc*, **196**, 967.
McCarthy, P., 1988: PhD thesis, University of California at Berkeley.
McCarthy, P., van Breugel, W., Kapahi, V.K., 1991: *Astrophys. J.*, **371**, 488.
McHardy, I.M., Abraham, R.G., Crawford, C.S., Ulrich, M.H., Mock, P.C., Vanderspeck, R.K., 1991: *Mon. Not. R. astr. Soc*, **249**, 742.
Morris,S.L., Stocke, J.T., Gioia, I.M., Schild, R.E., Wolter, A., Maccacaro, T., Ceca, R.D., 1991: *Astrophys. J.*, in press.
Murphy, D.W., 1988: PhD thesis, University of Manchester.
Neugebauer, G., Miley, G.K., Soifer, B.T., Clegg, P.E., 1986: *Astrophys. J.*, **308**, 815.
Orr, M.J.L., Browne, I.W.A, 1982: *Mon. Not. R. astr. Soc.*, **200**, 1067.
Ostriker, J.P., Vietri, M., 1985: *Nature*, **318**, 446.
Padovani, P., Urry, C.M., 1990: *Astrophys. J.*, **356**, 75.
Perley, R.A., 1982: *Astron. J.*, **87**, 859.
Rawlings, S., Saunders, R., 1991: *Nature*, **349**, 138.
Reid, M.J., Biretta, J.A., Junor, W., Muxlow, T.W.B., Spencer, R.E., 1991: *Astrophys. J.*, **336**, 112.
Rudy, R.J, 1984: *Astrophys. J.*, **284**, 33.
Saunders, R., Baldwin, J.E., Rawlings, S., Warner, P.J., Miller, L., 1989: *Mon. Not. R. astr. Soc.*, **238**, 777.
Scarrott, S.M., Rolph, C.D., Tadhunter, C.N., 1989: *Mon. Not. R. astr. Soc.*, **243**, 5P.
Scheuer, P.A.G., 1987: in *Superluminal Radio Sources*, eds J.A. Zensus and T.J. Pearson, Cambridge University Press.
Schwartz, D.A., Brissenden, R.J.V., Tuohy, I.R., Feigelson, E.D., Hertz, P.L., Remillard, R.A., 1989: in *BL Lac Objects*, ed L. Maraschi, T.Maccacaro, M.-H. Ulrich. (Springer-Verlag) P209.
di Serego Alighieri, S., Binette, L., Courvoisier, T.J.L., Fosbury, R.A.E., Tadhunter, C.N., 1988: *Nature*, **334**, 591.
Sitko, M.L., Zhu, Y., 1991: *Astrophys. J.*, **369**, 106.
Stickel, M., Padovani, P., Urry, C.M., Fried, J.W., Kühr, H., 1991: *Astrophys. J.*, **374**, 431.
Stocke, J.T., Morris, S.L., Gioia, I.M., Maccacaro, T., Schild, R., Wolter, A., Fleming, T.A., Henry, J.P., 1991: *Astrophys. J. Suppl.*, **76**, 813.
Tadhunter, C.N., Fosbury, R.A.E., Binetter, L., Danziger, I.J., Robinson, A., 1987: *Nature*, **325**, 504.
Ulrich, M.-H., 1989: in *BL Lac Objects*, ed L. Maraschi, T.Maccacaro, M.-H. Ulrich. (Springer-Verlag) P45.
Urry, C.M., Shaffer, R.A., 1984: *Astrophys. J.*, **280**, 549.
Urry, C.M., Padovani, P., 1991: *Astrophys. J.*, **371**, 60.
Urry, C.M., Padovani, P., Stickel, M., 1991: *Astrophys. J.*, in press.
Vagnetti, F., Giallongo, E., Cavaliere, A., 1991: *Astrophys. J.*, **368**, 366.
Whittle, M., 1989: in *Extranuclear Activity in Galaxies*, eds E.J.A. Meurs, R.A.E. Fosbury, ESO.
Wills. B.J., Browne, I.W.A., 1986: *Astrophys. J.*, **302**, 56.
Worrall, D.M., Wilkes, B.J., 1990: *Astrophys. J.*, **360**, 396.
Wolter, A., Gioia, A.M., Maccacaro, T., Morris, S.L., Stocke, J.T., 1991: *Astrophys. J.*, **369**, 314.
Yates, M.G., Longair, M.S., 1989: *Mon. Not. R. astr. Soc.*, **241**, 29.
Yee, H.K.C., Oke, J.B., 1978: *Astrophys. J.*, **226**, 753.

Discussion

Komberg: Why do FR1 radio galaxies live in clusters of galaxies while BL Lac objects do not?

Browne: I am not aware that a difference in cluster environments between BL Lacs and FR1 radio galaxies has been established. Clearly if it were, this would be a decisive argument against the BL Lac unified scheme.

Wilson: Could you please comment on the suggested negative evolution for BL Lac objects suggested by the X-ray samples.

Browne: I think there may be problems with the completeness of the X-ray selected BL Lac samples. With Maria Marcha I have been trying to estimate the probability that BL Lacs will be identified as such in the presence of the light from the underlying elliptical galaxy. For a reasonable distribution of optical to X-ray spectral index and host galaxy luminosities we estimate that, at the limit of the EMSS survey, the probability of the BL Lac being identified is $<$ 50and the redshift distributions and may lead to incorrect conclusions being drawn about evolution.

Preuss: There is a hotspot asymmetry in FR2 radio sources. Do you worry about this; does it fit into the unified scheme?

Browne: I believe you are referring to the trend for the brighter hotspot to lie in the lobe on the jet-side. I do not think this is a problem if one is willing to believe that some of the hotspot emission can be relativistically beamed. This seems a less wayout notion now that there is fairly clear evidence from the Laing/Garrington results for relativistic enhancement in the jets as they enter the radio lobes.

Wilkes: I have a comment. In work with Prajval Shastri described in a poster at this meeting we find a strong correlation in radio loud quasars between the X-ray slope and the core-to-lobe ratio R. This might indicate that the X-ray slope in the 0.1 - 3.5keV band is also an orientation indicator.

Pustil'nik: What is the agreement between estimates of Lorentz factor from statistical arguments and from direct measurements from superluminal motion?

Browne: The results seem reasonably consistent. Both indicate a need for a dispersion in Lorentz factors. For the FR2 scheme the agreement is helped by the inclusion of radio galaxies.

Unified Schemes for Radio-loud and Radio-quiet QSOs

Peter D. Barthel

Kapteyn Astronomical Institute, P.O. Box 800, 9700 AV Groningen, The Netherlands.

During this conference, we have heard some arguments contradicting a general QSR (radio-loud QSO) - RG (radio galaxy) unification, through the effects of 'obscurientation', as proposed by Barthel (1989). Before addressing the phenomena of axis orientation and anisotropic emission/obscuration in radio-quiet QSOs, I do wish to mention five new arguments in favour of a general unified theory for QSRs and (luminous) RGs. These will illustrate why I am still a faithful member of the unification church.

(1) Optical and near-IR imaging of the prototypical luminous (FR2) radio galaxy Cygnus A has revealed an obscured bright nucleus, consistent with a hidden QSO (Djorgovski et al. 1991).

(2) A correlation between radio jet power and optical narrow line luminosity was found to exist for luminous radio galaxies as well as QSRs, suggesting one common mechanism (Rawlings and Saunders 1991).

(3) Extended Lyman-alpha haloes, commonly present around luminous (high redshift) radio galaxies have also been detected around high z QSRs (Heckman et al. 1991). Their somewhat smaller angular extent is consistent with foreshortening due to projection effects.

(4) It was demonstrated that the combined population of QSRs and radio galaxies could be responsible for the complete X-ray background spectrum, with the soft X-rays being absorbed in the (obscured) radio galaxies (Setti and Woltjer 1989). Combining the populations might moreover decrease the need for radio source linear size evolution (Onuora 1991).

(5) Comparing luminosity functions, Padovani and Urry (Padovani, this volume) obtain good fits with core-dominated QSRs as the beamed population of luminous radio galaxies, and lobe-dominated QSRs as intermediate objects.

Generally speaking, unification models are appealing because they make astronomers' life easier. To understand the effects of anisotropic radiation (coupled to axis orientation) and dust obscuration is important for several reasons: anisotropic effects introduce an orientation bias in selecting objects, they lead to inferred luminosities which are incorrect, and they make us overestimate the AGN diversity. Understanding these effects permits a study of the physical interrelation between various types of AGN - and that is what we really want to do!

Radio-quiet QSOs outnumber their radio-loud counterparts by about a factor ten (Kellermann et al. 1989). It is challenging to try to fit the radio-quiet population into a general QSO unification scheme. Although the situation is certainly not clear-cut and homogeneous data bases are badly needed, some trends in the available data allow a first attempt, which I will describe in the following.

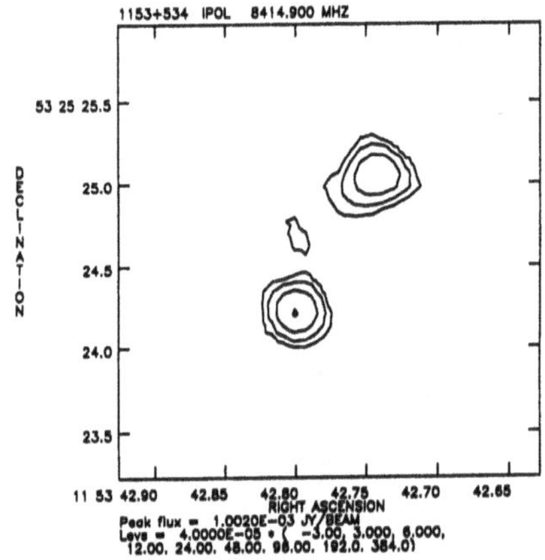

Fig. 1. Radio map of W1153+534 (VLA, 0.25 arcsec beam).

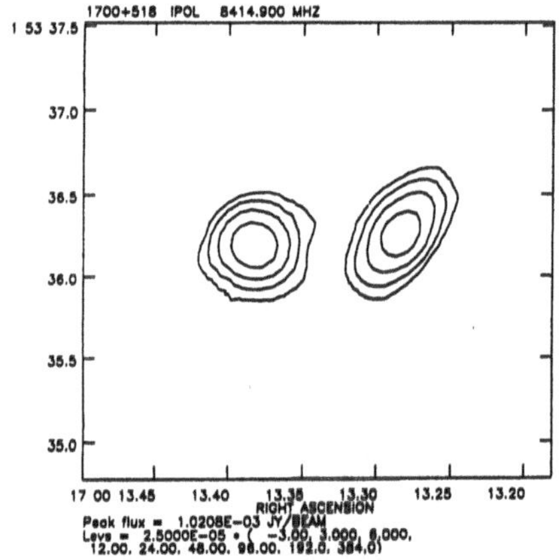

Fig. 2. Radio map of BAL QSO PG 1700 +518 (VLA, 0.25 arcsec beam).

First of all, it should be clear that radio-quiet QSOs are NOT misdirected QSRs. Studies of the (weak) radio emission in the radio-quiet QSO class (Strittmatter et al. 1980, Kellermann et al. 1989) have shown that this unification model, due to Scheuer and Readhead (1977) was untenable. Note that 'radio-quiet' rarely implies 'radio-silent': many radio-quiet QSOs are detectable at the (sub)mJy level. This permits a study of the radio properties (including polarization) of radio-quiet QSOs (Barthel and Coleman, work in progress). Confronting various properties of the two populations it is found that: (1) radio-quiet QSOs have radio-morphologies of subgalactic dimensions with low upper limits to their fractional radio polarization (see Fig. 1), quite different from the large-scale QSR radio-structures; (2) the underlying galaxy hosts differ in absolute brightness and morphology, as do the environments, at least at low redshift/luminosity (Malkan 1984, Yee and Green 1984, but see also Smith and Heckman 1990); (3) their optical emission lines and continua may differ (Barthel et al. 1990); (4) their absorption line properties differ, in the sense that about ten percent of the radio-quiet QSOs display Broad Absorption Line (BAL) systems in their optical spectra. It is this last property difference we will focus on.

A comparison of the absorption line properties in radio-loud and radio-quiet QSOs is being made by Tytler and Barthel (1991, see also Barthel

1991). These properties appear quite similar, including those for the so-called associated systems. A major difference however concerns the presence of BAL-systems. BAL absorption troughs, seen blueward of high-ionization heavy element emission lines such as CIV, SiIV and NV, are found in about ten percent of the QSO population. The inferred velocities of the absorbing gas with respect to the QSOs range up to c/5. Radio-loudness and the presence of a BAL-system appear mutually exclusive: no BAL has been seen in about one hundred QSRs with sufficiently good optical spectra (Barthel et al. 1990, Sargent et al. 1988, 1989) and no QSR has been found among some fifty (optically discovered) BAL QSOs (Stocke et al. 1984, Coleman, priv. comm.). I wish to stress here that (1) the often cited counterexample PKS 1157-014 is not a BAL QSR, but a QSR displaying damped Lyman-alpha absorption (Briggs et al. 1984), and (2) not all BAL QSOs are radio-silent. BAL QSOs make up about ten percent of the radio-quiet QSO population (Weymann et al. 1988). Prime question concerns the nature of BAL QSOs. Are we dealing with a stage in a QSO lifetime, a subclass among QSOs, or an aspect effect? Regarding this last possibility, could it be that all radio-quiet QSOs have intrinsic BALs with anisotropically distributed BAL clouds, such that our sightline cuts through an absorbing cloud in one out of ten QSOs?

Hartig and Baldwin (1986) and more recently Weymann et al. (1991) report fairly similar emission line properties for BAL QSOs and nonBAL QSOs. One significant difference appears to be an (UV) FeII emission excess in the BAL QSOs. The few BAL QSOs for which Barthel and Coleman (work in progress) obtained deep radio images display morphologies commonly seen among radio-quiet QSOs (e.g., Fig. 2).

The differences between BAL and nonBAL QSOs being minor, aspect effects seem to provide a logical explanation for the BAL phenomenon. Along these lines of thought (sight ..) the question emerges whether (extra-strength FeII) BAL QSOs can be unified with nonBAL QSOs through axis orientation.

What do we know about FeII in radio-loud QSOs – does FeII emission reflect radio axis orientation? On average, compact core-dominated QSRs display stronger FeII features than do extended lobe-dominated QSRs (Miley and Miller 1979, Steiner 1981, Joly 1991, Hooimeyer 1991, Jackson and Browne 1991). The group of compact QSRs is thought to represent otherwise normal extended QSRs, viewed from preferred direction (line of sight close to radio axis). This suggests that FeII - on average - is an orientation indicator. Support for this view comes from the discovery by Zheng and O'Brien (1990) that FeII strength in (radio-loud and -quiet) QSOs roughly anti-correlates with the Hβ emission line width. This correlation appears parallel to the Wills and Browne (1986) correlation between radio core dominance and Hβ line width in QSRs. Since the latter correlation is thought to reflect the preferred orientation of the core-dominated QSRs it

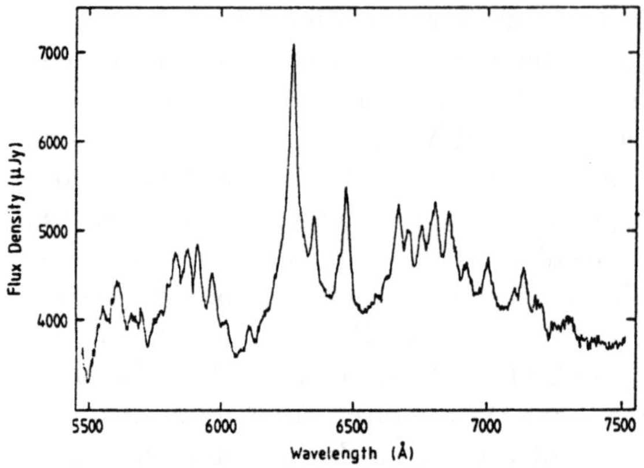

Fig. 3. **Optical spectrum of PG 1700+518 taken with Palomar 200-inch; redshifted Hβ at 6270 Angstrom, otherwise all FeII multiplets.**

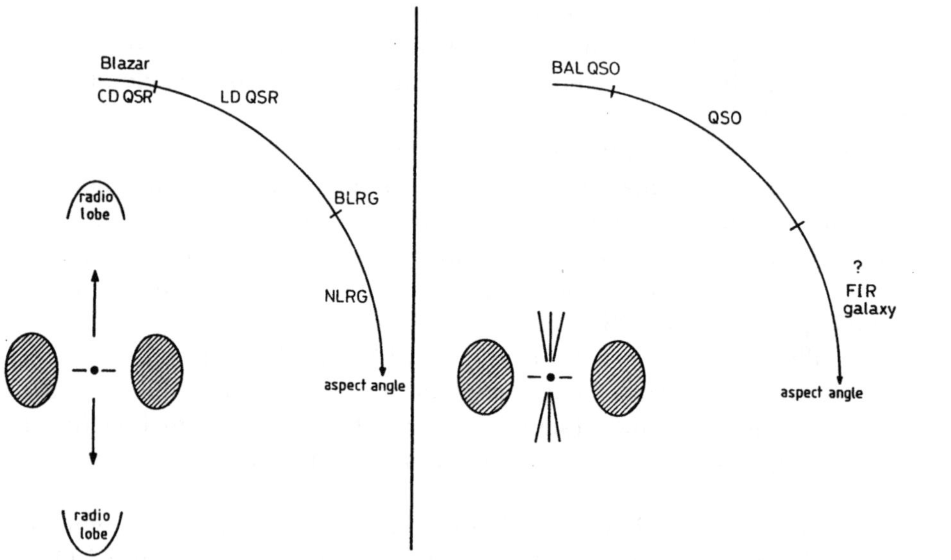

Fig. 4. **Unification models for radio-loud and radio-quiet QSOs.**

is tempting to attribute the former correlation to preferred orientation of the FeII-dominant QSOs in general. Do (extra-strength FeII) BAL QSOs have narrow Hβ lines, as predicted in this picture? Here unfortunately we lack data. Fig. 3 shows part of my optical spectrum (Palomar 200-inch) of PG 1700+518. Apart from many FeII complexes, indeed a narrow Hβ line (FWHM 2450 km/sec) is seen.

Summarizing, a case can be made that BAL QSO optical spectroscopic properties - FeII, Hβ - argue for identification with otherwise normal radio-quiet QSOs observed down a non-relativistic polar outflow cone. It is well possible that all radio-quiet QSOs have intrinsic BALs. Recalling that radio-loudness and BALs seem to represent a fundamental bimodality, I conclude by postulating that just two analogous scenario's (relativistic non-thermal versus non-relativistic thermal jet) may be able to explain the apparent dichotomy between radio-loud and radio-quiet QSOs. These scenarios are sketched in Fig. 4. I eagerly await homogeneous data to test these scenarios.

References

Barthel, P.D., 1989: *Astrophys. J.*, **336**, 606.

Barthel, P.D., 1991: in: *Proceedings of ESO Mini-Workshop on Quasar Absorption Lines*, ed. Shaver, Wampler, and Wolfe, Garching: ESO Sc. Rep. 9, p. 79.

Barthel, P.D., Tytler, D.R., Thompson, B., 1990: *Astron. Astrophys. Suppl.*, **82**, 339.

Briggs, F.H., Turnshek, D.A., Wolfe, A.M., 1984: *Astrophys. J.*, **287**, 549.

Djorgovski, S., Weir, N., Matthews, K., Graham, J.R., 1991: ApJL372,67.

Hartig, G.F., Baldwin, J.A., 1986: *Astrophys. J.*, **302**, 64.

Heckman, T.M., Lehnert, M.D., van Breugel, W., Miley, G.K., 1991: *Astrophys. J.*, **370**, 78.

Hooimeyer, J.R.A., 1991: *Ph.D. Thesis*, Leiden University.

Jackson, N., Browne, I.W.A., 1991: *Monthly Notices Roy. Astron. Soc.*, **250**, 422.

Joly, M., 1991: *Astron. Astrophys.*, **242**, 49.

Kellermann, K.I., Sramek, R., Schmidt, M., Shaffer, D.B., Green, R.F., 1989: *Astron. J.*, **98**, 1195.

Malkan, M.A., 1984: *Astrophys. J.*, **287**, 555.

Miley, G.K., Miller, J.S.: 1979, *Astrophys. J. (Letters)*, **228**, L55.

Onuora, L.I., 1991: *Astrophys. J.*, **377**, 36.

Rawlings., S., Saunders, R., 1991: *Nature* **349**, 138.

Sargent, W.L.W., Boksenberg, A., Steidel, C.C., 1988: *Astrophys. J. Suppl.*, **68**, 539.

Sargent, W.L.W., Steidel, C.C., Boksenberg, A., 1989: *Astrophys. J. Suppl.*, **69**, 703.

Smith, E.P., Heckman, T.M., 1990: *Astrophys. J.*, **348**, 38.

Scheuer, P.A.G., Readhead, A.C.S.: 1977, *Nature* **277**, 182.

Setti, G., Woltjer, L., 1989: AAL224,21.

Smith, E.P., Heckman, T.M., 1990: *Astrophys. J.*, **348**, 38.

Steiner, J.E.: 1981, *Astrophys. J.*, **250**, 469.

Stocke, J.T., Foltz, C.B., Weymann, R.J., Christiansen, W.A., 1984: *Astrophys. J.*, **280**, 76.

Strittmatter, P.A., Hill, P.,. Pauliny-Toth, I.I.K., Steppe, H., Witzel, A., 1980: *Astron. Astrophys. (Letters)*, **88**, L12.

Tytler, D.R., Barthel, P.D., 1991: in preparation.

Weymann, R.J., Morris, S.L. Anderson, S.F., 1988: in *Active Galactic Nuclei*, ed. Miller and Wiita (Berlin, Springer), p. 92.

Weymann, R.J., Morris. S.L., Foltz, C.B., Hewett, P.C., 1991: *Astrophys. J.*, **373**, 23.

Wills, B.J., Browne, I.W.A., 1986: *Astrophys. J.*, **302**, 56.

Yee, H.K.C., Green, R.F., 1984: *Astrophys. J.*, **280**, 79.

Zheng, Wei, O'Brien, P.T., 1990: *Astrophys. J.*, **353**, 433.

Luminosity Functions, Relativistic Beaming, and Unified Theories of AGN

Paolo Padovani[1] *and C.Megan Urry*[2]

[1] European Southern Observatory, Karl-Schwarzschild-Str. 2, 8046 Garching bei München, Germany
[2] Space Telescope Science Institute, 3700 San Martin Drive, Baltimore, MD 21218, USA

1. Introduction

If relativistically beamed emission dominates the observed flux from a class of objects, then there must be counterpart sources whose emission is beamed at larger angles to the line of sight. We evaluate such a scheme quantitatively testing the unification of BL Lacertae objects with Fanaroff-Riley type I (low-luminosity) radio galaxies, and of radio quasars with Fanaroff-Riley type II (high-luminosity) radio galaxies. Through a direct comparison of predicted and observed luminosity functions we show that both unification schemes are consistent with available data. The inferred ranges of Lorentz factors are in agreement with those derived from the (sparse) superluminal data. A full presentation of these results is given in Urry, Padovani, & Stickel, and Padovani & Urry [1].

2. Luminosity Functions and Relativistic Beaming

The possibility that radio emission from extragalactic sources is relativistically beamed has been considered for the past quarter century (see *Browne* [2] and *Barthel* [3]). If one accepts the existence of relativistic jets, the radiation must be highly anisotropic due to relativistic aberration. Sources beamed toward us — that is, with the relativistic velocity vector close to the line of sight — will appear much brighter, potentially by as much as three or four orders of magnitude, than in the case of isotropic emission, and still brighter than the same beamed source in the plane of the sky. Specifically, the apparent luminosity is enhanced by a factor δ^p, where $\delta = [\gamma(1-\beta\cos\theta)]^{-1}$ is the kinematic Doppler factor, $\gamma = (1-\beta^2)^{-1/2}$ is the Lorentz factor corresponding to velocity βc, θ is the angle to the line of sight, and $p \simeq n + \alpha$, where α is the spectral index ($F_\nu \propto \nu^{-\alpha}$) and n is roughly in the range 2 – 3, depending on the detailed jet physics [4].

Assuming that the jet speed is the same in every source and the directions are randomly distributed on the sky, the conditional probability of observing a luminosity $L = \delta^p \mathcal{L}$, given the intrinsic luminosity $\mathcal{L}$, can be shown to be $P(L|\mathcal{L}) = (1/\beta\gamma p)\ \mathcal{L}^{1/p}\ L^{-(p+1)/p}$. The observed luminosity function (LF) is the integral of the conditional probability times the intrinsic LF. In the case of an intrinsic LF of power law index B the observed LF will be a double power law of index B above the break and index $(p+1)/p \simeq 1.2 - 1.5$ below the break (see *Urry & Shafer* [5] for details).

This general character, a flat slope at the low-luminosity end due to the effects of beaming, and the same slope as the parent LF at the high-luminosity end, holds for most intrinsic LFs, including flatter parent LFs or more complex parent LFs [6] and for cases where there is a substantial contribution from unbeamed (isotropic) emission [7]. In the calculations that follow, we have assumed that the intrinsic luminosity of the beamed (jet) flux is proportional to the luminosity of the unbeamed flux: $\mathcal{L}_j = f\mathcal{L}_u$. Then the total luminosity, beamed plus unbeamed, is $L_t = (1 + f\delta^p)\mathcal{L}_u$. With this modification, the expression for $P(L|\mathcal{L})$ becomes slightly more complicated than previously derived [5]. If there exists a range of Lorentz factors in jets, then the flat-steep break will be less well-defined [6].

Our method is then the following: 1. we derive the observed LF of the supposed parent population; 2. we derive the observed LF of the beamed population; 3. we predict the LF of the beamed population from the parent LF; 4. we compare predictions and observations, adjusting the few free parameters (typically the Lorentz factor(s) and/or their distribution) to get the best agreement possible.

3. Unification of BL Lacs and FR Is

Arguably the most successful AGN unification scheme to date is the association of BL Lac objects with low-luminosity radio galaxies [2]. The most likely parent population, based on unbeamed characteristics like large-scale radio structure, host galaxy type, and emission line properties, consists of the Fanaroff-Riley type I (FR I) radio galaxies. We have previously applied the beaming formalism to BL Lac objects and FR I galaxies, using sizeable and complete X-ray samples, as well as the less complete optical data available, quantifying the proposed unification scheme and constraining the details of the beaming model [7], [8].

The radio data [9] are better suited for testing the beaming hypothesis for BL Lac objects because in the radio we have more and better data both for BL Lacs and FR Is. Perhaps because of this additional information,

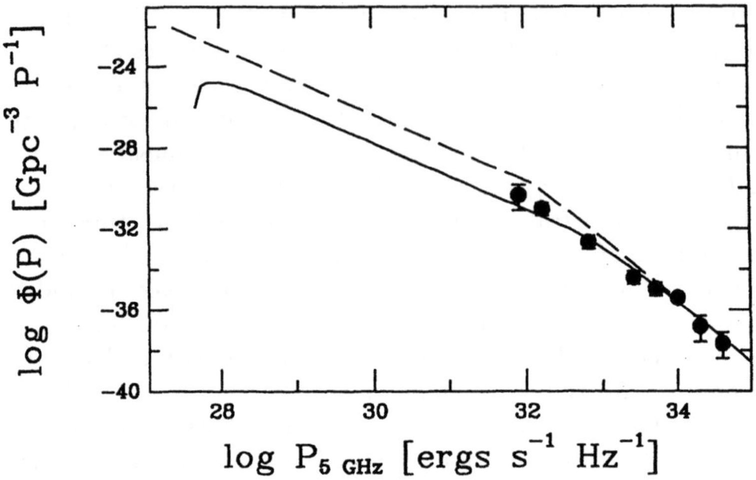

Fig. 1. The differential luminosity function of BL Lacertae objects (solid line) as predicted by our beaming model compared to the observational estimates (filled points) from *Stickel et al.* [9]. The differential luminosity function of FR I galaxies is also shown (dashed line)

it proved impossible to fit the observed radio LF of BL Lacs by beaming the radio LF of FR Is with a single Lorentz factor. Instead, we required a distribution of Lorentz factors in the range $5 \lesssim \gamma \lesssim 35$. The form of the distribution was weighted toward the lower values; e.g., for a power law of the form $n(\gamma) \propto \gamma^G$, the best-fit index was $G \sim -4$. The predicted (solid line) and observed (solid circles) radio LFs are compared in Fig. 1.

The mean Lorentz factor is $\langle\gamma\rangle \sim 7.5$, corresponding to a ratio between BL Lacs and FR Is of about 1.5%, i.e. about ten times smaller than in the X-ray case ($\sim$ 13%: [8]). The critical angle within which the object looks like a BL Lac (see Eq. [A1] of *Padovani & Urry* [8]) is approximately $\theta_c(\langle\gamma\rangle) \sim 10°$; for the extreme values of γ, $\theta_c(\gamma_1) \simeq 10°.2$ and $\theta_c(\gamma_2) \simeq 5°.5$.

Additional data are available from superluminal motion. All eleven BL Lac objects that have been observed more than once with VLBI techniques appear to show superluminal motion [10]. The apparent separation of radio components gives a lower limit to the Lorentz factor; for the BL Lacs the Lorentz factors are higher than ~ 4, and are distributed to at least $\gamma \sim 15$, with $\langle\gamma\rangle \sim 7$. These results agree roughly with the distribution in γ required to fit the luminosity function and core-halo ratios of radio-selected BL Lacs.

4. Unification of radio quasars and FR IIs

Turning to the more powerful radio sources, the idea that all radio-loud quasars are relativistically beamed is supported by recent studies suggesting that most, if not all, flat-spectrum radio quasars (FSRQs) are highly polarized and rapidly variable, i.e., blazar-like [11]. Moreover, superluminal motion is common in FSRQs, more so than in steep-spectrum radio quasars (SSRQs; [12]). A recent suggestion, discussed by *Peacock* [13] and explored by *Barthel* [14] (see also *Barthel* [3]) is that the parent population of all radio quasars consists of radio galaxies, with FSRQs being the most aligned objects and SSRQs at intermediate angles to the line of sight. The linear size distributions of high luminosity radio-galaxies and quasars (mostly SSRQs since Barthel used the low-frequency-selected 3CRR sample) support this hypothesis. Differences in the optical line emission properties were explained by invoking cylindrical obscuration with the same axis as the beamed radio emission. Here we test the scheme quantitatively incorporating the effects of beaming and including FSRQs.

We used the 2 Jy sample [15], a complete flux-limited sample including 233 sources with $F_{2.7GHz} \geq 2$ Jy, updating the redshifts and optical identifications using the latest version of the 1 Jy catalog [16]. The sources in the catalog were assigned to the following classes: FR I galaxies, FR II galaxies, SSRQs, FSRQs, BL Lacs, empty fields ($\lesssim 7\%$). The division between steep- and flat-spectrum sources was made at a spectral index $\alpha = 0.5$. The LF of FR II galaxies was derived from the 3CRR catalog [17], because a large fraction (nearly one-third) of FR II galaxies in the 2 Jy sample have uncertain redshift estimates and their evolutionary properties differ from those of radio galaxies in other samples. We checked carefully that neither the different selection frequency nor the different flux limit of the 3CRR as compared to the 2 Jy catalog will affect substantially the resulting LF.

The local LFs for FR IIs, SSRQs, and FSRQs were derived by first fitting a pure luminosity evolution model to each sample, testing the goodness of fit via a KS test, and then de-evolving the corresponding LFs. The resulting local LFs are shown in Fig. 2. Note that the LFs of SSRQs and FR II galaxies are similar in shape over the common range of powers, up to $P_{2.7} \sim 10^{34}$ ergs s^{-1} Hz^{-1}, but the SSRQs are lower in number density by a factor ~ 3. The ratio of the number densities of the two classes for $P_{2.7} \gtrsim 5 \times 10^{32}$ ergs s^{-1} Hz^{-1} is 3.1, in good agreement with the factor of 2.5 *Barthel* [14] found by simply dividing the number of 3CRR galaxies and quasars in the redshift range 0.5 – 1.

Even in this case it proved impossible to fit the observed LF of FSRQs with a single Lorentz factor. We required in fact a distribution in the range $5 \lesssim \gamma \lesssim 40$, again weighted toward low values: $n(\gamma) \propto \gamma^{-2}$, with a mean Lorentz factor $\langle\gamma\rangle \simeq 11$. The predicted (solid line) and observed (solid circles) radio LFs are compared in Fig. 2. The ratio between FSRQs and

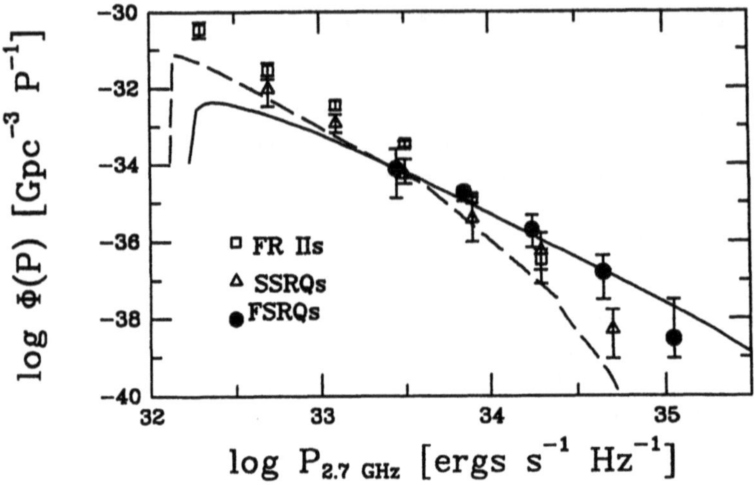

Fig. 2. Local luminosity functions for radio sources: filled circles represent flat spectrum radio quasars, open triangles represent steep spectrum radio quasars, while open squares represent FR II galaxies. The predictions of a beaming model are shown for flat spectrum radio quasars (solid line) and steep spectrum radio quasars (dashed line). The error bars have been derived by summing in quadrature the Poisson errors and the variations of the number density associated with a 1σ change in the evolutionary parameter τ (see *Padovani & Urry* [1] for details).

parents is of about 2%. The critical angles for the extreme values of γ are: $\theta_c(\gamma_1) \simeq 14°$ and $\theta_c(\gamma_2) \simeq 6°$.

VLBI observations of thirteen FSRQs in the 2 Jy sample are available in the literature. Every one of the thirteen FSRQs shows evidence of superluminal motion ($\beta_{app} > 1$), with a distribution of γ_{min} for the fastest component in the range 5 – 35, in rough agreement with the range necessary to explain the observed LF.

According to the beaming hypothesis, the steep-spectrum radio quasars are supposed to be at intermediate angles but their intrinsic properties — in particular, the value(s) of γ — must be identical to those of the FSRQs. Therefore, with these parameters already fixed, we can compare the predicted LF of SSRQs to the observed one. The method of calculation is similar to the one used in the previous section; however, since SSRQs are supposed to be misaligned objects, $\theta_{min} \neq 0$. Assuming there is no overlap between the two classes, we take $\theta_{min} = 14°$, the largest value of the critical angle for FSRQs. Note that in this case we have no free parameters.

The comparison between observations and the predictions of the beaming model for SSRQs is shown in Fig. 2. The agreement is again quite good. The critical angles for the extreme values of γ are: $\theta_c(\gamma_1) \simeq 38°$ and $\theta_c(\gamma_2) \simeq 14°$. Note that the same distribution of Lorentz factors produces a good fit to both the LFs of FSRQs and SSRQs. We stress that the model dependent value of $\simeq 38°$ we find for the angle separating SSRQs from FR II galaxies is in very good agreement with the angle derived from the number density ratio $\theta = \arccos(1 + 1/3.1)^{-1} \simeq 41°$. We remind the reader that the latter estimate is valid only when dealing with unbeamed luminosities or when the effect of beaming is not very strong because the objects are viewed off-axis, as in this case.

5. Conclusions

We have investigated the unification of BL Lac objects with FR I radio galaxies and of radio quasars with FR II radio galaxies, through direct comparisons of predicted and observed luminosity functions. The predictions of a beaming model are consistent with the observational data and the inferred ranges of Lorentz factors are in agreement with the (still sparse) superluminal data. More specifically, we have found:

- BL Lacs (radio data): $5 \lesssim \gamma \lesssim 35$, $n(\gamma) \propto \gamma^{-4}$, i.e., $\langle\gamma\rangle \simeq 7.5$; $\theta \lesssim 10°$;
- Radio quasars: $5 \lesssim \gamma \lesssim 40$, $n(\gamma) \propto \gamma^{-2}$, i.e., $\langle\gamma\rangle \simeq 11$;
  FSRQs: $\theta \lesssim 14°$; SSRQs: $14° \lesssim \theta \lesssim 40°$.

References

1. Urry, C.M., Padovani, P., Stickel, M., 1991: *Astrophys. J.*, in press.
   Padovani, P., Urry, C.M., 1991: *Astrophys. J.*, submitted.
2. Browne, I.W.A., 1992: this volume
3. Barthel, P.D., 1992: this volume
4. Lind, K.R., Blandford, R.D., 1985: *Astrophys. J.*, **295**, 358.
5. Urry, C.M., Shafer, R.A., 1984: *Astrophys. J.*, **280**, 569.
6. Urry, C.M., Padovani, P., 1991: *Astrophys. J.*, **371**, 60.
7. Padovani, P., Urry, C.M., 1991: *Astrophys. J.*, **368**, 373.
8. Padovani, P., Urry, C.M., 1990: *Astrophys. J.*, **356**, 75.
9. Stickel, M., Padovani, P., Urry, C.M., Fried, J.W., Kühr, H., 1991: *Astrophys. J.*, **374**, 431.
10. Mutel, R.L., 1990: in: *Parsec-Scale Radio Jets*, ed. J.A. Zensus, T.J. Pearson (Cambridge: Cambridge University Press), p. 98.
11. Fugmann, W., 1988: *Astron. Astrophys.*, **205**, 86. Impey, C.D., Tapia, S., 1990: *Astrophys. J.*, **354**, 124.
12. Hough, D.H., Readhead, A.C.S., 1987: *Astrophys. J. (Letters)*, **321**, L11.
13. Peacock, J.A., 1987: in: *Astrophysical Jets and Their Engines*, ed. W. Kundt (Dordrecht: Reidel), p. 185.
14. Barthel, P.D., 1989: *Astrophys. J.*, **336**, 606.
15. Wall, J.V., Peacock, J.A., 1985: *Monthly Notices Roy. Astron. Soc.*, **216**, 173.

16. Stickel, M., Kühr, H., 1991: in preparation.
17. Laing, R.A., Riley, J.M., Longair, M.S., 1983: *Monthly Notices Roy. Astron. Soc.*, **204**, 151.

Discussion

Camenzind: The beaming effect depends on p. What is the effect on the mean Lorentz factor?

Padovani: Our results have been derived for $p = 3 + \alpha$. If one takes $p = 2 + \alpha$ a good fit can still be obtained with a larger value of γ_2. Since the best fit distributions of γ are quite steep, the mean value of the Lorentz factor is essentially unchanged.

Bregman: VLBI studies indicate that the opening angle of the jet is not defined by the bulk Lorentz factor but it is likely to be quite different. How would this affect your results?

Padovani: If the opening angle is not defined by the bulk Lorentz factor, the ratio of beamed to parent objects is no longer related to the Lorentz factor. If that is the case, however, as it has been shown yesterday by Laura Maraschi, one ends up with quite large Lorentz factors ($\gamma \gtrsim 12$) for both the X-ray and radio emission of BL Lacs.

Maraschi: Just a comment to the remark of Joel Bregman. If you widen the opening angle of the jet, you need a larger Lorentz factor to have a similar intensity vs. angle distribution.

Meisenheimer: When I prepared the talk I gave yesterday, I came across PKS 0521-36, a famous BL Lac. It has an extended jet which is $\gtrsim 30\%$ polarized. That is hard to understand if the jet is more than 45° out of the plane of the sky.

Padovani: We find that BL Lac radio jets should be seen, on average, within 10° of the line of sight, so, if confirmed, the case you mentioned would not be consistent with our model.

Quasar Surveys

Paul Hewett

Institute of Astronomy, Madingley Road, Cambridge, CB3 0HA, UK.

Abstract. **Following a brief review of the status of quasar surveys, the effects of several properties of quasars and intervening hydrogen absorption on the detection of quasars in optical surveys are discussed. It is now possible to quantify precisely the details of survey selection functions although to date few surveys have been the subject of such a procedure. The techniques necessary for such a quantification are outlined and illustrated using data from a multi–colour survey.**

1. Introduction

The proceedings of the recent workshop held in Victoria, Canada (Ellingson and Crampton 1991) provides a comprehensive summary of the major quasar survey projects and their implications for our understanding of the evolution of the quasar luminosity function. To avoid duplication this contribution focuses on more general issues related to the implementation and interpretation of quasar surveys. Two topics are highlighted, spectroscopic and variability properties of quasars, not normally taken into account in the interpretation of survey results, and secondly, procedures necessary to provide a quantitative assessment of a survey's sensitivity to quasars as a function of redshift, magnitude and spectral energy distribution.

Section 2 gives a brief review of the status of quasar survey work and an outline of what is generally accepted concerning the form and evolution of the quasar luminosity function is presented in Section 3. After drawing attention in Section 4 to what is not well established, the affects of the dispersion in quasar spectral properties, long–term quasar variability and intervening absorption systems are discussed in Section 5. Procedures for determining the selection function for optical surveys are described in Section 6 using the high–redshift multi–colour survey of Warren, Hewett and Osmer (1991a,b) to illustrate some of the key issues. To conclude, a summary of the main points of the review is contained in Section 7. While the examples used throughout this contribution refer to surveys performed at optical wavelengths the principles apply equally to other wavebands, although in some instances the selection functions, *e.g.* at centimeter radio wavelengths, are simpler. Much of the discussion is particularly relevant to the determination of joint luminosity functions where samples of significant range in luminosity must be obtained at two different wavelengths,

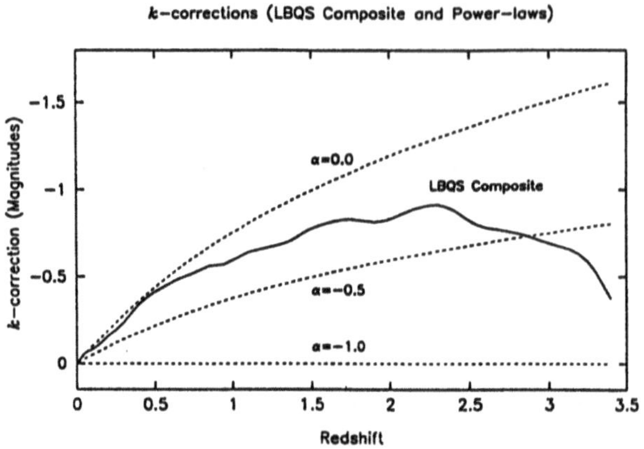

Fig. 1. k–corrections for the UK Schmidt Telescope B_J band as a function of redshift for three power–law spectral energy distributions commonly used to calculate k–corrections for quasar spectra. The curves are labeled with the slope α, $F(\nu) = \nu^{\alpha}$. The solid line is the k–correction for the composite spectrum of Francis *et al.* (1991).

e.g. radio–optical, infrared–optical. The discussion will be confined to high–luminosity objects, $M_B \leq -23$, using $H_0 = 50\,\mathrm{km\,s^{-1}\,Mpc^{-1}}$ and $q_0 = 0.5$.

2. Status

In a recent review Boyle (Boyle 1991) discussed the progress that has been made in obtaining large samples of spectroscopically confirmed quasars from surveys performed in the optical. Work in the latter half of the 1980s has brought about a significant improvement in the availability of quasar samples that have been identified in a relatively well–defined fashion and thus enable constraints to be placed on the change in the space density and luminosity function as a function of redshift. Figure 1 of Boyle's review illustrates the dramatic growth in the number of quasars discovered and shows that the absolute magnitude–redshift plane is relatively well covered for redshifts $0.2 < z < 3$ with a range of some five magnitudes accessible at fixed redshift. The combination of deep, narrow angle surveys (*e.g.* Koo and Kron 1988) and wide–field surveys to brighter magnitudes (*e.g.* Schmidt and Green 1983) results in there being few regions of the absolute magnitude–redshift plane that are not probed by existing surveys.

Further progress has since been made at all redshifts, most conspicuously at high redshifts, the detection of $z > 4$ quasars has now become almost routine (McMahon and Irwin 1991). The ability to generate large samples of quasars has resulted from a better understanding of discovery techniques coupled with the increased efficiency offered by improved single and multi–object spectrographs. Detection and confirmation of large numbers of quasars still requires substantial investment of telescope time but is no longer a limiting factor. The forthcoming edition of the Hewitt–

Burbidge catalogue will contain some 8000 confirmed quasars – numbers are no longer a problem!

Less positively the quality of the spectroscopic data is, by any objective standard, still extremely poor. The majority of published quasar spectra from surveys are of extremely low signal–to–noise ratio with many acquired at resolutions measured in tens of Angstroms. In a significant number of cases spectra are barely adequate to allow the identification of the strongest emission features and the measurement of even basic spectroscopic properties such as line equivalent–widths is not possible.

Photometric information is also of low precision and limited to few pass-bands. The plethora of photometric systems and the difficulty of deriving meaningful colour equations for objects with energy distributions containing strong features and which are not well approximated by blackbodies account for part of the problem. The propensity of astronomers working with slit-less spectroscopic material to prefer some form of continuum magnitude estimate while the ultra–violet excess and multi–colour surveys naturally present broad–band magnitude estimates is a further complication. The overriding factor however has been the difficulty of measuring magnitudes to ten per cent precision for many thousands of objects distributed over large areas of sky. Surveys concentrating on small areas of sky (*e.g.* Koo and Kron 1988 and Boyle *et al.* 1990) have been able to supply magnitudes to ten or twenty per cent precision in broad–band passes that approximate standard colour systems. The wide–field surveys on the other hand are woefully short of photometric information, for example, there are still no precise $U - B$ measures for the quasars in Schmidt and Green's Bright Quasar Survey.

More fundamentally the spectroscopic coverage in the quasar rest–frame is very restricted, a spectrum covering 2000Å in the observed frame providing a mere $\sim$ 700Å coverage in the rest–frame for a quasar with redshift $z = 2$. As a result, typical spectroscopic observations of quasars probe only narrow ranges of ionization energies for the strong emission lines and parameters specifying continuum shape are dominated by local effects rather than reflecting the desired large scale shape. It is sobering to note that the number of quasars for which the most basic of line properties exist, such as Lyman–α and Hα equivalent widths, is confined to a few tens of (unrepresentative) objects and the cases where precise relative–fluxes exist is even smaller.

3. Qualitative Picture

In the late 1960s and early 1970s Schmidt (1968) and others were able to show that the space density of the quasar population evolved rapidly for redshifts $0 < z < 2$. Marshall *et al.* (1984), Boyle *et al.* (1990) and others confirmed and refined this conclusion using much larger samples. The rate of evolution is so large that its effects were readily detectable in the samples

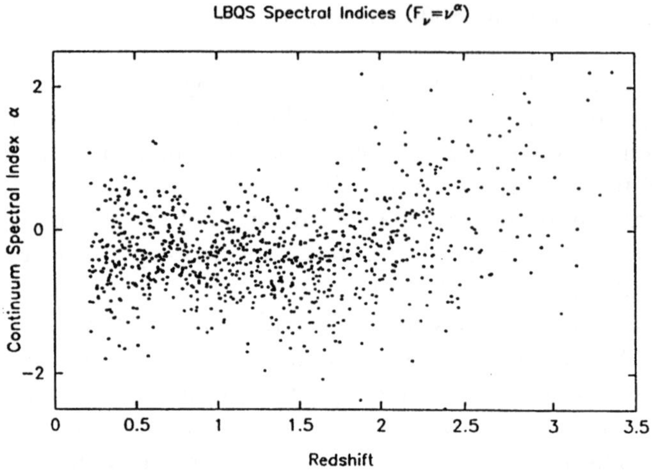

Fig. 2. **The distribution of power–law slopes, α, for quasars in the LBQS. There is a substantial spread in slopes at fixed redshift which exceeds the measurement errors by a large factor.**

of only a few tens of objects available at the end of the 1960s. A more quantitative appreciation of the nature of the evolution can be gained from inspection of Figure 2 in Boyle (1991), where the quasar luminosity function is plotted for different redshift intervals. At fixed space density the location of the luminosity function changes by a full magnitude for a change in redshift of $\Delta z = 0.5$, over the redshift interval $0 < z < 2$. In the now familiar "pure luminosity evolution" parameterization, the luminosity at fixed space density evolves as $L \propto (1+z)^{3.4}$ until a a redshift $z \sim 2$. In the most simplistic interpretation of such a parameterization, quasars were a factor $\sim$ 35 brighter at redshift $z = 2$ compared to those in the local universe. A feature to emerge particularly from the Boyle *et al.* work is the two–power law form of the luminosity function, the shape of which appears to be independent of redshift.

This change in the space density of quasars is not a subtle effect and agreement over the qualitative behaviour is hardly surprising given the strength of the evolution. Hartwick and Schade (1990) have made a detailed comparison of different surveys and conclude that the pure luminosity evolution parameterization advocated by Marshall *et al.* and Boyle *et al.* does reproduce the behaviour for redshifts $z < 2.2$, but inspection of Figure 2 in Hartwick and Schade (1990) shows the data is not entirely convincing, particularly with regard to an invariant two–power–law shape. Substantial disagreements still exist between surveys, Goldschmidt and Miller (1991) present results suggesting the Schmidt and Green (1983) survey may be a factor two incomplete. Increasing the counts at bright magnitudes as advocated by Goldschmidt and Miller has the effect of reducing the rate of evolution, and enhancing the number of high luminosity quasars making the steep decline in the number of quasars as a function of increasing luminosity less pronounced.

At redshifts $z > 2$ the situation is less clear, but observations do rule out the possibility that the rapid rate of evolution at lower redshifts continues much above $z = 2.5$. The data are still relatively sparse and increases or decreases in the space density of factors a few hard to establish. Hartwick and Schade (1990) conclude in common with several other authors that the space density remains essentially constant over the redshift range $2 < z < 3$.

Investigations of trends in the spectroscopic properties of the quasar population with redshift are severely hampered by the poor quality and restricted nature of the spectra as discussed in the previous section. Quantitative investigations are still in their infancy and at present all that can be said is there are no very strong changes in spectroscopic properties as a function of redshift.

4. Quantitative Picture – Limiting Factors

To derive the luminosity function from a sample of quasars it is necessary to calculate the absolute magnitude of each quasar at some fiducial wavelength, or, in ideal circumstances the bolometric luminosity. The aim is to construct a sample that may be compared directly to a quasar sample observed in some different redshift or absolute luminosity range. The form of the quasar luminosity function is such that only small changes in the depth of a survey results in large changes in the number of quasars detected. This is due to the very steep dependence of the space density on absolute magnitude. Taking the best–fit model derived by Boyle *et al.* (1990), the slope of the luminosity function at bright magnitudes results in a change of a factor two in number density for a change in absolute magnitude of only ~ 0.2 magnitudes. At fainter absolute magnitudes, below the break in the luminosity function, the dependence of space density on absolute magnitude is much weaker. However, the majority of quasar surveys probe, in part at least, the steep portion of the luminosity function. The derived quasar space density is thus critically dependent on the exact depth of a survey in apparent magnitude and also to the determination of absolute magnitudes from the measured apparent magnitudes. A feature of the reanalysis of published surveys by Hartwick and Schade (1990) is the use of consistent k–corrections and cosmological models which removes a number of apparent discrepancies between published analyses.

The importance of the k–correction is illustrated in Figure 1, where the k–corrections for three power–law quasar spectral energy distributions (SEDs), $F(\nu) = \nu^{\alpha}$ with $\alpha = 0.0$, -0.5 and -1.0, are shown together with that for the composite quasar spectrum of Francis *et al.* (1991). The Y–axis gives the value of the k–correction in magnitudes for the UK Schmidt Telescope B_J band employed in many photographic surveys at blue wave-

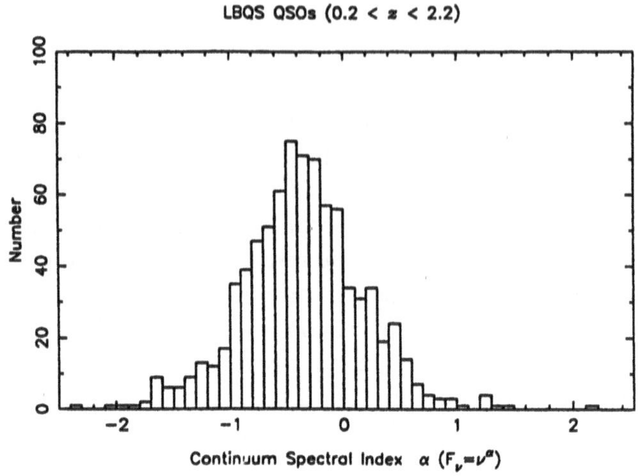

Fig. 3. Histogram of power–law indices, α, for LBQS quasars with redshifts $0.2 < z < 2.2$ where systematic trends in α due resulting from the changing rest–frame range visible in the spectra are small. Note the substantial spread in the values of α.

lengths. The size of the k–correction for different power–law SEDs changes significantly and thus space densities can vary by large factors depending on the form adopted for the quasar SEDs.

There is a considerable dispersion in the spectral energy distributions (SEDs) of quasars in the rest–frame ultraviolet and optical regions. Differences among objects are evident in the emission lines as well as the overall continuum slope and shape – see for example the relatively extended wavelength range, high signal–to–noise ratio spectra in Weymann *et al.* (1991) for an idea of the differences between broad absorption line and more "typical" quasars and also the dispersion within the two groups. On more extended wavelength scales McDowell *et al.* (1989) have stressed the variation in the strength of the ultraviolet component of quasar SEDs. However, it is usually assumed that the dispersion in continuum properties for quasars is small and that a mean k–correction is a good approximation for the quasar population as a whole. Evidence cited in favour of such a view includes measures of continuum slopes derived from extremely limited rest–frame wavelength ranges. An alternative view is suggested by Figure 2, which illustrates the distribution of power–law indices for quasars in the Large Bright Quasar Survey (LBQS) (Chaffee *et al.* (1991) and references therein). These continuum indices were measured during the generation of the composite quasar spectrum presented by Francis *et al.* (1991) and a fuller description of the relatively involved procedures used in the coaddition of the quasars and the measurement of the continuum slopes can be found in that paper.

A key problem in the description of quasar continua, including the simple power–law parameterization, is the identification of wavelength regions free of contamination from emission and absorption features. A further problem relates to the change in rest–frame wavelength coverage as a function of the quasar redshift which results in different continuum "windows" being used for objects at different redshifts. The measurements of Francis *et al.* are affected by both these problems and a number of systematic trends with redshift are evident in Figure 2. For redshifts $z < 2.2$ however the amplitude of the systematic trends with redshift are relatively small $\Delta\alpha \sim 0.3$. Figure 3 presents a frequency histogram of the continuum slopes for LBQS quasars with redshifts $0.2 < z < 2.2$. The error on an individual slope determination is ~ 0.25 and neither the systematic changes in slope with redshift nor the measurement errors can explain the observed range in continuum slopes, suggesting the bulk of the spread is due to the intrinsic dispersion among the quasar SEDs.

If quasar SEDs show a spread of even ~ 0.5 in their continuum slopes over wavelength intervals of ~ 2500Å there are significant consequences for our quantitative understanding of quasar evolution. Two populations of quasars with a continuum slope difference of $\Delta\alpha \sim 0.5$ observed at zero redshift will have observed–frame apparent magnitudes differing by 0.5 magnitudes when observed at redshift $z = 2$. This corresponds to probing 0.5 magnitudes deeper into the luminosity function for one population, resulting in an apparent increase of a factor ~ 4 in number density unless the correct k–corrections are applied for each population.

The evolution at low redshifts is so strong that the qualitative picture will not change if there is a spread in quasar SEDs but for redshifts significantly in excess of $z = 2$, where the very rapid evolution no longer holds, the potential effects of the dispersion in quasar SEDs can make the difference between interpretations where the space density increases or decreases. This is particularly true when observations probe regions of the luminosity function close to a change in slope or any other significant feature in the luminosity function. The spread in quasar SEDs precludes the unique identification of each quasar with a particular portion of the luminosity function and application of "mean" corrections will not in general solve the problem.

Intrinsic long–term variability of the quasar population is another factor usually ignored in surveys. Hook *et al.* (1991) have examined the variability properties for a large sample of optically selected quasars. Together with the work of Hawkins (1986) these results show that the majority of the quasar population exhibits magnitude variations of tens of per cent over timescales of 1-10 years. For ultraviolet excess and multi–colour surveys employing plates or CCD frames taken months or years apart the observed "colours" of many quasars are significantly modified by intrinsic variations in luminosity. The result is a blurring of supposedly well–defined colour criteria with the added complication that the amplitude of the effect may depend on both the luminosity and redshift of the quasars. An outline of the procedures necessary to take account of variability together with a graphic illustration of the effect can be found in Warren, Hewett and Osmer (1991c).

Modifications of quasar spectra caused by intervening material along our line of sight are important for rest–frame wavelengths $\lambda < 1216$Å. Our knowledge of the space density of Lyman–Limit absorbers, damped Lyman–α systems and the Lyman–α forest and their evolution with redshift has improved significantly in the last five years. The dispersion induced by these absorbers in the observed broadband colours that probe rest–frame wavelengths below 1216Å can amount to several magnitudes. Mean corrections are again of little use and a relatively sophisticated model of the distribution of intervening absorbers is necessary in order to simulate the effect of random lines of sight to quasars (Giallongo and Trevese 1990, Moller and Warren 1991).

5. Quantitative Picture - Requirements

The effects of just three aspects of quasars and intervening systems on the observed properties of quasars were outlined in the previous section. It follows that the sensitivity of most optical quasar surveys depends to some extent on these, and a number of other, factors. The effort involved in quantifying precisely the consequences of such effects on the sensitivity of a given survey is considerable and in many instances involves rather more work than identifying the quasars in the first place, which perhaps explains the apparent lack of enthusiasm astronomers have shown for the procedure. However, with the availability of digital data and reproducible selection procedures for the identification of quasar candidates such a quantification is possible.

The translation of quasar survey results into real constraints on theoretical models of the evolution of the space density and luminosity function consists of four main stages. The first involves the ability to generate synthetic quasar spectra as they would appear above the Earth's atmosphere. This phase must include quasars with the full range of SEDs - dispersion in continuum properties, emission line strengths *etc.* . It must also include a realistic model for the variability behaviour of the quasar population for time intervals equal to those over which the survey material is acquired - note that many existing surveys employ photographic plates obtained many years apart. A detailed model for the effects of intervening absorption on the quasar spectra is also required. In each case determination of the the "mean" effect is not sufficient, the distribution of observed properties in the population must be reproduced.

The second stage involves the precise specification of the observational experiment. This aspect of the procedure was first carried out by Schmidt, Schneider and Gunn (1986) when determining the effectiveness of their grism transit survey for high redshift quasars. In the case of multi–colour and ultraviolet excess surveys it must include magnitude calibration, specification of magnitude errors for each bandpass and as a function of apparent magnitude. Determination of filter bandpasses and any significant colour effects resulting from atmospheric extinction. Epoch differences between plate exposures to determine the effects of quasar variability on observed magnitudes and hence colours.

With these two stages complete it is possible to take a quasar of specified SED, models for the effects of intervening absorption and quasar variability and simulate how the object would appear in the observational data. To allow adequately for the substantial changes that can be introduced by such factors as variability, differing paths through intervening absorbers and the effects of magnitude errors it is necessary to produce many "realizations" of the individual object with a specified SED.

The third stage involves placing each realization of the simulated quasar into the *real* dataset and then applying the quasar candidate selection algorithm to find whether the object would have been detected. The process is repeated for each realization of the object, the number of occasions the synthetic object is identified as a quasar candidate divided by the total number of realizations gives the probability of detection. The use of the real dataset is essential as in most surveys the distribution of other objects, usually Galactic stars, influences whether an object is chosen as a candidate. For example, little is learned from the superficially impressive statement that a weak–emission line quasar is found to have a line with significance 3.5σ in some slitless spectroscopic survey. The relevant question is, given the distribution of "line detections" for the other (say) 10,000 stellar objects in the survey, would the 3.5σ object have been chosen as a quasar candidate. The third stage proceeds by taking the full range of quasar SEDs, luminosities and redshifts to which the survey may be sensitive and determining a three–dimensional array of detection probabilities.

In the final stage the range of models consistent with the observational data is determined. Taking a model prediction for the luminosity function, its evolution as a function of redshift and an intrinsic distribution of quasar SEDs - which may itself vary with parameters like redshift or luminosity - the number of quasars predicted to be found in the survey is determined by passing the model predictions through the array of detection probabilities. The goodness–of–fit can then be assessed using a two–dimensional Kolmogorov–Smirnov test in the *observed* apparent magnitude–redshift plane. This procedure avoids the instabilities that arise when attempts are made to derive the intrinsic distribution directly from observations that are in practice always of too low signal–to–noise ratio. It is in this final stage that surveyists can be less than helpful, in that their "favourite" model is shown to be consistent with the data, whereas the requirement is the determination of the range of models consistent with the data.

Figures 4, 5 and 6 illustrate some aspects of the discussion with specific examples taken from the Warren, Hewett and Osmer (1991a,b) survey. All three figures are two–colour projections of the six–dimensional magnitude

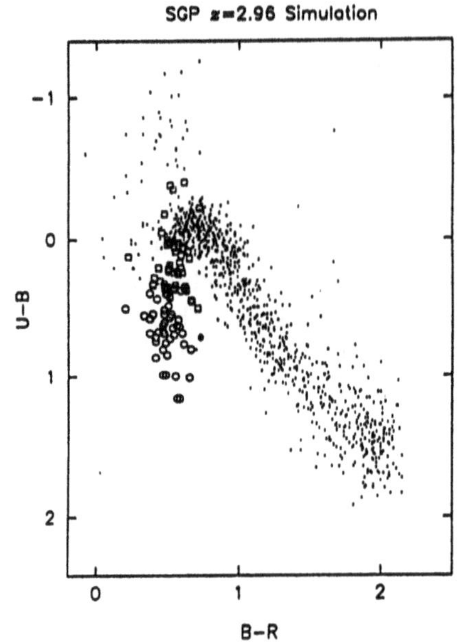

Fig. 4. $B - R$ *versus* $U - B$ **two–colour diagram for the south galactic pole field of the Warren *et al.* (1991a,b) quasar survey. Small points; stellar objects in the field. Solid circle; confirmed quasar, redshift $z = 2.96$. Open squares and circles; simulated objects with the same SED as the confirmed quasar. The significance of the spread in the positions of the simulated objects and the two different symbols are discussed in the text.**

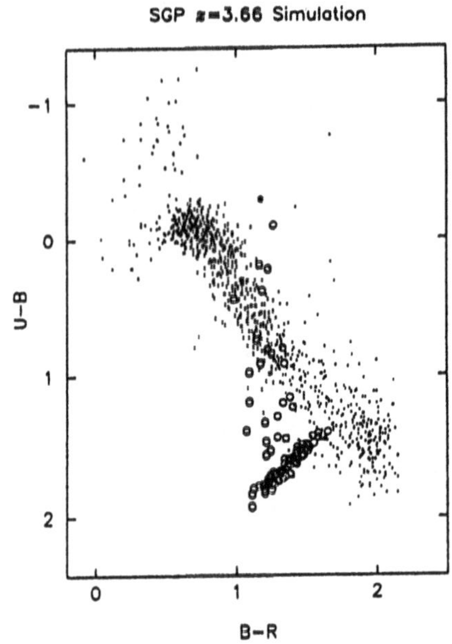

Fig. 5. As figure 4 but for confirmed quasar of redshift $z = 3.66$. The quasar is unusual in possessing no Lyman–Limit absorption system close to the quasar redshift. See text for details.

space used to select quasars by Warren *et al.*. The small dots show the colours of approximately 1000 stellar objects in one of the survey fields - they serve to define the position of the locus of main sequence stars in each figure. The solid circle in each figure is the location of a confirmed quasar. Over–plotted in each figure are the colours of 100 realizations of a simulation of a quasar of similar SED to that of the real quasar - *i.e.* the figures illustrate part of the third stage in the quantification procedure described above. The open circles represent realizations that were successfully identified as quasar candidates (using the full six–dimensional data set) while open squares those that alluded detection. Figure 4 shows a redshift $z = 2.96$ quasar and the location of the quasar is well reproduced by the simulations that would be detected as candidates. Much of the dispersion in the $U - B$ direction arises from the different redshifts of Lyman–Limit systems for each realization. The dispersion in $B - R$ arises from the effects of variability and to a lesser extent magnitude errors. Figure 5 shows the same information for an extremely unusual quasar that shows no Lyman–Limit system out to the atmospheric cutoff. The bulk of the simulations, corresponding to objects

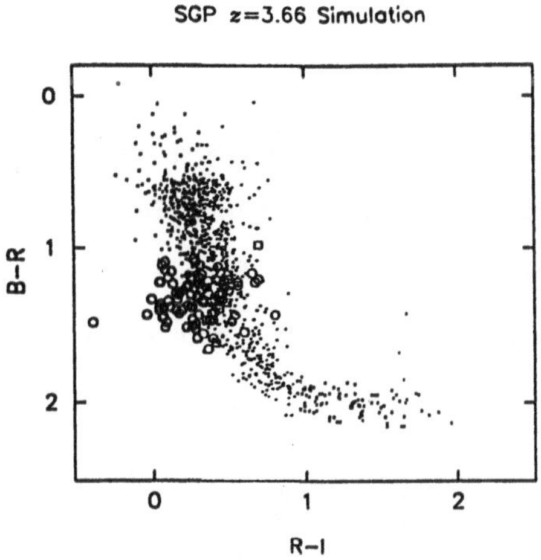

Fig. 6. *R − I versus B − R* **diagram for the** $z = 3.66$ **quasar of figure 5. Symbols as in figure 4. The simulations produce a good match to the quasar colours in the observed-frame colours not affected by rest-frame wavelengths shortward of Lyman-α. The text contains further details of the comparison.**

with Lyman–Limit systems much closer to the quasar redshift lie two magnitudes redder in $U - B$, but note how the simulations do reproduce a few objects with the colours of the detected quasar. Figure 6 shows a different two-colour projection for the same quasar but at wavelengths unaffected by the region shortward of Lyman-α, here the simulations fall close to the position of the quasar demonstrating that the simulations are appropriate to the quasar spectrum.

6. Summary

Quasar surveys have reached the stage where they can be relatively well targeted and produce large numbers of quasars. It is now possible to determine the selection function for surveys employing digital data and reproducible selection algorithms using the approach described in the previous section. The investment of resources to complete such comprehensive quantification investigations are considerable and to date observers have largely ignored the problem, appealing instead to comparisons with other surveys and qualitative assessments of completeness. If quantitative estimates of the space density, form of the luminosity function, distribution of quasar SEDs and their evolution with redshift are to be established such quantification is essential. Observers have also been keen to concentrate on the merits of their own surveys and to point out the complexity of selection criteria

I would like to thank Stephen Warren and Paul Francis for providing the data for the figures.

References

Boyle, B.J., 1991: *The quasar population at high redshifts* invited review at the Texas - ESO/CERN Symposium on Astrophysics, Brighton, UK

Boyle, B.J., Fong, R., Shanks, T., and Peterson, B.A., 1990: *Monthly Notices Roy. Astron. Soc.*, **243**, 1.

Chaffee, F.H., Foltz, C.B., Hewett, P.C., Francis, P.J., Weymann, R.J., Morris, S.L., Anderson, S.F., and MacAlpine, G.M., 1991: *Astron. J.*, **102**, 461.

Francis, P.J., Hewett, P.C., Foltz, C.B., Chaffee, F.H., Weymann, R.J., and Morris, S.L., 1991: *Astrophys. J.*, **373**, 465.

Giallongo, E., and Trevese, D., 1990: *Astrophys. J.*, **353**, 24.

Goldschmidt, P., and Miller, L., 1991: *The space distribution of quasars Astr. Soc. Pacific Conf. Series*, eds. Crampton, D. *et al.*

Hartwick, F.D.A., and Schade, D., 1990: *Ann. Rev. Astron. Astrophys.* **28**, 437.

Hawkins, M.R.S., 1986: *Monthly Notices Roy. Astron. Soc.*, **219**, 417.

Hook, I., Irwin, M.J., McMahon, R.G., and Boyle, B.J., 1991: *The space distribution of quasars Astr. Soc. Pacific Conf. Series*, eds. Crampton, D. *et al.*

Koo, D.C., and Kron, R.G., 1988: *Astrophys. J.*, **325**, 92.

McDowell, J.C., Elvis, M., Wilkes, B.J., Willner, S.P., Oey, M.S., Polomski, E., Bechtold, J., and Green, R.F., 1989: *Astrophys. J. (Letters)*, **345**, L 13.

Marshall, H.L., Avni, Y., Braccesi, A., Huchra, J.P., Tanenbaum, H., Zamorani, G., and Zitelli, V., 1984: *Astrophys. J.*, **283**, 50.

McMahon, R.G., and Irwin, M.J., 1991: *The space distribution of quasars Astr. Soc. Pacific Conf. Series*, eds. Crampton, D. *et al.*

Moller, P., and Warren, S.J., 1991: *The space distribution of quasars Astr. Soc. Pacific Conf. Series*, eds. Crampton, D. *et al.*

Schmidt, M., 1968: *Astrophys. J.*, **151**, 393.

Schmidt, M., and Green, R.F., 1983: *Astrophys. J.*, **269**, 352.

Schmidt, M., Schneider. D.P., and Gunn, J.E., 1986: *Astrophys. J.*, **306**, 411.

Warren, S.J., Hewett, P.C., and Osmer, P.S., 1991a: *Astrophys. J. Suppl.*, **76**, 1.

Warren, S.J., Hewett, P.C., and Osmer, P.S., 1991b: *Astrophys. J. Suppl.*, **76**, 23.

Warren, S.J., Hewett, P.C., and Osmer, P.S., 1991: *The space distribution of quasars Astr. Soc. Pacific Conf. Series*, eds. Crampton, D. *et al.*

Weymann, R.J., Morris, S.L., Foltz, C.B., and Hewett, P.C., 1991: *Astrophys. J.*, **373**, 23.

Discussion

Wilkes: How did you generate the composite spectrum? In particular did you take out the range of power–law slopes at a given redshift?

Hewett: it is not easy to answer that question succinctly, the problem occupied us for a considerable time. We thought it sufficiently important that full details are contained in the Francis *et al.* (1991) paper.

O'Dea: What fraction of the very luminous $z > 3$ quasars are radio–loud?

Hewett: there are relatively small numbers of $z > 3$ quasars in the LBQS but we have obtained VLA snapshots of some 250 LBQS quasars. Data reduction is still in progress, preliminary indications are that the radio–loud fraction is close to the canonical 10 per cent with indications that the fraction declines for objects at high–redshift (or equivalently luminosity).

Wandel: Is there a correlation between the continuum slope and luminosity of the LBQS quasars?

Hewett: there is a strong effect in the sense that objects at higher redshift are bluer, in agreement with several other studies. It is unlikely that we have sufficient dynamic range in luminosity at fixed redshift to enable to identify whether the correlation is due to the redshift or luminosity of the quasars.

Mannheim: What is the signal–to–noise–ratio of the LBQS composite spectrum?

Hewett: The signal–to–noise ratio varies with wavelength but it is approximately 400:1.

Appenzeller: How did you remove the absorption lines from intervening systems from the composite spectrum?

Hewett: We didn't! There are typically 300 quasars contributing to each wavelength bin and the effects of intervening absorbers, which are effectively randomly scattered in wavelength, is insignificant following the coaddition of the objects.

Quasars and Galaxy Formation

Martin J. Rees

Institute of Astronomy, Madingley Road, Cambridge CB3 0HA, U.K.

Abstract. I shall try to address four related questions:
1) **How much of a problem for theories of galaxy formation are the highest-z quasars, now being found in increasing numbers?**
2) **How quickly and efficiently can a black hole form and grow in the core of a young galaxy?**
3) **What can we learn from the clustering of quasars?**
4) **Can we detect "dead quasars" – massive black holes in the nuclei of nearby galaxies?**

The present text covers several somewhat disjoint points which are new, or which I think deserve emphasis – it is far from being comprehensive or balanced. I shall not elaborate on points already made by other speakers of this conference, nor dwell on topics which have been well-covered by other recent reviews.

1. Formation of the earliest quasars

To put the high-redshift objects into a proper cosmological context, it is useful to recall the relation between time and redshift in the standard models. For an Einstein-de Sitter model

$$t(z) = 13.6 h_{50}^{-1} (1+z)^{-3/2} yrs \qquad 1$$

The highest-z known quasars formed when the Universe was only 10^9 years old (when t was as little as $0.07t_{now}$). This is longer than the dynamical timescale within the 'luminous' part of a typical galaxy. But it is **not** long compared to the timescales for **extended halos:** the free-fall time from a radius r in a galactic halo is $\sim 10^9$ (r/100kpc) years, and the Universe must be at least twice as old as this before material at radius r can virialise. If the discs acquired their angular momentum from tidal torques there is, moreover, a well-known argument that the material now residing at 10kpc in a disc must have fallen in from 100kpc and then cooled down (Fall and

Efstathiou 1980). At these high redshifts we would not only expect galaxies to lack virialised halos – they would also not yet have developed discs.

The bigger redshifts now being discovered enable us to probe earlier epochs than could have been done in the early days of quasar research. The observed 'record' redshift has, year by year, gone up. But at the same time, the realization that galaxies (with their dark halos) are more extended and diffuse than was previously suspected has lowered theoretical estimates of the redshift of galaxy formation. In the 1970s, it was widely believed that 'galaxy formation' happened at earlier epochs than were directly observable. But it can now be claimed with some confidence that galaxy formation must still have been going on at eras what we can directly observe (corresponding to redshifts $z > 4$). Indeed the fact that quasars formed so early in cosmic history is an important constraint on models for galaxy formation, particularly 'top down' models in which large scale structures develop before individual galaxies; it is well known, for instance, that the simple adiabatic ('pancake') model dominated by neutrinos cannot, when the amplitude is normalized to fit clustering data, account for collapsing systems at such high redshifts. And this same question can be addressed quantitatively in the context of other specific model.

I shall discuss the incorporation of quasars into the standard Cold Dark Matter model for galaxy formation. This is just one of a number of alternative models currently 'in play', but it makes definite predictions about the fluctuation spectrum, and accounts for several features of galaxies and the way they are clustered. A feature of the CDM model is that structures built up hierarchically – from small scales to large – but that bound systems develop rather late: mergers are important even at redshifts of 1 or 2, and large galactic halos would generally have assembled rather recently. Only the most exceptionally high-amplitude fluctuations would have led to galactic-mass systems that had already virialised as early as z . 4. I should like to update and elaborate on an earlier (1988) paper by George Efstathiou and myself which considered whether the already-observed quasars with redshifts of order 5 were compatible with the CDM model.

Even the purely gravitational (non-dissipative) features of galaxy formation pose a severe challenge to numerical simulations. This is especially the case in the CDM model where, because of the very flat fluctuation spectrum, a very large dynamic range is required. For this reason, much can still be learnt from analytical studies. There have been many of these, going back to the classic paper by Press and Schechter (1974). Later authors, especially Bardeen et al. (1986), have discussed the statistics of gaussian fluctuations, and the quantitative details of the hierarchical built-up of collapsing and virialised systems.

The work of Ryden and Gunn (1987) is especially relevant to our present topic. These authors consider the situation that is schematically illustrated in Figure 1. They show how a massive galaxy can develop by 'nucleation'

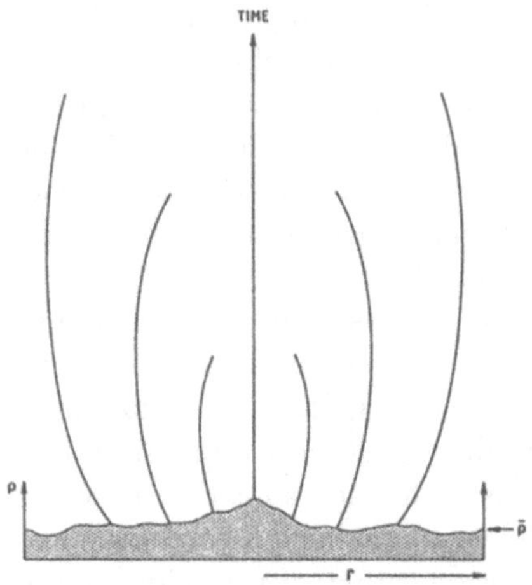

Fig. 1. The bottom part of the figure shows schematically a slice through the density field at an early cosmic epoch when the perturbations are still linear. The slice is centered on a high-amplitude peak of the initial fluctuation spectrum. The material immediately around the peak collapses quite early; the surrounding material (which 'feels an overdensity, but of smaller fractional amplitude) collapses later, forming an extended halo around the original core.

around a high-amplitude peak, which itself collapses at $z > 5$, but subsequently captures several hundred times its own mass. To quote an illustrative example form Ryden and Gunn, a $3\sigma(4\sigma)$ peak on a mass scale 10^9 $M_\odot$ will form a core with velocity dispersion 270 (450) km s^{-1} by a redshift $z = 8$ (10). It will subsequently capture the surrounding matter into a bound halo, by t_{now} the captured halo mass would be $2\ 10^{12}(4\ 10^{12})M_\odot$ (see Figure 2 and caption).

What would be the fate of the baryons, assumed to comprise 10 % of the total gravitating mass? The inner $3\ 10^9$ $M_\odot$ of baryons (associated with the inner $3\ 10^{10}$ $M_\odot$ of dark matter) can fall into a radius of $\sim$ 300 pc without losing angular momentum. The dynamical timescale at this radius is no more than $5\ 10^6$ years. Such models have been discussed by Hernquist (1990) and others – a common feature of most work on the subject is that the gas can settle into the centre on only a few dynamical timescales once it has become self-gravitating. Starbursts, etc. could further enhance the inflow by generating large-scale bulk motions and thereby maintaining a high viscosity.

In our earlier paper, Efstathiou and I emphasized that (i) it is unlikely that more than 1 % of the baryons in a galaxy would get sufficiently close to the centre to participate in the nuclear activity. (ii) the hosts of the first quasars would be the large galaxies, which arise from high amplitude peaks in the initial dark matter distribution; so it is appropriate to relate the density of quasars to that of present-day bright galaxies.

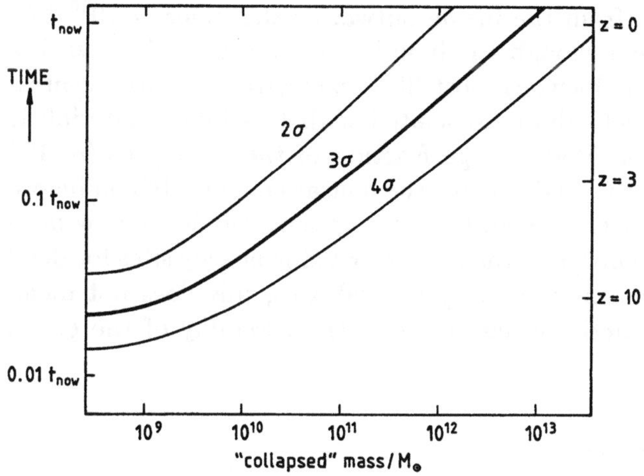

Fig. 2. **The accumulated virialised mass, plotted as a function of time (or redshift) that condenses around a 2, 3 or 4 peak on a mass scale of $10^9 M_\odot$. A CDM spectrum is assumed, normalized to the present epoch (adapted from Ryden and Gunn, 1987).**

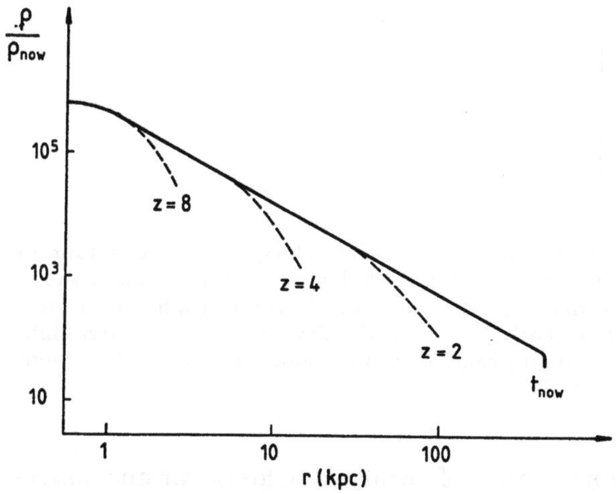

Fig. 3. **The density, as a multiple of the present-day critical density, at various stages in the development of a massive galactic halo. At early times only the core is virialised; the outer shells collapse later, and eventually give rise to a roughly isothermal ($\propto r^{-2}$) halo.**

At first sight, this suggested that the existence of any quasars with z = 5 might be an embarrassment. However, when we consider how these big galaxies would actually form, the problems seems less acute. Simulations show that large halos at the present epoch have different histories – some would have nucleated around a single peak, as depicted schematically in Figure 1; others may result from a relatively recent merger of systems that coagulated around separate high peaks in the initial density distribution.

But in general they form from the inside outwards: the inner $\sim 10^{10}$ $M_\odot$ may virialise at z > 8 even though the halo is not fully assembled until z ¡ 2 (See Figure 3). And the baryons most likely to aggregate into a central compact object are precisely those associated with this inner material: it is by no means improbable that a large fraction of **these** baryons ($\sim 10^9$ $M_\odot$) would participate in the AGN. Thus, the quasar could switch on before the halo was assembled. Even though the host systems would by now have accumulated $> 10^{12}$ $M_\odot$ halos, and the most promising nearby sites for dead quasars are therefore the centres of very big galaxies, this does not mean that the onset of quasar activity has to await the assembly of the entire halo.

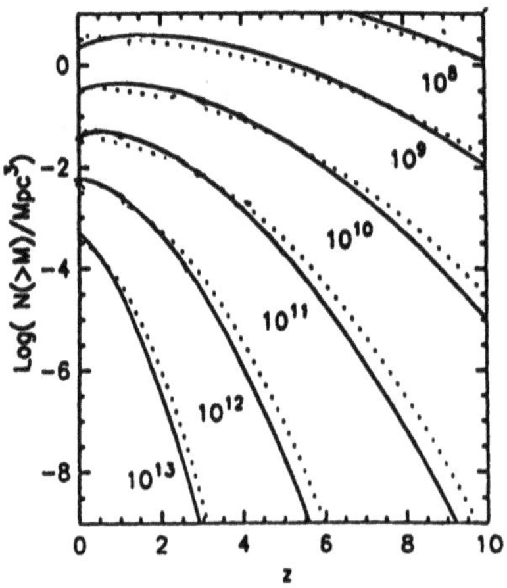

Fig. 4. This diagram, adapted from Efstathiou and Rees (1988), shows, as a function of epoch, the comoving number-density of virialised halos of different masses that is expected on the basis of a CDM cosmogony. The build-up of structure is hierarchical, so the larger-mass systems form more recently, and their density 'thins out towards higher redshifts because only exceptionally high peaks of initially gaussian distribution would already have collapsed at higher redshifts.

Figure 4 shows the comoving density of virialised halos of various masses, as a function of z. At high z, the curves drop precipitously, reflecting the extreme rareness of very high-amplitude peaks in a gaussian distribution. The drop occurs at a smaller redshift for the higher-mass systems. Even though the hosts of quasars may now be 10^{12} $M_\odot$ halos, the relevant curve in Fig. 4 is the one corresponding to 10^{10} $M_\odot$, because quasar activity can occur in a deep and compact potential well of only 10^{10} $M_\odot$. This may amount to only about 1 % of the final galaxy, but the baryons that participate in the quasar activity are predominantly those which were originally in this central ~ 1 %.

2. How rapidly can a black hole gain mass?

The foregoing discussion shows that, even in the CDM cosmogony, it would be possible for self-gravitating aggregates containing as much as 10^9 $M_\odot$ of baryons to form by z = 5, when the age of the universe (from equation (1) may be less than 10^9 years. The formation of such systems could be even earlier in other cosmogonies – indeed even if the dark matter is CDM, there is the possibility of a non-gaussian distribution of fluctuation amplitudes, triggered by textures, strings, fluctuations in the baryon-to-photon ratio, etc.

There are therefore many models in which gravitational potential mass of sufficient mass to create a quasar, and sufficient depth to retain baryons efficiently, would have formed by z = 5. The dynamical timescale within such regions would be les than 10^7 years; infall, and loss of angular momentum, may not take much longer than this. Moreover, the associated timescales shrink further as the baryons undergo further dissipation. here is no impediment, therefore, to the accumulation of 10^8 - 10^9 $M_\odot$ of baryons in a central region less than a few parsecs in size.

Even if a large mass of baryons accumulates in the central potential well, a quasar cannot switch on until this material has condensed into a relativistic (probably fully collapsed) object. The speed with which this can occur may·be constraint by other physical effects (e.g. radiation pressure). There is a simple and much quoted relation between the timescale for growth of a black hole the efficiency ε with which the gravitational energy is converted into radiation, and its luminosity:

$$M/\dot{M} = 4.4.10^7 \varepsilon_{0.1} (L/L_{Ed})^{-1} years \qquad 2$$

This formula is often used to estimate a maximum growth rate for a hole, on the assumption that $\varepsilon_{0.1} = 1$. While this may in some circumstances be a valid estimate, it is important to realise that the apparent simplicity of this formula is misleading because, when the central object is a black hole, the efficiency ε is itself very uncertain. (This contrasts with the case accretion onto a neutron star, where ε is determined by the gravitational binding energy at the surface). Because a black hole does not have a hard surface, the efficiency is less well-defined, being a function of the accretion rate $\dot{M}$ as well as of other parameters (angular momentum, etc.).

In the case of spherically-symmetric accretion, one can show explicitly how, if $\dot{M}$ is high, radiation emitted near the hole will be trapped and itself swallowed (see, for instance Begelman 1978). The 'trapping radius' moves out as $\dot{M}$ increases ; ε automatically drops as $\dot{M}$ rises, so that the hole can accept a arbitrarily large amount of mass without the emergent luminosity exceeding L_{Ed}.

[An extreme instance is the collapse of a supermassive star when it becomes subject to the well-known post-Newtonian instability. It collapses almost in free-fall. The radiation within it, which supplies almost all the internal pressure, is carried down the hole, because its outward diffusion relative to the flow is negligibly slow compared to the speed with which the bulk flow advects the radiation inwards].

It is not clear what happens when there is angular momentum. Abramowicz and his collaborators have presumed explicit solutions for thick discs, with narrow funnels along the axis, in which the binding energy of the orbit from which material is swallowed tends to zero, again allowing ε to become very small, and $\dot{M}$ very large, without the luminosity much exceeding L_E. It is, admittedly, possible that the excess mass inflow is always stemmed and the efficiency – the amount of radiation escaping per unit mass swallowed – cannot become very low. However, I hope enough has been said to emphasize that there is no **obvious** limit on the rate at which a black hole can grow: there are certainly specific models where there is **no** limit other than the rate at which material falls towards it.

There would clearly be a problem in explaining the high-z quasars if the growth rate were limited to the rate allowed by (2) for high efficiency – it would take longer than the (then) Hubble time for a hole that started off with only a stellar mass to grow by the requisite 20-odd powers of ε. As Turner (1991) has emphasized, optimum growth required low efficiency in the early stages of the hole's growth. But ε must be high during the quasar phase itself – otherwise the active lifetime will be $<< 4.10^7$ years. The number of quasar generations would then need to be correspondingly larger, and the mean remnant mass per galaxy would become too high to be consistent with the recent evidence on black holes in galactic nuclei (even if these are interpreted as upper limits).

I'd like to suggest that these requirements are naturally met in the type of picture discussed in the preceding section: when the hole mass is below 10^8 $M_\odot$ the rate of mass supply might well be highly supercritical, and ε will then be low enough to allow the hole to accept everything that falls towards it without L exceeding 1_{Ed}.

Once the inner part of a galaxy has virialised, the infall rate into the centre is likely to be constant. Therefore, a natural assumption is that $\dot{M}$ should be roughly constant (at least until infall and star formation are nearly complete. The three parts of Figure 5 illustrate this. When the hole mass is still small, the accretion rate is highly supercritical. But the hole can accept the inflowing material because ε adjusts to a sufficiently low value that L is below the appropriate L_{ed}. When M has risen to 10^8 $M_\odot$, the efficiency would rise, and the quasar phase would begin. The value of L/L_{Ed} would thereafter decline, because M continues to grow; at some stage, moreover, $\dot{M}$ will decrease because the infall stops or because the gas gets used up by star formation (Though it may be reactivated by, for instance, a merger).

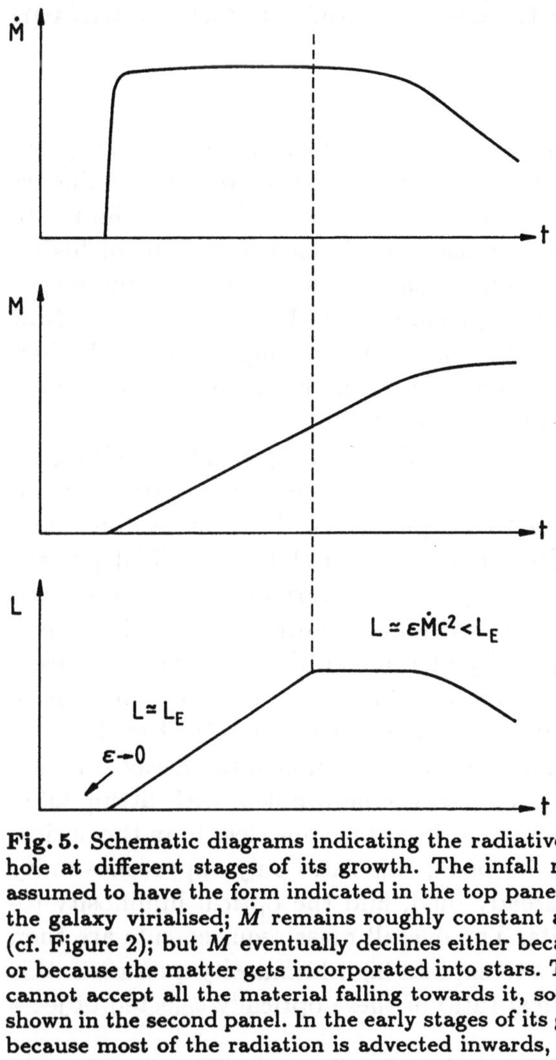

Fig. 5. Schematic diagrams indicating the radiative efficiency of accretion onto a black hole at different stages of its growth. The infall rate of material towards the hole is assumed to have the form indicated in the top panel. Infall starts when the inner part of the galaxy virialised; $\dot{M}$ remains roughly constant as progressively larger shells collapse (cf. Figure 2); but $\dot{M}$ eventually declines either because the galaxy completes its growth or because the matter gets incorporated into stars. There is no reason why the black hole cannot accept all the material falling towards it, so that its mass grows at a rate $\dot{M}$, as shown in the second panel. In the early stages of its growth, the radiative efficiency is low because most of the radiation is advected inwards, enabling the hole to accept the high influx of matter without exceeding L_{Ed}. At a certain stage (at the time indicated by the vertical dotted line) the mass has grown so large that the inflow rate $\dot{M}$ cannot supply a luminosity L, even if the efficiency ε is no longer reduced by trapping of radiation. The luminosity thereafter stays constant until $\dot{M}$ decreases, though the character of the emergent radiation may change, since this could be a function of L/L_{Ed}, which decreases, even for constant $\dot{M}$, as the mass M continues to grow.

An important feature of Figure 5 is that, at early stages in the growth, the mass increase with t is **linear**. This permits much faster growth than would be given by the exponential law obtained by regarding ε as a constant.

3. Some comments on starbursts and on the evolution of the AGN population

A surprising feature of quasar spectra is that their mean (rest-frame) properties display no obvious z-dependence (except for the greater prominence at higher z of the Gunn-Peterson discontinuity, and of Lyman limit absorbers). But the higher-z quasars must have formed in regions of higher mean density (and, especially, of higher mean gas density). Surrounding gas should therefore be spectrally more prominent at the earlier epochs. Now that larger samples are available, it would be interesting to check whether, for instance, the MgII lines are stronger in the higher-z objects, or if there is more evidence for 'fuzz' around such objects.

A more radical view is that many of the optically-identified QSOs do not involve supermassive compact objects, but are merely extremely intense 'starbursts' – see the paper by Filippenko in these proceedings, and the excellent earlier review by Heckman (1991). Relativistic VLBI jets are incompatible with this model; the rapid X-ray variability often observed is also very difficult to account for without a central relativistic object. However, in the case of most optically-identified objects, there is no direct evidence that these are less than a few kpc across, and the spectrum can be fitted by a suitably contrived star-burst, involving up to ~ 10 % of the total stellar content of an elliptical. In a pure starburst model, the X-ray emission must be attributed to supernova remnants. One problem with fitting QSO spectra is that the X-ray luminosity is at least ten times less than the optical luminosity when the burst is at its peak; this ratio only approaches unity when the brighter stars have already died, and the optical luminosity-to-mass ratio is only 100 solar units. The overall mass requirements are then embarrassingly high if one wishes to explain high-luminosity quasars in this way, especially if the X-ray/optical ratio of these optically-selected objects is indeed of order unity.

A superficially attractive feature of the starburst hypothesis is the surprisingly close match between the quasar luminosity function and the luminosity function of ellipticals, which Terlevich (1991) interprets by supposing that the bulges of all ellipticals formed in a sudden burst with a duration of order 10^7 years (this being the crossing-time for a typical bulge).

This similarity between the luminosity functions may, however, not be coincidental even in the black hole mode. The considerations of the previous section suggest that the amount of material accumulating in a central object would naturally depend on the density and on the depth of the potential well. Tentative corroboration comes from Dressler and Richstone's (1988) claim that the masses of the putative quiescent black holes in nearby galaxies are proportional to the bulge mass.

3.1 A conjecture on the evolution of the luminosity function

The luminosity function of AGNs and its z-dependence have been discussed by other speakers. Its main feature is a break or bend which occurs at a higher L at high z. There are compelling reasons for believing that the lifetime of an individual source (at least for those with high L) is $<< 10^9$ years. The luminosity function must therefore involve an integral over the life-cycle of each source, as well as being a function of other parameters. It is therefore surprising that any reasonably sharp feature survives in the integrated luminosity function.

How might we account for this? For black hole processes, the key parameter is L/L_{Ed} (This ratio obviously controls the importance of radiation pressure relative to gravity. It also controls another key parameter – the ratio of the cooling and inflow timescales – which may determine the efficiency, the ratio of thermal and non-thermal radiation, etc.z). It is therefore natural to identify the break in the luminosity function with a specific value of L/L_{Ed}.

But if this is so, we must conclude that **higher**-z quasars involve **bigger** black holes. This would be impossible if one attributed the z-dependence to the evolution of a single generation of objects. But it would be possible if there were many generations of short-lived AGNs, whose luminosity peaked soon after a black hole formed, and if conditions were more propitious for forming really big holes at high z. This proposal, though at first sight counter-intuitive, is actually not implausible. The optimum conditions for black hole formation are a deep potential well and a high density. (The latter is important because the cooling rate is then smaller relative to the collapse or expulsion timescale at high densities). These requirements may be best fulfilled, in combination by the deep and high-density potential well of the high-σ-perturbations which are the first to collapse (cf. Figure 2)

4. Cosmological implications

As more z>4 quasars are discovered, it becomes less problematic to maintain the view that UV from quasars is the primary ionization agent for the IGM, and accounts for the smallness of the Gunn-Peterson effect. (However, there is still no particular reason why quasars should be more important than bright stellar populations in young low-mass galaxies as sources of the overall UV background).

One potentially important difference between quasars and young galaxies as UV sources is that the former, being individually so much more luminous, would be more sparsely distributed, so that the ionization would occur inhomogeneously. Patchy or non-simultaneous ionization by the first quasars, separated by 50 Mpc, could create a large-scale environmental inhomogeneity (of non-gravitational origin) which modulates the efficiency of galaxy formation. This could be relevant to the surprisingly prominent large-scale features observed in the distribution of galaxies itself.

5. Dead quasars

One very important development in the last few years has been the strengthening evidence for massive black holes in nearby galaxies. Indeed, the evidence is consistent with the view that all large galaxies have central dark mass concentrations; moreover the mass seems to scale with that of the stellar bulge. I have recently reviewed this topic elsewhere (Rees 1990), so will merely recall the possibility that such holes could reveal their presence by the flares that occur when a star passes sufficiently close to the hole that it gets disrupted. The flare is energised by accretion onto the hole of some of the debris from the disrupted star; the remainder of the debris is ejected. The accretion process is very complicated – it is unsteady, the orbits are highly elliptical, and relativistic effects are important. (Moreover, because the orbital angular momentum of the disrupted star will general be misaligned with the hole's spin, Lense Thirring precession would destroy axisymmetry even if the debris had acquired circular orbits as a result of viscous effects.)

Because of these difficulties, the timescale for the flare is uncertain; so also is its spectrum, and the consequent bolometric correction. Further work on this problem is in progress. and more is certainly needed. All that can be said at the moment is that, for a characteristic timescale of a few months, the bolometric luminosity could attain a level of order the Eddington luminosity corresponding to the hole's mass – higher, therefore, than a supernova. Moreover, although the time-dependence is uncertain, there is no reason why it should mimic the light curve of a supernova. Any such event should therefore be recognizable, and should be detected by supernova searches, especially those using non-photographic techniques which do not discriminate against events in the high-surface-brightness central parts of galaxies.

The event rate in a galaxy harbouring a 10^8 $M_\odot$ hole depends, of course, on the properties of the compact system surrounding it: typical estimates are that there should be one disruption every 10^3 - 10^4 years. The Berkeley supernova search being carried out by Pennypacker and his collaborators, with an exposure of several thousand 'galaxy years' already, has not yet found any event of this kind. The significance of these limits cannot be fully quantified until the bolometric correction is better known. But it is clear that monitoring programmes of this kind would either soon reveal this new phenomenon, or else start to put significant limits on central black holes and/or the stellar distribution around them.

References

1. Bardeen, J.M., Bond, J.R., Kaiser, N., Szalay, A.S., 1986: *Astrophys. J.*, **304**, 15.
2. Begelman, M.C., 1978: *Monthly Notices Roy. Astron. Soc.*, **184**, 53
3. Dressler, A., Richstone, D.O., 1988: *Astrophys. J.*, **324**, 701.
4. Efstathiou, G.P., Rees, M.J., 1988: *Monthly Notices Roy. Astron. Soc.*, **230**, 5P.
5. Fall, S.M., Efstathiou, G.P, 1980: *Monthly Notices Roy. Astron. Soc.*, **193**, 189.
6. Heckman, T.M., 1991: in *Massive Stars and Starbursts*, ed. N. Walborn and C. Leitherer, Space Telescope Science Institute (in press)
7. Hernquist, L., 1990: Proc. 14th Texas Conference (Ann N.Y. Acad. Sci)
8. Press, W.H., Schechter, P., 1974: *Astrophys. J.*, **187**, 425.
9. Rees, M.J., 1990: *Science* **247**, 817.
10. Ryden, B., Gunn, J.E., 1987: *Astrophys. J.*, **318**, 15.
11. Terlevich, R., 1991: in *Relationships Between AGNs and Starburst Galaxies*, ed. A. Filippenko (in press)
12. Turner, E., 1991: *Astron. J.*, **101**, 5.

Discussion

Collin-Souffrin: How could you distinguish a supernova from a star being tidally disrupted and falling into the BH, in a galaxy nucleus?

Rees: It isn't yet possible to calculate the typical 'light curve' resulting from a stellar disruption, nor the spectrum and bolometric correction. All that I can say is that the total electromagnetic luminosity, integrated over the duration of the event, could be thousands of times larger than for a supernova. The events would not be specially standardized, as they would depend on the type of star swallowed, its initial impact parameter, and on the orientation of its orbital plane. But there is no reason to expect them to mimic a supernova especially closely. Incidentally, some very low-luminosity AGNs, especially those with no spatially-extended activity, could be interpreted as the fading stages of a stellar-disruption flare.

Wang: This concerns the recoil on the black hole. The black hole resides at the center of a dense (galactic) nucleus. Would the gravitational restoring force of the nucleus be sufficient to prevent the black hole from leaving the galaxy?

Rees: When two galaxies merge, then if each contained a massive black hole the two holes would spiral towards the centre of the merged galaxy and coalesce. The recoil due to gravitational wave asymmetry in the final coalescence has been estimated by Redmount and myself (and by others) to be up to a few hundred kilometres per second. This may not be enough to allow escape from the galaxy, in which case, as you imply, the merged hole would oscillate around the galactic centre until dynamical friction brought it to rest. However, in some cases the hole may achieve escape speed. Some supermassive black holes *may* therefore have broken loose from their host galaxies, and be hurtling through intergalactic space.

Wandel: Your suggestion that the break in the luminosity function is associated with a specific value of the intrinsic L/L_E ratio is demonstrated in a recent work (Caditz, Petrosian and Wandel, 1991, ApJLett, May 10). In this model the timescale for the evolution of the luminosity function is set by the intrinsic Eddington timescale. Can you comment on whether this scenario fits in the general scene luminosity-evolution of BH accretion you have presented.

Rees: Yes, your scheme is certainly one possibility. But it is hard to distinguish observationally between models where the Eddington ratio changes because the holes grow, or because the accretion rate fluctuates. And the holes may have a range of masses for other reasons. Knowledge of the masses of the dead remnants in nearby galaxies would help to pin down some of these uncertainties.

Bland-Hawthorn: At the 1989 Heidelberg meeting, there was much discussion of disk-disk mergers accounting for starburst, AGNs, QSOs, and even the complete Hubble diagram. Would it be possible to reproduce the quasar luminosity function by merging preferentially faint members of the Schechter luminosity function over a Hubble time? If so, would you still need the spheroidal component?

Rees: I haven't really thought about this possibility. Any acceptable theory certainly has to account for the remarkably steep drop in the density of all types of powerful AGNs between $z = 2$ and the present epoch. I don't think there can be much doubt that starbursts make a significant contribution to the total output at the fainter end of the luminosity function.

O'Dea: If the high redshift quasars are in protogalaxies, then they may be more compact and more depolarized than the low-z quasars.We find evidence that this is the case for the radio-loud quasars with sharply peaked radio spectra.

Rees: That is certainly interesting. As I emphasized in my talk, it would be very surprising if the properties of AGNs did not display some dependence on z, because the host galaxies (especially their outer parts) must have been very different at the highest redshifts.

Kundt: In your talk you have discussed the formation of black holes in the centers of galaxies. At this conference, however, a number of difficulties of the black-hole model have been mentioned: 1) The average gas spiral-in mass rate in our and neighbouring galaxies is of order $\leq M_\odot/yr$, but their nuclei contain less than some 10^7 $M_\odot$, perhaps only 10^3 $M_\odot$ for our galaxy (O. Gerhard, this meeting, W. Kundt, Astrophys. Space Sci. **172**, 285 and 109); 2) mass inflow rates in quasars are estimated to equal their outflow rates (S. Collin-Souffrin), like in the - very similar - stellar bipolar flows; 3) optical AGN spectra resemble those of supernova ejecta (R. Terlevich); 4) pair plasma of the needed high bulk Lorentz factors ($\gamma >> 10$) cannot escape from near the surface of a black hole. For these and other reasons, the SMC model has been proposed, (cf. NATO ASI C **208**, 1987). Question: What are your objections to it?

Rees: It would really take too long to respond adequately to all these points! Our understanding of the flow patterns around black holes, and of the balance between thermal and non-thermal emission mechanisms, is still at a very early stage. There are certainly, as you say, problems with getting very high bulk Lorentz factors for jets near black holes. But it isn't clear that these are demanded by the observations: the radiating particles may have been accelerated to high random gammas by a shock, and then may lose their energy before travelling very far. The most 'robust' prediction one can make about AGN models in general is that the activity should leave massive black holes as remnants in galactic centres – this should be true whether or not the black hole has already formed before the activity starts. A well-known argument first given by Soltan then implies that the remnant mass must be of order 10^7 solar masses per galaxy (the exact value depending on bolometric corrections, efficiencies, and the cube of the Hubble time). This is consistent with the present observations, even though some galaxies, including our own, have central masses which must be far below this average.

The Environments of Quasars

Josef W. Fried

Max-Planck-Institut für Astronomie, Königstuhl 17, D-6900 Heidelberg, Germany

1. Introduction

Overdensities of galaxies have been claimed not only near low ($z \approx 0.3$) and medium ($z \approx 0.6$), but also near high ($z \geq 1$) redshift quasars. These overdensities have been explained either by physical associations of galaxies and quasars, i.e. group or cluster environment, or - for the highest redshifts- by gravitational lensing of the background quasars by foreground galaxies, which would magnify the fluxes of the quasars and thereby raise them into flux limited samples.

In the following I will review the evidence for these claims and I will also present my own new data (a full analysis of these data will be given elsewhere) and compare these critically with the published data.

2. Low and medium redshift quasars

Several studies have shown that quasars with redshifts $z \leq 0.6$ reside in environments richer than the average galaxy density (Gehren et al. 1984, Green & Yee 1988, Tyson 1986, Hintzen et al. 1991, Ellingson, Yee and Green 1991a, EYG hereafter). However, there is little consensus about the amplitude of the excess; these claims range from groups to rich clusters. Rich clusters, however, have been claimed in only very few cases (3C 223.1, Oemler et al. 1972; 0016+16, Margon et al. 1983; 3C 275.1, Hintzen 1984; 3C 206, Ellingson et al. 1989) and therefore appear to be the exception rather than the rule. The most extensive imaging studies of the environments of low and medium redshift quasars have been done by Yee and Green (summarized by Green & Yee 1988). They concluded that quasars are located in groups or clusters; they also found an increase by a factor of ≈ 3 in the galaxy-quasar covariance amplitude between redshifts 0.4 and 0.6.

Clusters of Abell richness or any other large scale structure might actually escape detection because of the limited field of view of CCDs (a few core radii at $z \approx 0.3$). I have therefore studied the large scale environment of 2 quasars using a mosaique of twelve frames, covering a total of about 13 * 11□′. Galaxy isopleths down to $m = 22.5$ are shown for the quasar 1741+279 ($z = 0.372$) in figure 1; the total field corresponds to about 3.3 * 3 Mpc (H_0=75 km s^{-1}, q_0=0.5). There is clearly *no* pronounced increase in surface galaxy density towards the quasar (the result for the other observed quasar 2201+315 is similar). Clearly visible are large variations in galaxy surface density by factors of about 3 on angular scales of about 30″.

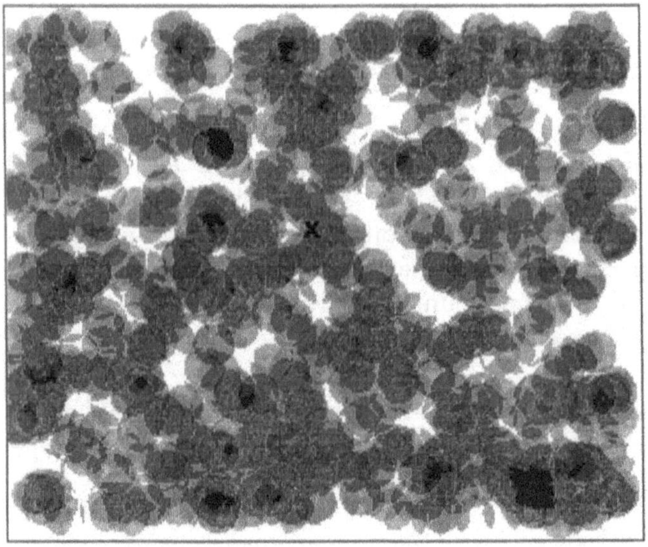

Fig. 1. Smoothed galaxy isopleths down to m=22.5 of the field around the quasar 1741+279, the position of which is marked. Field size 13 * 11□′.

I have also obtained deep ($m_{lim} \approx 23$) direct imaging data of a sample of 17 low redshift quasars ($z = 0.29 - 0.37$) and a sample of 18 medium redshift quasars ($z = 0.49 - 0.59$); the samples contain roughly equal numbers of radio quiet, flat and steep spectrum radio loud quasars. For the low redshift quasars, 2 adjacent CCD frames were obtained for each object, thus covering ≈ 1 Mpc around the quasars. For the medium redshift quasars a single frame covering approximately the same proper area around the quasars was obtained. The galaxy surface density as a function of distance from the quasars is shown in figure 1. There is a clear density increase towards the quasars.

The richness of environment is usually discussed in terms of the spatial correlation function which for galaxies is $\xi(r) = B_{gg}r^{-\gamma}$ with $\gamma = 1.77$. The two dimensional projection of this function on the sky is given by

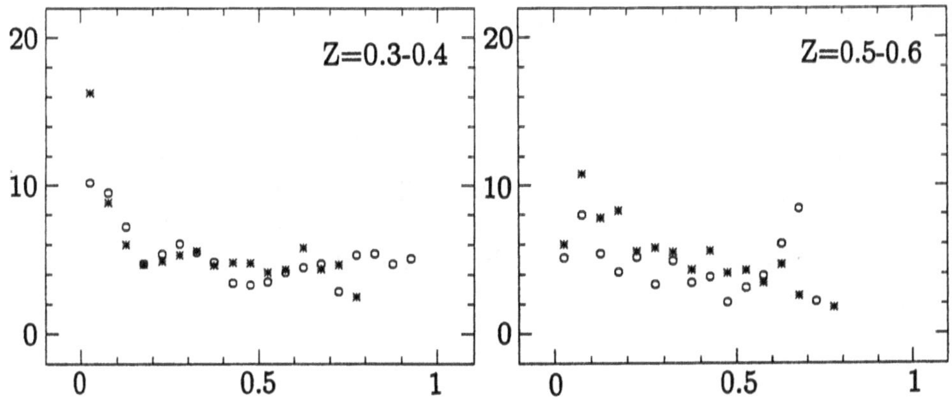

Fig. 2. Galaxy surface density per square arcmin as a function of radial distance in kpc from the quasars for the low and medium redshift quasar samples. Open circles: radio-quiet quasars, asterisks: radio-loud quasars.

$w(\theta) = A_{gg}\theta^{-(\gamma-1)}$. The conversion from the observable quantity A_{gg} to the physically meaningful quantity B_{gg} is described by a formula derived by Longair & Seldner (1979). It should be noted that this conversion is heavily model dependent; the numbers given below have been derived using model 3a of Yates et al. (1988). Typical B_{gg} values are 65 for a typical field galaxy (Seldner & Peebles 1978), and 360, 645 and 945 for Abell clusters of richness classes 0,1 and 2 respectively (Prestage & Peacock 1988, Yates et al. 1988).

For the radio-quiet quasars, I derive mean covariance amplitudes of $B_{gq} = 67$ for the low-redshift sample and $B_{gq} = 42$ for the medium redshift sample; this agrees very well with the results of EYG, who derived $B_{gq} = 72$; thus radio-quiet QSOs are found in environments typical for normal field galaxies.

For the radio-loud quasars, I find galaxy densities which are typical for groups or clusters of richness class 0; for the low-redshift sample I find a mean $B_{gq} = 188$, for the medium-redshift sample $B_{qq} = 301$. These data therefore provide independent evidence for the evolution of the environment of radio-loud quasars from $z = 0.3$ to $z = 0.6$ previously found by EYG. These authors find that this evolution pertains to the optically bright quasars only (similar results have been obtained for radio galaxies by Yates et al. 1988 and Hill & Lilly 1991). EYG explain their result by assuming that quasars in rich environments fade by ≈ 4 mag between $z = 0.6$ and $z = 0.3$, driven by changes in the velocity dispersion of the environment or the intracluster medium. It must be noted, however, that spectroscopic data of Ellingson, Green and Yee 1991 and Giraud 1991 show that the richest environments found by EYG are also the ones which are most heavily contaminated by foreground galaxies; some of the high B_{qg} values may need to be reduced. This would also explain the small velocity dispersions of

$400 - 500$ km s^{-1} found for the environments of the quasars, which would be unusually low for galaxy clusters.

Spectroscopic data for the galaxies close to the quasars are sparse because the faintness of these galaxies requires very long integration times. Such studies have been done by Stockton (1982), Heckman et al. (1984), Hintzen (1984), EYG and Giraud (1991). These studies show that a high fraction of the galaxies close to the quasars have redshifts within 1000 km s^{-1} of that of the quasar in the field; therefore these galaxies are probably truly associated with the quasars.

3. High redshift quasars

Webster et al. (1988) found an excess of galaxies towards optically selected quasars with redshifts between 0.5 and 3.3. In a subsample of 296 quasars they find 9 galaxies within 6 arcsec from the quasars, whereas they would expect only 5.5 ± 2.3 galaxies from the background. For another subsample of 334 quasars they find 7 certain and 2 probable quasar-galaxy associations (these revised numbers are from Webster & Hewett 1989). They interpret their result as being due to lensing; they have obtained redshifts for 5 galaxies, and the redshifts ($z = 0.1 - 0.3$) confirm that these galaxies are indeed foreground objects.

Images of high redshift quasars ($z = 1.0 - 1.5$) have been obtained by Tyson (1986); he found excess numbers of bright $R \leq 21$ galaxies within $r \leq 30''$ of the quasars. This excess pertained only to the flat spectrum ($\alpha > -0.5$) quasars in his sample. He interprets this result as being due to true quasar-galaxy associations, implying an evolution in the galaxy luminosities by more than 2 magnitudes. However, this excess was apparent only after a large correction had been made for the reduced probability of detection of faint objects near bright ones; furthermore, his data were obtained in white light and then matched to the R-band.

Hintzen, Romanishin and Valdes (1991) have observed a sample similar to the one of Tyson containing 16 radio loud quasars with redshifts $z = 0.9 - 1.5$. They could not confirm the result of Tyson (1986), but found excess numbers of galaxies at fainter magnitudes $R \leq 23$ and closer to the quasars $r \leq 15''$. In the R-band they found 32 galaxies where 19 ± 5 would have been expected from the background down to limiting magnitude $m_R = 23$. At $m_R = 22$ they found 15 galaxies compared with 9 ± 3 expected. They interpret the excess numbers of galaxies as being galaxy clusters associated with the quasars; they need to invoke a brightening of the galaxies by 1-3 magnitudes. Webster and Hewett (1989) constructed a Hubble diagram for these galaxies and found no correlation of apparent magnitude with redshift;

they concluded that either the evolution exactly cancels the cosmological effects (including K-dimming) or that the galaxies are actually foreground.

The strongest excess was found by Fugmann (1988); he found an 8-fold increase in galaxy surface density over background within 1 arcmin for a sample of 12 radio-loud flat-spectrum quasars with redshifts $z \geq 1.7$. Since these galaxies are almost certainly foreground galaxies, he explains this as being due to lensing of the quasars by these foreground galaxies. This lensing would magnify the quasars and thereby raise them into flux limited samples. I have observed a similar sample of 14 quasars; the data are summarized in figure 3, which shows the radial dependence of galaxy surface density down to 21.5 mag with angular distance from the quasars. Clearly, there is *no* pronounced excess close to the quasars; the data are compatible with constant galaxy surface density.

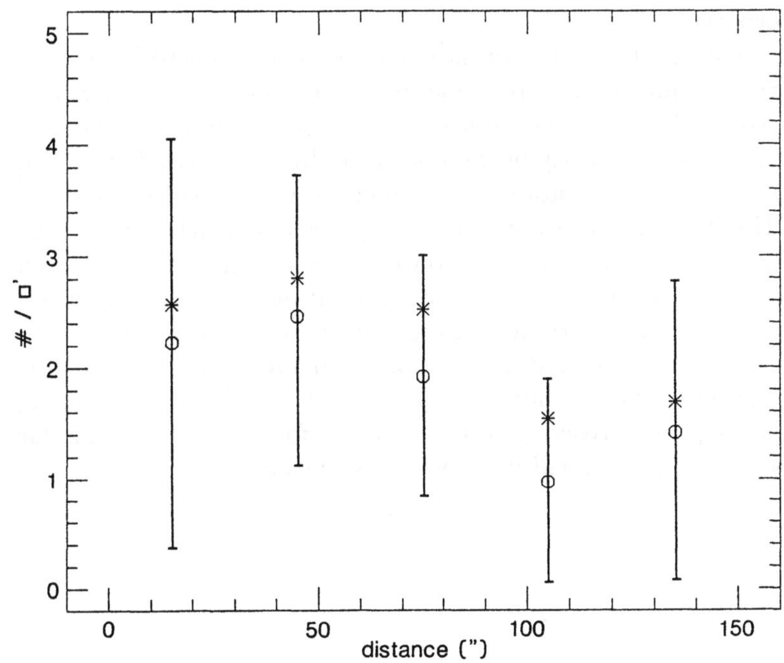

Fig. 3. Galaxy surface density per square arcmin as a function of angular distance for a sample of 14 radio-loud flat-spectrum quasars. Asterisks: all quasars, open circles: field of 0016+73, which is contaminated by foreground cluster, excluded. The data points are the sample means, the error bars are the standard deviations from the means.

Magain et al. (1989) analysed fields around high luminosity ($M \leq -29$) quasars with redshifts $z = 1 - 3.6$, which might by very good candidates for objects magnified by gravitational lensing. They found an excess over background by a factor of 1.8 towards the radio-loud flat-spectrum subsample at an estimated confidence level of 99.2%. It should be emphasized that their sample has been selected for high quasar luminosity and therefore differs greatly from the other samples discussed here.

4. Summary and conclusions

For the low and medium ($z \leq 0.6$) redshifts, there is convincing evidence for a density excess towards the quasars; redshift data for these galaxies are sparse, but support the interpretation of a true physical quasar-galaxy association. It is found that the radio-loud objects are located in environments as rich as groups or poor clusters; the radio-quiet objects live in poorer environments. There is also evidence that at redshifts ($z \approx 0.5$), the very bright radio-loud quasars live in environments 3 times richer than their low redshift counterparts.

For higher redshifts, the situation is much less clear. There is no consensus about the amplitude of a possible excess of galaxies near quasars. This may be partly due to differences in the samples studied (radio and optical properties and luminosity of the quasars), different filter bands and magnitude limits of the observations, and different galaxy background determinations. However, there are contradictory results which can not be presently understood. There is also no consensus about the interpretation of a possible excess of galaxies; both true physical quasar-galaxy associations (which would require enormous luminosity evolution of the associated galaxies at redshifts $z > 0.5$) and gravitational lensing have been put forward as explanations. These studies are necessarily statistical in nature; more data, especially spectroscopy of the galaxies found at small angular separations from the quasars, will improve the situation.

References

Ellingson, E., Yee, H.K.C., Green, R.F., Kinman, T.D., 1989: *Astron. J.*, **97**, 1539.
Ellingson, E., Yee, H.K.C., Green, R.F., 1991a: *Astrophys. J.*, **371**, 49 (EYG).
Ellingson, E., Green, R.F., Yee, H.K.C.,1991: preprint.
Fugmann, W., 1988: *Astron. Astrophys.*, **204**, 73.
Gehren, T., Fried, J.W., Wehinger, P., Wykoff, S., 1984: *Astrophys. J.*, **278**, 11.
Giraud, E., 1991: *ESO messenger* **63**, 67.
Green, R.F., Yee, H.K.C., 1988: in: *Proceedings of a workshop on optical surveys for quasars*, Osmer, P.S., Porter, A.C., Green, R.F, Foltz, C.B (eds.), Astronomical Society of the Pacific, San Francisco, p.233
Heckman, T.M., Bothun, G.D., Balick, B., Smith, E.P., 1984: *Astron. J.*, **89**, 958.
Hill, G.J., Lilly, S.J., 1991: *Astrophys. J.*, **367**, 1.
Hintzen, P., 1984: *Astrophys. J. Suppl.*, **55**, 533.
Hintzen, P., Romanishin, W., Valdes, F., 1991: *Astrophys. J.*, **366**, 7.

Longair, M.S., Seldner, M., 1979: *Monthly Notices Roy. Astron. Soc.*, **189**, 431.
Magain, P., Remy, M., Surdej, J.,Swings, J.-P., Smette, A., 1989: in: *Gravitational Lensing*, Mellier, Y., Fort, B., Soucail, G. (eds), Lecture Notes in Physics No.360, Springer Verlag, p.88
Margon, B., Downes, R.A., Spinrad, H., 1983: *Nature*, **301**, 221.
Oemler, A., Gunn, J.E., Oke, J.B., 1972: *Astrophys. J. (Letters)*, **176**, L47.
Prestage, R.M., Peacock, J.A., 1988: *Monthly Notices Roy. Astron. Soc.*, **230**, 131.
Seldner, M., Peebles, P.J.E., 1978: *Astrophys. J.*, **225**, 7.
Stockton, A., 1978: *Astrophys. J.*, **223**, 747.
Tyson, J.A., 1986: *Astron. J.*, **92**, 691.
Yates, M.G., Miller, L., Peacock, J.A., 1989: *Monthly Notices Roy. Astron. Soc.*, **240**, 129.
Webster, R.L., Hewett, P.C., Harding, M.E., Wegner, G.A.,1988: *Nature*, **336**, 358.
Webster R.L., Hewett P.C., 1989, in: *Gravitational Lensing*, Mellier, Y., Fort, B., Soucail, G. (eds), Lecture Notes in Physics No.360, Springer Verlag, p. 73

Structural Evolution of High-Redshift Radio-Quasars

Katherine Blundell, Paul Alexander, Richard McMahon, Philip Miller, Guy Pooley, Steve Rawlings, and Richard Saunders

Cavendish Laboratory, Cambridge CB3 0HE, England

Bright samples of radioquasars ($S_{178} > 10$Jy) have been studied (Barthel & Miley, hereafter B & M) and seem to indicate that quasars are more bent at high z. However, you need a much fainter comparison sample to try to separate the effects of z and radio luminosity.

So we defined our faint sample by:

- taking a 150 deg^2 patch of the 7C survey at 151MHz ($S_{151} > 0.1$Jy, resolution 1 arcmin, made with Cambridge Low Frequency Synthesis Telescope – see McGilchrist *et al.*, 1990);
- finding which of these 2000 7C sources appear associated with stellar objects by scanning POSS plates with the Cambridge APM;
- mapping the resultant 135 candidate radioquasars with the VLA — finally providing some 90 confirmed radioquasars.

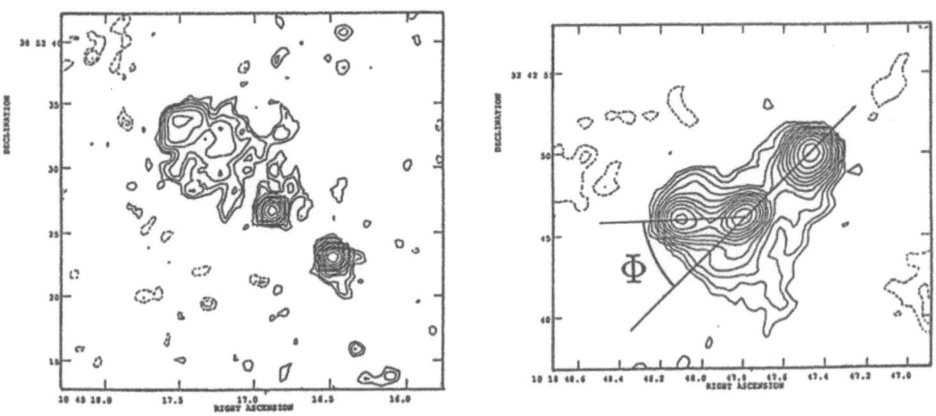

Fig. 1. Example VLA maps of 7C quasars

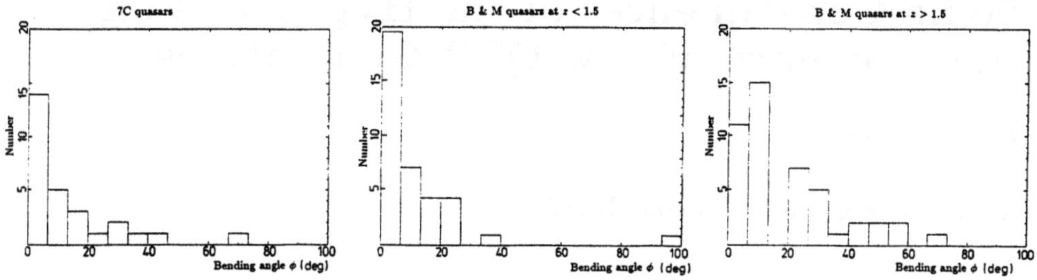

Fig. 2. left: 7C quasars; middle: B& M quasars at $z < 1.5$; right: B& M quasars at $z > 1.5$

We have measured various structural parameters, including a bending angle, ϕ, where possible (see fig. 1). In fig. 2 we compare ϕ distributions for B & M quasars at $z < 1.5$, B & M $z > 1.5$ and 7C.

The 7C distribution is more like that of B & M $z < 1.5$ than B & M $z > 1.5$: B & M $z > 1.5$ has median $\phi = 14°$; the other two samples both have median $\phi \simeq 7°$. Indeed, the Kolmogorov-Smirnov Test gives the probability p that the 7C and B & M $z < 1.5$ distributions are from the same populations as $> 99\%$, while $p = 3\%$ for 7C and B & M $z > 1.5$ and $p = 2\%$ for B & M $z < 1.5$ and B & M $z > 1.5$.

- Could this be because of some selection effect concerning inadequate resolution for the measurement of ϕ?
  - Unlikely, since the 7C quasar angular size distribution is similar to that of B & M $z > 1.5$.
- Could it be because the 7C quasars are nearby objects of low radio luminosity?
  - Unlikely, since the 7C quasar magnitude distribution shows the same characteristic break seen in UVX-selected samples, and so we expect the 7C sample to contain objects with $2 < z < 3$, but with radio luminosities ~ 30 times fainter than those of B & M $z > 1.5$.

The prima facie implication (subject to our obtaining a full set of redshifts for the 7C quasars) is that bending and distortion are linked primarily to high radio luminosity rather than to high redshift.

References

Barthel, P.D., Miley, G.K., 1988: *Nature* **333**, 319.

McGilchrist, M.M., Baldwin, J.E., Riley, J.M., Waldram, E.M., Warner, P.J., 1990: *Monthly Notices Roy. Astron. Soc.*, **246**, 110.

Some Consequences of the Hypothesis on the Existence of Few QSO Generations

B.V. Komberg

Astro Space Center, Moscow, Russia

1. Some assumptions:

1. There is QSO 0-generation that can be observed at $Z \geq 5$ (though their existence is rather possible than necessary). QSO 0-generation is low active nuclei of single, not bright, low massive distant primordial galaxies – protogalaxies (PG): $M > -19$, $m > 26$, $\rho \sim 10^{-3}/Mpc^3$, $N \sim 10^3/\sqcap^\circ$.
2. QSO I-st generation is very active nuclei of luminous and not too distant systems – merges-I, presenting product of merging of fainter members (PG) in groups and clusters. Because of $\tau_{merg} \sim R_{systems}/\sigma gal. \sim 10^9 years$ it becomes in general clear why QSO-I generation epoch of formation is $Z^x \sim 2-3$, while the epoch of galaxy formation is $Z^{xx} \sim 6-10$.
3. Very massive cD galaxies – mergers-II – can be formed in the central regions of rich clusters of galaxies due to the merging of less massive members (mergers-I) of the clusters. Active nuclei of these mergers-II could be observed as QSOs-II.

There are some indications that such merging is going at the present moment in some clusters, and during the process of "virialization" of the clusters the rate of merging decreases due to the increasing of members velocity dispersion.

4. Observed increasing of the ratio ρ_{QSG}/ρ_{QSS} at higher Z could be caused by earlier merging of E-groups – "E mergers" (their nuclei mostly are radio quiet – QSG rather than later merging of S-groups: their nuclei mostly are radio loud – QSS).

It is clear that the larger specific rotation momentum of the system the "merger" would be formed later and QSO would be observed at lower Z. The value of specific rotation momentum of the system has to be related to the morphology of its members: for larger fractions of disk galaxies the rotation momentum is larger. This fact leads to later formation of "mergers" with active nuclei, having higher activity in radio, because the probability of generation of stronger magnetic fields in the process of turbulent "dynamo" is higher (Multipole merging of small massive primordial galaxies led to massive "E-mergers". Massive "S-mergers" could be originate either as a result of multipole merging of abnormal high angular momentum of S-groups or due to merging of "E-merger"-pair.

5. Suggested evolution is directed from high luminosity FIR/star-burst galaxies via IR QSOs to normal QSO and later to galaxies with other kind of nucleus activity(to example: RG, SyG etc.).

2. Some consequences

1. QSO 0-generation is distributed in accordance with distribution of large scale inhomogeneities of matter at $Z > 5$.
   QSO-I generation is distributed at $Z \sim 1-5$ like clusters and groups of galaxies at Z smaller than 5.
   QSO-II generation is distributed at $Z < 0.5$ like rich clusters.
2. Far evolved QSO 0-generation in poorly populated regions (voids) could be observed as dwarf low luminosity galaxies. (Really, such a galaxy was found at $\lambda = 21cm$ in the known void: $M_V = -15$, $R_{21} = 10kpc$, $L_{21cm}/L_V = 1.5$, $M(HI) \sim 2 \cdot 10^8 M_\odot$ – the parameters are typical for low surface luminosity galaxies).
3. An epoch $Z^x \sim 2-3$ is specialized by the high merging rate via to formation of "mergers I". In this epoch is expected the highest volume density of FIR/star-burst galaxies and QSO I-generation. Radio quiet old ($\tau \geq 10^8 years$) QSO I-generation could be observed at $Z^x \sim 2-3$ as radio sources with special properties:
   $F_{GHz} < 0.1 - 1Jy$, compact double radio structure ($\Theta < 30''$, $l_r \leq 50kpc$) are placed inside optical image host galaxies, $\alpha_r > 1$.

References

Komberg, B., 1985: *Preprint Inst.Sp.Res.* (Russia, Moscow) **N971**.

Komberg, B., 1988: *Astrophys. J. (Letters)*, **327**, L1.

Komberg, B., 1989: *Astron.J. (Russia)* **66**, 710.

Radio-optical Studies of the BQS

Philip Miller

Mullard Radio Astronomy Observatory, Cavendish Laboratory, Madingley Road, Cambridge, CB3 0HE, UK

As part of a multi-wavelength investigation of the complete sample of 89 $z < 0.5$ quasars from the Bright Quasar Survey, we have made spectrophotometric observations of 39 objects to bring the completeness in available data for broad Hβ and narrow [OIII] to $\sim 85\%$. These data alone have allowed us to study the tightness and linearity of the correlations between optical continuum emission and line-emission from the BLR and NLR (Miller *et al.* 1991). We have made high-resolution ($\sim 0.4''$) maps of the sources detected at 5 GHz in the VLA A-array observations reported by Kellermann *et al.* (1989), who also observed these objects in D-array. The occurrence of radio-morphological types was found to be: 7 "No morph" — no available information; 9 "Non-det" — not detected, even in D-array; 25 "Quiet" — detected in D but not in A-array; 24 "Point" — most flux in unresolved point source; 8 "Single" — core of source shows some extension; 2 "Compact Double" — core resolved into sub-galactic components; 4 "Core–Halo" — unresolved core, no hotspots, much extended flux; 9 "Triple" — core and hotspots; 1 "Core–Jet" — the prototypical quasar 3C 273.

From fig. 1: the radio-quiet quasars ($P_{5GHz} < 10^{23}$ W Hz^{-1} sr^{-1}) show a clear P–$L_{\mathrm{[OIII]}}$ correlation, suggesting their radio-emission is dominated by a component of AGN origin; at intermediate radio-powers ($P_{5GHz} \sim 10^{23.5}$ W Hz^{-1} sr^{-1}) there is a scarcity of objects — seemingly distinct from the radio-loud and -quiet classes; and the radio-loud quasars ($P_{5GHz} > 10^{24}$ W Hz^{-1} sr^{-1}) — mostly of "triple" radio morphology — lie in a similar region of the plot to the 3C radioquasars.

Acknowledgements. We are grateful for the award of observing time on the University of Hawaii 2.2m telescope, to Dick Sramek for permission to use the calibrated VLA data of the BQS observations, and to Paddy Leahy for advice and assistance on AIPS reduction at the VLA.

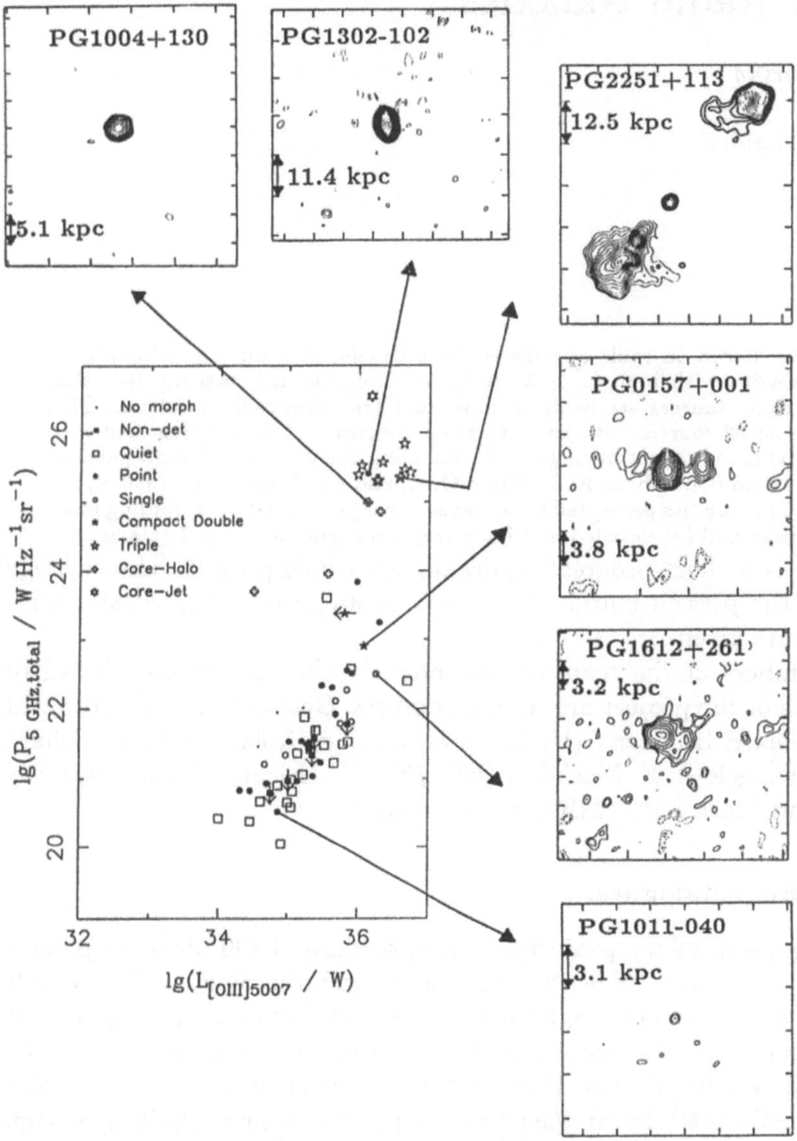

Fig. 1. Total 5 GHz radio luminosity versus the [OIII] narrow-line luminosity for the BQS $z < 0.5$ subsample. Symbols denote the morphology of the radiosource.

References

Kellermann, K.I., Sramek, R., Schmidt, M., Shaffer, D.B., Green, R.F., 1989: *Astron. J.*, **98**, 1195.

Miller, P., Rawlings, S., Saunders, R., Eales, S.A., 1991: *Monthly Notices Roy. Astron. Soc.*, submitted.

Distant Radio Galaxies

Huub Röttgering

Sterrewacht Leiden

Abstract. **Based on radio spectra we have developed a unique technique for finding galaxies with high ($z > 2$) redshifts. This method uses the fact that luminous radio sources are likely to have an ultra steep radio spectrum. Of a 4C sample of 32 sources with an ultra steep spectrum 25 % are identified with optically extended object having $z > 2$, including the two most distant galaxies known at z=3.8 (Chambers, K.C., Miley, G.K., and van Breugel, W. 1990 *Ap.J.*, 363, 21). This success prompted us to devise a project aimed at (i) finding more $z > 2$ galaxies and (2) detailed multi-spectral investigation of $z > 2$ galaxies.**

This poster is a short progress report on the whole project. The strategy is outlined, the present status of the various stages is presented and some conclusions are given.

The members of the team of astronomers who are involved in all or some aspects of the project are: Ken Chambers, George Miley (Sterrewacht Leiden), Richard Hunstead (U. Sydney), Jacques Roland (IAP), Richard Schilizzi (Dwingeloo), F. Machetto (ESA/STScI), Richard Saunders, Steve Rawlings and Mark Lacy (MRAO, Cambridge).

All sky radio catalogues.

The starting point of the project is a compendium of 800 ultra steep spectrum (USS, $\alpha \lesssim -1.1, \quad S \sim \nu^{\alpha}$) from various radio catalogues. The sample ranges over two orders of magnitude in flux and finding frequency so that the dependance of subsequent results on various radio properties can be disentangled. We have found that only a very small fraction (1-2 %) of a sample of small ($< 4'$), low frequency selected sources have a radio spectrum that is ultra–steep down to 38 MHz like 4C 41.17 (z=3.8).

VLA, Westerbork, and AT observations.

We observed USS sources from the compendium at the VLA, Westerbork or the AT to obtain radio positions, accurate fluxes and structures. An interesting result is that the relative number of diffuse (low luminosity) sources and the relative size of the sources appears to decrease with finding frequency. These both hint that the 38 MHz selected sample of USS sources contain fewer candidate high redshift objects than the 365 MHz selected samples of USS sources.

R–band CCD-imaging

We have carried out CCD imaging in R-band to a limiting magnitude $\lesssim 24$. We used the 2.2 ESO and the NTT for the southern sources and 2.5 m INT on Lapalma for the northern sources. Note that observations in both hemispheres are essential since the properties of the surveys we are using to select our samples are different in the northern and southern hemispheres.

Preliminary statistics of the identifications indicate that most of the radio sources that we have identified with galaxies show the so called "alignment effect" (the axis of the optical emission is reasonably well aligned with the principal axis of the radio emission, see Chambers, K.C., Miley, G.K., and van Breugel, W. 1987 *Ap. J.*, **329**, 604, McCarthy *et al.* 1987 *Ap.J.*, **319** L39).

Spectroscopy

We have carried out spectroscopy on about 70 of the sources in the sample. For the southern sources we used EFOSC on the 3.6, for the northern sources the FOS spectrograph on the WHT. We have found redshifts for 25 % of these objects, 10 sources are at $z > 2$ and 1 is at $z = 3.6$. It is not very likely that we have missed objects $z > 3$ with a spectrum similar to the 5 known $z > 3$, since the Lyα flux of these object are relatively large and would be redshifted to the region 5000 Å – 7000 Å where the sensitivity of the spectrographs we used is greater than at shorter wavelengths.

Questions

The use of USS radio sources is an efficient method to find distant galaxies. Spectroscopy of the samples of USS radio sources fainter then the 3C and 4C samples has led to the discovery of 10 galaxies $z > 2$ and one at 3.6. The three most distant galaxies known (z=3.6, 3.8, 3.8) are USS sources.

The alignment effect is present in most of the radio sources associated with galaxies from in our sample.

Our radio, imaging and spectroscopy data suggests a cutoff in the redshift distribution of powerful radiosources at z=3 – 3.5. Whether we have evidence for a decrease in the comoving density of powerful radio galaxies at about z=3 is not yet clear.

Polarization Properties and Unification Schemes

D.J. Saikia[1], *G.F. Holmes*[2], *S.T. Garrington*[2], *and R.A. Perley*[3]

[1] Tata Institute of Fundamental Research, National Centre for Radio Astrophysics, Poona University Campus, Post Bag No. 3, Ganeshkhind, Pune 411007, India
[2] University of Manchester, Nuffield Radio Astronomy Laboratories, Jodrell Bank, Macclesfield, Cheshire SK11 9DL, UK
[3] National Radio Astronomy Observatory, P.O. Box O, Socorro, NM 87801, USA

1. Introduction

The polarization properties of the cores of radio sources can be used as a probe of the physical conditions in the nuclear regions of active galaxies, and also to constrain unification schemes which attempt to understand apparently different kinds of active galaxies in a similar theoretical framework.

2. Rotation Measure

Rotation measure (RM) studies of the nuclear or radio core components are complicated by the effects of opacity and variability. Some have tried to find long-term average position angle differences in variable sources, while others have tried to circumvent the problem by observing at sufficiently close wavelengths so that the radiation is from the same electron population. The results of such measurements (Rudnick & Jones 1983; Saikia et al. 1985) suggested that core dominated sources have RMs < few tens of rad m^{-2}. However, more recent measurements at shorter wavelengths (e.g. O'Dea 1989) have shown that a number of core-dominated sources appear to have RMs of several hundred rad m^{-2}.

Our "quasi-simultaneous" observations of about a dozen sources with the VLA at $\lambda 20$, 18, 6 and 2 cm show that the RM of the core could be frequency dependent. An examination of the $\chi - \lambda^2$ relationship for these sources shows that such a relationship is not obeyed for the whole frequency range. In most cases the relationship appears to hold between $\lambda 20$ and 6 cm but there is a greater degree of rotation between $\lambda 6$ and 2 cm than would be predicted from an extrapolation of the longer wavelength data.

The long-wavelength |RM| ranges from ~ 0 to 130 rad m^{-2} with a median value of ~ 15 rad m^{-2}. The absolute values of the short wavelength RM ranges from ~20 to 290 rad m^{-2} with a median value of 130 rad m^{-2}. A possible explanation is that at shorter wavelengths, we are picking up radiation from a component which is not only more compact but perhaps buried deeper in the nuclear region, and hence giving rise to a higher RM.

These values suggest that there is no uniform dense medium in the vicinity of the nucleus capable of confining the broad-line clouds. Saikia et al. (1987) had earlier suggested that the low values of RM in these core-dominated sources are possibly due to their small angle to the line of sight so that the radio emission intercepts very little of the emission-line region, which could be in the form of a disk. In the unified schemes where the lobe-dominated quasars and radio galaxies are believed to be seen at progressively larger angles to the line of sight, one might expect a similar increase in the core RM if the radio emission from the nucleus intercepts a longer path length through the emission-line region for the sources seen at larger angles. It would be interesting to determine whether this is indeed the case. The core-polarization E-vector at λ6 cm tends to be orthogonal to the radio axis (e.g. Saikia & Shastri 1984)) for sources with $f_c < 0.5$. Here f_c is the fraction of emission from the core at an emitted frequency of 8 GHz and is used as a statistical measure of the orientation of the source relative to the line of sight. The lack of any such correlation for the core-dominated sources could be due to a combination of projection effects and aberration. Assuming that the polarization arises in nuclear jets with properties similar to the more extended jets in FRII radio sources, the correlation for those with $f_c < 0.5$ implies that the core RMs are less than about 100 rad m^{-2}. We are attempting to determine the precise RMs of a sample of lobe-dominated sources and examine what constraints this can place on the distribution of gas in the nuclear region for both radio galaxies and quasars were this scenario to be correct.

The median value of the degree of core polarization is ~ 2 per cent, similar to the values suggested earlier. Analysis of VLA 8 GHz data yield similar values.

3. One-sided Sources

For a sample of one-sided sources, we tried to determine their polarization properties to examine whether they might impose any constraints on their orientation relative to the line of sight. We made VLA scaled-array observations at λ 20 and 6 cm. The core RM and polarization values are similar to those discussed earlier.

The depolarization of the extended emission in the one-sided radio sources is similar to that of the approaching lobe, i.e. the one on the jet side, of the two-sided sources observed by Garrington et al. (1991). This is consistent with the apparent one-sidedness of these sources which have only one detected lobe, being due to relativistic motion of the outer lobes. The bulk velocity required to explain the degree of asymmetry is ~0.5c.

References

Garrington, S.T., et. al., 1991: *Monthly Notices Roy. Astron. Soc.*, **250**, 171
O'Dea, C.P., 1989: *Astron. Astrophys.*, **210**, 35.
Rudnick, L. & Jones, T.W., 1983: *Astron. J.*, **88**, 518.
Saikia, D.J. & Shastri, P., 1984: *Monthly Notices Roy. Astron. Soc.*, **211**, 47.
Saikia, D.J., Swarup, G. & Kodali, P.D., 1985: *Monthly Notices Roy. Astron. Soc.*, **216**, 385. 385.
Saikia, D.J., et. al.: *Monthly Notices Roy. Astron. Soc.*, **228**, 203.

Radio Variability of AGN: Shocks and Unified Models

Esko Valtaoja

Metsähovi Radio Research Station, Espoo, Finland

We have been conducting a large monitoring campaign of extragalactic radio sources at frequencies 22 GHz - 230 GHz since 1980 using radio telescopes at Metsähovi, Finland, at Crimean Astrophysical Observatory, and at SEST, Chile (Salonen et al. 1987, Tornikoski et al. 1991). Most of the time is spent on monitoring a core list of 60 bright flat-spectrum AGN, but we have also conducted several high radio frequency surveys. Our continuum observations seek answers to fundamental questions of the nature of radio-loud AGN: what mechanisms produce the observed flux and variations in it, what is the underlying radio structure, how are different classes of sources related to each other?

In seeking clues to the underlying structure and to the variability mechanisms of radio sources from continuum observations, the basic approach is to obtain well-sampled multifrequency flux curves during both quiescent and active phases of the sources. This enables us to separate the contributions from various components to the total flux and to follow the spectral evolution of the variable components (Valtaoja et al. 1988). With enough continuum data, it is even possible to derive some structural information (Valtaoja 1991).

The main results from our investigations thus far can be summarized as follows (Valtaoja et al. 1988, 1991a,b). All major radio outbursts exhibit the general characteristics expected from synchrotron-radiating shocks evolving in a relativistic jet (e.g., Marscher and Gear, 1985). In particular, a new outburst is caused by the appearance of a new component having a simple self-absorbed synchrotron spectrum ($\alpha_{thick} \simeq +2.5$ and $\alpha_{thin} \simeq -0.2$). The turnover frequency decreases monotonically in time, and the turnover flux shows the growth, plateau and decay stages in accordance with shock models. We have not found any significant systematic differences in these shock-induced variations when comparing radio galaxies, quasars and BL Lacs. We have also not yet encountered a single case of clearly different multifrequency behavior, which would require us to postulate an alternative mechanism for the outburst, such as gravitational microlensing or kinks in the jet. This need not mean that such mechanisms do not exist (and we are continuing our search for them with ever-improving database); however, we believe that one can quite categorically state that they are not major factors to radio variability.

Even if the basic variability mechanism is the same for all radio sources, one would still expect to find some differences in their continuum properties, depending for example on orientation. We have therefore also observed a number of complete samples of radio sources, searching for correlations and differences between the spectral properties, variability, and classification. Preliminary analysis of two samples has been completed (Tornikoski et al. 1991, Wirén et al. 1991). One sample consists of all Northern sources from the Kühr catalogue (Kühr et al. 1981) with S(>10 GHz) > 1 Jy and α(2.7-5 GHz) > -0.5 (113 sources), observed in Metsähovi at 22 and 37 GHz. The other complete sample from the Kühr catalogue consists of all sources with declinations between $-25°$ and $0°$, S(>22 GHz) > 1 Jy and α(2.7-5 GHz) > -0.5 (50 sources), observed at 100 and 230 GHz with SEST. For the analysis we have classified the sources as ordinary (low polarization) quasars, blazar-type (high polarization / OVV) quasars, and BL Lacs.

The main result, confirming the basic similarity of the radio emission mechanisms in these classes of AGN, is the lack of any outstanding differences. However, there are some small but statistically significant differences between the radio properties of flat-spectrum LPQs and HPQs, and correlations between their spectral shape and variability. These can be understood in terms of orientation-dependent models, where both the spectral steepness and the amount of rapid variability depend on the orientation of the beamed source relative to the line of sight.

Analysis of our data has also led us to suggest that the BL Lac objects have two physically different parent populations. The nearby 'BL Lacs' are beamed FR I radio galaxies, the distant 'BL Lacs' beamed quasars (i.e., FR II radio galaxies) (Valtaoja et al. 1991c).

References

Kühr, H., et al.: 1981, *Astron. Astrophys. Suppl.*, **45**, 367
Marscher, A.P., Gear, W.K.: 1985, *Astrophys. J.*, **298**, 114
Salonen, E., et al.: 1987, *Astron. Astrophys. Suppl.*, **70**, 409.
Tornikoski, M., et al.: 1991, in Variability of Blazars, eds. E. Valtaoja and M. Valtonen, Cambridge University Press (in press)
Valtaoja, E.: 1991, in Frontiers of VLBI, eds. H. Hirabayashi et al. (in press)
Valtaoja, E., et al.: 1988, *Astron. Astrophys.*, **203**, 1.
Valtaoja, E., et al.: 1991a,b, *Astron. Astrophys.*, (in press)
Valtaoja, E., et al.: 1991c, in Variability of Blazars (in press)
Wirén, S., et al.: 1991, in Variability of Blazars (in press)

Starburst–Black Hole Models for AGN

Robin Williams and Judith J. Perry

Institute of Astronomy, Cambridge, CB3 0HA, UK.

Abstract. **In 1985, Perry and Dyson (*MNRAS* 213, 665) suggested that radiative bowshocks around supernovae in the flowing interstellar medium of an AGN make clouds which emit high ionization broad lines (HIL); Collin-Souffrin et al. (*MNRAS* 232, 539 [1988]) included a disk to emit the low ionization lines. We describe further work on this integrated model. Perry (in Filippenko & Lo, A.S.P. Conf. Ser., *in press*) gives a detailed account of this programme and further references.**

1. Stellar evolution models

In the black hole model of AGN, gas must be fed to an accretion disc. If the fuel were evenly spread around the continuum source Thomson scattering would average out variations, so gas must be stored in clumps.

Stars can retain mass close to the central object and tight nuclear condensations of stars form naturally by galaxy interactions. Stellar mass loss then fuels the central object during *passive* stellar evolution (discussed before by Norman & Scoville *ApJ* **332,** 124 [1988]).

We have calculated fueling models with accurate stellar evolutionary tracks, for a *wide range* of initial mass functions and several star formation rates. We use these to determine the luminosity, the mass feeding rate and the supernova rate for the cluster. As an example, Fig. 1 shows a stellar cluster which forms in a single rapid burst; Fig. 2 is for star formation decaying exponentially with a 10^7 yr timescale. Stars form with a Salpeter IMF, between $1 \,\&\, 50\, M_\odot$. The curves are per solar mass of stars formed.

The luminosity of supernovae is always tiny compared to that of the background stars and the accretion luminosity dominates either of these. All luminosities are greatest at early times; the cluster will be most prominent at the early epochs when supernovae occur: this is the AGN phase.

Around $3 \times 10^9 M_\odot$ of stars must have formed in the initial burst for this cluster to reach 10^{47} ergs s^{-1}. The supernova rate is ~ 1 per year, as required by the HIL model of Perry & Dyson. As the gas supply comes from evolved stars, the HIL clouds will be enriched with C,N and especially O. The flows in the region will determine the clouds' abundances.

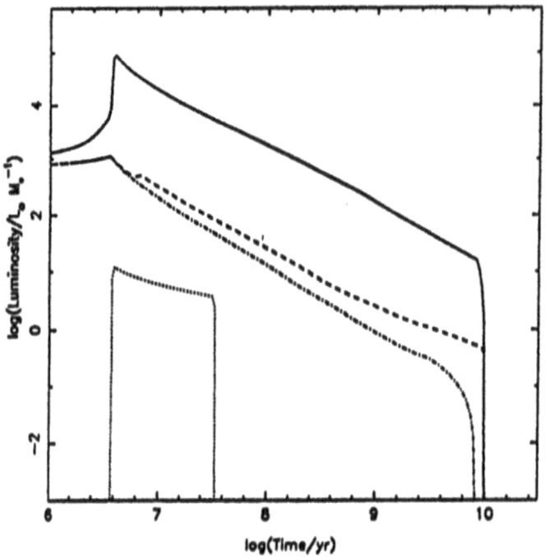

Fig. 1. Luminosity per solar mass of a 1–50 $M_\odot$ cluster with Salpeter IMF. The curves shown are: *Solid* accretion of all stellar mass loss with 10% of the rest-mass radiated, *Dashed* total stellar luminosity of the cluster, *Dash-dotted* luminosity of the main-sequence stars in the stellar cluster, *Dotted* luminosity of supernovae, if each produces a mechanical energy 10^{51} ergs which is efficiently radiated when the ejecta shock against the global flow. (The luminosity from normal supernovae would be 100 times smaller).

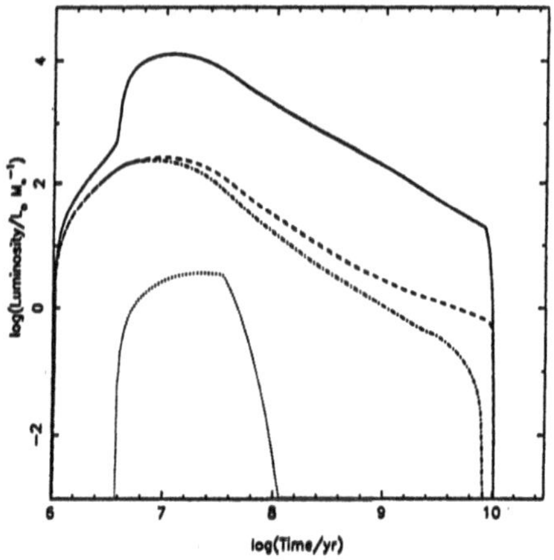

Fig. 2. As Fig. 1, but with star formation spread over $\sim 10^7$ yr.

2. Hydrodynamics of the broad line region

From the width of the high ionization lines, flow speeds are many tens of the Mach speed. Flow speeds of this order are also necessary to confine HIL clouds by ram-pressure (c.f. Perry & Dyson). However, Beltrametti & Perry (*A&A* **82,** 99 [1980]) showed that such speeds could be achieved for spherically symmetric systems, *only if the luminosity source extends outside the sonic surface of the flow.* This is not true for canonical models of AGN.

Axial symmetry is probably more realistic (because of the angular momentum of the fuel and anisotropic luminosity). We are carrying out hydrodynamical simulations in $2\frac{1}{2}$ dimensions in the hope of finding the source of these hypersonic flows.

3. Conclusion

Starburst clusters around massive black holes naturally produce many of the observed characteristics of AGN (Perry & Dyson). We hope to explain further details with realistic models of fueling and flow.

Variability as a Tool for Understanding Continuum Emission

Joel N. Bregman

Astronomy Department, University of Michigan, Ann Arbor, Michigan 48109, USA

Abstract. The continua from radio quiet AGN have been modeled with thermal emission from an accretion disk in the optical-ultraviolet region and dust reemission in the near to far infrared region. Variability studies support the dust emission model, but provide conflicting information for accretion disk models. Although the observed optical-ultraviolet time delays are shorter than expected from accretion disk models, periodic X- ray variations occur in NGC 6814, providing evidence for orbital motion around a central object.
In contrast to the radio quiet AGN, BL Lac objects have a nonthermal continuum probably produced by synchrotron emission within an inhomogeneous jet through which shocks propagate. This picture finds strong support from continuum variability programs, which are able to reproduce both flux density and polarization variations for a few objects. Although optical variations imply a smaller emitting region than the radio region, they are connected since optical variations are correlated with and precede radio variations, typically by 0.5-2 years. With increased effort, it is now feasible to spatially map the jet by determining time delays of fluctuations at a wide range of wavelengths. A concern with an underpinning of the model, incoherent synchrotron emission, is raised by the rapid optical-radio fluctuations in 0716+714, which suggests a radio brightness temperature so great that coherent emission processes may be necessary.

1. Introduction

A great frustration in the study of the nuclear emission from active galactic nuclei is that the vast majority of power emerges in spectral regions whose structure cannot be resolved with current astronomical techniques. This has forced workers to use various indirect methods aimed at revealing the salient properties of the central engine. One of these methods, used extensively in recent years, is variability measurements, which is used to deduce certain fundamental properties, such as sizes, shapes, and motion of thermal and nonthermal emitting plasma. Along with reviewing some results from these works, I will discuss a personal view of the working models for both radio quiet and radio loud AGN, followed by a discussion of key remaining issues and the impact that variability will hopefully provide. Since the analysis of the emission line region and their variability is addressed elsewhere at this meeting (Peterson, Boisson, this meeting), this review will focus on continuum emission.

2. The Continuum and Variability of Radio Quiet Quasars

Evidence is now strong that the continuum emission from the radio quiet quasars, and their lower luminosity counterparts, the Seyfert galaxies, is qualitatively different from the radio loud BL Lac objects. All radio quiet AGN have a strong "blue bump", where the power per logarithmic bandwidth rises from about 1 μm into the vacuum ultraviolet and possibly the extreme ultraviolet region (e.g. Malkan, this meeting). In a survey of such objects, it is found that the shape of this blue bump feature is similar in the rest frames of the sources (Malkan 1988; Sanders et al. 1989; Fig. 1). While most of the power of these AGN is emitted from this feature, typically one-third as much power is emitted in the infrared region. The continuum shape of the infrared region is also distinctive in that the power per logarithmic bandwidth has a broad peak in the 5-60 μm region, with a sharp dropoff at wavelengths beyond 100 μm.

In producing a model for the continuum emission in radio quiet objects, it is critical to notice that a few spectral features occur at the same location in the rest frame: the sharp falloff at wavelengths just beyond 100 μm, the local minimum in the power per logarithmic bandwidth near 1 μm, and the prominence of the blue bump in the ultraviolet region. Just as emission lines from different objects occur at the same rest wavelengths when adjusted for the redshift of the underlying galaxy, the alignment of these continuum features in the rest frame suggests that a thermal mechanism is responsible for the emerging continuum power. A model that was developed to reproduce these features involves an extended disk of gas and dust orbiting around a compact gravitating central object (Shakura and Sunyaev 1973; Czerny and Elvis 1987; Malkan, this meeting). In this picture, the central 10^{17} cm is dominated by a self- gravitating viscous accretion disk in which the temperature increases inward from 10^4 K to a maximum temperature of at least 3 10^4 K, and possibly as high as 5 10^5 K.

The presence of a hot disk of gas brings certain expectations, few of which are realized. While the absence of the anticipated thermal features is disappointing (continuum discontinuity; polarization; e.g. Laor, Netzer, and Piran 1990), it is possible to modify the model to accommodate observation. However, perhaps the most general and robust prediction of the model is that the hotter regions (3 10^4 K; peak flux near 0.1 μm) are smaller in diameter than the cooler regions (0.3-1 10^4 K; peak flux near 0.3-1 μm). Such size differences can be detected by variability studies, and fortunately, the blue bump does change intensity. If a disturbance, such as a flare, was to propagate from the inner to the outer region, a time delay in the outburst should be evident between the ultraviolet and optical region, with a value corresponding to or exceeding the difference in the light travel time between the radii of the two regions. Even if such events do not propagate,

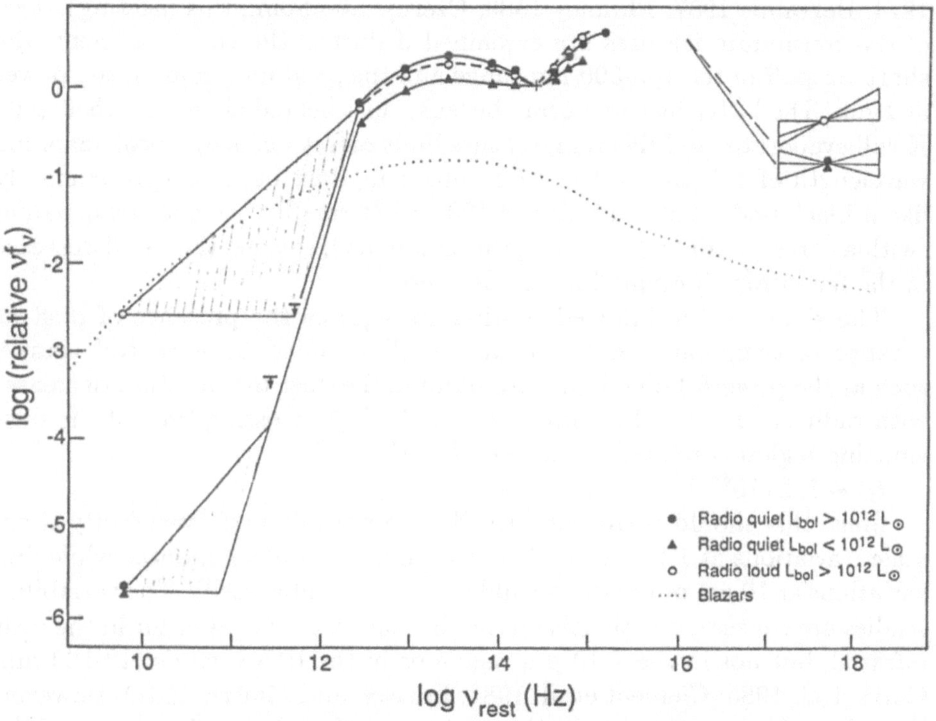

Fig. 1. The average continua of several types of AGN (from Sanders et al. 1989). The quasars show a local minimum in the power per logarithmic bandwidth near 1 μm, a prominent blue bump, and a sharp continuum falloff at wavelengths longward of 100 μm. It is argued that these features are due to thermal emission processes, features not present in the BL Lac continuum.

so that continuum fluctuations in both regions are independent, the characteristic timescale of variation should be proportional to the diameters of their respective regions.

Investigations of the temporal fluctuations in the blue bump region is still in a very early state and the predictions discussed above have not yet been well-tested. Malkan (this meeting) has discussed such observations for the best observed object, NGC 5548, and finds that optical and ultraviolet variations are correlated, but with a lag time less than 4 days. For NGC 4151, the lag time is less than 3 days, although detectable lags may exist for F 9 and 3C 273. Nevertheless, the upper limits to the lags are surprisingly short and may be inconsistent with the simple disk models. A major modification to his picture will be necessary if additional observations and more detailed modeling continue to point out behavior conflicting with thin accretion disks.

The 1-300 μm continuum of radio quiet AGN probably arises from dust with a distribution of temperatures in the range 50-1500 K (Penston et al.

1974; Barvainis 1987; Phinney 1989; Czerny, Stenholm, this meeting). Two of the continuum features are explained if dust is the emitting agent: the sharp dropoff in the 100-300 μm range and the local minimum in the power at 1 μm. The latter feature occurs because dust heated above the 1500-2000 K will evaporate, and this temperature limit corresponds to a peak emission wavelength of 1-2 μm. In the 1-60 μm range, dust behaves approximately like a black body, but longward of 100 μm, it rapidly becomes transparent (with a corresponding decrease in the emissivity), causing the rapid decrease in the far-infrared/submillimeter spectrum.

The shape of the infrared continuum requires the presence of dust at a range of temperatures. If the dust is illuminated by a central source, such as the powerful blue bump continuum, the dust temperature decreases with radius as r-1/2, which implies that the light crossing time of the dust emitting region is related to the wavelength as

$t_{lc} \sim 1(L/10^{45})^{1/2}(\lambda/1\mu m)^2$ months.

Since this should correspond roughly to the minimum variability timescale, variations at 1-2 μm should occur on a timescale of months while flux variations at 10 μm or greater should not be detectable easily. The variability studies are consistent with this picture in that flux changes occur in the near infrared, but not in the 2-10 μm region or in the IRAS bands (12-100 μm; Cutri et al. 1985; Clement et al. 1987; Edelson and Malkan 1987). However, there have been arguments in the past suggesting that at least part of the 1-40 μm emission in some radio quiet AGN is nonthermal (Edelson and Malkan 1986; Carleton et al. 1987; Wilkes, this meeting). Although there is little definitive support for this view (see Sanders et al. 1989), an observation that would help establish this argument would be infrared variability on a timescale less than a year at $\lambda >$f 5 μm.

Of the two remaining emission regions, the radio and the X-ray wavebands, the former is the only one that is clearly associated with a nonthermal process, synchrotron emission. However, there is little power in the radio region, and more thorough discussions of this topic are given elsewhere in this meeting, so it will not be discussed further (see Witzel, Marscher, this meeting). The X-ray continuum, which shows extremely rapid variability in many Seyfert galaxies, is produced in a very compact region around the central object (McHardy 1989; Pounds, this meeting). The origin of this emission remains a major issue in the field, although new observations are helping to build a framework for further understanding. The extremely soft X-ray emission in objects like Mkn 335 can have a very steep continuum, which is likely to be the high frequency extension of the blue bump (Wilkes and Elvis 1987; Turner and Pounds 1988). At higher energies (2-10 keV), the emission mechanism is clearly different, with the Compton process the most likely emission mechanism (Courvoisier; Walter et al., this meeting). These workers find a correlation of 2-10 keV slope with the ratio of scattered to seed photons (F_{x-ray}/F_{UV}), which is expected in the Compton model, but not in the pair cascade model. This promising line of study is certainly one that should be pursued for several more objects.

Some of the only evidence for the presence of an accretion disk comes from the X-ray variations in the object NGC 6814. There are reports of periodic variability in the X-ray continuum of NGC 6814 (Abramowitz, Massaro, this meeting), and although the data are fairly convincing, additional observations should be obtained. Similar behavior occurs in galactic X-ray binary sources, which are known to possess accretion disks. This evidence provides support for accretion disks in AGN as in turn implies the existence of a central gravitating object.

3. Blazar Continuum Variability

Unlike radio quiet quasars, blazars have a nonthermal component, which is established through observations that imply brightness temperatures vastly greater than the thermal mechanisms of their radio quiet counterparts (e.g. Bregman 1990; Marscher, this meeting). In particular, the nonthermal component is most clearly observed in the BL Lac class of objects, which show variability at all wavelengths (sometimes quite rapid), high polarization, superluminal motion of radio VLBI components, and the lack of either broad emission lines, a blue bump, or a local minimum in the power per logarithmic bandwidth at 1 μm (Miller 1989; Fig. 1). VLBI radio observations show the radio region to have a brightness temperature of 10^{12} K (e.g. Mutel et al. 1990) and IRAS variability studies show that the lower limit to the 60 μm brightness temperature is 10^6 K, at least three orders of magnitude above the evaporation temperature of dust (Edelson and Malkan 1987). Also, variability in the X-ray region is sometimes found to occur more rapidly than the light crossing time for a 10^9 $M_\odot$ black hole (Doxsey et al. 1983; Feigelson et al. 1986). Taken together, these observations imply both relativistic motion and nonthermal emission.

Another test for the presence of thermal emission in BL Lac objects would be a search for a continuum feature that occurs at the same wavelength in the rest frame. One of the few continuum features that are seen in some BL Lac objects is a sharp turnover in the spectrum toward higher frequency. This feature occurs in Red Quasars (really, a class of BL Lac objects; Rieke, Lebofsky, and Wisniewski 1982) in the infrared- optical region, and in certain BL Lac objects in the optical-ultraviolet region (e.g. BL Lacertae; Bregman et al. 1990). It is not yet clear how frequently this feature occurs because it would lie undetected if it occurred in the extreme ultraviolet region, at soft X-ray energies, or at mid-far infrared wavelengths. For the objects in which this feature is present, the frequency at which the exponential turnover occurs is not stationary in the rest frame of these objects. Consequently, it is unlikely that this rapid turnover is caused by thermal processes, and it is ascribed to a sharp upper energy limit to the nonthermal electron distribution (Beichman et al. 1981a, b; Bregman et al. 1981)

This recognition of two separate types of AGN continua, as exemplified by BL Lac and radio quiet quasars, might have occurred much earlier in the study of AGN if 3C 273 had not become the archetype quasar. The quasar 3C 273 and many other radio loud AGN consist of a combination of the thermal component found in radio-quiet quasars and the nonthermal continuum of BL Lac objects. Initially, it was natural for investigators to model the continuum of 3C 273 as a single phenomenon (usually, nonthermal emission was assumed). Multiwavelength studies of 3C 273 and a variety of other AGN have demonstrated the faults in these early studies. Perhaps if the first two AGN to be well-studied were a BL Lac object (e.g. OJ 287) and a classic Seyfert 1 galaxy, the division of AGN into two distinct classes might have been recognized initially.

A primary goal in studying the nonthermal continuum of BL Lac objects is to determine the spatial distribution of the emitting plasma and to identify important internal dynamical processes. I believe that the leading model is the "shock-in-jet" model, which incorporates a jet of inhomogeneous moving plasma (e.g. Ghisellini et al. 1985; Marscher 1991, also, this meeting). In the current models, the plasma properties (density, B, relativistic bulk motion) are approximately uniform along surfaces normal to the jet axis, but vary with position along the jet axis. The surface of the jet may be gently curving, such as a paraboloid, or a cone of constant opening angle; the opening angle is usually 1-20deg. The radiation process is incoherent synchrotron emission and the bulk motion of the plasma is usually relativistic, with values for the Lorentz factor of several.

To complete this model, a shock or series of shocks propagate outward and compress the magnetic field and plasma within turbulent cells, causing them to radiate more strongly. Immediately behind the shock, particle acceleration may be present, Compton losses will occur if the photon and particle density are sufficiently high, while synchrotron losses are significant if the cooling time is less than the time for the shock to cross the plasma element (Marscher and Gear 1985; Hughes, Aller, and Aller 1985; Marscher 1990). When the radiative timescales are long compared to the shock crossing time, the emitting parcel will evolve adiabatically. It is in this last case where the model was applied to radio variability measurements in the rare instances where individual radio outbursts can be identified. The best case is BL Lac, where Hughes, Aller, and Aller (1989a, b) find at least two isolated outbursts at 5 GHz, 8 GHz, and 15 GHz. For their model, only weak shocks are required and the orientation of the jet to the line of sight is determined by the relative polarized and unpolarized variations at the three wavelengths. Within the context of this model, they are able to map from the time variability domain to the spatial domain in a plasma jet. When this picture is compared to the VLBI observations, shocked regions can be identified with superluminal components.

At higher frequencies, the shock-in-jet model was applied for the outburst that occurred in 3C 273 (Robson et al. 1983; Marscher and Gear 1985), which was studied from the near infrared to radio millimeter region. The outburst first appeared in the near infrared and propagated to lower frequencies. At the initial frequencies of the outburst, the synchrotron loss time can be shorter than the shock crossing time, requiring the incorporation of radiative losses.

While such efforts are effective at testing the shock-in-jet model, they do not map out in a general way the distribution of material along the jet. To do so requires enough data to determine reliable time delays between different spectral regions. Variability data from the optical to the radio region can provide an estimate for the size of the emitting region as a function of wavelength. During 1983, when IRAS was operating, a few blazars were monitored irregularly in the optical, near infrared, infrared, mm, and cm region. Unfortunately, almost no variability was seen in some of the brightest objects (Impey and Neugebauer 1988), but the violently variable quasar 3C 446 did vary, as shown in Fig. 2 (Bregman et al. 1988). This figure suggests that the variability amplitude is larger and the timescale shorter at the higher frequencies. This issue can be explored further by comparing the variability timescales inferred from non- simultaneous monitoring data. The determination of a unique variability timescale is problematic if the power spectrum of the fluctuations is a power-law. However, one can hope to define a characteristic time as the timescale ($dt/d\ln F_\nu$) when the amplitude variation reaches some particular value ($\Delta F/F = 0.3$). With this definition, one finds the timescale to be about one week in the optical, about 1-2 months at 10 μm, about 6 months at 3 mm, and about 1.5 years at 4 cm, suggesting that the variability timescale increases with wavelength to the 0.3-0.4 power. However, this result is indicative of the possible situation rather than definitive, and more work along these lines would be welcome.

A well-defined timescale that is meaningful for the structure of the plasma jet is the time-delay between variations in two widely spaced wavebands, such as the optical and radio regions. This was investigated previously (e.g. Phomphrey et al. 1976) and recently reconsidered by Hufnagel and Bregman (1992) who analyzed the five best-studied objects at radio and optical wavelengths: OJ 287, BL Lacertae, 3C 446, 3C 345, and 0735+178. They find that the nature of the variability in the two wavelength regimes is qualitatively different, with the radio fluctuations described by shot noise and the optical fluctuations by a combination of flicker and shot noise. Since shot noise is the full integral of white noise, and flicker noise is the half integral of white noise, systems that display shot noise behavior have a longer effective "memory" than systems with flicker noise. In continuum emitting

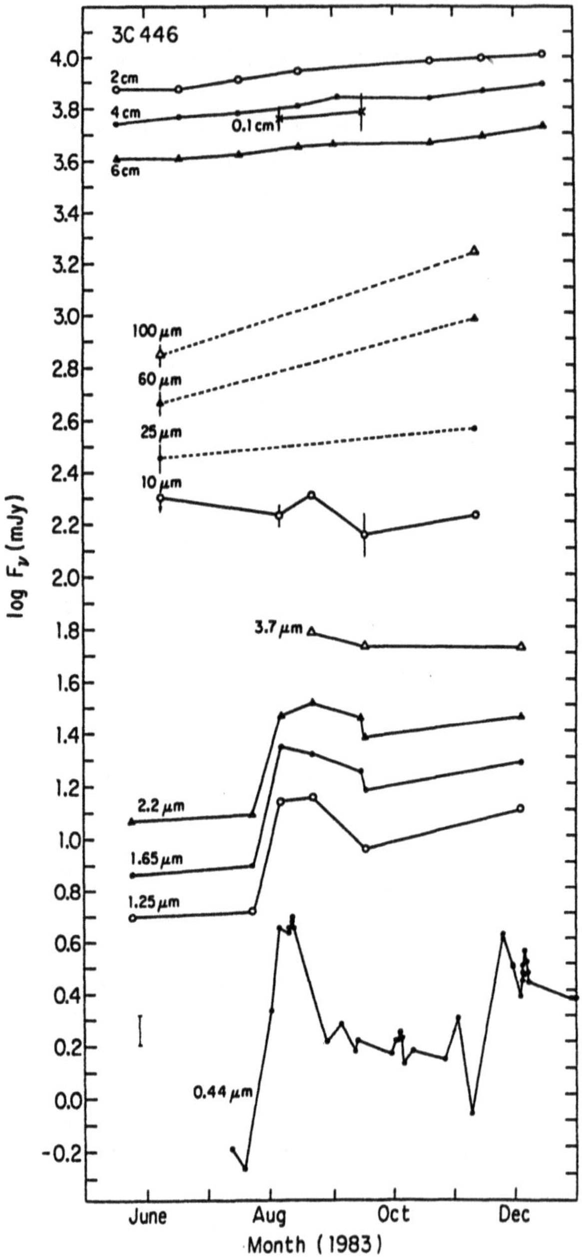

Fig. 2. **Variability of 3C 446 at optical through radio wavelengths (from Bregman et al. 1988) show greater amplitudes and shorter timescales at the shortest wavelength.**

regions, this shorter "memory" of the optical region may be a consequence of a briefer synchrotron cooling time.

In the measurement of an optical-radio time delay, the generic difference in the nature of the variability in the two wavebands introduces considerable difficulty because correlation techniques are best suited for two shifted but identical data streams. The discrete Fourier transform technique (Krolik and Edelson 1988) was used to investigate the presence of optical-radio time delays. Although the cross-correlations functions are weak, Hufnagel and Bregman (1992) find that in most sources, optical fluctuations precede those in the radio by typically 0.5-2 years.

Although many observations reviewed here are consistent with an inhomogeneous jet of plasma emitting incoherent synchrotron emission, a recent observation by Quirrenbach et al. (1991) may call into question a cornerstone upon which this model rests (also see Quirrenbach et al. 1989; Witzel, this meeting). These authors obtained very high-accuracy optical and radio measurements of the rapidly variable BL Lac object 0716+714 over a period of about one month. They find that both the optical and radio variability changes character from flickering to slower secular variations; the time delay between the optical and radio region is 3-4 days. If this interpretation is correct, then the variations found at 6 cm are intrinsic and occur in a region so small that a brightness temperature of $> 10^{17}$ K is implied. If this extraordinary photon density was to arise within an incoherent synchrotron emitting region, the Compton scattered emission at higher energies (i.e., X-rays) would greatly exceed the observations. To avoid this problem, coherent emission processes must be considered as the dominant radio emission process.

Before accepting a radical modification to the basic model, we need to examine whether rapid intrinsic radio variability is unavoidable. The type of rapid variations found at 0.3-5 GHz is common for compact flat-spectrum radio sources (Witzel et al. 1986; Heeschen et al. 1987) Much of the time, this variation can be attributed to interstellar scintillation effects, a process that becomes less effective at higher frequency. Some sources that were observed simultaneously over a range of radio frequencies, such as 0917+624 do not exhibit the anticipated frequency dependent variability from simple sources (e.g. a single point source; Quirrenbach et al. 1989). However, it is not yet clear whether the observed variability is inconsistent with more complicated (and possibly realistic) models, such as jets, which can have multiple emission components or tapered extended regions. For 0716+714, it might be possible to explain the radio variability due to external effects and the optical fluctuations as intrinsic if the two did not appear to be related. The cross-correlation between the two data streams is not strong, so their connection rests primarily upon the change in the nature of their variations. While this interpretation is sensible and suggested by the data, the implications are so great that it would be comforting to see additional observations.

For some BL Lac objects, the X-ray continuum is a distinct component and not an extension of the infrared through ultraviolet continuum. Compton scattering of synchrotron photons is often proposed for the origin of this emission, although there are other possible models, such as those that involve pair production. The simple prediction of the synchrotron-Compton model is that the seed photons (probably from the infrared-radio region) and the scattered photons fluctuate together and have the similar continuum shapes. There are some examples of sources in which the X-ray variability corresponds to the infrared-radio variations (e.g. 3C 345, 0735+178, BL Lac; Bregman et al. 1984, 1986, 1990), but only in BL Lac is the spectral shape of the X-ray emission determined simultaneously and found to be similar in shape to that of the presumed seed photons (Kawai et al. 1991). It is encouraging that such examples are found supporting the basic prediction, but there are also clear examples in which different behavior is observed, such as 3C 446, where changes in the 1 keV flux appear to be related to those in the optical/ultraviolet region.

4. Future Directions

Among the items and issues discussed above, I have chosen a few issues that I wish to see pursued. For the radio quiet quasar component, variability studies can help test models for accretion disks. In particular, for the blue bump region, it is necessary to examine the time delays of variation in the range 1 μm to 1 keV (rest frame). The short time delays seen thus far may be inconsistent with thin disk models, although a broad range of detailed model predictions still need to be calculated. Since the light travel time across a thin accretion disk would be hours, high accuracy observations need to be obtained with spacings of hours over a total time base of days to weeks. Such observations could be carried out for low redshift objects (Seyfert 1 galaxies) in the optical and ultraviolet regions, or for high redshift radio quiet quasars at optical and infrared wavelengths (time delays should be longer for these objects). A second promising line of study is that of periodic behavior in the X-ray region of NGC 6814. Since this periodic behavior is the first real evidence for rotation about a compact object, it must be studied in much greater detail to remove any doubts about its reality and to understand the phenomenon well enough to model. In addition, other objects of this type should be selected for study, possibly by examining whether any sources observed by previous X-ray telescopes suggest periodic behavior above the 2σ level.

For the study of the nonthermal component (BL Lac objects), the spatial mapping of the emitting jet, especially in the ultraviolet through submillimeter region is near the top of my "wish list". For this type of investigation, it will be necessary to resolve the temporal variations at every wavelength under consideration, an admittedly challenging task. This will require an enormous effort at the highest frequencies, where variations display a flicker-like behavior and timescales are short. One of the first efforts of this kind will be carried out during the coming year – the campaign to observe PKS 2155-304 from the X-ray through the radio region for about a month. However, more than a single campaign for a single object will be needed if we are to untangle the inner structure of jets.

The origin of the distinct X-ray component in nonthermal spectra has not been determined unambiguously. Although existing work supports the synchrotron-Compton model, these programs contain observations made at only a few times, and often, the X-ray spectral shape is not determined accurately. In future programs, it is important to select objects for which the X-ray continuum is a distinct component. Unfortunately, many well-known X-ray bright objects do not fit this criterion (e.g. Mkn 421) and it is often necessary to select sources with weaker X-ray fluxes, such as BL Lac.

Finally, the very short-term variations at optical and radio wavelengths that appear to be correlated are of great importance. These observations as interpreted require us to reconsider the fundamental radiation process of the model, incoherent synchrotron emission. Further observations, which include polarization information, are underway by Wagner and collaborators and we look forward to their results.

JNB would like to thank the meeting organizers for their efforts and support and to NASA for support through grant NAGW-2135.

References

Barvainis, R. 1987, *Astrophys. J.*, **320**, 537.
Beichman, C.A., Neugebauer, G., Soifer, B.T., Wootten, H.A., Roellig, T., and Harvey, P.M. 1981a, *Nature*, **293**, 711.
Beichman, C.A., Provdo, S.H., Neugebauer, G., Soifer, B.T., Matthews, K., and Wootten, H.A. 1981b, *Astrophys. J.*, **247**, 780.
Bregman, J.N. 1990, *Astron. Astrophys. Rev.*, **2**, 125.
Bregman, J.N., Lebofsky, M.J., Aller, M.F., Rieke, G.H., Aller, H.D., Hodge, P.E., Glassgold, A.E., and Huggins, P.J. 1981, *Nature*, **293**, 714.
Bregman, J.N., et al. 1984, *Astrophys. J.*, **276**, 454.
Bregman, J.N., et al. 1986, *Astrophys. J.*, **301**, 708.
Bregman, J.N., et al. 1988, *Astrophys. J.*, **331**, 746.
Bregman, J.N., et al. 1990, *Astrophys. J.*, **352**, 574.
Carleton, N.P., Elvis, M., Fabbiano, G., Willner, S.P., Lawrence, A., and Ward, M. 1987, *Astrophys. J.*, **318**, 595.
Clement, R., Sembay, S., Hanson, C.G., and Coe, M.J. 1988, *Monthly Notices Roy. Astron. Soc.*, **230**, 117.
Cutri, R.M., Wisniewski, W.Z., Rieke, G.H., and Lebofsky, M.J. 1985, *Astrophys. J.*, **296**, 423.
Czerny, B., and Elvis, M. 1987, *Astrophys. J.*, **321**, 305.
Doxsey, R., et al. 1983, *Astrophys. J. (Letters)*, **264**, L43.
Edelson, R.A., and Malkan, M.A. 1986, *Astrophys. J.*, **308**, 59.
Edelson, R.A., and Malkan, M.A. 1987, *Astrophys. J.*, **323**, 516.

Fiegelson, E.D., et al. 1986, *Astrophys. J.*, **302**, 337.
Ghisellini, G., Maraschi, L., Tanzi, E., Treves, A. 1985, *Astron. Astrophys.*, **146**, 204.
Heeschen, D.S., Krichbaum, T., Schalinski, C.J., and Witzel, A. 1987, *Astron. J.*, **94**, 1493.
Hufnagel, B.R., and Bregman, J.N. 1992, *Astrophys. J.*, in press.
Hughes, P.A., Aller, H.D., Aller, M.F. 1985, *Astrophys. J.*, **298**, 301.
Hughes, P.A., Aller, H.D., Aller, M.F. 1989a, *Astrophys. J.*, **341**, 54.
Hughes, P.A., Aller, H.D., Aller, M.F. 1989b, *Astrophys. J.*, **341**, 68.
Impey, C.D., and Neugebauer, G. 1988, *Astron. J.*, **95**, 307.
Kawai, N., et al. 1991, *Astrophys. J.*, in press.
Krolik, J.H., and Edelson, R.A. 1988, *Astrophys. J.*, **333**, 646.
Laor, A., Netzer, H., and Piran, T. 1990, *Monthly Notices Roy. Astron. Soc.*, **242**, 560.
Malkan, M.A., 1988, *Adv. Space Res.*, **8**, 49.
Marscher, A.P. 1990, in Parsec-Scale Radio Jets, ed. J.A. Zensus and T.J. Pearson (Cambridge Univ. Press), p. 236.
Marscher, A.P., and Gear, W.K. 1985, *Astrophys. J.*, **298**, 114.
McHardy, I.M. 1989, in BL Lac Objects, eds. Maraschi, L, Maccacaro, T., Ulrich, M.-H., Springer, Berlin, p. 189.
McHardy, I.M., Marscher, A.P., Gear, W.K., Muxlow, T., Lehto, H.J., and Abraham, R.G. 1990, *Monthly Notices Roy. Astron. Soc.*, **246**, 305.
Miller, J.S. 1989, in BL Lac Objects, eds. Maraschi, L, Maccacaro, T., Ulrich, M.-H., Springer, Berlin, p. 395.
Mutel, R.L., Phillips, R.B., Su, B., and Bucciferro, R.R. 1990, *Astrophys. J.*, **352**, 81.
O'Dell, S.L., et al. 1988, *Astrophys. J.*, **326**, 668
Penston, M.V., Penston, M.J., Selmes, R.A., Becklin, E.E., and Neugebauer, G. 1974, *Monthly Notices Roy. Astron. Soc.*, **169**, 357.
Phinney, E.S. 1989, in Theory of Accretion Disks, eds. F. Meyer, W. Duschl, and J. Frank, (Dordrecht: Kluwer), 1990, p. 457.
Phomphrey, R.B. 1976, *Astron. J.*, **81**, 489.
Quirrenbach, A., Witzel, A., Qian, S.J., Krichbaum, T., Hummel, C.A., and Alberdi, A. 1989, *Astron. Astrophys.*, **226**, L1.
Quirrenbach, A., Witzel, A., Wagner, S., Sanchez-Pons, F., Krichbaum, T.P., Wegner, R., Anton, K., Erkens, U., Haehnelt, M., Zensus, J.A., and Johnston, K.J. 1991, *Astrophys. J. (Letters)*, **372**, L71.
Rieke, G.H. 1985, in Astrophysics of Active Galaxies and Quasi-Stellar Objects, ed. J.S. Miller (University Science Books: Mill Valley), p. 236.
Rieke, G.H., Lebofsky, M.J., and Wisniewski, W.Z. 1982, *Astrophys. J.*, **263**, 73.
Robson, E.I., Gear, W.K., Clegg, P.E., Ade, P.A.R., Smith, M.G., Griffin, M.J., Nolt, I.G., Radostitz, J.V., and Howard, R.J. 1983, Nature, 305, 194.
Shakura, N.I., and Sunyaev, R.A. 1973, *Astron. Astrophys.*, **24**, 337.
Sanders, D.B., Phinney, E.S., Neugebauer, G., Soifer, B.T., and Matthews, K. 1989, *Astrophys. J.*, **347**, 29.
Turner, T.J., and Pounds, K.A. 1988, MNRAS, 232, 463.
Wilkes, B.J., and Elvis, M. 1987, *Astrophys. J.*, **323**, 243.
Witzel, A., Heeschen, D.S., Schalinski, C., and Krichbaum, Th. 1986, Mitt. Astr. Ges., 65, 239.

Discussion

H. Schulz: In Seyfert galaxies, one sometimes fails to see the local minimum in the power distribution at 1 μm. Has it been checked whether this can be attributed to stellar light?

J. Bregman: Studies indicate that stellar light, which peaks near 1 μm can contaminate the AGN spectrum and diminish the local minimum (Rieke 1985; Sanders et al. 1989).

R. Mushotzky: In your discussion of the far infrared to ultraviolet continua of radio-quiet AGN, material is distributed over scales ranging from pc to kpc, with the largest scales determined by the galactic distribution of dust and gas (the infrared emission). How does the disk know to arrange itself so that high luminosity AGN appear similar to objects 10^6 times less luminous?

J. Bregman: For the emission that originates in the accretion region near the central engine, one might expect a feedback mechanism (such as the Eddington limit) to create conditions that are similar from object to object. This could account for the near infrared through ultraviolet emission, but the 25-100 μm emission probably originates in the galactic ISM. It has been suggested that the AGN phenomenon occurs because a galaxy-galaxy encounters cause gas to orbit into the inner part of a galaxy. Then, the mass supply rate onto the central engine (i.e., the accretion disk luminosity) is related to the mass of dust available, leading to broad-band spectra with a characteristic shape.

Index of Contributors

MIX
Papier aus verantwortungsvollen Quellen
Paper from responsible sources
FSC® C105338

If you have any concerns about our products,
you can contact us on
ProductSafety@springernature.com

In case Publisher is established outside the EU,
the EU authorized representative is:
Springer Nature Customer Service Center GmbH
Europaplatz 3, 69115 Heidelberg, Germany

Printed by Libri Plureos GmbH
in Hamburg, Germany